Lecture Notes in Artificial Intelligence   2715

Edited by J. G. Carbonell and J. Siekmann

Subseries of Lecture Notes in Computer Science

T0189872

Springer
Berlin
Heidelberg
New York
Hong Kong
London
Milan
Paris
Tokyo

Taner Bilgiç   Bernard De Baets
Okyay Kaynak (Eds.)

Fuzzy Sets
and Systems –
IFSA 2003

10th International Fuzzy Systems Association World Congress
Istanbul, Turkey, June 30 – July 2, 2003
Proceedings

 Springer

Series Editors

Jaime G. Carbonell, Carnegie Mellon University, Pittsburgh, PA, USA
Jörg Siekmann, University of Saarland, Saarbrücken, Germany

Volume Editors

Taner Bilgiç
Boğaziçi University, Department of Industrial Engineering
Bebek 34342 Istanbul, Turkey
E-mail: taner@boun.edu.tr

Bernard De Baets
Ghent University, Dept. of Applied Mathematics, Biometrics and Process Control
Coupure links 653, 9000 Gent, Belgium
E-mail: bernard.debaets@ugent.be

Okyay Kaynak
Boğaziçi University, Department of Electrical and Electronics Engineering
Bebek 34342 Istanbul, Turkey
E-mail: kaynak@boun.edu.tr

Cataloging-in-Publication Data applied for

A catalog record for this book is available from the Library of Congress

Bibliographic information published by Die Deutsche Bibliothek
Die Deutsche Bibliothek lists this publication in the Deutsche Nationalbibliographie;
detailed bibliographic data is available in the Internet at <http://dnd.ddb.de>.

CR Subject Classification (1998): I.2, F.4.1, J.1, I, H.2

ISSN 0302-9743
ISBN 3-540-40383-3 Springer-Verlag Berlin Heidelberg New York

Springer-Verlag Berlin Heidelberg New York,
a member of BertelsmannSpringer Science+Business Media GmbH

http://www.springer.de

© Springer-Verlag Berlin Heidelberg 2003
Printed in Germany

Typesetting: Camera-ready by author, data conversion by Boller Mediendesign
Printed on acid-free paper   SPIN: 10928806   06/3142   5 4 3 2 1 0

Organization

The 10th IFSA World Congress (IFSA 2003) was organized by Boğaziçi University, Istanbul, in cooperation with the Soft Computational Intelligence Society, Turkey.

Executive Committee

Honorary Chair:	Lotfi A. Zadeh (USA)
Honorary Vice-chairs:	Michio Sugeno (Japan)
	I. Burhan Türksen (Canada)
Conference Chair:	Okyay Kaynak (Turkey)
Technical Program Chairs:	Taner Bilgiç (Turkey)
	Bernard De Baets (Belgium)
Organizing Committee:	Levent Akın (Turkey)
	Gökhan Aydın (Turkey)
	Eylem Koca (Turkey)
	Cem Say (Turkey)
	Uğur Yıldıran (Turkey)
Tutorials:	Nikola Kasabov (New Zealand)
Advisory Committee:	James Bezdek (USA)
	Z. Zenn Bien (Korea)
	Bernadette Bouchon-Meunier (France)
	Didier Dubois (France)
	Kaoru Hirota (Japan)
	Janusz Kacprzyk (Poland)
	László T. Kóczy (Hungary)
	Henri Prade (France)
	Sandra Sandri (Brazil)
	Ron Yager (USA)

Preface

This volume is a collection of papers presented at the 10th International Fuzzy Systems Association World Congress (IFSA 2003) during June 30–July 2 in Istanbul, Turkey. The IFSA World Congress is the main biennial event of IFSA. The 10th congress was organized by Boğaziçi University, Istanbul, in cooperation with the Soft Computational Intelligence Society, Turkey. The papers in this book are grouped together under five headings: invited papers, and the four area clusters of the congress (mathematical, methodological, application-oriented and cross-disciplinary). All areas were successful in attracting high-quality papers.

From 318 submitted papers, the technical program chairs together with the 12 area chairs selected 87 papers for publication as long papers in this volume. Another 155 papers were presented at IFSA 2003 as short papers, appearing in regular proceedings. We would like to thank all area chairs and reviewers for their conscientious reviews of all submissions under the tight time constraints imposed by the congress schedule.

We gratefully acknowledge the support of the Boğaziçi University Foundation and the Turkish Scientific and Technical Research Council (TÜBİTAK). The members of the organizing committee committed long hours of hard work for the success of the congress and the production of this volume. Special thanks go to Alexander Malinowski and Eylem Koca for their superb Web support.

April 2003

Taner Bilgiç
Bernard De Baets

Program Committee

Mathematical Areas
A1 Foundations: J. Fodor (Hungary)
A2 Pure Mathematics: S. Gottwald (Germany)
A3 Uncertainty Modelling: G. De Cooman (Belgium)

Methodological Areas
A4 Decision Making: M. Grabisch (France)
A5 Data Analysis and Data Mining: R. Kruse (Germany)
A6 Pattern Recognition and Image Processing: R. Krishnapuram (India)

Application Areas
A7 Control and Robotics: T. Fukuda (Japan)
A8 Information Systems: G. Chen (China)
A9 Business, Finance and Management: U. Kaymak
                                                   (The Netherlands)

Cross-disciplinary Areas
A10 Soft Computing: H. Takagi (Japan)
A11 Artificial Intelligence: L. Godo (Spain)
A12 Operations Research: K. Demirli (Canada)

Referees

Agell, N.	Bosc, P.	Díaz-Hermida, F.	Figueredo, J.M.C.
Al-Wedyan, H.	Branco, C.	da Silva, I.N.	Filev, D.P.
Angelov, P.P.	Bronevich, A.	De Baets, B.	Fodor, J.
Anthony, M.	Bugarín, A.	De Cock, M.	Fortemps, P.
Armengol, E.	Busquets, D.	De Cooman, G.	Foulloy, L.
Ashwin, T.V.	Cai, K.-Y.	De Mantaras, R.L.	Frisch, A.S.
B.-Meunier, B.	Calves, P.G.	De Meyer, H.	Fujimoto, K.
Baczynski, M.	Calvo, T.	De Tré, G.	Fuller, R.
Ballini, R.	Carlsson, C.	Delgado, M.	Furuhashi, T.
Baruch, I.S.	Castillo, O.	Demirli, K.	Gabriel, T.
Basak, J.	Castro, J.L.	Demisio, J.	Gebhardt, J.
Basir, O.	Cavalieri, S.	Denoeux, T.	Gil, M.A.
Benferhat, S.	Chakrabarty, K.	Deschrijver, G.	Glöckner, I.
Berthold, M.	Chekireb, H.	Detyniecki, M.	Gomide, F.
Berzal, F.	Chen, G.	Diamond, P.	Gottwald, S.
Bilgiç, T.	Chen, L.	Doignon, J.-P.	Grabisch, M.
Bloch, I.	Cheong, F.	Dombi, J.	Grauel, A.
Bodenhofer, U.	Coelho, L.S.	Dubois, D.	Greco, S.
Boel, R.	Cordon, O.	Ertugrul, S.	Höppner, F.
Bonarini, A.	Couso, I.	Esteva, F.	Hagiwara, M.
Bordogna, G.	Cubero, J.C.	Fargier, H.	Hajek, P.
Bors, A.G.	Döring, C.	Feng, G.	Hamzaoui, A.

Hayajneh, M.	Kummamuru, K.	Novak, V.	Surmann, H.
Hellendoorn, H.	Labreuche, C.	Pal, N.	Takagi, H.
Herrera, F.	Larsen, H.	Pap, E.	Tanaka, H.
Herrera-Viedma, E.	Lawry, J.	Pasi, G.	Tay, A.
Ho, N.C.	Lee, T.H.	Pedrycz, W.	Teixeira, R.
Hoffmann, F.	Levrat, E.	Pena, L.	Tikk, D.
Hong, T.-P.	Lim, C.-P.	Perfilieva, I.	Torra, V.
Hüllermeier, E.	Lin, C.-T.	Perny, P.	Trillas, E.
Hwang, H.	Lotlikar, R.	Perreria, R.A.M.	Troffaes, M.
Inou, H.	Maeda, Y.	Popov, A.	Tsoi, A.C.
Inuiguchi, M.	Maes, K.	Prade, H.	Utkin, L.
Ishibuchi, H.	Marichal, J.L.	Quaeghebeur, E.	Van De Ville, D.
Jaffray, J.Y.	Matthews, C.	Ralescu, A.	van den Berg, J.
Jenei, S.	Mattila, J.	Ramer, A.	van den Bergh, M.
Jeng, J.T.	Mauris, G.	Ramik, J.	van der Sluis, P.J.
Jimenez, F.	Mayer, H.	Reformat, M.	Van der Weken, D.
Johansson, S.	M.-Bautista, M.J.	Rifqi, M.	Vazirgiannis, M.
Joo, Y.-H.	Meghabghab, G.	Rocacher, D.	Vejnarova, J.
Kacprzyk, J.	Melaku, F.	Roubens, M.	Verdegay, J.L.
Kaymak, U.	Mesiar, R.	Ruan, D.	Vertan, C.
Keller, A.	Michels, K.	Ruiz, N.M.	Vincke, P.
Keller, J.	Miranda, E.	Runkler, T.	Walker, E.A.
Kerre, E.	Miranda, P.	Sànchez, D.	Wang, Z.
Kiguchi, K.	Mohamed, M.	Sabater, J.	Watada, J.
Klawonn, F.	Molhim, M.	Sabbadin, R.	Wei, Q.
Klose, A.	Moral, S.	Saffiotti, A.	Wets, G.
Kobayashi, F.	Mundici, D.	Sakawa, M.	Wettschereck, D.
Kolesarova, A.	Murofushi, T.	Sandri, S.	Willmott, S.
Kortelainen, J.	Nürnberger, A.	Scheffer, T.	Wolkenhauer, O.
Kothari, R.	Näther, W.	Segura, E.C.	Yager, R.R.
Kozma, R.	Naso, D.	Serrano, J.-M.	Yen, G.G.
Krishnapuram, R.	Nasraoui, O.	Sharma, S.K.	Yu, X.
Krogel, M.	Nauck, D.	Singh, R.	Zadrozny, S.
Kropp, J.	Navara, M.	Slowinski, R.	Zhang, J.J.
Kruse, R.	Nearchou, A.C.	Smets, P.	Zhang, Y.
Kubota, N.	Neogi, A.	Sousa, J.M.	
Kuchta, D.	Niskanen, V.A.	Sudkamp, T.	

Sponsoring Institutions

Boğaziçi University Foundation
Turkish Scientific and Technical Research Council (TÜBİTAK)

Table of Contents

Application Areas

Cross-Disciplinary Areas

A Perspective on the Philosophical Grounding of Fuzzy Theories

I. Burhan Türkşen

Director, Knowledge / Intelligence Systems Laboratory
Mechanical and Industrial Engineering
University of Toronto
Toronto, Ontario, M5S 3G8, CANADA
Tel: (416) 978-1298; Fax: (416) 946-7581
turksen@mie.utoronto.ca
http://www.mie.utoronto.ca/staff/profiles/turksen.html

Abstract. Philosophical grounding of theories are discussed within a framework of hierarchical levels of theoretical inquiry. Its application to classical and fuzzy set and logic theories are demonstrated within seven level of this hierarchy from ontological to application level.

1 Introduction

As fuzzy theorists and practitioners, we frequently find ourself confronting significant philosophical issues in our work. Indeed, if we are not doing so, we are probably and possibly missing out a lot. While different fuzzy theories and application approaches may be founded upon different set of philosophical presuppositions, all such theories rest upon *some epistemological and ontological* assumptions, whether explicitly acknowledged or not.

It is well known that Lotfi A. Zadeh has provided a continuous stream of novel and seminal ideas, from fuzzy sets, to fuzzy logic, to approximate reasoning to syllogistic reasoning, to computing with words and computing with perceptions. In this regard, we are greatly indebted to him for his continuing leadership. But very few of us have taken up some of his suggestions and clearly stated our particular stance in a systematic and constructive manner. I do not mean to state that we have not done significant progress over the last thirty eight years or so. We have, but we still have to do a lot more.

Part of the problem has been methodological. Despite the many significant developments, there have been few systematic or comprehensive attempts made to look at the complex and interweaving relationship among the philosophical and scientific issues in question. In this paper, we present a methodology with which one might explore the important philosophical bases of our fuzzy theories in a more structured and perhaps a more rigorous manner.

T. Bilgiç et al. (Eds.): IFSA 2003, LNAI 2715, pp. 1-15, 2003.

2   Underlying Philosophical Bases

An overview of a systematic approach to reviewing and observing philosophical issues of fundamental importance to fuzzy theory and its practice is presented below. This method involves an analysis of the stated or implied stances taken by any given fuzzy theory on a structured series of essential philosophical questions.

A hierarchy of levels of theoretical inquiry has been developed, and proposed which includes the Ontological, the Epistemological, Domain Specific Epistemological, and the Application Levels. Each level of this hierarchy poses its own fundamental philosophical questions. Each of these levels and questions in turn provides the philosophical "grounding" of subsequent ones.

3   Hierarchy of Levels of Theoretical Inquiry

A hierarchy of levels of theoretical inquiry and their questions, where the essential question thought pertinent to each level of inquiry being examined and which has been used as the framework for this approach is depicted in Table 1.

The bottom levels in this hierarchy are foundational to others: positions from Level 1 form the "grounding" or the conditions for the possibilities of positions on Level 2; those of 2 ground 3; etc, The Table 1 is to be read from bottom level up: from 1 to 7. Thus, on the application level 7, insights and theories are seen to rest upon a series of positions taken on each of the supporting levels 1 through 6.

4   Ontological Level

There are two sub-level in the Ontological Level. They are called Level 1 and Level 2. At the bottom, at Level 1 theoreticians of any sort must address the most fundamental of philosophical questions: "Is there any such thing as fuzziness independent or partially independent of us?" As well, "Is there fuzzy truth?" or, "Is there any absolute truth?" These questions about the existence of Reality as such are on what I refer to as the Ontological Level of Inquiry. It seems obvious that whether one answers yes or no to these questions will have profound implications for all other levels of the theory. Type of theories and science that we propose and construct in fact depend on whether we answer "yes" or "no" to these questions. In deed, if one answers yes to these questions, it is arguable that classical theories and science has to be re-assessed and must be rendered relevant on a new grounding. As such, this level is considered most fundamental or foundational. It is well known that Classical set and logic theorist's stance is "yes" that there is the absolute Truth and that there is a crisply defined Reality that exists independent of us. Whereas the stance that fuzzy theorist take is that there is no absolute truth and that there is a fuzzily defined Reality beginning with Zadeh's seminal paper (1965), i.e., that the Truth is a matter of degree and that the Reality is dependent on our perceptions (Zadeh, 1999).

Still within the realm of ontology at Level 2, a further, higher-level question then arises: "What is our position or relation to that Reality"? Are we originally separated or apart from it, or is it the very essence of our being *relational* in this respect? Some philosophical and scientific traditions take up stances very different from others on this still quite a fundamental level. If it is relational, then "What is the nature of that relation"?

The classical view is that our *relational* being to Reality is all or none. That is the elements of reality and their belonging to a set is "all or none". As well as the relation of these elements between sets is "all or none". The fuzzy view is that our *relational* being to reality is a matter of degree. That is the elements of reality in their belonging to a set as well as in their relation to each other between sets is a matter of degree, i.e., there are partial memberships in a set and partial degrees in participating in relation between sets. And the degree of truth of these membership values are also partial. This is compatible and in agreement with the level 1 stance that partial membership, partial participation in relations and partial truth are all perception based and expressed in our use of words and thus computing with words that capture imprecision in set membership and imprecision in combination of concepts via combination of sets.

5   General Epistemological Level

There are also two sub-levels here. Let us call them Level 3 and 4. Next level above the Ontological level is the General Epistemological Level of inquiry. A Level 3, the questions of general epistemology ask: "What is our access to truth or knowledge? Where is truth found in our paradigm? How or from what is it constituted?" These questions are addressed on this level in order to deal with the nature of human knowing in general. How do we acquire knowledge? Absolutely or partially? That is once we take a stance on description and/or verification, i.e., "Truth", being absolute or partial, then we have to explicitly state how we obtain it.

Based upon the stances adopted on Level 3 and still within the realm of epistemology will be questions of General Validity: "Given our General epistemological position on Level 3, "How do we *validate* our knowledge? How do we know it is true? What criteria do we use to assess its truth-value?" Again these questions are asked from the standpoint of the position and limits on human being or human knowledge in general.

6   Domain-Specific Epistemological Level

Here again there are two-sub-levels: Level 5 and Level 6.

The next major level of inquiry in the hierarchy is the Domain-Specific or Discipline-Specific Epistemological Level. At this Level 5, our questions take the form not of what human beings can know in general; but rather given our general epistemological and ontological position taken below on level 4. The more specific question is: "What can we know or hope to know or learn, within the discipline or setting (e.g., what can we know from within the experimental setting in terms of sampled training and testing data sets)?" Assertions on this level may include domain specific field experiments delimiting our statements, or positions (proposition) that attempt to define the proper limits or "horizons" to a given domain specific field.

As was the case with the General Epistemological Level, at Level 6 of the Domain-Specific Epistemological Level positions will also provide a basis for domain specific theories of validity, and thus of appropriate methodologies as well.

7   Application Level (Level 7)

Finally, it is only after implicitly or explicitly addressing all previous levels that we come to the Application Level of Inquiry proper. It is at this level of inquiry that our questions about system analysts and designers, managers, decision-makers, doctors, lawyers, engineers, etc., feel, behave, think, and interact, are raised, and our insights, theories, and different emphases are debated in our attempts to better understand them and support them.

Naturally, one might additionally propose various sub-levels within this complex hierarchy suggested in Table 1. But for the purpose of this section, the key point is that at the Application Level 7 issues should not be tackled in isolation from their philosophical underpinnings. As important as the theoretical, pragmatic and methodological controversies and different emphases that are within this level, changes of position that occur at lower levels of the hierarchy may even more profoundly shake the foundation of all that rest upon them. Such changes may be similar to a theoretical earthquake, which necessitates a great deal of subsequent rebuilding. In fact the introduction of the fuzzy set (Zadeh, 1965) has caused such a revolutionary (Grand) paradigm-shift (Khun, 1962) and subsequent rebuilding of fuzzy theory, approximate reasoning and computing with words.

8   Classical Vs Fuzzy Theory

Classical set and logic theory, at times known as Aristotelian theory, in contrast to Fuzzy set and logic theory, at times know as Zadehian theory, will be sketched out and articulated below in terms of the philosophical hierarchy discussed in the previous section. The purpose of this sketch at this juncture is *primarily methodological*; that is, it is presented more as a demonstration of the sort of analysis that may be facilitated by the hierarchical method described above in Section 2. It is not an exposition that reveals an exhaustive study of classical set and logic theory versus fuzzy set and logic theory. But it is believed that it forms a reasonable grounding.

Table 1. Hierarchy Of Levels Of Theoretical Inquiry And Their Questions.

APPLICATION LEVEL

7. How do people, decision-makers, feel, think, behave, and interact? How can we provide them with better decision-making tools?

DOMAIN-SPECIFIC EPISTEMOLOGICAL LEVEL

6. How do we validate knowledge appropriately in this domain specific field? What methodological approaches are appropriate to it?

5. What can we know or hope to learn within this domain-specific field or discipline? What are the limits or boundaries to it?

GENERAL EPISTEMOLOGICAL LEVEL

4. How do we validate our knowledge? How do we know it is true? What criteria do we use to assess its truth-value?

3. What is our access to truth and knowledge in general? Where is knowledge and its truth to be found? How or from what are they constituted?

ONTOLOGICAL LEVEL

2. What is our position or relation to that Reality (if we do assume that it exists on level 1 below)?

1. Is there any reality independent or partially independent of us? Does any absolute truth exist? Does fuzziness exists?

---

9   A Hierarchical Sketch of the Classical Set and Logic Theory Model

Classical theory is a well known *reductionist* theory that has helped us over more than two thousand years with many applications in explanations of physical and natural phenomena and electro-mechanical men made devices. In Table 2, we now give a sketch of classical theory.

Table 2. Positions Taken by Classical Set and Logic Theorists on the Hierarchy of Levels of Theoretical Inquiry.

Application Level	7. Emphasis on mechanistic Super additive systems theory of interactions, relations, equations, etc.
Domain Specific Epistemological Level	6. Validity and methodology dictated by meta-physical theories.e.g., principle of determinism and randomness. 5. Objective facts and truth accessible, but limited only by subjective distortions.
General Epistemological Level	4. Correspondence theory of Validity only accessible by Objective methods 3. Objectivist, empiricists
Ontological Level	2. sRo Cartesian dualism 1. Absolute Reality, crisp

On the Ontological Level, Level 1, of the proposed hierarchy, we must believe firmly in an *observer-independent* and absolute reality. The classical theory subscribes to the Cartesian "sRo" paradigm of describing our primary relation to that Reality on level 2. By the sRo paradigm, we refer to that ontological model that posits the subject, s, and the object (or world), o, as initially (at least in priciple) separate from each other but subsequently connected by some sort of relational, R, event. The prototypical relational event of this sort has been accepted as that of the primary cognitive act of "knowing"; i.e., subject and object come to be related through the subject's coming to *know* the object. The Cartesian split, between the subject and the object as well as between the mind and the body, appear to have been embraced by Classical thinkers throughout most of their theorizing. Such dualistic dichotomous thinking is well recognized and accepted by the defenders of the classical theory. Without reiterating all of their arguments here, Cartesian dualism, or sRo ontological model, inherit much of the terminology of the two-valued set and logic theory with its well known axioms that are exhibited in Table 3.

Briefly, every element belongs to a concept class, say A, either with full membership or none, i.e.,
$\mu_A: X \to \{0,1\}$, $\mu_A(x) = a \in \{0,1\}$, $x \in X$, where $\mu_A(x)$ is the membership assignment of an element $x \in X$ to a concept class A in a proposition. This is a *descriptive* assignment. Furthermore, this descriptive assignment, $D_{\{0,1\}}$, is verified or asserted to be absolutely True, T, or False, F, i.e., $\mu_V: \mu_A \to \{T,F\}$, where $V_{\{T,F\}}$, is the veristic assignment which is the atomic building block of two-valued logic expressions.

A particular consequence is the Law of Excluded Middle, $A \cup c(A)=X$ and Law of Contradiction $A \cap c(A)=\phi$. As a result of this, there exist the law of Conservation, i.e.,

$a \vee \bar{a} + a \wedge \bar{a} = 1$ and hence the "Principle of Invariance" in classical set and logic theory.

Table 3. Axioms of Classical & Set & Logic Theory, where A, B are crisp, two valued, sets and c(.) is the complement, X is the universal set and ϕ is the empty set.

Involution	$c(c(A)) = A$
Commutativity	$A \cup B = B \cup A$
	$A \cap B = B \cap A$
Associativity	$(A \cup B) \cup C = A \cup (B \cup C)$
	$(A \cap B) \cap C = A \cap (B \cap C)$
Distributivity	$A \cup (B \cap C) = (A \cup B) \cap (A \cup C)$
	$A \cap (B \cup C) = (A \cap B) \cup (A \cap C)$
Idempotence	$A \cup A = A$
	$A \cap A = A$
Absorption	$A \cup (A \cap B) = A$
	$A \cap (A \cup B) = A$
Absorption by X and ϕ	$A \cup X = X$
	$A \cap \phi = \phi$
Identity	$A \cup \phi = A$
	$A \cap X = A$
Law of contradiction	$A \cap c(A) = \phi$
Law of excluded middle	$A \cup c(A) = X$
De Morgan Laws	$c(A \cap B) = c(A) \cup c(B)$
	$c(A \cup B) = c(A) \cap c(B)$

On the Ontological Level, Levels 1 and 2, the acceptance of the descriptive assignment $D_{\{0,1\}}$ and the Veristic assignment, $V_{\{0,1\}}$ provide us the grounding for the formation of Two-valued "Truth Tables" and in turn for the derivation of the combination of concepts for any two crisp sets A and B to be "A AND B" = "$A \cap B$", "A OR B"="$A \cup B$" and "A IMP B" = $A \rightarrow B$ = $c(A) \cup B$, etc. This means that linguistic connectives "AND", "OR", "IMP", etc., are interpreted in a one-to-one correspondence, isomorphically to be equal to "$\cap$", "$\cup$", "$c(.)\cup$", respectively. That is the imprecise and varying meanings of linguistic connectives are precisitated in an absolute manner analogous to the absolute precisiation of the meaning of words interms of absolute crisp set representation. And all this is done for the sake of developing a "reductionist" theory of knowledge representation and reasoning with it!

On the General Epistemological Level 3, we have the foundation of the objectivism. Its stance is that real truth exists, is potentially accessible, and is to be found on the object side of the Cartesian split. The "subjective" elements, although understandable or symbolically meaningful, and of immense scientific interest, were nevertheless seen as basically unwanted distortions to be removed or eliminated (as much as possible) in order to get at the obscured "objective truth". On this General Epistemological Level, first at level 3, the observation based data are to be obtained from objects with the use of measurement agents such as sensors. Thus representations of objects are developed on measurement based data warehouses. This is done with the assumption of the fact that measurement based models so developed are to stand on the foundation developed on the Ontological level. They are consequently descriptive representations of model concepts that stand on two-valued set theory, i.e., $D_{\{0,1\}}$ which are verified with the two-valued logic theory, $V_{\{0,1\}}$.

Not surprisingly, on level 4 of the proposed hierarchy, dealing with General Theories of Validity, a "correspondence theory of truth" is accepted in most classical scientific investigations. According to this theory, a perception, or an observation, or a judgment is considered valid or true insofar as it may be shown to match with the objective or factual reality of the world around us. Hence on level 4, the "correspondence theory of truth" essentially is based on the two-valued set and truth framework, i.e., Description and Verification, $\{D_{\{0,1\}}, V_{\{0,1\}}\}$, which are accepted on the Ontological level. This means that models developed on the General Epistemological level are to be accepted as true depictions of real system behaviour. Furthermore, test data are to be used to validate results obtained from the model build on the level 3. Thus results are assumed to computationally determine as acceptable outcomes for given inputs of the test data.

Therefore on the General Epistemological level, we first have a model expressed in a general form as "A→B" as a descriptive model $D_{\{0,1\}}$ which is verified as $V_{\{0,1\}}$ within the frame work of a classical inference schema such as Modus Ponens which is stated as (A→B)oA = B such that the premise, $\{D_{\{0,1\}}, V_{\{0,1\}}\}$ for "A→B", combined with the premise $\{D_{\{0,1\}}, V_{\{0,1\}}\}$ for "A" results in a consequence which exactly matches the right hand side of the crisp rule, i.e., B, described and verified as $\{D_{\{0,1\}} V_{\{0,1\}}\}$, provided that the observation "A" matches exactly to the left hand side of the rule "A→B". The validation is based on a comparison of the actual output for a given test input data and model output for the same test input data. The error is usually accepted to be a true, $V_{\{0,1\}}$, verification based on a statistical risk which is to be a crisply evaluated assessment dependent on a crisp test of hypothesis.

On the Domain-Specific Epistemological level, the classical system development models that stand on observations which are measurement based data stored in data warehouses for a specific field of inquiry. Measurement data that is suppose to capture the system behaviour and its associated concepts that are represented with linguistic variables and linguistic terms are expressed in two-valued descriptive set theory $D_{\{0,1\}}$. Therefore system models so developed are assumed to be universally valid representations of domain specific object world that is captured by measurement based data only! That is, they are verified to be absolutely true, i.e., $V_{\{0,1\}}$, by measurements only. On level 5 of the hierarchy, dealing with the Domain-Specific Epistemology, for a particular investigation, the scientists appear to find limits to knowledge only on the *subjective side* of the Cartesian split. It is assumed that such subjective distortions may originate in the (system) analyst, but the "untruths" are all seen as coming from the s-or subject- side of the sRo split. The scientists seem to see few or no limits on the object side. That is, the object seems to be just what it is factually available; it is not ambiguous, or imprecise in itself, only in our misperceptions of it or in our measurement errors of it. It is interesting, to note, however, in the case of the

Domain-Specific Epistemology endorsed by classical thinking, such "objective facts of the world" (which by definition stand as true, regardless of our will, mood, or perspection on them) may include both external states in the world as well as certain "internal" (though nevertheless still "objective"), be abstract, states or realities. Classical science, articulated from the epistemological view-point, appears to have just discovered the internal abstract states, as it were, *out there* in the world. That is Classical science, in this perspective, does not accept that they have been created as concepts and have been expressed linguistically by subjects. Such internal, abstract, concepts are then considered to be universally valid features of the objective world. That is linguistic variables and their terms are assume to have crisp well defined meanings.

In terms of Level 6 of the proposed hierarchy, science deals with the question of validity and methodology in the domain specific field of study with reductionism and commitments to specific cause and effect hypothesis and metaphysical theories, e.g., drive a theory and investigate its derivatives. At times, this tended to arbitrarily restrict what was to be considered valid data and not within the particular domain of interest. For example, a good scientist of this persuasion should be actively searching for "derived derivatives" and attempt to find their origin in more primary theoretical precepts in accordance with the classical model.

On the Domain-Specific Epistemological Level, we find various developments of system models with application technologies known as statistical methods, such as multi-variate regression equations, programming methods such as linear and non-linear optimization algorithms or optimal control schemas developed on objective data that are obtained by measurement devises and depend on description and validation frameworks that are given as $\{D_{\{0,1\}}, V_{\{0,1\}}\}$.

As well the validation of the Domain-Specific models are assessed with domain-specific test data that are assumed to be standing on descriptive and verified framework of $\{D_{\{0,1\}}, V_{\{0,1\}}\}$. The validation of the domain specific models are executed with the classical inference schemas such as Modus Ponens as indicated above.

This may entail a re-computation of, say, regression, or programming or control models with test data. Results obtained from such models are assumed to be on $\{D_{\{0,1\}}, V_{\{0,1\}}\}$ framework based on some level of statistical risk.

This approach, however, may be seen by many to be quite problematic in that the theories and methods most properly belonging to the Application Level 7 appear here to be dictating what qualifies as good and valid data on the Domain Specific Epistemological Level 6. Such situation in which the position on lower levels of the hierarch are overtly dictated by higher ones is generally not recommended in this hierarchical approach.

Furthermore, the Application Level 7 theory determines or dictates what to be considered as valid knowledge in the field, the less likely it will be that anything not in conformance with the theory will ever be found, noticed, or admitted as evidence.

This in fact was the case for the rejection of fuzzy theory by orthodox, doctrinaire defenders of the classical theory. While such a hegemony of the Application Level 7 theory may appear to increase the coherence by guaranteeing a certain amount of conformity, it will also provide a basis for the systematic neglect of other potentially valid but contradictory observations such as the existence of gray between the black and white dichotomy.

10   A Hierarchical Sketch of Fuzzy Set and Two-Valued Logic Theory Model

Indeed several critical thinkers of the classical theory have argued that it restricts or reduces reality to be "objective"; and that classical theory leads inadvertently to many paradoxes in which the theory becomes over-structured, over-selective, and some-times even overshadows the real life data, e.g., Russell's paradox, Barber's paradox, Flackross paradox, etc. Such paradoxes ought to be demonstrations for the impor-tance of our being as conscious and aware as we can be of the philosophical presup-positions embodied in our application theories based on classical theory. This is par-ticularly so with respect to real-life experimental data, and thus the call of fuzzy theo-rists for more "experience-near" concepts and "expert-insight" approaches for the formation of fuzzy-expert system models. These models are either formed by expert interviews or by fuzzy data mining exercises more generally a combination of both.

The fuzzy theory is a non-reductionist theory that captures the gray information granules between black and white and helps us to cope with the complexity of our modern world in decision-making processes in a manner akin to human decision-making. In this theory, the paradoxes of two-valued classical theory are explained by admission of the gray information granules between black and white and hence al-lowing overlaps between classes and categories. In Table 4, we next give a sketch of the fuzzy theory in terms of the proposed hierarchy as we interpret them.

On the Ontological level, level 1, of the proposed hierarchy, we must believe firmly in an *observer-dependent and relative* reality that is perception based and communicated and computed with words and their numeric meaning representation with continuous memberships.

On level 2, the classical sRo Cartesian dualism is modified and extended to be "sR_1oR_2s" paradigm that describes a subject's, s's, primary relation to an object to be R_1 and in turn an object's, o's, relation to a subject, s, to be R_2. By the "sR_1oR_2s" paradigm, we refer to an ontological model that posits subject, s, and object, o, to be inter connected with relations R_1 and R_2. The prototypical relation R_1 is interpreted as the primary cognitive act of "knowing" based on perception of subjects articulated in a natural language, i.e., Computing With Words, CWW, followed by objects', o's, meaning representation in terms of continuous membership values that are next proc-essed by relation R_2 that generates a new cognitive interpretation in the act of "know-ing" based on meaning representation caused by relation R_2. Thus the Cartesian split between the subject and the object as well as mind and body are unified. Dualistic, dichotomous, thinking is discarded and overlap of categories are accepted. Consequently, "sR₁oR₂s" Ontological fuzzy theory terminology moves beyond the restrictions of two-valued set and logic theory and eliminates most of the axioms exhibited in Table 3. Instead it rests mainly on the limited set of axioms shown in Table 5 which are for t-norm and co-norm based structures. There are naturally further restrictions for pseudo t-norm and co-norm based structures which are not stated here.

Table 4. Position Taken by some of Fuzzy Set and Logic Theorists on the Hierarchy of Levels of Theoretical Inquiry

Application Level	7. Emphasis on humanistic non-linear systems theory of overlaps with fuzzy interactions, relations, equations, rules, etc.
Domain Specific Epistemological Level	6. Validity and methodology dictated by meta-linguistic theories of imprecision and uncertainty. 5. Subjective perception based facts and truth are accessible to capture imprecision and uncertainty.
General Epistemological Level	4. Correspondence theory of Validity with an integrated perspective of objective and subjective views. 3. Subbjectivist "experience-near", "expert-insight"
Ontological Level	2. sR₁oR₂s humanistic realism. 1. Relative Reality, imprecise and approximate defined by a continuous membership.

Table 5. Main Stream Axioms of General Fuzzy and Logic Theory

Involution	$c(c(A)) = A$
Commutativity	$A \cup B = B \cup A$ $A \cap B = B \cap A$
Associativity	$(A \cup B) \cup C = A \cup (B \cup C)$ $(A \cap B) \cap C = A \cap (B \cap C)$
Absorption by X and ∅	$A \cup X = X$ $A \cap \phi = \phi$
Identity	$A \cup \phi = A$ $A \cap X = A$
De Morgan Laws	$c(A \cap B) = c(A) \cup c(B)$ $c(A \cup B) = c(A) \cap c(B)$

Fundamental Assumption: Briefly, every element belongs to a concept class, say A, to a partial degree, i.e., μ_A: $X \to [0,1]$, $\mu_A(x)=a \in [0,1]$, $x \in X$, where $\mu_A(x)$ is the membership assignment of an element $x \in X$ to a concept class A in a proposition.

Corollary: Fuzzy set representations of concept classes can not be reduce to crisp sets; i.e., $\mu_A:X \to [0,1]$ is not reducable to $\mu_A:X \to \{0,1\}$.

Furthermore, the descriptive assignment $D_{[0,1]}$ is verified or asserted to be true, T, or false, F, i.e., $\mu_V:\mu_A \to \{T,F\}$ absolutely in *Descriptive fuzzy set theory*, where $V_{\{T,F\}}$, or $V_{\{0,1\}}$, is the veristic assignment which is the atomic building block of the two-valued logic. On the other hand, if the descriptive assignment $D_{\{0,1\}}$ or $D_{[0,1]}$ is verified or asserted to be partially true, i.e., $\mu_V:\mu_A \to [T,F]$ or $[0,1]$ in *Veristic fuzzy set theory* i.e., fuzzy set theory of truthood, which needs to be further verified or asserted to be absolutely True, T, or False, F, i.e., $\mu_{V'}$: $[\mu_V:\mu_A \to [0,1]] \to \{0,1\}$ where $V_{[0,1]}$ is a partial veristic truth assignment but $V'_{\{0,1\}}$ is a secondary absolute veristic assignment which is once again the atomic building block of the two-valued logic!

On the ontological level, levels 1 and 2, the acceptance of the descriptive assignments $D_{[0,1]}$ in *Descriptive fuzzy set theory* and the graded veristic assignment $V_{[0,1]}$ in *Veristic fuzzy set theory* with veristic assignment of $V_{\{0,1\}}$ for *Descriptive fuzzy set theory* and Veristic assignment of $V'_{\{0,1\}}$ for *Veristic fuzzy set theory* provide us the grounding for the formation of fuzzy-valued "Fuzzy Truth Tables" and in turn the derivation of the combination of concepts for any two fuzzy sets A and B, when they are represented by a Type 1 fuzzy sets, to be

"A AND B" = $\begin{cases} \text{FDCF(A AND B)} = A \cap B \\ \text{FCCF(A AND B)} = (A \cup B) \cap (c(A) \cup B) \cap (A \cup c(B)), \end{cases}$
and

"A OR B" = $\begin{cases} \text{FDCF(A OR B)} = (A \cap B) \cup (c(A) \cap B) \cup (A \cap c(B)) \\ \text{FCCF(A AND B)} = A \cup B, \end{cases}$
and

"A IMP B" = $\begin{cases} \text{FDCF(A IMP B)} = (A \cap B) \cup (c(A) \cap B) \cup (c(A) \cap c(B)) \\ \text{FCCF(A IMP B)} = c(A) \cup B, \end{cases}$

etc., in analogy to the two-valued set and logic theory where FDCF(.)=DNF(.) and FCCF(.)=CNF(.) *in form only*. Furthermore, as it is shown, (Türkşen, 1986-2002) FDCF(.)≠FCCF(.) and in particular we get FDCF(.)⊆FCCF(.) for certain classes of t-norms and t-conorms.

Particular consequences are that we receive:(1) FDCF(A OR NOT A)⊆FCCF(A OR NOT A) which is the realization of the law of **"Fuzzy Middle"** as opposed to the Law of Excluded Middle and (2) the Law of **"Fuzzy Contradiction"**, FDCF(A AND NOT A) ⊆FCCF(A AND NOT A) as opposed to the Law of Crisp Contradiction.

As a consequence, we obtain **"Fuzzy Laws of Conservation"** as:

$\mu[\text{FDCF}(A \text{ AND } c(A))] + \mu[\text{FCCF}(A \text{ OR } c(A))] = 1$; and

$\mu[\text{FDCF}(A \text{ OR } c(A))] + \mu[\text{FCCF}(A \text{ AND } c(A))] = 1$.

Hence we once again observe the "Principle of Invariance" re-established in Interval-Valued Type 2 fuzzy set theory, but as two distinct Laws of Conservation.

This means that linguistic connectives "AND", "OR", "IMP", etc., are *not* interpreted in a one-to-one correspondence, isomorphically, to be equal to "$\cap$", "$\cup$", "c(.)$\cup$", etc. That is the imprecise and varying meanings of linguistic connectives are not precisiated in an absolute manner and there is no absolute precisiation of the meaning of words nor is there an absolute precisiation of the meaning of connectives. This provides a framework for the representation of uncertainty in the combination of words and hence in reasoning with them. This particular interpretation and knowledge representation and reasoning forms a unique foundation for Type 2 fuzzy set theory in general and in particular for Interval-Valued Type 2 fuzzy set theory generated by the combination of linguistic concepts with linguistic connectives even if the initial meaning representation of words are to be reduced to Type 1 membership representation. More general representations start with Type 2 representation schema and then form Type 2 reasoning schemas to capture both imprecision and uncertainty.

On the General Epistemological Level 3, we have the foundation of an integrated subjectivist-objectivist perspective. Its stance is that real truth is relative and context dependent. It is potentially, partially and approximately accessible and it is to be found on the subject-object integrated interaction. On this level 3, the observation based data are obtained from subject-object interaction with perceptions as well as measurements. That is both human as well as electro-mechanical sensors provide data.

Membership functions can be obtained either with measurement theoretic experiments or by FCM based clustering techniques (Bilgic, 1995, Bilgic and Türkşen, 1995, Bezdek, 1981).

Thus representations of objects are developed on perceptions of humans and measurements of sensors for the use of human decision makers. This is done with the assumption of the fact that perception-measurement based models are developed to stand on the foundation proposed on the Ontological level, i.e., sR_1oR_2s paradigm. They are consequently descriptive representations of model concepts on fuzzy (infinite)-valued, i.e., $D_{[0,1]}$, sets which are verified with the two-valued logic theory, as $V_{\{0,1\}}$ for the *Descriptive fuzzy set theory*. On the other hand, descriptive propositions whether they be $D_{\{0,1\}}$ or $D_{[0,1]}$ if they are verified with fuzzy (infinite)-valued truthoods, as $V_{[0,1]}$ and then they are verified with the two-valued logic theory as $V'_{\{0,1\}}$ then we have a *Veristic fuzzy set theory*.

Next on the level 4, the correspondence theory of truth "is based on fuzzy valued sets, whether they be for *Descriptive fuzzy sets*, $D_{[0,1]}$, or *Veristic fuzzy sets*, $V_{[0,1]}$, paradigm and two-valued truth (verification) paradigm with either $V_{\{0,1\}}$ for *Descrip-*

tive fuzzy set paradigm or $V'_{\{0,1\}}$ for *Veristic fuzzy set paradigm* which are accepted on the Ontological Level. This means that models developed on the General Epistemological level are to be accepted as true (but approximate) depictions of a real system behaviour. Furthermore, test data are to be used to validate results obtained from the models build on the level 3. Thus results are assumed to be computationally but approximately determined as acceptable outcomes for given inputs of the test data,

Therefore on the General Epistemological level, we first have a model expressed in particular as Interval-Valued Type 2 fuzzy set as:

$$A\ IMP\ B = [\ FDCF(A\ IMP\ B),\ FCCF(A\ IMP\ B)\ ]$$

as a descriptive model, i.e., an Interval-Valued Type 2 rule, a premise. That is $\{\{D_{[0,1]}\ V_{\{0,1\}}\}\ IMP\ \{D_{[0,1]}\ V_{\{0,1\}}\}\}=\{D_{[0,1]}\ V_{\{0,1\}}\}$ which is within the framework of a fuzzy inference schema such as Generalized Modus Ponens, GMP, originally proposed by Zadeh as Compositional Rule of Inference, CRI, such that the first premise $\{D_{[0,1]}\ V_{\{0,1\}}\}$ for "A IMP B" combined with a second premise $\{D_{[0,1]}\ V_{\{0,1\}}\}$ for "A" result in a consequence $\{D_{[0,1]}\ V_{\{0,1\}}\}$ for B^*, where the fuzzy similarity of A' to A together with the t-norm and co-norm, that is chosen, result in B^*. The validation is based on a fuzzy comparison of the actual output for a given test input data and model output for the same test input data. The error is usually accepted to be a true, $V_{\{0,1\}}$, verification but based on a risk statistically but fuzzily evaluated assessment dependent on a fuzzy test of hypothesis. It should be noted that all of the proceeding exposition which is made for the *Descriptive fuzzy set paradigm*. A similar exposition is applicable to the *Veristic fuzzy set paradigm* as we have explained earlier!

On the Domain-Specific Epistemological Level, Level 5, we find various developments of system models with applications of technologies known as fuzzy statistical methods, such as fuzzy multi-variate regression equations, fuzzy linear and non-linear optimization algorithms, or fuzzy optimal control schemas developed on subjective-objective data that are obtained by expertise and measurement which are dependent on description and verification frameworks that are given as $\{D_{[0,1]}\ V_{\{0,1\}}\}$.

As well, the validation of the Domain-Specific models on Level 6 are assessed with domain-specific test data that are assumed to be standing on a descriptive and verified framework $\{D_{[0,1]}\ V_{\{0,1\}}\}$. The validation of the domain-specific models are executed with fuzzy inference schemas such as Generalized Modus Ponens as indicated above. They may entail re-computations of, say, fuzzy regression, or fuzzy programming, or fuzzy control models with, test data. Results obtained from such models are assumed to be on $\{D_{[0,1]}\ V_{\{0,1\}}\}$ framework for *descriptive fuzzy set models* based on some level of risk and on $\{V_{[0,1]}\ V'_{\{0,1\}}\}$ framework for *veristic fuzzy logic models*.

On Level 7 of the proposed hierarch, Application Level proper, we are all quite aware of the vast contributions made to fuzzy sets and systems field by Lotfi A. Zadeh as a foundation for numerous applications. The details of his numerous seminal ideas are well beyond the scope of this sketch. Let it suffice to point out here that his consistent emphasis on the foundations of fuzzy set and logic theories have provided a grounding for the Application Level 7 includes linguistic variables, their representation with fuzzy sets, i.e., their precisiation with membership functions, and reasoning with imprecise linguistic terms of linguistic variables that are precisiated with membership functions. This emphasis in turn has led to the notions of Computing With Words, CWW, and more recently Computing With Perceptions, CWP.

It is in these respects that many of the familiar revisions and alternatives to classical thinking, suggested by Black, Lukasiewicz, Kleene, etc., were preliminary break away strategies from the classical paradigm. With the grand paradigm shift caused by Zadeh's seminal work and continuous stream of visionary proposals, it is now clear that most of them reflect very different stances adopted at the more fundamental levels of our proposed hierarchy.

11 Conclusions

In order to expose the philosophical grounding of fuzzy theory we propose a hierarchical level of theoretical inquiry. We have shown that such a hierarchy exposes our stances with respect to crisp and fuzzy set and logic theories.

References

1. Bezdek, I.C.: Pattern Recognition with Fuzzy Objective Function Algorithms, Plenum Press, New York (1981)
2. Bilgic, T.: Measurement-Theoretic Frameworks for Fuzzy Set Theory with Applications to Preference Modeling, Ph.D. Thesis, University of Toronto, (supervisor, I.B. Türkşen) (1995)
3. Bilgic, T., Türkşen, I.B.: Measurement-Theoretic Justification of Connectives in Fuzzy Set Theory, Fuzzy Sets and Systems 76 (3), (1995) 289-308.
4. Resconi, G., Türkşen, I.B.: Canonical Forms of Fuzzy Truthoods by Meta-Theory Based Upon Modal Logic, Information Sciences, 131 (2001) 157-194
5. Türkşen, I.B.: Type 2 Representation and Reasoning for CWW, FSS, 127 (2002)17-36
6. Türkşen, I.B.: Type I and Type II Fuzzy System Modeling, FSS, 106, (1999)11-34
7. Türkşen, I.B.: Theories of Set and Logic with Crisp or Fuzzy Information Granules, J.of Advanced Computational Intelligence, 3, 4 (1999) 264-273
8. Türkşen, I.B.: Interval-Valued Fuzzy Sets and Compensatory AND, FSS, (1994) 87-100
9. Zadeh, L.A.: From Computing with Numbers to Computing With Words - From Manipulation of Measurements to Manipulation of Perceptions, IEEE, Trans. on Circuits and Systems, 45, (1999) 105-119
10. Zadeh, L.A.: Fuzzy Sets, Information and Control Systems, Vol.8, (1965) 338-353

Binary Operations on Fuzzy Sets:
Recent Advances

János Fodor*

Department of Biomathematics and Informatics
Faculty of Veterinary Science
Szent István University
István u. 2, H-1078 Budapest, Hungary
jfodor@univet.hu

Abstract. The main aim of this paper is to summarize recent advances on operations on fuzzy sets. First a detailed description of our present knowledge on left-continuous t-norms is presented. Then some new classes of associative operations (uninorms, nullnorms, t-operators) are reviewed. Finally we demonstrate the role of the evaluation scales (especially the case of totally ordered finite sets) in the choice of operations.

Keywords: associativity, left-continuous t-norm, nilpotent minimum, uninorm, nullnorm, t-operator, finite chain, unipolar and bipolar scale.

1 Introduction

When one considers fuzzy subsets of a universe, in order to generalize the Boolean set-theoretical operations like intersection, union and complement, it is quite natural to use *interpretations* of logic connectives $\wedge$, $\vee$ and $\neg$, respectively [30]. It is assumed that the conjunction $\wedge$ is interpreted by a *triangular norm* (*t-norm* for short), the disjunction $\vee$ is interpreted by a *triangular conorm* (shortly: *t-conorm*), and the negation $\neg$ by a *strong negation*.

Triangular norms and conorms are associative. Associativity of binary operations is a fundamental classical property, see [1] and further references there. With its help, a binary operation can be extended for more than two arguments in a unique way.

Nowadays it is needless to define t-norms and t-conorms in papers related to theoretical or practical aspects of fuzzy sets and logic: researchers have learned the basics and these notions have became part of their everyday scientific vocabulary. Nevertheless, from time to time it is necessary to summarize recent developments even in such a fundamental subject. This is the main aim of the present paper. Somewhat subjectively, we have selected topics where, on one hand, essential contributions have been made, and on the other hand, both theoreticians and practitioners may find it interesting and useful.

---

* Supported in part by FKFP 0051/2000 and by the Bilateral Scientific and Technological Cooperation Flanders–Hungary BIL00/51 (B-08/2000).

T. Bilgiç et al. (Eds.): IFSA 2003, LNAI 2715, pp. 16–29, 2003.

Therefore, we focus on three key issues. First, recent advances on left-continuous t-norms are summarized. The standard example of a left-continuous t-norm is the *nilpotent minimum* [41,11]. Starting from our more than ten years old algebraic ideas, their elegant geometric interpretations make it possible to understand more on *left-continuous t-norms with strong induced negations*, and construct a wide family of them. Studies on properties of fuzzy logics based on left-continuous t-norms have started recently. Secondly, we concentrate on associative operations that are more general than t-norms and t-conorms. These extensions are based on a flexible choice of the neutral (unit) element, or the null element of an associative operation. The resulted classes are known as *uninorms* and *nullnorms* (in other terminology: *t-operators*), respectively. Finally, we survey the role of the evaluation scales of membership values on the possible definitions (choices) of operations on fuzzy sets.

2   Left-Continuous Triangular Norms

Among numerous additional properties of t-norms, their continuity plays a key role. A t-norm T is *continuous* if for all convergent sequences $\{x_n\}_{n\in\mathbb{N}}$, $\{y_n\}_{n\in\mathbb{N}}$ we have

$$T\left(\lim_{n\to\infty} x_n, \lim_{n\to\infty} y_n\right) = \lim_{n\to\infty} T(x_n, y_n).$$

The structure of continuous t-norms is well known, see [30] for more details, especially Section 3.3 on *ordinal sums*.

In many cases, weaker forms of continuity are sufficient to consider. For t-norms, this property is *lower semicontinuity* [30, Section 1.3]. Since a t-norm T is non-decreasing and commutative, it is lower semicontinuous if and only if it is *left-continuous* in its first component. That is, if and only if for each $y \in [0,1]$ and for all non-decreasing sequences $\{x_n\}_{n\in\mathbb{N}}$ we have

$$\lim_{n\to\infty} T(x_n, y) = T\left(\lim_{n\to\infty} x_n, y\right).$$

If T is a left-continuous t-norm, the operation $I_T : [0,1]^2 \to [0,1]$ defined by

$$I_T(x,y) = \sup\{t \in [0,1] \mid T(x,t) \le y\} \tag{1}$$

is called the *residual implication* generated by T. An equivalent formulation of left-continuity of T is given by the following property ($x, y, z \in [0,1]$):

(R) $T(x,y) \le z$ if and only if $I_T(x,z) \ge y$.

We emphasize that the formula (1) can be computed for any t-norm T; however, the resulting operation I_T satisfies condition (R) if and only if the t-norm T is left-continuous. An interesting underlying algebraic structure of left-continuous t-norms is a commutative, residuated integral l-monoid, see [23] for more details.

The first known example of a left-continuous but non-continuous t-norm is the so-called *nilpotent minimum* [11] denoted as T^{nM} and defined by

$$T^{\mathrm{nM}}(x,y) = \begin{cases} 0 & \text{if } x+y \leq 1, \\ \min(x,y) & \text{otherwise.} \end{cases} \tag{2}$$

Note that the same operation appeared also in [41]. It can be understood as follows. We start from a t-norm (the minimum), and re-define its value below and along the diagonal $\{(x,y) \in [0,1] \mid x+y = 1\}$. So, the question is natural: if we consider a t-norm T and "annihilate" its original values below and along the mentioned diagonal, is the new operation always a t-norm? The general answer is "no" (although the contrary was "proved" in [41]). Therefore, the following problem was studied in [26].

Annihilation Let N be a strong negation (i.e., an involutive order reversing bijection of the closed unit interval). Let T be a t-norm. Define a binary operation $T_{(N)} : [0,1]^2 \to [0,1]$ as follows:

$$T_{(N)}(x,y) = \begin{cases} T(x,y) & \text{if } x > N(y) \\ 0 & \text{otherwise.} \end{cases} \tag{3}$$

We say that T *can be N-annihilated* when $T_{(N)}$ is also a t-norm. So, the question is: which t-norms can be N-annihilated?

A t-norm T is said to be a *trivial annihilation* (with respect to the strong negation N) if $N(x) = I_T(x,0)$ holds for all $x \in [0,1]$. It is easily seen that if a continuous t-norm T is a trivial annihilation then $T_{(N)} = T$.

Two t-norms T, T' are called N-*similar* if $T_{(N)} = T'_{(N)}$. Let T be a continuous non-Archimedean t-norm, and $\langle [a,b]; T_1 \rangle$ be a summand of T. We say that this summand is *in the center* (w.r.t. the strong negation N) if $a = N(b)$.

Theorem 1 ([26]). *(a) Let T be a continuous Archimedean t-norm. Then $T_{(N)}$ is a t-norm if and only if $T(x, N(x)) = 0$ holds for all $x \in [0,1]$.*

(b) Let T be a continuous non-Archimedean t-norm. Then $T_{(N)}$ is a t-norm if and only if

- *either T is N-similar to the minimum,*
- *or T is N-similar to a continuous t-norm which is defined by one trivial annihilation summand in the center.* □

Interestingly enough, the nilpotent minimum can be obtained as the limit of trivially annihilated continuous Archimedean t-norms, as the following result states.

Theorem 2 ([26]). *There exists a sequence of continuous Archimedean t-norms T_k ($k = 1, 2, \ldots$) such that*

$$\lim_{k \to \infty} T_k(x,y) = T^{\mathrm{nM}}(x,y) \qquad (x,y \in [0,1]).$$

Moreover, for all k, T_k is a trivial annihilation with respect to the standard negation.

The nilpotent minimum was slightly extended in [9] by allowing a weak negation instead of a strong one in the construction. Based on this extension, monoidal t-norm based logics (MTL) were studied also in [9], together with the involutive case (IMTL). Ordinal fuzzy logic, closely related to T^{nM}, and its application to preference modelling was considered in [6]. Properties and applications of the T^{nM}-based implication (called R_0 implication there) were published in [39]. Linked to [9], the equivalence of IMTL logic and NM logic (i.e., nilpotent minimum based logic) was established in [40].

Left-Continuous T-norms with Strong Induced Negations The following notions and some of the results were formulated in a slightly more general framework in [25]. We restrict ourselves to the case of left-continuous t-norms with strong induced negations; i.e., T is a left-continuous t-norm and the function $N_T(x) = I_T(x, 0)$ is a strong negation.

In [11] we proved the following result.

Theorem 3 ([11]). *Suppose that T is a t-norm such that (R) is satisfied, N is a strong negation. Then the following conditions are equivalent $(x, y, z \in [0, 1])$.*

(a) I_T has contrapositive symmetry with respect to N: $I_T(x, y) = I_T(N(y), N(x))$;
(b) $I_T(x, y) = N(T(x, N(y)))$;
(c) $T(x, y) \leq z$ if and only if $T(x, N(z)) \leq N(y)$.

In any of these cases we have

(d) $N(x) = I_T(x, 0)$,
(e) $T(x, y) = 0$ if and only if $x \leq N(y)$.   □

In [25], the same results were proved for operations T without having **(T4)** above. Moreover, in a sense, a converse statement was also established there: If T is a left-continuous t-norm such that $N_T(x) = I_T(x, 0)$ is a strong negation, then (a), (b) and (c) necessarily hold with $N = N_T$.

Already in [10], we studied the above algebraic property (c). Geometric interpretations of properties (b) and (c) were given in [25] under the names of *rotation invariance* and *self-quasi inverse property*, respectively. More exactly, we have the following definition.

Definition 1. Let $T : [0, 1]^2 \to [0, 1]$ be a symmetric and non-decreasing function, and let N be a strong negation. We say that T admits the *rotation invariance* property with respect to N if for all $x, y, z \in [0, 1]$ we have

$$T(x, y) \leq z \quad \text{if and only if} \quad T(y, N(z)) \leq N(x).$$

In addition, suppose T is left-continuous. We say that T admits the *self quasi-inverse* property w.r.t. N if for all $x, y, z \in [0, 1]$ we have

$$I_T(x, y) = z \quad \text{if and only if} \quad T(x, N(y)) = N(z). \quad □$$

For left-continuous t-norms, rotation invariance is exactly property (c) in Theorem 3, while self quasi-inverse property is just a slightly reformulated version of (b) there. Nevertheless, the following geometric interpretation was given in [25]. If N is a the standard negation and we consider the transformation $\sigma : [0,1]^3 \rightarrow [0,1]^3$ defined by $\sigma(x,y,z) = (y, N(z), N(x))$, then it can be understood as a rotation of the unit cube with angle of $2\pi/3$ around the line connecting the points $(0,0,1)$ and $(1,1,0)$. Thus, the formula $T(x,y) \leq z \iff T(y, N(z)) \leq N(x)$ expresses that the part of the unit cube above the graph of T remains invariant under σ. This is illustrated in the first part of the next figure.

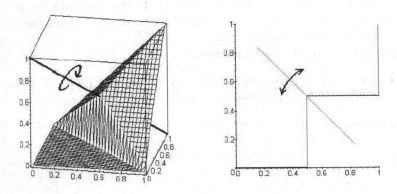

Fig. 1. Rotation invariance property (left).   Self quasi-inverse property (right).

The second part of this figure is about the self quasi-inverse property which can be described as follows (for quasi-inverses of decreasing functions see [42]). For a left-continuous t-norm T, we define a function $f_x : [0,1] \rightarrow [0,1]$ as follows: $f_x(y) = N_T(T(x,y))$. It was proved in [25] that f_x is its own quasi-inverse if and only if T admits the self quasi-inverse property. Assume that N is the standard negation. Then the geometric interpretation of the negation is the reflection of the graph with respect to the line $y = 1/2$. Then, if it is applied to the partial mapping $T(x,\cdot)$, extend discontinuities of $T(x,\cdot)$ with vertical line segments. Then the obtained graph is invariant under the reflection with respect to the diagonal $\{(x,y) \in [0,1] \mid x+y=1\}$ of the unit square.

Rotation Construction

Theorem 4 ([27]). *Let N be a strong negation, t its unique fixed point and T be a left-continuous t-norm without zero divisors. Let T_1 be the linear transformation of T into $[t,1]^2$. Let $I^+ = \ ]t,1]$, $I^- = [0,t]$, and define a function $T_{\mathbf{rot}} : [0,1]^2 \rightarrow [0,1]$ by*

$$T_{\text{rot}}(x,y) = \begin{cases} T(x,y) & \text{if } x,y \in I^+, \\ N(I_{T_1}(x,N(y))) & \text{if } x \in I^+ \text{ and } y \in I^-, \\ N(I_{T_1}(y,N(x))) & \text{if } x \in I^- \text{ and } y \in I^+, \\ 0 & \text{if } x,y \in I^-. \end{cases}$$

Then T_{rot} is a left-continuous t-norm, and its induced negation is N.

When we start from the standard negation, the construction works as follows: take any left-continuous t-norm without zero divisors, scale it down to the square $[1/2,1]^2$, and finally rotate it with angle of $2\pi/3$ in both directions around the line connecting the points $(0,0,1)$ and $(1,1,0)$. This is illustrated in Fig. 2.

Remark that there is another recent construction method of left-continuous t-norms (called rotation-annihilation) developed in [28].

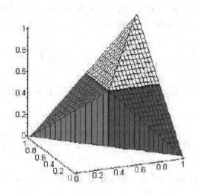

Fig. 2. T^{nM} as the rotation of the min, with the standard negation

3   New Classes of Associative Operators on the Unit Interval

In this section we give an overview of some extensions of t-norms and t-conorms on the unit interval. These extensions are based on a more general choice of either the neutral or the absorbing elements of the operations.

3.1   Uninorms

Uninorms were introduced in [43] as a generalization of t-norms and t-conorms. For uninorms, the neutral element is not forced to be either 0 or 1, but can be any value in the unit interval.

Definition 2 ([43]). A uninorm U is a commutative, associative and increasing binary operator with a neutral element $e \in [0, 1]$, i.e., for all $x \in [0, 1]$ we have $U(x, e) = x$. $\square$

It is interesting to notice that uninorms U with a neutral element in $]0, 1[$ are just those binary operators which make the structures $([0, 1], \sup, U)$ and $([0, 1], \inf, U)$ distributive semirings in the sense of Golan [18]. Further, in the theory of fuzzy measures and related integrals, uninorms play the role of pseudo-multiplication [36].

It is also known (see e.g. [21]) that in MYCIN-like expert systems *combining functions* are used to calculate the global degrees of suggested diagnoses. A careful study reveals that such combining functions are *representable uninorms* [5].

T-norms do not allow low values to be compensated by high values, while t-conorms do not allow high values to be compensated by low values. Uninorms may allow values separated by their neutral element to be aggregated in a compensating way. The structure of uninorms was studied by Fodor *et al.* [15].

For a uninorm U with neutral element $e \in]0, 1]$, the binary operator T_U defined by

$$T_U(x, y) = \frac{U(ex, ey)}{e}$$

is a t-norm; for a uninorm U with neutral element $e \in [0, 1[$, the binary operator S_U defined by

$$S_U(x, y) = \frac{U(e + (1 - e)x, e + (1 - e)y) - e}{1 - e}$$

is a t-conorm. The structure of a uninorm with neutral element $e \in]0, 1[$ (these are called *proper uninorms*) on the squares $[0, e]^2$ and $[e, 1]^2$ is therefore closely related to t-norms and t-conorms. For $e \in]0, 1[$, we denote by ϕ_e and ψ_e the linear transformations defined by $\phi_e(x) = \frac{x}{e}$ and $\psi_e(x) = \frac{x - e}{1 - e}$. To any uninorm U with neutral element $e \in]0, 1[$, there corresponds a t-norm T and a t-conorm S such that:

(i) for any $(x, y) \in [0, e]^2$: $U(x, y) = \phi_e^{-1}(T(\phi_e(x), \phi_e(y)))$;
(ii) for any $(x, y) \in [e, 1]^2$: $U(x, y) = \psi_e^{-1}(S(\psi_e(x), \psi_e(y)))$.

On the remaining part E of the unit square it satisfies

$$\min(x, y) \leq U(x, y) \leq \max(x, y),$$

and could therefore partially show a compensating behaviour, i.e., take values strictly between minimum and maximum. Note that any uninorm U is either *conjunctive*, i.e., $U(0, 1) = U(1, 0) = 0$, or *disjunctive*, i.e., $U(0, 1) = U(1, 0) = 1$.

Representable Uninorms In analogy to the representation of continuous Archimedean t-norms and t-conorms in terms of additive generators, Fodor *et al.* [15] have investigated the existence of uninorms with a similar representation in terms of a single-variable function. This search leads back to Dombi's class of *aggregative operators* [7]. This work is also closely related to that of Klement *et al.* on associative compensatory operators [29].

Consider $e \in]0,1[$ and a strictly increasing continuous $[0,1] \to \overline{\mathbb{R}}$ mapping h with $h(0) = -\infty$, $h(e) = 0$ and $h(1) = +\infty$. The binary operator U defined by

$$U(x,y) = h^{-1}(h(x) + h(y))$$

for any $(x,y) \in [0,1]^2 \setminus \{(0,1),(1,0)\}$, and either $U(0,1) = U(1,0) = 0$ or $U(0,1) = U(1,0) = 1$, is a uninorm with neutral element e. The class of uninorms that can be constructed in this way has been characterized [15]. Consider a uninorm U with neutral element $e \in]0,1[$, then there exists a strictly increasing continuous $[0,1] \to \overline{\mathbb{R}}$ mapping h with $h(0) = -\infty$, $h(e) = 0$ and $h(1) = +\infty$ such that

$$U(x,y) = h^{-1}(h(x) + h(y))$$

for any $(x,y) \in [0,1]^2 \setminus \{(0,1),(1,0)\}$ if and only if

(i) U is strictly increasing and continuous on $]0,1[^2$;
(ii) there exists an involutive negator N with fixpoint e such that

$$U(x,y) = N(U(N(x),N(y)))$$

for any $(x,y) \in [0,1]^2 \setminus \{(0,1),(1,0)\}$.

The uninorms characterized above are called *representable* uninorms. The mapping h is called an *additive generator* of U. The involutive negator corresponding to a representable uninorm U with additive generator h, as mentioned in condition (ii) above, is denoted N_U and is given by

$$N_U(x) = h^{-1}(-h(x)). \tag{4}$$

Clearly, any representable uninorm comes in a conjunctive and a disjunctive version, i.e., there always exist two representable uninorms that only differ in the points $(0,1)$ and $(1,0)$. Representable uninorms are almost continuous, i.e., continuous except in $(0,1)$ and $(1,0)$.

Continuous Uninorms on the Open Unit Square It is clear from [15] that a proper uninorm cannot be continuous on $[0,1]^2$. Therefore, Hu and Li [24] studied uninorms that are continuous on the *open* unit square $]0,1[^2$. Their results can be reinterpreted as follows.

Theorem 5. *A uninorm with neutral element $e \in]0,1[$ is continuous on $]0,1[^2$ if and only if one of the following two conditions is satisfied:*

(a) There exists $a \in [0,e[$ so that

$$U(x,y) = \begin{cases} U^*(x,y) & \text{if } x,y \in [a,1] \\ \min(x,y) & \text{otherwise,} \end{cases}$$

where U^* is a representable uninorm with neutral element $a + (1-a) \cdot e$.

(b) *There exists* $b \in]e, 1]$ *so that*

$$U(x,y) = \begin{cases} U^*(x,y) & \text{if } x,y \in [0,b] \\ \max(x,y) & \text{otherwise,} \end{cases}$$

where U^* *is a representable uninorm with neutral element* $b \cdot e$.   □

3.2 Nullnorms (T-operators)

In [2] we studied two functional equations for uninorms. One of those required to introduce a new family of associative binary operations on $[0,1]$ as follows.

Definition 3 ([2]). A *nullnorm* V is a commutative, associative and increasing binary operator with an absorbing element $a \in [0,1]$ (i.e., $V(x,a) = a$ for all $x \in [0,1]$), and that satisfies

- $V(x,0) = x$ for all $x \in [0,a]$,
- $V(y,1) = y$ for all $y \in [a,1]$.   □

On the other hand, in [4] the following notion was also defined, with the superfluous requirement of continuity [31].

Definition 4 ([4,31]). A *t-operator* is a two-place function $F : [0,1] \times [0,1] \rightarrow [0,1]$ which is associative, commutative, non-decreasing in each place and such that

- $F(0,0) = 0; F(1,1) = 1;$
- the sections $x \mapsto F(x,0)$ and $x \mapsto F(x,1)$ are continuous on $[0,1]$.   □

It turns out that Definitions 3 and 4 yield exactly the same operations; that is, t-operators coincide with nullnorms. When $a = 1$ we obtain t-conorms, while $a = 0$ gives back t-norms. The basic structure of nullnorms is similar to that of uninorms, as we state in the following theorem.

Theorem 6 ([2]). *A binary operation* V *on* $[0,1]$ *is a nullnorm with absorbing element* $a \in]0,1[$ *if and only if there exists a t-norm* T_V *and a t-conorm* S_V *such that*

$$U(x,y) = \begin{cases} a \cdot S_V(\frac{x}{a}, \frac{y}{a}) & \text{if } x,y \in [0,a], \\ a + (1-a) \cdot T_V\left(\frac{x-a}{1-a}, \frac{y-a}{1-a}\right) & \text{if } x,y \in [a,1], \\ a & \text{otherwise.} \end{cases}$$

Thus, nullnorms are generalizations of the well-known *median* operator. Further important properties of nullnorms (such as duality and self-duality, classification) can be found in [31].

Recall that classes of uninorms and t-operators satisfying some algebraic properties (such as modularity, distributivity, reversibility) as well as their structure on finite totally ordered scales were studied in [37,32,33,34].

4   Role of the Evaluation Scale

The usual evaluation scale in fuzzy sets and logics is a the closed unit interval, with its rich algebraic structure. When one considers a finite totally ordered scale instead, the picture is rather different.

Assume that $\mathcal{L} = \{x_0, x_1, \ldots, x_n, x_{n+1}\}$ is a totally ordered finite set of $n+2$ elements which are indexed increasingly, according to the asymmetric and negatively transitive relation $\prec$: $x_0 \prec x_1 \prec \ldots \prec x_n \prec x_{n+1}$. We use the notation $a = x_0$, $b = x_{n+1}$ in the sequel.

Now we briefly recall some results of [35] about counterparts of logical operations in $[0, 1]$.

First, a function $N : \mathcal{L} \to \mathcal{L}$ is a decreasing involution (called strong negation in case of $[0, 1]$) if and only if

$$N(x_i) = x_{n-i+1} \quad \text{for all } i \in \{0, 1, \ldots, n+1\}. \tag{5}$$

This unique involutive negation on $\mathcal{L}$ corresponds to the standard negation on the unit interval.

Second, Mayor and Torrens [35] determined all associative, commutative binary operations $T : \mathcal{L} \times \mathcal{L} \to \mathcal{L}$ which satisfy $T(b, b) = b$, and for all $x, y \in \mathcal{L}$

$$x \preceq y \quad \Longleftrightarrow \quad \text{there exists } z \in \mathcal{L} \text{ such that } x = T(y, z) \tag{6}$$

(note that $x \preceq y$ if and only if either $x \prec y$, or $x = y$).

We might call such a binary operation T a *smooth t-norm on* $\mathcal{L}$, for obvious reasons: it is not too difficult to justify that T satisfies all the four axioms of t-norms, and condition (6) is just equivalent with continuity of T when it is considered on $[0, 1]$.

Mayor and Torrens shown in [35] that the only Archimedean (i.e., for which $T(x, x) \prec x$ for all $x \in \mathcal{L}$) smooth t-norm on $\mathcal{L}$ is the following:

$$T(x_i, x_j) = \begin{cases} x_{i+j-(n+1)} & \text{if } i + j > n + 1, \\ x_0 & \text{otherwise} \end{cases}, \tag{7}$$

where $x_i, x_j \in \mathcal{L}$, $i, j \in \{0, 1, \ldots, n+1\}$. One can recognize that this T corresponds to the Łukasiewicz t-norm on $[0, 1]$. As a consequence, one essential difference between the finite case of $\mathcal{L}$ and the usual unit interval is that there is only one Archimedean smooth t-norm on a given $\mathcal{L}$, and it depends basically on the cardinality (i.e., the number of elements) of $\mathcal{L}$. No counterpart of any strict t-norm (like the product) on $[0, 1]$ exists on $\mathcal{L}$.

It can be proved that any smooth t-norm on $\mathcal{L}$ is the ordinal sum of such Archimedean t-norms. Moreover, there are exactly 2^n different smooth t-norms on $\mathcal{L}$ (where $|\mathcal{L}| = n + 2$) [35].

Smooth t-conorms can be obtained by duality with respect to the unique involution N in (5). Thus, the only Archimedean smooth t-conorm defined on $\mathcal{L}$ corresponds to the bounded sum well-known on $[0, 1]$.

As we mentioned before, equation (6) represents a condition which can be considered as a version of continuity of t-norms in $[0, 1]$. A drawback of (6) is that other types of binary operations (for example, t-conorms) on $\mathcal{L}$ need different definition.

Another concept of smoothness was introduced in [17]. A binary operation $M : \mathcal{L} \times \mathcal{L} \to \mathcal{L}$ may be considered 1-*smooth* if

$$M(x_i, x_j) = x_r,\ M(x_{i-1}, x_j) = x_p,\ M(x_i, x_{j-1}) = x_q \text{ imply } r - 1 \le p, q \le r. \tag{8}$$

It is really a kind of smoothness: if we move only one step (up or down) from x_i or x_j, while the other argument is unchanged, then the associated function value can move (up or down) at most one step too.

We proved in [13] that the class of associative, commutative operations T with $T(b, b) = b$ which satisfy condition (8) (i.e., which are 1-smooth) is wider than the class of smooth t-norms.

We also proved (see [13, Theorem 2]) that the condition (8) is equivalent to the well known *intermediate-value theorem*: any value x_ℓ between two function values $M(x_i, x_k)$ and $M(x_j, x_k)$ is attained at some point x_m between x_i and x_j. In addition, we determined the general form of smooth, nondecreasing and associative binary operations M on $\mathcal{L}$ satisfying also $M(a, a) = a$ and $M(b, b) = b$, given in the following theorem.

Theorem 7 ([13]). *Consider a function $M : \mathcal{L} \times \mathcal{L} \to \mathcal{L}$ with properties $M(a, a) = a$, $M(b, b) = b$. Let $\lambda = M(b, a)$, $\mu = M(a, b)$, and suppose $\lambda \preceq \mu$. If M is associative, increasing and smooth then M is of the following form:*

$$M(x, y) = \begin{cases} S_{\langle a, \lambda \rangle}(x, y) & \text{if } x, y \in \langle a, \lambda \rangle, \\ \lambda & \text{if } x \in \langle \lambda, b \rangle,\ y \in \langle a, \lambda \rangle, \\ y & \text{if } x \in \mathcal{L},\ y \in \langle \lambda, \mu \rangle, \\ \mu & \text{if } x \in \langle a, \mu \rangle,\ y \in \langle \mu, b \rangle, \\ T_{\langle \mu, b \rangle}(x, y) & \text{if } x, y \in \langle \mu, b \rangle \end{cases} \tag{9}$$

where $S_{\langle a, \lambda \rangle}$ is a smooth t-conorm on $\langle a, \lambda \rangle$ and $T_{\langle \mu, b \rangle}$ is a smooth t-norm on $\langle \mu, b \rangle$. □

The typical form of the operator M is given in Fig. 3, for sake of simplicity, on the unit square.

Denote $\mathcal{C}_{\langle a, b \rangle}$ the class of all associative, increasing, smooth binary operations on $\mathcal{L} = \langle a, b \rangle$ having a and b as idempotent elements (in other words: $\mathcal{C}_{\langle a, b \rangle}$ is the class of functions having form (9)). The following subclasses of $\mathcal{C}_{\langle a, b \rangle}$ were identified in [13]. Suppose $M \in \mathcal{C}_{\langle a, b \rangle}$.

1. If M is commutative then it is a nullnorm [2] on $\mathcal{L}$.
2. If M is idempotent then $M(x, y) = (\lambda \wedge x) \vee (\mu \wedge y) \vee (x \wedge y)$ $(x, y \in L)$. This operation was introduced and characterized in [12] as the noncommutative extension of the median (see [16]).
3. If M has an absorbing element then it is a nullnorm on $\mathcal{L}$.
4. If M has a neutral element then M is either a smooth t-norm or a smooth t-conorm (i.e., the neutral element is an extreme element of the scale).

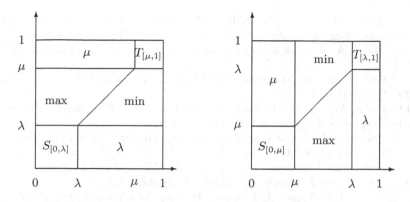

Fig. 3. M in case of $\lambda \preceq \mu$ (left) and in case of $\mu \preceq \lambda$ (right)

Recently, the use of unipolar and bipolar scales, and their effect on the definition of associative operations on such scales were also studied in the literature. The interested reader can find important results in [19,8,20].

5   Conclusion

Acknowledgement

The author is grateful to S. Jenei for placing Figures 1–2 at his disposal.

References

1. J. Aczél, *Lectures on Functional Equations and their Applications*, (Academic Press, New York, 1966).
2. T. Calvo, B. De Baets, J. Fodor, The functional equations of Frank and Alsina for uninorms and nullnorms, *Fuzzy Sets and Systems* **120** (2001) 385–394.
3. T. Calvo, G. Mayor, R. Mesiar, Eds. *Aggregation Operators: New Trends and Applications*, (Studies in Fuzziness and Soft Computing. Vol. 97, Physica-Verlag, Heidelberg, 2002).
4. T. Calvo, A. Fraile and G. Mayor, Algunes consideracions sobre connectius generalitzats, *Actes del VI Congrés Català de Lògica* (Barcelona, 1986), 45–46.
5. B. De Baets and J. Fodor, Van Melle's combining function in MYCIN is a representable uninorm: An alternative proof. *Fuzzy Sets and Systems* **104** (1999) 133–136.
6. B. De Baets, F. Esteva, J. Fodor and L. Godo, Systems of ordinal fuzzy logic with application to preference modelling, *Fuzzy Sets and Systems* **124** (2001) 353–359.
7. J. Dombi, Basic concepts for the theory of evaluation: the aggregative operator, *European J. Oper. Res.* **10** (1982), 282–293.
8. D. Dubois and H. Prade, On the use of aggregation operations in information fusion processes, *Fuzzy Sets and Systems* (to appear, 2003).

9. F. Esteva and L. Godo, Monoidal t-norm based logic: towards a logic for left-continuous t-norms, *Fuzzy Sets and Systems* **124** (2001) 271–288.
10. J.C. Fodor, A new look at fuzzy connectives, *Fuzzy Sets and Systems* **57** (1993) 141–148.
11. J.C. Fodor, Contrapositive symmetry of fuzzy implications, *Fuzzy Sets and Systems* **69** (1995) 141–156.
12. J.C. Fodor, An extension of Fung–Fu's theorem, *Int. J. Uncertainty, Fuzziness and Knowledge-Based Systems* **4** (1996) 235–243.
13. J. Fodor, Smooth associative operations on finite ordinal scales, *IEEE Transactions on Fuzzy Systems* **8** (2000) 791–795.
14. J.C. Fodor and M. Roubens, *Fuzzy Preference Modelling and Multicriteria Decision Support*, (Kluwer Academic Publishers, Dordrecht, 1994).
15. J.C. Fodor, R.R. Yager, A. Rybalov, Structure of uninorms, *Int. J. Uncertainty Fuzziness Knowledge-based Systems* **5** (1997) 411–427.
16. L.W. Fung and K.S. Fu, An axiomatic approach to rational decision making in a fuzzy environment, in: L.A. Zadeh *et al.*, Eds., *Fuzzy Sets and Their Applications to Cognitive and Decision Processes* (Academic Press, New York, 1975) pp. 227–256.
17. L. Godó and C. Sierra, A new approach to connective generation in the framework of expert systems using fuzzy logic, in: *Proc. of XVIIIth International Symposium on Multiple-Valued Logic*, Palma de Mallorca, Computer Society Press, Washington D.C. (1988), 157–162.
18. J. Golan, *The Theory of Semirings with Applications in Mathematics and Theoretical Computer Science* (Pitman Monographs and Surveys in Pure and Applied Mathematics, Vol. 54), (Longman Scientific and Technical, 1992).
19. M. Grabisch, Symmetric and asymmetric fuzzy integrals: the ordinal case, In: *Proc. 6th International Conference on Soft Computing* (Iizuka, Japan, October 2000).
20. M. Grabisch, B. De Baets and J. Fodor, The quest for rings on bipolar scales (submitted, 2003).
21. P. Hájek, T. Havránek and R. Jiroušek, *Uncertain Information Processing in Expert Systems* (CRC Press, 1992).
22. P. Hájek, *Metamathematics of Fuzzy Logic*, (Kluwer Academic Publishers, Dordrecht, 1998).
23. U. Höhle, Commutative, residuated l-monoids, in: U. Höhle and E.P. Klement, Eds., *Non-Classical Logics and their Applications to Fuzzy Subsets. A Handbook of the Mathematical Foundations of Fuzzy Set Theory.* (Kluwer Academic Publishers, Boston, 1995).
24. S.-k. Hu and Z.-f. Li, The structure of continuous uni-norms, *Fuzzy Sets and Systems* **124** (2001) 43–52.
25. S. Jenei, Geometry of left-continuous t-norms with strong induced negations, *Belg. J. Oper. Res. Statist. Comput. Sci.* **38** (1998) 5–16.
26. S. Jenei, New family of triangular norms via contrapositive symmetrization of residuated implications, *Fuzzy Sets and Systems* **110** (2000) 157–174.
27. S. Jenei, Structure of left-continuous t-norms with strong induced negations. (I) Rotation construction, *J. Appl. Non-Classical Logics* **10** (2000) 83–92.
28. S. Jenei, Structure of left-continuous triangular norms with strong induced negations. (II) Rotation-annihilation construction, *J. Appl. Non-Classical Logics*, **11** (2001) 351–366.
29. E.-P. Klement, R. Mesiar and E. Pap, On the relationship of associative compensatory operators to triangular norms and conorms, *Internat. J. Uncertain. Fuzziness Knowledge-Based Systems* **4** (1996), 129–144.

30. E.P. Klement, R. Mesiar, and E. Pap, *Triangular Norms*, (Kluwer Academic Publishers, Dordrecht, 2000).
31. M. Mas, G. Mayor, J. Torrens, t-operators, *Internat. J. Uncertainty Fuzziness Knowledge-Based Systems* **7** (1999) 31–50.
32. M. Mas, G. Mayor, J. Torrens, t-operators and uninorms in a finite totally ordered set, *Internat. J. Intell. Systems* **14** (1999) 909–922.
33. M. Mas, G. Mayor, J. Torrens, The modularity condition for uninorms and t-operators, *Fuzzy Sets and Systems* **126** (2002) 207–218.
34. M. Mas, G. Mayor, J. Torrens, The distributivity condition for uninorms and t-operators, *Fuzzy Sets and Systems* **128** (2002) 209–225.
35. G. Mayor and J. Torrens, On a class of operators for expert systems, *Int. J. of Intelligent Systems* **8** (1993) 771–778.
36. R. Mesiar, Choquet-like integrals, *J. Math. Anal. Appl.* **194** (1995) 477–488.
37. M. Monserrat and J. Torrens, On the reversibility of uninorms and t-operators, *Fuzzy Sets and Systems* **131** (2002) 303–314.
38. S. Ovchinnikov, On some bisymmetric functions on closed intervals, in: A. Sobrino and S. Barro (Eds.), *Estudios de lógica borrosa y sus aplicaciones* (Universidade de Santiago de Compostella, 1993), p. 353–364.
39. D. Pei, R_0 implication: characteristics and applications, *Fuzzy Sets and Sysytems* **131** (2002) 297–302.
40. D. Pei, On equivalent forms of fuzzy logic systems NM and IMTL, *Fuzzy Sets and Systems* (to appear, 2003).
41. P. Perny, *Modélisation, agrégation et exploitation des préférences floues dans une problématique de rangement*, (PhD thesis, Université Paris-Dauphine, Paris, 1992).
42. B. Schweizer and A. Sklar, *Probabilistic Metric Spaces*, (North-Holland, New York, 1983).
43. R.R. Yager, A. Rybalov, Uninorm aggregation operators, *Fuzzy Sets and Systems* **80** (1996) 111–120.
44. L.A. Zadeh, Fuzzy sets, *Inform. Control* **8** (1965) 338–353.

Multiple Criteria Choice, Ranking, and Sorting in the Presence of Ordinal Data and Interactive Points of View

Marc Roubens

University of Liege, Department of Mathematics,
12, Grande Traverse (B37),
B-4000 Liege, Belgium
M.Roubens@ulg.ac.be

Abstract. In this survey we use the discrete Choquet integral as a basic tool to solve ordinal multiple-attribute decision problems in the presence of interacting criteria . We consider ranking (all the alternatives from the decision set are ranked from the best one to the worst one), sorting (each alternative is assigned to one of the pre determined, ordered and disjoint classes) and choice (some alternatives are considered as the best ones or the worst ones).

1   Introduction

For each point of view, $i \in N$ every potential alternative from a finite set $A = \{\ldots, x, \ldots, y, \ldots\}$ is evaluated by a single decision-maker according to an ordinal performance scale defined on a totally ordered set with order $\preceq_i$. In other words, each alternative is completely identified by a vector x containing its partial evaluations x_i . Moreover the orderings $\preceq_i$ can be characterized by binary relations R_i, such that $R_i(x, y)$ equals 1 if $x_i \succeq_i y_i$ and 0 otherwise. Two different procedures may be used to merge the information included in the Boolean matrices R_i (see [3], [5])

(i) the pre-ranking methods : scores are first determined for each alternative $x \in A$ and for every point of view $i : S_i(x)$. These scores are aggregated to produce a global score $S(x)$. This way is followed to solve the sorting and ranking problems.

(ii) the pre-aggregation methods : a global valued binary relation R on A that aggregates the R_i's is first computed and in a second step either a a global score is calculated, either a choice function is proposed (as in game theory approach) taking into consideration some rationality concepts. This last way is used to propose a good or a bad choice to the decision-maker.

In the sorting and ranking approach, binary valued partial net scores $S_i(x)$ are computed. These scores are normalized in a meaningful way with the use of a discrete Choquet integral to produce a global score $C_v(x)$ for every alternative of

T. Bilgiç et al. (Eds.): IFSA 2003, LNAI 2715, pp. 30–38, 2003.
© Springer-Verlag Berlin Heidelberg 2003

the decision set A. The aggregator $\mathcal{C}_v$ depends on 2^{n-1} capacities $v(S), S \subset N$. These capacities cannot be directly evaluated but are tuned with the use of some prototypic elements that are ranked or sorted by the decision-maker.

In the choice approach, the $R_i(x,y)$ valuations are aggregated with a Choquet integral to obtain a global valued binary relation $R(x,y)$. Good and bad choices correspond to dominant and absorbent kernels in a valued digraph with vertex set A and arc family R. An efficient algorithm for computing these choices is proposed. Furthermore, formal correspondence between ordinal valued choices and kernels in ordinal valued digraphs is emphasized.

2 Ordinal Multiple-Attribute Sorting

2.1 Definition of the Sorting Problem

Let A be a set of potential alternatives, which are to be assigned to disjoint classes, and let $N = \{1, \ldots, n\}$ be a label set of points of view to satisfy. For each point of view $i \in N$, the alternatives are evaluated according to a s_i-point ordinal performance scale; that is a totally ordered set

$$X_i := \{g_1^i \prec_i g_2^i \prec_i \ldots \prec_i g_{s_i}^i\}.$$

We assume that each alternative $x \in A$ can be identified with the corresponding profile

$$(x_1, \ldots, x_n) \in \times_{i=1}^n X_i =: X,$$

where, for any $i \in N$, x_i represents the partial evaluation of x related to point of view i. In other words, each alternative is completely determined from its partial evaluations.

Now consider a partition of X into m non empty classes $\{\mathrm{Cl}_t\}_{t=1}^m$, which are increasingly ordered; that is, for any $r, s \in \{1, \ldots, m\}$, with $r > s$, the elements of Cl_r have a better comprehensive evaluation than the elements of Cl_s.

The sorting problem we actually face consists in partitioning the elements of A into the classes $\{\mathrm{Cl}_t\}_{t=1}^m$.

2.2 Normalized Scores as Criteria

For each point of view $i \in N$, the order $\preceq_i$ defined on X_i can be characterized by a valuation $R_i : A \times A \to \{0,1\}$ such that $R_i(x,y)$ is equal to 1, if $x_i \succeq_i y_i$ and to 0, otherwise. From each of these evaluations we determine a partial *net score* $S_i : A \to \mathbb{R}$ as follows:

$$S_i(x) := \sum_{y \in A} \{R_i(x,y) - R_i(y,x)\}, \quad (x \in A).$$

The integers $S_i(x)$ represent the number of times that x is preferred to any other alternative minus the number of times that any other alternative is preferred to x for point of view i. One can easily show that the partial net scores identify the corresponding partial evaluations; that is,

$$x_i \succeq_i y_i \Leftrightarrow S_i(x) \geq S_i(y).$$

This later aggregation makes sense since, contrary to the partial evaluations, the partial scores are *commensurable*; each partial score can be compared with any other partial score, even along a different point of view. The partial scores are defined according to the same interval scale. Positive linear transformations are meaningful with respect to such a scale (see [13]) and we can normalize these scores so that they range into the unit interval. We thus define *normalized partial scores* $S_1^N, \ldots, S_n^N$. In the sequel, we use the notation $S^N(x) = (S_1^N(x), \ldots, S_n^N(x))$.

2.3 Choquet Integral as a Discriminant Function

We aggregate the components of vector $S^N(x)$ by means of a Choquet integral (see [2], [7], [10], [11]) that allows to deal with interacting (dependant) points of view. We define a *global net score*

$$\mathcal{C}_v(S^N(x)) := \sum_{i=1}^{n} S_{(i)}^N(x)\{v(A_{(i)}) - v(A_{(i+1)})\}$$

where v represents a fuzzy measure on N; that is a monotone set function $v :$ $2^N \Rightarrow [0,1]$ fulfilling $v(\emptyset) = 0$ and $v(N) = 1$. The parentheses used for indices represent a permutation on N such that $S_{(1)}^N(x) \leq \cdots \leq S_{(n)}^N(x)$, and $A_{(i)}$ represents the subset $\{(i), \ldots, (n)\}$.

We observe that for additive measures $(v(S \cup T) = v(S) + v(T)$ whenever $S \cap T = \emptyset)$ the Choquet integral coincides with the usual weighted sum

$$\mathcal{C}_v(S^N(x)) = \sum_{i=1}^{n} v(i)S_i^N(x)$$

which is the natural extension of the *Borda score* as defined in voting theory.

If points of views cannot be considered as being independent, the importance of combinations $v(S)$, $S \in N$, has to be taken into account. The Choquet aggregator presents usual desirable properties. It is continuous, non-decreasing, located between min and max and its characterization ([9], [14]) clearly justifies the way the partial scores have been aggregated.

2.4 Assessment of Fuzzy Measures

Practically, the decision maker is asked to provide a set of prototypes $P \subseteq A$ and the assignment of each of these prototypes to a given class; that is a partition of P into prototypic classes $\{P_t\}_{t=1}^m$, where $P_t := P \cap \mathrm{Cl}_t$ for $t \in \{1, \ldots, m\}$. This information allows us to determine the fuzzy measure that is linked to any subset $S \subset N$. The global scores related to the prototypes determine intervals that are used to specify the limits (thresholds) of the m classes $\{\mathrm{Cl}_t\}_{t=1}^m$. Global scores are then calculated for all non-prototypic elements that are assigned to the given graded classes.

The basic condition that is used to strictly separate the classes Cl_t is the following

$$\mathcal{C}_v(S^N(x)) - \mathcal{C}_v(S^N(x')) \geq \varepsilon, \tag{1}$$

for each ordered pair of prototypic elements $(x, x') \in P_t \times P_{t-1}$ and each $t \in \{2, \ldots, m\}$, where ε is a given strictly positive threshold.

Due to the increasing monotonicity of the Choquet integral, the number of separation constraints is usually drastically reduced. On the basis of orders $\succeq_i$, $i \in N$, we can define a dominance relation D (partial order) on X as follows :

$$xDy \quad \text{iff} \quad x_i \succeq_i y_i, \text{ for all } i \in N.$$

It is then useful to determine, for each $t \in \{1, \ldots, m\}$, the set of non-dominating alternatives of P_t :

$$Nd_t := \{x \in P_t \text{ such that } \not\exists\, x' \in P_t \setminus \{x\} : xDx'\},$$

and the set of non-dominated alternatives of P_t :

$$ND_t := \{x \in P_t \text{ such that } \not\exists\, x' \in P_t \setminus \{x\} : x'Dx\},$$

and to consider constraint (1) only for each ordered pair $(x, x') \in Nd_t \times ND_{t-1}$ and each $t \in \{2, \ldots, m\}$.

The separation conditions (1) restricted to the prototypes of subset $Nd_t \cup ND_t$, $t \in \{1, \ldots, m\}$, put together with the boundary and monotonicity constraints on the fuzzy measure, form a linear program ([11]) whose unknowns are the capacities $v(S)$, $S \subset N$, that determine the fuzzy measure and where ε is a non-negative variable to be maximized in order to deliver well separated classes.

If there exists a k-additive fuzzy measure v^* (the Möbius transform m^* of v^* satisfies $m^*(S) = 0$ for S such that $\sharp S \succ k$, see [6]), k being kept as low as possible to respect the principle of parsimony, then any alternative $x \in A$ can be classified in the following way

- x is assigned to class Cl_t if $z_t \leq \mathcal{C}_{v^*}(S^N(x)) \leq Z_t$,
- x is assigned to class $\mathrm{Cl}_t \cup \mathrm{Cl}_{t-1}$ if $Z_{t-1} < \mathcal{C}_{v^*}(S^N(x)) < z_t$

where

$$z(t) := \min_{x \in Nd_t} \mathcal{C}_{v^*}(S^N(x)), \ Z(t) := \max_{x \in ND_t} \mathcal{C}_{v^*}(S^N(x)). \tag{2}$$

All the prototypic elements from P are well classified in a single class.

It may happen that the prototypic elements violate the axioms that are imposed to produce a discriminant function of Choquet type (see, for example,[9], [15]). In such a case we consider the following quadratic program

$$\min_{x \in \bigcup_{t\in\{1,\dots m\}} \{ND_t \cup ND_t\}} E(v,y) = \sum_x \{\mathcal{C}_v(S^N(x)) - y(x)\}^2$$

where the unknowns are

- the capacities $(v(S), S \subset N)$ – associated to boundary and monotonicity constraints – that determine the fuzzy measures,
- the components of vector y constrained by the global evaluation (classification) imposed by the decision-maker, i.e. $y(x) - y(x') > 0$ for every ordered pair $(x, x') \in Nd_t \times ND_{t-1}$, $t \in \{2, \dots, m\}$.

The intervals $I_t := \{z_t, Z_t\}$ defined according to (2) determine an interval order where $I_k \cap I_1 = \emptyset$, $k \neq 1$, is not necessarily verified and where the property to provide a semiorder $(Z_m \geq Z_{m-1} \geq \dots; z_m \geq z_{m-1} \geq \dots)$ might be violated.

The prototypic elements that induce violation of semiorder property are deleted from the set P and are considered as being *ill classified* with respect to the Choquet methodology and the remaining elements belonging to A are allocated either to well defined classes $(a \in \text{Cl}_k$, for some $k)$ either to the union of contiguous classes $(a \in \text{Cl}_k \cup \text{Cl}_{k+1} \cup \dots)$, that means classified with some ambiguity.

The prototypic elements are either ill-classified, either well defined and correctly or incorrectly classified, either ambiguously defined and correctly or incorrectly classified. The quality of the sorting based on piecewize linear discriminant functions (the Choquet integrals) depends on the distribution of the prototypes among these different classes.

This supervised classification has been implemented as a software tool called TOMASO (Technique for Ordinal Multiple criteria decision Aid in Sorting and Ordering) that can be freely downloaded on http://cassandra.ro.math.ulg.ac.be.

3   Ordinal Multiple-Attribute Ranking

In some situations, the prototypic elements can be ranked, i.e. a total ordering $\succ$ is provided by the decision maker on the elements belonging to P. All the alternatives of P are decomposed into disjoint classes reduced to single elements, where m corresponds to the total number of prototypes.

The methodology provided in section 2.4 applies in a straightforward way and the elements of P are either correctly classified (there exists a solution to the linear program) either the quadratic program produces some ill-classified items.

4   Choice Procedure

We now consider a binary relation R whose credibility is evaluated as follows :

$$R(x,y) = C_v[R_1(x,y), \ldots, R_i(x,y), \ldots, R_n(x,y)] \in [0,1], \text{ for all } x, y \in A.$$

In the sequel, we will only use the ordering of $R(x,y)$ and not their cardinality and we obtain a L-valued binary relation R (see [1]).

For all $x, y \in A$, $R(x,y)$ belongs to a finite set $L : \{c_0 = 0, c_1, \ldots, c_m = .5, \ldots, c_{2m} = 1\}$ that constitutes a $(2m + 1)$-element chain $c_0 \prec c_1 \prec \cdots \prec c_{2m}$. $R(x,y)$ may be understood as the level of credibility that "a is at least as good as b". The set L is built using the values of R taking into consideration an antitone unary contradiction operation $\neg$ such that $\neg c_i = c_{(2m-i)}$ for $i = 0, \ldots, 2m$.

If $R(x,y)$ is one of the elements of L, then automatically $\neg R(x,y)$ belongs to L. We call such a relation an L-valued binary relation.

We denote $L^{\succ m} : \{c_{m+1}, \ldots, c_{2m}\}$ and $L^{\prec m} : \{c_0, \ldots, c_{m-1}\}$.

If $R(x,y) \in L^{\succ m}$, we say that the proposition "$(x,y) \in R$" is L-true. If however $R(x,y) \in L^{\prec m}$, we say that the proposition is L-false. If $R(x,y) = c_m$, the median level (a fix point of the negation operator) then the proposition "$(x,y) \in R$" is L-undetermined.

In the classical case where R is a crisp binary relation, we define a digraph $G(A,R)$ with vertex set A and arc family R. A choice in $G(A,R)$ is a non empty set Y of A.

A dominant kernel is a choice that is stable in G i.e. $\forall x \neq y \in Y$, $(x,y) \notin R$, and dominant i.e. $\forall x \notin Y, \exists y \in Y$ such that $(y,x) \in R$.

R can be represented by a Boolean matrix and a kernel can be defined with the use of a subset characteristic row vector $Y(.) = (\ldots, Y(x), \ldots, Y(y), \ldots)$ where

$$Y(x) = \begin{cases} 1 \text{ if } x \in Y \\ 0 \text{ otherwise,} \end{cases} \text{ for all } x \in X.$$

A dominant kernel (if any) is a solution of the Boolean system of equations (see [14]) :

$$(Y \circ R)(x) = \vee_{y \neq x}(Y(x) \wedge R(y,x)) = \overline{Y(x)} = 1 - Y(x), \text{ for all } x \in X \quad (3)$$

$\vee$ and $\wedge$ represent respectively "disjunction" and "conjunction" for the 2-element Boolean lattice $B = \{0,1\}$ and $\circ$ represents the standard relational composition operator.

We now denote $G^L = G^L(A,R)$ a digraph with vertices set A and a valued arc family that corresponds to the L-valued binary relation R.

We define the level of stability qualification of subset Y of X as

$$\Delta^{sta}(Y) = \begin{cases} c_{2m} & \text{if } Y \text{ is a singleton} \\ \min_{\substack{y \neq x \\ y \in Y}} \min_{\substack{x \neq y \\ x \in Y}} \{\neg R(x,y)\} & \text{otherwise} \end{cases}$$

and the level of dominance qualification of Y as

$$\Delta^{dom}(Y) = \begin{cases} c_{2m} & \text{if } Y = A \\ \min_{x \notin Y} \max_{y \in Y} R(x,y) & \text{otherwise.} \end{cases}$$

Y is considered to be an L-good choice, i.e. L-stable and L-dominant, if $\Delta^{sta}(Y) \in L^{\succ m}$ and $\Delta^{dom}(Y) \in L^{\succ m}$. Its qualification corresponds to

$$Q^{good}(L) = \min(\Delta^{sta}(Y), \Delta^{dom}(Y)) \in L^{\succ m}.$$

We denote $C^{good}(G^L)$ the possibly empty set of L-good choices in G^L.

The determination of this set is an NP-complete problem even if, following a result of Kitainik [8], we do not have to enumerate the elements of the power set of A but only have to consider the kernels of the corresponding crisp strict median-level cut relation $R^{\succ m}$ associated to R, i.e. $(x,y) \in R^{\succ m}$ if $R(x,y) \in L^{\succ m}$.

As the dominant kernel in $G(X, R^{\succ m})$ is by definition a stable and dominant crisp subset of A, we consider the possibly empty set of kernels of $G^{\succ m} = G^{\succ m}(A, R^{\succ m})$ which we denote $C^{good}(G^{\succ m})$.

Kitainik proved that

$$C^{good}(G^L) \subseteq C^{good}(G^{\succ m}).$$

The determination of crisp kernels has been extensively described in the literature (see, for example [14]) and the definition of $C^{good}(G^L)$ is reduced to the enumeration of the elements of $C^{good}(G^{\succ m})$ and the calculation of their qualification.

A second approach (see [1]) to the problem of determining a good choice is to consider the valued extension of the Boolean system of equations (3).

If $\tilde{Y}(.) = (\dots, \tilde{Y}(x), \dots, \tilde{Y}(y), \dots)$, where $\tilde{Y}(x)$ belongs to L for every $x \in A$ is the characteristic vector of a fuzzy choice and indicates the credibility level of the assertion that "x is part of the choice $\tilde{Y}$", we have to solve the following system of equations :

$$(\tilde{Y} \circ R)(x) = \max_{y \neq x}[\min(\tilde{Y}(y), R(y,x))] = \neg \tilde{Y}(x), \quad \forall x, y \in A. \tag{4}$$

The set of solutions to the system of equations (4) is called $\tilde{Y}^{dom}(G^L)$.

In order to compare these fuzzy solutions to the solutions obtained in $C^{good}(G^L)$, we define the crisp choice

$$K_{\tilde{Y}} \subset A : \begin{cases} x \in K_{\tilde{Y}} \text{ if } \tilde{Y}(x) \in L^{\succ m} \\ x \notin K_{\tilde{Y}} \text{ otherwise} \end{cases} \tag{5}$$

and we consider a partial order on the elements of $\tilde{Y}^{dom}(G^L)$: $\tilde{Y}$ is sharper than $\tilde{Y}'$, noted $\tilde{Y}' \preceq \tilde{Y}$, iff $\forall x \in A$: either $\tilde{Y}(x) \leq \tilde{Y}'(x) \leq c_m$, either $c_m \leq \tilde{Y}'(x) \leq \tilde{Y}(x)$. The subset of the sharpest solutions in $\tilde{Y}^{dom}(G^L)$ is called $F^{dom}(G^L)$.

Bisdorff and Roubens have proved that the set of crisp choices constructed from $F^{dom}(G^L)$ using (5) and denoted $K(F^{dom}(G^L))$ coincides with $C^{dom}(G^L)$.

The decision maker might also be interested in bad choices. These choices correspond to absorbent kernels with a qualification greater than c_m. In the classical Boolean framework (see [14]) an (absorbent) kernel is a choice that is stable and absorbent i.e. $\forall x \notin Y, \exists\, Y \in Y$ such that $(x, y) \in R$. As $(x, y) \in R$ is equivalent to $(y, x) \in R^t$, where matrix R^t represents the transpose of matrix R, all the results obtained for dominant kernels can be immediately transposed for absorbent kernels and definitions like $\Delta^{bad}, Q^{bad}, F^{abs}, \tilde{Y}^{abs}$ are obviously and straightforwardly obtained from $\Delta^{good}, Q^{good}, F^{dom}, \tilde{Y}^{dom}$. We then compute good choices and bad choices and it might happen that some good choices are also bad choices with however different corresponding qualifications. These choices will be called *ambiguous choices*.

References

1. Bisdorff, R., Roubens, M. : On defining and computing fuzzy kernels from L-valued simple graphs. In : Da Ruan et al. (eds.) : Intelligent Systems and Soft Computing for Nuclear science and Industry, FLINS'96 Workshop. World Scientific Publishers, Singapore (1996), 113-123

2. Choquet, G. : Theory of Capacities. Annales de l'Institut Fourier **5** (1953) 131-295

3. Fodor, J., Roubens, M. : Fuzzy Preference Modelling and Multi-criteria Decision Support. Kluwer Academic publishers, Dordrecht Boston London (1994)

4. Fodor, J., Perny, P., Roubens, M. : Decision Making and Optimization. In : Ruspini, E., Bonissone, P., Pedrycz, W. (eds.) : Handbook of Fuzzy Computation. Institute of Physics Publications and Oxford University Press, Bristol (1998) F.5.1:1-14

5. Fodor, J., Orlovski, S.A., Perny, P., Roubens, M. : The use of fuzzy preference models in multiple criteria : choice, ranking and sorting. In : Dubois, D., Prade, H. (eds.) : Handbooks and of Fuzzy Sets, Vol. 5 (Operations Research and Statistics). Kluwer Academic Publishers, Dordrecht Boston London (1998) 69-101

6. Grabisch, M. : k-order additive discrete fuzzy measure and their representation. Fuzzy Sets and Systems **92** (1997) 167-189

7. Grabisch, M., Roubens, M. : Application of the Choquet Integral in Multicriteria Decision Making. In : Grabisch, M., Murofushi, T., Sugeno, M. (eds.) : Studies in Fuzzyness. Physica Verlag, Heidelberg (2000) 348-374

8. Kitainik, L. : Fuzzy Decision Procedures with Binary Relations : towards an unified Theory. Kluwer Academic Publishers, Dordrecht Boston London (1993)

9. Marichal, J.-L. : An axiomatic approach of the discrete Choquet integral as a tool to aggregate interacting criteria. IEEE Transactions on Fuzzy Systems **8** (2000) 800-807

10. Marichal, J.-L. : Aggregation of interacting criteria by means of the discrete Choquet integral. In : Calvo, T., Mayor, G., Mesiar, R. (eds.) : Aggregation operators : new trends and applications. Series : Studies in Fuzziness and Soft Computing, Vol. 97. Physica-Verlag, Heidelberg (2002) 224-244

11. Marichal, J.-L., Roubens, M. : On a sorting procedure in the presence of qualitative interacting points of view. In : Chojean, J., Leski, J. (eds.) : Fuzzy Sets and their Applications. Silesian University Press, Gliwice (2001) 217-230

38   Marc Roubens

12. Perny, P., Roubens, M. : Fuzzy Relational Preference Modelling. In : Dubois, D. and Prade, H. (eds.) : Handbooks of Fuzzy Sets, Vol. 5 (Operations Research and Statistics). Kluwer Academic Publishers, Dordrecht Boston London (1998) 3-30
13. Roberts, F. : Measurement Theory, with Applications to Decisionmaking, Utility and Social Science. Addison-Wesley Publishing Company, Reading, Massachusetts (1979)
14. Schmidt, G., Strhlein, T. : Relations and Graphs; Discrete mathematics for Computer Scientists. Springer-Verlag, Berlin Heidelberg New York (1991).
15. Wakker, P. : Additive Representations of Preferences : A new Foundation of Decision Analysis. Kluwer Academic Publishers, Dordrecht Boston London (1989)

Dual Interval Model and Its Application to Decision Making

Hideo Tanaka

Faculty of Human and Social Environment,
Hiroshima International University
555-36 Gakuendai, Kurose, Hiroshima 724-0695, JAPAN
h-tanaka@he.hirokoku-u.ac.jp

Abstract. In rough set approach, the rough approximations called lower
and upper ones have been discussed. This concept can be extended into
a new research field of data analysis. The proposed approach to data
modeling is to obtain dual mathematical models by using a similar concept to rough sets. The dual models called lower and upper models have
an inclusion relation. The lower and upper models are formulated by the
greatest lower and least upper bounds, respectively to obtain interval
models. The dual models are illustrated to be applied to AHP(Analytic
Hierarchy Process) which is the well-known tool for decision making.

1 Introduction

In the rough sets proposed by Z.Pawlak[1], lower and upper approximations
of the given objects are derived from our knowledge base which is a decision
table. Indiscernibility plays an important role in the rough set approach as the
granularity of knowledge.

In this paper, the rough approximations denoted as a pair of lower and upper
approximations are extended into dual models called lower and upper models in
data analysis. The proposed method can be described as two approximations to
a phenomenon under consideration such that

$$\text{Lower Model} \subseteq \text{Phenomenon} \subseteq \text{Upper Model}. \qquad (1)$$

Thus, the lower and upper models are obtained by the greatest lower bound
and the least upper bound, respectively. This property is illustrated by interval regression models. Interval regression models have an interval relationship
between inputs and outputs that can be regarded as an indiscernibility relationship. Therefore we can obtain a granular knowledge as an interval relationship.
In the other words, a phenomenon in an uncertain environment can be approximated by lower and upper models called dual models. It should be noted that
two models are dual if and only if the lower model can be defined from the upper
model and vice versa.

Using two concepts mentioned above, we have studied interval regression[2,
3, 4], interval AHP(Analytic Hierarchy Process)[5, 6], identification of dual possibility distributions[7, 8] and interval DEA(Data Envelopment Analysis)[9]. In

T. Bilgiç et al. (Eds.): IFSA 2003, LNAI 2715, pp. 39–51, 2003.
© Springer-Verlag Berlin Heidelberg 2003

these studies, we have obtained the lower and upper models as rough approximations such that (1) holds. There exists always an upper model for any data structure, while it is not assured to attain a solution for a lower model if the assumed model can not express the data structure . If we can not obtain the lower model, it might be caused by adopting a mathematical model not fitting to the given data. Interval regression analysis for interval outputs is formulated by dual mathematical models.

AHP[10, 11] is a useful method for multi-criteria decision making problems. The dual interval models are applied to AHP in order to formulate Interval AHP[5, 6] where we can obtain interval weights of priority to reflect the uncertainty of human judgment. The lower and upper models can be regarded as the certain and possible models, respectively.

2   Interval Regression Analysis

The given i-th input vector is denoted as $\mathbf{x}_j = (1, x_{j1}, \cdots, x_{jn})^t$ and the given j-th output is an interval denoted as $Y_j = (y_j, e_j)$ where $j = 1, \cdots, p$, y_j is a center of interval Y_j and e_j is a radius. e_j is called a width. An interval Y_j can be rewritten as $Y_j = [y_j - e_j, y_j + e_j]$.

An interval linear model is assumed to be

$$Y = A_0 + A_1 x_1 + \cdots + A_n x_n = \mathbf{A}\mathbf{x} \tag{2}$$

where $\mathbf{A} = (A_0, \cdots, A_n)$ is an interval coefficient vector, A_i is an interval denoted as $A_i = (a_i, c_i)$ and Y is an interval output. Using interval arithmetic, (2) can be written as

$$\begin{aligned} Y(\mathbf{x}_j) &= (a_0, c_0) + (a_1, c_1)x_{j1} + \cdots + (a_n, c_n)x_{jn} \\ &= (\mathbf{a}^t \mathbf{x}_j, \mathbf{c}^t |\mathbf{x}_j|) \end{aligned} \tag{3}$$

where $\mathbf{a} = (a_0, \cdots, a_n)^t$, $\mathbf{c} = (c_0, \cdots, c_n)^t$, $|\mathbf{x}_j| = (1, |x_{ji}|, \cdots, |x_{jn}|)^t$. It should be noted that the estimated interval output can be represented as the center $\mathbf{a}^t \mathbf{x}_j$ and the width $\mathbf{c}^t |\mathbf{x}_j|$.

The given input-output data are

$$(\mathbf{x}_j, Y_j) = (1, x_{j1}, \cdots, x_{jn}, Y_j), \ j = 1, \cdots, p \tag{4}$$

where $Y_j = (y_j, e_j)$. Since we consider the inclusion relation between the given interval output Y_j and the estimated interval $Y_j(\mathbf{x}_j)$, let us define the inclusion relation of two intervals. The inclusion relation of two intervals $A_1 = (a_1, c_1)$ and $A_2 = (a_2, c_2)$ is defined as follows.

$$A_1 \subseteq A_2 \Leftrightarrow \begin{cases} a_1 - c_1 \geq a_2 - c_2 \\ a_1 + c_1 \leq a_2 + c_2 \end{cases} \tag{5}$$

Using this inclusion relation, let us define the lower and upper models as follows.

<Lower Model>
The lowe model is denoted as

$$Y_*(\mathbf{x}_j) = A_{*0} + A_{*1}x_{j1} + \cdots + A_{*n}x_{jn} \tag{6}$$

where $A_{*i} = (a_{*i}, c_{*i})$, $i = 0, \cdots, n$. In the lower model, it is assumed that the estimated interval $Y_*(\mathbf{x}_j)$ should be included in the given output Y_j, that is

$$Y_*(\mathbf{x}_j) \subseteq Y_j. \tag{7}$$

Using the definition of inclusion relation (5), (7) can be rewritten as

$$\left. \begin{array}{l} y_j - e_j \leq \mathbf{a}_*^t \mathbf{x}_j - \mathbf{c}_*^t |\mathbf{x}_j| \\ \mathbf{a}_*^t \mathbf{x}_j + \mathbf{c}_*^t |\mathbf{x}_j| \leq y_j + e_j \end{array} \right\} \tag{8}$$

which can be regarded as the constraint conditions. Consider the optimization problem such that the sum of widths of the estimated intervals $Y_*(\mathbf{x}_j)$ is maximized subject to the constraint conditions (8). Therefore the object function is written as

$$\max_{\mathbf{a}_*, \mathbf{c}_*} J_* = \sum_{j=1}^{p} \mathbf{c}_*^t |\mathbf{x}_j|. \tag{9}$$

This optimization problem becomes an LP problem as follows.

$$\max_{\mathbf{a}_*, \mathbf{c}_*} \ J_* = \sum_{j=1}^{p} \mathbf{c}_*^t |\mathbf{x}_j| \tag{10}$$

$$s.t.) \ \ y_j - e_j \leq \mathbf{a}_*^t - \mathbf{c}_*^t |\mathbf{x}_j|$$
$$\mathbf{a}_*^t \mathbf{x}_j + \mathbf{c}_*^t |\mathbf{x}_j| \leq y_j + e_j$$
$$\mathbf{c}_* \geq 0 \ \ (j = 1, \cdots, p).$$

If there is a solution in (10), the lower model can be obtained by solving (10). The lower model is obtained as the greatest lower bound in the sense of inclusion relation. This model is corresponding to the lower approximation in rough sets.

<Upper Model>
The upper model is denoted as

$$Y^*(\mathbf{x}_j) = A_0^* + A_1^* x_{j1} + \cdots + A_n^* x_{jn} \tag{11}$$

where $A_i^* = (a_i^*, c_i^*)$, $i = 1, \cdots, p$. In the upper model, it is assumed that the estimated interval $Y^*(\mathbf{x}_j)$ should include the given output Y_j, that is

$$Y^*(\mathbf{x}_j) \supseteq Y_j \tag{12}$$

Using the definition of inclusion relation (5), (12) can be rewritten as

$$\left. \begin{array}{l} \mathbf{a}^{*t}\mathbf{x}_j - \mathbf{c}^{*t}|\mathbf{x}_j| \leq y_j - e_j \\ y_j + e_j \leq \mathbf{a}^{*t}\mathbf{x}_j + \mathbf{c}^{*t}|\mathbf{x}_j| \end{array} \right\} \tag{13}$$

which can be regarded as the constraint conditions. Consider the optimization problem such that the sum of widths of the estimated intervals $Y^*(\mathbf{x}_j)$ is minimized subject to the constraint (13). Therefore the objective function is written as

$$\min_{\mathbf{a}^*,\mathbf{c}^*} J^* = \sum_{j=1}^p \mathbf{c}^{*t}|\mathbf{x}_j|. \tag{14}$$

This optimization problem becomes an LP problem as follows.

$$\min_{\mathbf{a}^*,\mathbf{c}^*} \quad J^* = \sum_{j=1}^p \mathbf{c}^{*t}|\mathbf{x}_j| \tag{15}$$

$$s.t.) \quad \mathbf{a}^{*t}\mathbf{x}_j - \mathbf{c}^{*t}|\mathbf{x}_j| \le y_j - e_j$$
$$y_j + e_j \le \mathbf{a}^{*t} + \mathbf{c}^{*t}|\mathbf{x}_j|$$
$$\mathbf{c}^* \ge 0 \quad (j = 1, \cdots, p).$$

There exists always a solution in (15), because there is an admissible set of the constraint conditions (13) if a sufficient large positive vector is taken for $\mathbf{c}^*$. The upper model is obtained as the least upper bound in the sense of inclusion relation. This model is corresponding to the upper approximation in rough sets.

Since the lower and upper models can be obtained by solving LP problems (10) and (15), it is very easy to obtain two models, if there is a solution in the lower model (10). In order that there is an solution (a_i^*, c_i^*) in the lower model, some consistency between the given data structure and the assumed interval model is necessary.

It follows from the constraint conditions (7) and (12) that

$$Y_*(\mathbf{x}_j) \subseteq Y_j \subseteq Y^*(\mathbf{x}_j), \quad j = 1, \cdots, p \tag{16}$$

However, for the new sample vector $\mathbf{x}'$ it is not guaranteed that

$$Y_*(\mathbf{x}') \subseteq Y^*(\mathbf{x}'). \tag{17}$$

Therefore as the relation (17) for any $\mathbf{x}'$ can hold, let us consider the integrated model which the lower and upper models are combined into the following.

<Integrated Model>

$$\min_{\mathbf{a}^*,\mathbf{c}^*,\mathbf{a}_*,\mathbf{c}_*} \sum_{j=1}^p \mathbf{c}^{*t}\mathbf{x}_j - \sum_{j=1}^p \mathbf{c}_*^t\mathbf{x}_j \tag{18}$$

$$s.t.) \quad Y_*(\mathbf{x}_j) \supseteq Y_j$$
$$Y^*(\mathbf{x}_j) \subseteq Y_j$$
$$a_{*i} + c_{*i} \le a_i^* + c_i^*$$
$$a_i^* - c_i^* \le a_{*i} - c_{*i}$$
$$c_i^*, c_{*i} \ge 0, \quad i = 0, \cdots, n.$$

Since $A_i^* \supseteq A_{i*}$ $(i = 0, \cdots, n)$ are added to the constraint conditions in (18), for any $\mathbf{x}'$ (17) holds always. The given data structure can be approximated by the dual models $(Y_*(\mathbf{x}), Y^*(\mathbf{x}))$.

The solution of the lower model is not always guaranteed, because we fail to assume a proper regression model for the given data structure. In case of no solution for a linear system in the lower model, we can take the following polynominals.

$$Y(\mathbf{x}) = A_0 + \sum_i A_i x_i + \sum_{i,j} A_{ij} x_i x_j$$

$$+ \sum_{i,j,k} A_{ijk} x_i x_j x_k + \cdots. \tag{19}$$

Since a polynominal such as (19) can represent any function, the center of the estimated interval $Y(\mathbf{x})$ in the lower model can hit the center of the given output Y_j. Thus, one can select a polynominal as a lower model by increasing the number of terms of the polynominal (19) until a solution is found. It should be noted that (19) can be considered as a linear system with respect to parameters. Thus, we can use the proposed models with no difficulty to obtain the parameters in (19).

3   Application of Interval Regression Models to AHP

The Analytic Hierarchy Process(AHP)[10, 11] is the well-known and useful method to obtain priorities of each alternative in multiple criteria decision making problems. In the AHP, a decision maker is asked to estimate pairwise comparison ratios with respect to strength of preference between subjects of comparison. Thus the AHP is deeply related to human judgment. Then, inconsistency of rank and uncertainty of rank order have been discussed in [12, 13]. Even if pairwise comparison values with respect to preference are given as crisp values, the priority weights should be estimated as intervals because of a decision maker's uncertainty of judgments. This is our motivation to propose our approach for obtaining interval priorities.

It is assumed that the estimated weight vector is an interval vector to reflect the uncertainty of pairwise comparisons given by human judgment. The approach based on interval regression can be described as obtaining interval weights derived from inconsistent data and determining a partial order relation with respect to priorities. Considering all the possible ranges of weights obtained by the proposed method, a decision maker can afford to decide some crisp ordering with his or her sense. This approach might be suitable for handling the data given by human intuition.

3.1   Interval AHP for Crisp Data

The pairwise comparisons are denoted as a_{ij} for all i, j and the estimated interval weights are denoted as $W_i = [\underline{w_i}, \overline{w_i}]$ where $\underline{w_i}$ and $\overline{w_i}$ are the lower and upper bounds of the interval weight W_i. The estimated interval matrix can be defined as

$$\forall i, j \, (i \neq j) \; W_{ij} = \left[\frac{\underline{w_i}}{\overline{w_j}}, \frac{\overline{w_i}}{\underline{w_j}} \right] \tag{20}$$

which is the maximal range obtained from the estimated interval weights. It is assumed that $\underline{w_i} \geq \epsilon$ where ϵ is a very small positive number.

Let us consider the following definition[14] corresponding to the conventional normalization of weights, i.e. the sum of weights is one. Since the ratio model is considered, we need some normalization condition such as one in the conventional AHP.

Definition 1. An interval weight vector $(w_1, \ldots, w_n)$ is said to be normalized if and only if

$$\sum_i \overline{w_i} - \max_j \left(\overline{w_j} - \underline{w_j} \right) \geq 1 \tag{21}$$

$$\sum_i \underline{w_i} + \max_j \left(\overline{w_j} - \underline{w_j} \right) \leq 1. \tag{22}$$

In order to give some explanation of Definition 1, let us consider a numerical example. Assuming that $W_1 = [0.3, \, 0.6]$, $W_2 = [0.2, \, 0.4]$, $W_3 = [0.1, \, 0.2]$ which do not satisfy Definition 1, the value 0.3 in W_1 can not be taken, because there is no elements in W_2 and W_3 such that the sum of elements is equal to 1. Thus, the assumed interval weights should be embedded into, for example $W_1 = [0.5, \, 0.6]$, $W_2 = [0.2, \, 0.4]$, $W_3 = [0.1, \, 0.2]$ which satisfy Definition 1. Definition 1 allows to remove redundancy under the condition that the sum of some weight values in interval weights is equal to 1.

Given the pairwise comparisons a_{ij}, our problem is to find interval weights W_i according to the following conditions.

1. The given pairwise comparisons a_{ij} should be contained in the estimated interval comparisons W_{ij}. It yields that

$$a_{ij} \in W_{ij} \leftrightarrow \frac{\underline{w_i}}{\overline{w_j}} \leq a_{ij} \leq \frac{\overline{w_i}}{\underline{w_j}}$$

$$\leftrightarrow a_{ij}\overline{w_j} \geq \underline{w_i}, \; a_{ij}\underline{w_j} \leq \overline{w_i}.$$

$$\tag{23}$$

2. The estimated interval weights $W_i = [\underline{w_i}, \overline{w_i}]$ should satisfy the normalization condition described in Definition 1.
3. The estimated interval weights should be narrow as much as possible. Thus the following objective function can be considered.

$$\min_{\overline{w_i}, \underline{w_i}} \sum_i (\overline{w_i} - \underline{w_i}) \tag{24}$$

This problem yields the following LP problem called "Possibilistic AHP for Crisp data" which is abbreviated to PAHPC.

$< PAHPC >$

$$\min_{\overline{w_i}, \underline{w_i}} J = \sum_i (\overline{w_i} - \underline{w_i}) \tag{25}$$

$$subject\ to$$

$$\forall\, i, j\, (i \neq j) \qquad a_{ij} \overline{w_j} \geq \underline{w_i}$$

$$\forall\, i, j\, (i \neq j) \qquad a_{ij} \underline{w_j} \leq \overline{w_i}$$

$$\forall j \qquad \sum_{i \in \Omega-j} \overline{w_i} + \underline{w_j} \geq 1$$

$$\forall j \qquad \sum_{i \in \Omega-j} \underline{w_i} + \overline{w_j} \leq 1$$

$$\forall i \qquad \overline{w_i} \geq \underline{w_i}$$

$$\forall i \qquad \underline{w_i} \geq \epsilon$$

where Ω is a set $\{1, \cdots, n\}$.

Whatever the given comparison matrix is, we can obtain an optimal interval vector. The consistency between the given matrix and the model can be represented as the value of the objective function J in (24). If $J = 0$, it can be said that the given matrix is the perfect consistency.

The final preference can be obtained by interval arithmetic, but here we focus only on a priority interval vector. Thus, we introduce an interval preference relation denoted as $\succeq$.

Definition 2. The interval weight $W_i = \left[\underline{w_i},\ \overline{w_i}\right]$ is preferred to $W_j = \left[\underline{w_j},\ \overline{w_j}\right]$ if and only if

$$\underline{w_i} \geq \underline{w_j},\ \overline{w_i} \geq \overline{w_j} \tag{26}$$

Since the proposed method is a kind of regression analysis, we have the ratio model. Therefore we need the inequality that the number of data is larger than the number of decision variables. This requirment can be written as

$$\frac{n(n-1)}{2} \geq 2n \tag{27}$$

where n is the number of alternatives. With considering the above inequality, let us consider the numerical examples.

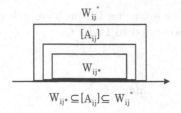

$$W_{ij*} \subseteq [A_{ij}] \subseteq W_{ij}^{\ *}$$

Fig. 1. Upper and lower approximations

Example 1

Let us consider the following matrix where each two rows satisfy the row dominance relation[12] defined as $a_{ik} \geq a_{jk}$ for all k.

$$A = \begin{pmatrix} 1 & 2 & 3 & 5 & 7 \\ \frac{1}{2} & 1 & 2 & 2 & 4 \\ \frac{1}{3} & \frac{1}{2} & 1 & 1 & 1 \\ \frac{1}{5} & \frac{1}{2} & 1 & 1 & 1 \\ \frac{1}{7} & \frac{1}{4} & 1 & 1 & 1 \end{pmatrix}$$

We obtained the following weights by solving PAHPC.

$$W_1 = 0.4528, \ W_2 = 0.2264,$$
$$W_3 = [0.1038, 0.1509]$$
$$W_4 = [0.0906, 0.1132]$$
$$W_5 = [0.0566, 0.1038]$$

while the solutions obtained by the eigenvector method are as follows.

$$W_1 = 0.4640, \ W_2 = 0.2413,$$
$$W_3 = 0.1120, \ W_4 = 0.0998,$$
$$W_5 = 0.0828$$

3.2   Interval AHP Models for Interval Data

Let us begin with interval scales in a reciprocal matrix denoted by the given interval pairwise comparisons $[A_{ij}] = [a_{ij}^L, a_{ij}^U]$ where a_{ij}^L and a_{ij}^U are the lower and upper bounds of the interval $[A_{ij}]$. The reciprocal property [13] is defined as :

$$a_{ij}^L = \frac{1}{a_{ji}^U}, a_{ij}^U = \frac{1}{a_{ji}^L} \tag{28}$$

where $[A_{ii}] = [1, 1]$.

Since we deal with interval data, we can consider two approximations as shown in Fig.1. According to these inclusion relations, the lower and upper approximations should satisfy the following constrain conditions

$$W_{ij*} \subseteq [A_{ij}] \, (Lower\,Approximation) \tag{29}$$
$$W_{ij}^* \supseteq [A_{ij}] \, (Upper\,Approximation) \tag{30}$$

where W_{ij*} and W_{ij}^* are the estimations of the lower and upper intervals of pairwise comparison ratios. Assume that the lower and upper interval weights are denoted as $W_{i*} = \left[\underline{w_{i_*}}, \overline{w_{i*}} \right]$ and $W_i^* = \left[\underline{w_i}^*, \overline{w_i}^* \right]$, respectively. Then according to (20), (29) and (30) can be rewritten as

$$
\begin{aligned}
W_{ij*} \subseteq [A_{ij}] &\leftrightarrow \left[\frac{\underline{w_{i_*}}}{\overline{w_{j}}_*}, \frac{\overline{w_{i*}}}{\underline{w_{j}}_*} \right] \subseteq [A_{ij}] \\
&\leftrightarrow a_{ij}^L \le \frac{\underline{w_{i_*}}}{\overline{w_{j}}_*}, \frac{\overline{w_{i*}}}{\underline{w_{j}}_*} \le a_{ij}^U \\
&\leftrightarrow a_{ij}^L \overline{w_{j}}_* \le \underline{w_{i_*}}, \; a_{ij}^U \underline{w_{j}}_* \ge \overline{w_{i*}}
\end{aligned}
\tag{31}
$$

$$
\begin{aligned}
W_{ij}^* \supseteq [A_{ij}] &\leftrightarrow \left[\frac{\underline{w_i}^*}{\overline{w_j}^*}, \frac{\overline{w_i}^*}{\underline{w_j}^*} \right] \supseteq [A_{ij}] \\
&\leftrightarrow \frac{\underline{w_i}^*}{\overline{w_j}^*} \le a_{ij}^L, \; a_{ij}^U \le \frac{\overline{w_i}^*}{\underline{w_j}^*} \\
&\leftrightarrow a_{ij}^L \overline{w_j}^* \ge \underline{w_i}^*, \; a_{ij}^U \underline{w_j}^* \le \overline{w_i}^*
\end{aligned}
\tag{32}
$$

Using the concepts of the greatest lower and least upper bounds, we can formulate the lower and upper models, respectively. The optimization problem to maximize the sum of widths of W_{ij*} subject to (31) can be reduced to the following LP problem.

$< Lower\,Model >$

$$\max_{\underline{w_{i_*}}, \overline{w_{i*}}} J_* = \sum_i \left(\overline{w_{i*}} - \underline{w_{i_*}} \right) \tag{33}$$

$subject\ to$

$$\forall i, j \, (i \ne j) \quad a_{ij}^L \overline{w_{j}}_* \le \underline{w_{i_*}}$$
$$\forall i, j \, (i \ne j) \quad a_{ij}^U \underline{w_{j}}_* \ge \overline{w_{i*}}$$
$$\forall j \quad \sum_{i \in \Omega - \{j\}} \underline{w_{i_*}} + \overline{w_{j}}_* \le 1$$
$$\forall j \quad \sum_{i \in \Omega - \{j\}} \overline{w_{i*}} + \underline{w_{j}}_* \ge 1$$
$$\forall i \quad \underline{w_{i_*}} \le \overline{w_{i*}}$$
$$\forall i \quad \underline{w_{i_*}} \ge \epsilon$$

where this is called the lower model.

Similarly, as to the upper model the optimization problem to minimize the sum of widths of W_{ij}^* subject to (32) can be reduced to the following LP problem.

< UpperModel >

$$\min_{\underline{w_i}^*, \overline{w_i}^*} \quad J^* = \sum_i \left(\overline{w_i}^* - \underline{w_i}^* \right) \tag{34}$$

subject to

$$\forall i, j \, (i \neq j) \qquad a_{ij}^L \overline{w_j}^* \geq \underline{w_i}^*$$

$$\forall i, j \, (i \neq j) \qquad a_{ij}^U \underline{w_j}^* \leq \overline{w_i}^*$$

$$\forall j \qquad \sum_{i \in \Omega - \{j\}} \overline{w_i}^* + \underline{w_j}^* \geq 1$$

$$\forall j \qquad \sum_{i \in \Omega - \{j\}} \underline{w_i}^* + \overline{w_j}^* \leq 1$$

$$\forall i \qquad \underline{w_i}^* \leq \overline{w_i}^*$$

$$\forall i \qquad \underline{w_i}^* \geq \epsilon$$

where this is called the upper model.

Example 2

Let us consider the following interval comparison matrix.

$$[A_{ij}] = \begin{pmatrix} 1 & [1,3] & [3,5] & [5,7] & [5,9] \\ [\frac{1}{3},1] & 1 & [1,4] & [1,5] & [1,4] \\ [\frac{1}{5},\frac{1}{3}] & [\frac{1}{4},1] & 1 & [\frac{1}{5},5] & [2,4] \\ [\frac{1}{7},\frac{1}{5}] & [\frac{1}{5},1] & [\frac{1}{5},5] & 1 & [1,2] \\ [\frac{1}{9},\frac{1}{5}] & [\frac{1}{4},1] & [\frac{1}{4},\frac{1}{2}] & [\frac{1}{2},1] & 1 \end{pmatrix}$$

The obtained interval priorities by solving the LP problems (33) and (34) are shown in Table 1. The results can be depicted in Fig.2 using the preference relation in Definition 2.

Table 1: The obtained interval priorities in the lower and upper models

Alter.	Lower-model	Upper-model
w_1	$[0.4225, 0.5343]$	$[0.2909, 0.4092]$
w_2	$[0.1781, 0.2817]$	$[0.1364, 0.2909]$
w_3	0.1408	$[0.0273, 0.1818]$
w_4	$[0.0763, 0.0845]$	$[0.0364, 0.1364]$
w_5	0.0704	$[0.0455, 0.1364]$

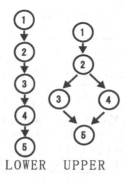

LOWER   UPPER

Fig. 2. Order relations of interval priorities

It should be noted from Table 1 that the following inclusion relations hold

$$W_{ij*} \subseteq [\Lambda_{ij}] \subseteq W_{ij}^*. \tag{35}$$

4   Comparison between Interval Regression and Rough Sets Concept

Let us briefly illustrate the elementary terms using in rough sets[1]. U denotes the universe of discourse, and R denotes an equivalence relation on U considered as an indiscernibility relation. Then the ordered pair $A = (U, R)$ is called an approximation space. Equivalence classes of the relation R are called elementary sets in A, denoted as $\{E_j, \ j = 1, \cdots, n\}$. Let a set $X \subset U$ be given. Then an upper approximation of X in A denoted as $A^*(X)$ means the least definable set containing X, and the lower approximation of X in A denoted as $A_*(X)$ means the greatest definable set contained in X. Thus, the inclusion relation $A_*(X) \subseteq A^*(X)$ is satisfied.

An accuracy measure of a set X in the approximation space $A = (U, R)$ is defined as

$$\alpha_A(X) = \frac{Card(A_*(X))}{Card(A^*(X))} \tag{36}$$

where $Card(A_*(X))$ is the cardinality of $A_*(X)$. This accuracy measure $\alpha_A(X)$ corresponds well to the measure of fitness $\varphi_Y(\mathbf{x})$ in interval regression analysis. When the classification $C(U) = \{X_1, \cdots, X_n\}$ is given, the accuracy of the classification $C(U)$ is defined as

$$\beta_A(U) = \frac{Card(\bigcup A_*(X_j))}{Card(\bigcup A^*(X_j))} \tag{37}$$

which is corresponding to φ_Y in interval regression.

Table 2: Comparison of the concept between interval regression and rough sets

Interval regression analysis	Rough sets
Upper estimation model : $Y^*(\mathbf{x})$	Upper approximation : $A^*(X)$
Lower estimation model : $Y_*(\mathbf{x})$	Lower approximation : $A_*(X)$
Spread of Y^* : $\mathbf{c}^{*t}|\mathbf{x}|$	Cardinality of $A^*(X)$: $Card(A^*(X))$
Spread of Y_* : $\mathbf{c}_*^t|\mathbf{x}|$	Cardinality of $A_*(X)$: $Card(A_*(X))$
Inclusion relation : $Y^*(\mathbf{x}_j) \supseteq Y_*(\mathbf{x}_j)$	Inclusion relation : $A^*(X) \supseteq A_*(X)$
Measure of fitness for j-th input : $\varphi_Y(\mathbf{x}_j)$	Accuracy measure of X_i : $\alpha_A(X_j)$
Measure of fitness for all data : φ_Y	Accuracy measure of classification : $\beta_A(U)$
(The higher, the better.)	(The higher, the better.)

Furthermore, the concept of adopting a polynominal (19) as an interval regression model corresponds to the refined elementary sets in rough sets. Table 2 shows the comparison of the concepts between interval regression and rough sets, where the measure of fitness for j-th input $\varphi_Y(\mathbf{x}_j)$ and the measure of fitness for all data φ_Y are defined as follows:

$$\varphi_Y(\mathbf{x}_j) = \frac{\mathbf{c}_*^t|\mathbf{x}_j|}{\mathbf{c}^{*t}|\mathbf{x}_j|} \tag{38}$$

$$\varphi_Y = \frac{\sum_j \varphi_Y(\mathbf{x}_j)}{p}. \tag{39}$$

The larger the value of φ_Y, the more the model is fitting to the data. Note that $0 \leq \varphi_Y \leq 1$ is the average ratio of lower spread to upper spread over the p data.

5   Concluding Remarks

It is shown that the rough set approach can be extended into a field of mathematical models. When the given data are intervals, we can approximate the data structure by the dual models with the inclusion relations like (1). It can be said that our obtained dual models are rough approximations which can be easily obtained by LP. In the case where the given data have uncertainty, our rough approximations are quite effective without complex calculations.

References

[1] Z. Pawlak(1991). *Rough Sets*, Kluwer Academic, Dordrecht.

[2] H. Tanaka and P. Guo(1999). *Possibilistic Data Analysis for Operations Research*, Physica-Verlag, Heidelberg.

[3] H. Tanaka and H. Lee(1998). Interval regression analysis by quadratic programming approach. In *IEEE Trans. on Fuzzy Systems*, volume 6, number 6, pages 473-481.

[4] H. Tanaka and H. Lee(1999). Interval regression analysis with polynominal and its similarity to rough sets concept. In *Fundamenta Informaticae*, IOS Press. volume 37, pages 71-87.

[5] K. Sugihara and H. Tanaka(2001). Interval evaluations in the analytic hierarchy process by possibility analysis. In *An Int. J. of Computational Intelligence*, volume 17, number 3, pages 567-579.

[6] K. Sugihara, H. Ishii and H. Tanaka(2000). On interval AHP. In *4th Asian Fuzzy Systems Symposium -Proceedings of AFSS2000-*, pages 251-254, Tsukuba, Japan, July 2000.

[7] P. Guo, H. Tanaka and H.J. Zimmermann(1999). Upper and lower possibility distributions of fuzzy decision variables in upper level decision problems. In *Int. J. of Fuzzy Sets and Systems* volume 111, pages 71-79.

[8] P. Guo and H. Tanaka(1998). Possibilistic data analysis and its application to portfolio selection problems. In *Fuzzy Economic Review*, volume 3/2, pages 3-23.

[9] T. Entani, Y. Maeda and H. Tanaka(2002). Dual models of interval DEA and its extension to interval data. In *European J. of Operational Research*, volume 136, pages 32-45.

[10] T.L. Saaty(1980). *The Analytic Hierarchy Process*, McGraw-Hill.

[11] T.L. Saaty(1998). *Multicriteria Decision Making, The Analytic Hierarchy Process*, RSW Publications.

[12] T.L. Saaty and L.G. Vergas(1984). Inconsistency and rank preservation. In *Journal of Mathematical Psychology*, volume 28, pages 205-214.

[13] T.L. Saaty and L.G. Vargas(1987). Uncertainty and rank order in the analytic hierarchy process. In *European Journal of Operational Research*, volume 32, pages 107-117.

[14] H. Tanaka, K. Sugihara and Y. Maeda(2001). Non-additive measures by interval probability functions. In *Proceedings of International Workshop on Rough Set Theory and Granular Computing*, Matsue, Japan, pages 63-67, May 2001.

Automatic Taxonomy Generation: Issues and Possibilities

Raghu Krishnapuram and Krishna Kummamuru

IBM India Research Lab, Block I, IIT,
Hauz Khas, New Delhi 110016. INDIA
{kraghura,kkummamu}@in.ibm.com

Abstract. Automatic taxonomy generation deals with organizing text documents in terms of an unknown labeled hierarchy. The main issues here are (i) how to identify documents that have similar content, (ii) how to discover the hierarchical structure of the topics and subtopics, and (iii) how to find appropriate labels for each of the topics and subtopics. In this paper, we review several approaches to automatic taxonomy generation to provide an insight into the issues involved. We also describe how fuzzy hierarchies can overcome some of the problems associated with traditional crisp taxonomies.

1 Introduction

The lack of a central structure and freedom from a strict syntax is responsible for making a vast amount of information available on the Web, but retrieving this information is not easy. Ranked lists returned by search engines are still a popular way of searching and browsing the Web today. However, this method is highly inefficient since the number of retrieved search results can be in the thousands for a typical query. Most users just view the top ten results and therefore might miss relevant information. Moreover, the criteria used for ranking may not reflect the needs of the user. A majority of the queries tend to be short [1], thus making them non-specific or imprecise [2]. The inherent ambiguity in interpreting a word or a phrase in the absence of its context means that a large percentage of the returned results can be irrelevant to the user.

Ranked lists have been found to be fairly effective for *navigational tasks* such as finding the URL of an organization. However, since the results are not summarized in terms of topics, they are not well suited for browsing tasks. One possible solution is to create a static hierarchical categorization of the entire Web and use these categories to organize the search results of a particular query. For example, the *dmoz* (*www.dmoz.org*) directory that categorizes Web sites is manually created by about 52 thousand editors. However, this solution is feasible only for small collections. For example, *dmoz* covers less than 5% of the Web. Secondly, even if we were to categorize the entire Web either manually or automatically, the categories may not be useful in organizing the search results of a particular query [3,4].

T. Bilgiç et al. (Eds.): IFSA 2003, LNAI 2715, pp. 52–63, 2003.

It has been observed that post-retrieval document clustering typically produces superior results [5]. Examples of search engines that return search results in terms of a hierarchy include *vivisimo* (*www.vivisimo.com*) and *kartoo* (*www.kartoo.com*). Some of the automatic taxonomy generation (ATG) approaches are *monothetic*, i.e., the cluster assignment is based on a signle feature, while others are *polythetic*. The approaches that produce hierarchical clusters can also be classified as either top-down or bottom-up, depending on how they build the clusters. The taxonomies can be generated by clustering either documents or words. Recently, there have been attempts at *co-clustering* or *simultaneous clustering*, where documents and words are clustered at the same time.

In this paper, we review various approaches to ATG to provide some insights into the issues involved. In Sections 2 and 3, we briefly describe a few representative algorithms based on document and word clustering, respectively. In Section 4, we discuss the co-clustering approach. In Section 5, we outline a few evaluation measures for taxonomies. In Section 6, we describe fuzzy hierarchies and argue that they can overcome many of the problems associated with crisp hierarchies. In Section 7, we summarize the issues in ATG and define some possible directions.

2   Approaches Based on Clustering Documents

2.1   Methods Based on Traditional Clustering

In the vector space model, documents are represented as M-dimensional vectors, where M is the size of the dictionary and each dimension represents a word. The elements of the vectors can be binary or real-valued. In the case of binary vectors, only the occurrence or nonoccurence of the word in the document is taken into account, whereas in the non-binary case, the frequency of occurrence is also taken into account. Once the documents are represented as numerical vectors, conventional clustering algorithms can be used for clustering. This approach to clustering documents is well-studied in the literature and can be used to generate one-level or hierarchical clusters of polythetic nature. The set of words that have a high frequency of occurrence within the cluster can be used as the 'label' for the cluster [5].

2.2   Grouper

Zamir and Etzioni [6] present an interface (called *Grouper*) to the HuskySearch meta search engine [7] that dynamically groups the search results into clusters labeled by phrases extracted from snippets. Grouper uses an algorithm called Suffix Tree Clustering (STC) for forming groups of "snippets" (or summaries) of Web pages. STC is based on standard techniques from the literature that allow the construction of "suffix trees" in time that is linear in the number of snippets (or document snippets), assuming that the number of words in each snippet is bounded by a constant. Each node in this tree captures a phrase (or a certain "suffix") of the snippet string, and has associated with it those snippets that contain it. These nodes are viewed as base clusters since they group together all documents that have one phrase in common. They then create a graph which has as its vertices the base clusters identified by the suffix tree. Vertices representing two base clusters are connected by an edge if at least half of the documents in each base cluster are common to both. They then run a connected component finding algorithm, and each connected component of the graph is identified as a group of documents addressing the same topic. (We can easily see that when the edge weights are not binary but fuzzy, the graph is a fuzzy relation, and finding the groups relates to identifying the complete α-cover [8].) Each topic is represented by the the phrases that occur in the connected component.

It is easy to see that, since the order of the words is taken into account when the suffix tree is constructed, this approach generates better summaries (polythetic labels) for the clusters. One problem with the STC approach is that the complexity of constructing the graph is exponential in the number of keywords. Thus, it becomes inefficient when the number of keywords is large.

2.3   Model-Based Hierarchical Clustering

Vaithyanathan and Dom [9,10] define a model for flat (i.e., one-level) clustering, and then generalize it to generate hierarchical clusters. The model assumes that the feature set T can be partitioned into two conditionally independent sets: a 'useful' set U, and a 'noise' set N. In addition to the selection of the optimal number of clusters in model selection, the model also includes the selection of the optimal feature set partition. Let D^T denote the set of documents in the original feature space corresponding to T, Ω the model structure, D^U the documents D^T projected onto the useful feature subspace, D^N the documents D^T projected onto the noisy feature subspace, Ψ the set of clusters indexed by k, and D_k^U the subset of D^U that has been assigned to cluster k. The marginal likelihood (ML) of the document collection can be expressed as:

$$P(D^T|\Omega) = P(D^N|\Omega) \prod_{k \in \Psi} P_k(D_k^U|\Omega) \qquad (1)$$

The ML needs to be maximized to find the optimal partition for the features, optimal parameters for the structure Ω, and the number of clusters. Vaithyanathan and Dom simplify their search by combining certain heuristics for searching the optimal feature partition with the multinomial Expectation-Maximization (EM) algorithm [11] to identify the structure. They extend the noise/useful concept to hierarchical clustering by interpreting the 'noise' set associated with a particular node to be the set of features that have a common distribution over child nodes of the given the node, and the 'useful' set to be the set of features that can be used to discriminate among the child nodes. Let U_k and N_k represent the set of useful and noise features associated with a cluster represented by node k of the hierarchy. Let $p(k)$ represent the parent node of node k in the hierarchy. Then, $\tilde{N}_k = N_k - N_{p(k)}$ represents the newly added 'noise' features at node k, i.e., the set of features that is uniquely common to the subtree rooted at node k. Thus, $\tilde{N}_k$ can be used to 'label' node k. It follows that $U_{p(k)} = U_k + \tilde{N}_k$. The marginal likelihood of the hierarchy (MLH) can be defined recursively by starting with (1) for the root node and then expanding each of the terms in the product at every non-leaf node. The algorithm proposed by Vaithyanathan and Dom is a bottom-up clustering algorithm that merges clusters to maximize the MLH.

The MLH approach is interesting from a modeling point of view. However, the computational complexity of determining the optimal solution is extremely high, unless simplifications are made. However, the ability to identify the unique words associated with each node in the hierarchy is a strong feature of the approach.

3   Approaches Based on Clustering Words

3.1   Thesaural Relationships among Words

The goal of word-based ATG is to organize words in terms of their thesaural relationships. There have been many attempts at automatically generating the-saural relationships from a corpus. Some of these methods are based on phrasal analysis [12,13]. They analyze the context (phrase) in which a term occurs to infer the relationships between various terms. There are other methods to establish relationships between words, e.g., by observing how often they co-occur in documents. [14,15,16]. In the following, we briefly describe some of those methods that perform hierarchical clustering.

3.2   Subsumption Algorithm

The Subsumption Algorithm builds concept hierarchies by finding pairs of concepts (x, y) in which x subsumes y [14]. Concept x is said to subsume concept y if $\Pr(x|y) > 0.8$ and $\Pr(y|x) < 1$. The threshold 0.8 was determined empirically. The algorithm considers an expanded set of terms obtained by Local Context Analysis (LCA) [17] and selects a few terms based on the relative frequency of their occurrence in the retrieved documents with respect to the entire corpus. Once these terms (concepts) are extracted, the algorithm computes the subsumption relationships between all pairs of concepts (x, y) and retains only those pairs in which x subsumes y. The hierarchy is then is built in a bottom-up fashion.

3.3   Dominating Set Approach

Lawrie et. al. [15] consider an approach based on a set of topic terms and a set of vocabulary terms to generate concept hierarchies. They propose a language model composed of all the conditional probabilities $\Pr_x(A|B)$ where $\Pr_x(A|B)$ is the probability of occurrence of A in the x-neighborhood of B. Here, A is a topic term and B is a vocabulary term. In the experiments reported in [15], the topic terms and the vocabulary terms are the same and are those that occur in at least two documents excluding numbers and stopwords. The ultimate goal is to find the set of terms that have maximal predictive power and coverage of the vocabulary. To achieve this goal, the language model is re-casted as a graph and the problem is posed as Dominating Set Problem (DSP). The authors propose a greedy approximation algorithm to solve DSP. The solution provides a set of terms that can be used as labels at the top level. To find the subtopics of a topic at the top level, the language model is constructed on the terms occurring in the neighborhood of the topic and the corresponding DSP is solved. The procedure can be applied recursively to obtain a hierarchy of topics.

3.4   Clustering Asymmetrically Related Data

In [16], the authors observe that relations between words are assymetric, and propose an algorithm called CAARD (Clustering Algorithm for Asymmetrically Related Data) to cluster such asymmetrically related data. They focus on the *inclusion* relation. The degree of inclusion of meaning of one word in another is estimated from a document corpus as follows. Let N be the number of documents in the corpus. Each concept is represented by a binary vector of dimension N. The i-th element of the vector is 1 if the concept is present in the ith document and is zero otherwise. Let $W = \{w_1, w_2, \ldots, w_M\}$ be the set of vectors representing the M concepts. The degree of inclusion of meaning w_i in w_j, denoted by $R(w_i, w_j)$ is computed as $|w_j \cdot w_i|/|w_i|$. The goal is to find a minimal subset S of W such that the elements of S are as distinct from each other as possible and each word in $W - S$ is is related to at least one of words in S to an extent that is greater than a predefined threshold, η. The subset S is referred to as the set of *leaders*.

Kummamuru and Krishnapuram propose a fast greedy algorithm to find the solution to the problem. The algorithm creates a possibilistic partition of the words by assigning each word to one or more of the leaders. The concept hierarchy is built by recursively applying CAARD to the words in each cluster until a terminating condition is reached. CAARD adaptively finds η at various levels so that the algorithm results in a hierarchy of a pre-specified size.

4   Approaches Based on Co-clustering

4.1   Co-clustering

As explained in Section 2.1, documents are traditionally represented as an M-dimensional vector in which the m-th element represents the occurrence (frequency) of m-th keyword. Similarly, keywords can be represented as an N-dimensional vector in which the l-th element represents the occurrence (frequency) of the word in l-th document. Using this representation for words, the keywords can also be clustered to find groups of related keywords. This type of clustering is known as distributional clustering [18,19]. Co-clustering deals with clustering keywords and documents simultaneously. The goal is to place all documents that have similar keyword occurrences in the same cluster, and at the same time place all keywords that have similar document occurrences in the same cluster. Thus, each co-cluster contains a cluster of documents and a cluster of keywords. We briefly describe few algorithms that aim to find co-clusters. Apart from those described below, there are also some approaches based on bipartite graph partitioning (for example, see [20]).

4.2 Fuzzy Co-clustering of Documents and Keywords (FCoDoK)

FCoDoK [21] is a variation of the FCCM algorithm [22] and aims to find co-clusters by maximizing what is known as *aggregation* among the clusters. Let $D_1, D_2, \ldots$, and D_N represent N documents and $W_1, W_2, \ldots$, and W_M, M words. Let C denote the number of co-clusters, u_{ci} denote the membership of D_i in co-cluster c and v_{cj} the membership of W_j in co-cluster c. The degree of aggregation for co-cluster c is given by $\sum_{i=1}^{N} \sum_{j=1}^{K} u_{ci} v_{cj} d_{ij}$, for $c = 1, \ldots, C$, where $D_i = [d_{i1}, \ldots, d_{iM}]$ and $d_{ij} = 1$ if the jth word occurs in ith document. The above objective function is maximized subject to the following conditions: $\sum_{c=1}^{C} u_{ci} = 1, u_{ci} \in [0,1], i = 1, \ldots, N$, and $\sum_{j=1}^{K} v_{cj} = 1, v_{cj} \in [0,1], c = 1, \ldots, C$. FCoDoK regularizes u_{ci} and v_{cj} by maximizing the Gini index, whereas FCCM does it by maximizing the entropy term. The objective function that is maximized is given by,

$$\sum_{c=1}^{C} \sum_{i=1}^{N} \sum_{j=1}^{K} u_{ci} v_{cj} d_{ij} - T_u \sum_{c=1}^{C} \sum_{i=1}^{N} u_{ci}^2 - T_v \sum_{c=1}^{C} \sum_{j=1}^{K} v_{cj}^2$$
$$+ \sum_{i=1}^{N} \lambda_i \left(\sum_{c=1}^{C} u_{ci} - 1 \right) + \sum_{c=1}^{C} \gamma_c \left(\sum_{j=1}^{K} v_{cj} - 1 \right).$$

This objective function is maximized using the Alternate Optimization technique [23]. Even though this algorithm generates a flat clustering, it can be used to generate hierarchies by using it recursively in a top-down approach.

4.3 Fuzzy Simultaneous Keyword Identification and Clustering of Documents (FSKWIC)

FSKWIC [24] performs clustering and feature weighing simultaneously by using a weighted cosine distance measure in which the weight for each feature depends on the cluster. Another version of the algorithm (referred to as SCAD) using a weighted Euclidean measure is given in [25]. The objective function that is minimized in FSKWIC is given by:

$$\sum_{c=1}^{C} \sum_{i=1}^{N} (u_{ci})^m \sum_{j=1}^{K} v_{cj} D_{cij} + \sum_{c=1}^{C} \delta_c \sum_{j=1}^{K} v_{cj}^2 + \sum_{c=1}^{C} \gamma_c \left(\sum_{j=1}^{K} v_{cj} - 1 \right),$$

where $D_{cij} = (\frac{1}{K} - d_{ij} \cdot p_{cj})$ and $\mathbf{p}_c = (p_{c1}, \ldots, p_{cK})$ is the centroid of cluster c. Note that the first term in the above expression reflects the sum of within cluster distances, the second term is due to the regularization and the third due to normalization of v_{cj}. The optimization is similar to Fuzzy C-Means algorithm except that values of v_{cj} are also updated in every iteration. In FSKWIC, the cluster centers $\mathbf{p}_c$ are normalized to unit length. By introducing a parameter δ_c with every regularization term, FSKWIC offers more flexibility when compared with FCoDoK. However, this comes at the expense of having to tune a larger number of parameters. A heuristic method for estimating the δ_c values in every iteration is given by the authors in the paper. Even though, FSKWIC was not intended to be a co-clustering algorithm, it can be thought of as a co-clustering algorithm since every document cluster can be associated with a co-cluster that consists of few keywords with small values of v_{cj}.

4.4   Rowset Partitioning and Submatrix Agglomeration (RPSA)

RPSA is a hierarchical clustering technique that builds a hierarchy of the dense submatrices of the document-word matrix [26]. The documents are represented by TFIDF values of the words contained in the documents. RPSA consists of two steps. The first step is a partitioning step that attempts to find small and dense submatrices of the document-word matrix. The second step is a agglomerative step in which the submatrices are hierarchically merged in a bottom-up fashion.

The crucial part in the partitioning step is to identify submatrices (leaf clusters) such that they have a minimum density λ. The method of identifying the submatrices is an extension of the ping-pong algorithm presented in [27] which deals with binary matrices. It starts with the row that has the maximum number of words as the leader of the first cluster. In every iteration, the algorithm finds the submatrix associated with the given leader, and identifies the leader for the subsequent cluster. The leader for the subsequent cluster would be the one that does not belong to and most dissimilar to any of the previous clusters, and contains large number of words. These iterations would go on till the algorithm covers a pre-specified fraction of the total number of documents. The remaining documents are assigned to the clusters with maximum overlap of words. In the agglomerative step, the dense submatrices are progressively merged based on a similarity measure.

5   Evaluation of Taxonomies

An important aspect of automatically generated taxonomies is their evaluation. Given a taxonomy, how do we evaluate its goodness? There are three possible scenarios. If the manually generated ground truth for an ideal hierarchy is available, then we can compare the generated hierarchy with the ideal one. When the ground truth is not available, we could measure the goodness of the hierarchy based on certain desirable properties. In the third scenario, we are simply interested in comparing two hierarchies generated by different algorithms to see how closely they match. We now briefly address these three aspects.

When the ground truth is available, two criteria that are commonly used are accuracy and mutual information. Accuracy for a one-level hierarchy generated by a clustering algorithm can be computed by assigning a class label for each cluster and then counting how many documents in the cluster are correctly labeled. If there are C concepts (classes) in the 1-level hierarchy, there are $C!$ ways of assigning the labels to clusters. We need to pick the assignment that gives the highest accuracy. Since this procedure is computationally expensive, often, the cluster label to be used is determined by majority voting. When the hierarchy has more than one level, this evaluation can be repeated at each level. However, it is not entirely clear how to combine the accuracies at various levels. The errors at higher levels of the hierarchy are more costly and the weighting scheme needs to reflect this fact.

Let $\mathcal{K} = \{k_1, k_2, \ldots, k_C\}$ and $\mathcal{K}' = \{k'_1, k'_2, \ldots, k'_C\}$ denote the set of C class labels and the set of cluster labels respectively. Mutual information is defined as [9,28,29]

$$MI(\mathcal{K}, \mathcal{K}') = \sum_{k_i \in \mathcal{K}} \sum_{k'_j \in \mathcal{K}'} p(k_i, k'_j) \log \frac{p(k_i, k'_j)}{p(k_i)p(k'_j)}. \tag{2}$$

The MI given in the above equation can be divided by the maximum of the entropies associated with $\mathcal{K}$ and $\mathcal{K}'$ so that the value always lies in [0,1].

When the ground truth is not available, one possible way to judge the goodness of the hierarchy is to measure how well the labels in the hierarchy *predict* the contents of the documents in the retrieved set. This can be done by computing the Expected Mutual Information Measure (EMIM) [15,30]. Let T denote the set of label words in the hierarchy and let V denote the set of non-stopwords that occur at least twice in the document collection. Then, $EMIM$ is given by

$$EMIM(T, V) = \sum_{t \in T} \sum_{v \in V} p(t, v) \log \frac{p(t, v)}{p(t)p(v)}. \tag{3}$$

Apart from $EMIM$, we can also consider reachability (i.e., percentage of documents covered by the hierarchy). Note that an automatically generated hierarchy is not guaranteed to cover all documents in the retrieved set, unless we include a *miscellaneous* or *other* label at each level. Another quantitative measure is the ease of access of all the relevant documents [15]. This can be measured by how many nodes of the hierarchy the user would have to expand to find all the relevant documents. This measure would favor hierarchies in which all relevant documents are concentrated under a small number of nodes.

To measure how well a given pair of hierarchies match, we could use how many parent-child pairs are common between the two hierarchies [14]. This measure can be normalized by the total number of parent-child pairs in one of the hierarchies. A better measure is to compute the edit distance between the two corresponding trees. The edit distance computes the minimum number of changes (additions, deletions and re-labeling of nodes) needed to transform one hierarchy into the other [31,32].

6 Fuzzy Taxonomies

6.1 Problems with Concept Hierarchies

The systems mentioned above present all clusters and their descriptions to the user. In the case of systems that generate concept hierarchies, the user is usually presented with the clusters at the top level which can refined recursively in a tree-like manner. While this interface is fairly effective in presenting an overview of the results, it can be quite rigid and cumbersome for retrieval. In other words, the hierarchy generated by a given algorithm will restrict the types of relevant documents that can be retrieved efficiently. For example, Let us consider a taxonomy that has the following concepts at a particular level: *books, electronics, art, computers, entertainment,* and *music*. It is not clear where the subconcept *books on computer art* should be placed. It could be under either *books* or *art* or *computers*. Similarly, the subconcept *art movie* could be under either *art* or *entertainment*. This problem can be partially alleviated by relaxing the constraint that each subconcept should be placed under one and only one concept. In other words, *books on computer art* can appear under *books* as well as under *art* and *computers*. Thus, the hierarchy is no longer a tree, but a directed acyclic graph (DAG). However, this solution still has some drawbacks. For example, CD players perhaps belong under *electronics*, but we may want to place them under *music* along with *compact disks* so that it would be easy for a music lover to look at some CD players while browsing CD's. However, *compact disc players* belong under *electronics* to a higher degree than they do under *music*. The problem is that most real-world concepts (and subconcepts) are not crisp and hence they defy crisp categorization. Fuzzy hierarchies can be used to overcome this drawback (see also [33]).

6.2 Fuzzy Concept Hierarchies

Let us consider an L-level crisp hierarchy, where level 0 represents the root, level 1 represents the first-level concepts (or topics), and so on. Let the number of concepts at level k be c_k. In a fuzzy concept hierarchy, each concept at level k of the hierarchy is a child of all the concepts at level $k-1$, albeit to different degrees. Let t_p^k denote the p-th concept at level k, and let $m_{(k,p)}(k-1,j)$ denote the degree to which t_p^k is a child of t_j^{k-1}. In other words, $m_{(k,p)}(k-1,j)$ is the membership of t_p^k in the fuzzy set of children of t_j^{k-1}. It is also the degree to which t_j^{k-1} is the parent of t_p^k. We can impose certain constraints on the memberships in order to achieve meaningful fuzzy hierarchies. For example, we can say:

$$\sum_{j=1}^{c_{k-1}} m_{(k,p)}(k-1,j) = 1; p = 1, \ldots, c_k; k = 1, \ldots, L.$$

If we define $m_{(k+1,l)}(k-1,j) = \sum_{p=1}^{c_k} m_{(k+1,l)}(k,p)m_{(k,p)}(k-1,j)$, then it can be easily verified that $\sum_{j=1}^{c_{k-1}} m_{(k+1,l)}(k-1,j) = \sum_{p=1}^{c_k} m_{(k+1,l)}(k,p) = 1$. Thus, the sum of the memberships of a grandchild across all grandparents is equal to one. This can be viewed as conservation of membership.

Another possibility is to impose a possibilistic constraint [34]. Let N_c denote the set of integers $\{1, 2, \ldots, c\}$.

$$\max_{j,j \in N_{c_{k-1}}} m_{(k,p)}(k-1,j) = 1; p = 1, \ldots, c_k; k = 1, \ldots, L.$$

This constraint essentially says that each concept has full membership in at least one of the parent concepts. We can always ensure that this condition is satisfied even when there are irrelevant or outlier concepts in the hierarchy by including a *miscellaneous* or *other* category. If we define $m_{(k+1,l)}(k-1,j) = \max_{p,p \in N_{c_k}} \min \left(m_{(k+1,l)}(k,p), m_{(k,p)}(k-1,j) \right)$, then, again we can easily verify that $\max_{j,j \in N_{c_{k-1}}} m_{(k+1,l)}(k-1,j) = \max_{p,j \in N_{c_k}} m_{(k+1,l)}(k,p) = 1$. Thus, a grandchild always has full membership in at least one of the grandparents.

Although fuzzy hierarchies make it simpler for the user to find relevant documents by removing the rigidity associated with crisp hierarchies, several issues related to the user interface need to be examined carefully. For example, when the user 'expands' a concept at a top level, should we show all concepts at the next level or only those with high memberships? Another interesting issue is how to generate fuzzy hierarchies in an efficient manner. It is easy to think of ways to modify the approaches presented in the earlier sections to produce fuzzy hierarchies.

7   Summary and Conclusions

In this paper, we presented an overview of several approaches to ATG. Many approaches presented here use soft assignments while creating the taxonomy. Since most real-world concepts are fuzzy, it makes sense to use a fuzzy approach to assign documents to the nodes in the hierarchy. Crisp assignments are rigid and make it harder for the user to access relevant documents. As explained in Section 6, we can also think of creating a fuzzy hierarchy, rather than a crisp one, to further ameliorate this problem.

In general, there will not be a unique taxonomy that is *correct* for a given collection of documents. However, in order to generate truly meaningful hierarchies, we also need to ensure that the labels at any given level are *comparable*, i.e., they refer to different aspects of a higher concept. For example, the concepts "Football" and "Baseball" are comparable because they are both sports, but the concepts "European" and "Botany" are incomparable. However, this requires very careful label selection based on domain or world knowledge and the context. Most information retrieval techniques use a list of *stopwords* (such as articles and prepositions) that are removed from further consideration. Unfortunately, what might be considered a stopword in one context may be meaningful in another context. For example, the word "can", when used as an auxiliary verb, may not carry much meaning, but it is a useful word if one is looking at various types of containers. Apart from the use of stopwords, there are also problems posed by synonyms and polysemy. Thus, the use of natural language processing, thesauri, and domain dictionaries/taxonomies is essential for deciphering the context and meaning of the text. There have been some attempts to incorporate natural language processing techniques such as part-of-speech (POS) patterns for the purpose of query disambiguation [35,36,37]. These techniques achieve disambiguation by identifying candidate documents (i.e., those that use the query words in the right *sense*) based on frequently occurring POS patterns. However, these systems do not provide a high-level summary of the contents of the document collection. The next leap in the quality of automatically generated taxonomies will perhaps occur when such techniques are integrated with fuzzy hierarchical clustering algorithms.

References

1. Franzen, K., Karlgren, J.: Verbosity and interface design. Technical Report T2000:04, Swedish Institute of Computer Science (SICS) (2000)
2. Sanderson, M.: Word sense disambiguation and information retrieval. In: Proceedings of SIGIR. (1994) 142–151
3. Salton, G.: Cluster search strategies and the optimization of retrieval effectiveness. Prentice Hall, Englewood Cliffs, N.J. (1971)
4. Griffiths, A., Luckhurst, H., Willett, P.: Using inter-document similarity information in document retrieval systems. Journal of the American Society for Information Sciences **37** (1986) 3–11
5. Hearst, M.A., Pedersen, J.O.: Reexamining the cluster hypothesis: Scatter/gather on retrieval results. In: Proceedings of SIGIR, Zürich, CH (1996) 76–84
6. Zamir, O., Etzioni, O.: Web document clustering: A feasibility demonstration. In: Research and Development in Information Retrieval. (1998) 46–54
7. Selberg, E., Etzioni, O.: Multi-service search and comparison using the MetaCrawler. In: Proceedings of the 4th International World-Wide Web Conference, Darmstadt, Germany (1995)
8. Klir, G.J., Yuan, B.: Fuzzy sets and Fuzzy logic. Prentice Hall, Englewood Cliffs, New Jersey (1995)
9. Vaithyanathan, S., Dom, B.: Model selection in unsupervised learning with applications to document clustering. In: The Sixth International Conference on Machine Learning (ICML- 1999). (1999) 423–433
10. Vaithyanathan, S., Dom, B.: Model-based hierarchical clustering. In: Proceedings of Sixth Conference on Uncertainty in Artificial Intelligence. (2000) 599–608
11. Nigam, K., McCallum, A.K., Thrun, S., Mitchell, T.M.: Learning to classify text from labeled and unlabeled documents. In: Proceedings of AAAI-98, 15th Conference of the American Association for Artificial Intelligence, Madison, US, AAAI Press, Menlo Park, US (1998) 792–799
12. Grefenstette, G.: Explorations in Automatic Thesaurus Discovery. Kluwer Academic Publishers (1994)
13. Hearst, M.A.: Automated discovery of WordNet relations. In Fellbaum, C., ed.: WordNet: an Electronic Lexical Database. MIT Press (1998)
14. Sanderson, M., W.B.Croft: Deriving concept hierarchies from text. In: Proceedings of SIGIR. (1999) 206–213
15. Lawrie, D., Croft, W.B., Rosenberg, A.: Finding topic words for hierarchical summarization. In: Proceedings of SIGIR, ACM Press (2001) 349–357

16. Krishna, K., Krishnapuram, R.: A clustering algorithm for asymmetrically related data with its applications to text mining. In: Proceedings of CIKM, Atlanta, USA (2001) 571–573
17. Xu, J., Croft, W.B.: Query expansion using local and global document analysis. In: Proceedings of SIGIR. (1996) 4–11
18. Baker, L.D., McCallum, A.K.: Distributional clustering of words for text classification. In: Proceedings of SIGIR, Melbourne, AU (1998) 96–103
19. Pereira, F.C.N., Tishby, N., Lee, L.: Distributional clustering of English words. In: Meeting of the Association for Computational Linguistics. (1993) 183–190
20. Dhillon, I.S.: Co-clustering documents and words using bipartite spectral graph partitioning. Technical Report TR2001-05, University of Texas, Austin (2001)
21. Kummamuru, K., Dhawale, A.K., Krishnapuram, R.: Fuzzy co-clustering of documents and keywords. In: Proceedings of FUZZIEEE, St. Louis, MO (2003)
22. Oh, C.H., Honda, K., Ichihashi, H.: Fuzzy clustering for categorical multivariate data. In: Proceedings of IFSA/NAFIPS, Vancouver, Canada (2001) 2154–2159
23. Bezdek, J.C., Hathaway, R.J.: Some notes on alternating optimization. In Pal, N.R., Sugeno, M., eds.: Advances in Soft Computing - AFSS 2002. Springer-Verlag (2002) 288–300
24. Frigui, H., Nasraoui, O.: Simultaneous categorization of text documents and identification of cluster-dependent keywords. In: Proceedings of FUZZIEEE, Honolulu, Hawaii (2002) 158–163
25. Frigui, H., Nasraoui, O.: Simultaneous clustering and attribute discrimination. In: Proceedings of FUZZIEEE, San Antonio (2000) 158–163
26. Mandhani, B., Joshi, S., Kummamuru, K.: A matrix density based algorithm to hierarchically co-cluster documents and words. In: Proceedings of WWW 2003 Conference, Budapest, Hungary (2003)
27. Oyanagi, S., Kubota, K., Nakase, A.: Application of matrix clustering to web log analysis and access prediction. In: Proceedings of WEBKDD, San Francisco (2001)
28. Liu, X., Gong, Y., Xu, W., Zhu, S.: Document clustering with cluster refinement and model selection capabilities. In: Proceedings of SIGIR, ACM Press (2002) 191–198
29. Cover, T.M., Thomas, J.A.: Elements of Information Theory. Wiley-Interscience (1991)
30. Van Rijsbergen, C.J.: Information Retrieval, 2nd edition. Dept. of Computer Science, University of Glasgow (1979)
31. Chawathe, S.S.: Comparing hierarchical data in external memory. In: Proceedings of the Twenty-fifth International Conference on Very Large Data Bases, Edinburgh, Scotland, U.K. (1999) 90–101
32. Shasha, D., Zhang, K.: Approximate Tree Pattern Matching. Oxford University Press (1995)
33. Lee, D.H., Kim, M.H.: Database summarization using fuzzy ISA hierarchies. IEEE Trans. On Systems Man And Cybernetics Part B- Cybernetics 27 (1997) 68–78
34. Krishnapuram, R., Keller, J.M.: A possibilistic approach to clustering. IEEE Transactions on Fuzzy Systems 1 (1993) 98–110
35. Grefenstette, G.: SQLET: Short query linguistic expansion techniques: Palliating one or two-word queries by providing intermediate structure to text. In: Proceedings of RIAO. (1997)
36. Anick, P.G., Tipirneni., S.: The paraphrase search assistant: Terminological feedback for iterative information seeking. In: Proceedings of SIGIR. (1999) 153–159
37. Allan, J., Raghvan, H.: Using part-of-speech patterns to reduce query ambiguity. In: Proceedings of SIGIR, Tampere, Finland (2002)

A Fuzziness Measure of Rough Sets*

Hsuan-Shih Lee

Department of Shipping and Transportation Management
National Taiwan Ocean University
Keelung 202, Taiwan
Republic of China

1 Introduction

The concept of the rough set is a new mathematical approach to imprecision, vagueness and uncertainty in data analysis beside fuzzy set theory. Rough sets were first introduced at the beginning of the eighties by Z. Pawlak [10,11,13] and belong to the family of concepts concerning the modeling and representing of incomplete knowledge [12,14].

The origin of the rough set philosophy is the assumption that with every object we associate some information. Objects can be something like patients, while the symptoms of the disease are the information employed to characterized patients. Objects are similar or indiscernible, if they are characterized by the same information. In contrast to the theory of fuzzy set which models the uncertainty with membership function, rough sets approach vague concept by charactering the vagueness with a pair of crisp sets, the lower and the upper approximation of the vague concept.

Dubios and Prade [6] combined fuzzy sets and rough sets in a fruitful way by defining rough fuzzy sets and fuzzy rough sets. Banerjee and Pal [1] have characterized a measure of roughness of a fuzzy set making use of the concept of rough fuzzy sets. They also suggested some possible applications of the measure in pattern recognition and image analysis problems. Rough sets and fuzzy sets are also studied by [2,3,8,9,16,17]. Recently, Chakrabarty et al. [4] gave a measure of fuzziness in rough set. In this paper, we give a different measure of fuzziness in rough sets, preventing the inconsistency problem of the measure proposed by Chakrabarty et al. By inconsistency, we mean that a rough set may have multiple fuzziness by the measure proposed by Chakrabarty et al.

2 Rough Sets

Let U be a nonempty set and R be an indiscernibility relation or equivalence relation on U. Then (U, R) is called a Pawlak approximation space. Let the

* This research work was partially supported by the National Science Council of the Republic of China under grant No. NSC90-2416-H-019-002-

T. Bilgiç et al. (Eds.): IFSA 2003, LNAI 2715, pp. 64–70, 2003.

concept X be a subset of U. Then the lower approximation of X in (U, R), denoted as $LA_R(X)$, is defined to be

$$LA_R(X) = \{x | [x]_R \subseteq X\}$$

and the upper approximation of X in (U, R), denoted as $UA_R(X)$, is defined to be

$$UA_R(X) = \{x | [x]_R \cap X \neq \phi\},$$

where $[x]_R$ is an equivalence class of R containing x. The equivalence classes of R and the empty set ϕ are called elementary or atomic sets in the approximation space (U, R). The union of one or more elementary sets is called a composed set. The family of all composed sets, including the empty set, is denoted by $Comp((U, R))$, which is a Boolean algebra and a subalgebra of Boolean algebra 2^U. Pawlak regards the group of subsets of U with the same upper and lower approximations in (U, R) as a rough set in (U, R). Using lower and upper approximations, an equivalence relation $\approx_R$ can be defined on the powerset of U:

$$X \approx_R Y \Leftrightarrow LA_R(X) = LA_R(Y) \text{ and } UA_R(X) = UA_R(Y),$$

where $X, Y \in 2^U$ and R is an equivalence relation on U.

This equivalence relation induces a partition on the power set 2^U. An equivalence class of such partition is called a P-rough set. The set of all P-rough set is denoted by $2^U / \approx_R$. More specifically, a P-rough set can be defined as follows:

Definition 1. *Given the approximation space (U, R) and two sets $A_1, A_2 \in Comp((U, R))$ with $A_1 \subseteq A_2$, a P-rough set is the family of subset of U described as follows:*

$$< A_1, A_2 >= \{X \in 2^U | LA_R(X) = A_1, UA_R(X) = A_2\}.$$

Equivalently, a P-rought set containing $X \in 2^U$ can be defined as:

$$[X]_{\approx_R} =$$
$$\{Y \in 2^U | LA_R(Y) = LA_R(X), UA_R(Y) = UA_R(X)\}.$$

In other words,

$$[X]_{\approx_R} =< LA_R(X), UA_R(X) > .$$

A member of $[X]_R$ is also referred to as a generator of the P-rough set [5].

P-rough set intersection $\sqcap$, union $\sqcup$ and complement $\neg$ are defined as follow:

$$[X]_{\approx_R} \sqcap [Y]_{\approx_R} = [Z]_{\approx_R},$$

with

$$LA_R(Z) = LA_R(X) \cap LA_R(Y),$$
$$UA_R(Z) = UA_R(X) \cap UA_R(Y);$$
$$[X]_{\approx_R} \sqcup [Y]_{\approx_R} = [Z]_{\approx_R},$$

with

$$LA_R(Z) = LA_R(X) \cup LA_R(Y),$$
$$UA_R(Z) = UA_R(X) \cup UA_R(Y);$$

and

$$\neg[X]_{\approx_R} = [Z]_{\approx_R},$$

with

$$LA_R(Z) = \overline{UA_R(X)}, UA_R(Z) = \overline{LA_R(X)}.$$

In general,

$$LA_R(X \cap Y) = LA_R(X) \cap LA_R(Y)$$
$$UA_R(X \cap Y) \subseteq UA_R(X) \cap UA_R(Y)$$
$$LA_R(X \cup Y) \supseteq LA_R(X) \cup LA_R(Y)$$
$$UA_R(X \cup Y) = UA_R(X) \cup UA_R(Y)$$

After Pawlak's initiative, Iwinski subsequently interpreted rough set in an algebraic way [7]. Let $\mathbb{B}$ be a complete subalgebra of the Boolean algebra 2^U. The pair $(U, \mathbb{B})$ is called a rough universe. Iwinski defined rough sets as follows:

Definition 2. *Given the rough universe* $(U, \mathbb{B})$, *the pair* (A_1, A_2) *is a rough set iff* $A_1, A_2 \in \mathbb{B}$ *and* $A_1 \subseteq A_2$.

We shall call (A_1, A_2) an I-rough set. In fact the views of Pawlak and Iwinski in rough sets are equivalent.

Theorem 1. *There is an one-to-one correspondence between approximation space* (U, R) *and rough universe* $(U, \mathbb{B})$.

3   Fuzziness of Rough Sets

First, the definition of fuzziness in rough sets proposed by Chakrabarty et al. [4] is reiterated.

Definition 3. *Let* $\tilde{A}$ *be a fuzzy set. Then the nearest ordinary set to* $\tilde{A}$ *is denoted by* $\underline{\tilde{A}}$ *and is given by*

$$\mu_{\underline{\tilde{A}}}(x) = \begin{cases} 0 & \text{if } \mu_{\tilde{A}}(x) < 0.5, \\ 1 & \text{if } \mu_{\tilde{A}}(x) > 0.5, \\ 0 \text{ or } 1 & \text{if } \mu_{\tilde{A}}(x) = 0.5. \end{cases}$$

Definition 4. *The index of fuzziness of a fuzzy set* $\tilde{A}$ *having n supporting points is defined as*

$$v_p(\tilde{A}) = (2/n^p)d(\tilde{A}, \underline{\tilde{A}}),$$

where $d(\tilde{A}, \underline{\tilde{A}})$ *denotes the distance between the fuzzy set* $\tilde{A}$ *and its nearest ordinary set* $\underline{\tilde{A}}$.

The value of p depends on the type of distance function used, e.g. $p = 1$ for a generalized Hamming distance whereas $p = 0.5$ for an Euclidean distance. When $p = 1$, $v_1(\tilde{A})$ is called the linear index of fuzziness of $\tilde{A}$. When $p = 0.5$, $v_{0.5}(\tilde{A})$ is called the quadratic index of fuzziness of $\tilde{A}$.

Let (U, R) be an approximation space and $X \subseteq U$. The rough set of X in (U, R) is $R(X) = (LA_R(X), UA_R(X))$. Given a subset X of U, Chakrabarty et al. [4] defined an induced fuzzy set $\tilde{F}_X^R$ based on the rough membership function: [15]

$$\mu_{\tilde{F}_X^R}(x) = \frac{|[x]_R \cap X|}{|[x]_R|}.$$

Chakrabarty et al. defined the fuzziness in an I-rough set as follows:

Definition 5. [4] The fuzziness of the I-rough set $R(X) = (LA_R(X), UA_R(X))$ is defined to be $v_k(\tilde{F}_X^R)$, where $\tilde{F}_X^R$ the induced fuzzy set of the rough set $R(X)$. The linear fuzziness in the rough set $R(X)$ is $v_1(\tilde{F}_X^R)$. The quadratic fuzziness in the rough set $R(X)$ is $v_{0.5}(\tilde{F}_X^R)$.

However, following the definition of the fuzziness of a rough set proposed by Chakrabarty et al., we may have different fuzziness for one rough set. To avoide such circumstances, we define the fuzziness of a rough set $R(X)$ as follows.

Definition 6. Let the fuzzy set of order k induced by the rough set $R(X)$ be denoted as $\tilde{G}_k^{R(X)}$. Then

$$\mu_{\tilde{G}_k^{R(X)}}(x) = \left(\frac{\sum_{u \in \{ \frac{|[x]_R \cap Y|}{|[x]_R|} | Y \in [X]_{\approx_R} \}} u^k}{|\{ \frac{|[x]_R \cap Y|}{|[x]_R|} | Y \in [X]_{\approx_R} \}|} \right)^{\frac{1}{k}}.$$

Following the definition, we have the following property:

Property 1. Given $A_1, A_2 \in Comp((U, R))$ and $A_1 \subseteq A_2$, $\tilde{G}_k^{R(X)} = \tilde{G}_k^{R(Y)}$ if $X, Y \in < A_1, A_2 >$

Definition 7. The fuzziness of the rough set $R(X)$ is defined to be the index of the fuzziness of the fuzzy set $\tilde{G}_k^{R(X)}$, $v_p(\tilde{G}_k^{R(X)})$.

Following Property 1, we define the fuzziness of a rough set (A_1, A_2) as follows.

Definition 8. Given $A_1, A_2 \in Comp((U, R))$ and $A_1 \subseteq A_2$, the fuzziness of the rough set (A_1, A_2) is $v_p(\tilde{G}_k^{R(X)})$, where $X \in < A_1, A_2 >$.

Property 2. For any X in an approximation space (U, R), the following holds:

$$\left\{ \frac{|[x]_R \cap Y|}{|[x]_R|} | Y \in [X]_{\approx_R} \right\} = \begin{cases} \{ \frac{j}{|[x]_R|} | 1 \leq j \leq \frac{|[x]_R| - 1}{|[x]_R|} \} & x \in UA_R(X) - LA_R(X) \\ \{1\} & x \in LA_R(X) \\ \{0\} & x \in U - UA_R(X). \end{cases}$$

Following property 2, we have the following:

Property 3. For any X in an approximation space (U, R),

$$|\{\frac{|[x]_R \cap Y|}{|[x]_R|}|Y \in [X]_{\approx_R}\}| = \begin{cases} |[x]_R| - 1 & x \in UA_R(X) - LA_R(X) \\ 1 & \text{otherwise.} \end{cases}$$

Following the definition 6 and properties above, we have:

Property 4. For any X in an approximation space (U, R), the following holds:

$$\mu_{\tilde{G}_k^{R(X)}}(x) = \begin{cases} (\frac{\sum_{j=1}^{|[x]_R|-1} j^k}{|[x]_R|^k(|[x]_R|-1)})^{\frac{1}{k}} & x \in UA_R(X) - LA_R(X) \\ 1 & x \in LA_R(X) \\ 0 & x \in U - UA_R(X) \end{cases}$$

When $k = 1$, the membership function coincides with the rough membership function proposed by Pawlak in [13]:

Property 5. For any X in an approximation space (U, R), the following holds:

$$\mu_{\tilde{G}_1^{R(X)}}(x) = \begin{cases} 0.5 & x \in UA_R(X) - LA_R(X) \\ 1 & x \in LA_R(X) \\ 0 & x \in U - UA_R(X). \end{cases}$$

Property 6. For any approximation space (U, R), we have

1. $\tilde{G}_k^{R(U)} = U$,
2. $\tilde{G}_k^{R(\phi)} = \phi$.

Property 7. For $X, Y \subseteq U$, if $X \subseteq Y$, then $\tilde{G}_k^{R(X)} \subseteq \tilde{G}_k^{R(Y)}$.

Property 8. For $X, Y \subseteq U$, the following holds:

1. $\tilde{G}_k^{R(X)} \cup \tilde{G}_k^{R(Y)} \subseteq \tilde{G}_k^{R(X \cup Y)}$
2. $\tilde{G}_k^{R(X)} \cup \tilde{G}_k^{R(Y)} = \tilde{G}_k^{R(X \cup Y)}$ if either $[X]_{\approx_R} = [Y]_{\approx_R}, X \subseteq Y$ or $[X]_{\approx_R} = [Y]_{\approx_R}, Y \subseteq X$.

Property 9. For $X, Y \subseteq U$, the following holds:

1. $\tilde{G}_k^{R(X)} \cap \tilde{G}_k^{R(Y)} \supseteq \tilde{G}_k^{R(X \cap Y)}$
2. $\tilde{G}_k^{R(X)} \cap \tilde{G}_k^{R(Y)} = \tilde{G}_k^{R(X \cap Y)}$ if $[X]_{\approx_R} = [Y]_{\approx_R}$.

Property 10. For any approximation space (U, R),

$$\tilde{G}_k^{R(X)} = \begin{cases} \overline{\tilde{G}_k^{R(\overline{X})}}, & \text{if } x \in UA_R(X) - LA_R(X) \\ \tilde{G}_k^{R(\overline{X})}, & \text{otherwise} \end{cases}$$

where $\overline{X}$ is the complement of X.

The linear index of fuzziness of $\tilde{G}_k^{R(X)}$ and the quadratic index of fuzziness of $\tilde{G}_k^{R(X)}$ can be computed according to the following:

Property 11. For any X in an approximation space (U, R), we have

$$v_1(\tilde{G}_k^{R(X)}) = \frac{2}{|U|} \sum_{x \in UA_R(X) - LA_R(X)} |[x]_R| \min((\frac{\sum_{j=1}^{|[x]_R|-1} j^k}{|[x]_R|^k(|[x]_R| - 1)})^{\frac{1}{k}},$$

$$1 - (\frac{\sum_{j=1}^{|[x]_R|-1} j^k}{|[x]_R|^k(|[x]_R| - 1)})^{\frac{1}{k}})$$

and

$$v_{0.5}(\tilde{G}_k^{R(X)}) = \frac{2}{\sqrt{|U|}}(\sum_{x \in UA_R(X) - LA_R(X)} |[x]_R|(\min((\frac{\sum_{j=1}^{|[x]_R|-1} j^k}{|[x]_R|^k(|[x]_R| - 1)})^{\frac{1}{k}},$$

$$1 - (\frac{\sum_{j=1}^{|[x]_R|-1} j^k}{|[x]_R|^k(|[x]_R| - 1)})^{\frac{1}{k}}))^2)^{\frac{1}{2}}.$$

Property 12.

$$v_p(\tilde{G}_k^{R(\overline{X})}) = v_p(\tilde{G}_k^{R(X)})$$

Property 13.

$$v_1(\tilde{G}_1^{R(X)}) = \frac{|UA_R(X) - LA_R(X)|}{|U|}$$

and

$$v_{0.5}(\tilde{G}_1^{R(X)}) = \sqrt{\frac{|UA_R(X) - LA_R(X)|}{|U|}}.$$

That is

$$v_{0.5}(\tilde{G}_1^{R(X)}) = \sqrt{v_1(\tilde{G}_1^{R(X)})}.$$

4   Conclusions

Rough set theory has been considered as a useful mean to model the vagueness and has been successfully applied in many fields. Every rough set is associated with some amount of fuzziness. In this paper, we have proposed a measure of fuzziness in rough set. Our measure prevents the inconsistency problem incurred in the measure proposed by Chakrabarty et al. Some properties of our measure also have been studied.

References

1. M. Banerjee, K.K. Pal, Roughness of a fuzzy set, Information Science 93 (1996) 235-246.
2. R. Biswas, On rough sets and fuzzy rough sets, Bull. Pol. Acad. Sci. Math. 42 (1994) 345-349.
3. R. Biswas, On rough fuzzy sets, Bull. Pol. Acad. Sci. Math. 42 (1994) 352-355.
4. K. Chakrabarty, Ranjit Biswas, and Sudarsan Nanda, Fuzziness in rough sets, Fuzzy Sets and Systems 110 (2000) 247-251.
5. S. Chanas, D. Kuchta, Further remarks on the relation between rough and fuzzy sets, Fuzzy Sets and Systems 47 (1992) 391-394.
6. D. Dubios, H. Prade, Rough fuzzy sets and fuzzy rough sets, Int. J. General Systems 17 (1990) 191-208.
7. T.B. Iwinski, Algebraic approach to rough sets, Bull. Polish Acad. Sci. Math. 35 (1987) 673-683.
8. A. Nakamura, Fuzzy rough sets, Note on Multiple-valued Logic in Japan 9 (8) (1988) 1-8.
9. S. Nanda, Fuzzy rough sets, Fuzzy Sets and Systems 45 (1992) 157-160.
10. Z. Pawlak, Rough sets, Report No. 431, Polish Academy of Sciences, Institute of Computer Science (1981).
11. Z. Pawlak, Rough sets, Internat. J. Inform. Comput. Sci. 11(5) (1982) 341-356.
12. Z. Pawlak, Rough classification, Internat. J. Man-Machine. Stud. 20 (1984) 469-483.
13. Z. Pawlak, Rough set and fuzzy sets, Fuzzy Sets and Systems 17 (1985) 99-102.
14. Z. Pawlak, Rough sets: A new approach to vagueness, in Fuzzy Logic for the Management of Uncertainty (L. A. Zadeh and J. Kacprzyk, Eds.), Wiley, New York (1992) 105-118.
15. Z. Pawlak and A. Skowron, Rough membership functions, in Fuzzy Logic for the Management of Uncertainty (L.A. Zadeh and J. Kacprzyk, Eds.), Wiley, New York (1994) 251-271.
16. Y.Y. Yao, Two views of the theory of rough sets in finite universes, International Journal of Approximate Reasoning 15 (1996) 291-317.
17. Y.Y. Yao, A comparative study of fuzzy sets and rough sets, Journal of Information Sciences 109 (1998) 227-242.

Fuzzy Closure Operators Induced by Similarity

Radim Bělohlávek[1,2]

[1] Dept. Computer Science, Palacký University, Tomkova 40, CZ-779 00, Olomouc,
Czech Republic
radim.belohlavek@upol.cz
[2] Inst. Res. Appl. Fuzzy Modeling, University of Ostrava, Bráfova 7, 701 03 Ostrava,
Czech Republic

Abstract. In fuzzy set theory, similarity phenomenon is approached
using so-called fuzzy equivalence relations. An important role in fuzzy
modeling is played by similarity-based closure (called also the extensional
hull). Intuitively, the degree to which an element x belongs to a similarity-
based closure of a fuzzy set A is the degree to which it is true that there
is an element y in A which is similar to x. In this paper, we give an
axiomatic characterization of the operation of a similarity-based closure
and discuss some consequences and related topics.

1 Introduction

The concept of similarity and related concepts of distance, nearness, proximity,
closeness etc. are among the basic concepts when modeling real-world phenomena. Of the most common approaches that allow us to quantify distance (or
nearness) of objects of interest is the concept of a metric space. Fuzzy set theory
offers another concept for modeling of similarity, so-called fuzzy equivalence. A
fuzzy equivalence induces so-called extensional hull operator. First, the extensional hull of A induced by a fuzzy equivalence E is the least fuzzy set compatible
with E which contains A. Second, the extensional hull of A may be interpreted
as an answer to a query A containing all elements similar to some element from
A. The concepts of a fuzzy equivalence and that of an extensional hull of a
fuzzy set are among the very important concepts having natural interpretation,
interesting properties, and immediate applications, see e.g. [5,7,8,9,10,13]. The
aim of this paper is investigate the concept of an extensional hull and to give
its complete characterization in terms of so-called fuzzy closure operators [2].
Moreover, we discuss the relationship of the concept of the extensional hull (i.e.
a similarity-based closure) and that of the metric-based closure. In Section 2 we
recall the necessary notions. The results of the paper and discussion is presented
in Section 3.

2 Preliminaries

We recall necessary notions form fuzzy logic and fuzzy sets. We will use complete
residuated lattices as the structures of truth values. Complete residuated lattices

T. Bilgiç et al. (Eds.): IFSA 2003, LNAI 2715, pp. 71–78, 2003.
© Springer-Verlag Berlin Heidelberg 2003

play a crucial role in fuzzy logic (see e.g. [6,7,8,9]). Recall that a complete residuated lattice is an algebra $\mathbf{L} = \langle L, \wedge, \vee, \otimes, \rightarrow, 0, 1 \rangle$ such that $\langle L, \wedge, \vee, 0, 1 \rangle$ is a complete lattice with the least element 0 and the greatest element 1; $\langle L, \otimes, 1 \rangle$ is a commutative monoid, i.e. $\otimes$ is commutative, associative, and $x \otimes 1 = x$ holds holds for each $x \in L$; and $\otimes, \rightarrow$ form an adjoint pair, i.e. $x \otimes y \leq z$ iff $x \leq y \rightarrow z$ holds for all $x, y, z \in L$. All properties of residuated lattices used in this paper can be found in the references. The most studied and applied set of truth values is the real interval $[0, 1]$ with $a \wedge b = \min(a, b)$, $a \vee b = \max(a, b)$, and with three well-known important pairs of adjoint operations: Łukasiewicz, minimum, and product. More generally, $\langle [0, 1], \min, \max, \otimes, \rightarrow, 0, 1 \rangle$ is a complete residuated lattice on $[0, 1]$ iff $\otimes$ is a left-continuous t-norm [8] and $a \rightarrow b = \max\{z \mid a \otimes z \leq b\}$. Another complete residuated lattice is the Boolean algebra **2** of classical logic with the support $2 = \{0, 1\}$.

"Standard" fuzzy set theory (i.e. $L = [0, 1]$) generalizes for general **L** in a straightforward way: An **L**-set (fuzzy set with truth degrees in **L**) [14,6] A in a universe set X is any map $A : X \rightarrow L$. By L^X we denote the set of all **L**-sets in X. Given $A, B \in L^X$, the subsethood degree $S(A, B)$ of A in B is defined by $S(A, B) = \bigwedge_{x \in X} A(x) \rightarrow B(x)$. We write $A \subseteq B$ if $S(A, B) = 1$. Analogously, the equality degree $(A \approx B)$ of A and B is defined by $(A \approx B) = \bigwedge_{x \in X}(A(x) \leftrightarrow B(x))$ where $\leftrightarrow$ is the so-called biresiduum defined by $a \leftrightarrow b = (a \rightarrow b) \wedge (b \rightarrow a)$. It is immediate that $E(A, B) = S(A, B) \wedge S(B, A)$. A binary **L**-relation E on a set X is called an **L**-equivalence (fuzzy equivalence, (**L**-)similarity) if

$$E(x, x) = 1 \tag{1}$$

$$E(x, y) = E(y, x) \tag{2}$$

$$E(x, y) \otimes E(y, z) \leq E(x, z). \tag{3}$$

An **L**-equivalence is called an **L**-equality if

$$E(x, y) = 1 \quad \text{implies} \quad x = y. \tag{4}$$

An **L**-set A in X is said to be compatible with an **L**-equivalence $\approx$ on X if $A(x) \otimes (x \approx y) \leq A(y)$ for each $x, y \in X$. The collection of all **L**-set in X compatible with $\approx$ will be denoted by $L^{\langle X, \approx \rangle}$.

3   Fuzzy Closure Induced by Similarity

Coming to similarity-based fuzzy closure The set $L^{\langle X, \approx \rangle}$ contains exactly those fuzzy sets in X that respect $\approx$. For a given fuzzy set A in X, the least fuzzy set $C(A)$ which both contains A and is compatible with $\approx$ is given by

$$C(A) = \bigcap \{B \mid B \in L^{\langle X, \approx \rangle}, A \subseteq B\}.$$

It is well-known that $C(A)$ may be described directly using $\approx$ by

$$C(A)(y) = \bigvee_{x \in X} A(x) \otimes (x \approx y). \tag{5}$$

If X is a set of elements of a database then a user-query may be given by listing some examples representing appropriate results for the query, each with a degree to which it is appropriate. That is, the query may be given by a fuzzy set A in X ($A(x)$ is the degree to which x is appropriate). Naturally, we expect the answer to the query A to contain those elements from X which are similar to some example from A. In other words, we expect that the degree Ans$(A)(y)$ to which an element y from X belongs to the answer Ans(A) is the truth degree of the fact "there is an element x in A such that x and y are similar". Basic rules of semantics of fuzzy logic tell that Ans$(A)(y)$ is just equal to C$(A)(y)$ defined in (5).

From the above examples it is clear that the operator of a similarity-based closure assigning a fuzzy set C(A) to a fuzzy set A is an important one. Our aim in the following is twofold. First, we discuss some relationships between similarity-based closure and metric closure. Second, we investigate abstract properties of the similarity-based closure operator and provide its complete axiomatization.

Similarity-based fuzzy closure and metric closure The concept of a (pseudo)metric is the one mostly applied when considering closeness of objects (usually of a geometric nature). There is an obvious question of what is the relationship between the notion of a metric and the notion of a **L**-similarity (i.e. a fuzzy equivalence) and that of the relationship between the well-established concept of a metric closure and that of a similarity-based closure.

We recall the following result (see e.g. [11]).

Theorem 1 (representation by additive generator). *A mapping* $\otimes$: $[0,1]^2 \to [0,1]$ *is a continuous Archimedean t-norm iff there is a continuous additive generator f such that*

$$a \otimes b = f^{(-1)}(f(x) + f(y)),$$

i.e. f is a strictly decreasing continuous mapping $f : [0,1] \to [0,\infty]$ with $f(1) = 0$ and $f^{(-1)}$ is the pseudoinverse of f defined by $f^{(-1)}(x) = f^{-1}(x)$ if $x \leq f(0)$ and $f^{(-1)}(x) = 0$ otherwise.

Recall that $\otimes$ is a continuous Archimedean t-norm if $\otimes$ is continuous as a real function and satisfies $a \otimes a < a$ for each $a \neq 0, 1$.

Łukasiewicz as well as product t-norms are both continuous and Archimedean. $f(x) = 1 - x$ and $f^{(-1)}(x) = \max(1 - x, 0)$ are an additive generator and its pseudoinverse of the Łukasiewicz t-norm; $f(x) = -\log(x)$ and $f^{(-1)}(x) = e^{-x}$ are an additive generator and its pseudoinverse of the product t-norm. Now, we have the following result. The following result shows a general relationship between pseudometrics and fuzzy equivalences with continuous Archimedean t-norms (the result follows by easy combination of results obtained in [4]).

Theorem 2 (similarity spaces vs. (pseudo)metric spaces). *Let $\otimes$ be a continuous Archimedean t-norm with an additive generator f, **L** be a residuated* *lattice on* $[0,1]$ *given by* $\otimes$, $\approx$ *be an* **L**-*equivalence on* X, δ *be a pseudometric on* X *in a generalized sense. Then (1)* $\delta_\approx : [0,1]^2 \to [0,\infty]$ *defined by*

$$\delta_\approx(x,y) = f(x \approx y)$$

is a pseudometric in a generalized sense which is a metric iff $\approx$ *is an* **L**-*equality;* *(2)* $\approx_\delta: [0,1]^2 \to [0,1]$ *defined by*

$$(x \approx_\delta y) = f^{(-1)}(\delta(x,y))$$

is an **L**-*equivalence on* X *which is an* **L**-*equality iff* δ *is a metric; (3)* $\approx$ *equals* $\approx_{\delta_\approx}$ *and if* $\delta(X,X) \subseteq [0,f(0)]$ *then* δ *equals* $\delta_{\approx_\delta}$.

Next, we discuss some basic relationships between the similarity-based fuzzy closure and metric closure. Recall that if $\mathbf{X},\delta$ is a (pseudo)metric space and $A \subseteq X$ then the set $C_\delta(A)$ defined by

$$C_\delta(A) = \{y \mid \text{for each } \varepsilon > 0 \text{ there is } x \in A : \delta(x,y) < \varepsilon\}$$

is called the closure of A in $\mathbf{X}$. Our aim is to discuss the relationship between C_δ and $C_\approx$ where $\approx$ is a fuzzy similarity corresponding (somehow) to δ and $C_\approx$ is the $\approx$-based operator defined by (5). If δ is a (generalized) pseudometric, then for any $a \in (0,\infty]$, the mapping $\delta_a : \langle x,y \rangle \mapsto \min(a,\delta(x,y))$ is a (generalized) pseudometric as well (easy to verify). Moreover, it is easy to show that $C_\delta = C_{\delta_a}$. Therefore, if we are interested in the metric closure only, we may restrict our attention to (generalized) (pseudo)metrics with $\delta(X,X) \subseteq [0,f(0)]$ where f is the generator of a given continuous Archimedean t-norm $\otimes$. Namely, $C_\delta = C_{\delta_{f(0)}}$ and, due to Theorem 2, there is a one-to-one correspondence between (generalized) (pseudo)metrics satisfying $\delta(X,X) \subseteq [0,f(0)]$ and **L**-similarities where **L** is given by the corresponding t-norm $\otimes$. The next theorem shows a way to describe C_δ using $C_\approx$.

Theorem 3. *Let* $\otimes$ *be a continuous Archimedean t-norm with a continuous additive generator* f, *let* δ *be a (generalized) (pseudo)metric and* $\approx$ *be an* **L**-*similarity corresponding to* δ, *i.e.* $\approx=\approx_\delta$ *and* $\delta = \delta_\approx$ *(cf. Theorem 2). Then*

$$C_\delta(A) = {}^1(C_\approx(A))$$

for each $A \subseteq X$. *Furthermore,*

$$x \in C_\delta(A) \text{ iff for each } \varepsilon < 1 : A \cap {}^\varepsilon C_\approx(\{{}^1/x\}) \neq \emptyset$$

for each $A \subseteq X$, $x \in X$.

Proof. First, we show $C_\delta(A) = {}^1(C_\approx(A))$: We have $y \in {}^1(C_\approx(A))$ iff $1 = \bigvee_{x \in A}(x \approx y) = \bigvee_{x \in A} f^{(-1)}(\delta(x,y))$. That is, for each $\eta > 0$ there is $x \in A$ such that $f^{(-1)}(\delta(x,y)) > 1 - \eta$ which is equivalent to saying that for each η with $f(0) \geq \eta > 0$ there is $x \in A$ such that $f^{(-1)}(\delta(x,y)) > 1 - \eta$. Since $f^{(-1)} : [0,f(0)] \to [0,1]$, is the inverse function to $f : [0,1] \to [0,f(0)]$, the latter condition is equivalent to saying that for each η with $f(0) \geq \eta > 0$ there is $x \in A$ such that $\delta(x,y) < f(1 - \eta)$. Now, since for $\varepsilon = f(1 - \eta)$ we have that $\varepsilon \to 0$ iff $f(1 - \eta) \to 0$, we further have that the latter condition holds iff for each $\varepsilon > 0$ there is $x \in A$ such that $\delta(x,y) < \varepsilon$ which is equivalent to $y \in C_\delta(A)$. The second part may be proved in a similar way.

Similarity-based closure as a special fuzzy closure operator A fuzzy set $C(A)$ is often called the extensional hull (or closure) of A. In what follows we consider the operator C from the point of view of closure operators of fuzzy sets as studied in [2,3], see also [5].

For **L**-sets $A, B \in L^X$, we put $\rho_{\mathbf{X}}(A,B) = \bigvee_{x,y \in X}(A(x) \otimes (x \approx y) \otimes B(y))$. $\rho_{\mathbf{X}}(A,B)$ is naturally interpreted as the truth degree of the fact that there are some x in A and y in B which are similar. One can easily see that $\rho_{\mathbf{X}}$ extends $\approx$ in that $(x \approx y) = \rho_{\mathbf{X}}(\{^1/x\}, \{^1/y\})$. $\rho_{\mathbf{X}}$ is a reflexive and symmetric relation on L^X which is not transitive in general. It is easy to see that if $A \in L^{\langle X, \approx \rangle}$ or $B \in L^{\langle X, \approx \rangle}$ then $\rho_{\mathbf{X}}(A,B) = \bigvee_{x \in X}(A(x) \otimes B(x))$, i.e. $\rho_{\mathbf{X}}(A,B)$ is the height of $A \otimes B$. An immediate verification shows that introducing $C_{\mathbf{X}} : L^X \to L^X$ by

$$C_{\mathbf{X}}(A)(x) = \rho_{\mathbf{X}}(\{^1/x\}, A). \tag{6}$$

we have $C_\approx(A)(x) = \bigvee_{y \in X} A(y) \otimes (x \approx y)$ which is the definition (5) of the similarity-based closure of A. We will freely use any of $C_{\mathbf{X}}$, $C_\approx$, and C to denote the operator we are dealing with. Recall the following definition and basic results from [2].

Definition 1. *Let $K \subseteq L$ be a $\leq$-filter (i.e. $K \neq \emptyset$, and $a \in K$, $a \leq b$ imply $b \in K$). An $\mathbf{L}_K$-closure operator on a set X is a mapping $C : L^X \to L^X$ satisfying*

$$A \subseteq C(A) \tag{7}$$
$$S(A_1, A_2) \leq S(C(A_1), C(A_2)) \quad \text{whenever } S(A_1, A_2) \in K \tag{8}$$
$$C(A) = C(C(A)) \tag{9}$$

for every $A, A_1, A_2 \in L^X$.

Definition 1 generalizes some earlier approaches to fuzzy closure operators [5], mainly in that it takes into account parital subsethood (sensitivity to to partial subsethood is parametrized by K). Particularly, for $L = [0,1]$, $\mathbf{L}_{\{1\}}$-closure operators are precisely fuzzy closure operators [5]. If $K = L$, we omit the subscript K and use the term **L**-closure operator. It is easily seen that for $\mathbf{L} = \mathbf{2}$ (classical logic), the notion of an $\mathbf{L}_K$-closure operator coincides with the notion of a closure operator. The next theorem gives a characterization of a system of closed fuzzy sets of **L**-closure operators (see [2]). For $C : L^X \to L^X$ denote $\mathcal{S}_C = \{A \mid A = C(A)\}$.

Theorem 4. *$\mathcal{S}$ is a system of all closed fuzzy sets of some **L**-closure operator C, i.e. $\mathcal{S} = \mathcal{S}_C$, iff $\mathcal{S}$ is closed under arbitrary intersections and a-shifts, i.e. for $A_i, A \in \mathcal{S}$, $a \in L$, we have $\cap_i A_i \in \mathcal{S}$ and $a \to A \in \mathcal{S}$.*

A system $\mathcal{S}$ of **L**-sets in X is called an **L**-closure system in X if it is closed under arbitrary intersections and a-shifts. In [2] it is shown that $C \mapsto \mathcal{S}_C$ and $\mathcal{S} \mapsto C_{\mathcal{S}}$, where $C_{\mathcal{S}}(A) = \bigcap \{B \in \mathcal{S} \mid A \subseteq B\}$, establish a bijective correspondence between **L**-closure operators and **L**-closure systems in X. We now proceed to show that similarity-based closures are exactly **L**-closure operators satisfying some additional properties.

Lemma 1. *Let* $\mathbf{X} = \langle X, \approx \rangle$ *be an* **L***-similarity space. Then the mapping* $C_{\mathbf{X}}$ *defined by (6) is an* **L***-closure operator satisfying, moreover,*

$$C_{\mathbf{X}}\left(\bigcup_{i \in I} A_i\right) = \bigcup_{i \in I} C_{\mathbf{X}}(A_i), \tag{10}$$

$$C_{\mathbf{X}}(\{\,^a/x\}) = a \otimes C_{\mathbf{X}}(\{\,^1/x\}), \tag{11}$$

$$C_{\mathbf{X}}(\{\,^1/x\})(y) = C_{\mathbf{X}}(\{\,^1/y\})(x) \tag{12}$$

for any $A_i \in L^X$ *($i \in I$),* $x, y \in X$, $a \in L$.

Proof. We have $A(x) = A(x) \otimes (x \approx x) \leq \bigvee_{y \in X} A(x) \otimes (x \approx y) = \rho_{\mathbf{X}}(\{\,^1/x\}, A)$, thus $A \subseteq C_{\mathbf{X}}(A)$, proving (7).

(9) is true iff for each $x \in X$ and every $A, B \in L^X$ we have $S(A, B) \leq C_{\mathbf{X}}(A)(x) \to C_{\mathbf{X}}(B)(x)$ which is equivalent to $C_{\mathbf{X}}(A)(x) \otimes S(A, B) \leq C_{\mathbf{X}}(B)(x)$ which is, indeed, true, proving (9). Because of the limited scope, we omit the rest of proof (it will be included in the full version of this paper).

Note that (11) implies $C_{\mathbf{X}}(\emptyset) = \emptyset$. Indeed, $C_{\mathbf{X}}(\emptyset)(x) = C_{\mathbf{X}}(\{\,^0/y\}) = 0 \otimes C_{\mathbf{X}}(\{\,^1/y\}) = 0$ for any $x, y \in X$.

Lemma 2. *Let* C *be an* **L***-closure operator on* X *that satisfies (10)–(12). For* $x, y \in X$ *put*

$$(x \approx_C y) = C(\{\,^1/x\})(y).$$

Then $\mathbf{X}_C = \langle X, \approx_C \rangle$ *is an* **L***-similarity space.*

Proof. We verify (1): $(x \approx_C x) = C(\{\,^1/x\})(x) \geq \{\,^1/x\}(x) = 1$, by (7). By (12), $(x \approx_C y) = C(\{\,^1/x\})(y) = C(\{\,^1/y\})(x) = (y \approx_C x)$ proving (2). Due to the limited scope, we omit the proof of transitivity of $\approx_C$.

Theorem 5. *The mappings sending* $\mathbf{X}$ *to* $C_{\mathbf{X}}$, *and* C *to* $\mathbf{X}_C$, *as defined in Lemmas 1 and 2, are mutually inverse mappings between the set of all* **L***-similarity spaces with support* X *and the set of all* **L***-closure operators on* X *satisfying (10)–(12).*

Proof. By Lemmas 1 and 2, we have to check that $\mathbf{X} = \mathbf{X}_{C_{\mathbf{X}}}$ and $C = C_{\mathbf{X}_C}$. We have $(x \approx_{C_{\mathbf{X}}} y) = C_{\mathbf{X}}(\{\,^1/x\})(y) = \rho_{\mathbf{X}}(\{\,^1/y\}, \{\,^1/x\}) = (x \approx y)$. Furthermore,

$$C(A)(x) = C(\bigcup_{y \in X} \{ A(y)/y \})(x) =$$

$$= \bigvee_{y \in X} C(\{ A(y)/y \})(x) = \bigvee_{y \in X} A(y) \otimes C(\{ 1/y \})(x) =$$

$$= \bigvee_{y \in X} A(y) \otimes (x \approx_C y) = C_{\mathbf{X}_C}(A)(x)$$

completing the proof.

In the following we find a suitable axiomatization of systems of closed elements of fuzzy closure operators which are induced by similarity spaces. Recall a well known fact that $A = C_{\approx}(A)$ iff A is compatible with $\approx$. Thus, $\mathcal{S}_{C_{\approx}} = L^{\langle X, \approx \rangle}$.

Theorem 6. *A system $\mathcal{S}$ of $\mathbf{L}$-sets in X is the system of closed sets of some similarity space (i.e. $\mathcal{S} = L^{\langle X, \approx \rangle}$ for some $\langle X, \approx \rangle$) iff it is an $\mathbf{L}$-closure system satisfying $\bigcup_{i \in I} A_i \in \mathcal{S}$, $a \otimes A \in \mathcal{S}$, and $A \to a \in \mathcal{S}$ for each $A_i, A \in \mathcal{S}$, $a \in L$.*

Proof. We given only a sketch of proof. Due to Theorem 4, we have to show that an $\mathbf{L}$-closure system $\mathcal{S}$ satisfies (10)–(12) iff $\bigcup_{i \in I} A_i \in \mathcal{S}$, $a \otimes A \in \mathcal{S}$, and $A \to a \in \mathcal{S}$ for each $A_i, A \in \mathcal{S}$, $a \in L$. For simplicity, we write only C instead of $C_{\mathcal{S}}$. For illustration, we show that (10) is equivalent to $\bigcup_{i \in I} A_i \in \mathcal{S}$, Assume (10); then for $A_i \in \mathcal{S}$ we have $\bigcup_i A_i = \bigcup C(A_i) = C(\bigcup_i A_i) \in \mathcal{S}$. Conversely, if $\bigcup_{i \in I} A_i \in \mathcal{S}$ for $A_i \in \mathcal{S}$, then from $C(A_i) \in \mathcal{S}$ we get $\bigcup_i C(A_i) \in \mathcal{S}$ and thus $\bigcup_i C(A_i) = C(\bigcup_i C(A_i)) \supseteq C(\bigcup_i A_i)$. Since we always have $\bigcup_i C(A_i) \subseteq C(\bigcup_i A_i)$, (10) follows. Complete proof can be found in the full version of this paper.

Corollary 1. *A system $\mathcal{S}$ of $\mathbf{L}$-sets in X is the system of all $\mathbf{L}$-sets in X compatible with some $\mathbf{L}$-equivalence $\approx$ on X (i.e. $\mathcal{S} = L^{\langle X, \approx \rangle}$) iff $\mathcal{S}$ satisfies*

$$\bigcap_{i \in I} A_i \in \mathcal{S}, \ \bigcup_{i \in I} A_i \in \mathcal{S}$$
$$a \otimes A \in \mathcal{S}, \ a \to A \in \mathcal{S}, \ A \to a \in \mathcal{S}.$$

for each $A_i, A \in \mathcal{S}$ and $a \in L$.

Proof. Directly by Theorem 4 and Theorem 6.

Note that using other methods, the result from Corollary 1 is obtained in [10]. Our arguments place this result into the appropriate context of closure operators.

Acknowledgement Supported by grant no. 201/02/P076 of the GA ČR and partly by grant no. B1137301 of GA AV ČR.

References

1. Bělohlávek R.: Similarity relations in concept lattices. *J. Logic and Computation* Vol. **10** No. 6(2000), 823–845.

2. Bělohlávek R.: Fuzzy closure operators. *J. Math. Anal. Appl.* **262**(2001), 473-489.
3. Bělohlávek R.: Fuzzy closure operators II. *Soft Computing* (to appear).
4. De Baets B., Mesiar R.: Pseudo-metrics and T-equivalences. *The Journal of Fuzzy Mathematics* Vol. **5**, No. 2(1997), 471–481.
5. Gerla G.: *Fuzzy Logic. Mathematical Tools for Approximate Reasoning.* Kluwer, Dordrecht, 2001.
6. Goguen J. A.: L-fuzzy sets. *J. Math. Anal. Appl.* **18**(1967), 145–174.
7. Gottwald S.: *A Treatise on Many-Valued Logics.* Research Studies Press, Baldock, Hertfordshire, England, 2001.
8. Hájek P.: *Metamathematics of Fuzzy Logic.* Kluwer, Dordrecht, 1998.
9. Höhle U.: On the fundamentals of fuzzy set theory. *J. Math. Anal. Appl.* **201**(1996), 786–826.
10. Klawonn F., Castro J. L.: Similarity in fuzzy reasoning. *Mathware & Soft Computing* **2**(1995), 197–228.
11. Klement E. P., Mesiar R., Pap E.: *Triangular Norms.* Kluwer, Dordrecht, 2000.
12. Klir G. J., Yuan B.: *Fuzzy Sets and Fuzzy Logic. Theory and Applications.* Prentice Hall, 1995.
13. Kruse R., Gebhardt J., Klawonn F.: *Fuzzy-Systeme.* B. G. Teubner, Stuttgart, 1995.
14. Zadeh L. A.: Fuzzy sets. *Inf. Control* **8**(3)(1965), 338–353.
15. Zadeh L. A.: Similarity relations and fuzzy orderings. *Information Sciences* **3**(1971), 159–176.

A New Approach to Teaching Fuzzy Logic System Design

Emine Inelmen[1], Erol Inelmen[2], Ahmad Ibrahim[3]

[1] Padova University,
Padova, Italy
eminemeral.inelmen@unipd.it
[2] Bogazici University,
Istanbul, Turkey
inelmen@boun.edu.tr
[3] DeVry Institute of Technology,
Toronto, Ontario, Canada
ahmad1@ieee.org

Abstract. In order for Fuzzy Systems to continue to flourish in all scientific research areas, not only engineering, a more efficient and effective way of transmitting the *relevant knowledge and skills* of this discipline is necessary. Conventional tutorials in this area follow a fixed outline starting from the basic principles (set operations, relations, etc) and ending with current applications such as pattern recognition, information classification and system control. We argue that an *iterative and concurrent top-down/down-top* approach to learning would be more effective, and suggest that a plan to discover concepts first and map them afterwards is viable. Several recent publications and a case study currently under development in the field of medicine are used as examples to show evidence for the future of this new approach.

1 Introduction

As can observed clearly in recent textbooks and journals, since its inception by Zadeh in 1965, Fuzzy Systems have had a very brilliant history of successful implementations [1]. At present, merging Fuzzy Systems with other emerging Artificial Intelligence disciplines -that have come to be called "Soft Computing" [2]- are opening a new horizon. Neural Networks and Genetic Algorithms have effectively permeated this new field of research [3].

Since we believe that conventional tutorials in this new area are not adequate for introducing the novice to this rapidly flourishing discipline, we suggest an alternative *"spot learning"* approach. This view is based on Kaynak and Sabanovic's proposal to develop a more general approach to education that could help to "bridge gaps" and generate new development [4]. Similarly Bissell suggests that the curriculum must be "inverted" in order the adapt to the needs of the learner [5]. Recently Fink *et al.* described the importance of developing a "problem based learning" approach to education [6].

This work provides a summary of the survey of the available literature in Fuzzy Systems and gives a roadmap for a novice [7]. The tutorial provided by Mathwork_is

T. Bilgiç et al. (Eds.): IFSA 2003, LNAI 2715, pp. 79–86, 2003.

used as the main reference and simple examples are developed in order to guide the learner [8]. A model based on research on nutrition for elderly, developed by the first author_to be used as a predictor for medical patients is proposed as an example of what can be accomplished with this new tool [9].

2   Fuzzy Systems

In order to understand the basics of Fuzzy Systems some preliminary concepts -set, membership, operations, relations, fuzziness- must be reviewed. Concepts such as terms, variables, rules, firing and tuning are necessary at the design stage [10]. Applications in pattern recognition, information classification and system control help to understand the philosophy behind Fuzzy Systems [11]. Ibrahim provides an introduction to Fuzzy Systems and an implementation in education, although textbooks are not the only source for learning [12]. The level of the publications in academic journals is very advanced. Some examples that would encourage learners to use fuzzy logic techniques follow.

Zhang addressed the problem of smoothly transform "digital image warping" using a new algorithm that computes the smooth in-betweening of two polygonal shapes based on fuzzy techniques. In this method, the vertex correspondence of the polygons under warping is established by maximizing the membership function of a fuzzy set of polygons. Normalized local coordinates are used in this interpolation. This approach can handle polygons with different location, orientation, size, and number of vertices. It can be extended to warping curved shapes. Examples of the use of the technique as compared with conventional approaches showing the improvements in the deformation between the source and target figures [13].

Karray et al. suggest that a soft-computing-based approach can be used in the design of the layout of temporary facilities. The design seeks the best arrangement of facilities within the available area considering the effective use of people, resources, equipment, space, and energy. The main objective is to obtain the closest relationship values between each pair of facilities in a construction site. To achieve this, an integrated approach incorporating the fuzzy set theory and genetic algorithms is used [14].

Miao et al. attempt to develop a dynamic fuzzy inference system -an extension of the fuzzy cognitive map- using causal relationships. Each concept can have its own value set, depending on how precisely it needs to be described in the network. This enables the tool to describe the strength of causes and the degree of the effects that are crucial to conducting meaningful inferences. The arcs define dynamic, causal relationships between concepts [16].

Cordon et al. propose an approach to design linguistic models that are accurate to a high degree, and may be suitably interpreted. The structure of the Knowledge Base of the Fuzzy Rule Base System used is hierarchical to make it more flexible. This allows the use of linguistic rules defined over linguistic partitions with different levels, and thus to improve the modeling of those problem subspaces where the former models had a bad performance [17].

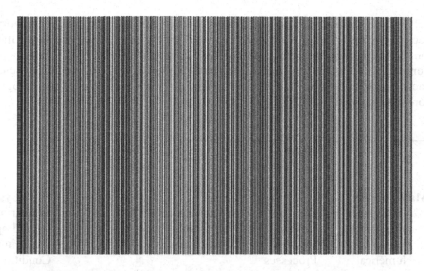

Fig. 1. Examples of image warping (from the source to the target figure [15]

Marin-Blazquez *et al.* present an effective and efficient approach for translating "fuzzy classification rules" that use approximate sets into rules that use descriptive sets and linguistic hedges of predefined meaning comprehensible to human users. Rules generated use approximative sets from training data, and then translate the resulting approximative rules into descriptive ones. The translated rules are functionally equivalent to the original approximative ones while reflecting their underlying preconceived meaning [18].

Caponetto *et al.* propose a methodology that applies artificial intelligence (AI) techniques to the modeling and control of some climate variables within a greenhouse. The nonlinear physical phenomena governing the dynamics of temperature and humidity in such systems are, in fact, difficult to model and control using traditional techniques. The paper proposes a framework for the development of *soft* computing-based controllers in modern greenhouses [19].

3 Education Today

Inelmen and Ibrahim have proposed a new control systems educational program. It is based on the concept of *inverted/integrated curriculum* approach where the students are presented with systems right at the beginning of the program [20]. Theoretical details are "grasped" progressively while working through guided projects (see Fig.1). A project includes all the stages in the design project from product definition to solution proposal.

Although the contents are being updated in accordance with the *emerging developments* in technology, the basic structure of the curriculum has always remained the same. Inelmen suggests that learners will be better motivated if a roadmap is delivered at the very beginning of an educational program [21]. Three levels of education (see Table 1) are given to better equip learners with the necessary professional tools that they need in an increasingly "innovation demanding" society.

This new approach can be implemented by changing the percentages of theory and practice courses. A unique research course along the timeline will boost the motivation to learn more. Keeping a close eye on "frontier research" in the *human* –social, economical, political- and *natural* –environment, technology, medicine- sciences, practitioners can make valuable contributions to the development of technology. For more details please see [22].

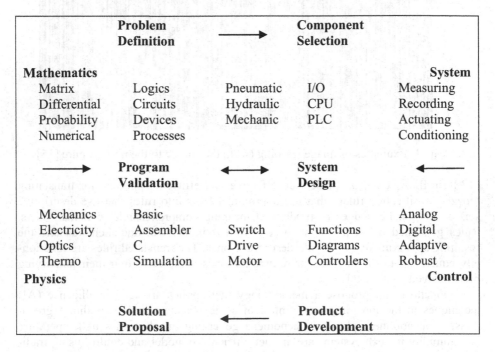

Fig. 2. Roadmap of the proposal for a control system education program

Table 1. Proposal for a new project based learning program.

Design	Source	Examples	Level
Practice	Catalogues	Components, Devices, Systems	Freshmen
Theory	Textbooks	Mathematics, Physics, Biology	Senior
Research	Journals	Laser, Fiber, Semiconductor	All

In the future education itself can be taken as a subject for developing a new fuzzy system. In such a case learning input and output variables (skills, expectations, character, motivation) can be related to existing conditions (resources, schedules, regulations, jobs, assignments, assessments) [11]. The character (values, beliefs, behavior, habits, feelings, mood, inspiration, attitude, attention) of all parties involved in the education is very important so the analysis of the model will be challenging.

4 Case Implementation

Although the Ruspini IEEE-video contains the basic Fuzzy Theory there are no examples for practical application [23]. A case in medicine suggested by the first author addresses the dietary patterns of the elderly living in different European communities [9], in relation to health and performance. The study is intended to address the following issues:

- to what extent do differences exist in the intake of energy, nutrients and foods between elderly living in different (European) communities.
- to what extent do differences exist in health and performance between the elderly living in different (European) communities. In the assessment: health complaints, use of medicines, self perceived health, activities of daily living, and physical activity are taken into account [24].

Comparisons are made between communities with relation to demographical and socio-economical variables, life style, social network and diet habits. Age, weight, height, status, income facilities, activity, diseases, medicines, diet are the input variables being considered. To start the modeling process: *given* a) the input and out variables membership degrees for different linguistic terms and b) set of rules for different combination of input variables, *find* for a set of input crisp values a) the implication and aggregation of rules and b) output crisp values. We are currently involved in the implementation of a working model.

5 Conclusion

We have proposed here a new approach to teaching fuzzy logic system design as given in Fig. 2. This new approach is an improvement on the problem-based learning approach that is becoming very popular nowadays. The content of this work has come about because of the interaction of three authors representing the expert, the learner and the knowledge provider.

Although Mathworks [8] provides an environment for learning using basic examples, further work is necessary to enhance the proposed approach to include the teaching of the theoretical side of fuzzy logic. The contributions of a larger number of experts in the fields can help in the development of a more robust educational environment that will lead to new developments in technology (See Fig. 3.).

This work will remain as a framework for the development of a teaching model in fuzzy logic system design. As new material and experience from actual projects emerge, the framework can be updated and enhanced reflecting the natural "learning path" of discovery. As with the case of machine learning, any new piece of information should fit into the suggested model. We believe that Soft Computing will be the future in Artificial Intelligence. Implementation of Neural Networks as suggested by Jang *et al.* will offer new possibilities in creating a better world [25]. The rich documentation provided by the IEEE remains still to be tapped.

			Fuzzy	*Systems*		
	Basics				**Properties**	
Universe	Elements	Set		Variables	Functions	Hedges
	Operations				**Rules**	
And	Or	Not		Fuzzy	Mapping	Defuzzy
	Implications				**Application**	
Relations	Families	Approx.		Control	Classif.	Recog.

Fig. 3. Concept Map for a Fuzzy Logic System Design (after [11])

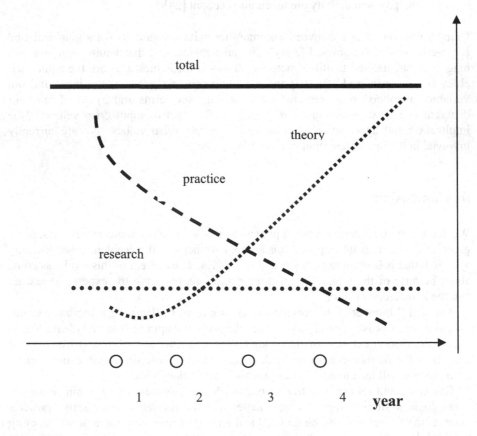

Fig. 4. Proposal for a Research Based Undergraduate Program in Engineering

Acknowledgement

The support given by Dr. Zenon J. Pudlowski is acknowledged. Science can only develop when organizations like UICEE create strong networks between researches in different fields and geographies. This contribution was made possible by the existence of such organizations.

References

1. Chen, G.; Pham T., Introduction to Fuzzy Sets, Fuzzy Logic, and Fuzzy Control Systems. Boca Raton, Fla.: CRC Press (2001).
2. Aliev, R. A.; Aliev, R. R., Soft Computing and its Applications, World Scientific, London (2001)
3. Wang, L., A Course in Fuzzy Systems and Control. Upper Saddle River, NJ: Prentice Hall PTR (1997).
4. Kaynak, O.; Sabanovic, A., Diffusion of New Technologies Through Appropriate Education and Training. Presented at the Diffusion of New Technologies Conference, St.Petersburg, June 13-17, 1994.
5. Bissell, C.C., Control Education: Time for Radical Change?, IEEE Control Systems, 10, (1999).44-49.
6. Fink, F.; Enemark, S.; Moesby, E., UICEE Centre for Problem-Based Learning (UCPBL) at Aalborg University. 6th Baltic Region Seminar on Engineering Education, UNESCO International Center for Engineering Education, Wismar, Germany, September 200, 34-38
7. Mordeson, J.N.; Premchand S., Fuzzy Graphs and Fuzzy Hypergraphs, Physica-Verlag, Heidelberg; New York (2000).
8. Mathworks (2002) In web http://www.mathworks.com/products/matlab/
9. Inelmen, Emine. Descriptive Analysis of the Prevalence Of Anemia in a Randomly Selected Sample of Elderly People Living at Home. Aging-Clinical and Experimental Research, 6:2 (1994) 81-89.
10. Yager, R.; Filev, D., Essentials of Fuzzy Modeling and Control. Wiley, New York (1994).
11. Yen, J. Langari; R., Fuzzy Logic: Intelligence, Control, and Information. Upper Saddle River, Prentice Hall, N.J. (1999).
12. Ibrahim, A., Assessment of Distance Education Quality Using Fuzzy Sets Model. Proceedings of the SEFI Annual Conference, Copenhagen, 12-14 September, 2001 (in CD).
13. Zhang, Y., A Fuzzy Approach to Digital Image Warping, IEEE Computer Graphics July 1996, Vol. 16, No. 4, 34-41.
14. Karray F.; Zaneldin E.; Hegazy T.; Shabeeb AHM; Elbeltagi E., Tools of Soft Computing as Applied to the Problem of Facilities Layout Planning. IEEE Transactions on Fuzzy Systems 8:4, (2000) 367-379
15. Parent, R. (2002). In web http://www.cis.ohio-state.edu/~parent/book/Full.html.
16. Miao Y, Liu ZQ, Siew CK, Miao CY., Dynamical Cognitive Network - An Extension of Fuzzy Cognitive Map. IEEE Transactions on Fuzzy Systems 9:5 (2001) 760-770
17. Cordon O, Herrera F, Zwir I., Linguistic Modelling by Hierarchical Systems of Linguistic Rules. IEEE Transactions on Fuzzy Systems 10:1(2002) 2-20
18. Marin-Blazquez JG, Shen Q., From Approximative to Descriptive Fuzzy Classifiers. IEEE Transactions on Fuzzy Systems 10:4 (2002) 484-497
19. Caponetto R, Fortuna L, Nunnari G, Occhipinti L, Xibilia MG., Soft Computing for Greenhouse Climate Control. IEEE Trans.on Fuzzy Systems 8:6 (2000) 753-760

20. Inelmen, Erol. ; Ibrahim, A.M., A Proposal for a novel Control Systems Undergraduate Program.   Proceedings of IASTED Modelling, Identification and Control (ed. M.H.Hamza), Innsbruck, (Austria), 19-22 February, 2001 494-499.
21. Inelmen Erol. Frontier Research' as a Novel Approach in the Engineering Curriculum of Tomorrow. 6th Baltic Region Seminar on Engineering Education, Wismar, Germany, 22 - 25 September, 2002. 109-111.
22. Ibrahim, A. M., Bringing Fuzzy Logic into Focus, IEEE Circuits & Devices, September 2001, 33-38.
23. Ruspini, E., Introduction to Fuzzy Set Theory and Fuzzy Logic Basic Concepts and Structures. (video) Piscataway: IEEE (1992).
24 EURO NUT (2002). In web: http://www.unu.edu/unupress/unupbooks/80633e/80633E09.htm
25. Jang, J. Sun, CT., and Mizutani, E., Neuro-fuzzy and soft computing, Upper Saddle River, NJ: Prentice Hall, (1997).

On the Transitivity of
Fuzzy Indifference Relations

Susana Díaz[1], Bernard De Baets[2], and Susana Montes[3]

[1] Department of Statistics and O.R., Faculty of Geology
University of Oviedo, Calvo Sotelo s/n, 33071 Oviedo, Spain
alu407@pinon.ccu.uniovi.es
[2] Department of Applied Mathematics, Biometrics and Process Control
Ghent University, Coupure links 653, B-9000 Gent, Belgium
Bernard.DeBaets@rug.ac.be
[3] Department of Statistics and O.R.
University of Oviedo, Nautical School, 33271 Gijón, Spain
smr@pinon.ccu.uniovi.es

Abstract. Transitivity is an essential property in preference modelling. In this work we study this property in the framework of fuzzy preference structures. In particular, we discuss the relationship between the transitivity of a fuzzy large preference relation R and the transitivity of the fuzzy indifference relation I obtained from R by some of the most important generators employed in the literature. We consider different types of transitivity for the fuzzy large preference relation R and identify the strongest type of transitivity of the corresponding fuzzy indifference relation I.

1 Introduction

In the context of preference modelling, the concept of transitivity arises as a natural property many relations must satisfy. In the classical setting, i.e. when working with crisp relations, the transitivity of a large preference relation R can be characterized by the transitivity of the corresponding indifference relation I and strict preference relation P and two additional relational inequalities involving P and I [12]. In case the relation R is complete, its transitivity is completely characterized by the transitivity of P and I only.

The above-mentioned characterization has also been studied in the fuzzy case, i.e. when working with fuzzy relations. In the well-defined context of additive fuzzy preference structures, a characterization of the transitivity of a fuzzy large preference relation R has been obtained when R is strongly complete [7]. Other studies require less restrictive completeness conditions (such as weak completeness) or no completeness condition at all [2,3,13]. However, in none of these studies a full characterization has been obtained.

In this paper, we focus on the propagation of the T-transitivity of a fuzzy large preference relation R to the corresponding fuzzy indifference relation I, constructed from R by means of some commutative conjunctor i. We will not

T. Bilgiç et al. (Eds.): IFSA 2003, LNAI 2715, pp. 87–94, 2003.
© Springer-Verlag Berlin Heidelberg 2003

only investigate whether I is also T-transitive, but identify the strongest type of transitivity I can exhibit, for different t-norms T and different commutative conjunctors i.

Our paper is organised as follows. In the next section, we give a brief introduction to crisp and fuzzy preference modelling. In particular, we explain how additive fuzzy preference structures can be constructed by means of an indifference generator. The next section is devoted to some important general results linked to the notion of dominating functions. In Section 4, we discuss in detail the case of ordinally irreducible t-norms.

2   Preference Modelling

2.1   Crisp Preference Structures

We briefly recall two equivalent relational representations of preferential information [12]. On the one hand, one can consider a large preference relation R, i.e. a *reflexive* binary relation on the set of alternatives A, with the following interpretation:

$$aRb \text{ if and only if } a \text{ is at least as good as } b.$$

On the other hand, R can be decomposed into disjoint parts: an irreflexive and asymmetric strict preference component P, a reflexive and symmetric indifference component I and an irreflexive and symmetric incomparability component J such that $P \cup P^t \cup I \cup J = A^2$ and $R = P \cup I$ (where ^{t} denotes the transpose of a relation). These components can be obtained by considering various intersections: $P = R \cap R^d$, $I = R \cap R^t$ and $J = R^c \cap R^d$ (where ^{c} denotes the complement of a relation and ^{d} denotes the dual of a relation, i.e. the complement of its transpose).

A binary relation Q on A is said to be *transitive* if

$$(aQb \wedge bQc) \Rightarrow aQc$$

for any $a, b, c \in A$. Transitivity can be stated equivalently as a relational inequality: $Q \circ Q \subseteq Q$. Using the latter notation, the characterization of the transitivity of a large preference relation R can be written as follows:

Theorem 1. [12] *For any reflexive binary relation R with corresponding preference structure (P, I, J) it holds that*

$$R \circ R \subseteq R \quad \Leftrightarrow \quad \begin{cases} P \circ P \subseteq P \\ I \circ I \subseteq I \\ P \circ I \subseteq P \\ I \circ P \subseteq P \end{cases}.$$

In case R is complete (i.e. aRb or bRa for any $a, b \in A$), the following simpler characterization holds. Note that in this case $J = \emptyset$.

Theorem 2. *For any complete binary relation R with corresponding preference structure $(P, I, \emptyset)$ it holds that*

$$R \circ R \subseteq R \quad \Leftrightarrow \quad \begin{cases} P \circ P \subseteq P \\ I \circ I \subseteq I \end{cases}.$$

2.2 Additive Fuzzy Preference Structures

In fuzzy preference modelling, a reflexive binary fuzzy relation R on A can also be decomposed into what is called an additive fuzzy preference structure, by means of an (indifference) generator i, which is defined as a symmetric (commutative) $[0, 1]^2 \rightarrow [0, 1]$ mapping located between the Łukasiewicz t-norm $T_{\mathbf{L}}$ and the minimum operator $T_{\mathbf{M}}$, i.e. $T_{\mathbf{L}} \leq i \leq T_{\mathbf{M}}$. More specifically, the strict preference relation P, the indifference relation I and the incomparability relation J are obtained as follows:

$$
\begin{aligned}
P(a, b) &= R(a, b) - i(R(a, b), R(b, a)) \\
I(a, b) &= i(R(a, b), R(b, a)) \\
J(a, b) &= i(R(a, b), R(b, a)) - (R(a, b) + R(b, a) - 1).
\end{aligned}
$$

An additive fuzzy preference structure (AFPS) (P, I, J) on A is then characterized as a triplet of binary fuzzy relations on A such that I is reflexive and symmetric and

$$P(a, b) + P(b, a) + I(a, b) + J(a, b) = 1,$$

whence the adjective 'additive'. The corresponding fuzzy large preference relation R is then given by $R(a, b) = P(a, b) + I(a, b)$.

The most popular type of transitivity of fuzzy relations is T-transitivity, with T a t-norm. A binary fuzzy relation Q on A is called T-transitive if it holds that

$$T(Q(a, b), Q(b, c)) \leq Q(a, c)$$

for any $a, b, c \in A$.

Recall that the sup-T composition of two binary fuzzy relations U and V on a universe A is the binary fuzzy relation $U \circ_T V$ on A defined by

$$U \circ_T V(x, z) = \sup_{y \in A} T(U(x, y), V(y, z)).$$

A trivial result is the fact that T-transitivity can be expressed equivalently as a relational inequality: $Q \circ_T Q \subseteq Q$.

As far as we know, the only generalization of Theorems 1 and 2 has been obtained in the case of a strongly complete fuzzy large preference relation R (i.e. $\max(R(a, b), R(b, a)) = 1$ for any $a, b \in A$). Note that in that case any generator i leads to the same AFPS and that again $J = \emptyset$.

Theorem 3. [7] *Consider a strongly complete binary fuzzy relation R with corresponding fuzzy preference structure $(P, I, \emptyset)$. For any t-norm $T \geq T_{\mathbf{L}}$ it holds that:*

$$R \circ_T R \subseteq R \quad \Leftrightarrow \quad \begin{cases} P \circ_{T_{\mathbf{M}}} P \subseteq P \\ I \circ_T I \subseteq I \\ P \circ_{T_{\mathbf{L}}} I \subseteq P \\ I \circ_{T_{\mathbf{L}}} P \subseteq P \end{cases}.$$

Note that although Theorem 3 is formally analogous to Theorem 1, it is only a generalization of Theorem 2.

3   General Propagation Results

Although we are primarily interested in the T-transitivity of a fuzzy large preference relation R, the class of t-norms is not always large enough to describe the strongest type of transitivity shown by a fuzzy indifference relation I defined from R. Actually, for introducing an adequate notion of transitivity, it suffices to consider a binary operation on $[0, 1]$ that generalizes Boolean conjunction. We therefore propose the following minimal definition.

Definition 1. *A conjunctor f is an increasing binary operation on $[0, 1]$ with neutral element 1.*

Note that the weakest conjunctor is given by the drastic product $T_{\mathbf{D}}$, while the strongest conjunctor is $T_{\mathbf{M}}$. We say that a binary fuzzy relation Q is f-*transitive*, with f a conjunctor, if

$$f(Q(a, b), Q(b, c)) \leq Q(a, c)$$

for any $a, b, c \in A$. Since all fuzzy relations we are dealing with are binary, we omit the adjective 'binary' from here on. Note that if $f_1 \leq f_2$, then f_2-transitivity implies f_1-transitivity. The composition of fuzzy relations can be generalized in the same way so that f-transitivity can again be written as $Q \circ_f Q \subseteq Q$. Of course, in order for this composition to be associative, the operation f should also be associative, but that is of no relevance to our study, at this moment.

   A second generalization we propose is the class of generators used for constructing fuzzy indifference relations. Strictly speaking, in order for the fuzzy indifference relation $I = i(R, R^t)$ to be part of an AFPS, i should be symmetric (commutative) and it should hold that $T_{\mathbf{L}} \leq i \leq T_{\mathbf{M}}$. As fuzzy indifference relations are also studied outside the framework of AFPS (see e.g. symmetric kernels in [1]) and the constraint $T_{\mathbf{L}} \leq i$ does not appear in any of the results in this paper, we will drop this constraint. Due to this, we would also loose that 1 is the neutral element, which is not desirable. On the other hand, generators are not necessarily monotone (increasing). This property, however, is needed for proving our results. In a nutshell, the foregoing discussion leads us to consider *commutative conjunctors* for generating fuzzy indifference relations. A generic commutative conjunctor will be denoted c; we will still refer to it as a generator.

A last generalization consists of considering an arbitrary fuzzy relation R, not necessarily being reflexive.

The following simple observation implies that any type of f-transitivity of R guarantees a kind of minimal transitivity of I.

Proposition 1. *Consider a commutative conjunctor c. For any fuzzy relation R it holds that*

$$R \text{ is } T_\mathbf{D}\text{-transitive} \quad \Rightarrow \quad I = c(R, R^t) \text{ is } T_\mathbf{D}\text{-transitive}.$$

On the other hand, we can identify the maximal potential transitivity of I. First of all, one easily verifies that the t-norm T in Theorem 3 can be replaced by an arbitrary commutative conjunctor $f \geq T_\mathbf{L}$. Moreover, the implication from right to left does not require the latter restrictions. A suitable counterexample then leads to the following claim.

Proposition 2. *Consider a conjunctor $f \geq T_\mathbf{L}$, a conjunctor g and a commutative conjunctor c. If for any fuzzy relation R it holds that*

$$R \text{ is } f\text{-transitive} \quad \Rightarrow \quad I = c(R, R^t) \text{ is } g\text{-transitive}$$

then $g \leq f$.

Of course, this proposition does not imply that this maximal transitivity is attained. We will show further on that in case of $T_\mathbf{M}$-transitivity this upper bound cannot always be reached.

A first important theorem in this generalized framework concerns the case of t-norm generators.

Theorem 4. *Consider a conjunctor f and a t-norm T such that $f \geq T$. For any fuzzy relation R it holds that*

$$R \text{ is } f\text{-transitive} \quad \Rightarrow \quad I = T(R, R^t) \text{ is } T\text{-transitive}.$$

The proof of this theorem mainly depends on the bisymmetry property of the t-norm T [9]:

$$(\forall (x, y, z, t) \in [0,1]^4)(T(T(x,y), T(z,t)) = T(T(x,z), T(y,t))),$$

which follows from its commutativity and associativity.

Theorem 4 and Proposition 2 then lead to the following proposition. It shows that in case of $T_\mathbf{M}$-transitive R, the fuzzy relation $I = T(R, R^t)$ can in general not be 'more' transitive than T-transitive.

Proposition 3. *Consider a t-norm T. The strongest conjunctor g such that for any $T_\mathbf{M}$-transitive fuzzy relation R it holds that $I = T(R, R^t)$ is g-transitive, is the t-norm T itself, i.e. $g = T$.*

A second important general result involves the concept of dominance. Although the original definition of dominance was given for t-norms (see e.g. [10]), we formulate it here for conjunctors.

Definition 2. *We say that a conjunctor f_1 dominates a conjunctor f_2, denoted $f_1 \gg f_2$, if*

$$(\forall (x,y,z,t) \in [0,1]^4)(f_1(f_2(x,y),f_2(z,t)) \geq f_2(f_1(x,z),f_1(y,t))).$$

It is well known that $T_{\mathbf{M}}$ dominates any other t-norm and that $T_{\mathbf{P}}$ dominates $T_{\mathbf{L}}$. Although these three t-norms belong to the parametric family of Frank t-norms, this family cannot be ordered by dominance. However, for the parametric family of Schweizer–Sklar t-norms this is effectively the case [10].

The following result is inspired by the works of De Baets and Mesiar [6] and Bodenhofer [1]. It concerns the preservation of f-transitivity, i.e. maximal transitivity of I.

Theorem 5. *Consider two commutative conjunctors f and c such that $f \ll c$. For any fuzzy relation R it holds that*

$$R \text{ is } f\text{-transitive} \quad \Rightarrow \quad I = c(R, R^t) \text{ is } f\text{-transitive.}$$

Recall that any binary operation f on $[0,1]$ can be isomorphically transformed by means of a $[0,1]$-automorphism ϕ in the following way:

$$f_\phi(x,y) = \phi^{-1}(f(\phi(x),\phi(y))).$$

The following proposition is very useful for generalizing a specific result to a whole class of operators.

Proposition 4. *Consider two conjunctors f and g, a commutative conjunctor c and a $[0,1]$-automorphism ϕ. The following statements are equivalent:*

(i) *For any fuzzy relation R it holds that:*

$$R \text{ is } f\text{-transitive} \quad \Rightarrow \quad I = c(R, R^t) \text{ is } g\text{-transitive.}$$

(ii) *For any fuzzy relation R' it holds that:*

$$R' \text{ is } f_\phi\text{-transitive} \quad \Rightarrow \quad I' = c_\phi(R', R'^t) \text{ is } g_\phi\text{-transitive.}$$

One easily verifies that dominance is preserved under ϕ-transformations. Indeed, for any two conjunctors f_1 and f_2 it holds that

$$f_1 \gg f_2 \quad \Leftrightarrow \quad (f_1)_\phi \gg (f_2)_\phi.$$

This allows us to generalize Theorem 5 using Proposition 4.

4   The Case of Ordinally Irreducible T-norms

In this section, we take a closer look at the propagation of T-transitivity, with T an ordinally irreducible t-norm, i.e. T is an idempotent t-norm (i.e. $T = T_{\mathbf{M}}$), a strict t-norm (i.e. $T = T_{\mathbf{P}\phi}$) or a nilpotent t-norm (i.e. $T = T_{\mathbf{L}\phi}$). We consider fuzzy indifference relations generated by means of one of these t-norms as well.

We have considered the 9 possible combinations, resulting from the decision not to complicate things by considering two different automorphisms. The results of our study are summarized in Table 1. This table contains the strongest type of transitivity exhibited by the fuzzy indifference relation I defined from any T-transitive fuzzy relation R by means of a t-norm c. It can be filled in using Theorems 4 and 5, and Propositions 2 and 3. For the case $T = T_{\mathbf{P}\phi}$ and $c = T_{\mathbf{L}\phi}$ a counterexample is available showing that this is really the strongest result possible.

Table 1. Transitivity of I

		Generator c		
		$T_{\mathbf{L}\phi}$	$T_{\mathbf{P}\phi}$	$T_{\mathbf{M}}$
T-Transitivity of R	$T_{\mathbf{L}\phi}$	$T_{\mathbf{L}\phi}$	$T_{\mathbf{L}\phi}$	$T_{\mathbf{L}\phi}$
	$T_{\mathbf{P}\phi}$	$T_{\mathbf{L}\phi}$	$T_{\mathbf{P}\phi}$	$T_{\mathbf{P}\phi}$
	$T_{\mathbf{M}}$	$T_{\mathbf{L}\phi}$	$T_{\mathbf{P}\phi}$	$T_{\mathbf{M}}$

Acknowledgement

The research reported on in this paper was partially supported by project MCYT BFM2001-3515.

References

1. U. Bodenhofer, *Representations and constructions of similarity-based fuzzy order-ings*, Fuzzy Sets and Systems, in press.
2. M. Dasgupta and R. Deb, *Transitivity and fuzzy preferences*, Social Choice and Welfare **13** (1996), 305–318.
3. M. Dasgupta and R. Deb, *Factoring fuzzy transitivity*, Fuzzy Sets and Systems **118** (2001), 489–502.
4. B. De Baets and J. Fodor, *Twenty years of fuzzy preference structures (1978-1997)*, JORBEL **37** (1997), 61–82.
5. B. De Baets and J. Fodor, *Generator triplets of additive fuzzy preference struc-tures*, Proc. Sixth Internat. Workshop on Relational Methods in Computer Science (Tilburg, The Netherlands), 2001, pp. 306–315.
6. B. De Baets and R. Mesiar, *T-partitions*, Fuzzy Sets and Systems **97** (1998), 211–223.

7. B. De Baets, B. Van De Walle and E. Kerre, *Fuzzy preference structure without incomparability*, Fuzzy Sets and Systems **76** (1995), 333–348.
8. S. Díaz and B. De Baets, *Transitive decompositions of weakly complete fuzzy preference relations*, Proc. EUROFUSE Workshop on Information Systems (Varenna, Italy), 2002, pp. 225–230.
9. J. Fodor and M. Roubens, *Fuzzy Preference Modelling and Multicriteria Decision Support*, Kluwer Academic Publishers, Dordrecht, 1994.
10. E.P. Klement, R. Mesiar and E. Pap, *Triangular Norms*, Kluwer Academic Publishers, Dordrecht, 2000.
11. R. Nelsen, *An Introduction to Copulas*, Lecture Notes in Statistics, Vol. 139, Springer-Verlag, New York, 1998.
12. M. Roubens and Ph. Vincke, *Preference modelling*, Lecture Notes in Economics and Mathematical Systems, Vol. **76**, Springer-Verlag, Berlin, 1985.
13. B. Van De Walle, *Het bestaan en de karakterisatie van vaagpreferentiestrukturen*, Ph.D. thesis (in Dutch), Ghent University, 1996.

About Z_f, the Set of Fuzzy Relative Integers, and the Definition of Fuzzy Bags on Z_f

Patrick Bosc, Daniel Rocacher

IRISA/ENSSAT
BP 447, 22305 LANNION Cédex, FRANCE
{bosc, rocacher}@enssat.fr

Abstract. A characterization of fuzzy bags with fuzzy integers (N_f) provides a general framework in which sets, bags, fuzzy sets and fuzzy bags are treated in a uniform way. In bag theory, the difference between two bags A and B is the relative complement of A intersection B to A. With fuzzy bags defined on N_f, this difference does not always exist and, in such a case, only approximations of the exact result can be defined. The problem comes from the fact that the fuzzy bag model considered so far is based on positive fuzzy integers. In this paper, we show that fuzzy relative integers (Z_f) offer a well-founded framework in which the difference of two fuzzy bags is always defined.

1 Introduction

A fuzzy bag is a collection which simultaneously deals with quantities and degrees of membership of the elements it contains. It provides a useful representation model for applications where quantifications and preferences play an important role: flexible querying of databases, fuzzy data mining and fuzzy information retrieval, among others. For example, in the field of databases, systems taking into account both flexible queries and bags lead to introduce fuzzy bags. Some of their operators have been defined by Yager [8] and a complementary study has been carried out in [5]. In [7] we have proposed a new approach in which fuzzy bags are characterized by a function from a universe U to conjunctive fuzzy integers (N_f). This approach provides a general framework in which sets, bags, fuzzy sets and fuzzy bags are treated in a uniform way and can then be composed.

In bag theory, the difference between two bags A and B is associated with a bag S corresponding to the elements which have to be 'added' to A∩B so that it equals A. Unfortunately, in the context of fuzzy bags, it is not always possible to find S, and only an approximation S1 of S can be defined. This operator is based on a specific difference between fuzzy integers called the optimistic difference.

This problem comes from the fact that the fuzzy bag model considered is based on positive fuzzy integers. In this paper, we show that, using fuzzy relative (or rational) integers (Z_f), the difference A-B of two fuzzy bags is always defined by one

T. Bilgiç et al. (Eds.): IFSA 2003, LNAI 2715, pp. 95–102, 2003.

equivalence class. The existence of negative multiplicities allows exact relative complementation.

The rest of this paper is organized as follows. The main definitions about fuzzy bags and the associated operators are briefly recalled in section 2. In section 3, we focus our study on the difference between two fuzzy bags and the underlying problems are pointed out. This leads, in section 4, to develop a complementary study which extends $\mathbb{N}_f$, the set of fuzzy integers, to $\mathbb{Z}_f$, the set of relative fuzzy integers. Last, section 5 deals with the way fuzzy bags are built on $\mathbb{Z}_f$ and we show how operators on fuzzy bags, especially the difference, can be specified.

2 Fuzzy Bags

A fuzzy bag is a bag in which each occurrence of each element is associated with a grade of membership. One way to describe a fuzzy bag is to enumerate its elements, for example: A = <1/a, 0.1/a, 0.1/a, 0.5/b> = {<1, 0.1, 0.1>/a, < 0.5>/b}.

In [1] we have presented two methods in order to well-defined fuzzy bags. The first one, which has the advantage of simplicity, is directly derived from one of the most important concepts of fuzzy sets: the α-cuts ; the second one, based on the concept of fuzzy integer, has the advantage of genericity.

In the first approach, we extend bag operators to fuzzy bag operators thanks to the α-cut concept, similarly to the extension of a set into a fuzzy set. We then define the α-cut of a fuzzy bag A as the crisp bag A_α which contains all the occurrences of the elements of a universe U whose grade of membership in A is greater than or equal to the degree α ($\alpha \in$]0, 1]). The number of occurrences of the element x in A_α is denoted: $\omega_{A_\alpha}(x)$. In order to preserve the compatibility between the bag and fuzzy bag structures, we define intersection and union of fuzzy bags satisfying the following properties : $(A \cap B)_\alpha = A_\alpha \cap B_\alpha$; $(A \cup B)_\alpha = A_\alpha \cup B_\alpha$. So, if A and B are two fuzzy bags, we have: $\omega_{(A \cap B)_\alpha}(x) = \min(\omega_{A_\alpha}(x), \omega_{B_\alpha}(x))$; $\omega_{(A \cup B)_\alpha}(x) = \max(\omega_{A_\alpha}(x), \omega_{B_\alpha}(x))$.

In the second approach, we present an extension of bags to fuzzy bags by using fuzzy cardinalities to define fuzzy numbers of occurrences in fuzzy bags.

The concept of fuzzy cardinality of a fuzzy set has been proposed by Zadeh [9]. The cardinality |A| of a fuzzy set A, called FGCount(A), is defined by: $\forall\ n \in \mathbb{N}$, $\mu_{|A|}(n) = \sup\{\alpha\ /\ |A_\alpha| \geq n\}$. Let us consider the fuzzy set A = {$1/x_1$, $0.1/x_2$, $0.1/x_3$}, then |A| = {1/0, 1/1, 0.1/2, 0.1/3}. The degree α associated with a number ω in the fuzzy cardinality of a fuzzy set A is interpreted as the extent to which A has at least ω elements. The fuzzy cardinality |A| has been defined as the convex hull of the fuzzy set of the cardinalities of all the α-cuts of A. |A| is a normalized and convex fuzzy set of integers. Properties of fuzzy cardinalities are presented, for instance, in [4].

Considering this notion of fuzzy cardinality, the occurrences of an element x in a fuzzy bag A can be characterized as a fuzzy integer denoted $\Omega_A(x)$. This fuzzy number is the fuzzy cardinality of the fuzzy set of occurrences of x in A. Thus, a fuzzy bag A, on a universe U, can be defined by a characteristic function Ω_A from U to $\mathbb{N}_f$, where $\mathbb{N}_f$ is the set of the fuzzy integers: Ω_A: U $\rightarrow$ $\mathbb{N}_f$. Using fuzzy integers A =

<1/a, 0.1/a, 0.1/a, 0.5/b> is represented by: A = {{1/0, 1/1, 0.1/2, 0.1/3}* a, {1/0, 0.5/1}* b} and Ω_A(a) = {1/0, 1/1, 0.1/2, 0.1/3}.

Through the application of the extension principle, crisp bag operations can be extended to fuzzy bags thanks to fuzzy numbers of occurrences which leads to: $\Omega_{A \cap B}$(x) = min(Ω_A(x), Ω_B(x)) ; $\Omega_{A \cup B}$(x) = max(Ω_A(x), Ω_B(x)). The Cartesian product and other algebraic operators have been developed in [7] but are not presented here due to the limitation of space.

3 Difference on Fuzzy Bags

In bag theory, the difference definition A–B is associated with the elements which have to be 'added' to A∩B so that it equals A: A–B = S ⟺ A = (A∩B) + S. Unfortunately, in the context of fuzzy bags, it is not always possible to find S, the relative complement of A∩B with respect to A. To solve this problem, the previous definition has to be relaxed [1]. This leads to the following specification: A)-(B = S1 = ∪{S_i : A ⊇ (B∩A) + S_i}. The fuzzy bag S1 approximates S, it is the greatest fuzzy bag S_i such as (A∩B) + S_i is contained in A. The α-cuts of S1 are such that for all α ∈]0, 1] and x ∈ X, ω_{S1_α}(x) is the greatest integer satisfying: $\omega_{A \cap B_\alpha}$(x) + ω_{S1_α}(x) ≤ ω_{A_α}(x).

Table 1. The different α-cuts of A = {<1, 0.8, 0.5, 0.2>/x}, B = {<1, 0.3>/x} and S1 = {<0.8, 0.2>/x}, the greatest fuzzy bag such that: (A ∩ B) + S1 ⊆ A.

α	ω_{A_α}(x)	ω_{B_α}(x)	s = ω_{A_α}(x)- ω_{B_α}(x)	ω_{S1_α}(x)
1	1	1	0	0
0.8	2	1	1	1
0.5	3	1	2	1
0.3	3	2	1	1
0.2	4	2	2	2

Using the extension with fuzzy numbers of occurrences, the difference between fuzzy bags boils down to find the solutions Ω_S(x) of the following equations:

∀ x ∈ X, Ω_A(x) = $\Omega_{A \cap B}$(x) + Ω_S(x).

It is known [2, 3, 6] that the solution X of the equation M + X = N, when it exists, is defined by the optimistic difference (denoted by: N)-(M) [1]. However, this solution X does not always exist, which leads to weaken the equation M + X = N into M + X ≤ N. Now there exits a set of solutions whose greatest element is given by X = N)-(M. X is the greatest fuzzy number such that: $\sup_{(x,y)/x+y=z}$ min(μ_M(x),μ_X(z)) ≤ μ_N(y) .

Consequently, the difference between two fuzzy bags is related to the optimistic difference between fuzzy integers. The solution obtained by solving the following equation on fuzzy numbers of occurrences: $\Omega_{A \cap B}$(x) + Ω_{S1}(x) ≤ Ω_A(x), is given by:

---

[1] $\mu_{N)-(M}$(y) = $\inf_{(x,z)/x+y=z}$ μ_M(x) $\wedge^{-1}$ μ_N(z) where $x_1 \wedge^{-1} x_2$ is defined as the greatest element t in [0, 1] such that min(x_1, t) ≤ x_2, i.e.: if $x_1 \le x_2$ then $x_1 \wedge^{-1} x_2$ = 1 else $x_1 \wedge^{-1} x_2 = x_2$

$\Omega_{S1}(x) = \max(0, \Omega_A(x))-(\Omega_{A \cap B}(x))$. In such a case, S1=A)-(B is defined as the greatest bag S_i that we can 'add' to A $\cap$ B so that A contains (A $\cap$ B) + S_i.

Example 3.1. A = {{1/0, 1/1, 0.8/2, 0.5/3, 0.2/4}* a } ; B = {{1/0, 1/1, 0.3/2}* a}. In this simple example A $\cap$ B = B, then: A)-(B = $\cup$ {S_i : B + S_i $\subseteq$ A}. A)-(B = {max(0, {1/0, 1/1, 0.8/2, 0.5/3, 0.2/4})-({1/0, 1/1, 0.3/2})* a} ={ {1/0, 0.8/2, 0.2/2}*a}.
We can check that A)-(B is the greatest fuzzy bag such that B + (A)-(B) = {{1/0, 1/1, 0.8/2, 0.3/3, 0.2/4}* a} is included in A.

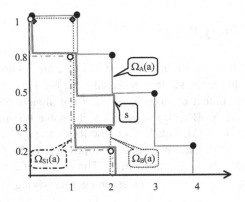

Fig. 1. $\Omega_{S1}(a) = \Omega_A(a))-(\Omega_B(a)$, an approximation of s ◆

After these backgrounds about fuzzy bags, we can tackle the problems remaining to solve. The difference x-y between two fuzzy integers does not always exist, when x is smaller than y, but, also, when y is smaller than x, as in the previous example. When we analyze fig. 1, we have the intuition that s appears to be the 'exact' representative of the difference between $\Omega_A(a)$ and $\Omega_B(a)$. Unfortunately s is not a fuzzy integer, it is the reason why an optimistic difference has been introduced in order to define S1 the best approximation (by inferior value) of the difference between A and B. The first problem is : as negative numbers $\mathbb{Z}$ have been introduced for defining unrestricted subtraction on integers, is it possible to introduce fuzzy rational (or relative) integers ($\mathbb{Z}_f$) in order to define unrestricted subtraction between fuzzy numbers ? The second question is then : what is the relationship between an entity such as s, in fig. 1, and a fuzzy relative integer ?

4 $\mathbb{Z}_f$, the Set of Fuzzy Relative Integers

Let a and b be two fuzzy integers, we define the equivalence relation $\mathcal{R}$ such that :

$\forall (a, b) \in \mathbb{N}_f \times \mathbb{N}_f$, (a, b) $\mathcal{R}$ (a', b') iff a + b' = a' + b
where + is defined via the sup-min convolution.

We denote $Z_f = (\mathbb{N}_f \times \mathbb{N}_f) / \mathcal{R}$ the set of all equivalence classes on $(\mathbb{N}_f \times \mathbb{N}_f)$ defined by $\mathcal{R}$. We can identify a fuzzy integer n to the class (n, 0) and a negative fuzzy integer to the class (0, n), latter being denoted by –n.

Example 4.1. a = {1/0, 1/1, 0.8/2, 0.5/3, 0.2/4} ; b = {1/0, 1/1, 0.3/2}.

The couple (a, b) = ({1/0, 1/1, 0.8/2, 0.5/3, 0.2/4}, {1/0, 1/1, 0.3/2}) $\in Z_f$.
(a, b) is one instance of an equivalence class which defines the difference between a and b: (a-b). Other instances of this class could be (a', b') = ({1/0, 0.8/1, 0.5/2, 0.2/3}, {1/0, 0.3/2}) or (a'', b'') = ({1/0, 1/1, 0.9/1, 0.8/2, 0.5/3, 0.2/4}, {1/0, 1/1, 0.9/3, 0.3/3})

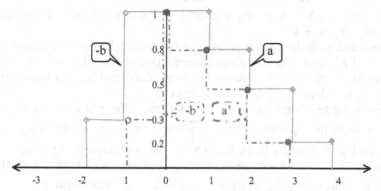

Fig. 2. Two equivalent graphic representations of the fuzzy rational (a-b) ◆

A fuzzy rational defined as : (a, b) = ({1/0, 1/1, 0.8/2, 0.5/3, 0.2/4}, {1/0, 1/1, 0.3/2}) can also be denoted by the following notation of fuzzy number : X={0.3/-2, 1/-1, 1/0, 1/1, 0.8/2, 0.5/3, 0.2/4}. The membership function of such a number X is a normalized convex fuzzy set defined on Z.

X can be decomposed into two separated parts, X^- and X^+, such that X^- is a non-decreasing function on]–∞, 0], and X^+ is a non-increasing function on [0, +∞[, and $X = X^- \cup X^+$.

4.1 Addition on Z_f

We now define the addition ⊕ on Z_f. Let $(X, Y) \in Z_f \times Z_f$, (X^+, X^-) an instance of X and , (Y^+, Y^-) an instance of Y. X ⊕ Y is defined as the class of $(X^+ + Y^+, X^- + Y^-)$. This addition is commutative, associative and has a neutral element 0_{Z_f} whose representative element is ({1/0}, {1/0}) or more generally {(a, a) / a $\in \mathbb{N}_f$}.

From our definition of the addition on Z_f we deduce : X ⊕ Y = $(X^- \cup X^+)$ ⊕ $(Y^- \cup Y^+)$ = $(X^- + Y^-) \cup (X^+ + Y^+)$ with + the addition defined via the extension principle. Using the addition defined via the extension principle on X and Y, the following property holds [2]: X+Y = $(X^- \cup X^+) + (Y^- \cup Y^+) = (X^- + Y^-) \cup (X^+ + Y^+) \cup (X^- + Y^-) \cup$

(X^-+Y^+). But, as $(X^++Y^-) \subseteq (X^- + Y^-) \cup (X^++Y^+)$ and $(X^-+Y^+) \subseteq (X^-+Y^-) \cup (X^++Y^+)$ we deduce : $X + Y = (X^-+Y^-) \cup (X^++Y^+)$

Consequently, the addition $\oplus$ is nothing but the addition + obtained by the extension principle.

4.2 About Difference on $\mathbb{Z}_f$

Each fuzzy rational $X = (X^+, X^-)$ has an opposite, denoted by $-X = (X^-, X^+)$, such that $X \oplus (-X) = (X^++ X^-, X^- + X^+) = 0_{\mathbb{Z}_f}$. The difference $X \ominus Y = X \oplus (-Y)$ is defined by the class of $(X^+ + Y^-, X^- + Y^+)$.

We now deal with the second problem: what is the relationship between an entity such as s in fig. 1 and a fuzzy rational integer ? Our intuition is that s (fig. 1) and the fuzzy rational (a, b) of fig. 2 represent the same entity: in other words we must show that s and (a, b) belong to the same equivalence class.

If a and b are two fuzzy integers, the optimistic difference a)-(b is the solution, when it exists, of the equation $a + x = b$, where x belongs to $\mathbb{N}_f$. If we represent a)-(b by the couple (a, b) and x by (x, 0), as $a + x = b$, we deduce (a, b) $\mathcal{R}$ (x, 0) and we conclude that the relative fuzzy integer (a, b) and the positive integer x are equivalent.

The interest to extend $\mathbb{N}_f$ to $\mathbb{Z}_f$ is to define a subtraction between any numbers. However, the entity s (fig. 1) comes from an approach by α-cuts and is not a fuzzy integer. So, in order to show that such an entity is equivalent to a fuzzy relative integer, we use an approach by α-cuts.

We define the α-cut of the fuzzy integer a as the set $\{x / \mu_a(x) \geq \alpha\}$. The elements of a_α can be associated with a creasing sequence of integers (0, 1, 2, ..., i) which will be denoted i. This integer i is such that $\mu_a(i)$ is the smallest degree satisfying $\mu_a(i)$) $\geq \alpha$. This definition generalizes the concept α-cut to entities such as s.

Table 2. The different α-cuts of the numbers a, b, (a-b), (a'-b') and s defined in table 1 and example 4.1.

α	a_α	b_α	$a_\alpha - b_\alpha$	a_α'	b_α'	$a_\alpha' - b_\alpha'$	s_α
1	1	1	1-1	0	0	0-0	0
0.8	2	1	2-1	1	0	1-0	1
0.5	3	1	3-1	2	0	2-0	2
0.3	3	2	3-2	2	1	2-1	1
0.2	4	2	4-2	3	1	3-1	2

Let (a, b) and (a', b') be two fuzzy relative integers (also denoted (a-b) and (a'-b')), we define the equivalence relation $\mathcal{R}'$ such that : $(a_\alpha, b_\alpha)\ \mathcal{R}'\ (a_\alpha', b_\alpha')$ iff $a_\alpha + b_\alpha' = a_\alpha' + b_\alpha$, $\forall \alpha \in [0, 1]$.

On the previous example we easily verify that (a, b) and (a', b') are equivalent with respects to $\mathcal{R}'$. Moreover, if we denote the different α-cuts of s with the sequence 0-

0, 1-0, 2-0, 1-0 and 2-0 (which represent (s^+_α, s^-_α)), we easily check that : (a_α, b_α) $\mathcal{R}'$ (s^+_α, s^-_α). Thus, (a-b), (a'-b') and s are equivalent, and, consequently, the entity s can be viewed as a relative fuzzy integer.

5 About Fuzzy Bags Based on $\mathbb{Z}_f$

In this section we quickly show how fuzzy bag operators can be defined on $\mathbb{Z}_f$.

There are conceptual difficulties to define fuzzy bags on $\mathbb{Z}_f$, that is a number of occurrences of an element can be associated with a fuzzy relative integers. What does it mean to assert that a collection contains negative occurrences ?

We extend the concept of a fuzzy bag by introducing the notion of positive or negative occurrence of an element. Such a fuzzy bag is a bag in which each signed occurrence of each element is associated with a grade of membership. A positive occurrence corresponds to an occurrence which has been added to the collection ; a negative occurrence corresponds to an occurrence which has to be suppressed from the collection. One way to describe this fuzzy bag is to enumerate the occurrences of its elements, for example: A = <1/a, 0.8/a, 0.2/a, 1/-a, 0.5/-a>. This bag is equivalent to <0.8/a, 0.2/a, 0,5/-a> because in the first description one occurrence of a has been added (at the level 1) and one occurrence of a (at the level 1) has to be suppressed.

Such a fuzzy bag A can be defined by a characteristic function Φ_A from a universe U to $\mathbb{Z}_f$, where $\mathbb{Z}_f$ is the set of the fuzzy relative integers : Φ_A: U $\to$ $\mathbb{Z}_f$.

Example 5.1. A = {{0.5/-2, 1/-1, 1/0, 1/1, 0.8/2, 0.2/3}* a} ; Φ_A (a) = {0.5/-2, 1/-1, 1/0, 1/1, 0.8/2, 0.2/3}.◆

In section 4 we have seen that a fuzzy relative integer X can be decomposed into a couple of positive fuzzy integers (X^+, X^-). Using this representation and the extension principle we define the extension of bag operations on $\mathbb{Z}_f$:

$$\Phi_{A\cap B}(x) = (\min (\Phi_A(x)^+, \Phi_B(x)^+) , \min (\Phi_A(x)^-, \Phi_B(x)^-))$$
$$\Phi_{A\cup B}(x) = (\max(\Phi_A(x)^+, \Phi_B(x)^+) , \max(\Phi_A(x)^-, \Phi_B(x)^-))$$
$$\Phi_{A+B}(x) = (\Phi_A(x)^+ + \Phi_B(x)^+, \Phi_A(x)^- + \Phi_B(x)^-).$$

The advantage and the aim of negative fuzzy numbers of occurrence is that they allow exact subtraction between fuzzy integers and exact relative complementation between fuzzy bags. As the difference between two fuzzy relative integers X and Y is defined by: X-Y = X+ (-Y) = (X^+ + Y^-, X^- + Y^+), we have:

$$\Phi_{A-B}(x) = (\Phi_A(x)^+ + \Phi_B(x)^-, \Phi_A(x)^- + \Phi_B(x)^+).$$

Applying this definition we get a fuzzy bag which is a representative of an equivalence class of fuzzy bags. The different representatives of an equivalence class of bags can be evaluated by working on the different representatives of each fuzzy numbers of occurrence of each element. The different representatives of a fuzzy relative integer can be evaluated by working on the different representatives of its different α-cuts which belong to $\mathbb{Z}$ (for example 0-1, also denoted −1, is a representative element of the class which contains 1-2, 2-3, 3-4, …).

6 Conclusion

Our main concern is here to go further our paper presented in [1]. In this study we have been confronted to a problem : in some case the relative complementation of a fuzzy bag A, with regards to a fuzzy bag B, does not exist (in the context of a model based on $\mathbb{N}_f$). The proposed solution was then to define approximations of this difference between fuzzy bags.

This paper offers a complementary study of the difference between fuzzy bags. The context of $\mathbb{N}_f$ does not permit the existence of any subtractions, so we propose the first ground for a new construction by extending $\mathbb{N}_f$ to $\mathbb{Z}_f$. In this context the difference A-B of two fuzzy bags is always defined by one equivalence class. The method applies for defining such an extension is very similar to this one used for extending $\mathbb{N}$ to $\mathbb{Z}$, and in that respect is not very original. However, inasmuch as very few attention as been paid to this kind of problem, up now, we think that this work sets up an interesting new contribution to the concept of fuzzy bags and brings out different perspectives.

As suggested in the introduction, fuzzy bags have interesting applications in the field of flexible querying of databases. In this context, one point that could be developed in further investigation, is the use of the difference operation for building up more complex operators such as the division.

References

1. Bosc P., Rocacher D., "About difference operations on fuzzy bags", *9th Inter. Conf. IPMU*, pp 1441-1546, 2002.
2. Dubois D., Prade H., "Inverse operations for fuzzy numbers", *Proc. IFAC symp. On fuzzy information, knowledge representation and decision analysis, Marseille,* pp 391-396, 1983.
3. Dubois D., Prade H., "Fuzzy set theoretic differences and inclusions and their use in the analysis of fuzzy equations", *Control and Cybernetics*, vol 13, n° 3, 1984.
4. Dubois D., Prade H., "Fuzzy cardinality and modeling of imprecise quantification", *Fuzzy Sets and Systems* , 16, 3, pp 199-230, 1985.
5. Miyamoto S., "Fuzzy Multisets and Fuzzy Clustering of Documents ", *10th Inter. Conf. on Fuzzy Systems*, (FUZZ IEEE'01), 2001.
6. Sanchez E., "Solutions of fuzzy equations with extended operations", *Fuzzy Set and Systems*, 12, pp 237-248, 1984.
7. Rocacher D., "On the use of fuzzy numbers in flexible querying", *9th IFSA World Cong. and 20th NAFIPS Inter. Conf.,* pp 2440-2445, 2001.
8. Yager R., "On the Theory of Bags", *Inter. J. of General Systems*, vol. 13, pp 23-27, 1986.
9. Zadeh L. A., "A computational approach to fuzzy quantifiers in natural languages", *Comp. Math. App., vol. 9, pp 149-184*, 1983.

The Difference Between 2 Multidimensional Fuzzy Bags: A New Perspective on Comparing Successful and Unsuccessful User's Web Behavior

George Meghabghab

Roane State
Dept of Computer Science Technology,
Oak Ridge, TN, 37830
gmeghab@hotmail.com

Abstract. In all studies of user's behavior on searching the web, researchers found that their search strategy or traversal process showed a pattern of moving back and forth between searching and browsing. These activities of searching and browsing are inclusive activities to complete a goal-oriented task to find the answer for the query. The moving back and forth, including backtracking to a former web page, constitute the measurable or physical states or actions that a user takes while searching the web. Is there a mathematical model that can help model actions of users to distinguish between 2 users while their repeated actions are accounted for since it is a part of their behavior or inherent essence to arrive at the goal of answering the questions regardless whether they are successful or unsuccessful? The idea of a set does not help account for repeated actions. It is known that in a set repeated objects are ignored. In some situations we want a structure in which a collections of objects in the same sense as a set but a redundancy counts. The structure of a bag as a framework can help study the behavior of users and uncover some intrinsic properties about users. This study continues in applying bags to user's mining of the web[10]. It considers queries that depend on more than one variable since multi variable queries have never been studied in the literature and the power of fuzzy bags in dealing which them. The queries are of the sort:" Find the number of web pages of users whose number of searches is less than average and hyperlink navigation is less than the average."

1 Introduction

In all studies of user's behavior on searching the web, researchers found that their search strategy or traversal process showed a pattern of moving back and forth between searching and browsing. These activities of searching and browsing are inclusive activities to complete a goal-oriented task to find the answer for the query [2,9]. The moving back and forth, including backtracking to a former web page, constitute the measurable or physical states or actions that a user takes while searching the web [7]. Is there a mathematical model that can help model actions of users to distinguish between 2 users while their repeated actions are accounted for since it is a part of their behavior or inherent essence to arrive at the goal of answering the questions regardless whether they are successful or unsuccessful? DiLascio [6]

T. Bilgiç et al. (Eds.): IFSA 2003, LNAI 2715, pp. 103-110, 2003.

introduced a set of triples <CS,PS,OR> to represent not only the user's cognitive state(CS) but also the current psychological state(PS) and the orientation (OR). They did not however compare the fuzzy management of user navigation in the case of success and the case of failure of users to a given query. This research will consider such an idea. The idea of a set does not help account for repeated actions. It is known that in a set repeated objects are ignored. In some situations we want a structure in which a collections of objects in the same sense as a set but a redundancy counts. Assume X is a set of elements that represent the actions of users while searching the web: a- users search or query a search engine using a single concept or a multiple concept query. Let S represent such an action, b- users browse one or more categories: let B be such an action, c-users scroll results and navigate hyperlinks: let H represent such an action, d- users backtrack between different web pages: let BT represent such an action, e- users select a web page looking for an answer for the query: let W represent such an action. Thus X as a set can be represented by X={S,B,H,BT,W}.

2   User's Actions While Searching the Web as a " Multidimensional Fuzzy Bag"

A fuzzy bag A by definition [5] is a characteristic function "E(x)" which gives the characteristic of an element x from a set $X=\{x_1,x_2,x_3,\ldots, x_n\}$:

$$A=\{\psi//x \mid \psi=E(x)\} \qquad (1)$$
A simplified version of (1) is:
$$A=\{\psi//x \mid E(x)\} \qquad (2)$$

Such a definition yields to the bag A being represented by $A=<a_1/x_1,a_2/x_2,a_3/x_3,..a_n/x_n>$ where $a_i=\psi//x_i$ From earlier definitions [5,11,12], one could summarize the followings: a set: $\{x|E(x)\}$ where E(x) returns a boolean, a fuzzy set: $\{\mu/x|E(x)\}$ where E(x) returns a degree, a crisp bag: $\{w*x|E(x)\}$ where E(x) returns an integer, a fuzzy bag: $\{\psi//x \mid E(x)\}$ where E(x) returns a bag of degrees. In the case of our user's behavior, the set X ={S,B,H,BT,W} will take on fuzzy attributes. To answer: "Find the number of web pages of users whose number of searches is less than average and hyperlink navigation is less than the average", a multi-dimensional fuzzy bag is considered.

2.1 The Association of n Fuzzy Bags

$$\text{Att, Bag}(X_1),\text{Att, Bag}(X_2)\ldots, \text{Att, Bag}(Xn)\rightarrow\text{Bag}(\text{Tuple}(\text{Att: }X_1,\text{Att: }X_2,\text{Att}:Xn)) \qquad (3)$$

In the case of 2 fuzzy bags:
$$(X_1,\text{Bag}(X_1)),(X_2,\text{Bag}(X_2))\rightarrow\text{Bag}(\text{Tuple}(X_1,X_2)) \qquad (4)$$

The Select operator [5] applies a "fuzzy predicate p" (less than average for example) to each occurrence of the fuzzy bag in (4) and results in a new fuzzy bag:
$$\text{Select}(\text{Bag},p):\text{Bag}(\text{Tuple}(X_1,X_2),(X_1\rightarrow\text{MBoolean}),(X_2\rightarrow\text{MBoolean})\rightarrow\text{Bag}(\text{Tuple}(X1,X2)) \qquad (5)$$

Where Mboolean is a multiset Boolean. By applying (5) to our query for successful users (see Appendix, table 1) yields:

$\psi_{(S,H)}$=1W$\rightarrow$<<1S,1H>,<1S,1H,<1S,1H>,<2S,3H>,<4S,4H>>,2W$\rightarrow$<<2S,2H>>, 3W$\rightarrow$< 2*<3S, 1H>>, 4W$\rightarrow$<<7S, 6H> ,<5S,3H>,<6S,1H>>, 9W$\rightarrow$<<8S,3H>>

The fuzzy bag A_S for successful users is made out of elements that can be written as (by applying (5)):

A_S={<<1,1>,<1,1>,<1,1>,<.9,.6>,<.7,.4>>//1W,<.9,.8>//2W,<<.8,1>,<.8,1>//3W,< <.4,.1>,<.6,.6>,<.5,1>>//4W,<.3,.6>//9W}. Figure 1 represents such a 2-D fuzzy bag. By applying (5) to unsuccessful users (see Appendix, table 2) yields:

$\psi_{(S,H)}$=1W$\rightarrow$<<2S,3H>,<1S,1H>,<3S,2H>>,4W$\rightarrow$<<3S,5H>,<5S,1H>>,5W$\rightarrow$ <<10S,7H>>,6W$\rightarrow$<<4S,5H>>,7W $\rightarrow$<<4S,4H>>.

The fuzzy bag A_F for unsuccessful users is made out of elements that can be written as (by applying(5)):
A_F={<<.9,.8>,<1,1>,<.8,.9>>//1W,<<.8,.6>,<.6,1>>//4W,<<.1,.4>>//5W, <<.7,.6>>//6W,<<.7,.7>>//7W}.Figure2 represents such a fuzzy bag.

These 2 degrees one for searches, and one for the hyperlinks, are commensurable, for example a degree 0.5 from the first predicate and a degree 0.5 from the second predicate do not have the same meaning. Consequently they cannot be combined, and as such any of the 2 dimensions cannot be reduced. Thus, the fuzzy bag has been coined a multi-dimensional fuzzy bag.What if fuzzy cardinalities ([13]) were used to define fuzzy numbers of occurrences of fuzzy bags? The concept of fuzzy cardinality of a fuzzy set called FGCount(A), is defined by:

$$\forall n: \mu |A| (n)=\sup\{\alpha/ |A| \geq n\}. \tag{6}$$

Then considering this notion of fuzzy cardinality, the occurrences of an element x in a fuzzy bag A can be characterized as a fuzzy integer denoted by $\Omega A (x)$. This fuzzy number is the fuzzy cardinality of the fuzzy set of occurrences of x in A. Thus a fuzzy bag A, on a universe U, can b defined by the characteristic function ΩA from U to Nf, where Nf is the set of fuzzy integers:

$$\Omega_A : U \rightarrow N_f \tag{7}$$

if A_S={<<1,1>,<1,1>,<1,1>,<.9,.6>,<.7,.4>>//1W,<.9,.8>//2W,<<.8,1>,<.8,1>//3W, <<.4,.1>,<.6,.6>,<.5,1>>//4W,<.3,.6>//9W},then:Ω_{AS}(1w)={<1,1>/0,<1,1>/1,<1,1>/2 ,<1,1>/3,<0.9,.6>/4,<.7,.4>/5},Ω_{AS}(2w)={<1,1>/0,<.9,.8>/1},Ω_{AS}(3w)={<1,1>/0,<.8,. 1>/1,<.8,.1>2},Ω_{AS}(4w)={<1,1>/0,<.6,.6>/1,.<.5,.1>/2,<.4,.1>/3},Ω_{AS}(9w)={<1,1>/0 ,<.3,.6>/1}.If
A_F={<<.9,.8>,<1,1>,<.8,.9>>//1W,<<.8,.6>,<.6,1>>//4W,<<.1,.4>>//5W,<<.7,.6>>// 6W, <<.7,.7>>//7W},then:
Ω_{AF}(1w)={<1,1>/0,<1,1>/1,<.9,.8>/2,<.8,.9>/3},Ω_{AF}(4w)={<1,1>/0,<.8,.6>/1,<.6,.1> /2},Ω_{AF}(5w)={<1,1>/0,<. 1,.4>/1},Ω_{AF}(6w)={<1,1>/0,<.7,.6>/1},Ω_{AF}(7w)={1/0,<.7,.7>/1}

$$\Omega_{A\wedge B}=\min(\Omega_A(x),\Omega_B(x)) \tag{8}$$
$$\Omega_{A\cup B}=\max(\Omega_A(x), \Omega_B(x)) \tag{9}$$
$$\Omega_{A+B}=\Omega_A(x) + \Omega_B(x) \tag{10}$$

2.2 Difference of 2 Multidimensional Fuzzy Bags

In the classical theory, the set difference can be defined without complementation by: $A-B=S \Leftrightarrow A=(A \cap B) \cup S$. In bag theory, this definition has an additive semantics. $A - B$ is associated with the elements, which have to be added to $(A \cap B)$ so that it equals:

$$A: A-B=S \Leftrightarrow A=(A \cap B)+S \tag{13}$$

Unfortunately, in the context of fuzzy bags, it is not always possible to find S, the relative complement of $(A \cap B)$ with respect to A. Let us consider 2 fuzzy bags $A_S=\{<.9,.9,.9,.8,.6>//1W\}$ and $A_F=\{<.9,.3>/1w\}$. $(A_S \cap A_F)=\{<.9,.3>//1W\}= A_F$ and there is no S such that $A_S = A_F + S$ because the occurrence of 1w with the degree of 0.3 cannot be upgraded to a higher levels thanks to the addition.

How does Yager's deal with the problem: The difference of 2 fuzzy bags [11,12] A and B is the smallest fuzzy bag S such that the addition of S and B contains A. In other words, the difference of 2 fuzzy bags is a fuzzy bag that contains the elements of A and does not contain the elements of B. The number of occurrences of the elements of the difference is the maximum value of the difference of the number of occurrences of x in A and the number of occurrences of x in B:

$$A - B = \{w*x \mid max\ (0,w_A\ (x) - w_B\ (x))\} \tag{14}$$

How does Connan deal with the difference of 2 fuzzy bags [5]: the difference of 2 fuzzy bags [5] A and B is based on the idea of what fuzzy bag "S" to be added to "B" to get a fuzzy bag that contains the fuzzy bag "A". To apply such a definition, we sort the elements of A and B in decreasing order. Every occurrence of an element of A is matched with an element of B which is greater or equal to that of A. Every element of A that cannot be put in correspondence to element of B is in the result of the difference:

$$(A - B) = \cap\{S \mid A \subseteq (B+S)\} \tag{15}$$

Applying (14) to A_S and A_F yields: $(A_S - A_F)=\{<max(0,3*<1,1>-1*<1,1>)$, $max(0,1*<.9,.6>-0*<.9,.6>)$, $max(0,\ 1*<.7,.4>-0*,<.7,.4>)//1W$, $<max(0,1*<.6,.6>-0*<.6,.6>)//4W,max(0,1*<.5,.1>-0*<.5,.1>)//4W,max(0,1*<.4,.1>-0*<.4,.1>)>//4W\}=\{<2*<1,1>,<.9,.6>,<.7,.4>>//1W,<<.6,.6>,<.5,.1>,<.4,.1>> //4W\}$ $A_S-A_F=\{<<1,1>,<1,1>,<.9,.6>,<.7,.4>>//1W,<<.6,.6>,<.5,1>,<.4,.1>>//4W \}$

Finding: The difference between A_S and A_F is a new fuzzy bag with 4 occurrences where 2 are made out of an element larger than average in the number of searches and larger than average in the number of hyperlinks, 1 occurrence made out of an element larger than average in the number of searches and smaller than average in the number of hyperlinks, and 1 occurrence made out of an element equal to average in the number of searches and smaller than average in the number of hyperlinks; and 3 occurrences where 1 occurrence is made out of an element equal to average in the number of searches and equal to average in the number of hyperlinks, and 2 occurrences made out of an element smaller than average in the number of searches and smaller than average in number of hyperlinks for 4 web pages. Figure 4 represents such a fuzzy bag.

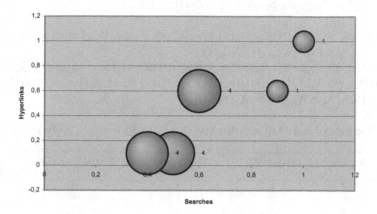

Fig 1. Difference between fuzzy bags (Yager's)

Applying (15) to A_S and A_F yields: $(A_S - A_F)=\{<<1,1>,<1,1>>//1W,<.4,.1>//4W\}$

Finding: The difference between A_S and A_F is a new fuzzy bag with 2 occurrences made out of an element larger than average in the number of searches and larger than average in the number of hyperlink navigations for 1 web page, and 1 occurrence is made out of an element smaller than average in the number of searches, and smaller than average in the number of hyperlink navigations for 4 web pages. Figure 5 represents such a difference.

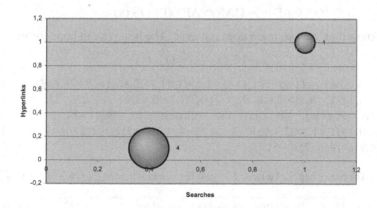

Fig 2. Difference between fuzzy bags (Connan's)

How does fuzzy cardinalities deal with the difference of 2 fuzzy numbers [4] in the case of a 2 dimensional fuzzy bag. As stated above, the condition in (13) has to be relaxed. Thus a fuzzy bag S_1 which approximates S such that it is the greatest fuzzy bag S_i that is contained in A:

$$S_1 = \cup \{ S_i : A \supseteq (A \cap B) + S_i \} \tag{16}$$

In other words, $\forall \alpha, \forall x : w_{(A \cap B)\alpha}(x) + w_{S1\alpha}(x) \leq w_{A\alpha}(x)$. Let us apply (16) to A_S and A_F: $w_{(AS \wedge AF)\alpha}(x) + w_{S1\alpha}(x) \leq w_{AS\alpha}(x)$.
$A_S \cap A_F = \{\{1,1>/0,<1,1>/1,<.9,.8>/2,<.8,.9>/3\}*1W, \{<1,1>/0,<.6,.6>/1,<.5,.1>/2\}*4W\}, A_S(1w) = \{<<1,1>,<1,1>,<1,1>,<.9,.6>,<.7,.4>>//1w\}, A_F(1w) = \{<<1,1>,<.9,.8><.8,.9>>//1W\}, (A_S \cap A_F)(1w) = A_F(1w)$

Table 1. The greatest fuzzy bag such that Max of $w_{(S1)\alpha}(1W) + w_{(AS \wedge AF)\alpha}(1w) \leq w_{(AS)\alpha}(1W)$

α	$w_{(AS)\alpha}(1W)$	$w_{(AF)\alpha}(1W)$	$w_{(AS)\alpha}(1W)-w_{(AF)\alpha}(1W)$	Max of $w_{(S1)\alpha}(1W)$	$w_{(AS \wedge AF)\alpha}(1w)$
<1,1>	3	1	2	2	1
<.9,.8>	3	2	1	1	2
<.9,.6>	4	2	2	2	3
<.8,.9>	4	3	1	1	3
<.7,.4>	5	3	2	2	3

Extending the results to fuzzy bags requires the following calculations:
$A_S \cap A_F$
$= \{<1,1>/0,<1,1>/1,<.9,.8>/2,<.8,.9>/3>//1w,<1,1>/0,<.6,.6>/1,<.5,.1>/2//4W\} \Rightarrow \Omega_{AS \wedge AF}(1w) = \{1/0,0.9/1,0.8/2,.7/3\}, A_S(1w) = \{<1,1>,<1,1>,<1,1>,<.9/.6>,<.7,.4>\} \Rightarrow \Omega_{AS}(1w) = \{<1,1>/0,<1,1>/1,<1,1>/2,<1,1>/3,<.9/.6>/4,<.7,.4>/5\}, \Omega_{AF}(1w) = \{<1,1>/0,<1,1>/1,<.9,.8>/2,<.8,.9>/3\} = \Omega_{AS \wedge AF}(1w) = \{<1,1>/0,<1,1>/1,<.9,.8>/2,<.8,.9>/3\}$
By extending the results to bags, we could say:

$$\forall x : \Omega AS (x) = \Omega AS \cap AF (x) + \Omega S (x) \tag{17}$$

We know that such solution does not exist. The best we can hope for is:

$$\forall x: \Omega_{AS \cap AF}(x) + \Omega_{S1}(x) \leq \Omega_{AS}(x) \tag{18}$$

$S_1(1w) = A_S(1w) - A_F(1w) = \{<1,1>/0,<1,1>/1,<1,1>/2,<1,1>/3,<.9/.6>/4,<.7,.4>/5\} - \{<1,1>/0,<1,1>/1,<.9,.8>/2,<.8,.9>/3\} = \{<1,1>/0,<1,1>/1,<.7,.4>/2\}$. If $S_1(1w)$ is added to $A_F(1w)$, it is the closest to $A_S(1w)$: $A_F(1w) + S_1(1w) = \{<1,1>/0,<1,1>/1,<1,1>/2,<.9,.8>/3,<.8,.9>/4,<.7,.4>/5\} \subseteq \{<1,1>/0,<1,1>/1,<1,1>/2,<1,1>/3,<.9/.6>/4,<.7,.4>/5\}$

Table 2. The fuzzy bag $S_1(1w)$ is the largest fuzzy bag where $w_{(S1)\alpha}(1w)(18) <= w_{(S1)\alpha}(1w)$

α	$w_{(AS)\alpha}(1w)$	$w_{(AF)\alpha}(1w)$	$w_{(AS)\alpha}(1w)- w_{(AF)\alpha}(1w)$	$w_{(S1)\alpha}(1w)(22)$	Max of $w_{(S1)\alpha}(1w)$
<1,1>	3	1	2	1	2
<.9,.8>	3	2	1	1	1
<.9,.6>	4	2	2	1	2
<.8,.9>	4	3	1	1	1
<.7,.4>	5	3	2	2	2

$A_S \cap A_F = \{<1,1>/0,<1,1>,<.9,.8>,<.8,.9>//1w,<.6,.6>,<.5,.1>//4w\}, A_S \cap A_F(4w)=$
$\{<.6,.6>/1,<.5,.1>\} \Rightarrow \Omega_{A_S \wedge A_F}(4w)=\{<1,1>/0,<.6,.6>/1,<.5,.1>/2\}, \Omega_{A_F}(4w)=$
$\{<1,1>/0,<.8,.6>/1,<.6,.1>/2\} \neq \Omega_{A_S \wedge A_F}(4w)=\{<1,1>/0,<.6,.6>/1,<.5,.1>/2\}, \Omega_{A_S}(4w)=$
$\{<1,1>/0,<.6,.6>/1,<.5,.1>/2,<.4,.1>/3\}, \Omega_{A_S}(4w)-\Omega_{A_F}(4w)= \{<1,1>/0, <.6,.6>/1,$
$<.5,.1>/2,<.4,.1>/3 \}-\{<1,1>/0,<.8,.6>/1,<.6,.1>/2\}=\{<1,1>/0,<.4, .1>/1\}, A_S-A_F=$
$\{<<1,1>,<.7,.4>>//1w,<<.4, .1>>//4w\}$

Finding: The difference between A_S and A_F according to cardinality [12] is a new fuzzy bag with 2 occurrences among which 1 occurrence is made out of an element larger than average in the number of searches and larger than average in the number of hyperlink navigations, and 1 occurrence is made out of an element larger than average in the number of searches and smaller than average in the number of hyperlink navigations for 1 web page; and 1 occurrence is made out of an element smaller than average in the number of searches, and smaller than average in the number of hyperlink navigations for 4 web pages. Figure 6 represents such a difference.

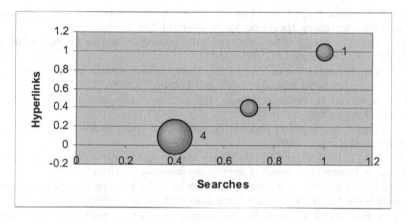

Fig 3. Difference between Fuzzy bags (Cardinality)

3   Discussion and Concluding Remarks

This study shows that bags are a viable "computational tool" to interpret users behavior. Although fuzzy bags since their creation by Yager [11,12] have not had many applications, more recently there is a resurgence of their application in object oriented data types [1], multi-criteria decision making problems [3], foundation of multi-sets [8], and databases [4]. Yager's theory on fuzzy bags [11,12] has the disadvantage of not generalizing their operators on fuzzy sets. Connan's thesis on fuzzy bags [5] anchors fuzzy bags in Zadeh's definition of fuzzy sets [13] and alpha cuts, but failed to implement it in the case of the difference between 2 fuzzy bags. In and as such is compatible with the operations on fuzzy sets. Connan's interpretation of fuzzy bags [5] has helped enrich user's behavior and offered a more flexible interpretation of the collected data of user's behavior. This study further enhances the use of bags to answer more questions about user's behavior. As such it becomes the only de facto technique when more complex queries are used. For example, if more than one attribute of the user's behavior was to be used, then better fuzzy bags operators will be needed to answer such a query. The results in figures 1,2, and 3 show a 2-dimensional bubble chart that were fuzzy on both attributes the searches and the number of hyperlink used. Compared to a study of how elementary school students behavior on a task was done [8]: "successful students looped searches and hyperlinks less than unsuccessful ones, examined a higher level of hyperlinks and homepages, and scrolled a lightly higher percentage of the screens and pages returned", this study showed a more complex comparison in nature and richer descriptive results between users who succeeded and those who failed through the use of queries.

References

1.   Albert, Algebraic Properties of bag data types, In the 17th International Conference on Very Large Data Bases, Barcelona, Spain, 1991 211-219.
2.   M.J. Bates, The design of browsing and berry picking techniques for the on-line search interface, Online Review 13 (1989) 407-431.
3.   R. Biswas, An application of Yager's Bag Theory in multi-criteria based decision making problems, International Journal of Intelligent Systems 14(1999) 1231-1238.
4.   D P. Bosc, and D. Rocacher, About difference operation on fuzzy bags, IPMM 02, Annecy, France (2002) 1541–1546.
5.   F.R. Connan, Interrogation Flexible de bases de donnees multimedias, These D'etat, University of Rennes I 1999.
6.   L.DiLascio, E. Fischetti, and V. Loia, Fuzzy Management of user actions during hypermedia navigation, International Journal of approximate reasoning 18 (1998) 271-303.
7.   R. Fidel, A visit to the information mall: Web searching behavior of high school students, Journal of the American Society for Information Science 50 (1999) 24-38.
8.   S. Grumbach, and T.M. Milo, Towards tractable algebras for multi-sets, ACM principles of Database Systems (1993).
9.   Y. Kafai, and M.J. Bates, Internet web-searching instruction in the elementary classroom: building a foundation for information literacy. School Library Media Quarterly 25 (1997) 103-111.
10. G. Meghabghab, and D. Windham, Mining User's Web behavior using "crisp and fuzzy bags": a new perspective and new findings, NAFIPS (2002), New Orleans, USA, 75-80.
11. R.R. Yager, On the Theory of Bags, International Journal of General Systems 13 (1986) 23-37.
12. R.R. Yager, Cardinality of Fuzzy Sets Via Bags, Mathematical Modeling 9(6) (1987) 441-446.
13. L.A. Zadeh, A computational approach to fuzzy quantifiers in natural languages, Comp. Math. App. 9 (1983) 149-184.

A Discussion of Indices for the Evaluation of Fuzzy Associations in Relational Databases

Didier Dubois[1], Henri Prade[1], and Thomas Sudkamp[2][*]

[1] IRIT-CNRS, Université Paul Sabatier, 31062 Toulouse, France,
{dubois,prade}@irit.fr
[2] Dept. of Computer Science, Wright State University, Dayton, Ohio 45385, USA,
tsudkamp@cs.wright.edu

Abstract. This paper investigates techniques to identify and evaluate associations in a relational database that are expressed by fuzzy if-then rules. Extensions of the classical confidence measure based on the α-cut decompositions of the fuzzy sets are proposed to address the problems associated with the normalization in scalar-valued generalizations of confidence. An analysis by α-level differentiates strongly and weakly supported associations and identifies robustness in an association. In addition, a method is proposed to assess the validity of a fuzzy association based on the ratio of examples to counterexamples.

1 Introduction

The proliferation of large databases provides both the impetus and the need for the development of algorithmic techniques for the identification of relationships among data. Data mining, the process of discovering relationships and patterns in data, consists of summarization and analysis using measures designed to identify the presence of associations among the objects in a database. In this paper we examine the use of fuzzy sets to represent associations in a relational database and the properties of measures designed to identify these relationships.

An association between attributes A and B is frequently represented in the form of a rule $A \Rightarrow B$ indicating that an element satisfying property A also satisfies B. The standard pair of indices used to measure the validity of an association rule are the support and the confidence,

$$ Sup(A \Rightarrow B) = \frac{|A \cap B|}{|U|} \qquad Con(A \Rightarrow B) = \frac{|A \cap B|}{|A|}, \qquad (1) $$

where $|\ |$ denotes the cardinality and U is the set of tuples in the database. The support measures the extent of the simultaneous occurrence of properties A and B in elements of the database while the confidence indicates the likelihood of an element with property A also having property B.

The introduction of fuzzy rules into data mining extends the representational capabilities of classical association rules. Two common types of fuzzy rules are

---

[*] Currently on leave at IRIT-CNRS, Université Paul Sabatier, Toulouse, France

T. Bilgiç et al. (Eds.): IFSA 2003, LNAI 2715, pp. 111–118, 2003.

certainty rules and *gradual rules* [5]. Certainty rules, which are based on Kleene-Dienes implication, have the form 'the more X is A, the more certainly Y is B' and provide a natural extension of crisp association rules. Gradual rules, based on residuated implication, have the form 'the more X is A, the more Y is B' and provide a set of increasingly restrictive constraints on the consequent based on the degree of satisfaction of the antecedent.

This paper examines the properties of measures proposed for the assessment of associations defined by fuzzy rules. It is sufficient to consider the relationship between two attributes $\mathcal{A}$ and $\mathcal{B}$ with domains $D_\mathcal{A} = \{a_1, \ldots a_{|D_\mathcal{A}|}\}$ and $D_\mathcal{B} = \{b_1, \ldots b_{|D_\mathcal{B}|}\}$, respectively. The tuples in the database R will be denoted $t_k = (a_{i_k}, b_{j_k})$, $k = 1, \ldots, r$, where a_{i_k} and b_{j_k} are elements from $D_\mathcal{A}$ and $D_\mathcal{B}$. When the index is immaterial, a tuple will be written $t = (a, b)$.

A subset A of $D_\mathcal{A}$ defines a subset $R(A)$ over R; a tuple (a, b) is in $R(A)$ if, and only if, $a \in A$. When A is fuzzy, $R(A)$ is a fuzzy set over the tuples of R. The membership value of the tuple $t = (a, b)$ in $R(A)$ is $A(a)$, the degree of membership of the attribute value a in the fuzzy set A.

Using the preceding notation, the confidence measure for an association $A \Rightarrow B$ of attributes in a relational database can be written

$$con(A, B) = |R(A) \cap R(B)| / |R(A)|. \tag{2}$$

When A and B are fuzzy sets, evaluating (2) requires selecting a T-norm for intersection and comparing the cardinalities of $R(A) \cap R(B)$ and $R(A)$.

In Section 2 we review extensions of the confidence measure that first combine membership values of corresponding tuples of $R(A)$ and $R(B)$ and then aggregate and normalize using cardinality. The presentation will focus on exhibiting the effect of the selection of the operations on the discriminability of the resulting measure. Section 3 reverses the order of the process; an initial step summarizes the database in terms of the membership values of the tuples. The summarization preserves the information necessary for computing confidence measures and provides the ability to evaluate associations based on the degree of relevance of the tuples. A combination of α-level confidence and tuple relevance is proposed as a criterion for supporting a fuzzy association. Section 4 assesses associations in terms of examples and counterexamples.

2   Scalar-Valued Measures

A common approach to extending the confidence measure to fuzzy sets is to replace the intersection and cardinality operations for crisp sets in (1) with corresponding fuzzy operations. Intersection is replaced with a T-norm, usually min, and cardinality with a scalar-valued cardinality. The simplest extension of cardinality to fuzzy sets is the Σ-count, which is the sum of the membership values of the elements in a fuzzy set. The generalization from crisp sets to fuzzy sets using the Σ-count produces the confidence measure

$$con_T(A, B) = \frac{\sum_{k=1}^{r} T(A(a_{i_k}), B(b_{j_k}))}{\sum_{k=1}^{r} A(a_{i_k})}, \tag{3}$$

where T is a T-norm and the sum is taken over the set of tuples in the database. A consequence of employing the Σ-count is that the summation permits a large number of tuples with small membership degrees to have the same effect as a single tuple with membership degree one, which may produce unintuitive results when assessing the association of variables.

An alternative approach is to give greater significance to tuples with higher membership degrees. Delgado et al. [3] proposed a scalar cardinality of fuzzy sets based on the weighted summation of the cardinalities of its α-cuts. This cardinality was used in [7] to reduce the impact of tuples with small membership degrees. The confidence measure using weighted cardinality is defined as

$$con_w(A, B) = \sum_{i=i}^{t-1}(\alpha_i - \alpha_{i+1})\frac{|(R(A) \cap R(B))|_{\alpha_i}}{|R(A)|_{\alpha_i}}, \qquad (4)$$

where $1 = \alpha_1, \ldots, \alpha_t = 0$ is an ordered listing of the union of the α-levels of $R(A) \cap R(B)$ and $R(A)$. The summation over the α-cuts places a greater emphasis on the elements with higher membership values, since a tuple with membership α_k occurs in each of the summands $k, k+1, \ldots, t$.

The following examples illustrate the influence of the distribution of membership values on the assessment of an association.

Example 1. The effect of multiple elements with small membership values can be seen using a modification of an example provided by Martín-Bautista et al [7]. The table below gives the number of tuples of each type, their membership values in A and B, and the result of the intersection.

Number	$A(a)$	$B(b)$	$\min(A(a), B(b))$
1	1	.01	.01
1	.01	1	.01
998	.01	.01	.01

Measuring confidence with the Σ-count yields $con_{\min}(A, B) = .91$. Thus a high measure of confidence in the association is produced, even though the only tuple that significantly satisfies A minimally satisfies B. With the weighted α-cuts, the contribution of the 999 tuples with membership .01 is reduced by the weighting. The result, $con_w(A, B) = .01$, is more in accord with an intuitive assessment of the validity of the rule.

The influence of small cardinalities on the confidence measure con_T is not limited to the T-norm min. Changing the B values from .01 to 1 in the final 999 tuples produces a like result regardless of the T-norm. Other probabilistic measures, such as conviction [2], produce the same anomalies when extended to fuzzy associations using the Σ-count. $\square$

Example 2. While minimum is the most common T-norm, its noncompensatory nature contributes to a loss of information when used for intersection. This phenomena can easily be exhibited by considering tuples whose A and B values are identical.

	Number	$A(a)$	$B(b)$	$\min(A(a), B(b))$
a)	3	.5	.5	.5
b)	3	.5	1	.5

Tuples in rows a) and b) differ only in value of the B attribute. The change
to the membership values affects neither the Σ-counts nor the α-cuts of $R(A)$
or $R(A) \cap R(B)$. Consequently, confidence measures that use min intersection or
are based on the intersection of α-cuts produce the same values for these sets of
tuples; $con_{\min}$ yields 1 for both the tuples in a) and b) and con_w yields .5. It is
also clear that the confidence measures will be independent of the B attributes
whenever their membership values satisfy $B(b_{j_k}) \geq A(a_{i_k})$ for all tuples t_k. This
type of behavior indicates that a compensatory T-norm may be more appropriate
when this type of data is expected. □

Additional scalar-valued measures of the validity of a fuzzy association have
been proposed by Hüllermeier [6] based on a reformulation of (1) in terms of
implication. Hüllermeier suggested the use of Kleene-Dienes implication for the
assessment of certainty rules and a residuated implication for gradual rules. The
resulting measures also normalize by the sum of the A membership values and
exhibit behavior similar to those illustrated above.

3   Summarization of Membership

In the preceding section, the analysis of an association first determined the
intersection of the values $A(a)$ and $B(b)$ for each tuple $t = (a, b)$ and then
aggregated the results using cardinality. In [1], Bosc et al. proposed the reversal
of this procedure. The first step consisted of summarizing the database using a
fuzzy cardinality. The information in the resulting fuzzy set was then used to
produce a confidence value. Moreover, the summarized data permitted a more
detailed analysis of relationships between the data. In this section we will adapt
the strategy of [1] to initially produce a summarization of the database.

The process begins with the selection of a small set of values that serve as
representative membership values for the fuzzy sets A and B. Let $1 = \sigma_1 > \sigma_2 >
\ldots > \sigma_r = 0$ and $1 = \gamma_1 > \gamma_2 > \ldots > \gamma_s = 0$ be the representative values for
A and B respectively. A two-dimensional 'membership map' of the database R
is constructed over the set $[1, r] \times [1, s]$ that summarizes the distribution of the
tuples in A and B. Each tuple $t = (a, b)$ is associated with a pair (σ_i, γ_j) based
on the proximity of $A(a)$ to σ_i and $B(b)$ to γ_j. The representative value σ_{i_t} for
an element a is selected as follows:

$$\sigma_{i_t} = \begin{cases} \sigma_1, & \text{if } (1 + \sigma_2)/2 < A(a) \leq 1; \\ \sigma_i, & \text{if } (\sigma_{i+1} + \sigma_i)/2 < A(a) \leq (\sigma_i + \sigma_{i-1})/2, \text{ for } i = 2, \ldots, r - 2; \\ \sigma_{r-1} & \text{if } 0 < A(a) \leq (\sigma_{r-1} + \sigma_{r-2})/2. \end{cases}$$

The value γ_{i_t} for b is selected in a similar manner. The entry $m(i, j)$ is the
number of tuples in the database that are associated with the pair (σ_i, γ_j).

This first level of processing provides two types of summarization. Organizing the tuples by their membership in the fuzzy sets A and B extracts the information required for the assessment of the association. The selection of the representative set of membership values reduces the size of the resulting map. If the fuzzy sets $R(A)$ and $R(B)$ have a small number of α-levels, the α-levels may be used as the representative values and the summarization has no resulting loss of membership information. In the examples that follow, the representative values σ_i will be the α-levels of the fuzzy sets.

The benefit of the summarization of the membership values is the ability to perform a more detailed analysis that considers the distribution of the values. In [1], fuzzy-valued cardinalities were used to store summarized membership information of both $R(A) \cap R(B)$ and $R(A)$. The fuzzy-valued cardinality, like the membership map, provides the information needed to compute the confidence of the α-cuts, $|(R(A) \cap R(B))_\alpha|/|R(A)_\alpha|$, to identify the types of tuples that contribute to the confidence measure.

The α-cut evaluation suggests a straightforward method for eliminating the adverse effect of small cardinalities. The standard criterion for the acceptance of an association is to employ a user-defined value θ as a threshold; a confidence value greater than θ indicates support for the rule. Using the α-cut analysis, the confidence threshold may be augmented by an α-cut requirement that designates a specific α-level α' on which the confidence threshold must be satisfied; that is, a rule is supported if the confidence of the α'-cut exceeds θ. The acceptance criteria in [1] used a threshold on the convex hull formed from the α-level confidences. The addition of either of these conditions would remove the difficulties presented in Example 1. For those tuples, the con_{min} α-cut confidence values are

σ-level	$con_{min}(A, B)$
1	0
.01	.91

and the association would not be accepted for any reasonable α-cut requirement.

Unfortunately, the designation of an α-cut may provide anomalous information when the support for the rule is not uniform over the α-levels.

Example 4. The variation of the α-cut confidence levels in the tuples

Number	$A(a)$	$B(b)$	$\min(A(a), B(b))$		α	con_α
1	1	1	1		1	1
n	.8	.7	.7		.8	$1/(n+1)$
m	.7	.7	.7		.7	1

indicates a lack of robustness in the support for the association; α-cut specifications α' of .9 and .7 support the association while .8 does not. Weighting the α-cuts produces the value $con_w = .9$, which indicates strong support for the association regardless of the lack of robustness indicated by the α-cuts.   □

Example 5. Another shortcoming of determining the support for an association based on α-cuts is that it does not consider the distribution of the tuples. The tuples in a) and b) both produce the α-cut confidence values shown on the right.

	Number	$A(a)$	$B(b)$	$\min(A(a), B(b))$
a)	n	1	1	1
	n	.6	.6	.6
	n	.1	.1	.1
b)	1	1	1	1
	$3n-1$	.1	.1	.1

α	con_α
1	1
.6	1
.1	1

In case a) the support for $A \Rightarrow B$ comes from tuples whose A membership values are distributed throughout the range $[0,1]$. This differs from b) in which only one tuple significantly supports the rule. However neither the selection of a particular α-cut, nor $con_{\min}$, nor con_w differentiates these sets of tuples.   □

To distinguish a confidence value produced by tuples that are strongly relevant to a rule from one generated by tuples that barely satisfy the antecedent, it is useful to consider the distribution of the tuples. The value

$$dst_{\sigma_i} = \begin{cases} \min\left(1, \dfrac{|A_{\sigma_i}|}{|supp(A)|(1-(\sigma_i+\sigma_{i+1})/2))}\,,\right), & \text{for } i = 1, \ldots, r-2 \\ 1, & \text{for } i = r-1 \end{cases} \qquad (5)$$

where $supp(A) = \{(a,b) \in R \mid A(a) > 0\}$, compares the number of tuples in $|A_{\sigma_i}|$ with the expected number from a uniform distribution across σ-levels. A σ-level that contains at least as many tuples as the expectation yields $dst_{\sigma_i} = 1$. Requiring dst_{σ_i} to be greater than the threshold θ, or another threshold for tuple distribution, incorporates strength of support into the confidence assessment.

The addition of the distribution criterion prevents the acceptance of the association for each of the problematic cases in Examples 1, 4, and 5. The σ-level and the associated dst_σ values are given below with $m = n = 100$ in Examples 4 and 5.

Ex. 1		Ex. 4		Ex. 5 a)		Ex. 5 b)	
σ	dst	σ	dst	σ	dst	σ	dst
1	.002	1	.002	1	1	1	.007
.01	1	.8	1	.6	1	.01	1
		.7	1	.1	1		

The lack of tuples at high σ-levels in all Examples except 5 a) would prevent the associations from receiving support.

4   Example-Counterexample Based Assessment

When A and B are crisp sets, $R(A) \cap R(B)$ is the set of tuples that are examples of the association $A \Rightarrow B$, $R(A) \cap \overline{R(B)}$ is the set of counterexamples, and $\overline{R(A)}$ is the set of tuples that are not relevant to the association. In the confidence measure, the impact of the set of counterexamples is limited to the normalization. By explicitly accumulating both the number of examples and counterexamples, the need for normalization and the resulting difficulties may be avoided.

Generalizing the notion of examples and counterexamples to tuples that have partial membership in sets A and B requires an examination of the meaning these terms. When A and B are fuzzy, a tuple with membership values $A(a) = 1$ and $B(b) = .9$ is intuitively closer to being an example of $A \Rightarrow B$ than a tuple with $A(a) = 1$ and $B(b) = .8$. Thus the property of being an example, like satisfying the properties A and B, is a matter of degree.

The degree to which a tuple $t = (a, b)$ is an example or a counterexample to $A \Rightarrow B$ is determined solely by the membership values $A(a)$ and $B(b)$. A straightforward approach for determining the size of these sets is to employ the Σ-count for cardinality and a T-norm for fuzzy set intersection. This produces the functions

$$ex_T(A, B) = \sum_{k=1}^{r} T(A(a_{i_k}), B(b_{j_k})), \quad cx_T(A, B) = \sum_{k=1}^{r} T(A(a_{i_k}), \overline{B}(b_{j_k})) \quad (6)$$

where the sum is taken over the set of tuples.

There are several potential difficulties with counting examples using min as in (6). The first is that the sum of the examples, counterexamples, and irrelevant instances may not be the number of elements in the universe. Moreover, the use of min for combining partial membership degrees may hide significant differences in the non-minimal value. The use of the product T-norm provides a uniform transition between the sets of pure examples and pure counterexamples. Moreover, with product intersection, the examples, counterexamples, and irrelevant cases of an association $A \Rightarrow B$ form a fuzzy partition of the database R.

When associations are assessed by examples and counterexamples, the decision to accept an association is based on a comparison of the size of the sets of examples and counterexamples. One possible method is to accept $A \Rightarrow B$ if

$$\frac{ex(A, B)}{cx(A, B)} \geq \frac{\theta}{1 - \theta}, \quad (7)$$

for the user-defined threshold θ. The ratio of examples to counterexamples in (7), which may be obtained from confidence measures by $con(A, B)/con(A, \overline{B})$, will be denoted $re(A, B)$. If $|ex(A, B)| > |cx(A, B) = 0|$, then $re(A, B)$ is assumed to be greater than any acceptance threshold.

The relative comparison differs from Kodratoff's proposal [8] of measuring the confidence of a crisp association as the difference of the number of examples and counterexamples normalized by the size of the database. In the latter, the confidence in an association with 90 examples and 10 counterexamples will be 10 times that of one with 9 examples and 1 counterexample, but both may be negligible depending upon the size of the database.

Example 6. The cardinalities of the sets of examples, counterexamples, and irrelevant cases for the tuples defined in Examples 1 and 2 are

Example	ex_{min}	cx_{min}	ir_{min}
1		10	10.97	988.02
2 a)	1.5	1.5	1.5
2 b)	1.5	0	1.5

Example	ex_{prod}	cx_{prod}	ir_{prod}
1		.1198	10.8702	989.01
2 a)	.75	.75	1.5
2 b)	1.5	0	1.5

Assessing an association using $re(A, B)$ avoids the undesirable properties exhibited in these cases. In Example 1, $re_{\min}(A, B) \approx 1$, indicating that $A \Rightarrow B$ has approximately the same number of examples as counterexamples and would not be accepted. For Example 2, $re_{\min}(A, B)$ is .5 for the tuples in case a) and exceeds the acceptance threshold for those in b), which differentiates these cases and supports the association only for the latter set of tuples. $\square$

The accumulation of values with small α-levels followed by scaling can cause difficulties with the assessment of examples and counterexamples as well as with confidence measures. When using the product T-norm, the 998 tuples with A and $\overline{B}$ membership values .01 and .99 cause the number of counterexamples to greatly exceed the examples and the association $A \Rightarrow \overline{B}$ would be accepted by any reasonable threshold. This situation may be addressed by examining the examples and counterexamples on α-levels as in the preceding section.

5   Conclusion

The assessment of a fuzzy association requires the consideration of the relevance of the tuples in addition to the degree to which they satisfy a confidence measure or an example-to-counterexample ratio. Analysis on α-levels exhibits the robustness of an association and permits the differentiation of associations based on the distribution of the supporting tuples.

References

1. P. Bosc, D. Dubois, O. Privert, H. Prade, and M. de Calmès. Fuzzy summarization of data using fuzzy cardinalities. In *Proceedings of the Ninth International Conference IPMU 2002*, pages 1553–1559, Annecy, France, 2002.
2. S. Brin, R. Motwani, J. D. Ullman, and S. Tsur. Dynamic itemset counting and implication rules for market basket data. *SIGMOD 1997, Proceedings ACM SIGMOD International Conference on Management of Data*, pages 255–264, Tuscon, 1997.
3. M. Delgado, D. Sánchez, and M. A. Vila. Fuzzy cardinality based evaluation of quantified sentences. *Inter. J. of Approximate Reasoning*, 23:23–66, 2000.
4. D. Dubois, H. Hüllermeier, and H. Prade. Toward the representation of implication-based fuzzy rules in terms of crisp rules. In *Proc. of the Joint 9th IFSA Congress and NAFIPS 20th Inter. Conf.*, pages 1592–1597, Vancouver, July 2001.
5. D. Dubois and H. Prade. What are fuzzy rules and how to use them. *Fuzzy Sets and Systems*, 84(2):169–186, 1996.
6. E. Hüllermeier. Fuzzy association rules semantic issues and quality measures. In B. Reusch, editor, *Computational Intelligence: Theory and Applications*, LNCS, pages 380–391. Springer-Verlag, Berlin, 2001.
7. M. J. Martín-Bautista, D. Sánchez, M. A. Vila, and H. Larsen. Measuring effectiveness in fuzzy information retrieval. In *Flexible Query Answering in Fuzzy Information Retrieval*, pages 396–402. Physica-Verlag, 2001.
8. Y. Kodratoff. Compating machine learning and knowledge discovery in databases: An application to knowledge discovery in texts. In G. Paliouras, V. Karkaletsis, and C. D. Spyropoulos, editors, *Machine Learning and Its Applications*, volume 2049, LNCS , pages 1–21. Springer, 2001.

On a Characterization of Fuzzy Bags

Miguel Delgado, María J. Martín-Bautista, Daniel Sánchez, and María A. Vila

Department of Computer Science and Artificial Intelligence, University of Granada
C/ Periodista Daniel Saucedo Aranda s/n, 18071 Granada, Spain
mdelgado@ugr.es, {mbautis,daniel,vila}@decsai.ugr.es

Abstract. Bags, also called multisets, were introduced by R. Yager as set-like algebraic structures where elements are allowed to be repeated. Since the original papers by Yager, different definitions of the concept of fuzzy bag, and the corresponding operators, are available in the literature, as well as some extensions of the union, intersection and difference operators of sets, and new algebraic operators.
In general, the current definitions of bag pose very interesting issues related to the ontological aspects and practical use of bags. In this paper we introduce a characterization of bags viewing them as the result of a count operation on the basis of a mathematical correspondence. We also discuss on the extension of our alternative characterization of bags to the fuzzy case. On these basis we introduce some operators on bags and fuzzy bags, and we compare them to existing approaches.

1 Introduction

Bags are set-like algebraical structures where an element can appear more than once [11]. They are also called *multisets* [8, 7, 2]. Bags have been extended to fuzzy bags in several different ways [11, 9, 4]. There are several areas where crisp and fuzzy bags can be useful, such as statistics [9], but they have been mainly applied in the database area [6, 11, 10] (see [1] for additional references).

In order to apply bags in practice, it is convenient to study their meaning and the processes that could generate them. In the context of computer science, bags can be useful in order to model situations of the real world, in particular those related to objects and their properties.

A usual situation in the real world is to find that several individually different objects verify the same properties in a certain context (space,time). In other words, we can say they are instances of the same class. Counting how many objects verify a certain property in a given context seems to be a natural application of bags in practice.

Following this idea, in this work we propose a characterization of bags based on two components: a correspondence that links properties to objects (more

---
[0] This work has been supported by the Spanish Ministry of Science and Technology under grant TIC2002-04021-C02-02.

T. Bilgiç et al. (Eds.): IFSA 2003, LNAI 2715, pp. 119–126, 2003.

generally, instances to classes) and a count of the number of objects that verify each property. The latter corresponds to the classical characterization of bags. In addition, an extension of this characterization to fuzzy bags is introduced.

2   A Characterization of Crisp Bags

2.1   Crisp Bags

Definition 1. *Let P and O be two finite universes (sets) we call "properties" and "objects" respectively. A bag B^f is a pair (f, β^f) where $f : P \to O$ is a correspondence and β^f is the following subset of $P \times \mathbb{N}$:*

$$\beta^f = \{ \ (p, card\,(f(p)))\,, \ p \in P \text{ and } f(p) \neq \emptyset\} \tag{1}$$

where $card(X)$ is the cardinality of the set X.

In this characterization, a bag B^f consists of two parts. The first one is the correspondence f, that can be seen as an information source about the relation between objects and properties. The second part, β^f, is a summary of the information in f obtained by means of a count operation. This summary corresponds to the classical view of bags. From now on, we shall refer to this kind of summaries as *sbags*.

Example 1. Let $O = \{John, Mary, Bill, Tom, Sue, Stan, Harry\}$ and P be the set of possible ages. Let $f_1, f_2, f_3, f_4 : P \to O$ be the correspondences in table 1 (data is extracted from [11]) with $f_i(p) \subseteq O \ \forall p \in P$.

p	21	27	17	35
$f_1(p)$	$\{John, Tom\}$	$\emptyset$	$\{Bill, Sue\}$	$\emptyset$
$f_2(p)$	$\{John, Tom, Stan\}$	$\emptyset$	$\{Bill, Sue\}$	$\{Harry\}$
$f_3(p)$	$\{Stan\}$	$\{Mary\}$	$\emptyset$	$\{Harry\}$
$f_4(p)$	$\{John, Stan\}$	$\emptyset$	$\{Bill\}$	$\emptyset$

Table 1. Several correspondences age-people

Examples of bags are $B^{f_i} = (f_i, \beta^{f_i})\ 1 \leq i \leq 4$ where

$$\beta^{f_1} = \{(21, 2), (17, 2)\}$$
$$\beta^{f_2} = \{(21, 3), (17, 2), (35, 1)\}$$
$$\beta^{f_3} = \{(21, 1), (27, 1), (35, 1)\}$$
$$\beta^{f_4} = \{(21, 2), (17, 1)\}$$

Following [11] we note $Count_{\beta^f}(p) = card\,(f(p))$ the number of objects that verify property p in the bag B^f. For instance, $Count_{\beta^{f_2}}(21) = 3$ and $Count_{\beta^{f_3}}(35) = 1$.

2.2   Operations and Relations

In the following, let $f, g : P \to O$ be two correspondences.

Definition 2 (Union, intersection, difference). *Let* $* \in \{\cap, \cup, \backslash\}$. *Then*

$$B^f * B^g = B^{f*g} = \left(f * g, \beta^{f*g}\right) \tag{2}$$

where $f * g : P \to O$ *such that* $(f * g)(p) = f(p) * g(p) \; \forall p \in P$.

Definition 2 is coherent with the characterization of bags given in definition 1. The union, intersection or difference operation is applied to the correspondences that define the bag. If we see the correspondences f and g as two information sources, these operations are different ways to aggregate that information. Finally, the aggregation is summarized as a sbag by means of a count operation.

Example 2. Table shows some operations between correspondences from example 1:

r	21	27	17	35
$(f_1 \cup f_2)(p)$	$\{John, Tom, Stan\}$	$\emptyset$	$\{Bill, Sue\}$	$\{Harry\}$
$(f_1 \cup f_3)(p)$	$\{John, Tom, Stan\}$	$\{Mary\}$	$\{Bill, Sue\}$	$\{Harry\}$
$(f_2 \cup f_3)(p)$	$\{John, Tom, Stan\}$	$\{Mary\}$	$\{Bill, Sue\}$	$\{Harry\}$
$(f_1 \cap f_2)(p)$	$\{John, Tom\}$	$\emptyset$	$\{Bill, Sue\}$	$\emptyset$
$(f_1 \cap f_3)(p)$	$\emptyset$	$\emptyset$	$\emptyset$	$\emptyset$
$(f_2 \cap f_3)(p)$	$\{Stan\}$	$\emptyset$	$\emptyset$	$\{Harry\}$
$(f_1 \backslash f_2)(p)$	$\emptyset$	$\emptyset$	$\emptyset$	$\emptyset$
$(f_1 \backslash f_3)(p)$	$\{John, Tom\}$	$\emptyset$	$\{Bill, Sue\}$	$\emptyset$
$(f_2 \backslash f_3)(p)$	$\{John, Tom\}$	$\emptyset$	$\{Bill, Sue\}$	$\emptyset$
$(f_3 \backslash f_2)(p)$	$\emptyset$	$\{Mary\}$	$\emptyset$	$\emptyset$

Table 2. Operations on correspondences from example 1

The corresponding sbags are:

$$\beta^{f_1 \cup f_2} = \{(21, 3), (17, 2), (35, 1)\}$$
$$\beta^{f_1 \cup f_3} = \{(21, 3), (27, 1), (17, 2), (35, 1)\}$$
$$\beta^{f_2 \cup f_3} = \{(21, 3), (27, 1), (17, 2), (35, 1)\}$$
$$\beta^{f_1 \cap f_2} = \{(21, 2), (17, 2)\}$$
$$\beta^{f_1 \cap f_3} = \emptyset$$
$$\beta^{f_2 \cap f_3} = \{(21, 1), (35, 1)\}$$
$$\beta^{f_1 \backslash f_2} = \emptyset$$
$$\beta^{f_1 \backslash f_3} = \{(21, 2), (17, 2)\}$$
$$\beta^{f_2 \backslash f_3} = \{(21, 2), (17, 2)\}$$
$$\beta^{f_3 \backslash f_2} = \{(27, 1)\}$$

Regarding relations between bags, we can consider different definitions by looking at the whole information or the count summary. From the point of view of the information (correspondences) from which a bag is obtained, we introduce the following definitions:

Definition 3 (Subbag). *A bag B^f is a subbag of B^g, noted $B^f \subseteq B^g$ when $f(p) \subseteq g(p)$ $\forall p \in P$.*

Definition 4 (Equality). *Two bags B^f and B^g are equal, noted $B^f = B^g$ when $B^f \subseteq B^g$ and $B^g \subseteq B^f$.*

The following definition introduces a relation between bags based on their corresponding sbags:

Definition 5 (Ranking by cardinality). *A bag B^f is*

- *smaller than B^g, noted $B^f < B^g$, when $count_{\beta^f}(p) < count_{\beta^g}(p)$ $\forall p \in P$*
- *equipotent to B^g, noted $B^f \equiv B^g$, when $count_{\beta^f}(p) = count_{\beta^g}(p)$ $\forall p \in P$*
- *greater than B^g, noted $B^f > B^g$, when $count_{\beta^f}(p) > count_{\beta^g}(p)$ $\forall p \in P$*

The relations "smaller" and "equipotent" in definition 5 are equivalent to the notions of inclusion (subbag) and equality of sbags as introduced in [11]. The following results are easy to show.

Proposition 1. *If $B^f \subseteq B^g$ then $B^f \leq B^g$.*

Corollary 1. *If $B^f = B^g$ then $B^f \equiv B^g$.*

That means the concepts of subbag and equality of bags given by definitions 3 and 4 are more restrictive than those proposed for sbags in [11]. The reciprocal is not true in general. For instance, in example 1 we can see that $B^{f_1} \subseteq B^{f_2}$ and consequently $B^{f_1} \leq B^{f_2}$. However, $B^{f_4} \leq B^{f_1}$ but $B^{f_4} \not\subseteq B^{f_1}$.

2.3   Discussion

The operations introduced in definition 2 verify the following properties.

Proposition 2. *Let $p \in P$ with $f(p) \cap g(p) = \emptyset$. Then*

1. *$count_{\beta^{f \cap g}}(p) = 0$*
2. *$count_{\beta^{f \cup g}}(p) = count_{\beta^f}(p) + count_{\beta^g}(p)$*
3. *$count_{\beta^{f \setminus g}}(p) = count_{\beta^f}(p)$*

Proof. Trivial.

Corollary 2. *Let $f(p) \cap g(p) = \emptyset$ $\forall p \in P$. Then*

1. *$\beta^{f \cap g} = \emptyset$*
2. *$\beta^{f \cup g} = \beta^f \oplus \beta^g$*
3. *$\beta^{f \setminus g} = \beta^f$*

where $\oplus$ is the addition of bags introduced in [11] to be

$$count_{A \oplus B}(p) = count_A(p) + count_B(p) \tag{3}$$

A particular case of the condition in corollary 2 is $f(P) \cap g(P) = \emptyset$. For instance, in example 2 we have $(f_1 \cap f_3)(p) = \emptyset$. We can see that $\beta^{f_1 \cap f_3} = \emptyset$, $\beta^{f_1 \cup f_3} = \beta^{f_1} \oplus \beta^{f_3}$, and $\beta^{f_1 \setminus f_3} = \beta^{f_1}$.

Proposition 3. *Let $p \in P$ with $f(p) \subseteq g(p)$. Then*

1. $count_{\beta^{f \cap g}}(p) = count_{\beta^f}(p)$
2. $count_{\beta^{f \cup g}}(p) = count_{\beta^g}(p)$
3. $count_{\beta^{g \setminus f}}(p) = count_{\beta^g}(p) - count_{\beta^f}(p)$
4. $count_{\beta^{f \setminus g}}(p) = 0$

Proof. Trivial.

Corollary 3. *Let $f(p) \subseteq g(p) \ \forall p \in P$. Then*

1. $\beta^{f \cap g} = \beta^f$
2. $\beta^{f \cup g} = \beta^g$
3. $\beta^{g \setminus f} = \beta^g \ominus \beta^f$
4. $\beta^{f \setminus g} = \emptyset$

where $A \ominus B$ is the removal of bag B from A introduced in [11] to be

$$count_{A \ominus B}(p) = \max\{count_A(p) - count_B(p), 0\} \tag{4}$$

The results of the first two cases in proposition 3 correspond to the usual definitions of intersection and union of bags [11]. However, the following proposition shows there is a more general case where this property holds.

Proposition 4. *Let $f, g : P \to O$ such that $f(p) \subseteq g(p)$ or $g(p) \subseteq f(p) \ \forall p \in P$. Then*

1. $\beta^{f \cap g} = \beta^f \ \textcircled{\cap} \ \beta^g$
2. $\beta^{f \cup g} = \beta^f \ \textcircled{\cup} \ \beta^g$

where $\textcircled{\cap}$ and $\textcircled{\cup}$ are respectively the intersection and union of bags introduced in [11] to be

$$count_{A \textcircled{\cap} B}(p) = \min\{count_A(p), count_B(p)\} \tag{5}$$

$$count_{A \textcircled{\cup} B}(p) = \max\{count_A(p), count_B(p)\} \tag{6}$$

Proof. Let $p \in P$ such that $f(p) \subseteq g(p)$. Then $\min\{count_{\beta^f}(p), count_{\beta^g}(p)\} = count_{\beta^f}(p) = count_{\beta^{f \cap g}}(p)$ and $\max\{count_{\beta^f}(p), count_{\beta^g}(p)\} = count_{\beta^g}(p) = count_{\beta^{f \cup g}}(p)$. The proof is similar for the case $g(p) \subseteq f(p)$.

124   Miguel Delgado et al.

As we have seen, the operations in definition 2 reduce to different usual operations defined on sbags in some particular cases. However, in the general case, these operations may have no counterpart in the existing literature.

Example 3. In example 2 we have

$$\beta^{f_1 \cup f_4} = \{(21,3),(17,2)\}$$
$$\beta^{f_1} \oplus \beta^{f_4} = \{(21,4),(17,3)\}$$
$$\beta^{f_1} \oslash \beta^{f_4} = \{(21,2),(17,2)\}$$
$$\beta^{f_1 \cap f_4} = \{(21,1),(17,1)\}$$
$$\beta^{f_1} \oslash \beta^{f_4} = \{(21,2),(17,1)\}$$
$$\beta^{f_1 \setminus f_4} = \{(21,1),(17,1)\}$$
$$\beta^{f_1} \ominus \beta^{f_4} = \{(17,1)\}$$

It is easy to verify that $\beta^{f_1 \cup f_4} \neq \beta^{f_1} \oplus \beta^{f_4}$ and $\beta^{f_1 \cup f_4} \neq \beta^{f_1} \oslash \beta^{f_4}$. In addition $\beta^{f_1 \cap f_4} \neq \beta^{f_1} \oslash \beta^{f_4}$ and $\beta^{f_1 \setminus f_4} \neq \beta^{f_1} \ominus \beta^{f_4}$.

3   A Characterization of Fuzzy Bags

3.1   Fuzzy Bags

Definition 6. *A fuzzy bag $B^{\tilde{f}}$ is a pair $\left(\tilde{f}, \beta^{\tilde{f}}\right)$ where $\tilde{f} : P \to O$ is a fuzzy correspondence (i.e., $\tilde{f}(p) \in \widetilde{\mathcal{P}}(O)$) and $\beta^{\tilde{f}}$ is the following subset of $P \times \widetilde{\mathcal{P}}(\mathcal{N})$:*

$$\beta^{\tilde{f}} = \left\{ \left(p, ED\left(\tilde{f}(p)\right)\right), \ p \in P \ and \ \tilde{f}(p) \neq \emptyset \right\} \tag{7}$$

where $ED(\tilde{f}(p))$ is the fuzzy cardinality of the fuzzy set $\tilde{f}(p)$ introduced in [5] to be

$$ED(\tilde{f}(p)) = \sum_{\alpha_i \in \Lambda(\tilde{f}(p)) \cup \{1\}} (\alpha_i - \alpha_{i+1})/|\tilde{f}(p)_{\alpha_i}| \tag{8}$$

with $\Lambda(F) = \{\alpha_1, \ldots, \alpha_k\}$, $\alpha_i > \alpha_{i+1}$, and considering $\alpha_{k+1} = 0$.

From definition 6, a crisp bag is a particular case of fuzzy bag where $\tilde{f}$ is crisp and the cardinality of $\tilde{f}(p)$ is the fuzzy set $ED(\tilde{f}(p)) = 1/|\tilde{f}(p)|$.

3.2   Operations and Relations

Definition 7 (Union). *The union of fuzzy bags $B^{\tilde{f}}$ and $B^{\tilde{g}}$ is the fuzzy bag*

$$B^{\tilde{f}} \cup B^{\tilde{g}} = B^{\tilde{f} \cup \tilde{g}} = \left(\tilde{f} \cup \tilde{g}, \beta^{\tilde{f} \cup \tilde{g}}\right) \tag{9}$$

where $\left(\tilde{f} \cup \tilde{g}\right)(p)(o) = \max\left\{\tilde{f}(p)(o), \tilde{g}(p)(o)\right\} \forall p \in P, o \in O.$

Definition 8 (Intersection). *The intersection of fuzzy bags $B^{\tilde{f}}$ and $B^{\tilde{g}}$ is the fuzzy bag*

$$B^{\tilde{f}} \cap B^{\tilde{g}} = B^{\tilde{f} \cap \tilde{g}} = \left(\tilde{f} \cap \tilde{g}, \beta^{\tilde{f} \cap \tilde{g}} \right) \tag{10}$$

where $\left(\tilde{f} \cap \tilde{g} \right)(p)(o) = \min \left\{ \tilde{f}(p)(o), \tilde{g}(p)(o) \right\} \ \forall p \in P, o \in O.$

Definition 9 (Difference). *The difference $B^{\tilde{f}} \backslash B^{\tilde{g}}$ is the fuzzy bag*

$$B^{\tilde{f}} \backslash B^{\tilde{g}} = B^{\tilde{f} \backslash \tilde{g}} = \left(\tilde{f} \backslash \tilde{g}, \beta^{\tilde{f} \backslash \tilde{g}} \right) \tag{11}$$

where $\left(\tilde{f} \backslash \tilde{g} \right)(p)(o) = \min \left\{ \tilde{f}(p)(o), (1 - \tilde{g}(p)(o)) \right\} \ \forall p \in P, o \in O.$

The following are some relations between fuzzy bags:

Definition 10 (Subbag). *A fuzzy bag $B^{\tilde{f}}$ is a subbag of $B^{\tilde{g}}$, noted $B^{\tilde{f}} \subseteq B^{\tilde{g}}$ when $\tilde{f}(p) \subseteq \tilde{g}(p) \ \forall p \in P$, i.e., $\tilde{f}(p)(o) \leq \tilde{g}(p)(o) \ \forall p \in P, \ o \in O.$*

Definition 11 (Equality). *Two fuzzy bags $B^{\tilde{f}}$ and $B^{\tilde{g}}$ are equal, noted $B^{\tilde{f}} = B^{\tilde{g}}$ when $B^{\tilde{f}} \subseteq B^{\tilde{g}}$ and $B^{\tilde{g}} \subseteq B^{\tilde{f}}.$*

In [5] a fuzzy ranking of the cardinalities of two fuzzy sets was introduced. Since cardinalities are fuzzy, it seems natural to provide a fuzzy ranking, that we define as a fuzzy subset of $\mathcal{R} = \{ <, =, > \}$. The fuzzy ranking of the cardinalities of fuzzy sets F and G is defined for every $* \in \mathcal{R}$ to be

$$R_{Prob}(|F| * |G|) = \sum_{\alpha_i : |F_{\alpha_i}| * |G_{\alpha_i}|} (\alpha_i - \alpha_{i+1}) \tag{12}$$

with $\alpha_i \in \Lambda(F) \cup \Lambda(G) \cup \{1\}$. Let us remark that

$$\sum_{* \in \mathcal{R}} R_{Prob}(|F| * |G|) = 1 \tag{13}$$

Definition 12 (Ranking by cardinality). *A fuzzy bag $B^{\tilde{f}}$ is*

- *smaller than $B^{\tilde{g}}$, noted $B^{\tilde{f}} < B^{\tilde{g}}$, when $R_{Prob}(|\tilde{f}(p)| < |\tilde{g}(p)|) = 1 \ \forall p \in P$*
- *equipotent to $B^{\tilde{g}}$, noted $B^{\tilde{f}} \equiv B^{\tilde{g}}$, when $R_{Prob}(|\tilde{f}(p)| = |\tilde{g}(p)|) = 1 \ \forall p \in P$*
- *greater than $B^{\tilde{g}}$, noted $B^{\tilde{f}} \geq B^{\tilde{g}}$, when $R_{Prob}(|\tilde{f}(p)| > |\tilde{g}(p)|) = 1 \ \forall p \in P$*

Notice that by equation (13) if $R_{Prob}(|\tilde{f}(p)| < |\tilde{g}(p)|) = 1$ then $R_{Prob}(|\tilde{f}(p)| = |\tilde{g}(p)|) = R_{Prob}(|\tilde{f}(p)| > |\tilde{g}(p)|) = 0$, so we are sure that the cardinality of $\tilde{f}(p)$ is smaller than the cardinality of $\tilde{g}(p)$. A similar conclusion holds for the cases "equipotent" and "greater".

4   Conclusions

A characterization of bags and fuzzy bags has been introduced, and some operations and relations between bags have been described. The characterization is based on a (fuzzy) correspondence that relates objects and properties, from which classical bags (sbags) are obtained as counts. This approach can be very useful in practice when we are dealing with counts of the number of objects that verify a certain property. That is the case for instance of queries in relational databases, where projections on non-key attributes yield a bag, since it is possible to define a correspondence between tuples of the original table and tuples in the result. Another example can be the answers to a certain question in a poll, where different people can give the same answer. Again, it is possible to define a correspondence between people and the corresponding answers.

Some questions will be dealt with in the future. Additional properties of operations on fuzzy bags are to be studied. We also plan to study the case when correspondences exist and are not available, but we know the summary. In such cases, we can only provide bounds for the counts of the union, intersection and difference of bags. We think that IC-bags [3] can be a very useful tool for this purpose.

References

[1] J. Albert. Algebraic properties of bag data types. In *Proceedings of the 17th Conference on Very Large Data Bases VLDB'91*, pages 211–218, 1991.

[2] W.D. Blizard. Multiset theory. *Notre Dame Journal of Formal Logic*, 30:36–66, 1989.

[3] K. Chakrabarty. On IC-bags. In *Proc. (CD-Rom) Int. Conf. On Computational Intelligence for Modelling, Control and Automation - CIMCA'2001*, volume 25, 2001.

[4] K. Chakrabarty, R. Biswas, and S. Nanda. Fuzzy shadows. *Fuzzy Sets and Systems*, 101(3):413–421, 1999.

[5] M. Delgado, M.J. Martín-Bautista, D. Sánchez, and M.A. Vila. A probabilistic definition of a nonconvex fuzzy cardinality. *Fuzzy Sets and Systems*, 126(2):41–54, 2002.

[6] A. Klausner and N. Goodman. Multirelations - semantics and languages. In *Proceedings of the 11th Conference on Very Large Data Bases VLDB'85*, pages 251–258, 1985.

[7] D. Knuth. *The Art of Computer Programming, Vol. 2: Seminumerical Algorithms*. Addison-Wesley, Reading, MA, 1981.

[8] J. Lake. Sets, fuzzy sets, multisets and functions. *Journal of the London Math. Society*, 2(12):323–326, 1976.

[9] Baowen Li. Fuzzy bags and applications. *Fuzzy Sets and Systems*, 34(1):61–71, 1990.

[10] I. Singh Mumick, H. Pirahesh, and R. Ramakrishnan. The magic of duplicates and aggregates. In *Proceedings of the 16th Conference on Very Large Data Bases VLDB'90*, pages 264–277, 1990.

[11] R.R. Yager. On the theory of bags. *Int. Journal of General Systems*, 13:23–37, 1986.

A New Proposal of Aggregation Functions: The Linguistic Summary

Ignacio Blanco[1], Daniel Sánchez[2], José M. Serrano[2], and María A. Vila[2]

[1] Universidad de Almería, Departamento de Lenguajes y Computación,
Camino de Sacramento s/n ,01071 Almería , SPAIN,
iblanco@ual.es
[2] Universidad de Granada, Depto. de Ciencias de la Computación e I.A.,
E.T.S.I de Informática
Periodista Daniel Saucedo Aranda s/n
18071 Granada SPAIN
daniel@decsai.ugr.es, jmserrano@decsai.ugr.es, vila@decsai.ugr.es
http://frontdb.decsai.ugr.es

Abstract. This paper presents a new way of giving the summary of a numerical attribute involved in a fuzzy query. It is based on the idea of offering a linguistic interpretation, therefore we propose to use a flat fuzzy number as summary. To obtain it, we optimize any index which measures the relation between the fuzzy bag (which is the answer to the fuzzy query) and the fuzzy number. Several indices should be considered: some of them are based on linguistic quantified sentences, other ones are founded on divergence measures. The method can be also used to summarize other related fuzzy sets such as fuzzy average, maximum, minimum etc.

1 Introduction

Aggregation functions are widely used in database querying. They also play a central role in data warehousing and data mining issues. Several "classical" aggregation functions are oriented to give a kind of summary for the bag of values which is the result of any database query. If we are dealing with numerical attributes the arithmetic average is the function usually employed for this purpose.

If the query involves any imprecise property, the result is a fuzzy bag of values. The current summarizing procedure consists of using the alpha cut representation, computing the attribute values average for each alpha cut and giving a "fuzzy set of means". The main drawbacks of this approach are that the result is not easily understandable and that using it in additional comparisons, arithmetic operations etc. is almost impossible. Consequently this approach does not seem to be useful for addressing problems such as nested fuzzy queries, data warehousing with imprecise dimension values etc.

For these reasons , we propose to use a flat fuzzy number as the summary of a fuzzy bag of numeric values. This fuzzy number has a direct linguistic interpretation and can be easily compared and operated. To obtain this "linguistic

T. Bilgiç et al. (Eds.): IFSA 2003, LNAI 2715, pp. 127–134, 2003.

summary", we propose to optimize any index which measures the relation between the considered fuzzy bag and the fuzzy number. Several indices should be considered: some of them are based in linguistic quantified sentences, other ones are founded in divergence measures.

We will present the problem in the next section, the following one is devoted to the mathematical formulation which leads to a constrained optimization problem. An optimization procedure is proposed in the next section where a real example is also offered. The paper finishes with some concluding remarks and the references.

2   Problem Presentation

Although the results we will present can be applied in many situations, the problem we face arises from a fuzzy query process. Frequently, the output of such a process consists of several pairs $\{(v_i, \alpha_i)\}$ where $i \in \{1, 2..n\}$; $\alpha_i \in [0, 1]$ and $v_i \in D_A$ being $D_A \subset R$ the domain of any numeric attribute. This kind of results appears when we ask for any numerical attribute value of tuples verifying an imprecise property. Some examples are the "salary of young people" or the "price of apartments located near to the beach".

It is important to remark that the pairs $\{(v_i, \alpha_i)\}$ are neither a set, nor a fuzzy set. In the output of a query like the examples above offered, there may be pairs (v_i, α_i) , (v_j, α_j) that verify $v_i = v_j$ and $\alpha_i = \alpha_j$. The mathematical structure corresponding to this situation is that of "fuzzy bag". This concept is a generalization of the fuzzy set and bag ones and it has been mainly developed by R. Yager ([YA96])

Let us consider now that the querying process has a summary objective. That is, the user does not want to know detailed values, but some kind of approximation which gives him a general idea about what is the value of an attribute for those items verifying some imprecise property. By continuing the above examples, we would ask for: "the approximate salary of young people" or for "the approximate price of apartments located near to the beach".

Additionally, it could be necessary to obtain a more specific aggregation function such as the average, the maximum or the minimum, and to consider queries such as: "the average salary of young people" or "the maximal price of apartments located near to the beach".

These last cases of specifics aggregations have been previously dealt with by E. Rudensteiner and L. Bic ([RU89]) and D.Dubois and H. Prade ([DP90]), by giving fuzzy sets as query solution. The central idea of these approaches is the following: starting from the fuzzy bag output $\widetilde{B} = \{(v_i, \alpha_i)\}$, we obtain the sequence $\{\beta_j\}, j \in \{1...m\}$; $\beta_j < \beta_{j+1}$; $\beta_m = 1$ of the different membership values which can appear in $\widetilde{B}$.

By obtaining the corresponding α-cut of $\widetilde{B}$ we can generate a crisp bag sequence $\{B_{\beta_j}\}$, and the application of the considered aggregation function $f(.)$ (average, maximum, minimum etc.) to each crisp bag gives us a sequence of pairs $\{(u_j, \beta_j)\}$ where $\forall j \in \{1, ..., m\}$ $u_j = f(\beta_j)$. This sequence can be easily transformed in a fuzzy set by using:

$$\forall v \in D_A \ \mu_f(v) = \begin{cases} sup\{\beta_j/u_j = v\} & \text{if } \exists j \in \{1,...n\} \ \ v = u_j \\ 0 & \text{otherwise} \end{cases}$$

However, in our opinion, offering a fuzzy bag or a fuzzy set as a solution for a summary query, may be meaningless, since a non expert user could be not able to understand it.

Additionally, the result of the query could be the input for another processes, such as to "nested queries", for example to ask for: "workers whose salary is greater than the average salary of young people". If the "average salary" is a fuzzy bag or a fuzzy set, solving this query is very difficult (if not impossible), since there is not a comparison procedure for general fuzzy sets.

For all these reasons we propose to give a fuzzy number as the result of a summary query either when we consider the whole original fuzzy bag, or we take any aggregation function which produces an intermediate fuzzy set.

It should be remarked our problem should be viewed a particular case that of the linguistic aproximation (LA) one. This problem has been mainly dealed with in the context of fuzzy control or fuzzy expert systems (see [[ku99],[Wha01]) and most of the proposed methods assume there is a set of label previously established as well as the item to be approximated is a fuzzy set. Since we start from different context (fuzzy querying processes) our approach is also quite different

3   Mathematical Formulation of the Problem

According to consideration above, our problem is focused in the following way:

"Given a fuzzy numerical bag $\widetilde{B}$ with support $[a, b] \in \mathbf{R}$ to obtain a fuzzy number $\widetilde{F}$ defined on the same support such that: "$\mathcal{M}(\widetilde{B}, \widetilde{F})is \ minimal$" where $\mathcal{M}$ stands for some kind of measure of distance, divergence, opposite to an approximation measure etc.. between $\widetilde{B}$ and $\widetilde{F}$.

Since the fuzzy number should actually be an approximation we may to consider it is a trapezoidal fuzzy number, which is easier to interpret and to compute. Therefore the problem can be formulated as follows:

$$\text{Minimize} \qquad \mathcal{M}(\widetilde{B}, \widetilde{F}(m_1, m_2, a_1, a_2)) \tag{1}$$

Subject to

$$h_k(m_1, m_2, a_1, a_2) \leq 0 \ \forall k \in \{1, ..l\}$$

where:

- $\widetilde{F}(m_1, m_2, a_1, a_2)$ is a trapezoidal fuzzy number with support $[m_1 - a_1, m_2 + a_2]$ and mode $[m_1, m_2]$
- $\forall k \in \{1, ..l\}$; $h_k(m_1, m_2, a_1, a_2) \leq 0$ stands for any constraint associated with the fuzzy number such as $a_1 \geq 0$ or $m_1 \leq m_2$. A more specific formulation of these constraints will offered in the follwing section 3.1.

It is clear that the optimization problem given in (1) has (m_1, m_2, a_1, a_2) as variables and we have to obtain them by solving it. However before to do it, we must to concrete the problem elements.

3.1   The Measure $\mathcal{M}$

$\mathcal{M}$ should be either a divergence measure or the opposite to a compatibility measure. We have considered both possibilities.

Divergence Measures The divergence between two fuzzy sets can be established in an axiomatic way and any fuzzy measure of this divergence weights their distance. A general study of this can be found in [MO02]. Among other divergence examples, this article presents the Hamming distance between fuzzy sets. We will use this measure by considering:

$$\mathcal{M}(\widetilde{B}, \widetilde{F}) = (\sum_{i=1}^{n} \|\alpha_i - \widetilde{F}(v_i)\|)/n$$

where $\widetilde{F}(.)$ denotes the membership function of $\widetilde{F}$

Compatibility Measures Since we have already defined a distance-based measure, a new approach is necessary if we want to consider truly different measures. Therefore we will consider that a good approximation to the fuzzy bag $\widetilde{B}$ could be the fuzzy number $\widetilde{F}$ which maximize the accomplishment degree of the quantified sentence

$$\text{"}Q \text{ elements of } \widetilde{B} \text{ are } \widetilde{F}\text{"} \tag{2}$$

where Q stands for any linguistic quantifier such as "most", "almost all", "all" etc. And we will define:

$$\mathcal{M}(\widetilde{B}, \widetilde{F}) = oposite\ of\{\text{accomplishment degree of (2)}\} \tag{3}$$

A wide study on the evaluation of quantified sentences and their relations with the different approaches to the fuzzy cardinal can be found in [DSV99].

The basic idea of this paper is that the accomplishment degree of the sentence: "Q of $\widetilde{D}$ are $\widetilde{A}$" can be obtained by means of the degree of matching between the quantifier ,(defined as a fuzzy set on [0,1]), and the "relative cardinal" of $\widetilde{A}$ with respect to $\widetilde{D}$, given also by a fuzzy set on [0,1]. The paper proposes two methods for computing this cardinal and the matching degree:

– *Possibilistic method*
 Let $M(\widetilde{A}) = \{\alpha \in [0,1] | \exists x_i \text{ such that } \widetilde{A}(x_i) = \alpha\}$, and:

$$M(\widetilde{A}/\widetilde{D}) = M(\widetilde{A} \cap \widetilde{D}) \cup M(\widetilde{D})$$

$$CR(\widetilde{A}/\widetilde{D}) = \left\{ \frac{|(\widetilde{A} \cap \widetilde{D})_\alpha|}{|\widetilde{D}_\alpha|} \text{ such that } \alpha \in M(\widetilde{A}/\widetilde{D}) \right\}$$

the relative cardinality of $\widetilde{A}$ with respect to $\widetilde{D}$ is defined as:

$$\forall c \in CR(\widetilde{A}/\widetilde{D}) : ES(\widetilde{A}/\widetilde{D}, c) = \max\left\{ \alpha \in M(\widetilde{A}/\widetilde{D}) | c = \frac{|(\widetilde{A} \cap \widetilde{D})_\alpha|}{|\widetilde{D}_\alpha|} \right\}$$

The matching degree with any quantifier Q is given by:

$$ZS_Q(\widetilde{A}/\widetilde{D}) = \max_{c \in CR(\widetilde{A}/\widetilde{D})} \min(ES(\widetilde{A}/\widetilde{D}, c), Q(c))$$

- *Probabilistic Method*
  Let us consider that the set $M(\widetilde{A}/\widetilde{D})$ defined in the above paragraph is ordered, that is $M(\widetilde{A}/\widetilde{D}) = \{\alpha_1, >, ..., > \alpha_{m+1}\}$ with $\alpha_1 = 1$ and $\alpha_{m+1} = 0$. Let

$$C(\widetilde{A}/\widetilde{D}, \alpha_i) = \frac{|(\widetilde{A} \cap \widetilde{D})_\alpha|}{|\widetilde{D}_\alpha|}$$

Then the relative cardinality of $\widetilde{A}$ with respect to $\widetilde{D}$, $ER(\widetilde{A}/\widetilde{D})$ is defined as:

$$\forall c \in CR(\widetilde{A}/\widetilde{D}) \ : \ ER(\widetilde{A}/\widetilde{D}, c) = \sum_{c = C(\widetilde{A}/\widetilde{D}, \alpha_i)} (\alpha_i - \alpha_{i+1})$$

The matching degree with any quantifier Q is given by:

$$GD_Q(\widetilde{A}/\widetilde{D}) = \sum_{c \in CR(\widetilde{A}/\widetilde{D})} ER(\widetilde{A}/\widetilde{D}, c) \times Q(c)$$

Verifying that both methods are usable in the fuzzy bag case is straightforward, since the involved processes are union, intersection, and computing the cardinal of an $\alpha-$cut are defined also for fuzzy bags.

Therefore, chosen any quantifier Q, we can use $ZS_Q(\widetilde{B}/\widetilde{F})$ or $GD_Q(\widetilde{B}/\widetilde{F})$ as the compatibility degree appearing in (3).

3.2   The Constraint System

The constraints defined on the four parameter of $\widetilde{F}$: (m_1, m_2, a_1, a_2) can be divided in two classes:

Those oriented to assure that $\widetilde{F}$ is a fuzzy number, which are:

$$\begin{aligned} -a_1 &\leq 0 \\ -a_2 &\leq 0 \\ m_1 - m_2 &\leq 0 \end{aligned} \tag{4}$$

And those oriented to fit the fuzzy number $\widetilde{F}(m_1, m_2, a_1, a_2)$ inside the $\widetilde{B}$ area. If $[a, b]$ is the minimal interval including the $\widetilde{B}$ support and $[c, d]$ the minimal interval including the $\widetilde{B}$ mode, we impose:

$$\begin{aligned} -m_1 + a_1 + a &\leq 0 \\ m_2 + a_2 - b &\leq 0 \\ m_1 - m_2 - (d - c) &\leq 0 \end{aligned} \tag{5}$$

Additionally, to control the imprecision of the obtained solution, we will impose that the fuzziness of $\widetilde{F}$ should be least than or equal to that of the $\widetilde{B}$:

$$\mathcal{F}(\widetilde{F}) - \mathcal{F}(\widetilde{B}) \leq 0 \tag{6}$$

We propose the Delgado et al.'s fuzziness measure (see [DVV98]) as fuzzy measure. The adaptation of this measure to the case of $\widetilde{B}$ is given by:

$$\mathcal{F}(\widetilde{B}) = \sum_{\beta_j \leq 1/2} (R(\beta_j) - L(\beta_j))(\beta_j - \beta_{j-1}) - \sum_{\beta_j > 1/2} (R(\beta_j) - L(\beta_j))(\beta_j - \beta_{j-1})$$

where the sequence $\{\beta_j\}$ has been defined in the section 2 and

$$\forall \alpha \in [0,1] \; ; \; R(\alpha) = \sup\{x|x \in \widetilde{B}_\alpha\} \text{ and } L(\alpha) = \inf\{x|x \in \widetilde{B}_\alpha\}$$

The following property can be also found in [DVV98].

Property 1. The fuzziness of any trapezoidal fuzzy number $\widetilde{F}(m_1, m_2, a_1, a_2)$ is given by $m_2 - m_1 + (a_2 - a_1)/2$

This property gives us the constraint (6):

$$1/2a_2 - 1/2a_1 + m_2 - m_1 - \mathcal{F}(\widetilde{B}) \leq 0 \qquad (7)$$

4   The Optimization Process

According to the formulation given above, the problem of obtaining a summary for a fuzzy query numerical answer can be focused in solving the following optimization problem:

Minimize $\qquad\qquad \mathcal{M}(\widetilde{B}, \widetilde{F}(m_1, m_2, a_1, a_2)) \qquad\qquad (8)$

Subject to

$$\begin{pmatrix} -1 & 0 & 0 & 0 \\ 0 & -1 & 0 & 0 \\ 0 & 0 & 1 & -1 \\ -1 & 0 & 1 & 0 \\ 0 & 1 & 0 & 1 \\ 0 & 0 & -1 & 1 \\ -1/2 & 1/2 & -1 & 1 \end{pmatrix} \begin{pmatrix} a_1 \\ a_2 \\ m_1 \\ m_2 \end{pmatrix} - \begin{pmatrix} 0 \\ 0 \\ 0 \\ -a \\ b \\ (d-c) \\ \mathcal{F}(\widetilde{B}) \end{pmatrix} \leq \begin{pmatrix} 0 \\ 0 \\ 0 \\ 0 \\ 0 \\ 0 \\ 0 \end{pmatrix}$$

Clearly, (8) is a constrained optimization problem, with a linear constraint system and with an objective function whose properties can vary depending on what alternative is chosen among those ones presented in the Section 3.1. Moreover, if we take some compatibility-based measure we have no formal expression for the objective of (8), although we can evaluate this objective at every point. Therefore, if we want to design an optimization process general enough to cover all possible objectives, it is necessary to use a direct search optimization method which only needs the objective function values.

For these reasons we have chosen a penalty method ([FC68]) to deal with the constraints and a direct search method to solve the unconstrained problems associated to the penalty approach, concretely we have used the Hooke-Jeeves search method. ([HJ66]).

This optimization procedure provides us with a way to get the trapezoidal fuzzy number which approximates the fuzzy bag, by considering the different kind of measures and even, in the case of compatibility-based measures, different linguistic quantifiers. The following example illustrates how this procedure can be used and its usefulness in order to provide a summarized answer.

Example 1. We have considered the marks obtained by a group of students in the subjects "Databases" and "Programming Languages". Our goal is now to know what is the mark in "Programming Languages" of those students which have attained a "good mark" in "Databases". We have applied the optimization procedure to the fuzzy bag direct result obtained from the query and the fuzzy sets average, maximum and minimum. The chosen objective has been the probabilistic compatibility measure and the linguistic quantifier "most", whose membership function is $\forall x \in [0,1]$; $Q(x) = x$. The results appears in the figure

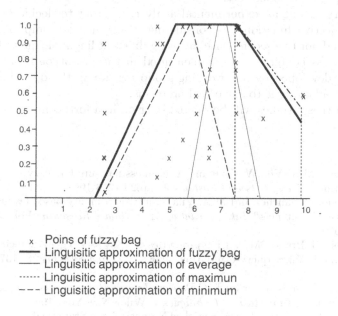

x   Poins of fuzzy bag
——— Linguisitic approximation of fuzzy bag
——— Linguisitic approximation of average
······ Linguisitic approximation of maximun
– – – Linguisitic approximation of minimum

Fig. 1. Results of example

Now it is possible to offer linguistic interpretations of the results such as:

- The marks in "Programming Languages" of students which are good in "Databases" are "more or less" between 5 and 7.5.
- The average of marks in "Programming Languages" of students which are good in "Databases" is more or less between 6.90 and 7.15.
- The maximum of marks in "Programming Languages" of students which are good in "Databases" is a bit greater than 7.5 and less than 10.
- The minimum of marks in "Programming Languages" of students which are good in "Databases" is around 5.5.

5   Concluding Remarks

We have presented a new way of giving the summary of a numerical attribute involved in a fuzzy query. This is based on the idea of offering a linguistic interpretation as result and it can be applied either to the initial fuzzy bag or to the fuzzy set that results from considering the average, maximum, minimum or, in general, every aggregation function.

A procedure to obtain this linguistic summary is also presented. It is founded in solving an constrained optimization problem whose objective function can adopt different formulations depending on what is the measure we choose.

A real example show us how this new approach can be more understandable for a non expert user, besides the possibility of using this result in an additional process such as a nested query.

However this is only an initial proposal. Deeper studies are necessary in order to improve the approach, concretely:

• We are carrying out an experimental analysis in order to decide on the more suitable objective function, by considering divergence or compatibility-based measures and, for this second case, by using different linguistic quantifiers.

• All this process is going to be implemented in the prototype of a fuzzy database system, developed by our working group for tuning the optimization algorithm and for adapting it to solve real problems.

The results of these studies will be offered later on in a forthcoming work.

References

[DVV98]   M. Delgado,M.A. Vila,W Voxman "A fuzziness measure for fuzzy numbers: Applications" *Fuzzy Sets and Systems 94* pp.205-216 1998.

[DSV99]   M.Delgado, D. Sanchez and M.A. Vila " Fuzzy cardinality based evaluation of quantified sentences" *Int. Journal of Approximate Reasoning* Vol. 23 pp. 23-66 2000.

[DP90]   D. Dubois, H. Prade, "Measuring properties of fuzzy sets: a general technique and its use in fuzzy query evaluation" *Fuzzy Sets and Systems* 38, 137-152, 1990.

[FC68]   Fiacco A. V. and McCormick G.P. *Non Linear Programming: Sequential Unconstrained Minimization Techniques* J. Wiley New York 1968.

[HJ66]   Hooke R., Jeeves T.A. "Direct Search of Numerical and Statistical Problems" *J. ACM* 8 212-229 1966.

[[ko99]   Kowalczyk R. "On numerical and linguistc quantifcation in linguistic approximation" *Proceeding of IEEE- SMC'99* pp. vol. 5 pp. 326-331 1999

[MO02]   S. Montes, I. Couso, P. Gil, C. Bertoluzza "Divergence measure between fuzzy sets" *Int. Journal of Approximate Reasoning* V. 30 pp. 91-105, 2002.

[RU89]   E.A. Rudensteiner, L. Bic, "Aggregates in possibilistic databases" *Proceedings of the Fifteenth Conference on Very Large Database* (VLDB'89), Amsterdam (Holland), 287-295, 1989.

[Wha01]   Whalen T., Scott B. "Empirical comparison of techniques for linguistic approximation" *Proceeding of 9th IFSA Congress* Vol. 1 pp.93-97 2001

[YA96]   R.R. Yager "On the Theory of Bags" *Int. Journal on General Systems* V. 13 pp. 23-37 1986

Fuzzy Quantifiers, Multiple Variable Binding, and Branching Quantification

Ingo Glöckner

AG Technische Informatik, Universität Bielefeld, 33501 Bielefeld, Germany
iglockner@web.de

Abstract. Lindström [1] introduced a very powerful notion of quantifiers, which permits multi-place quantification and the simultaneous binding of several variables. 'Branching' quantifification was found to be useful by linguists e.g. for modelling reciprocal constructions like "Most men and most women admire each other". Westerståhl [2] showed how to compute the three-place Lindström quantifier for "Q_1 A's and Q_2 B's R each other" from the binary quantifiers Q_1 and Q_2, assuming crisp quantifiers and arguments. In the paper, I generalize his method to approximate quantifiers like "many" and fuzzy arguments like "young". A consistent interpretation is achieved by extending the DFS theory of fuzzy quantification [3, 4], which rests on a system of formal adequacy criteria. The new analysis is important to linguistic data summarization because the full meaning of reciprocal summarizers (e.g. describing factors which are "correlated" or "associated" with each other), can only be captured by branching quantification.[1]

1 Introduction

The quantifiers found in natural language (NL) are by no means restricted to the absolute and proportional types usually considered in fuzzy set theory. The linguistic theory of quantification, i.e. the Theory of Generalized Quantifiers (TGQ) [5, 6], recognizes more than thirty different types of quantifiers, including quantifiers of exception like "all except about ten", cardinal comparatives like "many more than" and many others [6]. These quantifiers can be unary (like proper names in "Ronald is X") or multi-place; quantitative (like "about ten") or non-quantitative (like "all except Lotfi"); and they can be simplex or constructed, like "most married X's are Y's or Z's". However, it is not only the diversity of possible quantifiers in NL which poses difficulties to a systematic and comprehensive modelling of NL quantification. Even in simple cases like "most", the ways in which these quantifiers interact when combined in meaningful propositions can be complex and sometimes even puzzling. Consider "Most men and most women admire each other", for example, in which we find a reciprocal predicate, "admire each other". Barwise [7] argues that so-called branching quantification is needed to capture the meaning of propositions involving reciprocal predicates. Without branching quantifiers, the above example must be linearly phrased as either

a. $[\text{most } x : \text{men}(x)][\text{most } y : \text{women}(y)]\, \text{adm}(x, y)$
b. $[\text{most } y : \text{women}(y)][\text{most } x : \text{men}(x)]\, \text{adm}(x, y)\,.$

---

[1] The proofs of all theorems cited here are listed in [4].

T. Bilgiç et al. (Eds.): IFSA 2003, LNAI 2715, pp. 135–142, 2003.
© Springer-Verlag Berlin Heidelberg 2003

Neither interpretation captures the expected symmetry with respect to the men and women involved. In fact, we need a construction like

$$\left.\begin{array}{l}[Q_1\,x:\quad \mathrm{men}(x)]\\ [Q_2\,y:\mathrm{women}(y)]\end{array}\right\rangle \mathrm{adm}(x,y)$$

where $Q_1 = Q_2 =$ most operate in parallel and independently of each other. This branching use of quantifiers can be analysed in terms of Lindström quantifiers [1], i.e. multi-place quantifiers capable of binding several variables. We then have three arguments, and Q should bind x in $\mathrm{men}(x)$, y in $\mathrm{women}(y)$ and both x,y in $\mathrm{adm}(x,y)$. Thus, the above expression can be modelled by a Lindström quantifier of type $\langle 1,1,2\rangle$:

$$Q_{x,y,xy}(\mathrm{men}(x),\mathrm{women}(y),\mathrm{adm}(x,y))\,.$$

Obviously, the interpretation of Q depends on the meaning of "most" (majority of), i.e. $\mathbf{most}(Y_1,Y_2) = 1$ if $|Y_1 \cap Y_2| > \frac{1}{2}|Y_1|$ and 0 otherwise, where Y_1, Y_2 are crisp subsets of the given universe $E \neq \varnothing$. The quantifier Q, on the other hand, accepts the sets $A, B \in \mathcal{P}(E)$ (e.g. men and women), and the binary relation $R \in \mathcal{P}(E^2)$ (people admiring each other in the example). Barwise [7, p. 63] showed how to define Q in a special case; see also Westerståhl [2, p. 274, (D1)]. Hence suppose that Q_1 and Q_2, like "most", are nondecreasing in their second argument, i.e. $Q(Y_1, Y_2) \leq Q(Y_1, Y_2')$ whenever $Y_2 \subseteq Y_2'$. The complex quantifier Q can then be expressed as:

$$Q(A,B,R) = \begin{cases} 1 & : \exists U \times V \subseteq R : Q_1(A,U) = 1 \wedge Q_2(B,V) = 1 \\ 0 & : \text{else} \end{cases} \tag{1}$$

In the following, I will extend this analysis to approximate quantifiers and fuzzy arguments ("Many young and most old people respect each other"). To this end I describe the assumed formal framework, then incorporating Lindström quantifiers which bind several variables. Finally I apply the above analysis of branching quantifiers and its generalization by Westerståhl to the modelling of fuzzy branching quantification.

2   The Linguistic Theory of Fuzzy Quantification

TGQ rests on a simple but expressive model of two-valued quantifiers, which offers a uniform representation for the diversity of NL examples mentioned above. However, TGQ was not developed with fuzzy sets in mind, and its semantic analysis is essentially two-valued. To cover a broad range of fuzzy NL quantifiers, I hence developed the 'DFS theory' of fuzzy quantification [4, 3] which introduces the following basic notions.

Definition 1 *An n-ary fuzzy quantifier $\widetilde{Q}$ on a base set $E \neq \varnothing$ assigns a gradual interpretation $\widetilde{Q}(X_1,\ldots,X_n) \in [0,1]$ to all fuzzy subsets $X_1,\ldots,X_n \in \widetilde{\mathcal{P}}(E)$.*

$(\widetilde{\mathcal{P}}(E)$ is the fuzzy powerset). Fuzzy quantifiers are expressive operators, but hard to define because the usual cardinality is not applicable to the fuzzy sets they process. We hence need simplified specifications, powerful enough to embed all quantifiers of TGQ.

Definition 2 *An n-ary semi-fuzzy quantifier on a base set $E \neq \varnothing$ assigns a gradual result $Q(Y_1, \ldots, Y_n) \in [0,1]$ to all crisp subsets $Y_1, \ldots, Y_n \in \mathcal{P}(E)$.*

Semi-fuzzy quantifiers are much easier to define because the usual crisp cardinality is applicable to their arguments. An interpretation mechanism is used to associate these specifications with their matching fuzzy quantifiers:

Definition 3 *A quantifier fuzzification mechanism (QFM) $\mathcal{F}$ assigns to each semi-fuzzy quantifier Q a fuzzy quantifier $\mathcal{F}(Q)$ of the same arity and on the same base set.*

The resulting fuzzy quantifiers $\mathcal{F}(Q)$ can then be applied to fuzzy arguments. In order to ensure plausible results, the QFM should conform to all requirements of linguistic relevance. My research into various such properties converged into the following system of six basic postulates.

(Z-1) **Correct generalisation.** For all crisp arguments $Y_1, \ldots, Y_n \in \mathcal{P}(E)$, we require that $\mathcal{F}(Q)(Y_1, \ldots, Y_n) = Q(Y_1, \ldots, Y_n)$. (Combined with the other axioms, this condition can be restricted to $n \leq 1$). Rationale: $\mathcal{F}(Q)$ should properly generalize Q for crisp arguments.

(Z-2) **Membership assessment.** The two-valued quantifier defined by $\pi_e(Y) = 1$ if $e \in Y$ and $\pi_e(Y) = 0$ otherwise for crisp Y, has the obvious fuzzy counterpart $\tilde{\pi}_e(X) = \mu_X(e)$ for fuzzy subsets. We require that $\mathcal{F}(\pi_e) = \tilde{\pi}_e$. Rationale: Membership assessment (crisp or fuzzy) can be modelled through quantifiers. While π_e checks if e is present in its argument, $\tilde{\pi}_e$ returns the degree to which e is contained in its argument. It is natural to require that π_e be mapped to $\tilde{\pi}_e$, which serves the same purpose in the fuzzy case.

The fuzzy connectives which best match a QFM are given by a canonical construction.

Definition 4 *The induced truth function $\widetilde{\mathcal{F}}(f) : [0,1]^n \longrightarrow [0,1]$ of $f : \{0,1\}^n \longrightarrow [0,1]$ is defined by $\widetilde{\mathcal{F}}(f) = \mathcal{F}(f \circ \eta^{-1}) \circ \tilde{\eta}$, where $\eta(y_1, \ldots, y_n) = \{i : y_i = 1\}$ for all $y_1, \ldots, y_n \in \{0,1\}$ and $\mu_{\tilde{\eta}(x_1, \ldots, x_n)}(i) = x_i$ for all $x_i \in [0,1]$, $i \in \{1, \ldots, n\}$.*

Whenever $\mathcal{F}$ is understood, I abbreviate $\tilde{\vee} = \widetilde{\mathcal{F}}(\vee)$, $\tilde{\neg} = \widetilde{\mathcal{F}}(\neg)$ etc. These connectives are extended to fuzzy set operations in the usual ways. The desired criteria involving fuzzy complement and union can now be expressed as follows.

(Z-3) **Dualisation.** $\mathcal{F}$ preserves dualisation of quantifiers, i.e. $\mathcal{F}(Q')(X_1, \ldots, X_n) = \tilde{\neg}\mathcal{F}(Q)(X_1, \ldots, X_{n-1}, \tilde{\neg}X_n)$ for $X_1, \ldots, X_n \in \widetilde{\mathcal{P}}(E)$ given $Q'(Y_1, \ldots, Y_n) = \neg Q(Y_1, \ldots, Y_{n-1}, \neg Y_n)$ for all crisp arguments. Rationale: "All A's are B" and "It is not the case that some A's are not B's" should be equivalent.

(Z-4) **Union.** $\mathcal{F}$ must be compatible with unions of arguments, i.e. we should expect that $\mathcal{F}(Q')(X_1, \ldots, X_{n+1}) = \mathcal{F}(Q)(X_1, \ldots, X_{n-1}, X_n \tilde{\cup} X_{n+1})$ provided that $Q'(Y_1, \ldots, Y_{n+1}) = Q(Y_1, \ldots, Y_{n-1}, Y_n \cup Y_{n+1})$. Rationale: This postulate permits a compositional treatment of patterns like "Many A's are B's or C's"

(Z-5) **Monotonicity in arguments.** $\mathcal{F}$ must preserve monotonicity in arguments, i.e. if Q is nondecreasing/nonincreasing in the i-th argument, then $\mathcal{F}(Q)$ has the same property. (Combined with the other axioms, the condition can be restricted to nonincreasing Q). Rationale: The interpretation of "All men are tall" and "All young men are tall" must be systematically different and the former statement expresses the stricter condition.

The last criterion which I will state requires the extension of mappings $f : E \longrightarrow E'$ to fuzzy powerset mappings $f' : \widetilde{\mathcal{P}}(E) \longrightarrow \widetilde{\mathcal{P}}(E')$. The usual way of doing this is by applying the standard extension principle. In this case, the extension $f' = \hat{f}$ becomes $\mu_{\hat{f}(X)}(e') = \sup\{\mu_X(e) : e \in f^{-1}(e')\}$ for all $e' \in E'$. In order to define the sixth criterion, I must admit other choices which match the existential quantifiers $\mathcal{F}(\exists)$ of $\mathcal{F}$.

Definition 5 *The* induced extension principle *of* $\mathcal{F}$, *denoted* $\widehat{\mathcal{F}}$, *maps f to the extension* $\widehat{\mathcal{F}}(f)$ *defined* $\mu_{\widehat{\mathcal{F}}(f)(X)}(e') = \mathcal{F}(\pi_{e'} \circ \hat{f})$, *where* $\hat{f}(Y) = \{f(e) : e \in Y\}$ *for all* $Y \in \mathcal{P}(E)$.

(Z-6) **Functional application.** $\mathcal{F}$ must be compatible with 'functional application', i.e. $\mathcal{F}(Q')(X_1, \ldots, X_n) = \mathcal{F}(Q)(\widehat{\mathcal{F}}(f_1)(X_1), \ldots, \widehat{\mathcal{F}}(f_n)(X_n))$, where the semi-fuzzy quantifier Q' is defined by $Q'(Y_1, \ldots, Y_n) = Q(\hat{f}_1(Y_1), \ldots, \hat{f}_n(Y_n))$. Rationale: $\mathcal{F}$ must behave consistently over different domains E.

Definition 6 *A QFM $\mathcal{F}$ which satisfies (Z-1) to (Z-6) is called a* determiner fuzzification scheme *(DFS).*

(In linguistics, "most", "almost all" etc. are called 'determiners'). If $\mathcal{F}$ induces $\neg x = 1 - x$ and the standard extension principle, then it is called a *standard DFS*. Let us now consider some properties of these models. If $\mathcal{F}$ is a DFS, then

- $\mathcal{F}$ induces a reasonable set of fuzzy propositional connectives, i.e. $\widetilde{\neg}$ is a strong negation $\widetilde{\wedge}$ is a t-norm, $\widetilde{\vee}$ is an s-norm etc.
- $\mathcal{F}(\forall)$ is a T-quantifier and $\mathcal{F}(\exists)$ is an S-quantifier in the sense of Thiele [8].
- $\mathcal{F}$ is compatible with negations, e.g. "It is not the case that most A's are B's".
- $\mathcal{F}$ is compatible with the formation of antonyms, e.g. "Most A's are not B's".
- $\mathcal{F}$ is compatible with intersections of arguments, e.g. "Most A's are B's and C's".
- $\mathcal{F}$ is compatible with argument permutations. In particular, symmetry properties of a quantifier are preserved by applying $\mathcal{F}$.
- $\mathcal{F}$ is compatible with crisp adjectival restriction, e.g. "Many married A's are B's".

The models also account for some additional considerations of specifically linguistic interest. For a comprehensive discussion of semantical properties and a description of prototypical models, see [4]. These models include $\mathcal{M}_{\mathrm{CX}}$, a standard DFS which consistently generalises the Sugeno integral and hence the 'basic' FG-count approach to arbitrary n-place quantifiers. Due to its unique properties, this model is the preferred choice for all applications that need to capture NL semantics. Another interesting example $\mathcal{F}_{\mathrm{owa}}$, consistently generalises the Choquet integral and hence the 'basic' OWA approach. An efficient histogram-based method for implementing quantifiers in these models is described in [4].

3   Extension towards Multiple Variable Binding

A *Lindström quantifier* is a class Q of (relational) structures of type $t = \langle t_1, \ldots, t_n \rangle$, such that Q is closed under isomorphism [1, p. 186]. The cardinal $n \in \mathbb{N}$ specifies the number of arguments; the components $t_i \in \mathbb{N}$ specify the number of variables that the quantifier binds in its i-th argument position. For example, the existential quantifier, which accepts one argument and binds one variable, has type $t = \langle 1 \rangle$. The corresponding class $\mathcal{E}$ comprises all structures $\langle E, A \rangle$ where $E \neq \varnothing$ is a base set and $A \subseteq E$ is nonempty. In the introduction, we already met with a more complex quantifier Q of type $\langle 1, 1, 2 \rangle$. In this case, Q is the class of all structures $\langle E, A, B, R \rangle$ with $Q(A, B, R) = 1$, where $A, B \in \mathcal{P}(E)$, $R \in \mathcal{P}(E^2)$. To model quantifiers like "all except Lotfi", which depend on specific individuals, we must drop the assumption of isomorphism closure. Hence, in principle, a *generalized Lindström quantifier* is a class Q of relational structures of type $t = \langle t_1, \ldots, t_n \rangle$. However, it is convenient to stipulate the following alternative notions.

Definition 7 *A two-valued L-quantifier of type $t = \langle t_1, \ldots, t_n \rangle$ on a base set $E \neq \varnothing$ assigns a crisp quantification result $Q(Y_1, \ldots, Y_n) \in \{0, 1\}$ to each choice of crisp arguments $Y_i \in \mathcal{P}(E^{t_i})$, $i \in \{1, \ldots, n\}$. A full two-valued L-quantifier Q of type t assigns a two-valued L-quantifier Q_E of type t on E to each base set $E \neq \varnothing$.*

Hence 'full' L-quantifiers are in one-to-one correspondence with generalized Lindström quantifiers. The extension of L-quantifiers to gradual outputs should be obvious.

Definition 8 *A semi-fuzzy L-quantifier of type $t = \langle t_1, \ldots, t_n \rangle$ on $E \neq \varnothing$ assigns a gradual result $Q(Y_1, \ldots, Y_n) \in [0, 1]$ to all crisp $Y_i \in \mathcal{P}(E^{t_i})$, $i \in \{1, \ldots, n\}$.*

Thus, Q accepts crisp arguments of the indicated types, but it can express approximate quantification. Semi-fuzzy L-quantifiers establish a uniform specification medium for quantifiers with multiple variable binding. We further need operational quantifiers and fuzzification mechanisms which associate specifications and target quantifiers.

Definition 9 *A fuzzy L-quantifier of type t on $E \neq \varnothing$ assigns a gradual interpretation $\widetilde{Q}(X_1, \ldots, X_n) \in [0, 1]$ to all fuzzy arguments $X_i \in \widetilde{\mathcal{P}}(E^{t_i})$, $i \in \{1, \ldots, n\}$.*

Definition 10 *An L-QFM $\mathcal{F}$ assigns to each semi-fuzzy quantifier Q of some type t on $E \neq \varnothing$ a fuzzy L-quantifier $\mathcal{F}(Q)$ of the same type t and on the same base set E.*

Let me now associate with every L-QFM $\mathcal{F}$ a corresponding 'ordinary' QFM $\mathcal{F}_R$. For every n-ary semi-fuzzy quantifier $Q : \mathcal{P}(E)^n \longrightarrow [0, 1]$, on E, let Q' denote the n-ary quantifier on E^1 defined by $Q'(Y_1, \ldots, Y_n) = Q'(\widehat{\vartheta}(Y_1), \ldots, \widehat{\vartheta}(Y_n))$ for $Y_1, \ldots, Y_n \in \mathcal{P}(E^1)$, where $\vartheta : E^1 \longrightarrow E$ is the mapping $\vartheta((e)) = e$ for $(e) \in E^1$. Then let

$$\mathcal{F}_R(Q)(X_1, \ldots, X_n) = \mathcal{F}(Q')(\widehat{\widehat{\beta}}(X_1), \ldots, \widehat{\widehat{\beta}}(X_n))$$

for all $X_1, \ldots, X_n \in \widetilde{\mathcal{P}}(E)$, where $\widehat{\widehat{\beta}}$ is obtained from $\beta : E \longrightarrow E^1$ with $\beta(e) = (e)$ by applying the standard extension principle. The induced fuzzy connectives and extension principle of $\mathcal{F}$ are identified with the connectives and extension principle of the ordinary QFM $\mathcal{F}_R$. Based on these preparations, I can now develop criteria for plausible L-models of fuzzy quantification which parallel my requirements on QFMs. (The 'rationale' for these conditions is the same as above in each case).

(L-1) **Correct generalisation.** It is required that $\mathcal{F}(Q)(Y_1, \ldots, Y_n) = Q(Y_1, \ldots, Y_n)$ for all crisp arguments $Y_i \in \mathcal{P}(E^{t_i})$, $i \in \{1, \ldots, n\}$; combined with the other axioms, this condition can be restricted to quantifiers of types $t = \langle \rangle$ or $t = \langle 1 \rangle$.

(L-2) **Membership assessment.** Quantifiers for membership assessment of the special form $\pi_{(e)} : \mathcal{P}(E^1) \longrightarrow \{0, 1\}$ for some $e \in E$ also qualify as two-valued L-quantifiers of type $\langle 1 \rangle$ on E. These quantifiers should be mapped to their fuzzy counterparts $\tilde{\pi}_{(e)}$ of type $\langle 1 \rangle$ on E, i.e. we must have $\mathcal{F}(\pi_{(e)}) = \tilde{\pi}_{(e)}$.

(L-3) **Dualisation.** $\mathcal{F}$ preserves dualisation of quantifiers, i.e. $\mathcal{F}(Q')(X_1, \ldots, X_n) = \tilde{\neg} \mathcal{F}(Q)(X_1, \ldots, X_{n-1}, \tilde{\neg} X_n)$ for all fuzzy $X_i \in \tilde{\mathcal{P}}(E^{t_i})$ if $Q'(Y_1, \ldots, Y_n) = \tilde{\neg} Q(Y_1, \ldots, Y_{n-1}, \neg Y_n)$ for all crisp $Y_i \in \mathcal{P}(E^{t_i})$, $i \in \{1, \ldots, n\}$.

(L-4) **Union.** $\mathcal{F}$ must be compatible with unions of arguments, i.e. we should expect that $\mathcal{F}(Q')(X_1, \ldots, X_{n+1}) = \mathcal{F}(Q)(X_1, \ldots, X_{n-1}, X_n \tilde{\cup} X_{n+1})$ provided that $Q'(Y_1, \ldots, Y_{n+1}) = Q(Y_1, \ldots, Y_{n-1}, Y_n \cup Y_{n+1})$.

(L-5) **Monotonicity in arguments.** We require that $\mathcal{F}$ preserve monotonicity in arguments, i.e. if Q is nondecreasing/nonincreasing in the i-th argument, then $\mathcal{F}(Q)$ has the same property. (The condition can again be restricted to the case that Q is nonincreasing in its n-th argument).

(L-6) **Functional application.** Given a semi-fuzzy L-quantifier Q of type $t = \langle t_1, \ldots, t_n \rangle$ on E, another type $t' = \langle t'_1, \ldots, t'_n \rangle$ (same n), a set $E' \neq \varnothing$, and mappings $f_i : E'^{t'_i} \longrightarrow E^{t_i}$ for $i \in \{1, \ldots, n\}$, we can define a quantifier Q' of type t' on E' by $Q'(Y_1, \ldots, Y_n) = Q(\hat{f}_1(Y_1), \ldots, \hat{f}_n(Y_n))$ for all $Y_i \in \mathcal{P}(E'^{t'_i})$, $i \in \{1, \ldots, n\}$. It is required that $\mathcal{F}(Q')(X_1, \ldots, X_n) = \mathcal{F}(Q)(\hat{\mathcal{F}}(f)_1(X_1), \ldots, \hat{\mathcal{F}}(f)_n(X_n))$ for all fuzzy arguments $X_i \in \tilde{\mathcal{P}}(E'^{t'_i})$, $i \in \{1, \ldots, n\}$.

Definition 11 *An L-QFM which satisfies (L-1) to (L-6) is called an* L-DFS.

Theorem 1 *For every L-DFS $\mathcal{F}$, the corresponding QFM $\mathcal{F}_R$ is a DFS.*

Hence the generalized models are also suitable for carrying out 'ordinary' quantification. Now let $\mathcal{F}$ be an ordinary QFM and let Q be a semi-fuzzy L-quantifier of type $t = \langle t_1, \ldots, t_n \rangle$ on $E \neq \varnothing$. Let $m = \max\{t_1, \ldots, t_n\}$ and define $\zeta_i : E^{t_i} \longrightarrow E^m$ and $\kappa_i : E^m \longrightarrow E^{t_i}$ by

$$\zeta_i(e_1, \ldots, e_{t_i}) = (e_1, \ldots, e_{t_i-1}, e_{t_i}, e_{t_i}, \ldots, e_{t_i})$$
$$\kappa_i(e_1, \ldots, e_m) = (e_1, \ldots, e_{t_i})$$

for $i \in \{1, \ldots, n\}$. I introduce an n-ary semi-fuzzy quantifier Q' on E^m defined by

$$Q'(Y_1, \ldots, Y_n) = Q(\hat{\kappa}_1(Y_1 \cap \zeta_1(E^{t_1})), \ldots, \hat{\kappa}_n(Y_n \cap \zeta_n(E^{t_n}))),$$

for all $Y_1, \ldots, Y_n \in \mathcal{P}(E^m)$.

Definition 12 *For every QFM $\mathcal{F}$, the L-QFM $\mathcal{F}_L$ is defined by*

$$\mathcal{F}_L(Q)(X_1, \ldots, X_n) = \mathcal{F}(Q')(\hat{\check{\zeta}}_1(X_1), \ldots, \hat{\check{\zeta}}_n(X_n))$$

for all $X_i \in \widetilde{\mathcal{P}}(E^{t_i})$, $i \in \{1, \ldots, n\}$.

Theorem 2 *If $\mathcal{F}$ is a DFS, then $\mathcal{F}_{LR} = \mathcal{F}$, i.e. $\mathcal{F}_L$ properly generalizes $\mathcal{F}$.*

Theorem 3 *If $\mathcal{F}$ is a DFS, then $\mathcal{F}_L$ is an L-DFS, i.e. we obtain plausible models.*

Theorem 4 *If $\mathcal{F}$ is an L-DFS, then $\mathcal{F}_{RL} = \mathcal{F}$.*

Hence every L-DFS $\mathcal{F}'$ can now be expressed as $\mathcal{F}' = \mathcal{F}_L$. The canonical construction of $\mathcal{F}_L$ thus permits the re-use of $\mathcal{M}_{\mathrm{CX}}$ and $\mathcal{F}_{\mathrm{owa}}$ to handle fuzzy L-quantification.

4   Application to Fuzzy Branching Quantification

Let me now reconsider the motivating example, "Many young and most old people respect each other". In this case, we have semi-fuzzy quantifiers $Q_1 = $ **many**, defined by **many**$(Y_1, Y_2) = |Y_1 \cap Y_2|/|Y_1|$, say, and $Q_2 = $ **most**. Both quantifiers are nondecreasing in their second argument, i.e. we can adopt eq. (1). The modification to gradual truth values will be accomplished in the usual way, i.e. by replacing existential quantifiers with sup and conjunctions with min. The semi-fuzzy L-quantifier Q of type $\langle 1, 1, 2 \rangle$ constructed from Q_1, Q_2 then becomes

$$Q(A, B, R) = \sup\{\min(Q_1(A, U), Q_2(B, V)) : U \times V \subseteq R\}$$

for all $A, B \in \mathcal{P}(E)$ and $R \in \mathcal{P}(E^2)$. By applying $\mathcal{F}$, we then obtain the fuzzy L-quantifier $\mathcal{F}(Q)$ suitable for computing interpretations. In the example, we have fuzzy subsets **young**, **old** $\in \widetilde{\mathcal{P}}(E)$ of young and old people, and a fuzzy relation **rsp** $\in \widetilde{\mathcal{P}}(E^2)$ of people who respect each other. The interpretation of "Many young and most old people respect each other" is then given by $\mathcal{F}(Q)(\mathbf{young}, \mathbf{old}, \mathbf{rsp})$.

Finally let me describe how Westerståhl's generic method for interpreting branching quantifiers can be applied in the fuzzy case. Hence let Q_1, Q_2 be arbitrary semi-fuzzy quantifiers of arity $n = 2$. I introduce nondecreasing and nonincreasing approximations of the Q_i's, defined by $Q_i^+(Y_1, Y_2) = \sup\{Q_i(Y_1, L) : L \subseteq Y_2\}$ and $Q_i^-(Y_1, Y_2) = \sup\{Q_i(Y_1, U) : U \supseteq Y_2\}$, respectively. With the usual replacement of existential quantification with sup and conjunction with min, Westerståhls formula [2, p. 281, Def. 3.1] becomes:

$$Q(A, B, R) = \sup\{\min\{Q_1^+(A, U_1), Q_2^+(B, V_1), Q_1^-(A, U_2), Q_2^-(B, V_2)\} :$$
$$(U_1 \cap A) \times (V_1 \cap B) \subseteq R \cap (A \times B) \subseteq (U_2 \cap A) \times (V_2 \cap B)\}$$

for all $A, B \in \mathcal{P}(E)$ and $R \in \mathcal{P}(E^2)$. We can then apply $\mathcal{F}$ to fetch $\mathcal{F}(Q)$. As shown by Westerståhl [2, p.284], his method results in meaningful interpretations provided that (a) Q_1 and Q_2 are 'logical', i.e. $Q_i(Y_1, Y_2)$ can be expressed as a function of $|Y_1|$

and $|Y_1 \cap Y_2|$; and (b) the Q_i's satisfy $Q_i(Y_1, Y_2) \geq \min(Q_i(Y_1, L), Q_i(Y_1, U))$ for all $L \subseteq Y_2 \subseteq U$. The latter condition ensures that Q_1 and Q_2 can be recovered from their nondecreasing approximations Q_i^+ and their nonincreasing approximations Q_i^-, i.e. $Q_i = \min(Q_i^+, Q_i^-)$. This is the case when Q_1 and Q_2 are nondecreasing in their second argument ("many"), nonincreasing ("few"), or of unimodal shape ("about ten", "about one third"). An example with unimodal quantifiers, which demand the generic method, is "About fifty young and about sixty old persons respect each other".

5 Conclusion

In the paper, I proposed an extension of the DFS theory of fuzzy quantification with Lindström-like quantifiers. Westerståhl's method based on Lindström quantifiers which assigns a meaningful interpretation to branching NL quantification was then extended to approximate quantifiers and fuzzy arguments. The proposed analysis of reciprocal constructions in terms of fuzzy branching quantifiers is important to linguistic data summarization [9, 10]. Many summarizers of interest express mutual (or symmetric) relationships and can therefore be verbalized by a reciprocal construction. An ordinary summary like "$Q_1 X_1$'s are strongly correlated with $Q_2 X_2$'s" neglects the resulting groups of mutually correlated objects. The proposed analysis in terms of branching quantifiers, by contrast, permits me to support a novel type of summary specialized on groups of interrelated objects. Branching quantification, in this view, is a natural language technique for detecting such groups in the data. A possible summary involving a reciprocal predicate is "The intake of most vegetables and many health-related indicators are strongly associated with each other".

References

[1] Lindström, P.: First order predicate logic with generalized quantifiers. Theoria **32** (1966) 186–195
[2] Westerståhl, D.: Branching generalized quantifiers and natural language. In Gärdenfors, P., ed.: Generalized Quantifiers. Reidel (1987) 269–298
[3] Glöckner, I.: DFS – an axiomatic approach to fuzzy quantification. TR97-06, Technische Fakultät, Universität Bielefeld, P.O.-Box 100131, 33501 Bielefeld, Germany (1997)
[4] Glöckner, I.: Fundamentals of fuzzy quantification: Plausible models, constructive principles, and efficient implementation. TR2002-07, Technische Fakultät, Universität Bielefeld, P.O.-Box 100131, 33501 Bielefeld, Germany (2003)
[5] Barwise, J., Cooper, R.: Generalized quantifiers and natural language. Linguistics and Philosophy **4** (1981) 159–219
[6] Keenan, E., Stavi, J.: A semantic characterization of natural language determiners. Linguistics and Philosophy **9** (1986)
[7] Barwise, J.: On branching quantifiers in English. J. of Philosophical Logic **8** (1979) 47–80
[8] Thiele, H.: On T-quantifiers and S-quantifiers. In: The Twenty-Fourth International Symposium on Multiple-Valued Logic, Boston, MA (1994) 264–269
[9] Kacprzyk, J., Strykowski, P.: Linguistic summaries of sales data at a computer retailer: A case study. In: Proc. of IFSA '99. (1999) 29–33
[10] Yager, R.R.: On linguistic summaries of data. In Frawley, W., Piatetsky-Shapiro, G., eds.: Knowledge Discovery in Databases. AAAI/MIT Press (1991) 347–363

Modeling the Concept of Fuzzy Majority Opinion

Gabriella Pasi [1] and Ronald R. Yager [2]

[1] ITC-CNR, Via Bassini 15,
20133 Milano, Italy
gabriella.pasi@itc.cnr.it
[2] IonaCollege New Rochelle,
NY10801, USA
Ryager@Iona.edu

Abstract. In this paper we focus on the notion of majority in multi-agent (group) decision making. In group decision making the reduction of the decision makers opinions into an overall, representative opinion is usually performed through an aggregation process. In this paper we focus on the problem of defining an opinion synthesizing the majority of the experts; to this aim we propose a formalization of the concept of majority opinion as a fuzzy subset (*fuzzy majority opinion*). As we shall see this approach provides in addition to a value for a majority opinion also an indication of the strength of that value being used as the majority opinion. This fuzzy subset can be interpreted as a possibility distribution over the possible representatives of a majority opinion over a set of values.

1 Introduction

In this paper the context of multi-agent (group) decision making is considered. One of the main problems in this context is to define a decision strategy which takes into account the individual opinions of the decision makers and synthesize them into an overall representative opinion. The reduction of the individual values into a representative value is usually performed through an aggregation process, which can be guided by a linguistic quantifier (such as *most*) [1,3,4,6,7]. The concept of majority plays in this context a key role: what is often needed is an overall opinion which synthesizes the opinions of the *majority* of the decision makers. In this case the goal is to obtain a value which can be considered as the opinion of a majority, that is, a value that is similar for any large group of people, what we can call the *majority opinion*. In this paper we propose a formalization of the concept of majority as a vague concept. Based on this interpretation we propose a formalization of a *fuzzy majority opinion* as a fuzzy subset. As we shall see this approach provides in addition to a value for a majority opinion also an indication of the strength of that value being used as the majority opinion. The *fuzzy majority opinion* is modeled as a fuzzy subset on the values which represent a majority opinion over subsets of the sets of considered values. The definition of this fuzzy subset requires that we have both information about the similarity between the values provided, and some information about what

T. Bilgiç et al. (Eds.): IFSA 2003, LNAI 2715, pp. 143-150, 2003.

quantity constitutes the idea of a majority. This fuzzy subset can be interpreted as a possibility distribution over the possible representatives of a majority opinion over a set of values. In section 2 we formally define the concept of fuzzy majority opinion; we consider the case where the decision maker expresses her/his judgments as numeric values. In section 3 we consider the case in which the decision makers are allowed to linguistically express their opinions by means of labels on an ordinal scale.

2   The Concept of Fuzzy Majority Opinion

In group decision making, the usual approach to the definition of a consensual opinion is to compute a value which synthesizes the opinion of the experts involved in a decision problem. The aggregation process which computes this value can be guided by a linguistic quantifier, which expresses the quantity of the experts which one wants to take into account (e.g. most for the majority, a few for a minority). In the case of aggregation which reflects the opinion of a majority of the decision makers, the aggregated value is representative of the majority opinion. In this section we approach the problem of defining a *fuzzy majority opinion*. Under this interpretation the majority opinion is no longer represented as a value, but as a fuzzy subset. As we shall see this will provide in addition to a value for the majority opinion an indication of the strength of that value as a representative of the majority opinion.

In the following we shall let $A = \{a_1, ..., a_n\}$ be a bag of values which constitute the opinions of a group of people. Our definition of a fuzzy majority opinion requires that we have information about the similarity between the values provided. It requires also some information about what quantity constitutes the idea of a majority.

Here we shall assume the availability of a relation on the space from which the individual values are drawn indicating how similar two values are. In particular we assume a relationship Sim on the domain of A such that for any a_i and a_j $Sim[a_i, a_j] \in [0,1]$ and that Sim satisfies the properties $Sim(a_i, a_i) = 1$ and $Sim(a_i, a_j) = Sim(a_j, a_i)$. We note that the relationship Sim is not a formal similarity relationship as introduced by Zadeh [8], it lacks transitivity, but formally a proximity relationship [5]. However for linguistic convenience we shall refer to $Sim(x,y)$ as indicating the degree of similarity between x and y. Another tool we need is a formal definition of the quantity we consider to constitute a majority. The concept of a majority is a user and context dependent idea, however there are certain features common to any definition. We shall assume a user provided definition of a majority in terms of a fuzzy subset Q on the unit interval. In particular $Q: [0,1] \to [0,1]$ such that $Q(0) = 0$, $Q(1) = 1$ and $Q(x) \geq Q(y)$ if $x > y$. The monotonicity of Q implies that if $Q(y) = 1$ then for all $x > y$ we have $Q(x) = 1$. We shall call the point x*, the smallest value for which $Q(x) = 1$, the Point of Realization, POR.

The concept of majority always has some POR, typically POR < 1. It will be often useful to have definitions of Q that are strictly monotonic, if $x > y$ then $Q(x) > Q(y)$. This strict monotonicity requires $Q(x) < 1$ if $x < 1$ and hence POR = 1. This type of strictly monotonic definition of Q allows us to always be able to distinguish between sets of different cardinalities. One solution to this conflict between desiring strict monotonicity and POR ≠ 1 is to use a concept of "effective point of realization" EPOR. The quantifier displayed below illustrates this idea.

Figure 1. Implementing an EPOR

Here we allow our definition to be such that x* is our EPOR and we consider 0.99 to effectively denote complete satisfaction. In addition for x from x* to 1 we use a straight line such for $Q(x)$. We now proceed to use these ideas to introduce a concept of a majority opinion that will be a fuzzy subset that can be interpreted as a possibility distribution on the numeric majority opinions [2].

Let E be a crisp subset of A. Our first step is to determine the degree to which this is a subset containing a majority opinion. We shall say E contains a majority opinion if all the elements in E are similar and the cardinality of E satisfies our idea of being a majority of elements from A. We shall refer to a subset of values holding a majority opinion, as a **gang**. Thus a gang is a subset of E that contains a majority of elements having similar values. We let Majop(E) indicate the degree to which the values in E constitute a majority opinion, are a majority of elements from A with similar values.

We define $\text{Majop}(E) = Q(\dfrac{|E|}{n}) \wedge \text{Sim}(E)$ where $\text{Sim}(E) = \underset{a_i, a_j \in E}{\text{Min}} [\text{Sim}(a_i, a_j)]$.

We now define the concept of the *opinion* of the elements in E as $Op(E) = Ave(E)$

$= \dfrac{\underset{a_i \in E}{\sum} a_i}{|E|}$, the average value of the elements in E.

Using the concepts Op(E) and Majop(E) we can define a fuzzy subset F indicating the majority opinion of the set of elements in A as F = Majority Opinion =

$\underset{E \subseteq A}{\bigcup} \left\{ \dfrac{\text{Majop}(E)}{\text{Op}(E)} \right\}$. So for each subset E , the value Majop(E) indicates the degree to which the quantity OP(E) is a majority opinion.

With the fuzzy subset F corresponding to the fuzzy majority opinion we see the $\text{Max}_E[\text{Majop}(E)]$, the maximal membership grade in F, indicates the degree to which there exists a majority opinion. We provide an illustrative example.

Example: We assume that our values are drawn from a scale of 0 to 10. We assume the following simple similarity relation: $\text{Sim}(x, y)=1$ if $|x - y| \leq 2$, $\text{Sim}(x, y)=0.5(4 - |x-y|)$ if $2 \leq |x-y| \leq 4$ and $\text{Sim}(x, y) = 0$ if $|x - y| > 4$.

We assume the situation in which our concept majority, Q, is defined as shown in figure 2.

Figure 2. Definition of the quantity *majority*

Thus $Q(x) = 0$ if $x \leq 0.4$, $Q(x) = 5(x - 0.4)$ if $0.4 < x \leq 0.6$ and $Q(x) = 1$ if $x \geq 0.6$.

I. Consider the case where $A = \{1, 4, 5, 5, 6, 9\}$. Since $n = 6$ we have 2^6 possible subsets. However any subset having 2 or less elements has $Q(\frac{|E|}{6}) = 0$. In addition any subset having elements with a distance between any two of its members of four or more has $Sim(E) = 0$. Thus the following are the only subsets for which $Majop(E) \neq 0$: $E_1 = \{4, 5, 5, 6\}$, $E_2 = \{4, 5, 5\}$, $E_3 = \{4, 5, 6\}$ and $E_4 = \{5, 5, 6\}$.

E	AVE(E)	Q($|E|$/N)	Sim(E)	Majop(E)
E_1	5	1	1	1
E_2	4.7	0.5	1	0.5
E_3	5	0.5	1	0.5
E_4	5.3	0.5	1	0.5

Thus in this case our fuzzy majority F is $F = \{\frac{0.5}{4.7}, \frac{1}{5} \cdot \frac{0.5}{5.3}\}$, which we can see can be expressed as *about 5*.

II. Consider the case where $A = \{1, 1, 4.5, 6.5, 10, 10\}$. If we eliminate subsets with two or less elements and those which have elements at a distance of four from each other we get: $E_1 = \{1, 1, 4.5\}$, $E_2 = \{6.5, 10, 10\}$.

E	AVE(E)	Q($|E|$/N)	Sim(E)	Majop(E)
E_1	2.16	0.5	0.25	0.25
E_2	8.833	0.5	0.25	0.25

In this case we get as our majority opinion $F = \{\frac{0.25}{2.10}, \frac{0.25}{8.833}\}$. Here we see very little support for any fuzzy majority opinion. As matter of fact $Max_x F(x) = 0.25$, there is no subset constituting a majority of people with similar opinions.

We note that at a formal level F is a fuzzy subset of the real line such that

$$F(r) = \max_{\substack{\text{all } E \in A \text{ s.t.Ave}(E)=r}} (Q(\frac{|E|}{n}) \wedge Sim(|E|)$$

In describing F use can be made of the connection between fuzzy subsets and natural languages to allow, the expression of F in linguistic terms. We also again note that $\max_{E \in A}[\text{Majop}(E)]$ indicates the degree to which there exists a majority opinion.

3 Ordinal Environment

In the preceding we considered the situation in which the values to be aggregated where assumed to be numbers. Here we shall consider the problem of calculating the majority opinion in the case in which the individual opinions are assumed only to have an ordered nature. We let $S = \{s_1, s_2,..., s_m\}$ be an ordinal scale such that $s_i > s_j$ if $i > j$. We assume that the opinions to be aggregated $A = \{a_1, a_2, ..., a_n\}$ are drawn from S. A prototypical situation of this kind is the case in which opinions are expressed using linguistic terms such as *good*, *very good*, *perfect*.

In order to implement our method for obtaining a majority opinion we need some information as to what is a majority and information about the similarity of the objects in S. All we need for expressing this information is an ordinal scale. While this scale can be the scale S we need not require it be S. Here we shall assume a scale $T = (t_1, t_2,..., t_p)$ where $t_i > t_j$ if $i > j$. We emphasize that the assumption that T is not necessarily the same as S is less restrictive then assuming them to be the same. We note that we can provide a negation on this scale as $Neg(t_j) = t_{p+1-j}$. We also point out the assumption that this is an ordinal scale doesn't preclude us from using a numeric scale such as the unit interval. We also point out that while the information about the definition of similarity and the concept of majority do not have to be on the same scale as the opinions being aggregated the information about the definition of similarity and the concept of majority do have to be on the same scale.

We assume a relationship Sim on S such for any $s_i, s_j \in S$ we have $Sim(s_i, s_j) \in T$ such that $Sim(s_i, s_j) = Sim(s_j, s_i)$ and $Sim(s_j, s_j) = t_p$. In addition we assume a definition of majority, Q such that $Q: [0,1] \to T$ where $Q(0) = t_1$, $Q(1) = t_p$ and $Q(x) \geq Q(y)$ if $x > y$.

Using these tools we can build a concept of a majority opinion as a fuzzy subset in a manner analogous to the preceding. Here we again define our majority opinion as a fuzzy subset F such that $F = \bigcup_{E \subseteq A} \{\frac{\text{Majop}(E)}{\text{Op}(E)}\}$. For any subset E of A we define

$$\text{Majop}(E) = Q(\frac{|E|}{n}) \wedge Sim(E) \text{ where } Sim(E) = \min_{a_i, a_j \in E}[Sim(a_i, a_j)].$$

Here Majop determines the degree to which the subset E constitutes a majority of values that are compatible, similar. If we denote $E^* = \underset{a_i \in E}{\text{Max}}[a_i]$ and $E_* = \underset{a_i \in E}{\text{Min}}[a_i]$ then $\text{Sim}(E) = \text{Sim}(E^*, E_*)$.

The term Op(E) is the aggregated opinion of the elements in the subset E. In the preceding we used the average of the elements in E for OP(E), however here the ordered nature of the elements precludes our using this operation. In this ordinal case to calculate the aggregated opinion of the elements E we shall use the median of E, thus Op(E) = Med(E). Using the median we get as our majority opinion the fuzzy

subset $F = \underset{E \subseteq A}{\bigcup} \{\dfrac{\text{Majop(E)}}{\text{Med(E)}}\}$.

We now turn to some pragmatic issues related to this ordinal environment. First we note that the scale S on which we obtain the individual opinions generally arises naturally from the values presented by the experts. It is often assumed of a linguistic type.

The requirement that the scale used to measure the similarity of the expert opinions and that used to define the concept Q be the same arises because of the need to

perform the operation $Q(\dfrac{|E|}{n}) \wedge \text{Sim}(E)$.

This requirement can be somewhat relaxed in a pragmatic spirit. If in determining Sim(E) we can be satisfied in only establishing whether the elements in E are compatible with each other or not we can greatly reduce the information required with respect to the similarity. In this case we need only assign a value 1 or 0 (True or False) to Sim(E). Furthermore since $\text{Sim}(E) = \text{Sim}(E^*, E_*)$ all we need to determine is if the boundary elements in E are compatible or not. At a formal level the use of this binary type of measurement of similarity can be seen as defining the relationship Sim on a subscale of T, the subset scale being $\{t_1, t_p\}$. In particular if the elements in E are compatible $\text{Sim}(E) = t_p$ and if they are not compatible $\text{Sim}(E) = t_1$. Thus in the

case in which $\text{Sim}(E) = t_p$ we get $\text{Majop}(E) = Q(\dfrac{|E|}{n})$ and if elements in E are not

compatible, $\text{Sim}(E) = t_1$ then $\text{Majop}(E) = t_1$. It will be convenient to refer to these special elements in T as 0 and 1.

Using this more simplified measurement of the similarity we get

$$F = \underset{\text{Sim(E)}=t_p}{\bigcup} \{\dfrac{\text{Majop(E)}}{\text{Med(E)}}\}$$

Let us look more carefully at the collection of subsets of A with compatible elements, those with $\text{Sim}(E) = t_p$. We first note that if $E \subseteq E'$ then if the elements in E are not compatible then those in E' are not compatible. On the other hand if the elements in E' are compatible then all those in E are compatible.

We now introduce the idea of maximal compatibility set in this environment. Without loss of generality, we assume the elements in A have been indexed such that

$a_1 \leq a_2 \leq....\leq a_n$ if $i < j$. Let us start with a_1 and find the largest a_j such that $Sim(a_1, a_j)$ $= t_p = 1$ it is the farthest element in A still compatible with a_1. Let us denote $E_1 = \{a_j$ such that $j \geq 1$ and $Sim(a_1, a_j) = 1\}$. All the elements in E_1 are compatible and all the subsets of E_1 are made of compatible elements. More generally for any $a_i \in A$ let us denote $E_i = \{a_j \mid j \geq i, Sim(a_j, a_i) = 1\}$.

Consider a compatible set E_i and let G and H be any subsets of E_i then since all elements in G and H are compatible and since $Med(G) \in G$ and $Med(H) \in H$ then $Sim(Med(G), Med(H)) = 1$. In particular $Sim(Med(G), Med(E_i)) = 1$. Thus the similarity between the compatible set E_i and any of its subset in one.

Let C_A denote the collection of compatible subsets of A, $C_A = \bigcup_{i=1}^{n} 2^{E_i}$.

It is the union of the power sets of all of the E_i. We can further refine this definition. For any compatibility set E_i let us denote e_i^* as the maximal element, it is the element farthest from a_i, the seed of E_i. Let E_i and E_j be two compatible sets where $j > i$, $a_j > a_i$. We first note that $e_j^* \geq e_i^*$ that is the largest element in E_j must be at least as large as e_i^*. Further we note that if $e_j^* = e_i^*$ then $E_j \subseteq E_i$. We shall call E_i a maximal compatible set if $e_j^* > e_i^*$ for all $i < j$.

We shall denote the collection of indices corresponding to maximal compatible sets as **E**. We see that we can express $C_A = \bigcup_{i \in \mathbf{E}} 2^{E_i}$.

Thus assuming this binary similarity relationship we can express

$$F = \bigcup_{E \subset C_A} \{\frac{Q(\frac{|E|}{n})}{Med(E)}\} \text{ where } C_A = \bigcup_{i \in \mathbf{E}} 2^{E_i} \text{ and } \mathbf{E} \text{ is the collection of maximal}$$

compatible sets.

Let us now consider the definition of Q. First we recall that Q is monotonic. We can consider Q to be of the form shown as figure 3.

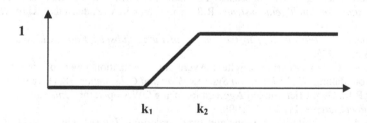

Figure 3. Basic form for Q

Thus there is some quantity of elements k_1 below which the degree of majority is zero and some quantity of elements k_2 for which we assume complete concept of majority. Using this we easily define Q in a natural manner: $Q(|E|) = 0$ if $|E| \leq k_1$, $Q(|E|) = \frac{|E| - k_1}{k_2 - k_1}$ if $k_1 < |E| < k_2$ and $Q(|E|) = 1$ if $|E| \geq k_2$.

Furthermore we note that if $B \subset E_i$ then $Q(|B|) \leq Q(|E_i|)$ this implies that for any maximal compatible set $Q(|E_i|)$ is at least as large as $Q(|B|)$ for its subsets. Furthermore since if $B \subset E_i$ then $Sim(Med(E_i), Med(B)) = 1$ we can effectively represent the majority opinion as

$$F = \bigcup_{i \in E} \{\frac{Q(E_i)}{Med(E_i)}\}$$

Thus all we need do is to find all the maximal compatibility sets, the degree to which each constitutes a majority of elements and obtain their respective medians.

4 Conclusions

In this paper a formalization of the concept of majority opinion has been defined as a fuzzy subset (*fuzzy majority opinion*). The notion of fuzzy majority opinion provides in addition to a value for a majority opinion also an indication of the strength of that value being used as the majority opinion.

References

1. Bordogna G, Fedrizzi M., Pasi G., A linguistic modeling of consensus in Group Decision Making based on OWA operators, *IEEE Trans. on System Man and Cybernetics*, 27(1), 1997.
2. Dubois, D. and Prade, H., *Possibility Theory: an Approach to Computerized Processing of Uncertainty*, Plenum Press: New York, 1988.
3. Herrera F., Herrera-Viedma E., Verdegay J.L., A Linguistic Decision Process in Group Decision Making, *Group Decision and Negotiation*, 5, 165-176, 1996.
4. Kacprzyk J., Fedrizzi M.and Nurmi H., Fuzzy Logic with Linguistic Quantifiers in Group Decision Making and Consensus Formation, in *An Introduction to Fuzzy Logic Applications in Intelligent Systems*, R.R. Yager and L.A. Zadeh eds., Kluwer, 263-280, 1992.
5. Kaufmann, A., *Introduction to the Theory of Fuzzy Subsets: Volume I*, Academic Press: New York, 1975.
6. Yager R. R., On Ordered Weighted Averaging aggregation Operators in Multi Criteria Decision Making, *IEEE Trans. on Systems, Man and Cybernetics*, 18(1), 183-190, 1988.
7. Yager R.R., Quantifier Guided Aggregation using OWA operators, *International Journal of Intelligent Systems*, 11, 49-73, 1996.
8. Zadeh, L. A., Similarity relations and fuzzy orderings, *Information Sciences*, 3, 177-200, 1971.

Modelling Fuzzy Quantified Statements under a Voting Model Interpretation of Fuzzy Sets

Félix Díaz-Hermida, Alberto Bugarín, Purificación Cariñena,
Manuel Mucientes, David E. Losada, and Senén Barro*

Dep. Electrónica e Computación. Univ. de Santiago de Compostela. Spain.
felixdh@usc.es, {alberto,puri,manuel}@dec.usc.es, dlosada@usc.es,
senen@dec.usc.es

Abstract. In this paper previous results by our group in the field of fuzzy quantification modelling and application are compiled. A general mechanism that is based on the voting-model interpretation of fuzzy sets is described. Application examples of quantified statements in the field of reasoning with fuzzy temporal rules, modelling of task-oriented vocabularies and information retrieval are presented.

Key words: Fuzzy quantification, semi-fuzzy quantifiers, theory of generalized quantifiers, extension of fuzzy operators, mobile robotics, information retrieval

1 Introduction

Modelling of fuzzy quantified statements is of great interest for tasks related with knowledge representation and reasoning, knowledge extraction, intelligent control, decision-making, fuzzy databases, information retrieval, etc. In spite of its importance for these and other fields, most of the approaches that are presented in the literature inherit the basic definitions of linguistic fuzzy quantifiers derived from the first approach [15], that solely represent a fuzzy quantity or proportion (absolute and relative quantifiers). Although this may be sufficient for some applications, this excludes dealing with more complex statements, such as those involving exception, comparative or non-quantitative quantifiers, therefore limiting expresiveness of fuzzy quantified statements. Furthermore, most of the approaches in literature fail to exhibit a plausible behaviour, since important properties for fuzzy quantification methods (correct generalisation, continuity, monotonicity, etc.) are not fulfilled [3,12].

Following [11,13] the evaluation of quantified sentences is presented under the approach of fuzzification of *semi-fuzzy quantifiers*. Under this approach the

* Authors wish to acknowledge support from the Spanish Ministry of Education and Culture through grant TIC2000-0873. D. E. Losada is supported by the "Ramón y Cajal" R&D program, which is funded in part by "Ministerio de Ciencia y Tecnología" and in part by FEDER funds.

T. Bilgiç et al. (Eds.): IFSA 2003, LNAI 2715, pp. 151–158, 2003.
© Springer-Verlag Berlin Heidelberg 2003

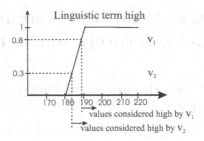

Fig. 1. Height values that are voted as "high" by two voters with different specificity.

problem of evaluating a fuzzy quantified statement is rewritten as the problem of applying an adequate fuzzification mechanism for transforming a semi-fuzzy quantifier into a fuzzy quantifier. Some *quantifier fuzzification mechanisms* that secure the fulfilment of most of the properties that are desirable for fuzzy quantifiers has been described in [11,13]. Our proposals do not fully fulfil the axiomatic framework presented in these works, but they show a very good behaviour [9], a clear and intuitive semantic interpretation that is based on voting models [2], and it also seems to be possible to modify these proposals for avoiding some counterintuitive behaviours [8].

2   Voting Model and Quantified Sentences Evaluation

Voting model [2] understands a fuzzy set $X \in \widetilde{\wp}(E)$ as the result of carrying out a random experiment that is summarized by its membership function μ_X. This interpretation assumes that a set of individuals (voters) take binary decisions about the elements of the referential that fulfil the property represented by X. In this way, membership function $\mu_X(e), e \in E$ indicates the probability that a randomly selected voter v state that e fulfils the property represented by X.

Under this construction, it can also be interpreted that vagueness arises from the different degrees of specificity of voters that are asked to decide which of the elements of the referential set are compatible with linguistic terms. This is similar to assuming that a uniform probability distribution on the specificity levels of the voters exists, or that all specificity levels are equally probable.

Figure 1 shows an example of this interpretation for the values on the referential set *height* that are considered to be *high* by two voters. In this figure, voter v_1 (with specificity level of 0.8) considers elements in $[188, \infty)$ as representatives of the property *high*; voter v_2 (with specificity level of 0.3) considers elements in $[183, \infty)$ to be high. It should be noted that it is natural to relate this interpretation with α-*cuts* (e.g. $high_{0.8} = [188, \infty)$ and $high_{0.3} = [183, \infty)$).

We now go on to make some definitions that make it possible to evaluate fuzzy quantified sentences. Later, we relate these definitions with voting models.

Definition 1 (Semi-fuzzy quantifier). *[11,13] An s-ary semi-fuzzy quantifier Q on a base set $E \neq \varnothing$ is a mapping $Q : \wp(E)^s \longrightarrow [0,1]$ which assigns a gradual result $Q(X_1, \ldots, X_s) \in [0,1]$ to each choice of crisp $X_1, \ldots, X_s \in \wp(E)$.*

Examples of semi-fuzzy quantifiers are:

$$\text{about } \_5_E(X_1, X_2) = T_{2,4,6,8}(|X_1 \cap X_2|)$$

$$\text{about80\%\_ormore}_E(X_1, X_2) = \begin{cases} S_{0.5,0.8}\left(\frac{|X_1 \cap X_2|}{|X_1|}\right) & X_1 \neq \varnothing \\ 1 & X_1 = \varnothing \end{cases}$$

where T represents a trapezoidal fuzzy set and S the Zadeh's S-function.

Fuzzy quantifiers are similar to semi-fuzzy quantifiers but taking values in the fuzzy powerset of E (i.e, a mapping $\widetilde{Q} : \widetilde{\wp}(E)^s \mapsto [0,1]$).

Semi-fuzzy quantifiers are much more intuitive and easier to define than fuzzy quantifiers, but they do not resolve the problem of evaluating fuzzy quantified sentences. In order to do so mechanisms are needed that enable us to transform a semi-fuzzy quantifier into a fuzzy quantifier; i.e., a function with domain on the universe of semi-fuzzy quantifiers and range on the universe of fuzzy quantifiers (quantifier fuzzification mechanisms (*QFM*) mentioned in the introduction):

$$F : (Q : \wp(E)^s \mapsto [0,1]) \mapsto \left(\widetilde{Q} : \widetilde{\wp}(E)^s \mapsto [0,1]\right) \tag{1}$$

Definition 2 (QFM related to P). *[9] Let $Q : \wp(E)^s \to [0,1]$ be a semi-fuzzy quantifier and $P(\alpha_1, \ldots, \alpha_s)$ a probability density function. We define the quantifier fuzzification mechanism related to P as:*

$$F^P(Q)(X_1, \ldots, X_s) = \int_0^1 \cdots \int_0^1 Q((X_1)_{\alpha_1}, \ldots, (X_s)_{\alpha_s}) P(\alpha_1, \ldots, \alpha_s) d\alpha_1, \ldots, d\alpha_s \tag{2}$$

$P(\alpha_1, \ldots, \alpha_s)$ is a probability density that describes the relationship that exists for every voter among the specificity levels used for defining the terms in the quantified statement. F^P represents the mean of the decisions of the voters.

According to different definitions of the probability density P different mechanisms can be defined. Among others, the following ones can be mentioned:

- **Maximum dependence model.** This scheme arises from assuming that voters are equally specific for all properties:

$$F^{MD}(Q)(X_1, \ldots, X_s) = \int_0^1 Q((X_1)_\alpha, \ldots, (X_s)_\alpha) d\alpha$$

  This expression, keeping appart differences related to representatives normalisation, generalizes one of the models defined in [6].

- **Independence model.** This scheme [7,9] arises from assuming that the specificity level of a voter for one property does not constraint his/her specificity level for other properties

$$F^I(Q)(X_1, \ldots, X_s) = \int_0^1 \cdots \int_0^1 Q((X_1)_{\alpha_1}, \ldots, (X_s)_{\alpha_s}) d\alpha_1 \ldots d\alpha_s$$

Example 1. Let us consider sentence: *"most tall men are blond"*, where quantifier $Q =$"most", and fuzzy sets $X_1 =$"tall" and $X_2 =$"blond" take the values:

$$X_1 = \{1, 0.9, 0.8, 0.7, 0.6, 0.5, 0.4\}, X_2 = \{0.9, 0.8, 0.7, 0.6, 0.5, 0.4, 0.3\}$$

$$Q(X_1, X_2) = \begin{cases} \max\left\{4\left(\frac{|X_1 \cap X_2|}{X_2}\right) - 3, 0\right\} & X_1 \neq \emptyset \\ 1 & X_1 = \emptyset \end{cases}$$

The results that are obtained for the two previously presented models are:

$$F^{MD}(Q)(X_1, X_2) = \int_0^1 Q\left((X_1)_\alpha, (X_2)_\alpha\right) d\alpha = 0.3$$

$$F^I(Q)(X_1, X_2) = \int_0^1 \int_0^1 D\left((X_1)_{\alpha_1}, (X_2)_{\alpha_2}\right) d\alpha_1 d\alpha_2 = 0.51$$

Previous models cannot be considered as "pure" quantifier fuzzification mechanism in the sense of definition 1, since the integral may not exist for non-finite referentials. Nevertheless, the finite case is sufficient for practical applications.

The methods that are derived from the probabilistic framework here presented do not fulfil all of the axioms that are stated in [11], but they do fulfil most of the important properties that are derived from them [9]. Furthermore, the simple and understandable semantic interpretation underlying the probabilistic framework is an interesting feature that should be pointed out. Some of the properties that are fulfiled by this framework are (due to lack of space, definitions and proofs are omitted): *correct generalization of crisp expressions, external negation, monotonicity, local monotonicity, specificity in quantifiers, induced operators,* etc. Moreover, the maximum dependence model guarantees the fulfilment of *internal meets*, whilst the independence model fulfils *antonymy* and *duality*.

In spite of their general good behaviour none of these probabilistic quantifier fuzzification mechanisms is fully coherent with the interpretation of fuzzy sets we are using. In particular, this happens when the fuzzy sets a quantifier takes as arguments are derived from linguistic terms associated to the same linguistic variable.

In figure 2 examples for explaining the reasons of this lack of coherence for the maximal dependence model are presented. In figure 2a assignation of the elements in the referential to the labels in the term set {*short,medium,tall*} is shown for a voter having an specificity level $\alpha_1 > 0.5$. We can see that some elements are assigned to no term. In figure 2b, assignation for a voter having an specificity level $\alpha_2 < 0.5$ is presented. For this case, some height values are simultaneously assigned to two labels. Both of these examples show situations that do not accommodate the reasonable voting model assumption that all voters should associate a single label to every element in the referential. Similar problems exists for the independence model. The following example shows the problems we have just mentioned when evaluating a fuzzy quantified sentence:

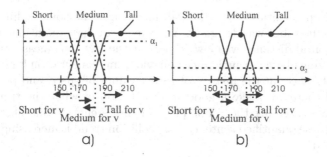

Fig. 2. Incoherence situations for the dependence model.

Example 2. Let us consider the following statement "all individuals are medium or tall", where semi-fuzzy quantifier $all_E : \wp(E) \to [0,1]$ is defined as:

$$all_E(X) = 1 \text{ iff } X = E, X \in \wp(E)$$

Let $E = \{e_1, \ldots, e_{10}\}$, and assume that $height(e_i) = 185, i = 1, \ldots, 10$. Intuitively, this sentence is true for the labels defined in figure 2. But if we calculate the union of *tall* and *medium* with *tconorm* max we obtain

$$F^{MD}(all_E)(medium\widetilde{\cup}tall) = F^I(all_E)(medium\widetilde{\cup}tall) = 0.5$$

In order to overcome the difficulties just mentioned, in [8] a preliminary definition has been by defining a probabilistic model over formulas involving linguistic terms.

3   Fields of Application for Fuzzy Quantified Statements

3.1   Fuzzy Temporal Rules in Intelligent Control (Temporal Fuzzy Control)

In the field of intelligent control, we have used fuzzy quantified statements within a more general framework known as Fuzzy Temporal Rules (FTRs) [5] in order to implement two behaviours in mobile robotics: wall-following and moving obstacles avoidance [14]. FTRs allow the control action to be taken after considering information on the state of the system (both the robot and its environment) within prior temporal instants and not only the current one, as is usual in fuzzy control. Quantification here plays the role of detecting whether certain situations of interest occur persistently on a given temporal window, in a partially persistent way or on a single temporal point. By means of the analysis of the evolution of state variables throughout temporal references a filtering of noisy sensorial inputs is also achieved, and therefore the control action is more reliable. This is a crucial aspect in robotic applications, where uncertainty about measures (obstacles and/or robot location, speed, ...) due to the ultrasound sensors limitations (crosstalking and other measurement errors) may be decisive.

For the implementation of the wall-following behaviour, a linear velocity controller for the robot was designed. An example of these quantified proposition is: "IF the frontal distance was low *in part of* the last four measurements...".

In the moving obstacles avoidance behaviour implementation FTRs are useful for taking into account the history of recent values of speed and position of an obstacle and therefore make a good interpretation of what the trend (passing before the robot, letting the robot pass, or indiferent) of this obstacle is. An examples of these quantified sentences is: "collision trend is increasing *throughout* the last second".

3.2   Modelling of Task-Oriented Vocabularies

Modelling of task-oriented vocabularies [10] allows us to increase expressiveness and aplicability of fuzzy knowledge based systems. Usual examples of use of these vocabularies range from flexible queries to data mining in databases, and also in the previously mentioned field of temporal fuzzy control [5]. By using similar approaches to the ones presented in the field of fuzzy quantification, the modeling of fuzzy specification operators (*"maximum"*, *"minimum"*, *"last"*, ...) and of fuzzy reduction operators (*"mean"*, ...) can be developed.

Let $f : \wp^s(E) \to \mathbb{R} \cup \{\theta\}$ be an arbitrary function, where θ indicates an undefined situation and the referential set E is supposed finite. On the basis of f we define the function $g^f : \wp^s(E) \times \mathbb{R} \cup \{\theta\} \to \{0,1\}$ as

$$g^f(X_1,\ldots,X_s)(r) = \begin{cases} 0 & f(X_1,\ldots,X_s) \neq r \\ 1 & f(X_1,\ldots,X_s) = r \end{cases}$$

and then, calculate the probability of r being the evaluation of f over $X_1,\ldots,X_s \in \widetilde{\wp}(E)$ for the probability density P as

$$P(r|X_1,\ldots,X_s) = \int_0^1 \cdots \int_0^1 g^f\left((X_1)_{\alpha_1},\ldots,(X_s)_{\alpha_s}\right)(r) P(\alpha_1,\ldots,\alpha_s) d\alpha_1 \ldots d\alpha_s$$

Example 3. Suppose that we wish to evaluate the sentence "the last temporal point at which the temperature was high during last minutes". Let $E = \{e_1, e_2, e_3, e_4, e_5, e_6, e_7\}$ be the temporal referential. Let

$$high(E) = \{0.7/e_1, 0.9/e_2, 0.2/e_3, 0/e_4, 1/e_5, 0.9/e_6, 0.7/e_7\}$$
$$last\_minutes(E) = \{0.3/e_1, 0.7/e_2, 1/e_3, 1/e_4, 1/e_5, 0.7/e_6, 0.3/e_7\}$$

be the sets "high temperatures" and "during last minutes". In the crisp case we could evaluate the sentence by formulating the function

$$f(X_1, X_2) = \begin{cases} max(X_1 \cap X_2) & X_1 \cap X_2 \neq \varnothing \\ \theta & X_1 \cap X_2 = \varnothing \end{cases}$$

where X_1 represents the "high temperatures" and X_2 "during last minutes". We define

$$g^f(X_1, X_2)(r) = \begin{cases} 0 & f(X_1, X_2) \neq r \\ 1 & f(X_1, X_2) = r \end{cases}$$

If we assume the independence profile for the extension of f we obtain

$$P\left(e_7|high, last\_minutes\right) = 0.21, P\left(e_6|high, last\_minutes\right) = 0.42$$
$$P\left(e_5|high, last\_minutes\right) = 0.37$$

3.3 Information Retrieval

The application of fuzzy set theory for Information Retrieval (IR) is not mature yet. Although a number of researchers have devoted their efforts to designing flexible retrieval models based on fuzzy sets (see e.g. the very interesting compendia [4]) there exists still a gap, especially at the experimentation level, that makes fuzzy approaches not very popular among the IR community [1]. Most of the works on fuzzy sets for IR did not pay much attention to supply efficient implementations and very few researchers attempted to test their theories against standard benchmarks of the field.

Our approach in this field is at a initial stage. It aims at extending fuzzy query languages and testing such models against standard evaluation datasets (at this moment we are employing a subset of the TREC collection containing more than 175.000 documents). Preliminary results are very promising. Our experiments show clearly that the introduction of quantifiers can lead to improvements in retrieval performance up to 20% of average precision[1] when compared to the vector-space model.

Our fuzzy IR model is based on the quantification models previously described. At the moment, best results were obtained using fuzzy quantifiers for the different statements of the query (title, description, narrative) and, then, aggregating these results by means of either a fuzzy quantifier or a tnorm.

More details on this experimentation including precision versus recall tables are being presented soon. We also aim to introduce more complex queries and more complex quantifiers. In these cases query results are expected to take advantage of the good theoretical behaviour of the probabilistic approach for fuzzy quantification and its capability for modelling different types of quantifiers.

4 Conclusions

In this paper a probabilistic fuzzy quantification model developed in our research group has been presented. Previous results regarding both its definition and applications are described. The main feature of this model is its clear underlying semantics and its flexibility.

References

1. R. Baeza-Yates and B. Ribeiro-Neto. *Modern Information Retrieval*. Addison Wesley, ACM press, 1999.

---

[1] Precision is an standard performance ratio that helps to measure the goodness of a ranking of documents.

2. J.F. Baldwin, J. Lawry, and T.P. Martin. Mass assignment theory of the probability of fuzzy events. *Fuzzy Sets and Systems*, 83:353–367, 1996.
3. S. Barro, A. Bugarín, P. Cariñena, and F. Díaz-Hermida. A framework for fuzzy quantification models analysis. *IEEE Transactions on Fuzzy Systems*, 11:89–99, 2003.
4. G. Bordogna and G. Pasi. Modeling vagueness in information retrieval. In M. Agosti, F. Crestani, and G. Pasi, editors, *ESSIR 2000, LNCS 1980*, pages 207–241. Springer-Verlag Berlin Heidelberg, 2000.
5. P. Cariñena, A. Bugarín, M. Mucientes, F. Díaz-Hermida, and S. Barro. *Technologies for Constructing Intelligent Systems*, volume 2, chapter Fuzzy Temporal Rules: A Rule-based Approach for Fuzzy Temporal Knowledge Representation and Reasoning, pages 237–250. Springer-Verlag, 2002.
6. M. Delgado, D. Sánchez, and M. A. Vila. Fuzzy cardinality based evaluation of quantified sentences. *International Journal of Approximate Reasoning*, 23(1):23–66, 2000.
7. F. Díaz-Hermida, A. Bugarín, P. Cariñena, and S. Barro. A framework for probabilistic approaches to fuzzy quantification. In *Proceedings 2001 EUSFLAT Conference*, pages 34–37, 2001.
8. F. Díaz-Hermida, A. Bugarín, P. Cariñena, and S. Barro. Un esquema probabilístico para el tratamiento de sentencias cuantificadas sobre fórmulas. In *Actas del XI Congreso Español Sobre Tecnologías y Lógica Fuzzy (ESTYLF 2002)*, pages 391–396, 2002.
9. F. Díaz-Hermida, A. Bugarín, P. Cariñena, and S. Barro. Voting model based evaluation of fuzzy quantified sentences: a general framework. Technical Report GSI-02-01, Intelligent Systems Group. Univ. Santiago de Compostela, 2002.
10. F. Díaz-Hermida, P. Cariñena, A. Bugarín, and S. Barro. Modelling of task-oriented vocabularies: an example in fuzzy temporal reasoning. In *Proceedings 2001 FUZZ-IEEE Conference*, pages 43–46, 2001.
11. I. Glöckner. DFS- an axiomatic approach to fuzzy quantification. TR97-06, Techn. Fakultät, Univ. Bielefeld, 1997.
12. I. Glöckner. A framework for evaluating approaches to fuzzy quantification. Technical Report TR99-03, Universität Bielefeld, May 1999.
13. I. Glöckner and A. Knoll. A formal theory of fuzzy natural language quantification and its role in granular computing. In W. Pedrycz, editor, *Granular computing: An emerging paradigm*, volume 70 of *Studies in Fuzziness and Soft Computing*, pages 215–256. Physica-Verlag, 2001.
14. M. Mucientes, R. Iglesias, C. V. Regueiro, A. Bugarín, P. Cariñena, and S. Barro. Fuzzy temporal rules for mobile robot guidance in dynamic environments. *IEEE Transactions on Systems, Man and Cybernetics, Part C*, 33(3):391–398, 2001.
15. L.A. Zadeh. A computational approach to fuzzy quantifiers in natural languages. *Comp. and Machs. with Appls.*, 8:149–184, 1983.

Arithmetic of Fuzzy Quantities Based on Vague Arithmetic Operations

Mustafa Demirci

Akdeniz University, Faculty of Sciences and Arts,
Department of Mathematics, 07058-Antalya, Turkey
demirci@akdeniz.edu.tr

Abstract. As a generalization of the standard fuzzy arithmetic operations, the present paper introduces the arithmetic operations among fuzzy quantities whenever the arithmetic operations are vaguely defined functions, or more precisely they are vague arithmetic operations. Furthermore, elementary properties of this kind of arithmetic of fuzzy quantities are established, and the relations between standard fuzzy arithmetic and vague arithmetic are pointed out, though the basic ideas of these two sorts of arithmetic are quite different from each other.
Keywords: Fuzzy arithmetic, Vague arithmetic, Fuzzy function, Fuzzy equivalence relation, Indistinguishability operator.

1 Introduction

As a practical implementation of vague algebraic notions, a new kind of fuzzy setting of arithmetic operations based on many-valued equivalence relations and strong fuzzy functions are introduced under the name of vague arithmetic operations in [5, 8]. In fuzzy arithmetic, the numbers are fuzzily defined objects and the arithmetic operations between them are functions in the classical sense [9]. In vague arithmetic, the numbers are certain objects, but the arithmetic operations between them are vaguely defined functions in the sense of [4, 5-6]. Fuzzy arithmetic operations on the set $\mathcal{F}(\mathbb{R})$ of all fuzzy quantities, where a fuzzy quantity is defined as a normalized fuzzy set of $\mathbb{R}$ [9], are defined as the extension of usual arithmetic operations on $\mathbb{R}$ to $\mathcal{F}(\mathbb{R})$. In this paper, as a generalization of fuzzy arithmetic operations, we introduce arithmetic operations on $\mathcal{F}(\mathbb{R})$ whenever the arithmetic operations on $\mathbb{R}$ are vaguely defined, or more precisely, they are vague arithmetic operations on $\mathbb{R}$ in the sense of [5, 8]. The purpose of this paper is to expose some elementary properties of these new kinds of arithmetic operations, and is to find out the relation between the standard fuzzy arithmetic and vague arithmetic, though the rudimentary ideas of these two sorts of arithmetic are completely different from each other.

2 Fuzzy Functions Based on Many-Valued Equivalence Relations

In this section, we first need to introduce the lattice-theoretic base of many-valued equivalence relations and fuzzy functions, namely integral, commutative

T. Bilgiç et al. (Eds.): IFSA 2003, LNAI 2715, pp. 159–166, 2003.

cl-monoid presented in [11]. A triple $M = (L, \leq, *)$ is called an integral, commutative cl-monoid iff the following conditions are satisfied: (i) $(L, \leq)$ is a complete lattice containing at least two distinct elements, where the join operation, the meet operation, the bottom element of L and the top element of L are denoted by $\bigvee$, $\bigwedge$, $\mathbf{0}$ and $\mathbf{1}$, respectively, (ii) $(L, *)$ is a commutative monoid with the unit $\mathbf{1}$, (iii) $*$ is distributive over arbitrary joins, i.e. for each subfamily $\{\alpha_i : i \in I\}$ of L and for each $\beta \in L$, $\beta * (\bigvee_{i \in I} \alpha_i) = \bigvee_{i \in I} (\beta * \alpha_i)$. For $L = [0, 1]$ and for a left continuous t-norm $*$, the triple $([0, 1], \leq, *)$ is an example for the integral, commutative cl-monoids. In this work, $M = (L, \leq, *)$ always stands for an integral, commutative cl-monoid. The letters X, Y, Z always denote nonempty usual sets. An L-fuzzy set of X is a mapping $\mu : X \to L$, and the set of all L-fuzzy sets of X is denoted by L^X. Each ordinary subset A of X can be represented by the particular L-fuzzy set $\mathbf{1}_A$ of X, given by $\mathbf{1}_A(x) = \begin{cases} \mathbf{1}, & if\ x \in A \\ \mathbf{0}, & if\ x \notin A \end{cases}$. The kernel $\ker(\mu)$ of an L-fuzzy set $\mu \in L^X$ is defined as the ordinary subset $\{x \in X : \mu(x) = 1\}$ of X [15]. An L-fuzzy set $\mu \in L^X$ is said to be normalized iff $\ker(\mu) \neq \emptyset$.

Definition 2.1 [6]. An M-equivalence relation on X is a map $E : X \times X \to L$ fulfilling the axioms: (E.1) $E(x, x) = \mathbf{1}, \forall x \in X$, (E.2) $E(x, y) = E(y, x), \forall x, y \in X$, (E.3) $E(x, y) * E(y, z) \leq E(x, z), \forall x, y, z \in X$.

An M-equivalence relation E on X is also called an M-valued similarity relation on X [10]. For the particular choices of the integral, commutative cl-monoid $M = (L, \leq, *)$, it is also named as an equality relation (an L-fuzzy equivalence relation) w.r.t. $*$ on X [12-13], a $*$-fuzzy equivalence relation [2], a $*$-indistinguishability operator [16], a similarity relation [17], a likeness relation [1] and a probabilistic relation [14]. An M-equivalence relation E on X is called a $*$-fuzzy equality on X iff it is separated [4-6], i.e. $E(x, y) = \mathbf{1} \Rightarrow x = y, \forall x, y \in X$. The classical $*$-fuzzy equality E_X^N on X, given by $E_X^N(x, y) = \begin{cases} \mathbf{1}, & if\ x = y \\ \mathbf{0}, & if\ x \neq y \end{cases}$, $\forall x, y \in X$, characterizes the equality relation on X in the classical sense [4-6]. For given M-equivalence relations ($*$-fuzzy equalities) E on X and F on Y, we may establish an M-equivalence relation (a $*$-fuzzy equality) $G(E, F)$ on the product set $X \times Y$ by the rule $G(E, F)((x, y), (x', y')) = E(x, x') * F(y, y'), \forall x, x' \in X$, $\forall y, y' \in Y$ [4-5]. For an M-equivalence relation E on X, an L-fuzzy set $\mu \in L^X$ is said to be extensional w.r.t. E if the inequality $\mu(x) * E(x, y) \leq \mu(y)$ is satisfied for each $x, y \in X$. The L-fuzzy set $[\mu]_E = \bigwedge\{\nu \mid \mu \leq \nu$ and ν is extensional w.r.t. $E\}$, called the extensional hull of μ w.r.t. E, is the smallest extensional L-fuzzy set of X w.r.t. E containing μ [12-13]. The extensional hull $[\mu]_E$ of $\mu \in L^X$ w.r.t. E, the extensional hull $[A]_E$ of an ordinary subset A of X w.r.t. E and the extensional hull $[y]_E$ of a point y of X w.r.t. E are explicitly given by $[\mu]_E(x) = \bigvee_{z \in X} \{\mu(z) * E(z, x)\}$, $[A]_E(x) = \bigvee_{z \in X} E(z, x)$ and $[y]_E(x) = E(y, x)$, $\forall x \in X$, respectively [12-13].

Definition 2.2 [6]. Let E and F be two M-equivalence relations on X and Y, respectively, and ρ an L-fuzzy relation from X to Y, i.e. $\rho \in L^{X \times Y}$.

(i) ρ is called a strong fuzzy function from X to Y w.r.t. E and F iff the conditions: (F.1) For each $x \in X$, $\exists y \in Y$ such that $\rho(x,y) = 1$, (F.2) $\rho(x,y) *$ $\rho(x',y') * E(x,x') \leq F(y,y')$, $\forall x, x' \in X$, $\forall y, y' \in Y$, are satisfied.

(ii) ρ is called a perfect fuzzy function from X to Y w.r.t. E and F iff the condition (F.1) and the conditions: (F.3) $\rho(x,y) * \rho(x,y') \leq F(y,y')$, (F.4) $\rho(x,y) * E(x,x') \leq \rho(x',y)$ and (F.5) $\rho(x,y) * F(y,y') \leq \rho(x,y')$, are satisfied for each $x, x' \in X$ and for each $y, y' \in Y$.

In the classical set-theory, for a given ordinary relation f from X to Y, it is well-known that f is defined as a function from X to Y iff f holds the conditions: (F.1') For each $x \in X$, $\exists y \in Y$ such that $(x,y) \in f$, (F.2') $(x,y) \in f$ and $(x',y') \in f$ and $x = x'$ implies $y = y'$ for each $x, x' \in X$ and for each $y, y' \in Y$. The conditions (F.1) and (F.2) in Definition 2.2 are nothing but the many-valued generalizations of (F.1') and (F.2'), respectively. The notion of perfect fuzzy function is a slightly modified form of fuzzy function in the sense of Klawonn [13], and its origin goes back to the work of Cerruti and Höhle [3]. For given M-equivalence relations E on X and F on Y, an ordinary function $f : X \to Y$ is said to be extensional w.r.t. E and F iff $E(x,y) \leq F(f(x), f(y))$, $\forall x, y \in X$ [6, 13]. It is shown in the pursuing results that the strong (perfect) fuzzy functions possess powerful representations by ordinary functions [5-6]:

Theorem 2.3. Let E and F be two M-equivalence relations on X and Y, respectively. For a given ordinary function $f : X \to Y$ extensional w.r.t. E and F, a fuzzy relation $\rho \in L^{X \times Y}$, which satisfies the conditions

$$\rho(x, f(x)) = 1 \quad \text{and} \quad \rho(x,y) \leq F(f(x), y), \tag{1}$$

$\forall x \in X, \forall y \in Y$, is a strong fuzzy function w.r.t. E and F. Conversely, for a given strong fuzzy function $\rho \in L^{X \times Y}$ w.r.t. E and F, there exists an ordinary function $f : X \to Y$ extensional w.r.t. E and F satisfying the conditions (1).

Theorem 2.4. Let E and F be two M-equivalence relations on X and Y, respectively. For a given ordinary function $f : X \to Y$ extensional w.r.t. E and F, the fuzzy relation $\rho \in L^{X \times Y}$, defined by the formula

$$\rho(x,y) = F(f(x), y), \quad \forall x \in X, \forall y \in Y, \tag{2}$$

is a perfect fuzzy function w.r.t. E and F. Conversely, for a given perfect fuzzy function $\rho \in L^{X \times Y}$ w.r.t. E and F, there exists an ordinary function $f : X \to Y$ extensional w.r.t. E and F fulfilling the equality (2).

In [5-6], using Theorem 2.3 and Theorem 2.4, it is proven that ρ is a perfect fuzzy function w.r.t. E and F iff ρ is a strong fuzzy function w.r.t. E and F satisfying the condition (F.5).

Definition 2.5 [6]. Let E and F be two M-equivalence relations on X and Y, respectively.

(i) For a given function $f : X \to Y$ extensional w.r.t. E and F, the perfect fuzzy function $\rho \in L^{X \times Y}$ w.r.t. E and F given by the formula (2) is called the $E - F$ vague description of f, and is denoted by $vag(f)$.

(ii) For a given strong fuzzy function $\rho \in L^{X \times Y}$ w.r.t. E and F, the ordinary function $f : X \to Y$ extensional w.r.t. E and F, which satisfies the conditions

(1), is called an ordinary description of ρ. The set of all ordinary descriptions of ρ is denoted by $ORD(\rho)$. For a given strong fuzzy function $\rho \in L^{X \times Y}$ w.r.t. E and F, if F is a $*$-fuzzy equality on Y, then ρ has a unique ordinary description, denoted by $ord(\rho)$, and so $ORD(\rho)$ consists of solely $ord(\rho)$ [5-6].

3   Arithmetic of Fuzzy Quantities on the Basis of Vague Arithmetic Operations

A normalized L-fuzzy set μ of the set $\mathbb{R}$ of all real numbers is called a fuzzy quantity [9], and the set of all fuzzy quantities will be denoted by $\mathcal{F}(\mathbb{R})$. Denoting one of the usual four arithmetic operations "+", "−", "." and "/" by Δ, let $\mathcal{F}^{\Delta}(\mathbb{R})$ represent the set $\mathcal{F}(\mathbb{R})$ for $\Delta \in \{+, -, .\}$ and the subset $\{\mu \in \mathcal{F}(\mathbb{R}) \mid \ker(\mu) \neq \{0\}\}$ of $\mathcal{F}(\mathbb{R})$ for $\Delta = /$. In the following definition, a generalized version of the fuzzy arithmetic operations for fuzzy quantities in [9] is now introduced on the basis of an integral, commutative cl-monoid:

Definition 3.1. For $\Delta \in \{+, -, ., /\}$, the map $\overline{\Delta} : \mathcal{F}(\mathbb{R}) \times \mathcal{F}^{\Delta}(\mathbb{R}) \to \mathcal{F}(\mathbb{R})$, $(\mu, \nu) \mapsto \overline{\Delta}(\mu, \nu)$, defined by

$$\overline{\Delta}(\mu, \nu)(x) = \bigvee_{x = z \Delta y} \{\mu(z) * \nu(y)\}, \quad \forall x \in \mathbb{R}, \forall \mu \in \mathcal{F}(\mathbb{R}), \forall \nu \in \mathcal{F}^{\Delta}(\mathbb{R}), \quad (3)$$

is called a fuzzy arithmetic operation on $\mathcal{F}(\mathbb{R})$. For $\Delta = +$ (resp., $\Delta = -$, $\Delta = .$ and $\Delta = /$) the fuzzy arithmetic operation $\overline{\Delta}$ on $\mathcal{F}(\mathbb{R})$ is called the fuzzy addition (resp., substraction, multiplication and division) operation on $\mathcal{F}(\mathbb{R})$.

Definition 3.2 [8]. Let U denote $\mathbb{R}^2$ for $\Delta \in \{+, -, .\}$ and $\mathbb{R} \times (\mathbb{R} - \{0\})$ for $\Delta = /$. Let E_U and $E_{\mathbb{R}}$ be M-equivalence relations on U and $\mathbb{R}$, respectively. Then a strong fuzzy function $\widetilde{\Delta}$ from U to $\mathbb{R}$ w.r.t. E_U and $E_{\mathbb{R}}$ is called an M-vague arithmetic operation on $\mathbb{R}$ w.r.t. E_U and $E_{\mathbb{R}}$ iff $\Delta \in ORD(\widetilde{\Delta})$. An M-vague arithmetic operation $\widetilde{\Delta}$ on $\mathbb{R}$ w.r.t. E_U and $E_{\mathbb{R}}$ is said to be perfect iff $\widetilde{\Delta}$ is a perfect fuzzy function, or equivalently $\widetilde{\Delta}(x, y, z) = E_{\mathbb{R}}(x \Delta y, z)$, $\forall (x, y) \in U$, $\forall z \in \mathbb{R}$. For $\Delta = +$ (resp., $\Delta = -$, $\Delta = .$ and $\Delta = /$), an M-vague arithmetic operation $\widetilde{\Delta}$ on $\mathbb{R}$ w.r.t. E_U and $E_{\mathbb{R}}$ is called an M-vague addition (resp., subtraction, multiplication and division) operation on $\mathbb{R}$ w.r.t. E_U and $E_{\mathbb{R}}$.

For a given M-vague arithmetic operation $\widetilde{\Delta}$ on $\mathbb{R}$, for the sake of simplicity, the notation $\widetilde{\Delta}(x, y, z)$ will be used instead of $\widetilde{\Delta}((x, y), z)$ for each $(x, y) \in U$ and for each $z \in \mathbb{R}$.

Fuzzy arithmetic carries the usual arithmetic operations on $\mathbb{R}$ to the arithmetic operations on $\mathcal{F}(\mathbb{R})$, although it does not establish the arithmetic operations on $\mathcal{F}(\mathbb{R})$, whenever the arithmetic operations on $\mathbb{R}$ are vaguely defined, or more precisely M-vague arithmetic operations on $\mathbb{R}$ are taken into consideration. In the following definition, the arithmetic operations on $\mathcal{F}(\mathbb{R})$ based on M-vague arithmetic operations on $\mathbb{R}$ will be introduced. For this purpose, for a given L-fuzzy relation $\rho \in L^{(X \times Y) \times Z}$, we first need to introduce the extension of ρ as an ordinary function $\overline{\rho} : L^X \times L^Y \to L^Z$, $(\mu, \nu) \mapsto \overline{\rho}(\mu, \nu)$, defined by

$$\overline{\rho}(\mu, \nu)(x) = \bigvee_{z \in X, y \in Y} \{\mu(z) * \nu(y) * \rho((z, y), x)\}, \quad \forall x \in Z, \forall \mu \in L^X, \forall \nu \in L^Y. \quad (4)$$

Definition 3.3. Let $\widetilde{\Delta}$ be an M-vague arithmetic operation on $\mathbb{R}$ w.r.t. E_U and $E_\mathbb{R}$. The ordinary function $\overline{(\widetilde{\Delta})} : \mathcal{F}(\mathbb{R}) \times \mathcal{F}^\Delta(\mathbb{R}) \to \mathcal{F}(\mathbb{R})$, obtained from $\widetilde{\Delta}$ by (4), i.e. for each $x \in \mathbb{R}$, for each $\mu \in \mathcal{F}(\mathbb{R})$ and $\nu \in \mathcal{F}^\Delta(\mathbb{R})$,

$$[\overline{(\widetilde{\Delta})}(\mu, \nu)](x) = \bigvee_{(z,y) \in U} \{\mu(z) * \nu(y) * \widetilde{\Delta}(z, y, x)\}, \tag{5}$$

is said to be an arithmetic operation on $\mathcal{F}(\mathbb{R})$ w.r.t. the M-vague arithmetic operation $\widetilde{\Delta}$ on $\mathbb{R}$. For $\Delta = +$ (resp., $\Delta = -$, $\Delta = .$ and $\Delta = /$), the arithmetic operation $\overline{(\widetilde{\Delta})}$ on $\mathcal{F}(\mathbb{R})$ w.r.t. the M-vague arithmetic operation $\widetilde{\Delta}$ on $\mathbb{R}$ is called the addition (resp., subtraction, multiplication and division) operation on $\mathcal{F}(\mathbb{R})$ w.r.t. $\widetilde{\Delta}$.

Definition 3.4. Let $\mathbb{R}^\Delta$ denote $\mathbb{R}$ for $\Delta \in \{+, -, .\}$ and $\mathbb{R} - \{0\}$ for $\Delta = /$. Let $\widetilde{\Delta}$ be an M-vague arithmetic operation on $\mathbb{R}$ w.r.t. E_U and $E_\mathbb{R}$. For given $x \in \mathbb{R}$ and $y \in \mathbb{R}^\Delta$, and for $\Delta = +$ (resp., $\Delta = -$, $\Delta = .$ and $\Delta = /$), the fuzzy quantity $\overline{(\widetilde{\Delta})}(\mathbf{1}_{\{x\}}, \mathbf{1}_{\{y\}}) \in \mathcal{F}(\mathbb{R})$, is called the vague addition of x and y (resp., the vague subtraction of y from x, the vague multiplication of x and y, the vague division of x by y) w.r.t. the M-vague addition (resp., subtraction, multiplication and division) operation $\widetilde{\Delta}$.

In Definition 3.4, taking the equality (5) into consideration, it is easily observed that $\overline{(\widetilde{\Delta})}(\mathbf{1}_{\{x\}}, \mathbf{1}_{\{y\}})(u) = \widetilde{\Delta}(x, y, u)$. Therefore, if we define the fuzzy quantity $\widetilde{\Delta}_{(x,y)} : \mathbb{R} \to L$ by $\widetilde{\Delta}_{(x,y)}(u) = \widetilde{\Delta}(x, y, u)$, $\forall x \in \mathbb{R}$, $\forall y \in \mathbb{R}^\Delta$, $\forall u \in \mathbb{R}$, we obviously have $\overline{(\widetilde{\Delta})}(\mathbf{1}_{\{x\}}, \mathbf{1}_{\{y\}}) = \widetilde{\Delta}_{(x,y)}$, $\forall x \in \mathbb{R}$, $\forall y \in \mathbb{R}^\Delta$.

In Definition 3.3, if the M-equivalence relations E_U and $E_\mathbb{R}$ and the M-vague arithmetic operation $\widetilde{\Delta}$ on $\mathbb{R}$ are particularly chosen as the classical $*$-fuzzy equality E_U^N on U, the classical $*$-fuzzy equality $E_\mathbb{R}^N$ on $\mathbb{R}$ and the characteristic function $\mathbf{1}_\Delta$ of Δ, i.e. $\widetilde{\Delta}(x, y, z) = \mathbf{1}_\Delta((x, y), z) = \begin{Bmatrix} 1, & if \ z = x\Delta y \\ 0, & if \ z \neq x\Delta y \end{Bmatrix}$, $\forall (x, y) \in U$, $\forall z \in \mathbb{R}$, respectively, then the equality (5) reduces to (3), i.e. $\overline{(\widetilde{\Delta})} = \overline{(\mathbf{1}_\Delta)} = \overline{\Delta}$, and so the arithmetic operation $\overline{(\widetilde{\Delta})}$ on $\mathcal{F}(\mathbb{R})$ w.r.t. the M-vague arithmetic operation $\mathbf{1}_\Delta$ on $\mathbb{R}$ w.r.t. the classical $*$-fuzzy equality E_U^N on U and the classical $*$-fuzzy equality $E_\mathbb{R}^N$ on $\mathbb{R}$ is exactly the fuzzy arithmetic operation $\overline{\Delta}$ on $\mathcal{F}(\mathbb{R})$. Therefore a fuzzy arithmetic operation on $\mathcal{F}(\mathbb{R})$ is nothing but a particular case of an arithmetic operation on $\mathcal{F}(\mathbb{R})$ w.r.t. an M-vague arithmetic operation on $\mathbb{R}$. In the subsequent results, some interesting properties of arithmetic operations on $\mathcal{F}(\mathbb{R})$ w.r.t. M-vague arithmetic operations on $\mathbb{R}$ are introduced:

Proposition 3.5. For a given L-fuzzy relation $\rho \in L^{(X \times Y) \times Z}$ and for a given M-equivalence relation E on $X \times Y$, for each $x \in X$, $y \in Y$, $z \in Z$, let us define the L-fuzzy relation $\rho_E \in L^{(X \times Y) \times Z}$ by

$$\rho_E((x, y), z) = \bigvee_{x' \in X, y' \in Y} \{E((x, y), (x', y')) * \rho((x', y'), z)\}, \tag{6}$$

Then we have the properties: (i) $\rho \leq \rho_E$, i.e. $\overline{\rho} \leq \overline{(\rho_E)}$, (ii) If ρ holds the condition (F.4) in Definition 2.2, then $\rho = \rho_E$, i.e. $\overline{\rho} = \overline{(\rho_E)}$.

Proposition 3.6. For given M-equivalence relation E on $X \times Y$ and M-equivalence relation ($*$-fuzzy equality) F on Z, let $\rho \in L^{(X \times Y) \times Z}$ be a strong fuzzy function from $X \times Y$ to Z w.r.t. E and F.

(i) The L-fuzzy relation $\rho_E \in L^{(X \times Y) \times Z}$, defined by (6), is a strong fuzzy function from $X \times Y$ to Z w.r.t. E and F such that $ORD(\rho) \subseteq ORD(\rho_E)$ ($ORD(\rho) = ORD(\rho_E)$, i.e. $ord(\rho) = ord(\rho_E)$).

(ii) If $\rho \in L^{(X \times Y) \times Z}$ holds the condition (F.5), or equivalently ρ is a perfect fuzzy function from $X \times Y$ to Z w.r.t. E and F, then $\rho = \rho_E$, i.e. $\overline{\rho} = \overline{(\rho_E)}$.

Proposition 3.7. For a given L-fuzzy relation $\rho \in L^{(X \times Y) \times Z}$ and for given M-equivalence relations E_1 on X, E_2 on Y and F on Z, let $\rho_{G(E_1, E_2)} \in L^{(X \times Y) \times Z}$ be the L-fuzzy relation given by (6). Then

$$\overline{\rho}([\mu]_{E_1}, [\nu]_{E_1}) = \overline{(\rho_{G(E_1, E_2)})}(\mu, \nu), \ \forall \mu \in L^X, \ \forall \nu \in L^Y.$$

Proof. Using the distributivity of $*$ over arbitrary joins, and considering the definition of $\rho_{G(E_1, E_2)}$, the required equality can be easily seen.

Corollary 3.8. For given M-equivalence relations E_1, E_3 on $\mathbb{R}$ and E_2 on $\mathbb{R}^\Delta$, let $\widetilde{\Delta}$ be an M-vague arithmetic operation on $\mathbb{R}$ w.r.t. $G(E_1, E_2)$ on U and E_3 on $\mathbb{R}$. Then the L-fuzzy relation $\widetilde{\Delta}_{G(E_1, E_2)} \in L^{U \times \mathbb{R}}$, obtained from $\widetilde{\Delta}$ by the formula (6), i.e. for each $x \in \mathbb{R}$, $y \in \mathbb{R}^\Delta$ and $z \in \mathbb{R}$,

$$\widetilde{\Delta}_{G(E_1, E_2)}(x, y, z) = \bigvee_{x' \in \mathbb{R}, y' \in \mathbb{R}^\Delta} E_1(x, x') * E_2(y, y') * \widetilde{\Delta}(x', y', z), \quad (7)$$

is an M-vague arithmetic operation on $\mathbb{R}$ w.r.t. $G(E_1, E_2)$ on U and E_3 on $\mathbb{R}$ such that $\Delta \in ORD(\widetilde{\Delta}_{G(E_1, E_2)})$, and holds the equality

$$(\overline{\widetilde{\Delta}})([\mu]_{E_1}, [\nu]_{E_2}) = \overline{(\widetilde{\Delta}_{G(E_1, E_2)})}(\mu, \nu), \ \forall \mu \in \mathcal{F}(\mathbb{R}) \forall \nu \in \mathcal{F}^\Delta(\mathbb{R}). \quad (8)$$

Corollary 3.9. For $\Delta \in \{+, -, ., /\}$ and for given M-equivalence relations E_1, E_3 on $\mathbb{R}$ and E_2 on $\mathbb{R}^\Delta$, let the usual arithmetic operation $\Delta : \mathbb{R} \times \mathbb{R}^\Delta \to \mathbb{R}$ be extensional w.r.t. $G(E_1, E_2)$ and E_3.

(i) The perfect M-vague arithmetic operation $vag(\Delta)$ on $\mathbb{R}$ w.r.t. $G(E_1, E_2)$ and E_3 satisfies the equality $(\overline{\widetilde{\Delta}})([\mu]_{E_1}, [\nu]_{E_2}) = (\overline{\widetilde{\Delta}})(\mu, \nu), \forall \mu \in \mathcal{F}(\mathbb{R}), \forall \nu \in \mathcal{F}^\Delta(\mathbb{R})$.

(ii) The L-fuzzy relation $(1_\Delta)_{G(E_1, E_2)} \in L^{U \times \mathbb{R}}$, obtained from 1_Δ by the equality (7), i.e. for each $x \in \mathbb{R}$, $y \in \mathbb{R}^\Delta$ and $z \in \mathbb{R}$,

$$(1_\Delta)_{G(E_1, E_2)}(x, y, z) = \bigvee_{\substack{x' \in \mathbb{R}, y' \in \mathbb{R}^\Delta \\ z = x' \Delta y'}} E_1(x, x') * E_2(y, y'), \quad (9)$$

is an M-vague arithmetic operation on $\mathbb{R}$ w.r.t. $G(E_1, E_2)$ on U and E_3 on $\mathbb{R}$ such that $\Delta \in ORD((1_\Delta)_{G(E_1, E_2)})$, and holds the equality

$$\overline{\Delta}([\mu]_{E_1}, [\nu]_{E_2}) = \overline{((1_\Delta)_{G(E_1, E_2)})}(\mu, \nu), \ \forall \mu \in \mathcal{F}(\mathbb{R}), \forall \nu \in \mathcal{F}^\Delta(\mathbb{R}). \quad (10)$$

Corollary 3.8 and Corollary 3.9 (i) give us a peculiar property of an arithmetic operation $(\overline{\widetilde{\Delta}})$ on $\mathcal{F}(\mathbb{R})$ w.r.t. an M-vague arithmetic operation $\widetilde{\Delta}$ on $\mathbb{R}$. In both Corollary 3.8 and Corollary 3.9, let us consider the particular L-fuzzy sets $\mu \in \mathcal{F}(\mathbb{R})$, $\nu \in \mathcal{F}^{\Delta}(\mathbb{R})$ defined as the extensional hull $[\ker(\mu)]_{E_1}$ of $\ker(\mu)$ w.r.t. the M-equivalence relation E_1 on $\mathbb{R}$ and the extensional hull $[\ker(\nu)]_{E_2}$ of $\ker(\nu)$ w.r.t. the M-equivalence relation E_2 on $\mathbb{R}^{\Delta}$, respectively. Corollary 3.8 shows that, for a given M-vague arithmetic operation $\tilde{\Delta}$ on $\mathbb{R}$ w.r.t. $G(E_1, E_2)$ on U and E_3 on $\mathbb{R}$, the shapes of the L-fuzzy sets μ and ν do not effect the calculation of the fuzzy quantity $\overline{(\tilde{\Delta})}(\mu, \nu)$, and it is sufficient to know only $\ker(\mu)$ and $\ker(\nu)$. In other words, exploiting the equality (8), $\overline{(\tilde{\Delta})}(\mu, \nu)$ can be simply evaluated in terms of $\ker(\mu)$ and $\ker(\nu)$ by the equality

$$\overline{(\tilde{\Delta})}(\mu, \nu) = \overline{(\tilde{\Delta}_{G(E_1, E_2)})}(1_{\ker(\mu)}, 1_{\ker(\nu)}). \tag{11}$$

As a direct consequence of the equality (11), for a given M-vague arithmetic operation $\tilde{\Delta}$ on $\mathbb{R}$ w.r.t. $G(E_1, E_2)$ on U and E_3 on $\mathbb{R}$, for each $x \in \mathbb{R}$ and $y \in \mathbb{R}^{\Delta}$, considering the extensional hull $[x]_{E_1}$ of x w.r.t. the M-equivalence relation E_1 on $\mathbb{R}$ and the extensional hull $[y]_{E_2}$ of y w.r.t. the M-equivalence relation E_2 on $\mathbb{R}^{\Delta}$, we observe that

$$\overline{(\tilde{\Delta})}([x]_{E_1}, [y]_{E_2}) = [\tilde{\Delta}_{G(E_1, E_2)}]_{(x,y)}. \tag{12}$$

The equality (12) means that the addition (resp., the subtraction, the multiplication and the division) of the fuzzy quantities $[x]_{E_1}$ and $[y]_{E_2}$ w.r.t. the M-vague addition (resp., subtraction, multiplication and division) operation $\tilde{\Delta}$ is nothing but the vague addition (resp., the vague subtraction, the vague multiplication and the vague division) of the real numbers x and y w.r.t. the M-vague addition (resp., subtraction, multiplication and division) operation $\tilde{\Delta}_{G(E_1, E_2)}$. Corollary 3.9 (i) proves that the M-vague arithmetic operation $\tilde{\Delta}_{G(E_1, E_2)}$ on $\mathbb{R}$ defined by (7) can be simply taken as the M-vague arithmetic operation $\tilde{\Delta}$ itself in the equality (8) whenever $\tilde{\Delta}$ is a perfect M-vague arithmetic operation on $\mathbb{R}$ w.r.t. $G(E_1, E_2)$ and E_3, or equivalently $\tilde{\Delta} = vag(\Delta)$. Therefore, if $\tilde{\Delta} = vag(\Delta)$, $\tilde{\Delta}_{G(E_1, E_2)}$ can be replaced by $\tilde{\Delta}$ itself in the equalities (11) and (12). Finally, Corollary 3.9 (ii) establishes a connection between the fuzzy arithmetic operations and the vague arithmetic operations. For given M-equivalence relations E_1, E_3 on $\mathbb{R}$ and E_2 on $\mathbb{R}^{\Delta}$, and for a given usual arithmetic operation $\Delta \in \{+, -, ., /\}$ extensional w.r.t. $G(E_1, E_2)$ and E_3, considering the M-vague arithmetic operation $(1_{\Delta})_{G(E_1, E_2)}$ on $\mathbb{R}$, defined by (9), the equality (10) directly yields that, for each $x \in \mathbb{R}$ and $y \in \mathbb{R}^{\Delta}$, $\overline{\Delta}([x]_{E_1}, [y]_{E_2}) = [(1_{\Delta})_{G(E_1, E_2)}]_{(x,y)}$. This equality shows that, for each $x \in \mathbb{R}$ and $y \in \mathbb{R}^{\Delta}$, the fuzzy addition (resp., the fuzzy subtraction, the fuzzy multiplication and the fuzzy division) of the fuzzy quantities $[x]_{E_1}$ and $[y]_{E_2}$ is exactly same as the vague addition (resp., the vague subtraction, the vague multiplication and the vague division) of the real numbers x and y w.r.t. $(1_{\Delta})_{G(E_1, E_2)}$.

Acknowledgement. This study has been supported by Turkish Academy of Sciences in the framework of the Young Scientist Award Program (MD/TUBA-GEBIP/2002-1-8).

References

1. Bezdek, J.C. and Harris, J.O.: Fuzzy Partitions and Relations: An Axiomatic Basis for Clustering. Fuzzy Sets and Systems 1 (1978) 111-127.
2. Boixader, D. Jacas, J. and Recasens, J.: Fuzzy Equivalence Relations: Advanced Material. In: Dubois, D., Prade., H. (eds.): Fundamentals of Fuzzy Sets. The Handbooks of Fuzzy Sets, Vol. 7. Kluwer Acad. Publishers, Boston (2000) 261-290.
3. Cerruti, U. and Höhle, U.: An Approach to Uncertainity Using Algebras Over a Monoidal Closed Category. Suppl. Rend. Circ. Matem. Palermo Ser. II 12 (1986) 47-63.
4. Demirci, M.: Fuzzy Functions and Their Applications. J. Math. Anal. Appl. 252 (2000) 495-517.
5. Demirci, M.: Fundamentals of M-Vague Algebra and M-Vague Arithmetic Operations. Int. J. Uncertain. Fuzz. 10 (1) (2002) 25-75.
6. Demirci, M.: Foundations of Fuzzy Functions and Vague Algebra Based on Many-Valued Equivalence Relations, Part I: Fuzzy Functions and Their Applications. Int. J. General Systems 32 (2) (2003) 123-155.
7. Demirci, M.: Foundations of Fuzzy Functions and Vague Algebra Based on Many-Valued Equivalence Relations, Part II: Vague Algebraic Notions. Int. J. General Systems 32 (2) (2003) 157-175.
8. Demirci, M.: Foundations of Fuzzy Functions and Vague Algebra Based on Many-Valued Equivalence Relations, Part III: Constructions of Vague Algebraic Notions and Vague Arithmetic Operations. Int. J. General Systems 32 (2) (2003) 177-201.
9. Dubois, D. and Prade, H.: Possibility Theory: An Approach to Computerized Processing of Uncertainty. Plenum Press, New York (1988).
10. Höhle, U.: Quotients with Respect to Similarity Relations. Fuzzy Sets and Systems 27 (1988) 31-44.
11. Höhle, U.: Commutative, Residuated l-monoids. In: Höhle, U., Klement, E.P. (eds.): Non-Classical Logics and Their Applications to Fuzzy Subsets. The Handbooks of Fuzzy Sets, Vol. 32. Kluwer Acad. Publishers, Dordrecht (1995) 53-106.
12. Klawonn, F. and Castro, J.L.: Similarity in Fuzzy Reasoning. Mathware and Soft Computing 2 (1995) 197-228.
13. Klawonn, F.: Fuzzy Points, Fuzzy Relations and Fuzzy Functions. In: Novák, V., Perfilieva, I. (eds.): Discovering the World with Fuzzy Logic, Vol. 57. Physica-Verlag, Heidelberg (2000) 431-453.
14. Menger, K.: Probabilistic Theory of Relations. Proc. Nat. Acad. Sci. USA 37 (1951) 178-180.
15. Novăk, V.: Fuzzy Sets and Their Applications. Adam Hilger, Bristol (1986).
16. Valverde, L.: On the Structure of F-Indistinguishability Operators. Fuzzy Sets and Systems 17 (1985) 313-328.
17. Zadeh, L. A.: Similarity Relations and Fuzzy Orderings. Inform. Sci. 3 (1971) 177-200.

Level-Sets as Decomposition of the Topological Space SpecA

Paavo Kukkurainen

Lappeenranta University of Technology, P.O. Box 20
53851 Lappeenranta, Finland
{paavo.kukkurainen}@lut.fi
http://www.it.lut.fi

Abstract. Let $SpecA$ be the set of prime ideals of an MV-algebra A. Then $SpecA$ is a topological space with the usual spectral topology. All closed sets of $SpecA$ are $W(I) = \{p \in SpecA \mid I \subset p\}$, where I is a subset (often an ideal but not here) of A ([1]). We prove that $SpecA = \bigcup W(a_i) = \bigcup \mathcal{A}_{W(a_i)}$ is a decomposition of closed sets and $W(a_i)$-level sets of a fuzzy set $\mathcal{A}$ for some $a_i \in A$ iff the set $S^* = \{a_i^* \mid a_i^* \in A\}$ is an orthogonal subset of A i.e. $a_i^* \wedge a_j^* = 0$, $i \neq j$, $a_k^* \neq 0$. If A is complete there is a decomposition $SpecA = \bigcup \mathcal{A}_{W((a_i))}$, where every (a_i) is a principal ideal of A generated by a_i.

Keywords : MV-algebra, ideal, topological space, decomposition, fuzzy level-set.

1 Introduction

L. P. Belluce and S. Sessa considered in ([1]) $SpecA$ (see Abstract) as a topological space with open sets $V(I) = \{p \in SpecA \mid I \not\subset p\}$, where I is an ideal or a subset of an MV-algebra. In this paper we assume that I is a subset. So all closed sets consist of sets $W(I) = \{p \in SpecA \mid I \subset p\}$. In ([1]), it is proved that if $SpecA = \bigcup T_i$ is a decomposition of closed sets, then $T_i = \bigcup(T_M \mid M \in MaxA \cap T_i)$ for each index i, where $MaxA$ is the set of maximal ideals of A and $T_M = \{p \in SpecA \mid p \subset M\}$. In fact, $SpecA = \bigcup(T_M \mid M \in MaxA)$. We construct a decomposition of closed sets $SpecA = \bigcup W(a_i)$, $a_i \in A$ and show that $W(a_i) = \mathcal{A}_{W(a_i)}$ and $W((a_i)) = \mathcal{A}_{W((a_i))}$(see Abstract). In this connection it is utilized the work of B. Seselja and A. Tepavcevic ([4]).

2 Preliminaries

A poset is a partially ordered set $P = (P, \leq)$ with a binary relation $\leq$. A lattice is a poset any two of whose elements x and y have a least upper bound $x \vee y$ and a greatest lower bound $x \wedge y$. A lattice order $\leq$ is defined in the following way: $x \leq y$ if and only if $x \vee y = y$ if and only if $x \wedge y = x$ for any elements x and y of a lattice L. A lattice A is complete when each of its subsets has the greatest lower bound and the least upper bound in A ([2]).

T. Bilgiç et al. (Eds.): IFSA 2003, LNAI 2715, pp. 167–171, 2003.

The following concepts are found, for example, from ([1]) or ([3]). Let A be a nonempty set with binary operations $+$, $\cdot$ and an unary operation *. Assume that A contains two distinct elements 0 and 1 and the following conditions are valid: $(A, +, 0)$, $(A, \cdot, 1)$ are commutative monoids with identity 0 and 1, respectively, $(x+y)^* = x^* \cdot y^*$, $(x \cdot y)^* = x^* + y^*$, $x^{**} = x$, $0^* = 1$ and $x + x^* \cdot y = y + y^* \cdot x$. Then a system $A = (A, +, \cdot, ^*, 0, 1)$ is called an MV-algebra. Define $x \vee y = x + x^* \cdot y$ and $x \wedge y = (x^* \vee y^*)^*$ on A. Then the MV-algebra $A = (A, +, \cdot, ^*, 0, 1)$ induces a distributive lattice $(A, \vee, \wedge, 0, 1)$ with the least element 0 and the greatest element 1. An ideal of the MV-algebra A is a subset I of A which satisfies the following conditions: (i) $0 \in I$, (ii) $x \in I$, $y \in A$ and $y \leq x$ implies $y \in I$, and (iii) $x + y \in I$ for every x, $y \in I$. An ideal of a lattice L with 0 and 1 is a subset I of L which satisfies the conditions (i) and (ii), and the condition: $x \vee y \in I$ for every x, $y \in I$. Moreover, by ([2]), a nonempty subset $I \subset L$ is an ideal of L exactly when $x \vee y \in I$ if and only if $x \in I$ and $y \in I$. Every ideal of the MV-algebra is also an ideal of the induced lattice but not conversely. An ideal I of the MV-algebra A (the lattice L) is proper if $I \neq A$ $(I \neq L)$, and prime if it is proper and satisfies the condition: $x \wedge y \in I$ implies $x \in I$ or $y \in I$. Moreover, an MV-ideal or a lattice (with 0, 1) ideal I is proper iff $1 \notin I$.

Next fuzzy concepts are represented in ([4]): Let A be a nonempty set and $P = (P, \leq)$ a poset (a partially ordered set). Then a poset-valued mapping $\mathcal{A} : A \longrightarrow P$ is a P-(fuzzy)set on A. For every $p \in P$, $\mathcal{A}_p = \{x \in A \mid \mathcal{A}(x) \geq p\}$ is a p-level set (shortly a level-set) of $\mathcal{A}$.

3   Closed Sets and Level-Sets

The following propositon is represented by B. Seselja and A. Tepavcevic ([4]):

Proposition 3.1 *Let A be a nonempty set, and P a family of its subsets i.e. $P \subset \mathcal{P}(A)$, such that*

(i) $\bigcup P = A$,
(ii) *for every $x \in A$, $\bigcap(p \in P \mid x \in p) \in P$.*

Let $\mathcal{A} : A \longrightarrow P$ be defined with

$$\mathcal{A}(x) = \bigcap(p \in P \mid x \in p). \tag{1}$$

Then, $\mathcal{A}$ is a P-set, where $(P, \leq)$ is a poset ordered under $p \leq q$ if and only if $q \subset p$ $(p, q \in P)$, and for every $p \in P$,

$$p = \mathcal{A}_p. \tag{2}$$

Let Γ be any subset of $\mathcal{P}(A)$ such that Γ contains $\{0\}$. Because an MV-ideal is prime iff it is a prime ideal of the induced lattice ([3]), and in Chapter 4, we apply the next Proposition 3.2 only to prime MV-ideals, it is used the notation $SpecA$ meaninig also the set of prime ideals of a lattice A with 0 and 1 although A is not necessary an MV-algebra.

Proposition 3.2 *Let A be a lattice with the least element 0 and the greatest element 1, and $P = (W(I) \mid I \in \Gamma) \subset \mathcal{P}(SpecA)$. Then for every $p \in SpecA$,*

$$\bigcap(W(I) \in P \mid p \in W(I)) = W(\bigcup I) \in P. \tag{3}$$

Let $\mathcal{A} : SpecA \longrightarrow P$ be defined with

$$\mathcal{A}(p) = \bigcap(W(I) \in P \mid p \in W(I)). \tag{4}$$

Then $\mathcal{A}$ is a P-set, where $(P, \leq)$ is a poset ordered under $W(I_1) \leq W(I_2)$ if and only if $W(I_2) \subset W(I_1)$, and for every $W(I) \in P$,

$$W(I) = \mathcal{A}_{W(I)}. \tag{5}$$

Proof. The proof is based on Proposition 3.1. For any $p \in SpecA$ we consider the sets $I \in \Gamma$ such that $p \in W(I)$. Always $\bigcup W(I) \subset SpecA$. Let $q \in SpecA$. Since $\{0\} \subset q$ (i.e. $0 \in q$) for every $q \in SpecA$, then $q \in W(0) \subset \bigcup W(I)$. Therefore $\bigcup W(I) = SpecA$. Next we show that $\bigcap(W(I) \in P \mid p \in W(I)) = W(\bigcup I) \in P$. We conclude $q \in \bigcap W(I)$ iff $q \in W(I)$ for every I iff $I \subset q$ for every I iff $\bigcup I \subset q$ iff $q \in W(\bigcup I) = \{q \in SpecA \mid \bigcup I \subset q\}$.

The following proposition is represented by M. H. Stone ([5]). However, notations and formulation in this paper are different from the original one.

Proposition 3.3 *Let A be a distributive lattice. If I and J are distinct ideals of A, then there is a prime ideal p of A which contains I but not J.*

Let $b \in A$. We recall that a principal ideal (b) of a lattice A is an ideal of A generated by b, $(b) = \{x \in A \mid x \leq b\}$ ([3]).

Proposition 3.4 *Let (b) be a principal ideal of a lattice A generated by b. Then $(b) \subset I$ if and only if $b \in I$ for any ideal I of A.*

Proof. Let $(b) \subset I$. Since $b \leq b$, $b \in (b) \subset I$. Let $b \in I$ and $x \in (b)$. Since I is an ideal and $x \leq b$, we conclude that $x \in I$. Therefore $(b) \subset I$.

Proposition 3.5 *In any distributive lattice the following conditions are equivalent:*

(i) $W((b)) = W((c))$
(ii) $(b) = (c)$
(iii) $b = c$
(iv) $W(b) = W(c)$

Proof. Let $(b) \neq (c)$. By Proposition 3.3, there is a prime ideal p such that $(b) \subset p$ and $(c) \not\subset p$ which implies $p \in W((b))$ and $p \notin W((c))$. Hence $W((b)) \neq W((c))$ and (i) implies (ii). Let $(b) = (c)$. Then $b \in (b) = (c)$ and so $b \leq c$. Also $c \leq b$. Therefore $b = c$ and (ii) implies (iii). Let $b = c$. We obtain $p \in W(b)$ iff $c = b \in p$ iff $p \in W(c)$, and (iii) implies (iv). Finally, we show that $W((b)) = W(b)$. By Proposition 3.4, we conclude $p \in W((b))$ iff $(b) \subset p$ iff $b \in p$ iff $p \in W(b)$.

We still apply Proposition 3.2 in the case where A is a complete MV-algebra (a complete lattice).

Proposition 3.6 *Let A be a complete lattice with the least element 0 and the greatest element 1, and $P = (W((b)) \mid b \in A)$. Then for every $p \in SpecA$,*

$$\bigcap(W((b)) \in P \mid p \in W((b))) = W(\bigcup(b)) \in P. \tag{6}$$

Let $\mathcal{A} : SpecA \longrightarrow P$ be defined with

$$\mathcal{A}(p) = \bigcap(W((b)) \in P \mid p \in W((b)). \tag{7}$$

Then $\mathcal{A}$ is a P-set, where $(P, \leq)$ is a poset ordered under $W((b_1)) \leq W((b_2))$ if and only if $W((b_2)) \subset W((b_1))$, and for every $W((b)) \in P$,

$$W((b)) = \mathcal{A}_{W((b))}. \tag{8}$$

Proof. Because A is complete there exits supremum $\bigvee b \in A$, $b \in A$. First we show that $\bigcup(b)$, $b \in A$, is a principal ideal of A generated by $\bigvee b$. There is $b_0 \in A$ such that $x \in \bigcup(b)$ iff $x \in (b_0)$ iff $x \leq b_0$. Since $x \leq b_0 \leq \bigvee b$ for any $b_0 \in A$, it follows that $x \in (\bigvee b)$. Therefore $\bigcup(b) \subset (\bigvee b)$. Conversely, since $\bigvee b \in A$, $(\bigvee b) \subset \bigcup(b)$. Hence $\bigcup(b) = (\bigvee b)$ and $W(\bigcup(b)) = W((\bigvee b)) \in P$. Next we state that 0 is included in the set of the principal ideals (b), since $(0) = \{0\}$. If every I consists of a single element (b), $b \in A$ in Proposition 3.2, we have $W((b)) = \mathcal{A}_{W((b))}$.

4   Level Sets as Decomposition

The following definitions are represented in ([1]):

Let A be an MV-algebra. A nonempty set $S \subset A$ is orthogonal if $0 \notin S$ and $x, y \in S$, $x \neq y$ implies $x \wedge y = 0$.

A decomposition of a topological space T is an union $T = \bigcup T_i$, where every T_i is a nonempty subspace of T and $T_i \bigcap T_j = \emptyset$, $i \neq j$.

Proposition 4.1 *Let A be an MV-algebra. Then*

(i) *The sets $W(a_i)$, $a_i \in A$, $a_i \neq 1$ constitute a decomposition of a topological space $SpecA$ if and only if the set $S^* = \{a_i^* \mid a_i^* \in A\}$ is an orthogonal subset of A.*

(ii) *If A is complete then the sets $W((a_i))$, $a_i \in A$, $a_i \neq 1$ constitute a decomposition of a topological space $SpecA$ if and only if the set $S^* = \{a_i^* \mid a_i^* \in A\}$ is an orthogonal subset of A.*

Proof. As in the proof of Proposition 3.2 we can show that $\bigcup W(a_i) = SpecA$ $(a_i \neq 1)$. Further, for any (prime) ideal p we conclude $p \in W(a_i \vee a_j)$ iff $a_i \vee a_j \in p$ iff $a_i \in p$ and $a_j \in p$ iff $p \in W(a_i) \bigcap W(a_j)$. Let $i \neq j$. Since p is proper as a prime ideal, $1 \notin p$, and we obtain $W(a_i \vee a_j) = W(a_i) \bigcap W(a_j) = \emptyset =$

$\{p \in SpecA \mid 1 \in p\} = W(1)$. By Proposition 3.5, $a_i \vee a_j = 1$ i.e. $a_i^* \wedge a_j^* = 0$. Moreover, $a_k^* \neq 0$ for any k, since $a_k \neq 1$. The assertion (i) is proved. At the end of the proof of Proposition 3.5 it is proved that $W((a_i)) = W(a_i)$. So we first state as in the proof of Proposition 3.2 that $\bigcup W((a_i)) = SpecA$ and then $W(1) = \emptyset = W((a_i)) \bigcap W((a_j)) = W(a_i) \bigcap W(a_j) = W(a_i \vee a_j)$. By Proposition 3.5, $a_i \vee a_j = 1$ iff $a_i^* \wedge a_j^* = 0$. Also the assertion (ii) is proved.

Combining Proposition 3.2 (in the case where every I consists of a single element a_i) and Propositions 3.6 and 4.1 we summarize:

Corollary 4.2 *The following assertions are equivalent:*

(i) $SpecA = \bigcup W(a_i) = \bigcup \mathcal{A}_{W(a_i)}$ *is a decomposition of a topological space SpecA.*

(ii) *If A is complete, $SpecA = \bigcup W((a_i)) = \bigcup \mathcal{A}_{W((a_i))}$ is a decomposition of a topological space SpecA.*

(iii) $S^* = \{a_i^* \mid a_i^* \in A\}$ *is an orthogonal subset of A.*

References

1. L.P. Belluce and S. Sessa, Orthogonal decompositions of MV-spaces. Mathware and Soft Computing vol 4 (1997), 5-22.
2. G. Birkhoff, Lattice Theory, volume XXV. AMS, Province, Rhode Island, third edition, 1995. Eight printing.
3. R. O. Cignoli and I. M. L. D'Ottaviano and D. Mundici, Algebraic Foundations of Many-valued Reasoning, volume VII, Kluwer Academic Publishers, Dordrecht, The Netherlands, 2000.
4. B. Seselja and A. Tepavcevic, On a construction of codes by P-fuzzy sets, Zb. Rad. Prirod.-Mat. Fak. Ser. Mat. vol 20 (1990), 71-80.
5. M. H. Stone, Topological representations of distributive lattices and Brouwerian logics, Casopis pro pestovani matematiky a fysiky, vol 67 (1937-1938), 1-25.

Axiomatization of Any Residuated Fuzzy Logic Defined by a Continuous T-norm

Francesc Esteva[1], Lluis Godo[1], and Franco Montagna[2]

[1] IIIA - CSIC, 08193 Bellaterra, Spain
{esteva,godo}@iiia.csic.es
[2] Dept. Mathematics, University of Siena, 53100 Siena, Italy
montagna@unisi.it

Abstract. In this paper we axiomatize the subvarieties of the variety of BL-algebras generated by single BL-chains on [0, 1]. From a logical point of view, this corresponds to find the axiomatization of every residuated many-valued calculus defined by a continuous t-norm and its residuum.

1 Introduction

There is a common agreement in recent and important monographs [8, 13, 7] on considering the core of fuzzy logic in narrow sense [14] as the family of residuated many-valued logical calculi with truth values on the real unit interval [0, 1], and with min, max, a t-norm $\star$ and its residuum $\rightarrow$ as basic truth functions (interpreting the lattice meet and joint connectives (additive "and" and "or"), and a strong conjunction (multiplicative "and") and its adjoint implication, respectively).

In [8] Hájek introduced the so-called Basic Fuzzy Logic, BL for short, together with the corresponding variety of BL-algebras, to cope with the 1-tautologies common to all many-valued calculi in [0, 1] defined by a continuous t-norm and its residuum. The claim that BL would be the logic of continuous t-norms was proved soon after [9, 4]. BL is a common sublogic of three welknown fuzzy logics: Lukasiewicz's infinitely-valued logic, Gödel's infinitely-valued logic and Product logic, corresponding to the three basic t-norms, i.e. Lukasiewicz, minimum and product t-norms. Hájek also proved in [8] that there exists a one-to-one correspondence between subvarieties of BL-algebras and axiomatic extensions of the BL logic, through a natural translation between algebraic equations and logical axioms.

Some subvarieties of **BL** are already well-known, in particular, the subvarieties of the three basic varieties **L**, **G** and **Π** (the algebraic counterparts of Lukasiewicz, Gödel and Product logics), which are fully described and equationally characterized in the literature. Another example is the subvariety **SBL** of pseudo-complemented BL-algebras, which is studied in [5]. The corresponding logic, also noted SBL, was proved to be the logic of the pseudo-complemented continuous t-norms, i.e. t-norms whose associated negation $\neg x = x \rightarrow 0$ verifies $\min(x, \neg x) = 0$, or equivalently $\neg$ is Gödel negation. But so far, the most

T. Bilgiç et al. (Eds.): IFSA 2003, LNAI 2715, pp. 172–179, 2003.
© Springer-Verlag Berlin Heidelberg 2003

general study of subvarieties of **BL** is to be found in the paper by Aglianó and Montagna [1]. Basic results of this paper are the decomposition of BL-chains as ordinal sums of Wajsberg hoops (algebraic structures already studied by Blok and Ferreirim in [3]), the characterization of the *generic* BL-chains (chains that generate the full variety **BL**) and the equational definition of the subvarieties generated by BL-chains that are finite ordinal sums of Wajsberg hoops.

The goal of the present paper is, for any continuous t-norm $\star$, the study and axiomatic definition of the residuated fuzzy logic, extension of BL, which is complete with respect to the BL-chain over [0,1] defined by the t-norm $\star$ and its residuum, denoted $[0,1]_\star$. Relevant related work in this direction are Hániková's papers [10, 11]. In this paper we attack the problem from an algebraic point of view, that is, our goal is the equational characterizations of the subvarieties $Var(\star)$ of **BL** generated by each BL-chain $[0,1]_\star$. In more details, after providing some background in the next section, we show in Section 3 that each standard BL-chain $[0,1]_\star$ has a canonical form generating the same subvariety. This is used in Section 4 to show that any subvariety generated by a standard BL-chains is *finitely axiomatizable*. Moreover we provide an algorithm to find the equations. From a logical point of view, this means that we provide, for each continuous t-norm, an effective method to find the axiomatic extension of BL defining the logic of the given continuous t-norm. Proofs are not included here but they will be included in the long version of this paper [6].

2   Preliminaries

We suppose that the notions of continuous t-norm, its residuum and their basic properties are known to the reader, as well as the notion of ordinal sum of t-norms and the decomposition of a continuous t-norm as an ordinal sum of components isomorphic to Lukasiewicz, Minimum or Product t-norms (from Mostert and Shields theorem) (see e.g. [12]).

A BL-algebra is an algebraic structure $\mathbf{L} = (L, \vee, \wedge, \star, \rightarrow, 0, 1)$ such that the following conditions are satisfied:

(BL1) $(L, \vee, \wedge, 0, 1)$ is a lattice with maximum 1 and minimum 0;
(BL2) $(L, \star, 1)$ is a commutative semigroup with unit 1;
(BL3) $\star$ and $\rightarrow$ form and adjoint pair, i.e. they satisfy $z \leq (x \rightarrow y)$ iff $x \star z \leq y$, for all x, y, z;
(BL4) $x \wedge y = x \star (x \rightarrow y)$;
(BL5) $(x \rightarrow y) \vee (y \rightarrow x) = 1$.

Hájek proved ([8, Lemma 2.3.16]) that each BL-algebra is a subdirect product of a set of linearly ordered BL-algebras, called BL-chains from now on. Hence, each subvariety of **BL** is characterized by the family of BL-chains it contains. The structure of BL-chains has been studied in [9] and in [4], where a representation theorem proves that each saturated BL-chain is an ordinal sum of MV, Gödel and Product chains, in an analogous way to Mostert and Shields's theorem for continuous t-norms.

BL-chains on [0,1], i.e. algebraic structures $[0,1]_\star = ([0,1], \max, \min, \star, \rightarrow, 0, 1)$ where $\star$ is a continuous t-norm and $\rightarrow$ is its residuum, are known as *standard* BL-algebras. More generally, in this paper a BL-chain $\mathcal{A}$ will be called *t-norm algebra* if $\mathcal{A}$ is isomorphic to $[0,1]_\star$ for some continuous t-norm $\star$. The class of standard BL-chains (or t-norm algebras) generates the whole variety **BL** [9, 4]. We will denote by $\mathcal{G}, \mathcal{L}$ and Π the standard BL-chains induced by the Minimum, Lukasiewicz and Product t-norms respectively.

On the other hand Aglianó and Montagna give in [1] another decomposition of BL-chains as ordinal sums of *Wajsberg hoops*. Basically, hoops[1] are also residuated structures $(H, \star, \rightarrow, 1)$ satisfying (BL4)[2] and possibly without bottom. A hoop is Wajsberg if it satisfies the equation $(x \rightarrow y) \rightarrow y = (y \rightarrow x) \rightarrow x$. A hoop is *cancellative* if $x \star z \le y \star z$ implies $x \le y$ for all x, y, z. Cancellative hoops are also Wajsberg. Ordinal sums of hoops differ from the ordinal sum of BL-chains in the fact that the top elements of all hoop components are identified with the top of the ordinal sum. In the decomposition of a standard BL-chain as an ordinal sum of Wajsberg hoops, there is always a first component, which is necessarily bounded, and any component appearing in the decomposition is (isomorphic to) one of following hoops:

2, the hoop defined on the set of two idempotent elements $\{0, 1\}$, coinciding with the 2-element Boolean algebra.
$\mathcal{L}$, the hoop defined on [0,1] by Lukasiewicz's t-norm and its residuum, and coinciding with the corresponding standard BL-algebra. (This is obviously a bounded hoop.)
$\mathcal{C}$, the cancellative hoop defined on the semi-open interval $(0, 1]$ by the product t-norm and its residuum.

As a matter of example, let $\star$ be a continuous t-norm whose decomposition as ordinal sum of BL-chains is (isomorphic to) $\Pi \oplus \mathcal{G} \oplus \mathcal{L}$. Then the ordinal sum decomposition of $\star$ in terms of Wajsberg hoops is the following[3]: $(\mathbf{2} \oplus \mathcal{C}) \oplus (\bigoplus_{x \in [0,1]} \mathbf{2}) \oplus \mathcal{L}$. Observe that, as hoops, a product BL-chain component Π is isomorphic to $\mathbf{2} \oplus \mathcal{C}$, a Gödel component $\mathcal{G}$ is isomorphic to a sum of continuum many copies of **2** while the $\mathcal{L}$ component remains unchanged.

The subclass of BL-chains which are *finite* ordinal sums of Wajsberg hoops **2**, $\mathcal{C}$ and $\mathcal{L}$ will be denoted by *Fin*.

Definition 1. *For a continuous t-norm $\star$, $Fin(\star)$ denotes the set of all finite ordinal sums $\mathcal{W}_0 \oplus \ldots \oplus \mathcal{W}_n$ of Wajsberg hoops such that the following conditions hold:*

- *Each $\mathcal{W}_i$ is isomorphic either to **2**, or to $\mathcal{C}$ or to $\mathcal{L}$.*
- *$\mathcal{W}_0$ is either **2** or $\mathcal{L}$.*

---

[1] The interested reader is referred to [3] for details.
[2] Where the lattice meet $\wedge$ is definable from the lattice ordering defined by $x \le y$ iff $x \rightarrow y = 1$.
[3] We will use the same symbol $\oplus$ for the ordinal sum of Wajsberg hoops and of BL-chains.

- *There are components $\mathcal{A}_0 < \ldots < \mathcal{A}_n$ of $\star$ such that $\mathcal{A}_0$ is the first component of $\star$, and for every i, if $\mathcal{W}_i$ is isomorphic to $\mathcal{L}$, then $\mathcal{A}_i$ is isomorphic to $\mathcal{L}$, if $\mathcal{W}_i$ is isomorphic to $\mathcal{C}$ then $\mathcal{A}_i$ is isomorphic either to $\mathcal{C}$ or to $\mathcal{L}$, and if $\mathcal{W}_i$ is isomorphic to $\mathbf{2}$ then $\mathcal{A}_i$ is isomorphic either to $\mathbf{2}$ or to $\mathcal{L}$.*

This definition plays an important role in the paper. Notice that, for any continuous t-norm $\star$, the set $Fin(\star)$ fully determines the variety generated by $\star$ in the following sense.

Theorem 1. *For all continuous t-norms $\star$ and $\circ$ we have:*

(1) $Var(\star) = Var(Fin(\star))$,
(2) $Var(\star) \subseteq Var(\circ)$ iff $Fin(\star) \subseteq Fin(\circ)$.

3   Canonical T-norm Algebras

From Theorem 1, it is possible to show that each standard algebra $\star$ admits a *canonical* form (which is again another standard algebra) $\star'$ such that:

(i) $Fin(\star) = Fin(\star')$, hence it generates the same subvariety, and
(ii) it can be expressed as a *finite* ordinal sum whose components may be either basic components (i.e. $\mathbf{2}, \mathcal{C}, \mathcal{L}$ components) or complex components consisting in turn of ordinal sums of infinite copies of $\mathbf{2}$, Π and $\mathcal{L}$ components.

In the sequel, Π^∞ will denote the ordinal sum of ω copies of Π, and $\mathcal{L}^\infty$ the ordinal sum of ω copies of $\mathcal{L}$. (Recall that $\mathcal{G}$ denotes in fact the ordinal sum of continuously many copies of $\mathbf{2}$.)

In [1] it is proved that any BL-chain whose decomposition as ordinal sum of Wajsberg hoops begins with $\mathcal{L}$ and contains infinitely many $\mathcal{L}$ components is *generic*, i.e. it generates the whole variety **BL**. We will take $\mathcal{L}^\infty$ as the *canonical* form for all generic t-norm algebras. Analogously, any BL-chain whose decomposition as ordinal sum of Wajsberg hoops begins with $\mathbf{2}$ and contains infinitely many components $\mathcal{L}$ generates the whole variety **SBL**. The simplest t-norm algebras generating **SBL** are $\mathcal{G} \oplus \mathcal{L}^\infty$ and $\Pi \oplus \mathcal{L}^\infty$. We will take the latter as the *canonical* form for all those generating **SBL**.

From these cases, and the fact that two BL-chains which are finite ordinal sums of components $\mathcal{L}$, $\mathbf{2}$ and $\mathcal{C}$ define the same variety iff they have the same components, we can define what canonical continuous t-norms are.

Definition 2. *A t-norm algebra $\star$ is said to be* canonical *iff either $\star = \mathcal{L}^\infty$, or $\star = \Pi \oplus \mathcal{L}^\infty$, or $\star$ is a finite ordinal sum of components of the form $\mathcal{L}$, Π, $\mathcal{G}$ and Π^∞, where each component $\mathcal{G}$ is not preceded and not followed by another $\mathcal{G}$, and each component Π^∞ is not preceded and not followed by $\mathcal{G}$, or by Π or by another Π^∞.*

The main result of this section is the following theorem providing the canonical representative for each continuous t-norm:

Theorem 2. *Each t-norm algebra has a unique corresponding canonical form, which is given by the following procedure:*

(i) *if its (hoop or BL-chain) decomposition begins with $\mathcal{L}$ and has infinitely many components $\mathcal{L}$, the canonical form is $\mathcal{L}^\infty$.*

(ii) *if its hoop decomposition begins with $\mathbf{2}$ and has infinitely many components $\mathcal{L}$, the canonical form is $\Pi \oplus \mathcal{L}^\infty$.*

(iii) *otherwise, if the t-norm has finitely many components $\mathcal{L}$, let its BL-chain decomposition be $A_1 \oplus \mathcal{L}_1 \oplus \ldots \oplus \mathcal{L}_n \oplus A_n$. (Notice that in each non-empty A_i there can only be $\mathcal{G}$ and Π components, and moreover there are no consecutive $\mathcal{G}$ components.) Then the canonical form is the ordinal sum resulting from replacing only those A_i which contain infinitely-many components Π by Π^∞.*

As an example, by applying the procedure described in the last theorem to the t-norm algebra $\star = \mathcal{L} \oplus \Pi^\infty \oplus \mathcal{G} \oplus \Pi^\infty \oplus \mathcal{L}$, one obtains $\star' = \mathcal{L} \oplus \Pi^\infty \oplus \mathcal{L}$ as the canonical form of $\star$.

As a consequence of this last theorem, any canonical t-norm can be coded as a finite string, hence there are only countably many different canonical t-norms, hence there are only countably many different varieties generated by single continuous t-norms [4]. Therefore, the set of different residuated fuzzy logics which are standard complete with respect to a t-norm BL-chain (one logic for each continuous t-norm) is countable as well.

4   Axiomatizations of Varieties Generated by a Standard BL-algebra

We start by introducing some definitions and previous results. Consider the following terms:

$$e_{\mathcal{L}}(x) : (x \to x^2) \vee ((x \to x^3) \to x^2)$$
$$e_{\mathcal{C}}(x) : x \to x^2$$
$$e_{\mathbf{2}}(x) : (x \to x^3) \to x^2$$

where expressions of the form x^m stand for abbreviations of $x \star \ldots \star x$, m times. Notice that:

– the equation $e_{\mathcal{L}}(x) = 1$ is valid in $\mathbf{2}$ and in cancellative hoops and it is not valid in any MV chain with more than two elements.

– the equation $e_{\mathcal{C}}(x) = 1$ is valid in $\mathbf{2}$ and it is not valid either in any MV chain with more than two elements or in non-trivial cancellative hoop.

– the equation $e_{\mathbf{2}}(x) = 1$ is valid in any cancellative hoop and it is not valid either in $\mathbf{2}$ or in any MV chain.

This leads us to define an equation associated to each BL-chain of *Fin*. Namely, let $\mathcal{A} = \bigoplus_{i=0,n} \mathcal{A}_i \in Fin$, and for each $i = 0, \ldots, n$ let $e_i^{\mathcal{A}}$ be $e_{\mathcal{L}}$ if $\mathcal{A}_i$ is an MV algebra with more than two elements, be $e_{\mathcal{C}}$ if $\mathcal{A}_i$ is a non-trivial cancellative hoop, and be $e_{\mathbf{2}}$ if $\mathcal{A}_i$ is $\mathbf{2}$. Then we define the following equation:

---

[4] This result can also be derived from results of [11].

$$(e_{\mathcal{A}}): \quad [(\bigwedge_{i=0...n-1}((x_{i+1} \to x_i) \to x_i)) \,\&\, (\neg\neg x_0 \to x_0) \to (\bigvee_{i=0...n} x_i)] \vee (\bigvee_{i=0...n} e_i^{\mathcal{A}}(x_i)) = 1 .$$

By construction, the equation $(e_{\mathcal{A}})$ is not valid in $\mathcal{A}$. Moreover, we have a stronger result.

Lemma 1. *Let $\star$ be a continuous t-norm algebra, and let $\mathcal{A} \in Fin$. Then the equation $(e_{\mathcal{A}})$ is valid in all $\mathcal{B} \in Fin(\star)$ iff $\mathcal{A} \notin Fin(\star)$.*

The key guiding idea is that a t-norm algebra $\star$ can be characterized by the equations corresponding to those BL-chains in Fin which are not in $Fin(\star)$, and more important, the fact that in any case there are only finitely-many different such equations.

To this end, for each t-norm algebra $\star$, $\star^{\perp}$ will denote the set $Fin \setminus Fin(\star)$, and $Min(\star^{\perp})$ the set of minimal elements of $\star^{\perp}$ with respect to the relation $\preceq$ in Fin defined as: $\mathcal{D} \preceq \mathcal{E}$ iff $\mathcal{D} \in \mathbf{ISP}_u(\mathcal{E})^5$. When $\mathcal{D} \preceq \mathcal{E}$ we say that the algebra $\mathcal{D}$ embeds into the algebra $\mathcal{E}$. For instance we have $\mathbf{2} \preceq \mathcal{L}$ while $\mathbf{2} \oplus \mathcal{C} \npreceq \mathcal{L}$. Some interesting examples of $Min(\star^{\perp})$ are the following ones:

1. If $\star = \mathcal{L}^{\infty}$ then $\star^{\perp} = Min(\star^{\perp}) = \emptyset$, and if $\star = \Pi \oplus \mathcal{L}^{\infty}$ then $Min(\star^{\perp}) = \{\mathcal{L}\}$.
2. If $\star = \Pi^{\infty}$ then $Min(\star^{\perp}) = \{\mathcal{L}, \mathbf{2} \oplus \mathcal{L}\}$.
3. $Min(\mathcal{L}^{\perp}) = \{\mathbf{2} \oplus \mathbf{2}, \mathbf{2} \oplus \mathcal{C}\}$, $Min(\mathcal{G}^{\perp}) = \{\mathcal{L}, \mathbf{2} \oplus \mathcal{C}\}$, $Min(\Pi^{\perp}) = \{\mathcal{L}, \mathbf{2} \oplus \mathbf{2}, \mathbf{2} \oplus \mathcal{C} \oplus \mathcal{C}\}$.

The above claims are formally written down in the following main theorem.

Theorem 3. *Let $\star$ be any standard BL-chain. Then:*

(i) $Var(\star)$ is axiomatized by $AX(\star) = \{e_{\mathcal{B}} : \mathcal{B} \in \star^{\perp}\}$.
(ii) $Var(\star)$ is axiomatized by $AX_0(\star) = \{e_{\mathcal{B}} : \mathcal{B} \in Min(\star^{\perp})\}$.
(iii) $Min(\star^{\perp})$ is finite

Thus, to find an axiomatization of the variety $Var(\star)$ generated by a standard BL-algebra $\star$ it is enough to find the set $Min(\star^{\perp})$ of minimal elements of $\star^{\perp}$. The algorithm below describes a procedure to find such minimal elements, once we assume $\star$ is already in its canonical form. The algorithm makes use of the notion of *maximal embeddability* of one algebra into another one.

Definition 3. *Let $\mathcal{A}$ be a canonical t-norm BL-chain, with $\mathcal{A} = \mathcal{A}_1 \oplus ... \oplus \mathcal{A}_n$, where $\mathcal{A}_i \in \{\mathbf{2}, \mathcal{C}, \mathcal{G}, \Pi^{\infty}, \mathcal{L}\}$, for $i = 1, ..., n$. Let $\mathcal{B} \in Fin$. Then:*

(i) we say that $\mathcal{B}$ maximally embeds in $\mathcal{A}$, written $\mathcal{B} \hookrightarrow \mathcal{A}$, if $\mathcal{B} \in Fin(\mathcal{A}_1 \oplus ... \oplus \mathcal{A}_n)$ but $\mathcal{B} \notin Fin(\mathcal{A}_1 \oplus ... \oplus \mathcal{A}_{n-1})$, in the case $n > 1$. In the case $n = 1$, we say that $\mathcal{B}$ maximally embeds in $\mathcal{A}$ if simply $\mathcal{B} \in Fin(\mathcal{A})$.
(ii) the degree of "maximal embeddability" of $\mathcal{B}$ in $\mathcal{A}$ is defined as follows:

$$g(\mathcal{B} \hookrightarrow \mathcal{A}) = \begin{cases} k, & \text{if } \mathcal{B} \in Fin(\mathcal{A}) \text{ and } \mathcal{B} \hookrightarrow \mathcal{A}_1 \oplus ... \oplus \mathcal{A}_k, \text{ with } k \leq n \\ n+1, & \text{if } \mathcal{B} \notin Fin(\mathcal{A}) \end{cases}$$

---

[5] Symbols $\mathbf{I}, \mathbf{S}, \mathbf{P}_u$ have the usual meaning in universal algebra, i.e. isomorphic images, subalgebras and ultraproducts, respectively.

Now we are ready to describe a general procedure of finding the elements of $Min(\star^\perp)$ for any acanonical t-norm algebra $\star$. We use $\mathcal{U}_\emptyset$ to denote the empty ordinal sum, and by convention we take $g(\mathcal{U}_\emptyset \leftrightarrow \star) = 0$ for any $\star$.

procedure $find\_Min^\perp(\star)$
% input: $\star = \star_1 \oplus ... \oplus \star_n$, canonical t-norm BL-chain, where
% $\star_1 \in \{\Pi, \mathcal{L}, \mathcal{G}, \Pi^\infty\}$ and $\star_i \in \{\Pi, \mathcal{L}, \mathcal{G}, \Pi^\infty\}$ for $i > 1$
% output: $minimal\_list$ – list in which minimal elements of $\star^\perp$ are stored
% auxiliary list: $open\_list$ – list containing nodes to be expanded
    $n = length(\star)$; $open\_list = [\mathcal{U}_\emptyset]$; $minimal\_list = [\ ]$;
    do while $open\_list \neq [\ ]$
        $\mathcal{U} = first(open\_list)$; $k = g(\mathcal{U} \leftrightarrow \star)$;
        if $k = 0$ then $expanded\_nodes = \{\mathcal{U} \oplus \mathbf{2}, \mathcal{U} \oplus \mathcal{L}\}$;
        if $0 < k < n$ then $expanded\_nodes = \{\mathcal{U} \oplus \mathbf{2}, \mathcal{U} \oplus \mathcal{C}, \mathcal{U} \oplus \mathcal{L}\}$;
        if $k = n$ then $expanded\_nodes = \{\mathcal{U} \oplus \mathbf{2}, \mathcal{U} \oplus \mathcal{C}\}$;
        for all $\mathcal{U}' \in expanded\_nodes$ do
            $l = g(\mathcal{U}' \leftrightarrow \star)$;
            if $l = 1$ or $(1 < l \leq n$ and $\star_l \neq \mathcal{G}$ and $\star_l \neq \Pi^\infty)$ then
                $open\_list = update(open\_list, \mathcal{U}')$;
            if $l = n + 1$ then $minimal\_list = update(minimal\_list, \mathcal{U}')$;
        end for
        $open\_list = remove(open\_list, \mathcal{U})$;
    end do
end procedure
function $update(list, \mathcal{U})$
% inputs: $list$ to be possibly updated with node $\mathcal{U}$
% output: $list$ after being updated
    for all $\mathcal{W} \in list$ do
        if $g(\mathcal{W} \leftrightarrow \mathcal{A}) = g(\mathcal{U} \leftrightarrow \mathcal{A})$ then do
            if $\mathcal{W} \preceq \mathcal{U}$ then return $list$;
            if $\mathcal{U} \prec \mathcal{W}$ then $list = remove(list, \mathcal{W})$;
        end do
    end for
    $list = append(list, \mathcal{U})$; return $list$;
end function

As an example, let us consider a continuous t-norm $\star$ isomorphic to $\mathcal{G} \oplus \mathcal{L} \oplus \Pi^\infty \oplus \mathcal{L}$. The above procedures yield the expanded tree of Figure 1, where one can get $Min(\star^\perp) = \{\mathcal{L}, \mathbf{2} \oplus \mathcal{C} \oplus \mathcal{L} \oplus \mathbf{2}, \mathbf{2} \oplus \mathcal{C} \oplus \mathcal{L} \oplus \mathcal{C}\}$.

5   Conclusions and Future Work

In this paper we have solved the problem of finding the equational definition of subvarieties of **BL** generated by a single standard BL-chain. However the problem of axiomatizing the subvarieties generated by *families* of standard BL-chains remains. Although there are already some ideas towards the solution, the problem deserves further research.

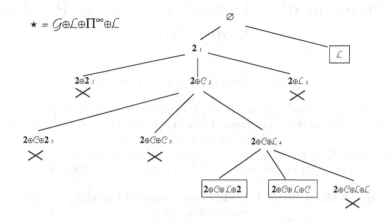

$\star = \mathcal{G} \oplus \mathcal{L} \oplus \Pi^{\infty} \oplus \mathcal{L}$

Fig. 1. Expanded tree for $\star = \mathcal{G} \oplus \mathcal{L} \oplus \Pi^{\infty} \oplus \mathcal{L}$.

Acknowledgments The Spanish researchers acknowledge partial support by a grant of the Italian GNSAGA-INDAM for a stay at the University of Siena in July 2002 and to the project LOGFAC (TIC2001-1577-C03-01).

References

[1] Aglianó P., Montagna F.: *Varieties of BL-algebras I: general properties*. Journal of Pure and Applied Algebra, to appear.
[2] Aglianó P., Ferreirim I.M.A., Montagna F.: *Basic hoops: an algebraic study of continuous t-norms*. Studia Logica, to appear.
[3] Blok W.J., Ferreirim I.M.A.: *On the structure of hoops*, Algebra Universalis 43 (2000) 233-257.
[4] Cignoli R., Esteva F., Godo L., Torrens A.: *Basic logic is the logic of continuous t-norms and their residua*. Soft Comp. 4 (2000) 106-112.
[5] Esteva F., Godo L., Hájek P., Navara M.: *Residuated Fuzzy Logic with an involutive negation* Archive of Mathematical Logic 39 (2000), 103-124.
[6] Esteva F., Godo L., Montagna F.: *Equational characterization of the subvarieties of* **BL** *generated by t-norm algebras*. Submitted.
[7] Gottwald S. *A traitise on Multiple-valued Logics*. Studies in logic and computation. Research Studies Press, Baldock, 2001.
[8] P. Hájek. *Metamathematics of Fuzzy Logic*. Kluwer, 1998.
[9] Hájek P.: *Basic logic and BL-algebras*. Soft Computing 2 (1998) 124-128.
[10] Haniková Z.: *Standard algebras for fuzzy propositional calculi*, Fuzzy Sets and Systems, vol. 123, n.3 (2001), 309-320.
[11] Haniková Z.: *A note on propositional tautologies of individual continuous t-norms*. Neural Network World vol. 12 n.5 (2002), 453-460.
[12] Klement P., Mesiar R., and Pap L.: *Triangular Norms*. Kluwer, 2000
[13] Novák V., Perfilieva I. and Močkoř J.: *Mathematical Principles of Fuzzy Logic*, Kluwer 1999.
[14] Zadeh L.A.: Preface. In *Fuzzy Logic Technology and Applications*, (R. J. Marks-II Ed.), IEEE Technical Activities Board (1994).

Extension of Łukasiewicz Logic by Product Connective

Rostislav Horčík[1]* and Petr Cintula[2]**

[1] Center for Machine Perception, Dept. of Cybernetics, Faculty of Elec. Eng.
Czech Technical University in Prague Technická 2, 166 27 Prague 6, Czech Republic
xhorcik@cmp.felk.cvut.cz
[2] Institute of Computer Science, Academy of Sciences of the Czech Republic
Pod vodárenskou věží 2, 182 07 Prague 8, Czech Republic
and
Dept. of Mathematics, FNSPE, Czech Technical University in Prague.
cintula@cs.cas.cz

1 Introduction

Formulas in Łukasiewicz logic [10,7] can be represented by piecewise linear functions with integer coefficients (Mc-Naughton functions). If we want to work with polynomials in the framework of Łukasiewicz logic, we have to extend the set of its connectives by a new connectives representing product. The main aim of this paper is to define and develop such a logic. We call this logic PŁ logic. We also study extension of PŁ logic by Baaz's $\triangle$ so-called PŁ$_\triangle$ logic. Furthermore we examine the Pavelka style extension and predicate versions of all defined logics.

All investigated logics lie between Łukasiewicz logic and RŁΠ logic [3,5]. Relations between them are depicted here:

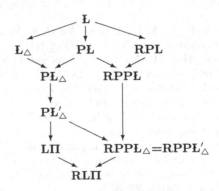

---

* The work of the first author was supported by the Czech Ministry of Education under project MSM 212300013, by the Grant Agency of the Czech Republic under project GACR 201/02/1540, by the Grant Agency of the Czech Technical University in Prague under project CTU 0208613, and by Net CEEPUS SK-042.
** The work of the second author was supported by grant IAA1030004 of the Grant Agency of the Academy of Sciences of the Czech Republic.

T. Bilgiç et al. (Eds.): IFSA 2003, LNAI 2715, pp. 180–188, 2003.

2   Preliminaries

The Łukasiewicz logic (Ł) is a well-known logic [10]. The language of Łukasiewicz logic contains a set of propositional variables, an implication $\to$ and the
constant $\bar{0}$. Further connectives are defined as follows:

$\neg\varphi$	is $\varphi \to \bar{0}$	$\varphi \oplus \psi$	is $\neg\varphi \to \psi$,
$\varphi \ominus \psi$	is $\varphi \otimes \neg\psi$	$\varphi \otimes \psi$	is $\neg(\neg\varphi \oplus \neg\psi)$,
$\varphi \wedge \psi$	is $\varphi \otimes (\varphi \to \psi)$	$\varphi \vee \psi$	is $(\varphi \ominus \psi) \oplus \psi$,
$\varphi \equiv \psi$	is $(\varphi \to \psi) \otimes (\psi \to \varphi)$	$\bar{1}$	is $\neg\bar{0}$.

The following are the *axioms* of Ł:

(1) $\varphi \to (\psi \to \varphi)$
(2) $(\varphi \to \psi) \to ((\psi \to \chi) \to (\varphi \to \chi))$
(3) $(\neg\varphi \to \neg\psi) \to (\psi \to \varphi)$
(4) $((\varphi \to \psi) \to \psi) \to ((\psi \to \varphi) \to \varphi)$

The *deduction rule* is modus ponens. The notions of *theory, proof, model,* and
theorem are defined as usual. It is shown in [7] that Ł is the schematic extension
of Hájek's Basic logic BL by the axiom $\neg\neg\varphi \to \varphi$.

In [7], Hájek also studies the extension of Ł by a unary connective $\triangle$. The
axioms of the Łukasiewicz logic with $\triangle$ ($Ł_\triangle$) are those of Ł plus:

(Ł$\triangle$1) $\triangle\varphi \vee \neg\triangle\varphi$
(Ł$\triangle$2) $\triangle(\varphi \vee \psi) \to (\triangle\varphi \vee \triangle\psi)$
(Ł$\triangle$3) $\triangle\varphi \to \varphi$
(Ł$\triangle$4) $\triangle\varphi \to \triangle\triangle\varphi$
(Ł$\triangle$5) $\triangle(\varphi \to \psi) \to (\triangle\varphi \to \triangle\psi)$

Deduction rules are modus ponens and *necessitation*: from φ derive $\triangle\varphi$.
Corresponding semantics models for Łukasiewicz logic are so-called MV-algebras.
By abuse of language, we use the same symbols to denote logical connectives and
the corresponding algebraic operations.

Definition 1. *An MV-algebra is a structure* $\mathbf{L} = (L, \oplus, \neg, \mathbf{0})$ *such that,* $(L, \oplus, \mathbf{0})$
is a commutative monoid and and letting $x \ominus y = \neg(\neg x \oplus y)$, *and* $1 = \neg 0$ *the
following conditions are satisfied:*

(MV1) $x \oplus 1 = 1$
(MV2) $\neg\neg x = x$
(MV3) $(x \ominus y) \oplus y = (y \ominus x) \oplus x$

The corresponding algebraic structures for $Ł_\triangle$ logic are MV$_\triangle$-algebras [7]. Let
us also recall the definition of *o-groups* and *o-rings* (see also [6]).

Definition 2. *A linearly ordered Abelian group (o-group for short) is a structure*
$(G, +, 0, -, \leq)$ *such that* $(G, +, 0, -)$ *is an Abelian group,* $(G, \leq)$ *is a linearly
ordered lattice, and if* $x \leq y$, *then* $x + z \leq y + z$ *for all* $z \in G$.

Definition 3. *A linearly ordered commutative ring with strong unit (o-ring for
short) is a structure* $(R, +, -, \times, 0, 1, \leq)$ *such that* $(R, +, 0, -, \leq)$ *is an o-group,*
$(R, +, -, \times, 0, 1)$ *is a commutative ring with strong unit, and the following is
satisfied: if* $x \geq 0$ *and* $y \geq 0$, *then* $x \times y \geq 0$.

3   PŁ Logic

In this section we introduce the PŁ logic (PŁ for short), an extension of Łukasiewicz logic by adding a new binary connective $\odot$. This connective plays the role of multiplication. Thus the basic connectives of PŁ are $\rightarrow, \odot, \bar{0}$. Additional connectives $\oplus, \otimes, \ominus, \neg, \wedge, \vee, \equiv, \bar{1}$ are defined as in Łukasiewicz logic.

Definition 4. *The axioms of PŁ logic are the axioms of Łukasiewicz logic plus:*

(P1) $(\chi \odot \varphi) \ominus (\chi \odot \psi) \equiv \chi \odot (\varphi \ominus \psi)$
(P2) $\varphi \odot (\psi \odot \chi) \equiv (\varphi \odot \psi) \odot \chi$
(P3) $\varphi \rightarrow \varphi \odot \bar{1}$
(P4) $\varphi \odot \psi \rightarrow \varphi$
(P5) $\varphi \odot \psi \rightarrow \psi \odot \varphi$

Deduction rule is modus ponens.

It is obvious that all theorems of Łukasiewicz logic are also theorems of PŁ. There is a useful theorem of PŁ $(\varphi_1 \equiv \psi_1) \otimes (\varphi_2 \equiv \psi_2) \rightarrow (\varphi_1 \odot \varphi_2 \equiv \psi_1 \odot \psi_2)$ stating that the connective $\equiv$ is a congruence w.r.t. the product $\odot$ (the fact that $\equiv$ is a congruence w.r.t. other connectives is known from [7]).

Since we have the same deduction rules in PŁ as in BL, we obtain also the same deduction theorem for PŁ. For details see [7, Theorem 2.2.18].

Theorem 1 (Deduction theorem). *Let T be a theory and φ, ψ be formulas. $T \cup \{\varphi\} \vdash \psi$ iff there is an n such that $T \vdash \varphi^n \rightarrow \psi$, where $\varphi^n = \varphi \otimes \ldots \otimes \varphi$.*

Now we proceed with a definition of the algebra corresponding to PŁ logic.

Definition 5. *A PŁ-algebra is a structure $\mathbf{L} = (L, \oplus, \neg, \odot, 0, 1)$, where the reduct $(L, \oplus, \neg, 0, 1)$ is an MV-algebra, $(L, \odot, 1)$ is commutative monoid, and the following identity holds: $(a \odot b) \ominus (a \odot c) = a \odot (b \ominus c)$, where $a \ominus b = \neg(\neg a \oplus b)$.*

PŁ-algebras are a special subclass of so-called product MV-algebras introduced by Dvurečenskij and Di Nola in their paper [4]. They do not require $\mathbf{1}$ to be a neutral element for product $\odot$ and commutativity of $\odot$.

Example 1. If $([0, 1], \oplus, \neg, 0, 1)$ is the standard MV-algebra (i.e. $x \oplus y = min(1, x + y)$ and $\neg x = 1 - x$) and $\odot$ is the usual algebraic product of reals then $[\mathbf{0}, \mathbf{1}]_{\mathbf{S}} = ([0, 1], \oplus, \neg, \odot, 0, 1)$ is called a standard PŁ-algebra.

Theorem 2. *Every PŁ-algebra is a subdirect product of linear PŁ-algebras.*

There is a correspondence between linearly ordered PŁ-algebras and o-rings like between linearly ordered MV-algebras and o-groups.

Definition 6. *Let $\mathcal{F} = (F, +, -, \times, 0, 1, \leq)$ be an o-ring. Let $L = \{x \in F | 0 \leq x \leq 1\}$. Define for all $x, y \in L$, $x \oplus y = \min\{1, (x + y)\}$ and $\neg x = 1 - x$. By $\odot$ we denote the operation $\times$ restricted to L. Then the algebra $(L, \oplus, \odot, \neg, 0, 1)$ is called the interval algebra of $\mathcal{F}$.*

Theorem 3. *An algebra* **L** *is a linearly ordered PŁ-algebra if and only if* **L** *is isomorphic to the interval algebra of some o-ring.*

Theorem 4 (Strong Completeness). *Let T be a theory and φ be a formula. Then the following are equivalent:*

- $T \vdash \varphi$
- $e(\varphi) = \mathbf{1_L}$ *for each PŁ-algebra* **L** *and each* **L**-model e *of* T
- $e(\varphi) = \mathbf{1_L}$ *for each linearly ordered PŁ-algebra* **L** *and each* **L**-model e *of* T

Thanks to latter theorem and standard completeness of Łukasiewicz logic we obtain that PŁ is a conservative extension of Łukasiewicz logic.

Standard completeness of PŁ logic seems to be related to the Henriksen and Isbell's result stating that the variety generated by standard ring of reals is not finitely axiomatizable (see [8,9]). However we are not able to prove that this result implies that the variety generated by standard PŁ-algebra is not finitely axiomatizable. Thus the problem of standard completeness of PŁ seems to be open for us.

4   PŁ$_\triangle$ Logic

In this section, we extend the language of PŁ logic by unary connective $\triangle$ and introduce PŁ$_\triangle$ logic. This connective was introduced in Gödel logic by Baaz (see [1]) and generalized to BL by Hájek (see [7]). Since it can be shown that PŁ$_\triangle$ logic is not standard complete, we introduce an extension PŁ$'_\triangle$ which enjoys the standard completeness theorem.

Definition 7. *The axioms of PŁ$_\triangle$ logic are the axioms of Ł$_\triangle$ plus (P1)–(P5). The PŁ$'_\triangle$ logic is an extension of PŁ$_\triangle$ by $\triangle\neg(\varphi \odot \varphi) \rightarrow \triangle\neg\varphi$.*

Definition 8. *A PŁ$_\triangle$-algebra is a structure* $\mathbf{L} = (L, \oplus, \neg, \odot, \triangle, 0, 1)$ *where* $(L, \oplus, \neg, \triangle, 0, 1)$ *is an MV$_\triangle$-algebra and* $(L, \oplus, \neg, \odot, 0, 1)$ *is a PŁ-algebra. A PŁ$'_\triangle$-algebra is a PŁ$_\triangle$-algebra where it holds $\triangle\neg(\varphi \odot \varphi) = \triangle\neg\varphi$.*

Example 2. We may extend the definition of the standard PŁ-algebra $[0,1]_\mathbf{S}$ by $\triangle$ by setting $\triangle(x) = \mathbf{1}$ iff $x = \mathbf{1}$ and $\triangle(x) = \mathbf{0}$ otherwise. Then we obtain the standard PŁ$_\triangle$-algebra which we denote $[0,1]_{\mathbf{S}\triangle} = ([0,1], \oplus, \neg, \odot, \triangle, 0, 1)$.

Theorem 5. *Every PŁ$_\triangle$-algebra is a subdirect product of linearly ordered PŁ$_\triangle$-algebras.*

Theorem 6 (Strong Completeness). *Let $\mathcal{C}$ be either PŁ$_\triangle$ or PŁ$'_\triangle$, T be a theory over $\mathcal{C}$ and φ be a formula. Then the following are equivalent:*

- $T \vdash \varphi$
- $e(\varphi) = \mathbf{1_L}$ *for each $\mathcal{C}$-algebra* **L** *and each* **L**-model e *of* T
- $e(\varphi) = \mathbf{1_L}$ *for each linearly ordered $\mathcal{C}$-algebra* **L** *and each* **L**-model e *of* T

It can be shown that the difference between $PL_\triangle$-algebras and $PL'_\triangle$-algebras is that $PL'_\triangle$-algebras contain only trivial zero-divisors, i.e. a linearly ordered $PL_\triangle$-algebra is $PL'_\triangle$-algebra iff $x \odot y = \mathbf{0}$ implies $x = \mathbf{0}$ or $y = \mathbf{0}$. It has several important consequences: the standard $PL_\triangle$-algebra is $PL'_\triangle$-algebra, $PL'_\triangle$ logic is strictly stronger than $PL_\triangle$ logic, and the following theorem which leads to the standard completeness of $PL'_\triangle$.

Theorem 7. *Let* **L** *be linearly ordered PL-algebra. Then* **L** *is a PL-reduct of some $PL'_\triangle$-algebra if and only if* **L** *is isomorphic to the interval algebra of some linearly ordered domain of integrity.*

Theorem 8 (Strong Standard Completeness). *Let T be a finite theory over $PL'_\triangle$ and φ a formula. Then $T \vdash \varphi$ iff $e(\varphi) = 1$ for each $[\mathbf{0},\mathbf{1}]_{\mathbf{S}\triangle}$-model e of T.*

As corollary we get that $PL_\triangle$ logic is a conservative extension of PL logic. Furthermore $\text{Ł}\Pi$, $\text{Ł}\Pi\frac{1}{2}$, and $R\text{Ł}\Pi$ are conservative extensions of $PL'_\triangle$.[1]

5 Pavelka Style Extension

In this section we add rational constants into the language of our logics together with book-keeping axioms. Rational Pavelka logic (RPL) has very interesting properties and has been very widely studied. RPL was introduced in Pavelka's series of papers [12] and simplified to its modern form in [7,11].

Definition 9. *Let $\mathcal{L}$ be one of the PL, $PL_\triangle$, $PL'_\triangle$. The language of $RP\mathcal{L}$ arises from the language of $\mathcal{L}$ by adding a truth constant $\overline{r}$ for each $r \in \mathbb{Q} \cap [0,1]$. The notion of evaluation extends by the condition $e(\overline{r}) = r$. The axioms of $RP\mathcal{L}$ logic are the axioms of $\mathcal{L}$ plus the book-keeping axioms for each rational $r, s \in [0,1]$:*

$$(\overline{r} \oplus \overline{s}) \equiv \overline{\min(1, r+s)} \qquad (\overline{r} \odot \overline{s}) \equiv \overline{r \cdot s}$$
$$\neg \overline{r} \equiv \overline{1-r} \qquad \qquad \triangle \overline{r} \equiv \overline{\triangle r}$$

The deduction rule of $RPPL$ is modus ponens. The deduction rules of $RPPL_\triangle$ and $RPPL'_\triangle$ are modus ponens, neccesitation and the following infinitary deduction rule IR: from $\overline{r} \to \varphi$ for all $r < 1$, derive φ.

Since we added the infinitary deduction rule IR, we have to change the notion of the proof. Let T be a theory, then the set $C_{RP\mathcal{L}}(T)$ of all provable formulas in T is the smallest set containing T, axioms of $RP\mathcal{L}$ and closed under all deduction rules. For simplicity we denote $\varphi \in C_{RP\mathcal{L}}(T)$ as $T \vdash \varphi$.

Theorem 9 (Deduction theorem). *Let φ and ψ be formulas.*

 – *Let T be a theory over $RPPL$. Then $T \cup \{\varphi\} \vdash \psi$ iff there in n such that $T \vdash \varphi^n \to \psi$.*

---

[1] For the definition and details about $\text{Ł}\Pi$, $\text{Ł}\Pi\frac{1}{2}$, and $R\text{Ł}\Pi$ see [5].

− Let T be a theory either over $RPPL_\triangle$ or over $RPPL'_\triangle$. Then $T \cup \{\varphi\} \vdash \psi$ iff $T \vdash \triangle\varphi \to \psi$.

Definition 10. Let $\mathcal{L}$ be one of the PL, $PL_\triangle$, $PL'_\triangle$, T be a theory over $RP\mathcal{L}$ and φ be a formula.

(1) The truth degree of φ over T is $\|\varphi\|_T = \inf\{e(\varphi) \mid e$ is a model of $T\}$.

(2) The provability degree of φ over T is $|\varphi|_T = \sup\{r \mid T \vdash \bar{r} \to \varphi\}$.

Theorem 10 (Pavelka's style completeness). Let $\mathcal{L}$ be one of the PL, $PL_\triangle$, $PL'_\triangle$ and T a theory over $RP\mathcal{L}$ and φ be a formula. Then $|\varphi|_T = \|\varphi\|_T$.

The proof of Pavelka's style completeness for RPPŁ could be obtained as a corollary of [11, Corollary 4.6] or [7, Theorem 3.3.19] but it is not the case of RPPŁ$_\triangle$ and RPPŁ$'_\triangle$. Now we prove the strong standard completeness for RPPŁ$_\triangle$ and RPPŁ$'_\triangle$ and show that these logics coincide.

Theorem 11. Let T be a theory over $RPPL_\triangle$ or $RPPL'_\triangle$ and φ be a formula. Then $T \vdash \varphi$ iff $e(\varphi) = 1$ for all standard models e.

As corollary we get that RPPŁ$_\triangle$ and RPPŁ$'_\triangle$ logics coincide. RPPŁ$_\triangle$ is a conservative extension of PŁ$'_\triangle$. RPPŁ$_\triangle$ is a conservative extension of RPŁ. RŁΠ is a conservative extension of RPPŁ$_\triangle$.[2]

6 The Predicate Logics

This section deals with predicate versions of the PŁ, PŁ$_\triangle$ and PŁ$'_\triangle$ logics. The basic definitions of predicate language, term, (atomic) formula, (safe) **L**-structure **M**, **M**-evaluation v; truth value of (atomic) formula φ under **M**-evaluation v in **L**-structure **M** ($\|\varphi\|^\mathbf{L}_{\mathbf{M},v}$), truth value of formula φ in **L**-structure **M** ($\|\varphi\|^\mathbf{L}_\mathbf{M}$) and theory T (of closed formulas) and (safe) **L** model **M** of theory T are analogous to the definitions of corresponding concepts in [7]. The differences are mainly due to the differences in our languages. We also suppose that the reader is familiar with the notion of free or bounded occurrence of an object variable in a formula and the notion of substitutable term into the formula φ.

Definition 11. Let $\mathcal{C}$ be either PL, $PL_\triangle$ or $PL'_\triangle$, and I be a predicate language. The logic $\mathcal{C}\forall$ is given by the following axioms and the deduction rules:

− the formulas resulting from the axioms of $\mathcal{C}$ by the substitution of the propositional variables by the formulas of I
− $(\forall x)\varphi(x) \to \varphi(t)$ where t is substitutable for x in φ
− $(\forall x)(\chi \to \varphi) \to (\chi \to (\forall x)\varphi)$, where x is not free in χ
− deduction rules are modus ponens, generalization (from $\varphi(x)$ derive $(\forall x)$ $\varphi(x)$), and necessitation of $\triangle$ in case of $PL_\triangle$ or $PL'_\triangle$ logic

---

[2] For the definition and details about RŁΠ see [5].

Theorem 12 (Strong Completeness). *Let I be a predicate language, C either PŁ, PŁ$_\triangle$ or PŁ$'_\triangle$, T a theory over $C\forall$, φ a closed formula. Then $T \vdash \varphi$ iff $\|\varphi\|^L_M = 1$ for each linearly ordered C-algebra L and each safe L-model M of T.*

The question, whether PŁ$\forall$ is a conservative extension of the Łukasiewicz predicate logics, seems to be open. The standard completeness of the PŁ$\forall$, PŁ$_\triangle\forall$, and PŁ$'_\triangle\forall$ is a related problem. This question can be answered. Here we assume that the reader is familiar with basic concepts of undecidability and arithmetical hierarchy (the preliminary [7, Section 6.1] is satisfactory for our needs). In the following we assume that our predicate language is at most countable.

Theorem 13. *The set of $[0,1]_S$-tautologies of PŁ$\forall$ is Π_2-complete. The set of $[0,1]_{S\triangle}$-tautologies of PŁ$_\triangle\forall$ is not arithmetical.*

As corollary we obtain that PŁ$\forall$, PŁ$_\triangle\forall$, and PŁ$'_\triangle\forall$ have not the standard completeness property. Another corollary is that all our logics (understood as the set of $[0,1]_S$ or $[0,1]_{S\triangle}$-tautologies) are undecidable. Further it can be shown that PŁ$_\triangle\forall$ and PŁ$'_\triangle\forall$ logics (as set of theorems) are undecidable.

6.1 Pavelka Style Extension of Predicate Logics

In this section we build Rational Pavelka's style extension of the predicate logics defined in the previous subsection. Our approach will be analogous to the one from Section 5. As we proved in that section, adding of infinitary rule IR, turns the rational extensions of PŁ$_\triangle$ and PŁ$'_\triangle$ into the same logics. Thus in this section we will develop only RPPŁ$\forall$ and RPPŁ$_\triangle\forall$ logics. Again in this subsection we restrict ourselves to the standard algebras of the truth values ($[0,1]_S$ and $[0,1]_{S\triangle}$).

Definition 12. *Let C be either PŁ or PŁ$_\triangle$.*

- *We extend the set of logical symbols by truth constant $\bar{r}$ for each $r \in \mathbb{Q} \cap [0,1]$.*
- *We extend the definition of a formula by a clause that $\bar{r}$ is a formula.*
- *We extend the definition of truth value by a condition $\|\bar{r}\|^L_{M,v} = r$.*
- *The logic RPC$\forall$ results from $C\forall$ by adding the book-keeping axioms.*
- *The deduction rules of RPPŁ$\forall$ are modus ponens and generalization, RPPŁ$_\triangle\forall$ has additional deduction rules neccesitation and IR.*

Definition 13. *Let C be either RPPŁ or RPPŁ$_\triangle$. Let T be a theory over $C\forall$ and φ be a formula.*

(1) *The truth degree of φ over T is $\|\varphi\|_T = \inf\{\|\varphi\|_M \mid M \text{ is a model of } T\}$.*
(2) *The provability degree of φ over T is $|\varphi|_T = \sup\{r \mid T \vdash \bar{r} \to \varphi\}$.*

Theorem 14 (Pavelka's style completeness). *Let C be either RPPŁ or RPPŁ$_\triangle$. Let T be a theory over $C\forall$ and φ a formula. Then $\|\varphi\|_T = |\varphi|_T$.*

Theorem 15 (Strong standard completeness). *Let φ be a formula and T a theory over $RPPL_\triangle \forall$. Then $T \vdash \varphi$ iff $\|\varphi\|_{\mathbf{M}} = 1$ for all standard models $\mathbf{M}$.*

As corollary we get that RŁΠ∀ is a conservative extension of $RPPL_\triangle \forall$.[3]

At the end of this section we prove a very important consequence of the standard completeness of $RPPL_\triangle \forall$. We define a small modification of the language of $RPPL_\triangle \forall$: the logic TT results from $RPPL_\triangle \forall$ by omitting those truth constants, which can not be expressed in the form $\frac{k}{2^n}$, $k, n \in \mathbb{N}$. For TT we can prove the analogy of Theorem 15 for TT thus we know that $RPPL_\triangle \forall$ is a conservative extension of TT.

Takeuti and Titani's logic was introduced by Takeuti and Titani in their work [13]. It is a predicate fuzzy logic based on Gentzen's system of intuitionistic predicate logic. The connectives used by this logic are just the connectives of TT logic. This logic has two additional deduction rules and 46 axioms. We will not present the axiomatic system and we only recall that this logic is sound and complete w.r.t. standard $PL_\triangle$-algebra (cf. [13, Theorem 1.4.3]). This leads us to the following conclusion:

Theorem 16. *Takeuti and Titani logic coincides with the logic TT. Furthermore $RPPL_\triangle \forall$ logic is the conservative extension of Takeuti and Titani logic.*

This theorem allows to translate the very interesting results from the Takeuti and Titani's logic into our much more simpler (in syntactical sense) logical system of the TT or $RPPL_\triangle \forall$ logic.

References

1. M. Baaz: *Infinite-valued Gödel logic with 0-1 projector and relativisations.* In Gödel'96: Logical foundations of mathematics, computer science and physics. Ed. P. Hájek, Lecture notes in logic 6:23-33, 1996.
2. R. Cignoli, I.M.L. D'Ottaviano, D. Mundici: *Algebraic Foundations of Many-Valued Reasoning.* Kluwer, Dordrecht, 1999.
3. P. Cintula: *The ŁΠ and ŁΠ$\frac{1}{2}$ propositional and predicate logics.* Fuzzy Sets and Systems 124/3:21–34, 2001.
4. A. Di Nola, A. Dvurečenskij: *Product MV-algebras.* Multiple-Valued Logic 6:193–215, No.1-2, 2001.
5. F. Esteva, L. Godo, F. Montagna: *The ŁΠ and ŁΠ$\frac{1}{2}$ logics: two complete fuzzy systems joining Łukasiewicz and product logics.* Archive for Mathematical Logic 40:39–67, 2001.
6. L. Fuchs: *Partially Ordered Algebraic Systems.* Pergamon Press, Oxford, 1963.
7. P. Hájek: *Metamathematics of Fuzzy Logic.* Kluwer, Dordrecht, 1998.
8. M. Henriksen, J. R. Isbell: *Lattice-ordered Rings and Function Rings.* Pacific J. Math. 12:533–565, 1962.
9. J. R. Isbell: *Notes on Ordered Rings.* Algebra Univ. 1:393–399, 1972.
10. J. Łukasiewicz: *O logice trojwartosciowej (On three-valued logic)* Ruch filozoficzny 5:170-171, 1920.

---

[3] For the definition and details about RŁΠ∀ see [3].

11. V. Novák, I. Perfilieva, J. Močkoř: *Mathematical Principles of Fuzzy Logic*. Kluwer, Norwell, 1999.
12. J. Pavelka: *On Fuzzy Logic I,II,III*. Zeitschr. f. math. Logic und Grundlagen der Math. 25:45–52,119–134,447–464, 1979.
13. G. Takeuti, S. Titani: *Fuzzy Logic and Fuzzy Set Theory*. Archive for Mathematical Logic 32:1–32, 1992.

Formulas of Łukasiewicz's Logic Represented by Hyperplanes

Antonio Di Nola and Ada Lettieri

[1] Department of Mathematics and Informatics
University of Salerno
dinola@cds.unina.it
[2] Dep. di Costruzioni e Metodi Matematici in Architettura
University of Naples Federico II
lettieri@unina.it

Abstract. Formulas of n variables of Łukasiewicz's sentential calculus can be represented, via McNaughton's theorem, by piecewise linear functions, with integer coefficients, from hypercube $[0,1]^n$ to $[0,1]$, called *McNaughton functions*. McNaughton functions, which are truncated functions to $[0,1]$ of the restriction to $[0,1]^n$ of single hyperplanes, are called *simple McNaughton functions*.

In the present work we describe a class of formulas that can be represented by simple McNaughton functions.

1 Introduction and Preliminaries

Łukasiewicz's logic introduced in [7] is a calculus among others introduced as a many valued generalization of classical logic. Let us denote it by $\mathcal{L}_\omega$. As the truth-values of sentences of Łukasiewicz logic can be interpreted as degrees of truth, in the last decades the studies on such a logic received a renowed impulse, viewing this calculus as a variant of Fuzzy Logic [6]. Chang provided an algebraic proof of the completeness of Łukasiewicz logic, [2, 3]. In [2] Chang introduced a class of algebraic structures, known as *MV*-algebras. Such algebras play the role with respect to Łukasiewicz' s logic analogous to the one played by Boolean Algebras with respect to the classical logic.

An *MV*-algebra is an algebraic structure $A = (A, 0, 1, {}^*, \odot, \oplus)$ satisfying the following axioms:

(1) $(x \oplus y) \oplus z = x \oplus (y \oplus z)$;
(2) $x \oplus y = y \oplus x$;
(3) $x \oplus 0 = x$;
(4) $x \oplus 1 = 1$;
(5) $0^* = 1$;
(6) $1^* = 0$;
(7) $x \odot y = (x^* \oplus y^*)^*$;
(8) $(x^* \oplus y)^* \oplus y = (y* \oplus x)^* \oplus x$.

T. Bilgiç et al. (Eds.): IFSA 2003, LNAI 2715, pp. 189–194, 2003.

From (8), with $y = 0$ it follows $(x^*)^* = x$ and with $y = 1$, $x^* \oplus x = 1$. On A two new operations $\vee$ and $\wedge$ are defined as follows: $x \vee y = (x^* \oplus y)^* \oplus y$ and $x \wedge y = (x^* \odot y)^* \odot y$. The structure $(A, \vee, \wedge, 0, 1)$ is a bounded distributive lattice. We shall write $x \leq y$ iff $x \wedge y = x$. Boolean algebras are just the MV-algebras obeying the additional equation $x \odot x = x$. Let $B(A) = \{x \in A / x \odot x = x\}$ be the set of all idempotent elements of A. Then, $B(A)$ is a subalgebra of A, which is also a Boolean algebra. Indeed, it is the greatest Boolean subalgebra of A.

Throughout th paper, R and Z denote the set of real and integer numbers, respectively. Besides we will denote by $\mathbf{x}$ the n-tuple $(x_1, \ldots, x_n) \in R^n$ and, for every $\mathbf{x} \in R^n$, $\mathbf{x}^{r_1, \ldots, r_t}$ shall denote the the (n-t)-tuple, obtained from $\mathbf{x} = (x_1, \ldots, x_n)$, by deleting the r_i-components of $\mathbf{x}$, for $i \in \{1, \ldots, t\}$. For $\mathbf{x}, \mathbf{y} \in R^n$, we set $\mathbf{x} \bullet \mathbf{y} = x_1 y_1 + \ldots + x_n y_n$, where $+$ is the usual addition in R and xy is the usual product in R of x and y. $\bullet$ operation shall be considered more binding than the addition.

In [9] D. Mundici proved an equivalence functor Γ between the category of abelian ℓ-groups with a strong unit and the category of MV-algebras. Let $(G, +, \mathbf{0})$ be an abelian ℓ-group with a strong unit u, then the univers of the MV-algebra $\Gamma((G, u))$ is the set $\{x \in G : 0 \leq x \leq u\}$ and the operations are defined as follows:

$$0 = \mathbf{0};$$
$$1 = u;$$
$$x \oplus y = u \wedge (x + y);$$
$$x \odot y = 0 \vee (x + y - 1);$$
$$x^* = 1 - x.$$

Particularly, denoting by $(R, +, 0)$ the ordered group of the real numbers, we obtain the MV-algebra $\Gamma(R, 1)$, having as support the real interval $[0, 1]$. Thus the basic MV-algebraic operations on $[0, 1]$, are

$$x \oplus y = \min(1, x + y);$$
$$x \odot y = \max(0, x + y - 1);$$
$$x^* = 1 - x.$$

We refer to this MV-algebra by $[0, 1]$.

Let A be an MV-algebra, $x \in A$ and n a nonegative integer. In the sequel we will denote by $n.x$ the element of A, inductively defined by $0.x = 0$, $n.x = (n-1).x \oplus x$; whereas we will denote by nx the element of $\Gamma^{-1}(A)$, inductively defined by $0x = 0$, $nx = (n-1)x + x$. Moreover we consider the $*$ operation more binding than any other operation, and $\odot$ operation more binding than $\oplus$.

For all unexplained notions regarding MV-algebras we refer to [4]. Remarkable results on MV-algebras, due to Chang [3] are the algebraic completeness theorem and the subdirect representation theorem.

Theorem 1. *An equation holds in $[0, 1]$ if and only if it holds in every MV-algebra.*

Theorem 2. *Every non trivial MV-algebra is a subdirect product of totally ordered MV-algebras.*

A real valued function $f : [0,1]^n \to [0,1]$ is a *McNaughton function* (of n variables) if and only if the following conditions are satisfied:

(9) f is continuous;
(10) f is piecewise linear, that is there exist finitely many linear polynomials $p_1, \ldots, p_{k(f)}$ such that $p_i(\mathbf{x}) = \mathbf{a}_i \bullet \mathbf{x} + b_i$ and for any $\mathbf{x} \in [0,1]^n$ there is an index $j \in \{1, \ldots, k(f)\}$ for which $f(\mathbf{x}) = p_j(\mathbf{x})$;
(11) For each $i \in \{1, \ldots, k(f)\}$ all the coefficients $a_{i1}, \ldots, a_{in}, b_i$ are integers.

For every positive integer n, denote by $\mathcal{M}_n$ the class of all McNaughton functions of n variables. It is easy to check that for every n the following pointwise operations defined on the class $\mathcal{M}_n$, by:

$$f^*(\mathbf{x}) = 1 - f(\mathbf{x}) \qquad \text{and} \qquad (f \oplus g)(\mathbf{x}) = \min(1, f(\mathbf{x}) + g(\mathbf{x}))$$

bear the structure $(\mathcal{M}_n, {}^*, \oplus, \mathbf{0}, \mathbf{1})$ as an MV-algebra, where $\mathbf{0}$ and $\mathbf{1}$ are the 0-constant function and the 1-constant function, respectively.

As in [4], for every real-valued function g, defined on $[0,1]^n$, $g^\#$ shall denote the function defined by $g^\#(\mathbf{x}) = 0 \vee (g(\mathbf{x}) \wedge 1)$.

An n variable *McNaughton function* f, defined on $[0,1]^n$, will be called *simple* if there is an n variable real polynomial $g(\mathbf{x}) = \mathbf{a} \bullet \mathbf{x} + b$, a_k and b integers, for every $k = 1, \ldots, n$ and $f(\mathbf{x}) = g^\#(\mathbf{x})$, for every $\mathbf{x} \in [0,1]^n$.

A formula ϕ with variables $v_1, \ldots, v_n$ uniquely determines a function $f_\phi : [0,1]^n \to [0,1]$ as follows:

(12) If $\phi = v_i$, then $f_\phi(x_1, \ldots, x_n) = x_i$;
(13) If $\phi = \neg\psi$, then $f_\phi = 1 - f_\psi$;
(14) If $\phi = \psi \to \chi$, then $f_\phi = \min(1, 1 - f_\psi + f_\chi)$,

where $\neg$ and $\to$ are the negation and implication symbols of $\mathcal{L}_\omega$, respectively.

Theorem 3. *The class of functions determined by formulas of $\mathcal{L}_\omega$ coincides with the class of Mc Naughton functions.*

As a corollary of the above Mc Naughton's theorem, [8], we get that, for every n, the free MV-algebra L_n over n generators is isomorphic to $\mathcal{M}_n$.

In Sections 2 and 3 of this paper we shall describe a class $\mathfrak{S}$ of formulas which can be represented by simple Mc Naughton functions. About this question see also [1].

2   Simple Formulas of One Variable

Let $\mathbb{S}_1^\emptyset$ denote the class of formulas of one variable of $\mathcal{L}_\omega$ defined as follows:
for every $n \in N$ and for every $\ell \in N_0$, such that $0 \leq \ell \leq n - 1$, we set:

$\pi_0^n(v) = n.v,$
$\pi_1^n(v) = \bigoplus_{i=1}^{n-1} F_{0i}(v),$

$$\ldots\ldots\ldots ,$$
$$\pi_\ell^n(v) = \bigoplus_{i=\ell}^{n-1} F_{0,1,\ldots,\ell-1,i}(v),$$
$$\ldots\ldots\ldots ,$$
$$\pi_{n-1}^n(v) = F_{0,1,\ldots,n-2,n-1}(v),$$

where v is a propositional variable and the formulas $F_{0,1,\ldots,\ell-1,i}(v)$ are defined as follows:

for every integer $i > 0$, $F_{0,i}(v) = v \odot (i.v)$,
for every integer $i > 1$, $F_{0,1,i}(v) = (F_{0,1}(v) \oplus \ldots \oplus F_{0,i-1}(v)) \odot F_{0,i}(v)$,
and, by induction,
for every integer i such that $i > \ell$,
$\quad F_{0,1,\ldots,\ell,i}(v) = (F_{0,1,\ldots,\ell-1,\ell}(v) \oplus \ldots \oplus F_{0,1,\ldots,\ell-1,i-1}(v)) \odot F_{0,1,\ldots,\ell-1,i}(v)$.

Let $\mathbb{S}_1^{\{1\}}$ be the set of formulas obtained with formulas from $\mathbb{S}_1^{\emptyset}$ substituting the single propositional variable v by $\neg v$. Finally set $\mathfrak{S}_1 = \mathbb{S}_1^{\emptyset} \cup \mathbb{S}_1^{\{1\}}$. The elements of the set $\mathfrak{S}_1$ shall be called *simple* formulas with one variable.

We shall denote the Mc Naughton functions corresponding to formulas from $\mathbb{S}_1^{\emptyset}$ by the same symbols $\pi_\ell^n(v)$, up to the substitution of the propositional variable v by the real variable $x \in [0,1]$, that is:

$$f_{\pi_\ell^n(v)}(x) = \pi_\ell^n(x);$$

and the Mc Naughton functions corresponding to formulas from $\mathbb{S}_1^{\{1\}}$ by the same symbols $\pi_\ell^n(\neg v)$, up to the substitution of $\neg v$ by $x^* \in [0,1]$, that is:

$$f_{\pi_\ell^n(\neg v)}(x) = \pi_\ell^n(x^*).$$

In [5] the authors described the behaviour of such functions in the case they are defined on an MV-algebra A, belonging to a subvariety of MV generated by only one finite chain.

Now we are going to describe the behaviour of the polynomials $\pi_\ell^n(x)$ when they are defined on the MV-algebra $[0,1]$.

For every positive real number a, we will denote by $[a]$ the integer part of a, that is

$$[a] = max\{n \in N_0/n \le a\}. \tag{1}$$

The next proposition characterizes the functions determinated by one variable *simple* formulas as one variable *simple McNaughton* functions.

Proposition 4. *Let $f : [0,1] \to [0,1]$ be a function. Then the following conditions are equivalent:*

(a) $f(x) = (px + q)^{\#}$, for every $x \in [0,1]$ and $p,q \in Z$.
(b)

$$f(x) = \begin{cases} \pi_\ell^n(x), & if\ p \ge 0 \\ \pi_m^n(x^*) & if\ p < 0. \end{cases} \tag{2}$$

for suitable $n \in N$ and $\ell, m \in Z$.

3   Simple Formulas of n Variables

Until further notice a_k, $k = 1, \ldots, n$ will denote positive integer numbers, and b an element of Z. Then we set:

$\lambda(\mathbf{a}^1, b) = max\{-1, b - \sum_{k=2}^{n} a_k\}$
and
$\mu(a_1, b) = min\{b, a_1 - 1\}$,
where $\mathbf{a} = (a_1, \ldots, a_n)$.

It is immediate that if $0 \le b < \sum_{k=1}^{n} a_k$, then $\lambda(\mathbf{a}^1, b) \le \mu(a_1, b)$.

Under the above notations we define, by induction, the class $\mathbb{S}_n^{\emptyset}$ of formulas with n variables as follows:

if $n = 1$, set $S_b^{a_1}(v_1) = \pi_b^{a_1}(v_1)$;
if $n > 1$,

$$S_b^{\mathbf{a}}(\mathbf{v}) = \begin{cases} 0 & \text{if } b \ge \sum_{k=1}^{n} a_k, \\ 1 & \text{if } b < 0, \\ \bigoplus_{\lambda(\mathbf{a}^1,b) \le i \le \mu(a_1,b)} \pi_i^{a_1}(v_1) \odot S_{b-1-i}^{\mathbf{a}^1}(\mathbf{v}^1) & \text{if } 0 \le b < \sum_{k=1}^{n} a_k; \end{cases} \tag{3}$$

where each component v_k of $\mathbf{v}$ is a propositional variable. Set now $I_n = \{1, 2, \ldots, n\}$ and denote by $\mathbb{P}(I_n)$ the power set of I_n. For every $A \in \mathbb{P}(I_n)$, let $\mathbf{v}_A$ be the vector, obtained from $\mathbf{v}$, by substituting, for each $k \in A$, the component v_k by $\neg v_k$ and let $\mathbb{S}_n^A$ be the set of formulas obtained with formulas from $\mathbb{S}_n^{\emptyset}$ by substitution the vector $\mathbf{v}$ by $\mathbf{v}_A$.

Formulas from $\mathfrak{S}_n = \bigcup_{A \in \mathbb{P}(I_n)} \mathbb{S}_n^A$ shall be called *simple* formulas with n variables and formulas from $\mathfrak{S} = \bigcup_n \mathfrak{S}_n$ shall be called *simple*.

As in the previous section we shall denote the McNaughton functions corresponding to formulas from $\mathbb{S}_n^{\emptyset}$ by the same symbols $S_b^{\mathbf{a}}(\mathbf{v})$, up to the substitution of the propositional variables $\mathbf{v}$, by the real variables $\mathbf{x} = (x_1, \ldots, x_n) \in [0, 1]^n$, that is

$$f_{S_b^{\mathbf{a}}(\mathbf{v})}(\mathbf{x}) = S_b^{\mathbf{a}}(\mathbf{x});$$

and, for every $A \in \mathbb{P}(I_n)$, the McNaughton functions corresponding to formulas from $\mathbb{S}_n^A$ by the same symbols $S_b^{\mathbf{a}}(\mathbf{v}_A)$, up to the substitution of the vector $\mathbf{v}_A$, by the real vector $\mathbf{x}_A$, where, for each $k \in A$, the k-th component is x_k^*, that is:

$$f_{S_b^{\mathbf{a}}(\mathbf{v}_A)}(\mathbf{x}) = S_b^{\mathbf{a}}(\mathbf{x}_A).$$

Immediately by definition they follow:

($S1$) $S_b^{\mathbf{a}}(\mathbf{0}) = 0$ for every $b \ge 0$,
($S2$) $S_b^{\mathbf{a}}(\mathbf{1}) = 1$ for every $b \le \sum_{k=1}^{n} a_k$,
($S3$) $S_b^{\mathbf{a}}(\mathbf{x})$ is increasing with respect to each variable x_k.

As well as, in the previous section, the one variable simple formulas of $\mathcal{L}_\omega$, $\pi_\ell^n(x)$, have been represented by one variable *simple McNaughton* functions, in a similar way, here, we will show that the n variable polynomials $S_b^a(x)$ can be represented by *simple McNaughton* functions of n variables. Indeed we have:

Theorem 5. *Let ϕ be a formula of $\mathcal{L}_\omega$. Then the following holds:*

If ϕ is simple, then $f_\phi(\mathbf{x})$ is a simple McNaughton function.

References

[1] S. Aguzzoli, *The Complexity of McNaughton Functions of One Variable* , Advances in Applied Mathematics, **21** (1998) 58-77.
[2] C. C Chang, *Algebraic Analysis of infinite valued logic* , Trans. Amer. Math. Soc. **88** (1958), 467-490.
[3] C. C Chang, *A new proof of the completeness of the Łukasiewicz axioms*, Trans. Amer. Math. Soc. **93** (1959), 74-90.
[4] R. Cignoli, I. M. L. D'Ottaviano, D. Mundici, *Algebraic Foundations of Many-Valued Reasoning* Trends in Logic, Volume 7, Kluwer, Dordrecht, 1999.
[5] A. Di Nola, G. Georgescu, A. Lettieri, *Extending Probabilities to States of MV-algebras*, Collegium Logicum, Annals of the Kurt Goedel Society, 3–30, 1999.
[6] P. Hajek, *Metamathematics of Fuzzy Logic* ,Trends in Logic,Kluwer, Dordrecht, (1998).
[7] J. Łukasiewicz, A. Tarski, *Untersuchungen uber den Aussagenkalkul, Comptes Rendus* des séances de la Société des Sciences et des Lettres de Varsovie, Classe III, **23**, 30–50, (1930).
[8] R. McNaughton, *A theorem about infinite-valued sentential logic*, Journal of Symbolic Logic, **16** (1951) 1-13.
[9] D. Mundici, *Interpretation of AFC^* -algebras in Łukasiewicz sentential calculus*, J. Funct. Analysis **65**, (1986), 15–63.

Fuzzifying the Thoughts of Animats

Iztok Lebar Bajec, Nikolaj Zimic, and Miha Mraz

Faculty of Computer and Information Science, University of Ljubljana,
Ljubljana SI-1000, Slovenia,
iztok.bajec@fri.uni-lj.si,
http://lrss.fri.uni-lj.si

Abstract. In this article we present a fuzzy logic based method for the construction of thoughts of artificial animals (animats). Due to the substantial increase of the processing power of personal computers in the last decade there was a notable progress in the field of animat construction and simulation. Regardless of the achieved results, the coding of the animat's behaviour is very inaccurate and can, to someone not familiar with common physics variables like speed, acceleration, banking, etc., seem like pure black magic. Our leading hypothesis is, that by using linguistic programming based on common sense, unclear and even partially contradictory knowledge of dynamics, we can achieve comparable, if not better, simulation results. We begin the article with the basics of animats, continue with their fuzzyfication and end with the presentation and comparison of simulation results.

1 Introduction to Animats

The research field of the construction of simple artificial life was started by J. von Neumann in the mid 20th century, when he first set ground for the basics of cellular automata. The main field of his research was the modelling of the basic characteristics of living organisms, with self-replication being one of them. Von Neumann, using a few sets of crisp rules, constructed a cellular automaton, which allowed specific structures to accurately self-replicate in a finite number of time steps.

The boost in the computing power of personal computers in the last decade introduced a new field of research. This field is dedicated to the modelling and analysis of the behaviour of groups of artificial organisms. Nowadays we commonly address such artificial organisms with the term animat, which was first introduced by Wilson [7]. Representative examples of group behaviours that can be found in nature are flocks of birds, herds of sheep, schools of fish, packs of wolves, swarms of bees, ant foraging, etc.

In this article we focus our attention on the boid - a special type of animat - which was first introduced by Reynolds [5]. Observing the behaviour of a group of boids, a strong resemblance to the characteristic behaviour of a flock of birds can be sensed. Reynolds also states that the boid model can be used to simulate behaviours of herds and schools. The behaviour of every boid is based on a set of

T. Bilgiç et al. (Eds.): IFSA 2003, LNAI 2715, pp. 195–202, 2003.
© Springer-Verlag Berlin Heidelberg 2003

geometrical expressions. On their basis every boid (every flock member) chooses the direction and speed of flight that allows him to be in the flock. From the point of view of a boid the processing of this decision takes place in discrete time steps, and from the point of view of the flock it takes place simultaneously. If the boid is by definition a model of an independent flock member, we argue the use of geometrical expressions as a source of the boid's decision. We find it contradictory from at least three points of view:

- it is hardly imaginable that animals have the ability to sense crisp accurate numerical data (such as distance) from the environment,
- it is hardly imaginable that animals have the ability to execute accurate numerical or geometrical calculations - the basis of the Reynolds model,
- the crisp numerically geometrical concept of the boid model was constructed empirically and not as a result of observation and description of the dynamics of a real flock, and is as such incomprehensible to a lay public.

In the article we first present a formal definition of the boid model and explain its background. We continue with the implementation of fuzzy logic in the decision process, and in this way generalize and upgrade the existing definitions with logic.

2   Definition of Boids

In this section we present a formal definition of the boid model. For better understanding of the definition we first present the Moore automaton [1] and the animat [2], [7] formal definitions.

Definition 1. *A Moore automaton is defined as a five-tuple $< X, Q, Y, \delta, \lambda >$, where X, Q and Y are finite non-empty sets representing the input alphabet, the internal states and the output alphabet respectively. δ is a mapping called the transition function and λ is a mapping called the output function:*

$$\delta : X \times Q \to Q, \tag{1}$$

$$\lambda : Q \to Y. \tag{2}$$

At any discrete time step t the automaton is in a state $q(t) \in Q$ emitting the output $\lambda(q(t)) \in Y$. If an input $x(t) \in X$ is applied to the automaton, in the next discrete time step $t+1$ the automaton instantly assumes the state $q(t+1) = \delta(x(t), q(t))$ and emits the output $\lambda(q(t+1))$.

Definition 2. *An animat is a special Moore automaton $A = < X, Q, Y, \delta, \lambda >$, where the transition function δ is defined by a sequence of three levels of functions. These are specified by two function vectors $\boldsymbol{P}$ and $\boldsymbol{S}$ and a mapping B named the perception functions vector, steering functions vector and behaviour function respectively. Formally these are defined as:*

$$P_i : X \times Q \to \mathcal{P}(X), \ i = 1, ..., k, \tag{3}$$

$$S_j : \mathcal{P}(X)^k \times Q \to F, \ j = 1, ..., l, \tag{4}$$

$$B : F^l \times Q \to Q. \tag{5}$$

Let us sketch the appropriate picture informally. At any discrete time step t the input $x(t) \in X$ is the current state of the world, which is a finite non-empty set of animats. The three processing levels of the transition function try to imitate the behaviour of an animal. More precisely, they represent the perception, goal selection and action selection processes. The first level using the perception functions P_i (3) selects from the input $x(t) \in X$ only according to the specific perception function P_i relevant information - the observed animat's neighbours, for example - thus obtaining a vector of neighbour states. These are the input for the steering functions S_j (4), which calculate a vector of steering forces. Finally the steering forces are combined by the behaviour function B (5), and the animat's next discrete time step state $q(t+1)$ is generated.

Definition 3. *A Boid B is an animat, where the perception functions vector is defined as $P = (P_f)$, the steering functions vector is defined as $S = (S_s, S_a, S_c)$ and the behaviour function is defined as B_{pa}. The animat's state at a discrete time step t is $q(t)$ as defined in eq.(6), where $p(t) \subset \mathbb{R}^d (d = 2, 3)$ is the boid's position in space, $v(t) \in \mathbb{R}^d (d = 2, 3)$ is the boid's velocity, r is the boid's radii of perception, m is the boid's mass, $maxf$ is the boid's maximal force and $maxs$ is the boid's maximal speed.*

$$q(t) = (p(t), v(t), r, m, maxf, maxs), \ q(t) \in Q, \tag{6}$$

$$P_f(x(t), q_c(t)) = \{(s_i, q_{B_i}(t)) : B_i \in x(t), D(q_c(t), q_{B_i}(t)) < r\}, \tag{7}$$

$$s_i = 1 - \left(\frac{D(q_c(t), q_{B_i}(t))}{r} \right)^2. \tag{8}$$

The perception function P_f is defined with eq.(7). At any discrete time step t it, based on the input $x(t) \in X$ and the observed boid's state $q_c(t)$, returns a set of pairs $(s_i, q_{B_i}(t))$, where $q_{B_i}(t)$ is the internal state of boid B_i, whose distance from the observed boid is less then r and s_i is the level of importance of boid B_i. The importance s_i decreases with the square of distance. The distance metric $D(q_c(t), q_{B_i}(t))$ is Euclidean distance and is in the case of $d = 2$ given with eq.(9).

$$D(q_c(t), q_{B_i}(t)) = \sqrt{(q_c(t).p.x - q_{B_i}(t).p.x)^2 + (q_c(t).p.y - q_{B_i}(t).p.y)^2}. \tag{9}$$

The steering functions vector S is represented by three steering functions named separation S_s, alignment S_a and cohesion S_c. Stated briefly as rules, and in order of decreasing precedence, they are [5]:

− *separation*: avoid collisions with nearby neighbours - attempt to keep an "appropriate" distance from nearby neighbours,

- *alignment*: attempt to match the speed and direction of flight with nearby neighbours,
- *cohesion*: attempt to stay close to nearby neighbours - attempt to move into the centre of the nearby neighbours.

The boid's next discrete time step state $q(t+1) \in Q$ is calculated by combining these three urges - steering forces. The formal definitions of the S_s, S_a, S_c steering functions as well as of the behaviour function B_{pa} can be found in [2], [5] and will be omitted in this paper.

3   Fuzzification of Boids

In this section we will implement one of the boid's urges in fuzzy logic and through simulation show its advantages. Simulation showed [2] that the urge of alignment has the biggest influence on the boid's behaviour when it is a member of a flock. What is more, it showed that this urge is also the most suitable for the fuzzy logic implementation. As already mentioned, the foremost purpose of this urge is to match the speed and direction of flight with those of the nearby neighbours.

According to the definition of the animat, the result of the steering function is a steering force, which is, according to the definition of the boid, the force required to achieve the desired next discrete time step state. The alignment steering function S_a is defined with eq.(10), where N_f is the set of neighbour states returned by the perception function P_f:

$$S_a(N_f, q_c(t)) = \left(\frac{1}{|N_f|} \sum_{(s_i, q_{B_i}) \in N_f} (q_c(t).v + s_i(q_{B_i}(t).v - q_c(t).v)) \right) - q_c(t).v.$$

(10)

The alignment steering function is concerned only with the speed and direction of flight of the nearby neighbours and ignores their positions. The vectors $q_c(t).v$ and $q_{B_i}(t).v$ therefore represent the observed boid's and the neighbour's velocity vectors. The velocity vector gives the relative position changes per coordinate axis in the Cartesian coordinate system and thus describes the boid's speed and direction of flight at time step t. The main distinction of the alignment urge is its predictive collision avoidance [5]. This is mainly caused by the reason that if a boid does a good job aligning with its nearby neighbours, it is unlikely that it will collide with any of them in the near future.

Let us represent s_i - the level of importance of boid B_i - with a linguistic variable Imp composed of three fuzzy sets (LOW, MEDIUM and HIGH). To continue, we shall use a linguistic variable $OrDiff$ composed of three fuzzy sets (LEFT, SAME and RIGHT) to represent the relative difference of the direction of flight between the observed boid and boid B_i. Finally, we shall use a linguistic variable $SpdDiff$ composed of three fuzzy sets (SLOWER, SAME and FASTER) to represent the relative difference of the speed of flight between the observed boid and boid B_i.

Let us declare the linguistic variables $OrChng$ and $SpdChng$, which can be decomposed into the same set of terms as $OrDiff$ and $SpdDiff$, and represent the desired orientation and speed changes respectively. Then the following rules give the fuzzy alignment steering function:

```
if (OrDiff is SAME) then (OrChng is SAME),
if (Imp is LOW) and (OrDiff is RIGHT) then (OrChng is SAME),
if (Imp is MEDIUM) and (OrDiff is RIGHT) then (OrChng is RIGHT),
if (Imp is HIGH) and (OrDiff is RIGHT) then (OrChng is RIGHT),
if (Imp is LOW) and (OrDiff is LEFT) then (OrChng is SAME),
if (Imp is MEDIUM) and (OrDiff is LEFT) then (OrChng is LEFT),
if (Imp is HIGH) and (OrDiff is LEFT) then (OrChng is LEFT),
if (SpdDiff is SAME) then (SpdChng is SAME),
if (Imp is LOW) and (SpdDiff is FASTER) then (SpdChng is SAME),
if (Imp is MEDIUM) and (SpdDiff is FASTER) then (SpdChng is FASTER),
if (Imp is HIGH) and (SpdDiff is FASTER) then (SpdChng is FASTER),
if (Imp is LOW) and (SpdDiff is SLOWER) then (SpdChng is SAME),
if (Imp is MEDIUM) and (SpdDiff is SLOWER) then (SpdChng is SLOWER),
if (Imp is HIGH) and (SpdDiff is SLOWER) then (SpdChng is SLOWER).
```

4   The Results of Experiments

The two graphs in Fig. 1 at point (x, y) give the alignment steering force in the case when the observed boid is at location (x, y) travelling away from the centre with speed $maxs$ and its only neighbour is at location $(0,0)$ travelling in the positive y direction with speed $maxs$. The left graph stands for the crisp implementation of the function, whereas the right stands for our fuzzy logic implementation. It can be seen that even with a simple set of fuzzy logic rules, for which we did not use fitting or other forms of automatic generation, we get a remarkably similar mapping. It can also be noticed that with the fuzzy implementation distant neighbours have less impact on the change of direction and speed of the observed boid (see outer perimeter in Fig. 1). The similarity of the two implementations is even more evident in the metrics, which will be presented later on.

To test the quality of the alignment steering function based on fuzzy logic we run a simple experiment. The experiment included fifty boids in an uninteresting environment; an environment without obstacles. Every boid had random initial position, speed and direction. The other parameters of their states were fixed and equal for all boids. We ran 2000 steps of the simulation, where at each frame we measured the cumulative number of collisions, the number of flocks, average flock speed, average flock speed variation, average flock direction and average flock direction variation. In the paper we will, for reasons of limited space, present only the graphs of the most interesting metrics, the others will be only commented upon.

The cumulative number of collisions is the same in both cases, although the collisions in the case of the fuzzy logic implementation happen earlier in the

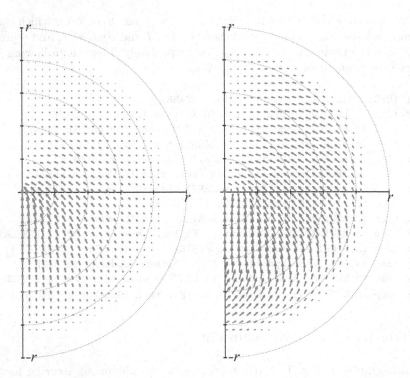

Fig. 1. Graph of the alignment steering function (left) crisp, (right) fuzzy implementation.

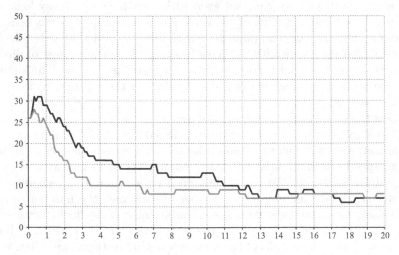

Fig. 2. Number of flocks: (black) crisp, (grey) fuzzy.

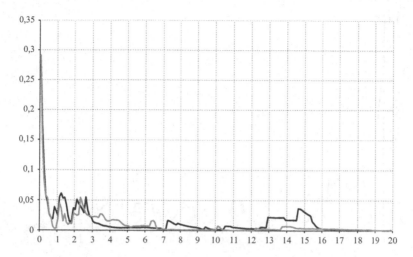

Fig. 3. Average flock speed variation (black) crisp, (grey) fuzzy logic.

simulation. The graph in Fig. 2 shows the number of flocks. As we can see, the fuzzy logic implementation in this case generates the flocks faster than the crisp implementation. Both versions shows similar signs of being unable to keep the flocks together, which would, in an uninteresting environment, not be expected. Nevertheless this deficiency is less evident in the fuzzy logic implementation. Therefore we can conclude that due to the almost identical tendencies of both implementations they behave almost identically where the fuzzy logic implementation has a slight lead over the crisp implementation.

The graph in Fig. 3 shows the average flock speed variation. As we can see the graph also shows similar tendencies of both implementations. Nevertheless it looks as if the fuzzy logic implementation gives better results since the last part of the graph has no turbulences (see frames 120-160).

The graph in Fig. 4 shows the average flock direction variation and similarly as Fig. 3 shows almost identical tendencies of both implementations. The earlier mentioned superiority of the fuzzy logic implementation is even more evident in this graph (see frames 75-185).

5   Acknowledgements

The work presented in this paper was done at the Computer Structures and Systems Laboratory, Faculty of Computer and Information Science, Ljubljana, Slovenia and is part of the Ph.D. thesis being prepared by I. Lebar Bajec.

6   Conclusion

In this paper we explore the use of fuzzy logic as a tool for the construction of artificial animals (animats). We limited our research to the construction of a

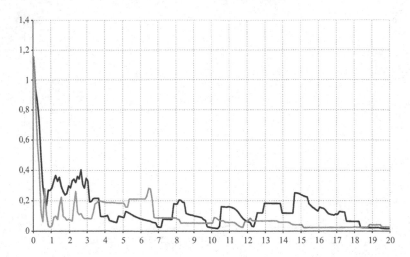

Fig. 4. Average flock direction variation (black) crisp, (grey) fuzzy logic.

boid - a special type of animat. In our case, analogous to other fields of modelling [3], [4], [6], the fuzzy logic approach results as a more suitable and user friendlier than the traditional crisp numerical approaches. We introduced a set of simple linguistic rules that describe the boid's urge of alignment when in a flock. The behaviour of a group of boids that uses these rules is comparable to the behaviour of a group of boids that uses the original geometrical function. This proves that a flock member can base its decisions purely on unclear evaluations of its environment and linguistic rules even without the knowledge of the Newton's laws of motion.

References

1. Kohavi Z.: Switching and Finite Automata Theory. McGraw-Hill, Inc., (1978).
2. Lebar Bajec I.: Computer model of bird flocking. MSc Thesis, University of Ljubljana, Faculty of Computer and Information Science, Ljubljana, Slovenia, (2002).
3. Mraz M., Lapanja I., Zimic N., Virant J.: Notes on fuzzy cellular automata. Journal of Chinese Institute of Industrial Engineers, Vol.17, No.5, 469-476, (2000).
4. Mraz M., Zimic N., Virant J.: Intelligent bush fire spread prediction using fuzzy cellular automata, Journal of Intelligent and Fuzzy Systems, Vol.7, 203-207, (1999).
5. Reynolds C.W.: Flocks, Herds, and Schools: A Distributed Behavioral Model, Computer Graphics (SIGGRAPH87 Conference Proceedings), Vol.21, No.4, 25-34, (1987).
6. Virant J.: Design considerations of time in fuzzy systems. Kluwer Academic Publishers, (2000).
7. Wilson S.W.: Knowledge Growth in an Artificial Animal, Proceedings of ICGA'85, Pittsburgh, PA, 16-23, (1985).

Approximating Fuzzy Control Strategies via CRI

Siegfried Gottwald[1], Vilem Novák[2], and Irina Perfilieva[2]

[1] Leipzig University, Institute for Logic, Leipzig, Germany
gottwald@uni-leipzig.de
[2] Ostrava University, IRAFM, Ostrava, Czech Republic
{Vilem.Novak/Irina.Perfilieva}@osu.cz

Abstract. We start from the observation that ZADEH's compositional rule of inference (CRI) is a strategy to determine approximately a roughly given control function.
From this point of view the problem to solve a system of fuzzy relation equations also becomes the problem to determine approximately such a control strategy. This gives a natural interpretation for approximate solutions of unsolvable systems of relation equations.
Therefore we discuss and generalize some approaches and results about approximate solutions of systems of relation equations.
Finally we discuss here how a choice of the t-norm which is involved in the sup-t-composition modifies the solvability behavior.

Key words: fuzzy control, approximation strategies, compositional rule of inference, fuzzy relation equations, solvability behavior, sup-t-composition

1 Introduction

The standard mathematical understanding of approximation is that by an approximation process some mathematical object A, e.g. some function, is approximated, i.e. determined within some (usually previously unspecified) error bounds.

Additionally one assumes that the approximating object B for A is of some predetermined kind, e.g. a polynomial function.

So one may approximate some transcendental function, e.g. the trajectory of some non-linear process, by a piecewise linear function, or by a polynomial function of some bounded degree. Similarly one approximates e.g. in the Runge-Kutta methods the solution of a differential equation by a piecewise linear function, or one uses splines to approximate a difficult surface in 3-space by planar pieces.

2 CRI Is an Approximation Strategy

In the context of fuzzy control the object which has to be determined, viz. the control function Φ, additionally is described only roughly, i.e. given only by its behavior in some (fuzzy) points of the state space.

T. Bilgiç et al. (Eds.): IFSA 2003, LNAI 2715, pp. 203–210, 2003.

The way to roughly describe the control function is to give (in the simplest case of one input variable) a list

$$\text{IF } \alpha \text{ is } A_i, \text{ THEN } \beta \text{ is } B_i, \quad i = 1, \dots, n, \tag{1}$$

of linguistic control rules connecting fuzzy subsets A_i of the input space $\mathcal{X}$ with fuzzy subsets B_i of the output space $\mathcal{Y}$, understood as indicating that one likes to have

$$\Phi^*(A_i) = B_i, \quad i = 1, \dots, n \tag{2}$$

for a suitable "fuzzified" version $\Phi^* : \mathbb{F}(\mathcal{X}) \to \mathbb{F}(\mathcal{Y})$ of the control function $\Phi : \mathcal{X} \to \mathcal{Y}$.

And the additional approximation idea explained in ZADEH's compositional rule of inference (CRI) is that one likes to approximate Φ^* by a fuzzy function $\Psi^* : \mathbb{F}(\mathcal{X}) \to \mathbb{F}(\mathcal{Y})$ which is determined for all $A \in \mathbb{F}(\mathcal{X})$ by the equation

$$\Psi^*(A) = A \circ R \tag{3}$$

which refers to some suitable fuzzy relation $R \in \mathbb{F}(\mathcal{X} \times \mathcal{Y})$, and understands $\circ$ as sup-t-composition.

Formally this means that the equations (2) become transformed into the system

$$A_i \circ R = B_i, \quad i = 1, \dots, n \tag{4}$$

of relation equations which has to be solved w.r.t. the unknown fuzzy relation R.

This approximation idea fits well with the fact that one often is satisfied with pseudo-solutions of (4), to use the terminology introduced in [3], and particularly with the MA-pseudo-solution R_{MA} of MAMDANI/ASSILIAN [7], or the S-pseudo-solution $\widehat{R}$ of SANCHEZ [10].

Neither R_{MA} nor $\widehat{R}$ needs to be a solution of (4), however both these pseudo-solutions determine approximations Ψ^* to the (fuzzified) control function Φ^*.

3   Approximate Solutions of Fuzzy Relation Equations

The authors of this paper used in their previous papers, e.g. in [1,3], the notion of approximate solution only naively in the previously explained sense of a fuzzy relation which roughly describes the intended control behavior which some list (1) of (linguistic) control rules describes.

A precise qualitative definition of a notion of approximate solution was given by WU [11] and used e.g. by KLIR/YUAN [4,5]. In this approach an approximate solution $\widetilde{R}$ of (4) is defined as a fuzzy relation which satisfies the following two conditions:

1. There are fuzzy sets A_i', B_i' such that for all $i = 1, \dots, n$ one has $A_i \subseteq A_i'$ and $B_i' \subseteq B_i$ as well as $\widetilde{R}'' A_i' = B_i'$.

2. If there exist fuzzy sets A_i^*, B_i^* for $i = 1, \ldots, n$ and a fuzzy relation R^* such that $R^{*\prime\prime} A_i^* = B_i^*$ and $A_i \subseteq A_i^* \subseteq A_i{}', B_i{}' \subseteq B_i^* \subseteq B_i$ for all $i = 1, \ldots, n$ then one has $A_i^* = A_i{}'$ and $B_i^* = B_i{}'$ for all $i = 1, \ldots, n$.

These conditions formalize the ideas that (1) an approximate solution $\widetilde{R}$ should be a solution of a system of relation equations with input-output data $(A_i{}', B_i{}')$ which may (slightly) differ from the original input-output data (A_i, B_i) and which (2) has the additional property that no system of relational equations with input-output data which "strongly better" approximate the original ones is solvable.

4   Some Generalizations of WU's Approach

It is obvious that the two conditions (1), (2) of WU are independent.

What is, however, not obvious at all – and even rather arbitrary – is that condition (1) also says that the approximating input-output data $(A_i{}', B_i{}')$ should approximate the original input data *from above*[1] and the original output data *from below*. To overcome this artificial restriction we redefine the crucial notion of approximate solution here in the following way.

Before we give the generalizing definition we coin the name of an *approximating system* for (4) and understand by it any system

$$C_i \circ R = D_i, \qquad i = 1, \ldots, n \tag{5}$$

of relation equations with the same number of equations.

Definition 1 *A ul-approximate solution of a system (4) of relation equations is a solution of a ul-approximating system for (4), i.e. of an approximating system (5) for (4) which satisfies*

$$A_i \subseteq C_i \quad and \quad B_i \supseteq D_i, \qquad for\ i = 1, \ldots, n. \tag{6}$$

An lu-approximate solution of a system (4) of relation equations is a solution of an lu-approximating system (5) for (4), i.e. of a system which satisfies

$$A_i \supseteq C_i \quad and \quad B_i \subseteq D_i, \qquad for\ i = 1, \ldots, n. \tag{7}$$

An l*-approximate solution of a system (4) of relation equations is a solution of an l*-approximating system (5) for (4), i.e. of a system which satisfies

$$A_i \supseteq C_i \quad and \quad B_i = D_i, \qquad for\ i = 1, \ldots, n. \tag{8}$$

In a similar way one may define the notions of ll-approximate solution, of uu-approximate solution, of u*-approximate solution, of *l-approximate solution, and of *u-approximate solution.

---
[1]   The ordering we refer to here and later on in this discussion is the usual inclusion for fuzzy sets.

Then one immediately has e.g.

Corollary 1 (*i*) *Each *l-approximate solution of* (4) *is also an ul-approximate solution and an ll-approximate solution of* (4).

(*ii*) *Each u*-approximate solution of* (4) *is also an ul-approximate solution and an uu-approximate solution of* (4).

Proposition 2 *For each system* (4) *of relation equations its S-pseudo-solution $\widehat{R}$ is an *l-approximate solution.*

This generalizes a result of KLIR/YUAN [4], cf. also [5].

Proposition 3 *For each system* (4) *of relation equations with normal input data its MS-pseudo-solution R_{MA} is an *u-approximate solution.*

Together with Corollary 1 these two Propositions say that each system of relation equations has approximate solutions of any one of the types we introduced in this section.

5   Optimality of Approximate Solutions

All the previous results do not give any information about some kind of "quality" of the approximate solutions or the approximating systems. This is to some extent related to the fact that up to now we disregarded in our modified terminology WU's condition (2) which is a kind of optimality condition.

Definition 2 *An inclusion based approximate solution $\widetilde{R}$ of a system* (4) *is called* optimal *iff there does not exist a solvable system $R''C_i' = D_i'$ of relation equations whose input-output data (C_i', D_i') approximate the original input-output data of* (4) *strongly better than the input-output data (C_i, D_i) of the system which determines $\widetilde{R}$.*

Proposition 4 *If an inclusion based *l-approximate solution $\widetilde{R}$ is optimal, then it is also optimal as an ul-approximate solution and as a ll-approximate solution.*

Of course, similar results holds true also for (l*-, u*- and) *u-approximate solutions.

6   Some Optimality Results

The problem arises immediately whether the two standard pseudo-solutions $\widehat{R}$ and R_{MA} are optimal (inclusion based) approximate solutions. For the S-pseudo-solution $\widehat{R}$ as an ul-approximate solution this optimality was shown in [4,5].

We have that $\widehat{R}$ is even an $\subseteq$-optimal *l-approximate solution. The proofs of this and the other results of this Section can be found in [9].

Proposition 5 *The fuzzy relation $\widehat{R}$ is always an $\subseteq$-optimal $*l$-approximate solution.*

For the MA-pseudo-solution the situation is different.

Proposition 6 *There exist systems (4) of relation equations for which their MA-pseudo-solution R_{MA} is an $*u$-approximate solution which is not optimal.*

A closer inspection of the proof of Proposition 5 shows that the crucial difference of the previous optimality result for $\widehat{R}$ to the present situation of R_{MA} is that in the former case the solvable approximating system has its own (largest) solution $\widehat{S}$. But in the present situation a solvable approximating system may fail to have his MA-pseudo-solution R_{MA} as a solution.

However, this remark leads us to a partial optimality result w.r.t. the MA-pseudo-solution.

Definition 3 *Let us call a system (4) of relation equations MA-solvable iff its MA-pseudo-solution R_{MA} is a solution of this system.*

Then we have the following result.

Proposition 7 *If a system (4) of relation equations has an MA-solvable $*u$-approximating system*

$$A_i \circ R = B_i^*, \qquad i = 1, \ldots, n \qquad (9)$$

such that for the MA-pseudo-solution R_{MA} of (4) one has

$$B_i \subseteq B_i^* \subseteq A_i \circ R_{\mathrm{MA}}, \qquad i = 1, \ldots, n,$$

then one has

$$B_i^* = A_i \circ R_{\mathrm{MA}} \qquad \text{for all} \quad i = 1, \ldots, n.$$

Corollary 8 *If all input sets of (4) are normal then the system*

$$R'' A_i = A_i \circ R_{\mathrm{MA}}, \qquad i = 1, \ldots, n, \qquad (10)$$

is the smallest MA-solvable $*u$-supersystem for (4).

Corollary 9 *Let $\widehat{R}$ be the S-pseudo-solution of (4), let be $\widehat{B_i} = A_i \circ \widehat{R}$ for $i = 1, \ldots, n$, and suppose that the modified system*

$$A_i \circ R = \widehat{B_i}, \qquad i = 1, \ldots, n, \qquad (11)$$

is MA-solvable. Then the iterated pseudo-solution $R_{\mathrm{MA}}[\widehat{R}[B_k]'' A_k]$, introduced in [3], is an optimal $*l$-approximate solution of (4).

This last Proposition can be further generalized. To do this assume that $\mathbb{S}$ is some *pseudo-solution strategy*, i.e. some mapping from the class of families $(A_i, B_i)_{1 \leq i \leq n}$ of input-output data pairs into the class of fuzzy relations, which yields for any given system (4) of relation equations an $\mathbb{S}$-pseudo-solution $R_{\mathbb{S}}$. Of course the system (4) will be called $\mathbb{S}$-*solvable* iff $R_{\mathbb{S}}$ is a solution of the system (4).

Definition 4 *We shall say that the $\mathbb{S}$-pseudo-solution $R_{\mathbb{S}}$ depends isotonically (w.r.t. inclusion) on the output data of the system (4) of relation equations iff the condition*

$$\text{if} \quad B_i \subseteq B_i' \quad \text{for all} \quad i = 1, \ldots, n \quad \text{then} \quad R_{\mathbb{S}} \subseteq R_{\mathbb{S}}'.$$

holds true for the $\mathbb{S}$-pseudo-solutions $R_{\mathbb{S}}$ of the system (4) and $R_{\mathbb{S}}'$ of an "output-modified" system $R'' A_i = A_i \circ R = B_i'$, $i = 1, \ldots, n$.

Definition 5 *Furthermore we understand by an $\mathbb{S}$-optimal *u-approximate solution of the system (4) the $\mathbb{S}$-pseudo-solution of an $\mathbb{S}$-solvable *u-approximating system of (4) which has the additional property that no strongly better *u-approximating system of (4) is $\mathbb{S}$-solvable.*

Proposition 10 *Suppose that the $\mathbb{S}$-pseudo-solution depends isotonically (w. r.t. inclusion) on the output data of the systems of relation equations. Assume furthermore that for the $\mathbb{S}$-pseudo-solution $R_{\mathbb{S}}$ of (4) one always has $B_i \subseteq R_{\mathbb{S}}'' A_i$ (or that one always has $R_{\mathbb{S}}'' A_i \subseteq B_i$) for all $i = 1, \ldots, n$. Then the $\mathbb{S}$-pseudo-solution $R_{\mathbb{S}}$ of (4) is an $\mathbb{S}$-optimal *u-approximate (or: *l-approximate) solution of the system (4).*

It is immediately clear that Corollary 8 is the particular case of the MA-pseudo-solution strategy. But also Proposition 5 is a particular case of this Proposition: the case of the S-pseudo-solution strategy (having in mind that S-solvability and solvability are equivalent notions).

7   Modifying the Approximation Behavior via T-norm Changes

Now we make some remarks which take into account the change of the basic t-norm our considerations are based upon. Again we omit the proofs which are not hard to find.

Let us use for t-norms t_1 and t_2 their ordering $\leqq$ as functions given by

$$t_1 \leqq t_2 \quad \text{iff} \quad \text{always } t_1(x, y) \leq t_2(x, y),$$

and let us call a t-norm t_1 *weaker* than a t-norm t_2 iff $t_1 \leqq t_2$ holds true.

Related to two left continuous t-norms t_k for $k = 1, 2$ let us simplify our notation and write $\&_k$ and $\to_k$ for the corresponding t-norm based connectives $\&_{t_k}, \to_{t_k}$ of the formalized language. Let us furthermore write

$$A \rhd_k B \text{ for } A \rhd_{t_k} B, \qquad \widehat{R}^k \text{ for } \widehat{R}^{(t_k)},$$

$$A \times_k B \text{ for } A \times_{t_k} B, \qquad R^k{}_{\text{MA}} \text{ for } R^{(t_k)}{}_{\text{MA}}.$$

It is well known, and proved e.g. in [2], that for the truth degree functions seq_k of the R-implications related with the (left continuous) t-norms t_i for $k = 1, 2$ one has

$$t_1 \leqq t_2 \quad \text{iff} \quad \text{seq}_2 \leqq \text{seq}_1,$$

which immediately gives the next result.

Proposition 11 *Let two t-norms t_1, t_2 be given. Then one has for the MA-pseudo-solutions $R^k{}_{\text{MA}}$ and the S-pseudo-solutions $\widehat{R}^k$ of a given system (4) of relation equations the relationships*

$$t_1 \leqq t_2 \Rightarrow R^1{}_{\text{MA}} \subseteq R^2{}_{\text{MA}}, \tag{12}$$

$$t_1 \leqq t_2 \Rightarrow \widehat{R}^2 \subseteq \widehat{R}^1. \tag{13}$$

Then it is a routine matter to prove also the

Proposition 12 *For any two t-norms t_1, t_2 with $t_1 \leqq t_2$ one has for any system (4) of relation equations and for each t-norms t the inclusion relations*

$$A_i \circ_t \widehat{R}^2 \subseteq A_i \circ_t \widehat{R}^1 \subseteq B_i \subseteq A_i \circ_t R^1_{\text{MA}} \subseteq A_i \circ_t R^2_{\text{MA}}. \tag{14}$$

which means that the two standard pseudo-solutions w.r.t. the weaker t-norm give better approximate solutions for the system (4) of relation equations.

To evaluate this result, the reader has to have in mind that in (14) the t_k-related pseudo-solutions are used together with sup-t-composition for an arbitrary t-norm t. The standard situation is, however, to consider t_k-related pseudo-solutions together with sup-t_k-composition. For this standard situation we get

Corollary 13 *For any two t-norms t_1, t_2 with $t_1 \leqq t_2$ one has for any system (4) of relation equations w.r.t. their MA-pseudo-solutions*

$$B_i \subseteq A_i \circ_{t_1} R^1_{\text{MA}} \subseteq A_i \circ_{t_2} R^2_{\text{MA}},$$

which means that weaker t-norms provide better MA-pseudo-solvability.

A similar result w.r.t. the S-pseudo-solutions is unknown: and it is an open problem whether such a result may hold at all. For, from (14) one gets for t-norms $t_1 \leqq t_2$ only

$$A_i \circ_{t_1} \widehat{R}^2 \subseteq A_i \circ_{t_1} \widehat{R}^1 \quad \text{and} \quad A_i \circ_{t_1} \widehat{R}^2 \subseteq A_i \circ_{t_2} \widehat{R}^2.$$

For the particular case of MA-solvable systems of relation equations this result gives immediately

Corollary 14 *If a system (4) of relation equations is MA-solvable w.r.t. some t-norm t then it is also MA-solvable w.r.t. every weaker t-norm.*

And generally one has also the following evaluation for the most preferred choice of t-norms.

Corollary 15 *For the largest t-norm* min *the MA-pseudo-solution R_{MA} gives the worst possible MA-approximation quality among all t-norms.*

References

1. GOTTWALD, S. (1993): *Fuzzy Sets and Fuzzy Logic*. The Foundations of Application – from a Mathematical Point of View. Vieweg: Braunschweig/Wiesbaden and Teknea: Toulouse.
2. GOTTWALD, S. (2001): *A Treatise on Many-Valued Logics*. Research Stud. Press, Baldock.
3. GOTTWALD, S., NOVAK, V. and I. PERFILIEVA (2002): Fuzzy control and t-norm-based fuzzy logic. Some recent results. In: *Proc. 9th Internat. Conf. IPMU 2002*, vol. 2, ESIA – Université de Savoie: Annecy 2002, 1087–1094.
4. KLIR, G. and B. YUAN (1994): Approximate solutions of systems of fuzzy relation equations. In: *FUZZ-IEEE '94*. Proc. 3rd Internat. Conf. Fuzzy Systems, June 26-29, 1994, Orlando/FL, 1452–1457.
5. KLIR, G. and B. YUAN (1995): *Fuzzy Sets and Fuzzy Logic*. Theory and Applications. Prentice Hall: Upper Saddle River.
6. KLAWONN, F. (2001): Fuzzy points, fuzzy relations and fuzzy functions. In: *Discovering the World with Fuzzy Logic* (V. Novák, I. Perfilieva eds.) Advances in Soft Computing, Physica-Verlag: Heidelberg 2000, 431–453.
7. MAMDANI, A. and S. ASSILIAN (1975): An experiment in linguistic synthesis with a fuzzy logic controller. *Internat. J. Man-Machine Studies* **7**, 1–13.
8. NOVÁK, V., PERFILIEVA, I. and J. MOČKOŘ (1999): *Mathematical Principles of Fuzzy Logic*. Kluwer Acad. Publ., Boston.
9. PERFILIEVA, I. and S. GOTTWALD (200x): Solvability and approximate solvability of fuzzy relation equations (submitted).
10. SANCHEZ, E. (1976): Resolution of composite fuzzy relation equations. *Information and Control* **30**, 38–48.
11. WU WANGMING (1986): Fuzzy reasoning and fuzzy relation equations, *Fuzzy Sets Systems* **20**, 67–78.

Inequalities in Fuzzy Probability Calculus

Saskia Janssens[1], Bernard De Baets[1], and Hans De Meyer[2]

[1] Department of Applied Mathematics, Biometrics and Process Control
Ghent University, Coupure links 653, B-9000 Gent, Belgium
{Saskia.Janssens,Bernard.DeBaets}@rug.ac.be
[2] Department of Applied Mathematics and Computer Science
Ghent University, Krijgslaan 281 (S9), B-9000 Gent, Belgium
Hans.DeMeyer@rug.ac.be

Abstract. We present all Bell-type inequalities concerning at most four random events of which not more than two are intersected at the same time. Reformulating these inequalities in the context of fuzzy probability calculus leads to related inequalities on commutative conjunctors, and in particular on triangular norms and commutative copulas. For the most important parametric families of t-norms, we identify the parameter values for which each of the inequalities is fulfilled. Some of the inequalities hold for any commutative copula, while all of them are preserved under ordinal sums.

1 Introduction

Pykacz and D'Hooghe [6] recently studied which of the numerous Bell-type inequalities that are necessarily satisfied by Kolmogorovian probabilities may be violated in various models of fuzzy probability calculus. They showed that the most popular model of fuzzy probability calculus based on min and max cannot be distinguished from the Kolmogorovian model by any of the inequalities studied by Pitowsky [5]. They also proved that if we consider fuzzy set intersection pointwisely generated by a Frank t-norm T_λ^F, then the borderline between models of fuzzy probability calculus that can be distinguished from Kolmogorovian ones and models that cannot be distinguished (by the same set of inequalities) is situated at $\lambda = 9 + 4\sqrt{5}$.

The Bell-type inequalities are not only of interest to fuzzy probability calculus, but they also appear in other applications of fuzzy logic. One particular inequality for instance has shown to be of primordial importance in the design of transitivity-preserving fuzzification schemes for cardinality-based similarity measures [1].

Our paper is organized as follows. In the following section, we discuss various models of fuzzy intersection based on triangular norms and copulas. In Section 3, we generate all Bell-type inequalities in which the number of random events is smaller than or equal to four and in which at most two events are intersected at the same time. To that end, we have used the cdd package of Fukuda (URL http://www.ifor.math.ethz.ch/~fukuda/cddhome/cdd.html), which is an

T. Bilgiç et al. (Eds.): IFSA 2003, LNAI 2715, pp. 211–218, 2003.
© Springer-Verlag Berlin Heidelberg 2003

efficient implementation of the double description method [3]. In Section 4, a straightforward interpretation of these inequalities in fuzzy probability calculus leads to similar inequalities on commutative conjunctors. Finally, in Section 5 we present some general results for commutative copulas and ordinal sums, as well as specific results for the most important parametric families of t-norms. Specific attention is given to the product t-norm $T_1^{\mathbf{F}} = T_{\mathbf{P}}$.

2   Conjunctors, Triangular Norms, and Copulas

As usual, we define the intersection of two fuzzy sets A and B pointwisely, i.e. $A \cap B(x) = f(A(x), B(x))$, by means of an appropriate function f that generalizes Boolean conjunction. Since in this paper we will intersect at most two fuzzy sets at the same time, it suffices to consider as suitable f a commutative conjunctor.

Definition 1. *A* conjunctor *is a binary operation f on $[0,1]$, i.e. a function $f : [0,1]^2 \rightarrow [0,1]$, such that for all $x_1, x_2, y_1, y_2 \in [0,1]$ with $x_1 \leq x_2$ and $y_1 \leq y_2$, it holds that:*

(i) $f(x_1, y_1) \leq f(x_2, y_2)$;
(ii) $f(x_1, 1) = f(1, x_1) = x_1$.

A commutative conjunctor is generically denoted by I. Two interesting classes of commutative conjunctors are the class of triangular norms (t-norms for short) and the class of commutative copulas. Triangular norms were introduced by Menger in 1942 and permit to define a kind of triangle inequality in the setting of probabilistic metric spaces. Copulas were introduced by Sklar in 1959 [7] and are used for combining marginal probability distributions into joint probability distributions. We adopt here the definitions and notations from [2,4].

Definition 2. *A* t-norm *is an associative and commutative conjunctor, and is generically denoted by T.*

Definition 3. *A* copula *is a binary operation C on $[0,1]$, i.e. a function $C : [0,1]^2 \rightarrow [0,1]$, such that for all $x_1, x_2, y_1, y_2 \in [0,1]$ with $x_1 \leq x_2$ and $y_1 \leq y_2$, it holds that:*

(i) $C(x_1, y_1) + C(x_2, y_2) \geq C(x_1, y_2) + C(x_2, y_1)$;
(ii) $C(x_1, 0) = C(0, x_1) = 0$;
(iii) $C(x_1, 1) = C(1, x_1) = x_1$.

Condition (i) is called the property of *moderate growth* and implies that any copula is 1-Lipschitz. Among the four main t-norms, the minimum operator $T_{\mathbf{M}}$, the Łukasiewicz t-norm $T_{\mathbf{L}}$ and the algebraic product $T_{\mathbf{P}}$ are (associative and commutative) copulas. The drastic product $T_{\mathbf{D}}$ is not a copula.

Proposition 1. *For each copula C it holds that $T_{\mathbf{L}} \leq C \leq T_{\mathbf{M}}$.*

Table 1 lists the parametric families of t-norms used in this paper and indicates for which parameter values the corresponding t-norm is a copula.

Table 1. Parametric families of t-norms and their subfamilies of copulas.

T-norm family	$T_\lambda(x,y) =$	Copulas for
Frank	$\begin{cases} T_{\mathbf{M}}(x,y) & \text{if } \lambda = 0 \\ T_{\mathbf{P}}(x,y) & \text{if } \lambda = 1 \\ T_{\mathbf{L}}(x,y) & \text{if } \lambda = +\infty \\ \log_\lambda[1 + \frac{(\lambda^x-1)(\lambda^y-1)}{(\lambda-1)}] & \text{if } \lambda \in]0,+\infty[\,\backslash\{1\} \end{cases}$	$\lambda \in \mathbb{R}^+$
Hamacher	$\begin{cases} T_{\mathbf{D}}(x,y) & \text{if } \lambda = +\infty \\ 0 & \text{if } \lambda = x = y = 0 \\ \frac{xy}{\lambda+(1-\lambda)(x+y-xy)} & \text{otherwise} \end{cases}$	$\lambda \in [0,2]$
Schweizer–Sklar	$\begin{cases} T_{\mathbf{M}}(x,y) & \text{if } \lambda = -\infty \\ T_{\mathbf{P}}(x,y) & \text{if } \lambda = 0 \\ T_{\mathbf{D}}(x,y) & \text{if } \lambda = +\infty \\ \max(x^\lambda + y^\lambda - 1, 0)^{\frac{1}{\lambda}} & \text{otherwise} \end{cases}$	$\lambda \in [-\infty, 1]$
Sugeno–Weber	$\begin{cases} T_{\mathbf{D}}(x,y) & \text{if } \lambda = -1 \\ T_{\mathbf{P}}(x,y) & \text{if } \lambda = +\infty \\ \max(\frac{x+y-1+\lambda xy}{1+\lambda}, 0) & \text{otherwise} \end{cases}$	$\lambda \in [0,+\infty]$
Dombi	$\begin{cases} T_{\mathbf{D}}(x,y) & \text{if } \lambda = 0 \\ T_{\mathbf{M}}(x,y) & \text{if } \lambda = +\infty \\ \frac{1}{1+((\frac{1-x}{x})^\lambda + (\frac{1-y}{y})^\lambda)^{\frac{1}{\lambda}}} & \text{otherwise} \end{cases}$	$\lambda \in [1,+\infty]$
Aczel–Alsina	$\begin{cases} T_{\mathbf{D}}(x,y) & \text{if } \lambda = 0 \\ T_{\mathbf{M}}(x,y) & \text{if } \lambda = +\infty \\ \exp^{-((-\log x)^\lambda + (-\log y)^\lambda)^{\frac{1}{\lambda}}} & \text{otherwise} \end{cases}$	$\lambda \in [1,+\infty]$
Yager	$\begin{cases} T_{\mathbf{D}}(x,y) & \text{if } \lambda = 0 \\ T_{\mathbf{M}}(x,y) & \text{if } \lambda = +\infty \\ \max(1 - ((1-x)^\lambda + (1-y)^\lambda)^{\frac{1}{\lambda}}, 0) & \text{otherwise} \end{cases}$	$\lambda \in [1,+\infty]$

3   Bell-Type Inequalities in Probability Theory

The probability of occurrence of a single random event A_i is denoted by $p_i = P(A_i)$ and the probability of the intersection of a pair of random events is denoted by $p_{ij} = P(A_i \cap A_j)$. Since p_i and p_{ij} are probabilities, the following inequalities hold:

$$0 \leq p_{ij} \leq p_i \leq 1,$$
$$0 \leq p_{ij} \leq p_j \leq 1,$$
$$p_i + p_j - p_{ij} \leq 1. \tag{1}$$

In the case of experiments concerning four random events in which at most two events are intersected at the same time and only considering four possible intersections, Pitowsky [5] found the following set of inequalities:

$$0 \leq p_i - p_{ij} - p_{ik} + p_{jk}, \tag{2}$$

$$p_i + p_j + p_k - p_{ij} - p_{ik} - p_{jk} \leq 1 \,, \tag{3}$$
$$-1 \leq -p_i - p_k + p_{ik} + p_{il} + p_{jk} - p_{jl} \leq 0 \,, \tag{4}$$

for different $i, j, k, l \in \{1, 2, 3, 4\}$. Inequalities (2) and (3) are called the Bell–Wigner inequalities. Inequalities (4) are only considered for the following indices:

$$-1 \leq -p_1 - p_3 + p_{13} + p_{14} + p_{23} - p_{24} \leq 0 \,,$$
$$-1 \leq -p_1 - p_4 + p_{14} + p_{13} + p_{24} - p_{23} \leq 0 \,,$$
$$-1 \leq -p_2 - p_3 + p_{23} + p_{24} + p_{13} - p_{14} \leq 0 \,,$$
$$-1 \leq -p_2 - p_4 + p_{24} + p_{23} + p_{14} - p_{13} \leq 0 \,,$$

and are referred to as the Clauser–Horne inequalities.

Next to the above inequalities, we have generated the remaining Bell-type inequalities (taking into account five and six possible intersections) using the cdd package of Fukuda :

$$0 \leq p_i + p_j + p_{ij} - p_{ik} - p_{il} - p_{jl} - p_{jk} + p_{kl} \,, \tag{5}$$
$$p_i + p_j + p_k + p_l - p_{ij} - p_{ik} - p_{il} - p_{jk} - p_{jl} - p_{kl} \leq 1 \,, \tag{6}$$
$$2p_i + 2p_j + 2p_k + 2p_l - p_{ij} - p_{ik} - p_{il} - p_{jk} - p_{jl} - p_{kl} \leq 3 \,, \tag{7}$$
$$-p_i + p_{ij} + p_{ik} + p_{il} - p_{jk} - p_{jl} - p_{kl} \leq 0 \,, \tag{8}$$
$$p_i + p_j + p_k - 2p_l - p_{ij} - p_{ik} + p_{il} - p_{jk} + p_{jl} + p_{kl} \leq 1 \,. \tag{9}$$

4   Bell-Type Inequalities for Commutative Conjunctors

We can rewrite the above-mentioned Bell-type inequalities in the context of fuzzy probability calculus. Let A_i and A_j be fuzzy sets in a finite universe with cardinality n. Modelling fuzzy set intersection pointwisely by means of a commutative conjunctor I, the inequality

$$0 \leq P(A_i) + P(A_j) - P(A_i \cap A_j) \leq 1$$

is equivalent to

$$0 \leq \sum_u \frac{A_i(u)}{n} + \sum_u \frac{A_j(u)}{n} - \sum_u \frac{I(A_i(u), A_j(u))}{n} \leq 1 \,.$$

A necessary and sufficient condition for the above double inequality to hold in any finite universe (any finite n) clearly is the double inequality

$$0 \leq x + y - I(x, y) \leq 1 \,,$$

for any $x, y \in [0, 1]$. In this way, we can establish equivalent conditions (inequalities) on the commutative conjunctor I for each of the Bell-type inequalities. To simplify the discussion of these inequalities we introduce a unique code I_i^j for

Table 2. Bell-type inequalities for commutative conjunctors.

code	Bell-type inequalities
I_2^1	$0 \leq x + y - I(x,y) \leq 1$
I_3^2	$0 \leq x - I(x,y) - I(x,z) + I(y,z)$
I_3^3	$0 \leq x + y + z - I(x,y) - I(x,z) - I(y,z) \leq 1$
I_4^4	$0 \leq x + t - I(x,z) - I(x,t) - I(y,t) + I(y,z) \leq 1$
I_4^5	$0 \leq x + t - I(x,y) - I(x,z) + I(x,t)$ $+ I(y,z) - I(y,t) - I(z,t)$
I_4^6	$x + y + z + t - I(x,y) - I(x,z) - I(x,t) - I(y,z)$ $- I(y,t) - I(z,t) \leq 1$
I_4^7	$2x + 2y + 2z + 2t - I(x,y) - I(x,z) - I(x,t)$ $- I(y,z) - I(y,t) - I(z,t) \leq 3$
I_4^8	$0 \leq x - I(x,y) - I(x,z) - I(x,t)$ $+ I(y,z) + I(y,t) + I(z,t)$
I_4^9	$x + y + z - 2t - I(x,y) - I(x,z) + I(x,t)$ $- I(y,z) + I(y,t) + I(z,t) \leq 1$

each inequality where i denotes the number of events involved and j is a sequential number. These inequalities are summarized in Table 2 and will be referred to as Bell-type inequalities for commutative conjunctors.

The Bell-type inequalities are of particular interest in the context of cardinalities of fuzzy sets. Consider for instance the inequality

$$\#(A \bigtriangleup B) + \#(B \bigtriangleup C) \geq \#(A \bigtriangleup C) \qquad (10)$$

which is valid for any ordinary sets A, B and C in a universe X of arbitrary dimension n. This inequality can be rewritten as

$$\#A + \#B - 2\#(A \cap B) + \#B + \#C - 2\#(B \cap C) \geq \#A + \#C - 2\#(A \cap C),$$

or also

$$2(\#B - \#(A \cap B) - \#(B \cap C) + \#(A \cap C)) \geq 0.$$

Modelling fuzzy set intersection by means of a commutative conjunctor I, the above inequality will remain true for fuzzy sets in finite universes if and only if I satisfies

$$2(x - I(x,y) - I(x,z) + I(y,z)) \geq 0, \qquad (11)$$

i.e. if and only if I satisfies inequality I_3^2, which holds for all commutative copulas (see Section 5).

If we are working in a setting where $\#(A \setminus B) = \#A - \#(A \cap B)$ and $\#(A \Delta B) = \#A + \#B - 2\#(A \cap B)$ hold for cardinalities of fuzzy sets, we can also conclude that inequality (10) will hold if inequality (11) is satisfied.

5   Inequalities for Commutative Copulas, Ordinal Sums, and Parametric Families of T-norms

We have studied the Bell-type inequalities for each of the parametric families of t-norms mentioned in Table 1. Before that, we mention some general results.

Theorem 1. *The following inequalities are fulfilled for any commutative copula C:*

$$I_2^1 : \ 0 \le x + y - C(x,y) \le 1,$$
$$I_3^2 : \ 0 \le x - C(x,y) - C(x,z) + C(y,z),$$
$$I_4^4 : \ 0 \le x + t - C(x,z) - C(x,t) - C(y,t) + C(y,z) \le 1,$$
$$I_4^5 : \ 0 \le x + t - C(x,y) - C(x,z) + C(x,t) + C(y,z) - C(y,t) - C(z,t).$$

The following theorem expresses that ordinal sums preserve the Bell-type inequalities. We first recall the definition of an ordinal sum in a slightly more general form.

Definition 4. *Consider a family $(I_\alpha)_{\alpha \in A}$ of commutative conjunctors and a family $(]a_\alpha, e_\alpha[)_{\alpha \in A}$ of non-empty, pairwise disjoint open subintervals of $[0,1]$. The commutative conjunctor I defined by*

$$I(x,y) = \begin{cases} a_\alpha + (e_\alpha - a_\alpha) I(\frac{x - a_\alpha}{e_\alpha - a_\alpha}, \frac{y - a_\alpha}{e_\alpha - a_\alpha}) & \text{if } (x,y) \in [a_\alpha, e_\alpha]^2, \\ \min(x,y) & \text{otherwise.} \end{cases}$$

is called the ordinal sum of the summands $\langle a_\alpha, e_\alpha, I_\alpha \rangle$, $\alpha \in A$.

Theorem 2. *The ordinal sum of a family of commutative conjunctors that all satisfy the same Bell-type inequality also satisfies that inequality.*

The proof of these two theorems falls outside the scope of this paper and will be published elsewhere. Next to these results, we have identified for each parametric family of t-norms and each Bell-type inequality the interval of parameter values for which the inequality is fulfilled. These results are summarized in Table 3. It would lead us to far to give a detailed account of these lengthy calculations, but we can say that all of them are based upon the investigation of the first-order derivatives in order to find extrema. We had to do this for each family individually. Note that except when T_D occurs as limit case, all other members of the considered families are differentiable. For all differentiable members T of these families, the Bell-type inequalities (except for I_3^2 and I_4^4) can be shown to be equivalent to

Table 3. Conditions on the parameter λ.

Family	$I_2^1, I_3^2, I_4^4, I_4^5$	$I_3^3, I_4^8, \ I_4^9$	I_4^6	I_4^7
Frank	$\mathbb{R}^+$	$[0, 9 + 4\sqrt{5}]$	$[0, 9.2946]$	$[0, 9.2946]$
Hamacher	$[0, 2]$	$[0, 2.9386]$	$[0, 2.6529]$	$[0, 2.2220]$
Schweizer-Sklar	$[-\infty, 1]$	$[-\infty, \frac{1}{2}]$	$[-\infty, 0.3435]$	$[-\infty, \frac{1}{2}]$
Sugeno-Weber	$[0, +\infty]$	$[3, +\infty]$	$[8, +\infty]$	$[2, +\infty]$
Dombi	$[1, +\infty]$	$[\frac{1}{2}, +\infty]$	$[\frac{\ln 2}{\ln(3 + \frac{4}{3}\sqrt{2})}, +\infty]$	$[\frac{\ln 2}{\ln 3}, +\infty]$
Aczel-Alsina	$[1, +\infty]$	$[\frac{\ln 2}{\ln 3 - \ln 2}, +\infty]$	$[0.73, +\infty]$	$[0.82, +\infty]$
Yager	$[1, +\infty]$	$[\frac{\ln 2}{2\ln 2 - \ln 3}, +\infty]$	$[\frac{\ln 2}{\ln 7 - \ln 6}, +\infty]$	$[\frac{\ln 2}{\ln 3 - \ln 2}, +\infty]$

$$I_2^1 : 2x - T(x, x) \leq 1\,,$$
$$I_3^3 : 3x - 3T(x, x) \leq 1\,,$$
$$I_4^5 : -2x + 4T(x, y) - T(x, x) - T(y, y) \leq 0\,,$$
$$I_4^6 : 4x - 6T(x, x) \leq 1\,,$$
$$I_4^7 : 8x - 6T(x, x) \leq 3\,,$$
$$I_4^8 : -x + 3T(x, y) - 3T(y, y) \leq 0\,,$$
$$I_4^9 : 3x - 2y - 3T(x, x) + 3T(x, y) \leq 1\,.$$

Note that we are not able to show that these equivalences hold in general for any differentiable t-norm.

A closer look at the inequalities of type $c_1 x - c_2 T(x, x) \leq c_3$, with constants $c_1, c_2, c_3 \geq 0$, such as the inequality $3x - 3T(x, x) \leq 1$, suggests the following general form, $n \geq 2$:

$$nx - \binom{n}{2} T(x, x) \leq 1\,. \tag{12}$$

For $n = 2$, we obtain the inequality $2x - T(x, x) \leq 1$, i.e. I_2^1. Increasing n to 3, we obtain the inequality $3x - 3T(x, x) \leq 1$, i.e. I_3^3, and for $n = 4$, we obtain the inequality $4x - 6T(x, x) \leq 1$, i.e. I_4^6. Also for n greater than 4, we obtain inequalities of this form.

We can prove that the only Frank t-norms for which inequality (12) is fulfilled for all $n \geq 2$ are the t-norms between the algebraic product T_P and the minimum operator T_M (i.e. with $\lambda \in [0, 1]$). These are also the only Frank t-norms for which all Bell-type inequalities are fulfilled.

6   Conclusions

In this paper, we have described all Bell-type inequalities concerning four random events in which at most two events are intersected at the same time. We have reformulated these inequalities as inequalities on commutative conjunctors. General results concerning copulas and ordinal sums have been complemented by a complete study of the validity of each of the inequalities for the most important parametric families of t-norms.

References

1. B. De Baets and H. De Meyer, *Transitivity-preserving fuzzification schemes for cardinality-based similarity measures*, European J. Oper. Res., to appear.
2. E.P. Klement, R. Mesiar and E. Pap, *Triangular Norms*, Kluwer Academic Publishers, Dordrecht, 2000.
3. T. Motzkin, H. Raiffa, G. Thompson and R. Thrall, *The double description method*, in: H.W. Kuhn and A.W.Tucker, eds., Contributions to the Theory of Games, Vol. 2, Princeton University Press, Princeton, RI, 1953.
4. R. Nelsen, *An Introduction to Copulas*, Lecture Notes in Statistics, Vol. **139** (Springer-Verlag, New York, 1998).
5. I. Pitowsky, *Quantum Probability – Quantum Logic*, Lecture Notes in Physics **321**, Springer, Berlin, New York, 1989.
6. J. Pykacz and B. D'hooghe, *Bell-type inequalities in fuzzy probability calculus*, Internat. J. of Uncertainty, Fuzziness and Knowledge-based Systems **9** (2001), 263–275.
7. A. Sklar, *Fonctions de répartition à n dimensions et leurs marges*, Publ. Inst. Statist. Univ. Paris **8** (1959), 229–231.

Fuzziness and Uncertainty within the Framework of Context Model

Van-Nam Huynh[1,3], Mina Ryoke[2], Yoshiteru Nakamori[3], and Tu Bao Ho[3]

[1] Department of Computer Science, Quinhon University
170 An Duong Vuong, Quinhon, VIETNAM
[2] Graduate School of Business Sciences, University of Tsukuba
Otsuka 3-29-1, Bunkyo, Tokyo 112-0012, JAPAN
ryoke@gssm.otsuka.tsukuba.ac.jp
[3] Japan Advanced Institute of Science and Technology
Tatsunokuchi, Ishikawa, 923-1292, JAPAN
{huynh,nakamori,bao}@jaist.ac.jp

Abstract. In this paper, we will briefly show that the notion of context model introduced by Gebhardt and Kruse (1993) can be considered as a unifying framework for representing fuzziness and uncertainty. Firstly, from a decision making point of view, the Dempster- Shafer theory of evidence will be reinterpreted within the framework of context model. Secondly, from a concept analysis point of view, the context model will be semantically considered as a data model for constructing membership functions of fuzzy concepts in connection with likelihood as well as random set views on the interpretation of membership grades. Furthermore, an interpretation of mass assignments of fuzzy concepts within the context model is also established.

Keywords: Context model, uncertainty modelling, fuzzy set, decision making

1 Introduction

In [9] Gebhardt and Kruse have introduced the notion of context model as an integrating model of vagueness and uncertainty. The motivation for the context model arises from the intention to develop a common formal framework that supports a better understanding and comparison of existing models of partial ignorance to reduce the rivalry between well-known approaches. Particularly, the authors presented basic ideas keyed to the interpretation of Bayes theory and the Dempster-Shafer theory within the context model. Furthermore, a direct comparison between these two approaches based on the well-known decision-making problems within the context model were also examined in their paper.

In this paper, we will briefly show that the notion of context model introduced by Gebhardt and Kruse [9] can be considered as a unifying framework for representing fuzziness and uncertainty. Particularly, from a decision analysis point of view, the Dempster- Shafer theory of evidence can be reinterpreted within the

T. Bilgiç et al. (Eds.): IFSA 2003, LNAI 2715, pp. 219–228, 2003.

framework of the context model. On the other hand, from a concept analysis point of view, the context model can be semantically considered as a data model for constructing membership functions of fuzzy concepts in connection with likelihood as well as random set views on the interpretation of membership grades. Furthermore, a probabilistic-based semantics of fuzzy concepts via the notion of mass assignments introduced by Baldwin [1,2] can be also established within the context model.

2 Basics of the Context Model

First, let us recall briefly the interpretation of data and the kinds of imperfection within the context model [9]. By this approach, data characterizes the state of an object (*obj*) with respect to underlying relevant frame conditions (*cond*). In this sense, we assume that it is possible to characterize *obj* by an element state(*obj, cond*) of a well-defined set dom(*obj*) of distinguishable object states. dom(*obj*) is usually called the *universe of discourse* or *frame of discernment* of *obj* with respect to *cond*. Then we are interested in the problem that the original characterization of state(*obj, cond*) is not available due to a lack of information about *obj* and *cond*. Generally, *cond* merely permits us to use statements like "state(*obj, cond*) $\in$ char(*obj, cond*)", where char(*obj, cond*) $\subseteq$ dom(*obj*) and called an *imprecise characterization of obj* with respect to *cond*. The second kind of imperfect knowledge in context model is *conflict*. This kind of imperfection is induced by information about preferences between the elements of char(*obj, cond*) that interprets for the existence of contexts. The combined occurence of imprecision and conflict in data reflects *vagueness* in the context model, and state(*obj, cond*) is described by a so-called *vague characteristic* of *obj* with respect to *cond*. We should note that while the lack of information about relevant conditions causes the imprecision in describing char(*obj, cond*), the existence of contexts provides the information about preferences between the elements of char(*obj, cond*) to be the original characterization of state(*obj, cond*).

Formally, a context model is defined as a triple $\langle D, C, A_C(D) \rangle$, where D is a nonempty *universe of discourse*, C is a nonempty *finite set of contexts*, and the set $A_C(D) = \{a | a : C \to 2^D\}$ which is called the set of all vague characteristics of D with respect to C. Let $a \in A_C(D)$, a is said to be *contradictory* (respectively, *consistent*) if and only if $\exists c \in C, a(c) = \emptyset$ (respectively, $\bigcap_{c \in C} a(c) \neq \emptyset$). For $a_1, a_2 \in A_C(D)$, then a_1 is said to be *more specific* than a_2 iff $(\forall c \in C)(a_1(c) \subseteq a_2(2))$. In this paper we confine ourselves to only vague characteristics that are not contradictory in the context model.

If there is a finite measure $P_C : 2^C \to \mathbb{R}^+$ that fulfills $(\forall c \in C)(P_C(\{c\}) > 0)$, then $a \in A_C(D)$ is called a *valuated vague characteristic* of D with respect to P_C. Then we call a quadruple $\langle D, C, A_C(D), P_C \rangle$ a valuated context model. Let a be a vague characteristic in the valuated context model. For each $X \in 2^D$, we define the acceptance degree $\mathsf{Acc}_a(X)$ that evaluates the proposition "state(*obj, cond*) $\in$ X" is true. Due to inherent imprecision of a, it does not allow us to uniquely determine acceptance degrees $\mathsf{Acc}_a(X)$, $X \in 2^D$. However, as shown in [9], we can calculate lower and upper bounds for them as follows:

$$\underline{\text{Acc}}_a(X) = P_C(\{c \in C | \emptyset \neq a(c) \subseteq X\}) \qquad (1)$$

$$\overline{\text{Acc}}_a(X) = P_C(\{c \in C | a(c) \cap X \neq \emptyset\}) \qquad (2)$$

More details on the context model as well as its applications could be found in, e.g. [9,10,11,16].

3   Dempster-Shafer Theory within the Context Model

We first recall in this section necessary notions from the Dempster-Shafer theory of evidence (DS theory, for short). The theory aims at providing a mechanism for representing and reasoning with uncertain, imprecise and incomplete information. It is based on Dempster's original work [4] on the modeling of uncertainty in terms of upper and lower probabilities induced by a multivalued mapping.

A multivalued mapping F from space Q into space S associates to each element q of Q a subset $F(q)$ of S. The domain of F, denoted by $\text{Dom}(F)$, is defined by

$$\text{Dom}(F) = \{q \in Q | F(q) \neq \emptyset\}$$

From a multivalued mapping F, a probability measure P on Q can be propagated to S in such a way that for any subset T of S the lower and upper bounds of probabilities of T are defined as

$$P_*(T) = \frac{P(F^-(T))}{P(F^-(S))} \qquad (3)$$

$$P^*(T) = \frac{P(F^+(T))}{P(F^+(S))} \qquad (4)$$

where
$$F^-(T) = \{q \in Q | q \in \text{Dom}(F) \wedge F(q) \subseteq T\}$$
$$F^+(T) = \{q \in Q | F(q) \cap T \neq \emptyset\}$$

Clearly, $F^+(S) = F^-(S) = \text{Dom}(F)$, and P_*, P^* are well defined only when $P(\text{Dom}(F)) \neq 0$. Furthermore, Dempster also observed that, in the case that S is finite, these lower and upper probabilities are completely determined by the quantities

$$P(F^{-1}(T)), \text{ for } T \in 2^S$$

where for each $T \in 2^S$,

$$F^{-1}(T) = \{q \in Q | F(q) = T\}$$

As such Dempster implicitly gave the prototype of a mass function also called *basic probability assignment*. Shafer's contribution has been to explicitly define the basic probability assignment and to use it to represent evidence directly. Simultaneously, Shafer has reinterpreted Dempster's lower and upper probabilities as degrees of belief and plausibility respectively, and abandoned the idea that they arise as lower and upper bounds over classes of Bayesian probabilities [20].

Formally, the definitions of these measures are given as follows:

1. A function $Bel : 2^S \to [0,1]$ is called a *belief measure* over S if $Bel(\emptyset) = 0, Bel(S) = 1$ and

$$Bel(\bigcup_{i=1}^n A_i) \geq \sum_{\emptyset \neq I \subseteq \{1,\ldots,n\}} (-1)^{|I|+1} Bel(\bigcap_{i \in I} A_i)$$

for any finite family $\{A_i\}_{i=1}^n$ in 2^S.

2. A function $Pl : 2^S \to [0,1]$ is called a *plausibility measure* if $Pl(\emptyset) = 0, Pl(S) = 1$ and

$$Pl(\bigcap_{i=1}^n A_i) \leq \sum_{\emptyset \neq I \subseteq \{1,\ldots,n\}} (-1)^{|I|+1} Pl(\bigcup_{i \in I} A_i)$$

for any finite family $\{A_i\}_{i=1}^n$ in 2^S.

It should be noted that belief and plausibility measures form a dual pair, namely

$$Pl(A) = 1 - Bel(\overline{A}), \text{ for any } A \in 2^S$$

In the case of a finite universe S, a function $m : 2^S \to [0,1]$ is called a *basic probability assignment* if $m(\emptyset) = 0$ and

$$\sum_{A \in 2^S} m(A) = 1$$

A subset $A \in 2^S$ with $m(A) > 0$ is called a *focal element* of m. The difference between $m(A)$ and $Bel(A)$ is that while $m(A)$ is our belief committed to the subset A excluding any of its subsets, $Bel(A)$ is our degree of belief in A as well as all of its subsets. Consequently, $Pl(A)$ represents the degree to which the evidence fails to refute A. Furthermore, the belief and plausibility measures are in an one-to-one correspondence with basic probability assignments. Namely, given a basic probability assignment m, the corresponding belief measure Bel and its dual plausibility measure Pl are determined by

$$Bel(A) = \sum_{\emptyset \neq B \subseteq A} m(B)$$

$$Pl(A) = \sum_{B \cap A \neq \emptyset} m(B)$$

Conversely, given a belief measure Bel, the corresponding basic probability assignment m is determined by

$$m(A) = \sum_{B \subseteq A} (-1)^{|A \setminus B|} Bel(B)$$

As we already observed above, both Dempster-Shafer theory and the context model are closely related to the theory of multivalued mappings. In fact, each vague characteristic in the context model is formally a multivalued mapping from the set of contexts into the universe of discourse.

Let $\mathcal{C} = \langle D, C, A_C(D), P_C \rangle$ be a valuated context model. Here, for the sake of discussing essential remarks regarding the interpretation of the Dempster-Shafer theory within the context model, we assume that P_C is a probability measure on C. Let a be a vague characteristic in $\mathcal{C}$ considered now as a multivalued mapping from C into D. Then a induces lower and upper probabilities, in the sense of Dempster, on 2^D as respectively defined in (3) and (4). Namely, for any $X \in 2^D$,

$$P(a)_*(X) = \frac{P_C(a^-(X))}{P_C(a^-(D))}$$

$$P(a)^*(X) = \frac{P_C(a^+(X))}{P_C(a^+(D))}$$

In the case where a is non-contradictory, we have $\mathrm{Dom}(a) = C$. Then, these probabilities coincide with lower and upper acceptance degrees as defined in (1) and (2) respectively. That is, for any $X \in 2^D$,

$$P(a)_*(X) = \underline{\mathrm{Acc}}_a(X)$$

$$P(a)^*(X) = \overline{\mathrm{Acc}}_a(X)$$

Furthermore, Gebhardt and Kruse also defined the so-called *mass distribution* m_a of a as follows

$$m_a(X) = P_C(a^{-1}(X)), \text{ for any } X \in 2^D$$

Then, for any $X \in 2^D$, we have

$$\underline{\mathrm{Acc}}_a(X) = \sum_{A \in a(C): \emptyset \neq A \subseteq X} m_a(A)$$

$$\overline{\mathrm{Acc}}_a(X) = \sum_{A \in a(C): A \cap X \neq \emptyset} m_a(A)$$

As such the mass distribution m_a induced from a in the context model $\mathcal{C}$ can be considered as the counterpart of a basic probability assignment in the DS theory.

It should also be noticed that to deal with the problem of synthesis of vague evidence linguistically provided by the experts in some situations of decision analysis, the notion of context-dependent vague characteristics as well as an extension of context model called fuzzy context model have been introduced in [14]. It is shown that each context-dependent vague characteristic within fuzzy context model directly induces a uncertainty measure of type 2 interpreted as "vague" belief function, which is inferred from vague evidence expressed linguistically.

4   Context Model for Fuzzy Concept Analysis

In the connection with formal concept analysis, it is of interest to note that in the case where C (the set of contexts) is a single-element set, say $C = \{c\}$, a context model formally becomes a formal context in the sense of Wille (see Ganter and Wille [8]) as follows. Let $\langle D, C, A_C(D) \rangle$ be a context model such that $|C| = 1$. Then the triple (O, A, R), where $O = D$, $A = A_C(D)$ and $R \subseteq O \times A$ such that $(o, a) \in R$ iff $o \in a(c)$, is a formal context. Thus, a context model can be considered as a collection of formal contexts. In [12] we have considered and introduced the notion of fuzzy concepts within a context model and the membership functions associated with these fuzzy concepts. It is shown that fuzzy concepts can be interpreted exactly as the collections of α-cuts of their membership functions.

We may agree that vague concepts are used as verbal descriptions about characteristics of objects with a tolerance of imprecise in human reasoning. Generally, people often use statements like "att(obj) is A", where A is a linguistic term that qualitatively describes an attribute of the object denoted by att(obj). In the terms of fuzzy sets, we may know att(ob) but must determine to what degree obj is considered to be compatible with A. In connection with this, as noted by Resconi and Turksen [19], the specific meaning of a vague concept in a proposition is usually evaluated in different ways for different assessments of an entity by different agents, contexts, etc. For example, consider a sentence such as:"John is tall", where "tall" is a linguistic term of a linguistic variable, the height of people (Zadeh [21]). Assume that the domain $D = [0, 3m]$ which is associated with the base variable of the linguistic variable $height$. As mentioned above, we may know John's height but must determine to what degree he is considered "tall". Next consider a set of possible worlds W in which each world evaluates the sentence as either $true$ or $false$. That is each world in W responds either as true or false when presented with the sentence "John is tall". These worlds may be contexts, agents, persons, etc. This implicitly shows that each world w_i in W determines a subset of D given as being compatible with the linguistic term $tall$. In the other words, this subset represents w_i's view of the vague concept "tall". At this point we see that the context model can be semantically considered as a data model for constructing membership functions of vague concepts.

Let us consider a context model $\mathcal{C} = \langle D, C, A_C(D) \rangle$, where D is a domain of an attribute att which is applied to objects of concern, C is a non-empty finite set of contexts, and $A_C(D)$ is a set of linguistic terms associated with the domain D considered now as vague characteristics in the context model. For example, consider $D = [0, 3m]$ which is interpreted as the domain of the attribute $height$ for people, C is a set of contexts such as Japanese, American, Swede, etc., and $A_C(D) = \{$ very short, short, medium, tall, more or less tall, ...$\}$. Each context determines a subset of D given as being compatible with a given linguistic term. Formally, each linguistic term can be considered as a mapping from C to 2^D. Furthermore, we can also associate with the context model a weighting function or a probability distribution Ω defined on C. As such we obtain a valuated context model $\mathcal{C} = \langle D, C, A_C(D), \Omega \rangle$.

By this context model, each linguistic term $a \in A_C(D)$ can be semantically represented by the fuzzy set A as follows

$$\mu_A(x) = \sum_{c \in C} \Omega(c) \mu_{a(c)}(x) \tag{5}$$

where $\mu_{a(c)}$ is the characteristic function of $a(c)$. Intuitively, while each subset $a(c)$, for $c \in C$, represents the c's view of the vague concept a, the fuzzy set A is the result of a weighted combined view of the vague concept. It is worthwhile to note that the formulation of membership function as in (5) is essentially comparable to likelihood as well as random set views on the interpretation of membership grades. Under such a formulation, we can then formulate the set-theoretic operations such as complement, intersection and union defined on fuzzy sets within the framework of context model.

Note that Kruse et al. [16] considered the same set of contexts for many domains of concern. While this assumption is acceptable in the framework of fuzzy data analysis where the characteristics (attributes) of observed objects are considered simultaneously in the same contexts, it may not be longer suitable for fuzzy concept analysis. For example, let us consider two attributes *Height* and *Income* of a set of people. Then, a set of contexts used for formulating of vague concepts of the attribute *Height* may be given as above; while another set of contexts for formulating of vague concepts of the attribute *Income* (like *high*, *low*, etc.) may be given as a set of kinds of employees or a set of residential areas of employees. Based on the meta-theory developed by Resconi et al. in 1990s [18], we have proposed in [13] a model of modal logic for fuzzy concept analysis from a context model. By this approach, we can integrate context models by using a model of modal logic, and then develop a method of calculating the expression for the membership functions of composed and/or complex fuzzy concepts based on values $\{0, 1\}$ corresponding to the truth values $\{F, T\}$ assigned to a given sentence as the response of a context considered as a possible world. It is of interest to note that fuzzy intersection and fuzzy union operators by this model are truth-functional, and, moreover, they are a well-known dual pair of *product t-norm* T_P and *probabilistic sum t-conorm* S_P [15].

5   Fuzzy Sets by Context Model and Mass Assignments

In this section we establish a mass assignment interpretation of fuzzy concepts within the context model. The mass assignment for a fuzzy concept was introduced in Baldwin [1,2] and can be interpreted as a probability distribution over possible definitions of the concept. These varying definitions may be provided by a population of voters where each is asked to give a crisp definition of the concept.

Let F be a fuzzy subset of a finite universe U such that the range of the membership function μ_F is $\{y_1, \ldots, y_n\}$, where $y_i > y_{i+1} > 0$, for $i = 1, \ldots, n-1$.

Then the mass assignment of F, denoted by m_F, is a probability distribution on 2^U satisfying $m_F(\emptyset) = 1 - y_1, m_F(F_i) = y_i - y_{i+1}$, for $i = 1, \ldots, n-1$, and $m_F(F_n) = y_n$, where $F_i = \{u \in U | \mu_F(u) \geq y_i\}$, for $i = 1, \ldots, n$. $\{F_i\}_{i=1}^n$ are referred to as the focal elements of m_F. The mass assignment of a fuzzy concept is then considered as providing a probabilistic based semantics for membership function of the fuzzy concept. The mass assignment theory of fuzzy sets have been applied in some fields such as induction of decision trees [3] and computing with words [17].

Given a context model $\mathcal{C} = \langle D, C, A_C(D), \Omega \rangle$. Assume $a \in A_C(D)$ and μ_A denotes the fuzzy set induced from a as defined by (5) in the preceding section. The weighting function Ω can be extended to 2^C as a probability measure by

$$\Omega(X) = \sum_{c \in X} \Omega(c), \text{ for any } X \in 2^C$$

Denote $\{\omega_1, \ldots, \omega_k\}$ the range of Ω defined on 2^C such that $\omega_i > \omega_{i+1} > 0$, for $i = 1, \ldots, k-1$. Clearly, $\omega_1 = 1$.

Set $C_i = \{X \in 2^C | \Omega(X) = \omega_i\}$, for $i = 1, \ldots, k$. We now define $\{A_i\}_{i=1}^k$ inductively as follows

$$A_1 = \bigcap_{c \in C} a(c)$$

$$A_i = A_{i-1} \cup \bigcup_{X \in C_i} \bigcap_{c \in X} a(c), \text{ for } i > 1$$

Let s be the least number such that $A_s \neq \emptyset$.

Obviously, $A_s \subset A_{s+1} \subset \ldots \subset A_k$. If a is consistent then we have $s = 1$. In this case let us define $m : 2^D \longrightarrow [0,1]$ by

$$m(E) = \begin{cases} \omega_i - \omega_{i+1} & \text{if } E = A_i \\ 0 & \text{otherwise} \end{cases}$$

where, by convention, $\omega_{k+1} = 0$.

In the case where $s > 1$, i.e. that a is not consistent, we define $m : 2^D \longrightarrow [0,1]$ by

$$m(E) = \begin{cases} 1 - \omega_s & \text{if } E = \emptyset \\ \omega_i - \omega_{i+1} & \text{if } E = A_i \text{ and } i > s \\ 0 & \text{otherwise} \end{cases}$$

Clearly, in both cases m is a probability distribution over 2^D with $\{A_i\}_{i=s}^k$ is a nested family of focal elenments of m. Furthermore, we have

$$m = m_{\mu_A}$$

where m_{μ_A} denotes the mass assignment of the fuzzy set μ_A in the sense of Baldwin as defined above.

On the other hand, for each $a \in A_C(D)$, it naturally generates a mass distribution m_a over 2^D defined as follows

$$m_a(E) = \Omega(\{c \in C | a(c) = E\}), \text{for any } E \in 2^D$$

In this case, if the mass assignment for a fuzzy concept could be interpreted as a probability distribution over possible definitions of the concept, it would seem desirable that the natural mass distribution m_a coincides with the mass assignment of the fuzzy set μ_A induced by a up to a permutation of C. However, this is not generally the case. Actually, due to the additive property imposed on Ω, we have $m_a = m_{\mu_A}$ if and only if the family $\{a(c)|c \in C\}$ forms a nested family of subsets in D.

6 Conclusions

A unifying framework for representing fuzziness and uncertainty based on the notion of context model has been provided in this paper. It should be emphasized that the notion of context model can be also extended to deal with the uncertainty of type 2 that arises in situations of decision analysis involving the elicitation of degrees of belief from experts with vague knowledge expressed linguistically [14]. Actually, the context model can also give a unifying interpretation for the notions of rough sets and fuzzy sets so that we can semantically relate one of these notions to each other. This problem as well as further development of the proposal proposed in this paper are being the subject of our further work.

References

1. J.F. Baldwin, The management of fuzzy and probabilistic uncertainties for knowledge based systems, in: S.A. Shapiro (Ed.), *The Encyclopaedia of AI*, Wiley, New York, 1992, pp. 528–537.
2. J.F. Baldwin, J Lawry & T.P. Martin, A mass assignment theory of the probability of fuzzy events, *Fuzzy Sets and Systems* **83** (1996) 353–367.
3. J.F. Baldwin, J Lawry & T.P. Martin, Mass assignment based induction of decision trees on words, *Proceedings of IPMU'98*, 1996, pp. 524–531.
4. A.P. Dempster, Upper and lower probabilities induced by a multivalued mapping, *Annals of Mathematics and Statistics* **38** (1967) 325–339.
5. T. Denœux, Modeling vague beliefs using fuzzy-valued belief structures, *Fuzzy Sets and Systems* **116** (2000) 167–199.
6. T. Denœux, Reasoning with imprecise belief structures, *International Journal of Approximate Reasoning* **20** (1999) 79–111.
7. D. Dubois & H. Prade, *Possibility Theory – An Approach to Computerized Processing of Uncertainty*, Plenum Press, New York, 1987.
8. B. Ganter & R. Wille, *Formal Concept Analysis: Mathematical Foundations*, Springer-Verlag, Berlin Heidelberg, 1999.
9. J. Gebhardt & R. Kruse, The context model: An integrating view of vagueness and uncertainty, *International Journal of Approximate Reasoning* **9** (1993) 283–314.
10. J. Gebhardt & R. Kruse, Parallel combination of information sources, in D.M. Gabbay & P. Smets (Eds.), *Handbook of Defeasible Reasoning and Uncertainty Management Systems*, Vol. 3 (Kluwer, Doordrecht, The Netherlands, 1998) 393–439.
11. J. Gebhardt, Learning from data – Possibilistic graphical models, in D.M. Gabbay & P. Smets (Eds.), *Handbook of Defeasible Reasoning and Uncertainty Management Systems*, Vol. 4 (Kluwer, Doordrecht, The Netherlands, 2000) 314–389.

12. V.N. Huynh & Y. Nakamori, Fuzzy concept formation based on context model, in: N. Baba et al. (Eds.), *Knowledge-Based Intelligent Information Engineering Systems & Allied Technologies* (IOS Press, Amsterdam, 2001), pp. 687–691.

13. V.N. Huynh, Y. Nakamori, T.B. Ho & G. Resconi, A context model for constructing membership functions of fuzzy concepts based on modal logic, in: T. Eiter & K.-D. Schewe (Eds.), *Foundations of Information and Knowledge Systems*, LNCS 2284, Springer-Verlag, Berlin Heidelberg, 2002, pp. 93–104.

14. V.N. Huynh, G. Resconi & Y. Nakamori, An extension of context model for modelling of uncertainty of type 2, in E. Damiani et al. (Eds.), *Knowledge-Based Intelligent Information Engineering Systems & Allied Technologies* (IOS Press, Amsterdam, 2002) 1068–1072.

15. E.P. Klement, Some mathematical aspects of fuzzy sets: Triangular norms, fuzzy logics, and generalized measures, *Fuzzy Sets and Systems* **90** (1997) 133–140.

16. R. Kruse, J. Gebhardt & F. Klawonn, Numerical and logical approaches to fuzzy set theory by the context model, in: R. Lowen and M. Roubens (Eds.), *Fuzzy Logic: State of the Art*, Kluwer Academic Publishers, Dordrecht, 1993, pp. 365–376.

17. J. Lawry, A methodology for computing with words, *International Journal of Approximate Reasoning* **28** (2001) 51–89.

18. G. Resconi, G. Klir & U. St. Clair, Hierarchically uncertainty metatheory based upon modal logic, *International Journal of General Systems* **21** (1992) 23–50.

19. G. Resconi & I.B. Turksen, Canonical forms of fuzzy truthoods by meta-theory based upon modal logic, *Information Sciences* **131** (2001) 157–194.

20. G. Shafer, *A Mathematical Theory of Evidence* (Princeton University Press, Princeton, 1976).

21. L.A. Zadeh, The concept of linguistic variable and its application to approximate reasoning, *Information Sciences,* I: **8** (1975) 199–249; II: **8** (1975) 310–357.

Uncertainty in Noise Mapping: Comparing a Probabilistic and a Fuzzy Set Approach

Tom De Muer and Dick Botteldooren

Dept. of Information Technology, Ghent University
Sint-Pietersnieuwstraat 41, B-9000 Gent(Belgium)
{tom.demuer;dick.botteldooren}@intec.rug.ac.be

Abstract. Complex georeferenced simulation resulting in immission maps loose credibility because of lacking or poor quality input data and model approximations. For modeling the resulting imperfection in the maps, several techniques can be used. In this paper we briefly compare a probabilistic approach implemented using Monte Carlo and a Fuzzy Approach. The theoretical foundations are highlighted and their consequences for the simulations are outlined. An experiment is set up to compare practical results of both techniques. Despite numerical differences in results, both techniques prove usable while the Fuzzy Approach has a clear advantage in calculation speed compared to a Monte Carlo Approach.

1 Introduction

Recently the European Commission decided to require Member States to draw noise maps for large cities and important infrastructure [1]. Interim models for source characterization and propagation were put forward and research aimed at defining the new generation of models was started. In this process very little attention is paid however to the quality of the resulting noise maps. This quality is not solely determined by the quality of the models but is for a very large part dependent on the quality of the input data. Indeed, the noise prediction models have become very accurate during the last years but it is very difficult to gather all the information needed to feed the complex noise prediction models. This problem becomes worse when larger regions need to be predicted. Some of the model parameters involved may even be unknown or it may only be possible to determine a range of possible values resulting in *hard uncertainties* [9]. Specific for noise mapping, assumptions are made about the meteorological conditions, ground parameters, ... which have their impact on the uncertainty/precision of the developed emission and propagation models. [5]

Based on these observations it was decided to compare available techniques that allow to accurately track the propagation of imperfect information from the input data through the models to the resulting noise map. The problem is actually more general than the noise mapping application envisaged in a first instance. All processes using geographic referenced data extracted from a GIS and involving a numerically intensive propagation or diffusion model (possibly

T. Bilgiç et al. (Eds.): IFSA 2003, LNAI 2715, pp. 229–236, 2003.

influenced by meteorological data as well) to obtain impact maps, suffer from the same data problems and can be treated in a similar way.

Because the conceptual simplicity of the approach, many have used a probabilistic model for the imperfection, implemented using a Monte Carlo technique. In this approach it is assumed that a probability distribution for each input variable is known. Existing simulation software can then be run on an instance drawn at random of all variables involved. In this paper we compare a fuzzy set based technique with this probabilistic Monte Carlo approach. The technique is not only conceptually better suited to describe many of the hard uncertainties [9] (of which no distribution can be built) involved, but also reduces computation times considerable. The focus of this paper will be on comparing results of both approaches in an experimental and realistic, many source environment.

2   Noise Model

The model for the prediction of the noise level is based on a ray tracing technique. For each point source in the region of interest rays are traced through the geometric environment of houses and terrain. When one of these rays hit a receiver the attenuation along the ray path is computed and multiplied with the sound power of the source. The contribution of each ray hitting the receiver is cumulated to obtain the total noise level. The most important sound source considered in this paper is road traffic. Noise emission (or sound power) by road traffic depends on vehicle fleet, traffic velocity, start-stop conditions, and road surface. The fleet daily averaged emission is related to these parameters using parametric models extracted from field measurements. The noise emission of all traffic in a street is represented by a line of point sources. The attenuation and source power are computed for each octave band center frequency which is assumed representative for the whole octave frequency band.

3   Analysis of the Uncertainty Involved

A lack of quality and imperfection of models and input data can either be caused by uncertainty or imprecision. Uncertain information can be characterized by the partial knowledge of the true value of a statement. Imprecise information is linked to approximate information or not exact information [8]. Uncertain information is commonly modeled by fuzzy set theory and imprecision by probability theory [8]. Between the two extreme cases of uncertain but exact and certain but imprecise information, a continuous spectrum of gradations exists.

In noise mapping, the propagation model can be biased because influences that become important at certain location are omitted in the model [5] . This bias is clearly a cause of uncertainty. Noise producing activities that are missing or not quantified in the underlying georeferenced (GIS) data are an additional cause of uncertainty. Examples of these are construction works, air fans, ... For this lacking information not even a probability distribution is available and is labeled a *hard uncertainty*. For other variables only partial knowledge in the form of upper and/or lower limits is available (e.g. only an upper limit of the traffic intensity of a local road is known), which also results in a (hard) uncertainty. The ground and façade reflection coefficients are an example of imprecise information.

4   Possibility and Probability Framework

4.1   Probability

A probabilistic number is described by a function mapping from the real axis to [0,1] with the constraint that the integral of the distribution is exactly 1. Operations on the probability distribution like addition and multiplication can be calculated with the use of the cumulative distribution. The probability density function is then derived from the cumulative. As an example consider multiplication.

$$
\begin{aligned}
F_{A \cdot B}(x) &= \int_{-\infty}^{+\infty} Prob[A \le x/b \wedge b < B \le b + db] db \\
&= \int_{-\infty}^{+\infty} F_A(x/y) f_B(b) db
\end{aligned}
\tag{1}
$$

where f_X is the probability density function and F_X the cumulative distribution. Above calculations assume that there is no correlation between the two variables involved in the binary operation. Correlation between the variables complicates the equation considerably.

4.2   Possibility

A fuzzy number is defined as a mapping from the real axis onto the interval [0,1] with the restriction that the supremum of the image is 1. Fuzzy numbers are based on the possibility theory put forward by Zadeh [10]. Operations on fuzzy numbers are based on the extension principle that allows extending mathematical relationships between non-fuzzy variables to relations between fuzzy variables. Again giving the product as an example:

$$
\mu_{A \cdot B}(y) \equiv \mu_C(y) = \sup_x \min \{\mu_A(y/x), \mu_B(x)\}
\tag{2}
$$

where μ_X is the membership of the fuzzy number X. Although the underlying conceptual framework is quite different, the similarity between eq.1 and eq.2 is striking.

Both probability and possibility mappings result in a distribution that can be approximated by suitable analytical curves or piecewise linear approximation. Comparable simplification of the numerical implementation of the calculus results from this. In the case of possibility convex fuzzy numbers relieve the computational burden considerably of eq. 2 which theoretically involves an optimisation. This is done by using α-cuts and basing the fuzzy calculus on the interval calculus of the α-cuts.

5 Meta Models

The term metamodel is used to refer to the model that results from the introduction of uncertainty/imprecision calculus in the classical, crisp model on which it is based. Analyzing the propagation of uncertainty/imprecision through a complex model can be done at different levels. We introduce the term *macro level meta model* as a model which uses the original model as a submodel. The macro meta level approach leaves the original model untouched and instead builts a layer around it. The crisp input parameters are sampled from their imperfect counterparts. This sample set is then fed to the original model. For each sample the original model needs to be rerun. The distribution of the crisp outputs obtained through the multiple runs forms the resulting uncertain/imprecise output variable. Because the original numerical model is not altered implementation is easy. Calculation is however very long and the number of inputs to sample poses limits on the applicability.

A *micro level meta model* is a meta model which changes the internal working of the original model by replacing all basic operational blocks with the counterparts that handle imperfect data. For this the original model needs to be altered. A practical problem can be the full access to this micro level of the software implementing the model. Relational operators introduce another problem. At the micro level, the flow of the numerical simulation depends on decisions based on relations between calculated values and crisp constants as defined in the original crisp model. There are three ways to extend this decision process to the corresponding framework:

- The distribution (fuzzy set or probability) can be mapped to a crisp value and the decision can than be based on the resulting crisp number. The advantage of this approach is that it is easy to understand and to implement although it is difficult to define a generally suited mapping function.
- The second option is to use an extended relational operator and calculate a probability/possibility degree of the relation. While this is a well-established technique, it is not straightforward to choose a general applicable threshold, above which the decision is to be taken.
- Finally, both calculation paths could be taken and their result aggregated taking into account the respective probability/possibility. The choice of aggregation method is not easy, but the main disadvantage of this approach is that both paths have to be followed. This quickly leads to an exponential growth in the number of calculations.

A macro level implementation demands offline storage of the result of each simulation run. When output is of higher dimension (maps, volume data) as is the case in noise mapping this may be a problem. A micro level implementation requires more online storage because during the simulation each real number is replaced by a distribution.

6   Discussion

The choice of meta model level and the representation of the imperfect information can theoretically be made independently. Figure 1 shows the design space with the two axes. The most commonly used combinations are a Micro Fuzzy Approach (MFA) and a Macro Monte Carlo Probabilistic Approach (MMCA). Other approaches could be a Macro Fuzzy Approach (MAFA) and a Micro Probability Approach (MPA). The choice of which one to use depends on several factors.

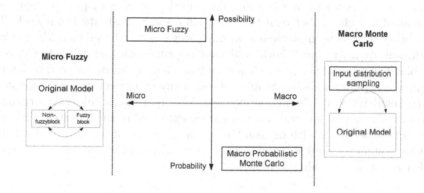

Fig. 1. Design space of meta models

A first factor to consider is the nature of imperfection that is of main interest as this determines the choice between probability and possibility (fuzzy set theory). However transformation from one framework into another is possible using methods described in [3] although interpretation problems may arise during this conversion as exemplified in [3] and [6]. Transformation is required when some information is uncertain and other is imprecise but only a single framework is used.

Imprecise and uncertain information also propagates differently through the respective frameworks. To illustrate the effect numerically, subsequent multiplication with and addition of the same number are considered. Taking the Shannon entropy as a measure for imperfection of a probability distribution [7] and a related entropy measure for the possibilistic distribution [4] (The possibilities π_i in eq. 4 are sorted in descending order):

$$H_{\text{Shannon}} = -\sum_{i=1}^{n} p_i \log_2(p_i) \tag{3}$$

$$H_{\text{Klir}} = \sum_{i=2}^{n} \pi_i \log_2\left(\frac{i}{i-1}\right) - \sum_{i=1}^{n-1} (\pi_i - \pi_{i+1}) \log\left(1 - i \sum_{j=i+1}^{n} \frac{\pi_j}{j(j-1)}\right) \tag{4}$$

Figure 2 shows the entropy ratio for the possibilistic versus the probabilistic operations after N operations. The figures show that the information propagation is not only different between frameworks but also differs between operators and depends on the form of the distributions. Moreover, correlation (probability approach) between variables plays an important role. For addition, the entropy ratio is close to one for perfect correlation, which indicates that possibility (fuzzy set) and probability grow in a similar way under this operation. For multiplication interpretation is more difficult. The figures serve as an illustration that special care is needed while interpreting modeling results between the different frameworks.

The choice of micro or macro implementation level is in practice not independant of the choice between the possibility and probability framework. The probability approach, implemented at macro level can be based on a Monte Carlo simulation. At micro level, both analytical approximation and Monte Carlo are suitable. The possibility approach is very fast when it can be reduced to interval calculus as explained above. At micro level, many transformation are monotonic and therefore interval calculus can be used. Most non-monotonic transformations are relatively easy so interval calculus can be extended to them as well. At macro level, propagation algorithms result in complex non-monotonic transformation for which interval calculus requires a complex optimisation procedure. Therefore MAFA is not an appealing option.

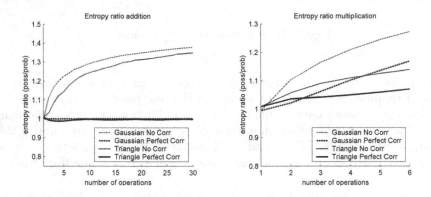

Fig. 2. Entropy ratio (possibility/probability) evolution of addition and multiplication

7   Experiment

An experiment was set up to highlight the differences between two different approaches: a Micro Fuzzy Approach and a Macro Monte Carlo Probabilistic Approach. The simulated region was about one square km of a city. Geographical data (location of buildings and streets) was extracted from a Geographical

Information System (GIS). Traffic on the roads was measured at a few locations and estimated on the bases of these measurements at comparable places. For the emission model, information with a high level of imperfection are the traffic intensity, traffic speed distribution and individual vehicle emission level. For the propagation model these were the façade and ground reflection.

The Fuzzy Approach (MFA) assumed gaussian shaped normalized possibility distribution cut off at three times the standard deviation. The α-cuts of the fuzzy numbers were at $0.0, 0.05, 0.31, 0.99$. The Macro Monte Carlo Probabilistic Approach (MMCA) also used gaussian shaped probability density function but appropriately normalized and assumed statistical independent. Geometric imprecision on geographical data was only considered in the Monte Carlo Approach as this is not possible to include in the fuzzy approach. The MFA took 28 minutes of CPU-time while the MMCA took 4 days to complete, involving 500 individual simulations of each 11 minutes. Figure 3 shows, on the left, the difference between the mean value of MMCA-simulated sound level and the sound level having the highest possibility simulated using the MFA. This difference amounts up to 4 dB(A) at specific locations. Figure 3 shows the width of the possibility distribution at half height (MFA) and the standard deviation of the probability distributions (MMCA) at individual locations in the map.

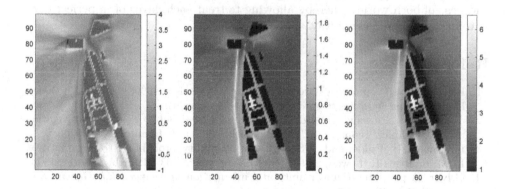

Fig. 3. Difference between the sound level with highest possibility (MFA) and the mean level (MMCA) (left). Standard deviation of the MMCA (mid) and width of the fuzzy numbers at possibility 0.5 (right)

The same geographical areas light up in both maps of figure 3, indicating highly imperfect results. In general, the imperfection corresponds better between the frameworks close to the source than further away. This is due to the fact that near the source the uncertainty/imprecision on the emission model is seen and this model is of low order compared to the propagation model. The propagation model [2] is complex and involves many operations and parameters.

A more important phenomenon is that in fuzzy the uncertainty builds up with distance to the source region (central area of the maps) because more sources with about equal strength start to contribute. The uncertainty is then the combined uncertainty of all these sources with their propagation uncertainty. In the probabilistic framework the imprecision tends to average out due to absence of correlation.

8   Conclusion

Two different frameworks dealing with imperfect information are compared. They model different aspects of imperfect information. We illustrated that both aspects are involved in complex geographically referenced impact modeling (e.g. noise mapping). Using a framework-transformation on the input, a single meta model can be used. The results for a propagation problem correspond well near the sources. Further away the results diverge and the fuzzy approach reports more imperfection than the probabilistic approach with statistical independent variables. The most important advantage of the fuzzy approach is its speed. Using a fuzzy approach the cost of computing the uncertainty information is only a factor 3 of the original, where a Monte Carlo approach is several orders slower. This makes the fuzzy approach very usable in practice.

The noise mapping application may benefit from an approach that combines the best of both worlds, thereby allowing to treat each flavor of imperfect data and models in an optimal way.

References

1. Future noise policy, Green Paper, Brussels 1996; COM(96)540def, http://europa.eu.int/en/record/green/gp9611/noise.htm.
2. J. Kragh B. Plovsing. Nord2000. Comprehensive outdoor sound propagation model. Technical report, DELTA, www.delta.dk. Draft Report AV 1849/00.
3. D. Dubois, H. Prade, and S. Sandri. On possibility/probability transformations. In M. Roubens R. Lowen, editor, *Fuzzy Logic*, pages 103–112. Kluwer A.P., 1993.
4. G.Klir. A principle of uncertainty and information invariance. *I.J. General Systems*, (2):249–275, 1990.
5. J. Kragh. News and needs in outdoor noise prediction. In *Proc. of Internoise 2001, The Hague*, pages 2573–2582, 2001.
6. R. Krishnapuram S. Medasani, J. Kim. An overview of membership function generation techniques for pattern recognition. *I.J.Appr.Reas.*, (19):391–417, 1998.
7. C.E. Shannon. A mathematical theory of communication. *The Bell System Technical Journal*, 27:379–423,623–656, july,october 1948.
8. P. Smets. Imperfect information: Imprecision–uncertainty. In *Uncertainty Management in Information Systems: From Needs to Solution*. Kluwer A.P., Boston, 1996.
9. R.A. Young. *Uncertainty and the Environment, Implications for decision making and environmental policy*. Edward E. Publishing, Inc., Northampton, USA, 2001.
10. L.A. Zadeh. Fuzzy sets as a basis for theory of possibility. *Fuzzy Sets and Systems*, (1):3–28, 1978.

Trapezoidal Approximations of Fuzzy Numbers

Przemysław Grzegorzewski and Edyta Mrówka

Systems Research Institute, Polish Academy of Sciences,
Newelska 6, 01-447 Warsaw, Poland,
pgrzeg@ibspan.waw.pl, mrowka@ibspan.waw.pl

Abstract. The problem of the trapezoidal approximation of fuzzy numbers is discussed. A set of criteria for approximation operators is formulated. A new nearest trapezoidal approximation operator preserving expected interval is suggested.

1 Introduction

Fuzzy numbers play a significant role among all fuzzy sets since the predominant carrier of information are numbers. When operating with fuzzy numbers, the result of our calculations strongly depend on the shape of their membership functions. Less regular membership functions lead to more complicated calculations. On the other hand, fuzzy numbers with simpler shape of membership functions often have more intuitive and more natural interpretation.

All these reasons cause a natural need of simple approximations of fuzzy numbers that are easy to handle and have a natural interpretation. For the sake of simplicity the trapezoidal or triangular fuzzy numbers are most common in current applications. As noted by Trillas: "the problems that arise with vague predicates are less concerned with precision and are more of a qualitative type; thus they are generally written as linearly as possible. Normally it is sufficient to use a trapezoidal representation, as it makes it possible to define them with no more than four parameters" (see [10]).

In the present paper we discuss the problem of the trapezoidal approximation of fuzzy numbers. Since it could be done in many ways one has to specify the constraints for each approximation algorithm. In Sec. 3 we formulate a set of criteria for trapezoidal approximation of fuzzy numbers which the approximation operator can or should exhibit. Some of the criteria are similar to those specified for defuzzification operators (see [11], [12]) or for interval approximation operators (see [1], [5], [6]). However, there are some points that have no counterpart in the defuzzification strategies and relate exclusively to problems typical for this framework.

In Sec. 4 we propose an approximation operator which produces a trapezoidal fuzzy number that is the closest to given original fuzzy number among all trapezoidal fuzzy numbers having identical expected interval as the original one. As it is shown in Sec. 5 this operator possesses many desired properties, like continuity, invariance to transformations, correlation invariance, ordering and information invariance.

T. Bilgiç et al. (Eds.): IFSA 2003, LNAI 2715, pp. 237–244, 2003.

2   Fuzzy Numbers

A fuzzy subset A of the real line $\mathbb{R}$ with membership function $\mu_A : \mathbb{R} \to [0,1]$ is called a fuzzy number if it is normal, fuzzy convex, its membership function μ_A is upper semicontinuous and its support is a closed interval (see [2]). A space of all fuzzy numbers will be denoted by $\mathbb{F}(\mathbb{R})$.

A useful tool for dealing with fuzzy numbers are their α−cuts. The α−cut of a fuzzy number A is a crisp set $A_\alpha = \{x \in R : \mu_A(x) \geq \alpha\}$, for $\alpha \in (0,1]$, while for $\alpha = 0$ we assume $A_0 = supp A = cl(\{x \in \mathbb{R} : \mu_A(x) > 0\})$, where cl is the closure operator. According to the definition of a fuzzy number it is seen at once that every α−cut of a fuzzy number is a closed interval. Hence we have $A_\alpha = [A_L(\alpha), A_U(\alpha)]$, where

$$A_L(\alpha) = \inf\{x \in \mathbb{R} : \mu_A(x) \geq \alpha\}, \tag{1}$$
$$A_U(\alpha) = \sup\{x \in \mathbb{R} : \mu_A(x) \geq \alpha\}.$$

For two arbitrary fuzzy numbers A and B with α−cuts $[A_L(\alpha), A_U(\alpha)]$ and $[B_L(\alpha), B_U(\alpha)]$, respectively, the quantity

$$d(A, B) = \sqrt{\int_0^1 (A_L(\alpha) - B_L(\alpha))^2 d\alpha + \int_0^1 (A_U(\alpha) - B_U(\alpha))^2 d\alpha} \tag{2}$$

is the distance between A and B (for more details we refer the reader to [4]).

Another important notion connected with fuzzy numbers is an expected interval $EI(A)$ of a fuzzy number A, introduced independently by Dubois and Prade [3] and Heilpern [7]. It is given by

$$EI(A) = \left[\int_0^1 A_L(\alpha) d\alpha, \int_0^1 A_U(\alpha) d\alpha \right]. \tag{3}$$

The most often used fuzzy numbers are, so called, *trapezoidal fuzzy numbers* with a membership function of a form

$$\mu_B(x) = \begin{cases} 0 & \text{if} \quad x < t_1 \\ \frac{x - t_1}{t_2 - t_1} & \text{if} \quad t_1 \leq x < t_2 \\ 1 & \text{if} \quad t_2 \leq x \leq t_3 \\ \frac{t_4 - x}{t_4 - t_3} & \text{if} \quad t_3 < x \leq t_4 \\ 0 & \text{if} \quad t_4 < x. \end{cases} \tag{4}$$

A family of all trapezoidal fuzzy number will be denoted by $\mathbb{F}^T(\mathbb{R})$. By (3) and (4) the expected interval of the trapezoidal fuzzy number is given by

$$EI(B) = \left[\frac{t_1 + t_2}{2}, \frac{t_3 + t_4}{2} \right]. \tag{5}$$

3   Criteria for Approximation

Suppose we want to approximate a fuzzy number by a trapezoidal fuzzy number. Thus we have to use an operator $T : \mathbb{F}(\mathbb{R}) \to \mathbb{F}^T(\mathbb{R})$ which transforms all fuzzy numbers into a family of trapezoidal fuzzy numbers, i.e. $T : A \longmapsto T(A)$. Since we can do this in many ways we propose a number of criteria which the approximation operator should or just can possess.

α–cut invariance

We say that an approximation operator T is α_0–invariant if

$$(T(A))_{\alpha_0} = A_{\alpha_0}. \tag{6}$$

E.g., 0-invariant operator preserves the support of a fuzzy number, 1-invariant preserves its core, while 0.5-invariant operator preserves a set of values that belong to A to the same extent as they belong to its complement.

Translation invariance

We say that an approximation operator T is invariant to translations if

$$T(A + z) = T(A) + z \quad \forall z \in \mathbb{R}. \tag{7}$$

It means that the relative position of the approximation remains constant when the membership function is moved to the left or to the right.

Scale invariance

We say that an approximation operator T is scale invariant if

$$T(\lambda \cdot A) = \lambda \cdot T(A) \quad \forall \lambda \in \mathbb{R} \setminus \{0\}. \tag{8}$$

It is worth noting that for $\lambda = -1$ we get, so called, symmetry constraint, which means that the relative position of the approximation does not vary if the orientation of the support interval changes.

Monotony

The criterion of monotony states that for any two fuzzy numbers A and B holds:

$$\text{if} \quad A \subseteq B \quad \text{then} \quad T(A) \subseteq T(B). \tag{9}$$

Identity

This criterion states that the approximation of a trapezoidal fuzzy number is equivalent to that number, i.e.

$$\text{if} \quad A \in \mathbb{F}^T(\mathbb{R}) \quad \text{then} \quad T(A) = A. \tag{10}$$

Nearness criterion

We say that an approximation operator T fulfills the nearness criterion if for any fuzzy number its output value is the nearest trapezoidal fuzzy number to A with respect to metric d. In other words, for any $A \in \mathbb{F}^T(\mathbb{R})$ we have

$$d(A, T(A)) \le d(A, B) \quad \forall B \in \mathbb{F}^T(\mathbb{R}). \tag{11}$$

Continuity

The continuity constraint means that if two original fuzzy numbers are close then their approximations should also be close. Or, in other words, that a small deviation in the degree of membership function should not result in a big change in the approximation. Hence we say that an approximation operator T is continuous if for any $A, B \in \mathbb{F}(\mathbb{R})$ we have

$$\forall(\varepsilon > 0) \; \exists(\delta > 0) \quad d(A, B) < \delta \Rightarrow d(T(A), T(B)) < \varepsilon, \tag{12}$$

where d denotes a metric defined in $\mathbb{F}(\mathbb{R})$.

Fuzzy operations criterion

This criterion states that any max-min operation (based on the extension principle) will not increase the deviation when approximations of the original fuzzy number are used in the operation, i.e. for any $A, B \in \mathbb{F}(\mathbb{R})$ and for any max-min operator $* : \mathbb{F}(\mathbb{R}) \to \mathbb{F}(\mathbb{R})$ we have $\forall(\varepsilon > 0)$

$$\text{if} \quad d(A, T(A)) < \varepsilon, d(B, T(B)) < \varepsilon \quad \text{then} \quad d(A * B, T(A) * T(B)) < \varepsilon, \tag{13}$$

where d denotes a metric defined in $\mathbb{F}(\mathbb{R})$. Since all fuzzy arithmetic operations are of that type, this criterion enable to perform operations with assurance that the validity of the final approximate result will be at least equal to the validity of the approximation made.

Order invariance

It is known that fuzzy numbers are not linearly ordered and no ranking method is the best one. However, a reasonable approximation operator should preserve an ordering accepted for given situation. Let $A \succ B$ means that A is "greater" than B with respect to ordering $\succ$. Then we say that an approximation operator T is order invariant with respect to ordering $\succ$ if for any $A, B \in \mathbb{F}(\mathbb{R})$ we have

$$A \succ B \Leftrightarrow T(A) \succ T(B). \tag{14}$$

Correlation invariance

In many applications we are interested in finding a correlation between fuzzy numbers which describe the relationship between these fuzzy numbers. We say that an approximation operator T is correlation invariant if for any $A, B \in \mathbb{F}(\mathbb{R})$ we have

$$\rho(A, B) = \rho(T(A), T(B)), \tag{15}$$

where $\rho(A, B)$ denotes a correlation measure of two fuzzy numbers.

Information invariance

When we approximate one model with another one, we want to convert information of one type to another, preserving its amount. This expresses the spirit of, so called, the principle of information invariance. Therefore, it seems desirable that a trapezoidal approximation $T(A)$ of a fuzzy number A should contain the same amount of information as the initial fuzzy number A. Here one can consider different measures of uncertainty, information, specifity or nonspecifity, etc. If we denote such a measure by I then we can write this natural requirement in a following way

$$I(A) = I(T(A)). \tag{16}$$

4   Trapezoidal Approximation of a Fuzzy Number

In this section we propose an approximation operator $T : \mathbb{F}(\mathbb{R}) \to \mathbb{F}^T(\mathbb{R})$ which produces a trapezoidal fuzzy number that is the closest to given original fuzzy number among all trapezoidal fuzzy numbers having identical expected interval as the original one.

Suppose A is a fuzzy number and $[A_L(\alpha), A_U(\alpha)]$ is its α–cut. Given A we try to find a trapezoidal fuzzy number $T(A)$ nearest to A with respect to metric (2). By (4) a trapezoidal fuzzy number is completely described by four real numbers $t_1 \le t_2 \le t_3 \le t_4$ and the α–cut of $T(A)$ is equal to $[t_1+(t_2 - t_1)\alpha, t_4-(t_4-t_3)\alpha]$. It is easily seen that in order to minimize $d(A, T(A))$ it suffices to minimize function $d^2(A, T(A))$ which in our case reduces to

$$d(^2 A, T(A)) = \int_0^1 [A_L(\alpha) - (t_1 + (t_2 - t_1)\,\alpha)]^2\,d\alpha \\ + \int_0^1 [A_U(\alpha) - (t_4 - (t_4 - t_3)\alpha)]^2\,d\alpha. \tag{17}$$

However, we want to find a trapezoidal fuzzy number which is not only closest to given fuzzy number but which also preserves the expected interval of that fuzzy number, i.e. $EI(T(A)) = EI(A)$. Thus by (3) and (5) our problem is to find such real numbers $t_1 \le t_2 \le t_3 \le t_4$ that minimize (17) with respect to conditions

$$\frac{t_1+t_2}{2} - \int_0^1 A_L(\alpha)d\alpha = 0, \\ \frac{t_3+t_4}{2} - \int_0^1 A_U(\alpha)d\alpha = 0. \tag{18}$$

Using well known method of the Lagrange multipliers our problem reduces to finding such real numbers $t_1 \le t_2 \le t_3 \le t_4$ that minimize

$$H(t_1, t_2, t_3, t_4) = \int_0^1 [A_L(\alpha) - (t_1 + (t_2 - t_1)\,\alpha)]^2\,d\alpha \\ + \int_0^1 [A_U(\alpha) - (t_4 - (t_4 - t_3)\alpha)]^2\,d\alpha \\ + \lambda_1 \left(\frac{t_2+t_1}{2} - \int_0^1 A_L(\alpha)\,d\alpha\right) + \lambda_2 \left(\frac{t_4+t_3}{2} - \int_0^1 A_U(\alpha)\,d\alpha\right), \tag{19}$$

where λ_1 and λ_2 are real numbers.

Thus we have to find partial derivatives and to solve

$$\begin{cases} \frac{\partial H}{\partial t_1} = 2 \int_0^1 [A_L(\alpha) - (t_1 + (t_2 - t_1)\,\alpha)]\,(\alpha - 1)\,d\alpha + \frac{1}{2}\lambda_1 = 0 \\ \frac{\partial H}{\partial t_2} = 2 \int_0^1 [A_L(\alpha) - (t_1 + (t_2 - t_1)\,\alpha)]\,(-\alpha)\,d\alpha + \frac{1}{2}\lambda_1 = 0 \\ \frac{\partial H}{\partial t_3} = 2 \int_0^1 [A_U(\alpha) - (t_4 - (t_4 - t_3)\alpha)]\,(-\alpha)\,d\alpha + \frac{1}{2}\lambda_2 = 0 \\ \frac{\partial H}{\partial t_4} = 2 \int_0^1 [A_U(\alpha) - (t_4 - (t_4 - t_3)\alpha)]\,(\alpha - 1)\,d\alpha + \frac{1}{2}\lambda_2 = 0. \end{cases} \tag{20}$$

The solution is

$$t_1 = t_1(A) = -6 \int_0^1 \alpha A_L(\alpha)\, d\alpha + 4 \int_0^1 A_L(\alpha)\, d\alpha,$$
$$t_2 = t_2(A) = 6 \int_0^1 \alpha A_L(\alpha)\, d\alpha - 2 \int_0^1 A_L(\alpha)\, d\alpha,$$
$$t_3 = t_3(A) = 6 \int_0^1 \alpha A_U(\alpha)\, d\alpha - 2 \int_0^1 A_U(\alpha)\, d\alpha,$$
$$t_4 = t_4(A) = -6 \int_0^1 \alpha A_U(\alpha)\, d\alpha + 4 \int_0^1 A_U(\alpha)\, d\alpha. \tag{21}$$

Moreover, $\det \left[\frac{\partial^2 H}{\partial t_i \partial t_j} \right]_{i,j=1,2,3,4} = \frac{1}{9} > 0$, $\det \left[\frac{\partial^2 H}{\partial t_i \partial t_j} \right]_{i,j=1,2,3} = \frac{2}{9} > 0$,

$\det \left[\frac{\partial^2 H}{\partial t_i \partial t_j} \right]_{i,j=1,2} = \frac{1}{3} > 0$, $\det \left[\frac{\partial^2 H}{\partial t_1^2} \right] = \frac{2}{3} > 0$, and hence t_1, t_2, t_3, t_4 given by (21), actually minimize $d(A, T(A))$. Therefore a trapezoidal fuzzy number $T(A)$, with t_1, t_2, t_3, t_4 given by (21), is indeed the nearest trapezoidal fuzzy number to fuzzy number A (with respect to metric d) preserving the expected interval.

5   Properties

In this section we consider some properties of the approximation operator suggested in Sec. 4. It can be shown that

Proposition 1. *The approximation operator T described by (21)*
  a) is invariant to translations,
  b) is scale invariant,
  c) is monotonic,
  d) fulfills the identity criterion,
  e) fulfills the nearness criterion with respect to metric (2) in subfamily of all trapezoidal fuzzy numbers with fixed expected interval,
  f) is continuous,
  g) fulfills fuzzy operations criterion.

Now let us consider the problem of order invariance. We can rank fuzzy numbers in many ways. Recently, Jimenez [10] proposed a ranking method based on the comparison of expected intervals. Let $EI(A) = [E_*(A), E^*(A)]$ and $EI(B) = [E_*(B), E^*(B)]$ denote expected intervals of fuzzy numbers A and B, respectively. Then we get the preference fuzzy relation $M(A, B)$ with the following membership function

$$\mu_M(A, B) = \begin{cases} 0 & \text{if } E^*(A) - E_*(B) < 0 \\ 1 & \text{if } E_*(A) - E^*(B) > 0 \\ \frac{E^*(A) - E_*(B)}{E^*(A) - E_*(B) - (E_*(A) - E^*(B))} & \text{otherwise} \end{cases}, \tag{22}$$

where $\mu_M(A, B)$ is the degree of preference of A over B. It was shown that the preference relation M is a complete fuzzy order. It is seen that

Proposition 2. *The approximation operator T described by (21) is order invariant with respect to preference relation M, i.e.*

$$M(T(A), T(B)) = M(A, B) \quad \forall A, B \in \mathbb{F}(\mathbb{R}). \tag{23}$$

In many applications the correlation between fuzzy numbers is of interest. Hung and Wu [8] defined a correlation coefficient by means of expected interval. Namely, they defined the correlation coefficient between A and B as

$$\rho(A, B) = \frac{E_*(A)E_*(B) + E^*(A)E^*(B)}{\sqrt{(E_*(A))^2 + (E^*(A))^2}\sqrt{(E_*(B))^2 + (E^*(B))^2}}. \tag{24}$$

This correlation coefficient shows not only the degree of relationship between the fuzzy numbers but also whether these fuzzy numbers are positively or negatively related. By (24) we get immediately

Proposition 3. *The approximation operator T described by (21) is correlation invariant with respect to the correlation coefficient (24), i.e.*

$$\rho(T(A), T(B)) = \rho(A, B) \quad \forall A, B \in \mathbb{F}(\mathbb{R}). \tag{25}$$

Now let us check whether our approximation operator fulfills the principle of information invariance. A concept of information is intimately connected with the concept of uncertainty. Among numerous uncertainty measures there are nonspecifity measures. One of the simplest and the most natural nonspecifity measure is, so called, the width of a fuzzy number (see [1]), defined as

$$w(A) = \int_{-\infty}^{\infty} \mu_A(x)dx. \tag{26}$$

It can be proved that our approximation operator preserves the width of a fuzzy number. Hence we get

Proposition 4. *The approximation operator T described by (21) is information invariant with respect to nonspecifity measure (26), i.e.*

$$w(A) = w(T(A)) \quad \forall A \in \mathbb{F}(\mathbb{R}). \tag{27}$$

6   Conclusions

In the present paper we have discussed the problem of the trapezoidal approximation of fuzzy numbers. We have formulated a list of criteria which would be desirable for approximation operators to possess. These requirements have different origin: they are motivated by geometrical considerations, by some theoretical concepts and foundations, by operations typical for application of fuzzy numbers, etc. Which properties are more important depend, of course, on particular situation. These constraints can be used for direct operator derivation. However, in any case a satisfactory approximation operator should be easy to implement, computational inexpensive and should have convincing interpretation.

We have also suggested a new approach to trapezoidal approximation of fuzzy numbers. The proposed operator, called the nearest trapezoidal approximation operator preserving expected interval, possess many desired properties: it is simple and natural, it is continuous and invariant to translations and scale transformations. Moreover, it preserves ordering, correlation and information amount. It is worth noting that our approach might be generalized for one-sided fuzzy numbers as well.

References

1. Chanas S., On the interval approximation of a fuzzy number, Fuzzy Sets and Systems 122 (2001), 353–356.
2. Dubois D., Prade H., Operations on fuzzy numbers, Int. J. Syst. Sci. 9 (1978), 613–626.
3. Dubois D., Prade H., The mean value of a fuzzy number, Fuzzy Sets and Systems 24 (1987), 279–300.
4. Grzegorzewski P., Metrics and orders in space of fuzzy numbers, Fuzzy Sets and Systems 97 (1998), 83–94.
5. Grzegorzewski P., Interval approximation of a fuzzy number and the principle of information invariance, In: Proceedings of the 9th International Conference on Information Processing and Management of Uncertainty IPMU'2002, Annecy, 1-5 July 2002, pp. 347–354.
6. Grzegorzewski P., Nearest interval approximation of a fuzzy number, Fuzzy Sets and Systems 130 (2002), 321–330.
7. Heilpern S., The expected value of a fuzzy number, Fuzzy Sets and Systems 47 (1992), 81–86.
8. Hung W., Wu J., A note on the correlation of fuzzy numbers by expected interval, International Journal of Uncertainty, Fuzziness and Knowledge–Based Systems 9 (2001), 517–523.
9. Jimenez M., Ranking fuzzy numbers through the comaparison of its expected intervals, International Journal of Uncertainty, Fuzziness and Knowledge–Based Systems 4 (1996), 379–388.
10. Jimenez M., Rivas J.A., Fuzzy number approximation, International Journal of Uncertainty, Fuzziness and Knowledge-Based Systems, 6 (1998), 68–78.
11. van Leewijck W., Kerre E.E., Defuzzification: criteria and classification, Fuzzy Sets and Systems 108 (1999), 159–178.
12. Runkler T.A., Glesner M., A set of axioms for defuzzification strategies – towards a theory of rational defuzzification operators, in: Proc. 2nd IEEE Internat. Conf. on Fuzzy Systems, San Francisco, 1993, pp. 1161–1166.

A Discrete-Time Portfolio Selection with Uncertainty of Stock Prices

Yuji Yoshida[1], Masami Yasuda[2], Jun-ichi Nakagami[2], and Masami Kurano[3]

[1] Faculty of Economics and Business Administration, the University of Kitakyushu
4-2-1 Kitagata, Kokuraminami, Kitakyushu 802-8577, Japan
yoshida@kitakyu-u.ac.jp
[2] Faculty of Science, Chiba University
1-33 Yayoi-cho,Inage-ku,Chiba 263-8522, Japan
{yasuda, nakagami}@math.s.chiba-u.ac.jp
[3] Faculty of Education, Chiba University
1-33 Yayoi-cho,Inage-ku,Chiba 263-8522, Japan
kurano@math.e.chiba-u.ac.jp

Abstract. A mathematical model for dynamic portfolio model with uncertainty is discussed. To consider this uncertainty modelling, the randomness and fuzziness are evaluated simultaneously cooperating with both of probabilistic expectation and mean values with λ-weighting functions. By dynamic programming approach, we will derive an optimality equation for the optimal consumption and wealth problem in a fuzzy stochastic process and then an optimal portfolio is given. It is shown that the optimal total expected utility is a solution of the optimality equation under a reasonable assumption.

1   Introduction and Notations

In a financial market, the portfolio is one of the most important tools for the asset management. Static portfolio models have been studied by many authors [6,1,2,11], using linear programming on the basis of Markovitz's model. On the other hand, dynamic approaches to the portfolio have been studied by martingale method and dynamic programming ([7]). Recently, dynamic financial engineering has been developing on the basis of the Black-Scholes log-normal stochastic differential models. However it is not easy to predict the future actual prices. When we sell or buy stocks by Internet in a financial market, there sometimes exists a difference between the actual prices and the theoretical value which derived from Black-Scholes methods. The difficulty comes from not only randomness of financial stochastic systems but also uncertainty which we cannot represent by only probability theory. When the market are changing rapidly, the losses/errors often become bigger between the decision maker's expected price and the actual price. These losses/errors occur, and one of the reasons is because the investors can not know all information about the stock markets at current time. When we deal with systems like financial markets, fuzzy logic works well that the markets contain the uncertain factors which are different from probabilistic essence and

T. Bilgiç et al. (Eds.): IFSA 2003, LNAI 2715, pp. 245–252, 2003.

in which there exists a difficulty to identify actual price values exactly. In this paper, probability is applied as the uncertainty such that something occurs or not with probability, and fuzziness is applied as the vagueness such that we cannot specify the exact values because of a lack of knowledge regarding the present stock market. In this paper, we consider a dynamic portfolio model with fuzzy prices in a discrete-time financial market.

By introducing fuzzy logic to the discrete-time stochastic processes for the financial market, we present a new model with uncertainty of both randomness and fuzziness, which is a reasonable and natural extension of the original discrete-time stochastic processes in Black-Scholes model. Applying fuzzy logic to the portfolio is valid for not only maximizing the investor's consumption and wealth but also avoiding the worst scenarios which will occur in uncertain environments, and it brings us a robust asset management in uncertain environments. To valuate the dynamic portfolio model, we need to deal with optimal trading strategies in discrete-time stochastic processes (Ross [10] and so on). In this paper, we present an optimal portfolio model regarding discrete-time stochastic processes with fuzziness, and we discuss the optimal trading strategies for dynamic portfolio model in the financial markets. In order to describe the portfolio model with fuzziness, we need to extend real-valued random variables in probability theory to *fuzzy random variables*, which are random variables with fuzzy number values. We introduce a *fuzzy stochastic process* by fuzzy random variables to define prices in dynamic portfolio model, and we evaluate the randomness and fuzziness by probabilistic expectation and mean values with λ-weighting functions from the viewpoint of Yoshida et al. [14,15].

We consider a portfolio model with a bond and n stocks, where n is a positive integer. In the remainder of this section, we describe bond price processes and stock price processes. We deal with a model where an investor's actions do not have any impact on the stocks, so-called *small investors hypothesis* ([4]). Let $\mathbb{T} := \{0, 1, 2, \cdots, T\}$ be the time space with an expiration date T, and $\mathbb{R}$ denotes the set of all real numbers. Let $(\Omega, \mathcal{M}, P)$ be a probability space, where $\mathcal{M}$ is a σ-field of Ω and P is a non-atomic probability measure. Take a probability space $\Omega := (\mathbb{R}^{n+1})^{T+1}$ by the product of $\mathbb{R}$. Let a positive number r_t be an *interest rate* of a bond price at time t for $t = 1, 2, \cdots, T$, and put a *bond price process* $\{S_t^0\}_{t=0}^T$ by $S_0^0 = 1$ and

$$S_t^0 := \prod_{s=1}^t (1 + r_s) \quad \text{for } t = 1, 2, \cdots, T. \tag{1}$$

We define *stock price processes* $\{S_t^i\}_{t=0}^T$ for stock $i = 1, 2, \cdots, n$ as follows: An initial stock price S_0^i is a positive constant and stock prices at positive time t are given by

$$S_t^i := S_0^i \prod_{s=1}^t (1 + Y_s^i) \quad \text{for } t = 1, 2, \cdots, T, \tag{2}$$

where $\{Y_t^i\}_{t=1}^T$ is a uniform integrable sequence of independent identically distributed real random variables on $[-1, \infty)$. A sequence of σ-fields $\{\mathcal{M}_t\}_{t=0}^T$

on Ω is given as follows: $\mathcal{M}_0$ is the completion of $\{\emptyset, \Omega\}$ and $\mathcal{M}_t$ for $t = 1, 2, \cdots, T$ denote the complete σ-fields generated by random variables $\{Y_s^i \mid i = 1, 2, \cdots, n; s = 1, 2, \cdots, t\}$. In this paper, we present a portfolio model where stock price processes S_t^i take fuzzy values using fuzzy random variables, whose mathematical notations are introduced in the next section.

2 Fuzzy Stochastic Processes

Fuzzy random variables, which take values in fuzzy numbers, were first studied by Kwakernaak [5] and have been discussed by many authors [9]. First we introduce fuzzy numbers. A fuzzy number is denoted by its membership function $\tilde{a} : \mathbb{R} \mapsto [0, 1]$ which is normal, upper-semicontinuous, fuzzy convex and has a compact support (Zadeh [16]). $\mathcal{R}$ denotes the set of all fuzzy numbers. In this paper, we identify fuzzy numbers with their corresponding membership functions. The α-cut of a fuzzy number $\tilde{a}(\in \mathcal{R})$ is given by $\tilde{a}_\alpha := \{x \in \mathbb{R} \mid \tilde{a}(x) \geq \alpha\}$ $(\alpha \in (0, 1])$ and $\tilde{a}_0 := \mathrm{cl}\{x \in \mathbb{R} \mid \tilde{a}(x) > 0\}$, where cl denotes the closure of an interval. We write the closed intervals as $\tilde{a}_\alpha := [\tilde{a}_\alpha^-, \tilde{a}_\alpha^+]$ for $\alpha \in [0, 1]$. Hence we introduce a partial order $\succeq$, so called the *fuzzy max order*, on fuzzy numbers $\mathcal{R}$ ([3]): Let $\tilde{a}, \tilde{b} \in \mathcal{R}$ be fuzzy numbers. Then, $\tilde{a} \succeq \tilde{b}$ means that $\tilde{a}_\alpha^- \geq \tilde{b}_\alpha^-$ and $\tilde{a}_\alpha^+ \geq \tilde{b}_\alpha^+$ for all $\alpha \in [0, 1]$. Here $(\mathcal{R}, \succeq)$ becomes a lattice.

A fuzzy-number-valued map $\tilde{X} : \Omega \mapsto \mathcal{R}$ is called a *fuzzy random variable* if the maps $\omega \mapsto \tilde{X}_\alpha^-(\omega)$ and $\omega \mapsto \tilde{X}_\alpha^+(\omega)$ are measurable for all $\alpha \in [0, 1]$, where $\tilde{X}_\alpha(\omega) = [\tilde{X}_\alpha^-(\omega), \tilde{X}_\alpha^+(\omega)] = \{x \in \mathbb{R} \mid \tilde{X}(\omega)(x) \geq \alpha\}$ ([12]). Next we need to introduce expectations of fuzzy random variables in order to describe a portfolio model in the next section. A fuzzy random variable $\tilde{X}$ is called integrably bounded if both $\omega \mapsto \tilde{X}_\alpha^-(\omega)$ and $\omega \mapsto \tilde{X}_\alpha^+(\omega)$ are integrable for all $\alpha \in [0, 1]$. Let $\tilde{X}$ be an integrably bounded fuzzy random variable. The expectation $E(\tilde{X})$ of the fuzzy random variable $\tilde{X}$ is defined by a fuzzy number

$$E(\tilde{X})(x) := \sup_{\alpha \in [0,1]} \min\{\alpha, 1_{E(\tilde{X})_\alpha}(x)\}, \quad x \in \mathbb{R}, \tag{3}$$

where $E(\tilde{X})_\alpha := [\int_\Omega \tilde{X}_\alpha^-(\omega)\,dP(\omega), \int_\Omega \tilde{X}_\alpha^+(\omega)\,dP(\omega)]$ for $\alpha \in [0, 1]$ ([9]). A sequence of integrably bounded fuzzy random variables $\{\tilde{X}_t\}_{t=0}^T$ is also called a *fuzzy stochastic process*.

3 Consumption Processes and Wealth Processes with Uncertainty of Stock Prices

In this section, we introduce a multi-period portfolio model where stock prices have uncertainty and we discuss its properties. Let $i = 1, 2, \cdots, n$, and let $\{\delta_t^i\}_{t=0}^T$ be a stochastic process such that $0 < \delta_t^i(\omega) < S_t^i(\omega)$ for all $\omega \in \Omega$. We give a fuzzy stochastic process $\{\tilde{S}_t^i\}_{t=0}^T$ of the stock prices by the following fuzzy random variables:

$$\tilde{S}_t^i(\omega)(x) := L((x - S_t^i(\omega))/\delta_t^i(\omega)) \tag{4}$$

for $t \in \mathbb{T}$, $\omega \in \Omega$ and $x \in \mathbb{R}$, where $L(x) := \max\{1 - |x|, 0\}$ $(x \in \mathbb{R})$ is the triangle type shape function and $\{S_t^i\}_{t=0}^T$ is defined by (2). Hence, $\delta_t^i(\omega)$ is a spread of triangular fuzzy numbers and it corresponds to the amount of fuzziness in the stock price process $\{\tilde{S}_t^i\}_{t=0}^T$. Now we introduce the following assumption to reduce the complexity of computation ([13]).

Assumption S. Let $i = 1, 2, \cdots, n$. The stochastic process $\{\delta_t^i\}_{t=0}^T$ is represented by

$$\delta_t^i(\omega) := \eta^i S_t^i(\omega) \quad \text{for } t \in \mathbb{T} \text{ and } \omega \in \Omega,$$

where η^i is a constant satisfying $0 < \eta^i < 1$.

One of the most difficulties is estimation of the actual volatilities of the stocks ([10, Sect.7.5.1]). Therefore, it is reasonable that the size of fuzziness $\delta_t^i(\omega)$ depends on the volatility. In this model, we represent by η^i the fuzziness of the appreciate rate and the volatility for the stock price $\tilde{S}_t^i$, and we call η^i a *fuzzy factor* of the stock price process ([13]). Thus, we subjectively estimate the fuzzy quantity of the stock i with $\delta_t^i = \eta^i S_t^i$ in Assumption S. From now on, we suppose that Assumption S holds. We also represent the bond price process $\{\tilde{S}_t^0\}_{t=0}^T$ by the following crisp number

$$\tilde{S}_t^0(x) := 1_{\{S_t^0\}}(x), \quad x \in \mathbb{R} \tag{5}$$

for $t \in \mathbb{T}$, where $1_{\{.\}}$ denotes the characteristic function of a singleton.

Next we introduce trading strategies, consumption and wealth for the bond and the stocks. We assume that the securities are perfectly divisible, and we consider a case where their short sales are allowed. A *trading strategy* $\pi = \{\pi_t\}_{t=0}^T = \{(\pi_t^0, \pi_t^1, \cdots, \pi_t^n)\}_{t=0}^T$ is a $\mathbb{R}^{n+1}$-valued $\mathcal{M}_t$-predictable process such that $\sum_{t=0}^T E(|\pi_t^0|) < \infty$, $\sum_{t=0}^T E(|\pi_t^i|S_t^i) < \infty$ and $\sum_{t=0}^{T-1} E(|\pi_{t+1}^i|S_t^i) < \infty$ for all $i = 1, 2, \cdots, n$. Here π_t^0 means *the amount of the bond* $\tilde{S}_t^0$ and π_t^i means *the amount of the stock* $\tilde{S}_t^i$ *at time* t. In this paper, we deal with an optimization problem regarding consumption and wealth in a *fuzzy wealth process* $\{\tilde{V}_t\}_{t=0}^T$ defined by fuzzy random variables

$$\tilde{V}_t(\omega) := \pi_t^0(\omega)\tilde{S}_t^0(\omega) + \sum_{i=1}^n \pi_t^i(\omega)\tilde{S}_t^i(\omega) \tag{6}$$

for $t \in \mathbb{T}$ and $\omega \in \Omega$. A *consumption process* $\{\tilde{C}_t\}_{t=0}^{T-1}$ is defined by

$$\tilde{C}_t(\omega) := (\pi_t^0(\omega) - \pi_{t+1}^0(\omega))\tilde{S}_t^0(\omega) + \sum_{i=1}^n (\pi_t^i(\omega) - \pi_{t+1}^i(\omega))\tilde{S}_t^i(\omega) \tag{7}$$

for $t = 0, 1, \cdots, T-1$ and $\omega \in \Omega$. Let γ be an $\mathcal{M}_T$-adapted real random variable which is independent to $\mathcal{M}_{T-1}$. Then we put a crisp random variable

$$\tilde{C}_T(\omega) := 1_{\{\gamma(\omega)\}} \tag{8}$$

for $\omega \in \Omega$, and it is called a *terminal consumption*. A trading strategy π is called *admissible* if

$$\tilde{C}_t(\omega) \succeq \tilde{0} \tag{9}$$

for all $t \in \mathbb{T}$ and $\omega \in \Omega$, where $\tilde{0} = 1_{\{0\}}$ is the crisp number zero and $\succeq$ is the fuzzy max order. The condition (9) means that compensation money or supplementary money is not allowed at any time even in cases of the worst scenarios. In this paper, we adopt the admissible condition (9) in stead of the self-financing ([8]).

Hence we consider utility estimation of consumption and wealth in the portfolio model. Let $\overline{\mathbb{R}} := [-\infty, \infty)$. A map $U_1 : \mathbb{T} \times \mathbb{R} \mapsto \overline{\mathbb{R}}$ is called a *consumption utility function* if $U_1(t, \cdot)$ is continuous, increasing and strictly concave on $(0, \infty)$ such that $\lim_{c \to \infty} U_1(t, c) = \infty$, $\lim_{c \downarrow 0} U_1(t, c) = -\infty$ and $U_1(t, c) = -\infty$ if $c \leq 0$ for all $t \in \mathbb{T}$. Further, a map $U_2 : \mathbb{R} \mapsto \overline{\mathbb{R}}$ is also called a time-invariant *wealth utility function* if U_2 is continuous, increasing and strictly concave on $(0, \infty)$ such that $\lim_{w \to \infty} U_2(w) = \infty$ $\lim_{w \downarrow 0} U_2(w) = -\infty$ and $U_2(w) = -\infty$ if $w \leq 0$. Since the consumption process $\tilde{C}_t$ and the fuzzy wealth process $\tilde{V}_t$ take fuzzy values, for their estimation we introduce their fuzzy utilities $\tilde{U}_1(t, \cdot) : \mathcal{R} \mapsto \mathcal{R}$ and $\tilde{U}_2 : \mathcal{R} \mapsto \mathcal{R}$ by

$$\tilde{U}_1(t, \tilde{a})(y) := \sup_{x : U_1(t,x)=y} \tilde{a}(x), \quad y \in \overline{\mathbb{R}} \quad \text{for } t \in \mathbb{T} \text{ and } \tilde{a} \in \mathcal{R}; \tag{10}$$

$$\tilde{U}_2(\tilde{a})(y) := \sup_{x : U_2(x)=y} \tilde{a}(x), \quad y \in \overline{\mathbb{R}} \quad \text{for } \tilde{a} \in \mathcal{R}. \tag{11}$$

Let $\boldsymbol{x} = (x^0, x^1, \cdots, x^n)$ be initial securities prices and let w be an initial wealth such that $x^i > 0$ for all $i = 0, 1, \cdots, n$ and $w > 0$, where x^0 is an initial price for bond and x^i is an initial price for stock i. Let $\pi = \{(\pi_t^0, \pi_t^1, \cdots, \pi_t^n)\}_{t=0}^T$ be an admissible trading strategy satisfying $\sum_{i=0}^n x^i \pi_0^i = w$. Then, for a terminal consumption $\tilde{C}_T$ satisfying $\tilde{C}_T \preceq \tilde{V}_T$, the sum of the expected utilities of the consumption process $\{\tilde{C}_t\}_{t=0}^T$ and the terminal wealth $\tilde{W}_T := \tilde{V}_T - \tilde{C}_T$ is given by

$$E_{\boldsymbol{x},w} \left(\sum_{t=0}^T \tilde{U}_1(t, \tilde{C}_t) + \tilde{U}_2(\tilde{W}_T) \right), \tag{12}$$

where $E_{\boldsymbol{x},w}(\cdot)$ is the probabilistic expectation with a trading strategy π satisfying $\sum_{i=0}^n x^i \pi_0^i = w$ for initial securities prices $\boldsymbol{x}$ and an initial wealth w. We note that (12) is a fuzzy number by (3). Therefore, we need an estimation method of fuzzy numbers because the fuzzy max order $\succeq$ in Section 2 is a partial order but not linear order on $\mathcal{R}$.

Now we introduce mean values of fuzzy numbers from Yoshida et al. [14]. Let $\mathcal{I}$ be the set of all non-empty bounded closed intervals. Let $g : \mathcal{I} \mapsto \mathbb{R}$ be a map such that

$$g([x, y]) := \lambda x + (1 - \lambda)y, \quad [x, y] \in \mathcal{I}, \tag{13}$$

where λ is a constant satisfying $0 \leq \lambda \leq 1$. Hence, g is called a λ-*weighting function*, and λ is called a *pessimistic-optimistic index* and means the pessimistic degree in the investor's decision making. Then, the mean value of a fuzzy number $\tilde{a} \in \mathcal{R}$ is

$$\tilde{E}(\tilde{a}) := \int_0^1 g(\tilde{a}_\alpha) \, d\alpha. \tag{14}$$

Then, *the mean value of the total expected utilities* of the consumption process $\{\tilde{C}_t\}_{t=0}^T$ and the wealth $\tilde{W}_T$ at the terminal time T is given by

$$J(\boldsymbol{x}, w, \pi, \tilde{C}_T) := \tilde{E} \left(E_{\boldsymbol{x}, w} \left(\sum_{t=0}^T \tilde{U}_1(t, \tilde{C}_t) + \tilde{U}_2(\tilde{W}_T) \right) \right) \tag{15}$$

for an initial securities price $\boldsymbol{x}$, an initial wealth w and an admissible trading strategy π satisfying $\sum_{i=0}^n x^i \pi_0^i = w$.

4 The Optimal Consumption and Wealth under Uncertainty

In this section, we discuss the following optimal portfolio problem by dynamic programming ([7,8]).

Problem P (Optimal Consumption and Wealth Problem). Let $\lambda \in [0,1]$ and let $\boldsymbol{x}$ be initial securities prices. Maximize the total expected utility $J(\boldsymbol{x}, w, \pi, \tilde{C}_T)$ by admissible trading strategies π and admissible terminal consumptions $\tilde{C}_T$, where $J(\boldsymbol{x}, w, \pi, \tilde{C}_T)$ is defined by (15).

Define the *optimal total expected utility* by

$$J(\boldsymbol{x}, w) := \sup_{\pi, \tilde{C}_T} J(\boldsymbol{x}, w, \pi, \tilde{C}_T) \tag{16}$$

for an initial securities price $\boldsymbol{x}$ and an initial wealth w, where π and $\tilde{C}_T$ are taken over admissible trading strategies and admissible terminal consumptions satisfying the initial condition $\sum_{i=0}^n x^i \pi_0^i = w$. Let a family of admissible trading strategies and terminal consumptions by $\mathcal{A}(t) := \{(\pi_t, \pi_{t+1}, \cdots .\pi_T, \tilde{C}_T) \mid \tilde{C}_s(\omega) \succeq \tilde{0}$ for almost all $\omega \in \Omega$ and $s = t, \cdots, T; \tilde{C}_T \preceq \tilde{V}_T\}$ for $t \in \mathbb{T}$. To discuss the optimality equation for Problem P, we define the mean value of the total expected utility after time t by

$$I_{\pi, \tilde{C}_T}(t) := \tilde{E} \left(E_\pi \left(\sum_{s=t}^T \tilde{U}_1(s, \tilde{C}_s) + \tilde{U}_2(\tilde{W}_T) \mid \mathcal{M}_t \right) \right) \tag{17}$$

for trading strategies $\pi = (\pi_0, \pi_1, \cdots .\pi_T)$ satisfying $(\pi_t, \pi_{t+1}, \cdots .\pi_T, \tilde{C}_T) \in \mathcal{A}(t)$, where $E_\pi(\cdot \mid \mathcal{M}_t)$ is the probabilistic conditional expectation with the trading strategy π at current initial time t. We obtain the following results about the total expected utility $I_{\pi, \tilde{C}_T}(t)$.

Lemma 1. *Let $t \in \mathbb{T}$. The mean value of the total expected utility (17) is reduced to*

$$I_{\pi,\tilde{C}_T}(t) = E_\pi \left(\sum_{s=t}^{T} C_s + R_T \mid \mathcal{M}_t \right), \tag{18}$$

where real random variables C_s and R_T are defined by

$$C_s := \int_0^1 (\lambda U_1(s, \tilde{C}_{s,\alpha}^-) + (1-\lambda)U_1(s, \tilde{C}_{s,\alpha}^+))\, d\alpha, \quad s = 0, 1, \cdots, T; \tag{19}$$

$$R_T := \int_0^1 (\lambda U_2(\tilde{W}_{T,\alpha}^-) + (1-\lambda)U_2(\tilde{W}_{T,\alpha}^+))\, d\alpha. \tag{20}$$

Let $t \in \mathbb{T}$. Similarly to (15), we put the mean value of the total expected utility after time t by

$$I(t, \boldsymbol{x}, \boldsymbol{u}, \pi, \tilde{C}_T) := E_{t,\boldsymbol{x},\boldsymbol{u},\pi}(I_{\pi,\tilde{C}_T}(t)) \tag{21}$$

for securities prices $\boldsymbol{x}$, current initial trading strategies $\boldsymbol{u} = (u^0, u^1, \cdots, u^n) \in \mathbb{R}^{n+1}$, an admissible trading strategy $\pi = (\pi^0, \pi^1, \cdots, \pi^n)$, an admissible terminal consumption $\tilde{C}_T$ such that $\pi_t^i = u^i$ for $i = 0, 1, \cdots, n$ holds at time t, where $E_{t,\boldsymbol{x},\boldsymbol{u},\pi}(\cdot)$ is the probabilistic expectation with a trading strategy π satisfying $\pi_t^i = u^i$ $(i = 0, 1, \cdots, n)$ for current initial securities prices $\boldsymbol{x}$ and current initial trading strategies $\boldsymbol{u}$ at time t. Then, Lemma 1 implies

$$I(t, \boldsymbol{x}, \boldsymbol{u}, \pi, \tilde{C}_T) = E_{t,\boldsymbol{x},\boldsymbol{u},\pi} \left(\sum_{s=t}^{T} C_s + R_T \right). \tag{22}$$

Now we define the *optimal total expected utility after time t* by

$$v_t(\boldsymbol{x}, \boldsymbol{u}) := \max_{(\pi_t, \cdots \pi_T, \tilde{C}_T) \in \mathcal{A}(t)} I(t, \boldsymbol{x}, \boldsymbol{u}, \pi, \tilde{C}_T) \tag{23}$$

for $t = 0, 1, \cdots, T$, where C_s and R_T are defined by (19) and (20). Now, we obtain the following optimality equations by dynamic programming.

Theorem 1 (Optimality equation).

(i) *The optimal total expected utility is a solution of the following backward recursive equation:*

$$v_t(\boldsymbol{x}, \boldsymbol{u}) = \max_{\pi_{t+1}} E_{t,\boldsymbol{x},\boldsymbol{u},\pi}(C_t + v_{t+1}(Z_{t+1}, \pi_{t+1})) \tag{24}$$

for $t = 0, 1, \cdots, T-1$, securities prices $\boldsymbol{x} = (x^0, x^1, \cdots, x^n)$ and current initial trading strategies $\boldsymbol{u}$; and at the terminal time T it holds that

$$v_T(\boldsymbol{x}, \boldsymbol{u}) = \max_{\tilde{C}_T} E_{\boldsymbol{x},\boldsymbol{u}}(C_T + R_T), \tag{25}$$

where Z_{t+1} is given by

$$Z_{t+1} := (x^0(1 + r_{t+1}), x^1(1 + Y_{t+1}^1), x^2(1 + Y_{t+1}^2), \cdots, x^n(1 + Y_{t+1}^n)). \tag{26}$$

(ii) *Let π^* and $\tilde{C}_T^*$ be an admissible trading strategy and an admissible consumption attaining the maxima in (24) and (25). Then it holds that*

$$v_0(\boldsymbol{x}, \boldsymbol{u}) = I(0, \boldsymbol{x}, \boldsymbol{u}, \pi^*, \tilde{C}_T^*) \qquad (27)$$

for initial securities prices $\boldsymbol{x} = (x^0, x^1, \cdots, x^n)$ and current initial trading strategies $\boldsymbol{u} = (u^0, u^1, \cdots, u^n)$ satisfying $\pi_0^{*i} = u^i$ for $i = 0, 1, \cdots, n$. Further, π^* and $\tilde{C}_T^*$ are optimal for Problem P: Let $\boldsymbol{x} = (x^0, x^1, \cdots, x^n)$ be an initial securities price, and let w be an initial wealth. Then it holds that

$$J(\boldsymbol{x}, w) = \max_{\boldsymbol{u}\, :\, \boldsymbol{x}\boldsymbol{u}' = w} v_0(\boldsymbol{x}, \boldsymbol{u}), \qquad (28)$$

where $\boldsymbol{x}\boldsymbol{u}' = \sum_{i=0}^n x^i u^i = w$ with the transpose $\boldsymbol{u}'$ of $\boldsymbol{u} = (u^0, u^1, \cdots, u^n)$.

References

1. Inuiguchi, M., Ramík, J.: Possibility linear programming: a brief review of fuzzy mathematical programming and a comparison with stochastic programming in portfolio selection problem. Fuzzy Sets and Systems **111** (2000) 3-28.
2. Inuiguchi, M., Tanino, T.: Portfolio selection under independent possibilistic information. Fuzzy Sets and Systems **115** (2000) 83-92.
3. Klir, G.J., Yuan, B.: Fuzzy Sets and Fuzzy Logic: Theory and Applications. Prentice-Hall, London (1995).
4. Korn, R., Korn, E.: Options Pricing and Portfolio Optimization Modern Models of Financial Mathematics. Amer.Math.Soc., Providence (2001).
5. Kwakernaak, H.: Fuzzy random variables – I. Definitions and theorem. Inform. Sci. **15** (1978) 1-29.
6. Markowits, H.: Mean-Variance Analysis in Portfolio Choice and Capital Markets. Blackwell, Oxford (1990).
7. Merton, R.C.: Continuous-Time Finance. Blackwell, Cambridge, MA (1990).
8. Pliska, S.R.: Introduction to Mathematical Finance: Discrete-Time Models. Blackwell Publ., New York (1997).
9. Puri, M.L. .Ralescu, D.A: Fuzzy random variables. Math. Anal. Appl. **114** (1986) 409-422.
10. Ross, S.M.: An Introduction to Mathematical Finance. Cambridge Univ. Press, Cambridge (1999).
11. Tanaka, H., Guo P., Türksen, L.B.: Portfolio selection based on fuzzy probabilities and possibilistic distribution. Fuzzy Sets and Systems **111** (2000) 397-397.
12. Wang, G., Zhang, Y.: The theory of fuzzy stochastic processes. Fuzzy Sets and Systems **51** (1992) 161-178.
13. Yoshida, Y.: The valuation of European options in uncertain environment. European J. Oper. Res. to appear.
14. Yoshida, Y., Yasuda, M., Nakagami, J., Kurano, M.: A new evaluation of mean value for fuzzy numbers and its application to American put option under uncertainty. preprint.
15. Yoshida, Y., Yasuda, M., Nakagami, J., Kurano, M.: A discrete-time consumption and wealth model under uncertainty: A mean value evaluation approach. preprint.
16. Zadeh, L.A.: Fuzzy sets. Inform. and Control **8** (1965) 338-353.

A Fuzzy Approach to Stochastic Dominance of Random Variables

Bart De Schuymer[1], Hans De Meyer[1], and Bernard De Baets[2]

[1] Department of Applied Mathematics and Computer Science
Ghent University, Krijgslaan 281 (S9), B-9000 Gent, Belgium
{Bart.DeSchuymer,Hans.DeMeyer}@rug.ac.be
[2] Department of Applied Mathematics, Biometrics and Process Control
Ghent University, Coupure links 653, B-9000 Gent, Belgium
Bernard.DeBaets@rug.ac.be

Abstract. The discrete dice model essentially amounts to comparing discrete uniform probability distributions and generates reciprocal relations that exhibit a particular type of transitivity called dice-transitivity. In this contribution, this comparison method is extended and applied to general probability distributions and the generated reciprocal relations are shown to generalize the concept of stochastic dominance. For a variety of parametrized probability distributions, we analyse the transitivity properties of these reciprocal relations within the framework of cycle-transitivity. The relationship between normal probability distributions and the different types of stochastic transitivity is emphasized.

1 Introduction

Recently, the present authors have introduced a discrete model, suitable for comparing discrete uniform probability distributions and leading to a broad class of reciprocal relations [3]. Let us recall that a reciprocal relation Q on a set of alternatives A is a mapping from A^2 to $[0,1]$, such that for all $a, b \in A$ it holds that $Q(a,b) + Q(b,a) = 1$.

Reciprocal relations quite naturally appear in the framework of fuzzy preference modelling [1]. The dice model yields a way of representing fuzzy preferences and indifferences. Here the name dice is reserved for an m-dimensional multiset of (possibly equal) integers, each number having equal likelihood of showing up when the hypothetical dice is randomly thrown. Furthermore, two dice can be compared by considering the winning probability of one dice with respect to the other. In the following definition, V denotes the space of all possible dice (multisets in $\mathbb{N}$).

Definition 1. [3] *Consider the mappings* $P : V^2 \to [0,1]$ *and* $I : V^2 \to [0,1]$ *defined by:*

$$P(A,B) = P\{A \text{ wins from } B\} = \frac{\#\{(a,b) \mid a > b, a \in A, b \in B\}}{(\#A)(\#B)}, \quad (1)$$

T. Bilgiç et al. (Eds.): IFSA 2003, LNAI 2715, pp. 253–260, 2003.

and

$$I(A,B) = P\{A \text{ and } B \text{ end in a draw}\} = \frac{\#\{(a,b) \mid a = b, a \in A, b \in B\}}{(\#A)(\#B)}. \quad (2)$$

For the mapping $D : V^2 \to [0,1]$ defined by

$$D(A,B) = P(A,B) + \frac{1}{2}I(A,B), \quad (3)$$

it holds that

$$D(A,B) + D(B,A) = 1. \quad (4)$$

We say that an n-dimensional reciprocal relation $Q = [q_{ij}]$ is generated by a collection $(A_1, A_2, \ldots, A_n)$ of n dice, if it holds that $q_{ij} = D(A_i, A_j)$ for all i, j. The collection of dice is called a dice model for Q.

For two dice A and B, it can be stated that $A >_s B$ (A statistically wins from B) if $Q(A,B) > 1/2$ and $A =_s B$ (A is statistically indifferent to B) if $Q(A,B) = 1/2$. One of the main features of the dice model is that it can generate reciprocal relations that show a cyclic behaviour, which in dice terminology means that it may occur for some dice A, B, C that $A >_s B, B >_s C$ and $C >_s A$, in other words, there is no winning dice among A, B and C. Clearly, the possible occurence of cycles implies that the relation $>_s$ derived from the reciprocal relation Q generated by a dice model is in general not transitive. However, by generalizing the well-known concept of T-transitivity (with T a triangular norm) to that of cycle-transitivity, we have been able to prove that the reciprocal relation generated by a dice model exhibits a particular type of cycle-transitivity, called dice-transitivity [2,3].

In probabilistic terms, the dice model is a discrete model since a dice is characterized by a discrete uniform probability distribution on a discrete (multi)set of integers, whereas the reciprocal relation Q generated by a collection of independent discrete uniform probability distributions yields a natural way of comparing these distributions.

In the present paper, our principal aim is to generalize the discrete dice model so that we can also deal with general continuous probability distributions and in particular also discrete probability distributions other than uniform distributions. Then we will investigate the transitivity properties of the reciprocal relations generated by collections of various types of parametrized probability distributions, and the specific relationship between collections of normal probability distributions and stochastic transitivity will be emphasized.

2   Generalizations of the Dice Model

We first propose the following generalization of the dice model to the case of general discrete probability distributions. From here on, a random variable will also be denoted r.v.

Definition 2. *Let* $(X_1, X_2, \ldots, X_n)$ *denote an n-dimensional discrete r.v.,* X_i *taking values in the discrete set* C_i *of integers, with joint probability mass function* $p_{X_1, X_2, \ldots, X_n}(k_1, k_2, \ldots, k_n)$. *Then the relation* $Q = [q_{ij}]$ *with elements*

$$q_{ij} = \sum_{k \in C_i} \sum_{l \in C_j, l < k} p_{X_i, X_j}(k, l) + \frac{1}{2} \sum_{k \in C_i \cap C_j} p_{X_i, X_j}(k, k) \tag{5}$$

is a reciprocal relation.

It should be noted that $p_{X_i, X_j}(k_i, k_j)$ denotes the two-dimensional marginal probability mass function and that the r.v. $X_1, X_2, \ldots, X_n$ are not necessarily independent. If they are independent, the joint probability mass function and the two-dimensional marginal mass functions are all factorizable in one-dimensional marginal distributions, and if moreover these marginal distributions are uniform, then the classical dice model (Definition 1) is recovered.

Another generalization of the dice model arises when considering (absolutely) continuous random variables.

Definition 3. *Let* $(X_1, X_2, \ldots, X_n)$ *denote an n-dimensional continuous r.v. with joint probability density function* $f_{X_1, X_2, \ldots, X_n}(x_1, x_2, \ldots, x_n)$. *Then the relation* $Q = [q_{ij}]$ *with elements*

$$q_{ij} = \int_{-\infty}^{+\infty} dx_i \int_{-\infty}^{x_i} f_{X_i, X_j}(x_i, x_j) dx_j , \tag{6}$$

is a reciprocal relation.

Note that in the transition from the discrete to the continuous case, the second contribution to q_{ij} in (5) has disappeared since in the latter case the probability of having $X_i = X_j$ is zero.

In the discussion of these generalized models, we will maintain the terminology related to the original dice model. It suffices to imagine a collection of dice as a collection of real random variables. Comparing two dice X_i and X_j from the collection is done by considering q_{ij}. If $q_{ij} > 1/2$, we still say that dice X_i wins from dice X_j, which is denoted by $X_i >_s X_j$, and if $q_{ij} = 1/2$ we say that both dice are statistically indifferent, which is denoted by $X_i =_s X_j$.

An alternative concept for comparing two probability distributions is that of stochastic dominance [4] which is particularly popular in financial mathematics.

Definition 4. *A r.v. X with cumulative distribution function* F_X *stochastically dominates by first degree (FSD) a r.v. Y with cumulative distribution function* F_Y, *denoted as* $X \succeq_1 Y$, *if for all real t it holds that* $F_X(t) \leq F_Y(t)$, *and the strict inequality holds for at least one t.*

The condition for first degree stochastic dominance is rather severe, as it requires that the graph of the function F_X lies beneath the graph of F_Y. The need to relax this condition has led to other types of stochastic dominance, such as second-degree and third-degree stochastic dominance. We don't go into more details, since we just want to emphasize the following relationship between first-degree stochastic dominance and the winning probabilities in the dice model.

Theorem 1. *For two independent r.v. X and Y it holds that $X \succeq_1 Y$ implies $X \geq_s Y$.*

The relation $\geq_s$ therefore generalizes the stochastic dominance relation $\succeq_1$, whereas the reciprocal relation generated by a dice model clearly is a fuzzy version of the crisp relation $\geq_s$ (i.e. the $\frac{1}{2}$-cut of Q). Hence, we can interpret the reciprocal relation Q also as a fuzzy alternative to stochastic dominance.

3   Cycle-Transitivity

Further on we will use the concept of cycle-transitivity to describe the transitivity properties of certain reciprocal relations generated by r.v. For a reciprocal relation $Q = [q_{ij}]$, we define for all i, j, k the following quantities $\alpha_{ijk} = \min(q_{ij}, q_{jk}, q_{ki})$, $\beta_{ijk} = \mathrm{median}(q_{ij}, q_{jk}, q_{ki})$ and $\gamma_{ijk} = \max(q_{ij}, q_{jk}, q_{ki})$. Obviously, $\alpha_{ijk} \leq \beta_{ijk} \leq \gamma_{ijk}$.

Definition 5. *[2] A reciprocal relation $Q = [q_{ij}]$ is called cycle-transitive w.r.t. the upper bound $U : [0,1]^3 \to \mathbb{R}$, if for all i, j, k it holds that:*

$$\alpha_{ijk} + \beta_{ijk} + \gamma_{ijk} - 1 \leq U(\alpha_{ijk}, \beta_{ijk}, \gamma_{ijk}). \tag{7}$$

For the function U it must hold that $U(0,0,1) \geq 0$, $U(0,1,1) \geq 1$ and for all $0 \leq a \leq b \leq c \leq 1$ that $U(a,b,c) + U(1-c, 1-b, 1-a) \geq 1$.

Note that for each triplet of dice there are six conditions of type (7) of which only two are independent. Also, U should only be defined in points (a, b, c) with $0 \leq a \leq b \leq c \leq 1$. Due to the reciprocity of Q, inequality (7) is equivalent to the statement that for all i, j, k it must hold that:

$$L(\alpha_{ijk}, \beta_{ijk}, \gamma_{ijk}) \leq \alpha_{ijk} + \beta_{ijk} + \gamma_{ijk} - 1, \tag{8}$$

where the lower bound L is defined by:

$$L(\alpha_{ijk}, \beta_{ijk}, \gamma_{ijk}) = 1 - U(1 - \gamma_{ijk}, 1 - \beta_{ijk}, 1 - \alpha_{ijk}). \tag{9}$$

This definition implies that if a reciprocal relation Q is cycle-transitive w.r.t. U_1 and $U_1(a,b,c) \leq U_2(a,b,c)$ for all $0 \leq a \leq b \leq c \leq 1$, then Q is cycle-transitive w.r.t. U_2. When the lower bound equals the upper bound, i.e. $L(a,b,c) = U(a,b,c)$ for all $0 \leq a \leq b \leq c \leq 1$ (in which case the inequalities in (7) and (8) become equalities), we say that the function U is self-dual.

Cycle-transitivity includes as special cases many other well-known types of transitivity, such as T-transitivity with T a Frank t-norm, and stochastic transitivity. Given a t-norm T, a fuzzy relation $Q = [q_{ij}]$ is called T-transitive if for all i, j, k it holds that $T(q_{ij}, q_{jk}) \leq q_{ik}$.

Theorem 2. [2] *A reciprocal relation Q is T-transitive for a given Frank t-norm T (with dual t-conorm S), if and only if Q is cycle-transitive w.r.t. the upper bound U_T defined by*

$$U_T(\alpha, \beta, \gamma) = \alpha + \beta - T(\alpha, \beta) = S(\alpha, \beta). \tag{10}$$

Well-known examples of Frank t-norms are the minimum operator $(T_\mathbf{M})$, i.e. $T_\mathbf{M}(x, y) = \min(x, y)$, the algebraic product $(T_\mathbf{P})$, i.e. $T_\mathbf{P}(x, y) = xy$ and the Łukasiewicz t-norm $(T_\mathbf{L})$, i.e. $T_\mathbf{L}(x, y) = \max(x + y - 1, 0)$. $T_\mathbf{M}$-transitivity implies $T_\mathbf{P}$-transitivity and $T_\mathbf{P}$-transitivity implies $T_\mathbf{L}$-transitivity.

A reciprocal relation $Q = [q_{ij}]$ is called strongly (moderately, weakly) stochastic transitive if for all i, j, k for which $q_{ij} \geq 1/2$ and $q_{jk} \geq 1/2$, it holds that $q_{ik} \geq \max(q_{ij}, q_{jk})$, $(q_{ik} \geq \min(q_{ij}, q_{jk})$, $q_{ik} \geq 1/2)$.

A reciprocal relation Q is strongly (moderately, weakly) stochastic transitive, if and only if Q is cycle-transitive w.r.t. the upper bound U_{ss} (U_{ms}, U_{ws}), with $U_{ss}(\alpha, \beta, \gamma) = \beta$, $U_{ms}(\alpha, \beta, \gamma) = \gamma$ and $U_{ws}(\alpha, \beta, \gamma) = \beta + \gamma - 1/2$ when $\beta \geq 1/2$ and $\alpha \neq 1/2$, $U_{ss}(\alpha, \beta, \gamma) = U_{ms}(\alpha, \beta, \gamma) = U_{ws}(\alpha, \beta, \gamma) = 1$ when $\beta < 1/2$ and $U_{ss}(\alpha, \beta, \gamma) = U_{ms}(\alpha, \beta, \gamma) = U_{ws}(\alpha, \beta, \gamma) = 1/2$ when $\alpha = 1/2$.

In a previous paper we have shown that:

Theorem 3. [3] *The reciprocal relation generated by a (classical) dice model (independent uniform probability distributions on discrete sets) is cycle-transitive w.r.t. the upper bound $U_\mathbf{D}$ defined by*

$$U_\mathbf{D}(\alpha, \beta, \gamma) = \beta + \gamma - \beta\gamma. \tag{11}$$

Cycle-transitivity w.r.t. $U_\mathbf{D}$ is called dice-transitivity.

The converse, however, is only true for 3-dimensional relations:

Theorem 4. [3] *A 3-dimensional dice-transitive reciprocal relation Q, with q_{12}, q_{23}, q_{31} rational numbers, can be generated by a (classical) dice model.*

Dice-transitivity can be situated between $T_\mathbf{P}$-transitivity and $T_\mathbf{L}$-transitivity, and also between moderate stochastic transitivity and $T_\mathbf{L}$-transitivity.

4   General Transitivity Properties

The main result of this section is the fact that dice-transitivity remains a property of reciprocal relations generated by a collection of random variables, provided that the random variables are independent.

Theorem 5. *The reciprocal relation generated by a collection of independent random variables is dice-transitive.*

As before, the converse is only true for 3-dimensional relations, namely:

Theorem 6. *A 3-dimensional dice-transitive reciprocal relation can be generated by a generalized dice model in which the random variables are independent.*

Dice-transitivity is a very weak form of transitivity, as it is a type of transitivity that does not exclude cyclic behaviour. Nevertheless, T_L-transitivity is still a weaker type of transitivity, and in many applications the minimum kind of transitivity that one should require. Hence the question arises whether a generalized dice model could generate all T_L-transitive reciprocal relations. It turns out that by dropping the condition of independence the question can be affirmatively answered.

Theorem 7. *The reciprocal relation generated by a collection of random variables is T_L-transitive.*

Theorem 8. *A 3-dimensional T_L-transitive reciprocal relation can be generated by a generalized dice model.*

At the other end of the transitivity scale, in the region of 'strong' transitivity, we obtain the following general result.

Theorem 9. *If the cumulative distribution functions F_{X_i} of the independent r.v. $X_1, X_2, \ldots, X_n$ of a dice model are derived by translation from the same cumulative distibution function F_X, i.e. $F_{X_i}(x) = F_X(x - t_i)$ for all i with t_i given arbitrary real constants, then the reciprocal relation generated by this dice model is strongly stochastic transitive.*

As an example, let us consider a dice model in which the random variables X_i are all normally distributed with the same variance σ^2. More explicitly, if $X_i \sim N(\mu_i, \sigma^2)$ and the X_i are independent, then the generated reciprocal relation is strongly stochastic transitive. The same conclusion can be drawn when the normal distributions are replaced by Laplace distributions with constant (but arbitrary) variance.

In the next section we will investigate the type of cycle-transitivity related to different types of probability distributions.

5   Some Specific Dice Models

We first analyse the transitivity of the reciprocal relation generated by a dice model in which the randon variables are distributed according to certain single-parameter families of probability distributions.

Theorem 10. *The reciprocal relation generated by a collection of independent r.v. that are all exponentially distributed (i.e. $f_{X_i}(x) = \lambda_i \exp(-\lambda_i x)$ for $x \geq 0$) or all geometrically distributed (i.e. $p_{X_i}(k) = p_i(1 - p_i)^{k-1}$), is cycle-transitive w.r.t. the upper bound U_E defined by*

$$U_E(\alpha, \beta, \gamma) = \alpha\beta + \alpha\gamma + \beta\gamma - 2\alpha\beta\gamma. \tag{12}$$

It is not at all surprising that the exponential distribution and the geometric distribution share the same transitivity property, since the latter distribution can be regarded as a discretization of the former.

As examples of two-parameter distributions we consider the uniform distribution on a compact interval and the Laplace distribution.

Theorem 11. *The reciprocal relation generated by a collection of independent r.v. that are all uniformly distributed over an interval of length λ (i.e. $f_{X_i}(x) = 1/\lambda$ for $x \in [a_i, a_i + \lambda]$ and 0 elsewhere), with arbitrary (but fixed) $\lambda > 0$, is cycle-transitive w.r.t. the upper bound $U_{\mathbf{U}}$ defined by*

$$U_{\mathbf{U}}(\alpha, \beta, \gamma) = \begin{cases} \alpha + \beta - \dfrac{1}{2}[\min(0, 1 - \sqrt{2\alpha} - \sqrt{2\beta})]^2 & , \beta < 1/2 \,, \\[2mm] \beta + \gamma - 1 + \dfrac{1}{2}[\min(0, 1 - \sqrt{2(1-\gamma)} - \sqrt{2(1-\beta)})]^2 & , \beta \geq 1/2 \,. \end{cases}$$

$$(13)$$

Theorem 12. *The reciprocal relation generated by a collection of independent r.v. that are all Laplace-distributed with the same variance (i.e. $f_{X_i}(x) = \exp(-|x - a_i|/b)/(2b))$ with arbitrary but fixed $b > 0$, is cycle-transitive w.r.t. the upper bound $U_{\mathbf{Lap}}$ defined by*

$$U_{\mathbf{Lap}}(\alpha, \beta, \gamma) = \begin{cases} \alpha + \beta - F(F^{-1}(\alpha) + F^{-1}(\beta)) & , \beta < 1/2 \,, \\[2mm] \beta + \gamma - 1 + F(F^{-1}(1 - \gamma) + F^{-1}(1 - \beta)) & , \beta \geq 1/2 \,, \end{cases}$$

$$(14)$$

with

$$F : \mathbb{R}^+ \to [0, 1/2] : x \to \frac{1}{2}\left(1 + \frac{x}{2}\right) e^{-x} \,.$$

Obviously, $U_{\mathbf{U}} \leq U_{\mathbf{ss}}$ and $U_{\mathbf{Lap}} \leq U_{\mathbf{ss}}$. This is in agreement with the general result of Theorem 9, since in both cases the distributions only differ from one another by a translation. One can easily verify that upper bounds (12), (13) and (14) are self-dual.

One can also easily verify that the reciprocal relation generated by a collection of independent random variables that are Laplace-distributed and that have the same expectation and possibly different variance have the property $\alpha_{ijk} = \beta_{ijk} = \gamma_{ijk} = 1/2$ for all i, j, k.

In the foregoing cases of two-parameter distributions, we have fixed one of the parameters and just exploited the remaining degree of freedom. However, if both parameters were allowed to take different values for the random variables in the collection, it becomes very difficult to find closed-form expressions for the corresponding upper bounds in the definition of cycle-transitivity. There seems to be one exception to this, namely when we consider the class of normal probability distributions.

6 Dice Models with Normal Random Variables

We will consider three different types of collections of normal distributions.

Theorem 13. *For the reciprocal relation generated by a collection of independent r.v. that are all normally distributed and that have the same variance σ^2 (i.e. $f_{X_i}(x) = (\sqrt{2\pi\sigma^2})^{-1} \exp(-1/2\left((x - \mu_i)/\sigma\right)^2))$ it holds for all i, j, k, that*

$$\Phi^{-1}(\alpha_{ijk}) + \Phi^{-1}(\beta_{ijk}) + \Phi^{-1}(\gamma_{ijk}) = 0 \,,$$

$$(15)$$

which implies that this relation is cycle-transitive w.r.t. the upper bound $U_{\mathbf{N}}$ defined by

$$U_{\mathbf{N}}(\alpha, \beta, \gamma) = \alpha + \beta - \Phi(\Phi^{-1}(\alpha) + \Phi^{-1}(\beta)). \tag{16}$$

Herein, Φ denotes the c.d.f. of the standard normal distribution.

One can show that on account of (15) it holds that $L_{\mathbf{N}}(a, b, c) = U_{\mathbf{N}}(a, b, c)$ for all $0 \le a \le b \le c \le 1$. Clearly, $U_{\mathbf{N}} \le U_{\mathbf{ss}}$, which is again in agreement with Theorem 9.

Theorem 14. *The reciprocal relation generated by a collection of independent normal r.v., i.e. $X_i \sim N(\mu_i, \sigma_i^2)$, is moderately stochastic transitive.*

By dropping the condition of independence, we know that we can expect an even weaker type of transitivity of the generated reciprocal relation. In fact, it can be shown that:

Theorem 15. *The reciprocal relation generated by a collection $(X_1, X_2, \ldots, X_n)$ of r.v. that are distributed according to an n-variate normal distribution, i.e. with joint probability density function*

$$f_{\underline{X}}(\underline{x}) = \frac{1}{\sqrt{(2\pi)^n \det \Sigma}} e^{-\frac{1}{2}(\underline{x}-\underline{\mu})\Sigma^{-1}(\underline{x}-\underline{\mu})^t}, \tag{17}$$

where the underlined quantities stand for row-vectors, ^{t} denotes the transposition and Σ represents the covariance matrix, is weakly stochastic transitive.

References

1. B. De Baets and J. Fodor, Twenty years of fuzzy preference structures, *Belgian Journal of Operations Research, Statistics and Computer Science* **37** (1997) 61–81.
2. H. De Meyer, B. De Baets and S. Jenei, Cyclic evaluation of transitivity of reciprocal relations, *Social Choice and Welfare*, submitted.
3. B. De Schuymer, H. De Meyer, B. De Baets and S. Jenei, On the cycle-transitivity of the dice model, *Theory and Decision*, submitted.
4. H. Levy, *Stochastic Dominance*, Kluwer Academic Publishers, Norwell, MA, 1998.

Extracting Strict Orders
from Fuzzy Preference Relations

Koen Maes and Bernard De Baets

Department of Applied Mathematics, Biometrics and Process Control
Ghent University, Coupure links 653, B-9000 Gent, Belgium
{Koen.Maes,Bernard.DeBaets}@rug.ac.be

Abstract. Since crisp relations are too poor to represent the human way of reasoning, we often use fuzzy relations in preference modelling and multicriteria decision support. At the final stage, the decision maker expects a crisp answer to his problems. Many solutions have been proposed to perform this fuzziness dissolution step. In this paper, interval-valued preference structures are used to achieve this goal.

1 Introduction

Fuzzy relations are a natural generalization of crisp relations. Binary fuzzy relations express "the degree to which the elements in question are in relation, or in other words, the strength of the link between any two elements" [6]. There exist many ways of transforming the decision maker's opinion into a binary fuzzy relation [8]. Once we have this fuzzy relation at our disposal, we need to reduce it to a crisp preference structure in order to draw some conclusions from it. One way of realizing this is by considering the fuzzy pre-order closure (i.e. reflexive min-transitive closure) of the fuzzy relation [1]. When taking the α-cuts of the transformed relation, we obtain a hierarchical structure of crisp pre-orders, in which, with decreasing α, incomparability disappears at the cost of increasing indifference [9]. In this paper, however, we will extract a crisp preference structure from the original fuzzy relation by means of auxiliary interval-valued preference structures.

In Section 2 we recall some basic notions concerning additive fuzzy preference structures [5]. Special attention is paid to the Frank t-norm family. In Section 3 we weaken the results of Bilgiç [2]: if a reflexive binary fuzzy relation fulfils his TS-property, one does not necessarily obtain a strict order. In Section 4, we introduce a new definition of interval-valued preference structures and propose two ways of extracting crisp preference structures from them. Furthermore we reveal under which conditions the extracted crisp preference relation represents an order. Finally, we apply our results to reflexive inclusion measures in Section 5.

Before we start our study, we recall some definitions. For a t-norm T, a t-conorm S and an involutive negator N, we say that $\langle T, S, N \rangle$ is a De Morgan triplet if

$$N(S(x,y)) = T(N(x), N(y))$$

T. Bilgiç et al. (Eds.): IFSA 2003, LNAI 2715, pp. 261–268, 2003.

is satisfied for all $(x, y) \in [0, 1]^2$. If the negator N is determined by the $[0, 1]$-automorfism ϕ, we write $N \equiv \mathcal{N}_\phi$, with $\mathcal{N}$ the standard negator. In this paper the Frank t-norm family plays a very profound role. For $s \in]0, 1[\cup]1, \infty[$, the Frank t-norm T^s is defined by

$$T^s(x, y) = \log_s \left(1 + \frac{(s^x - 1)(s^y - 1)}{s - 1} \right).$$

The limit cases are given by T_M ($s \to 0$), T_P ($s \to 1$) and T_L ($s \to \infty$). The dual t-conorm $(T^s)^{\mathcal{N}}$ is usually denoted by S^s.

2   Additive Fuzzy Preference Structures

Let ϕ be an automorphism of the unit interval. Consider the continuous De Morgan triplet $\langle T_\phi^\infty, S_\phi^\infty, \mathcal{N}_\phi \rangle$. The minimal characterization of a ϕ-fuzzy preference structure is a straightforward extension of its classical counterpart [4].

Theorem 1. [4] *A triplet (P, I, J) of binary fuzzy relations on the set of alternatives $\mathcal{A}$ is a ϕ-fuzzy preference structure (ϕ-FPS) on $\mathcal{A}$ if and only if I is irreflexive and symmetric, and for any $(a, b) \in \mathcal{A}^2$:*

$$\phi(P(a, b)) + \phi(P(b, a)) + \phi(I(a, b)) + \phi(J(a, b)) = 1.$$

In case ϕ is the identity mapping we talk about a fuzzy preference structure (FPS). Fuzzy preference modelling has already witnessed quite some historical development (see [4]). The most recent approach is to construct a fuzzy preference structure by means of a generator [5].

Definition 1. *A generator i is a symmetric function $i : [0, 1]^2 \to [0, 1]$, such that $T^\infty \leq i \leq T^0$.*

The following proposition shows that a generator i indeed generates a fuzzy preference structure.

Proposition 1. [5] *Given a reflexive binary fuzzy relation R on the set of alternatives $\mathcal{A}$ and a generator i, the triplet (P, I, J), defined by*

$$P(a, b) := p(R(a, b), R(b, a)) = R(a, b) - i(R(a, b), R(b, a)),$$
$$I(a, b) := i(R(a, b), R(b, a)),$$
$$J(a, b) := j(R(a, b), R(b, a)) = i(R(a, b), R(b, a)) - (R(a, b) + R(b, a) - 1),$$

is a FPS on $\mathcal{A}$.

The triplet (p, i, j) is also called a generator triplet of the FPS (P, I, J).

Definition 2. [5] *A generator triplet (p, i, j) is called monotone if*

  (i) p is increasing in the first and decreasing in the second argument;
 (ii) i is increasing in both arguments;
(iii) j is decreasing in both arguments.

The monotonicity of a generator triplet is then easily characterized in terms of its generator i.

Theorem 2. [5] *A generator triplet* (p, i, j) *is monotone if and only if* i *is increasing and* 1-*Lipschitz.*

In accordance with the axiomatic study of fuzzy preference structures by Fodor and Roubens [6], the Frank t-norm family plays a very important role in the study of generator triplets.

Theorem 3. [5] *If the generator* i *of a generator triplet* (p, i, j) *is a* t-*norm then the following statements are equivalent:*

(i) *the mapping* $p(x, 1 - y)$ *is a t-norm;*
(ii) i *is a Frank t-norm* T^s, $s \in [0, \infty]$.

In the latter case we get $p(x, y) = T^{1/s}(x, 1 - y)$ *and* $j(x, y) = T^s(1 - x, 1 - y)$.

3   Extracting Strict Orders: Based on Normal Forms

Suppose that we dispose of a reflexive binary fuzzy relation R, representing the decision maker's subjective weak preference on a set of alternatives $\mathcal{A}$. How can we create from this fuzzy relation R some (strict) order representing as good as possible the decision maker's opinion? Remark that, for reasons of simplicity, we will often denote $R(a, b)$ (resp. $R(b, a)$) as x (resp y).

In order to incorporate some higher order vagueness into the decision process, Bilgiç [2] constructs an interval-valued preference structure by using disjunctive and conjunctive normal forms. The desired strict order is then defined on the basis of relative positions of the constructed intervals. Inspired by the disjunctive and conjunctive Boolean normal form expressions of a Boolean preference structure

$$P(a, b) = x \wedge y' = (x \vee y) \wedge (x \vee y') \wedge (x' \vee y'),$$
$$I(a, b) = x \wedge y = (x \vee y) \wedge (x \vee y') \wedge (x' \vee y),$$
$$J(a, b) = x' \wedge y' = (x \vee y') \wedge (x' \vee y) \wedge (x' \vee y').$$

Bilgiç defines the following interval-valued fuzzy preference structure

$$\mathbb{P}[a, b] = [P_D(a, b), P_C(a, b)] = [T(x, y^N), T(S(x, y), S(x, y^N), S(x^N, y^N))],$$
$$\mathbb{I}[a, b] = [I_D(a, b), I_C(a, b)] = [T(x, y), T(S(x, y), S(x, y^N), S(x^N, y))],$$
$$\mathbb{J}[a, b] = [J_D(a, b), J_C(a, b)] = [T(x^N, y^N), T(S(x, y^N), S(x^N, y), S(x^N, y^N))].$$

Note that for this construction to make sense, the disjunctive fuzzy normal form must be smaller than or equal to the conjunctive one. This particular problem has been discussed in [7]. From this interval-valued fuzzy preference structure, Bilgiç defines a crisp asymmetric relation $\succ$ in the following way [2]:

$$a \succ b \quad \Leftrightarrow \quad P_D(a, b) > P_C(b, a).$$

Furthermore, he claims that, for $\langle T, S, N \rangle = \langle T_\phi^\infty, S_\phi^\infty, \mathcal{N}_\phi \rangle$ or $\langle T, S, N \rangle = \langle T_\phi^1, S_\phi^1, \mathcal{N}_\phi \rangle$, the relation $\succ$ is transitive if and only if for all $a, b, c \in \mathcal{A}$:

$$\wedge \begin{cases} T(R(a,b), R(b,a)^N) \geq S(R(b,a), R(a,b)^N) \\ T(R(b,c), R(c,b)^N) \geq S(R(c,b), R(b,c)^N) \end{cases}$$

$$\Downarrow \qquad\qquad (TS)$$

$$T(R(a,c), R(c,a)^N) \geq S(R(c,a), R(a,c)^N).$$

Unfortunately, this claim is false. Consider $\langle T^1, S^1, \mathcal{N} \rangle$ and let R_1 be a reflexive binary fuzzy relation on $\mathcal{A} = \{a, b, c\}$ with $R_1(a,b) = R_1(b,c) = R_1(c,a) = 0.65$ and $R_1(b,a) = R_1(c,b) = R_1(a,c) = 0.4$. It is easily verified that R_1 fulfils the TS-property, while $\succ$ is not transitive. Conversely, consider the reflexive binary fuzzy relation R_2 on $\mathcal{A}$ with $R_2(a,b) = R_2(b,c) = 1$, $R_2(b,a) = R_2(c,b) = 0$, $R_2(a,c) = 0.65$ and $R_2(c,a) = 0.4$. Although in this case $\succ$ is transitive, the TS-property does not hold for R_2.

4   Extracting Strict Orders: Based on a Pair of Generators

The most obvious way of constructing some interval-valued preference structure $(\mathbb{P}, \mathbb{I}, \mathbb{J})$ on a set of alternatives $\mathcal{A}$ is given by

$$\mathbb{P}[a,b] = [x - i_2(x,y), x - i_1(x,y)],$$
$$\mathbb{I}[a,b] = [i_1(x,y), i_2(x,y)],$$
$$\mathbb{J}[a,b] = [i_1(x,y) - (x+y-1), i_2(x,y) - (x+y-1)],$$

where i_1 and i_2 are two generators such that $i_1(x,y) \leq i_2(x,y)$, for all $(x,y) \in [0,1]^2$. Remark that Bilgiç's construction does not fit into this framework. It is impossible to find two generators $i_1 \leq i_2$, such that $P_C(a,b) = x - i_1(x,y)$, $I_C(a,b) = i_2(x,y)$ or $J_C(a,b) = i_2(x,y) - (x+y-1)$ holds for all reflexive binary fuzzy relations R. We now want to construct a crisp preference structure $(\succ, \simeq, \|)$ based on the relative positions of the intervals of $(\mathbb{P}, \mathbb{I}, \mathbb{J})$. Because of the various possibilities that occur it is difficult to decide when to call two alternatives indifferent resp. incomparable or when to decide that one alternative is better than another. We propose here two different approaches.

4.1   First Approach

We fix the asymmetric crisp relation $\succ$ in the following way:

$$a \succ b \iff x - i_2(x,y) > y - i_1(x,y). \qquad (1)$$

It means that we prefer a to b if and only if $\mathbb{P}[a,b]$ is situated strictly to the right of $\mathbb{P}[b,a]$. To obtain a strict order, the relation $\succ$ should be transitive. We are looking for a condition on R such that $\succ$ is transitive for any choice of i_1 and i_2.

Theorem 4. *Consider a reflexive binary fuzzy relation R on the set of alternatives $\mathcal{A}$. If for all $(a, b, c) \in \mathcal{A}^3$ it holds that*

$$[R(a,b) > R(b,a)] \;\wedge\; [R(b,c) > R(c,b)]$$

$$\Downarrow \qquad\qquad\qquad\qquad \text{(T1)}$$

$$\vee \begin{cases} R(a,c) \geq R(a,b) > R(b,a) \geq R(c,a) \\ R(a,c) \geq R(b,c) > R(c,b) \geq R(c,a), \end{cases}$$

then the relation $\succ$, based on two increasing generators i_1 and i_2 fulfilling the 1-Lipschitz property, is transitive.

It is immediately clear that any min-transitive fuzzy relation R also satisfies condition T1. On the other hand, condition T1 closely resembles and implies the T0-property[1]: for all $(a, b, c) \in \mathcal{A}^3$ it holds that

$$[R(a,b) > R(b,a)] \;\wedge\; [R(b,c) > R(c,b)] \;\Rightarrow\; [R(a,c) > R(c,a)]. \qquad \text{(T0)}$$

Although in the crisp case both conditions coincide, this is no longer true in the fuzzy case. If the binary fuzzy relation R only fulfils the T0-property, the associated crisp relation $\succ$ is not always transitive. This is easily illustrated by taking $i_1 = T^\infty$, $i_2 = T^0$, $R(a,b) = R(b,c) = 0.7$, $R(b,a) = R(c,b) = R(c,a) = 0.3$ and $R(a,c) = 0.6$.

We now want to introduce the indifference $\simeq$ and incomparability $\|$ components of our crisp preference structure. In case $a \nsucc b$ and $b \nsucc a$ we say that a is indifferent to b if $\mathbb{I}[a, b]$ is situated totally to the right of the three other intervals:

$$a \simeq b \;\Leftrightarrow\; |x - y| \leq i_2(x,y) - i_1(x,y) < x + y - 1 \;\wedge\; \max(x,y) < 2\,i_1(x,y).$$

In all other cases we say that $a \| b$. By considering $\succ \cup \simeq$, we obtain a crisp relation $\succeq$ fulfilling the T0-property.

4.2 Second Approach

We start by defining the indifference relation $\simeq$ and decide when two alternatives are definitely incomparable. Once this is done we conclude where to put the strict preferences. Finally, whatever remains undecided is declared incomparable. Using the same principles as in the previous paragraph, we obtain the following definitions

$$a \simeq b \Longleftrightarrow i_2(x,y) - i_1(x,y) < x + y - 1 \;\wedge\; \max(x,y) < 2\,i_1(x,y),$$
$$a \succ b \Longleftrightarrow [x + y - 1 \leq i_2(x,y) - i_1(x,y) \;\vee\; 2\,i_1(x,y) \leq x] \;\wedge$$
$$[1 - x - y \leq i_2(x,y) - i_1(x,y) \;\vee\; 2\,i_1(x,y) \leq 2\,x + y - 1] \;\wedge$$
$$x - y > i_2(x,y) - i_1(x,y).$$

---

[1] Also known as weak transitivity [2].

Indifference will occur when $\mathbb{I}[a,b]$ is situated strictly to the right of $\mathbb{P}[a,b]$, $\mathbb{P}[b,a]$ and $\mathbb{J}[a,b]$. In case $a \not\simeq b$ we obtain $a \succ b$ if $\mathbb{P}[a,b]$ is situated strictly to the right of $\mathbb{P}[b,a]$ and if $\mathbb{J}[a,b]$ is not strictly to the right of the three other intervals. Two alternatives a and b are incomparable if $a \not\simeq b$, $a \not\succ b$ and $b \not\succ a$. As in the first approach, we would like the strict preference relation $\succ$ to be transitive. Preferably, in the case of a min-transitive R, the relation $\succ$ should be transitive as well.

Theorem 5. *Consider two increasing generators $i_1 \leq i_2$ fulfilling the 1-Lipschitz property. It then holds that:*

1. *If $i_1(x,y) < T^1(x,y)$, for all $(x,y) \in]0,1[^2$, or $T^1(x,y) < i_1(x,y)$, for all $(x,y) \in]0,1[^2$, and R is min-transitive, then the relation $\succ$ is not always transitive.*
2. *If $i_1 = T^1$ and R fulfils property T1, then the relation $\succ$ is transitive.*

From Theorem 3 we know that the Frank t-norm family plays an important role in the construction of FPS. Due to the previous theorem, we get the following interesting result.

Corollary 1. *Consider a Frank t-norm i_1 and an increasing generator i_2 fulfilling the 1-Lipschitz property such that $i_1 \leq i_2$. It then holds that:*

1. *If $i_1 \neq T^1$ and R is min-transitive, then the relation $\succ$ is not always transitive.*
2. *If $i_1 = T^1$ and R fulfils property T1, then the relation $\succ$ is transitive.*

In practical situations the following observations can be useful.

Theorem 6. *Let R be a reflexive binary fuzzy relation on a set of alternatives $\mathcal{A}$ and let $(\mathbb{P}, \mathbb{I}, \mathbb{J})$ be an interval-valued preference structure generated from R. For any $a,b \in \mathcal{A}$ it then holds that*

1. *If $\min(R(a,b), R(b,a)) > 2/3$, then the indifference interval $\mathbb{I}[a,b]$ is situated totally to the right of $\mathbb{P}[a,b]$, $\mathbb{P}[b,a]$ and $\mathbb{J}[a,b]$.*
2. *If $\max(R(a,b), R(b,a)) < 1/3$, then the incomparability interval $\mathbb{J}[a,b]$ lies totally to the right of $\mathbb{P}[a,b]$, $\mathbb{P}[b,a]$ and $\mathbb{I}[a,b]$.*

For the second approach, this means that $a \simeq b$ in case $\min(x,y) > 2/3$ and $a \parallel b$ whenever $\max(x,y) < 1/3$.

5   Reflexive Inclusion Measures in Multicriteria Decision Support

Generally speaking, inclusion measures express the degree to which a subset A of a finite universe X is contained in another subset B. They can be considered as particular binary fuzzy relations on $\mathcal{P}(X)$. Consider such an inclusion measure I. Given a set of alternatives $\mathcal{A}$ and a finite set of criteria $\mathcal{C}$, we will now construct a fuzzy weak preference relation R on $\mathcal{A}$ by means of the inclusion measure I.

For $a \in \mathcal{A}$, let C_a denote the set of criteria satisfied by a. The fuzzy relation $R(a, b) := I(C_b, C_a)$ then defines a fuzzy weak preference relation on the set of alternatives $\mathcal{A}$.

We will investigate the reflexive inclusion measures introduced in [3] and defined by:

$$H(A, B) = \frac{\alpha |B \setminus A| + \beta |A \cap B| + \gamma |X \setminus (A \cup B)|}{|A \setminus B| + \alpha |B \setminus A| + \beta |A \cap B| + \gamma |X \setminus (A \cup B)|}, \qquad (2)$$

where $(\alpha, \beta, \gamma) \in \{0, 1\}^3$ and $\alpha + \beta + \gamma \neq 0$. Most of these reflexive inclusion measures fulfil the T0-property.

Theorem 7. *Consider the reflexive inclusion measure H.*

1. *If $\alpha = 1$, then H fulfils the T0-property.*
2. *If $\alpha = 0$ and for all $(A, B, C) \in \mathcal{P}(X)^3$ it holds that*

$$\beta |A \cap B| + \gamma |X \setminus (A \cup B)| \neq 0 \quad \wedge \quad \beta |B \cap C| + \gamma |X \setminus (B \cup C)| \neq 0$$

$$\Downarrow$$

$$\beta |A \cap C| + \gamma |X \setminus (A \cup C)| \neq 0,$$

then H also fulfils the T0-property.

Similarly to the previous section, we want to construct a crisp preference structure $(\succ, \simeq, \parallel)$ on $\mathcal{A}$, where $a \succ b$ means that C_b is rather well contained in C_a and $a \simeq b$ that C_a is as much contained in C_b as C_b is contained in C_a. Within the class of reflexive inclusion measures fulfilling the T0 property we are now looking for an inclusion measure that immediately yields a transitive relation $\succ$. Consider the reflexive inclusion measure

$$H_3(A, B) := \frac{|B \setminus A|}{|A \setminus B| + |B \setminus A|}, \qquad (\alpha = 1, \ \beta = \gamma = 0).$$

Remark that in the limit case $A = B$ we define $H_3(A, A) := 1$. When using both approaches of the foregoing section in order to construct a crisp preference structure $(\succ, \simeq, \parallel)$ from the fuzzy relation R obtained by means of H_3, we get the following remarkable result.

Theorem 8. *The inclusion measure H_3 has the T1-property. Furthermore, both approaches for extracting a strict order from the fuzzy relation R constructed by means of H_3 yield the same result.*

6 Conclusions and Further Research

In order to extract a crisp preference structure $(\succ, \simeq, \parallel)$ from a reflexive binary fuzzy relation R, we have constructed an auxiliary interval-valued preference structure based on a pair of generators (i_1, i_2), $i_1 \leq i_2$. Next, we have proposed two different ways of constructing a crisp preference structure from it. For both approaches we have derived some sufficient conditions on R such that $\succ$ is transitive. Finally, we have applied these concepts to fuzzy weak preference relations modelled by means of some particular reflexive inclusion measures.

There is still quite a lot of work to be done on the topic. When starting with the construction of the symmetrical part of $(\succ, \simeq, \parallel)$, can we use indifference intervals with lower bound $i_1 \neq T^1$ for the extraction of transitive strict orders $\succ$? How can we adapt R in a minimal way so that $\succ$ is transitive? Do there exist other inclusion measures for which $\succ$ immediately is transitive? For which fuzzy relations R does the crisp relation $\succ$ have the Ferrers property?

References

1. W. Bandler and L. Kohout, *Special properties, closures and interiors of crisp and fuzzy relations*, Fuzzy Sets and Systems **26** (1988), 317–331.
2. T. Bilgiç, *Interval-valued preference structures*, European Journal of Operational Research **105** (1998), 162–183.
3. B. De Baets, H. De Meyer, and H. Naessens, *On rational cardinality-based inclusion measures*, Fuzzy Sets and Systems **128** (2002), 169–183.
4. B. De Baets and J. Fodor, *Twenty years of fuzzy preference structures (1978-1997)*, Belgian Journal of Operations Research, Statistics and Computer Science **37** (1997), 61–82.
5. _____, *Generator triplets of additive fuzzy preference structures*, Proc. Sixth Internat. Workshop on Relational Methods in Computer Science (Tilburg, The Netherlands), 2001, pp. 306–315.
6. J. Fodor and M. Roubens, *Fuzzy Preference Modelling and Multicriteria Decision Support*, Kluwer Academic Publishers, 1994.
7. K. Maes, B. De Baets, and J. Fodor, *Towards n-ary disjunctive and conjunctive fuzzy normal forms*, Proc. Seventh Meeting of the EURO Working Group on Fuzzy Sets (Varenna, Italy), 2002, pp. 297–302.
8. I.B. Türkşen and T. Bilgiç, *Interval valued strict preference with zadeh triples*, Fuzzy Sets and Systems **78** (1996), 183–195.
9. B. Van De Walle, B. De Baets, and E. Kerre, *Fuzzy multi-criteria analysis of cutting techniques in a nuclear reactor dismantling project*, Fuzzy Sets and Systems **74** (1995), 115–126.

T-Ferrers Relations versus T-biorders

Susana Díaz[1], Bernard De Baets[2], and Susana Montes[3]

[1] Department of Statistics and O.R., Faculty of Geology
University of Oviedo, Calvo Sotelo s/n, 33071 Oviedo, Spain
alu407@pinon.ccu.uniovi.es
[2] Department of Applied Mathematics, Biometrics and Process Control
Ghent University, Coupure links 653, B-9000 Gent, Belgium
Bernard.DeBaets@rug.ac.be
[3] Department of Statistics and O.R.
University of Oviedo, Nautical School, 33271 Gijón, Spain
smr@pinon.ccu.uniovi.es

Abstract. In this paper, we study the Ferrers property of relations in the context of fuzzy preference modelling. A logical approach leads us to the notion of T-Ferrers relations, while a relational approach brings us to T-biorders. We characterize the t-norms for which both notions coincide. We also describe the kind of completeness exhibited by reflexive T-Ferrers relations or reflexive T-biorders. Finally, we investigate the relationship between the T-Ferrers properties of a reflexive fuzzy relation R and the corresponding strict preference relation P, and the relationship between R and P being T-biorders.

1 Introduction

In classical relational calculus, graph theory, preference modelling [18], and many other disciplines, the Ferrers property is an interesting property of binary relations. Recall that a binary relation Q on a universe A is said to have the *Ferrers property*, or in short, to be a *Ferrers relation*, or even to be *Ferrers*, if it holds that

$$(aQb \wedge cQd) \Rightarrow (aQd \vee cQb) \tag{1}$$

for any $a, b, c, d \in A$. In this paper, we are concerned with the Ferrers property in the context of fuzzy preference modelling. As usual in fuzzy set theory, there is no unique way of generalizing a given concept. We will explore various possibilities and complement the results known from the literature with new points of view.

We briefly recall two equivalent relational representations of preferential information. On the one hand, one can consider a large preference relation R, i.e. a *reflexive* binary relation on the set of alternatives A, with the following interpretation:

aRb if and only if a is at least as good as b.

T. Bilgiç et al. (Eds.): IFSA 2003, LNAI 2715, pp. 269–276, 2003.
© Springer-Verlag Berlin Heidelberg 2003

On the other hand, R can be decomposed into disjoint parts: an irreflexive and asymmetric strict preference component P, a reflexive and symmetric indifference component I and an irreflexive and symmetric incomparability component J such that $P \cup P^t \cup I \cup J = A^2$ and $R = P \cup I$ (where ^{t} denotes the transpose of a relation). These components can be obtained by considering various intersections: $P = R \cap R^d$, $I = R \cap R^t$ and $J = R^c \cap R^d$ (where ^{c} denotes the complement of a relation and ^{d} denotes the dual of a relation, i.e. the complement of its transpose).

In fuzzy preference modelling, a reflexive binary fuzzy relation R on A can also be decomposed into what is called an additive fuzzy preference structure, by means of an (indifference) generator i, which is defined as a symmetric $[0,1]^2 \to [0,1]$ mapping located between the Łukasiewicz t-norm $T_{\mathbf{L}}$ and the minimum operator $T_{\mathbf{M}}$, i.e. $T_{\mathbf{L}} \le i \le T_{\mathbf{M}}$. More specifically, the strict preference relation P, the indifference relation I and the incomparability relation J are obtained as follows:

$$P(a,b) = R(a,b) - i(R(a,b), R(b,a))$$

$$I(a,b) = i(R(a,b), R(b,a))$$

$$J(a,b) = i(R(a,b), R(b,a)) - (R(a,b) + R(b,a) - 1).$$

An additive fuzzy preference structure (AFPS) (P, I, J) on A is then characterized as a triplet of binary fuzzy relations on A such that I is reflexive and symmetric and
$$P(a,b) + P(b,a) + I(a,b) + J(a,b) = 1,$$
whence the adjective 'additive'. The corresponding fuzzy large preference relation R is then given by $R(a,b) = P(a,b) + I(a,b)$.

An important observation is the fact that the logical formulation (1) of the Ferrers property of a binary relation Q can be stated equivalently in terms of a relational inequality:

$$Q \text{ is Ferrers } \Leftrightarrow Q \circ Q^d \circ Q \subseteq Q. \qquad (2)$$

A relation Q satisfying the above inequality is also called a *biorder*, hence Ferrers relations and biorders are one and the same thing.

One easily verifies that a reflexive Ferrers relation R is necessarily *complete* (i.e. aRb or bRa for any a, b), which is equivalent to saying that the corresponding incomparability relation J is the empty relation. This is due to its reflexivity. Reflexive biorders are also known as *interval orders*. The Ferrers property of a reflexive relation R implies the Ferrers property of its strict preference relation P. The converse also holds if the preference structure is one without incomparability. Preference structures without incomparability ($J = \emptyset$) therefore play an important role in the study of the Ferrers property.

In the fuzzy case, a fuzzy preference structure (P, I, J) without incomparability ($J = \emptyset$) is characterized by a *weakly complete* large preference relation R (i.e. $R(a,b) + R(b,a) \ge 1$ for any $a, b \in A$), from which it can be reconstructed by means of the generator $i = T_{\mathbf{L}}$. Conversely, the AFPS generated from a weakly complete large preference relation R by means of $i = T_{\mathbf{L}}$ is always without incomparability. Note that in case of a *strongly complete* large preference relation R (i.e. $\max(R(a,b), R(b,a)) = 1$ for any $a, b \in A$), the resulting AFPS is independent of the generator used.

It should be clear that there exist many ways of expressing the same thing in the crisp case. This will be no longer true in the fuzzy case, and each of the alternative expressions may lead to a different concept, with its own advantages and disadvantages. The purpose of this paper is to unravel some of these approaches.

2   The Ferrers Property versus Biorders

2.1   Definitions

As mentioned before, for crisp relations the notions of a Ferrers relation and of a biorder are one and the same thing. In this section, we investigate what remains true of this equivalence in the fuzzy case. To that end, we first introduce some appropriate definitions. We will stay within the usual setting of t-norms. We first recall the well-known notion of T-transitivity in order to be able to draw some analogy.

Definition 1. [10] *Consider a t-norm T. A binary fuzzy relation Q on A is called T-transitive if it holds that*

$$T(Q(a,b), Q(b,c)) \leq Q(a,c) \tag{3}$$

for any $a, b, c \in A$.

Recall that the sup-T composition of two binary fuzzy relations U and V on a universe A is the binary fuzzy relation $U \circ_T V$ on A defined by

$$U \circ_T V(x,z) = \sup_{y \in A} T(U(x,y), V(y,z)).$$

A sufficient condition for the sup-T composition of fuzzy relations to be associative is the left-continuity of the t-norm T. When working on finite universes, as is the case in fuzzy preference modelling, this condition becomes superfluous. A trivial result is the fact that T-transitivity can equivalently be expressed as a relational inequality.

Proposition 1. [10] *Consider a t-norm T. A binary fuzzy relation Q on A is T-transitive if and only if*

$$Q \circ_T Q \subseteq Q. \tag{4}$$

The Ferrers property can be generalized to fuzzy relations in a similar way.

Definition 2. [10] *Consider a t-norm T and a t-conorm S. A binary fuzzy relation Q on A is called (T, S)-Ferrers if it holds that*

$$T(Q(a,b), Q(c,d)) \leq S(Q(a,d), Q(c,b)) \tag{5}$$

for any $a, b, c, d \in A$.

A particular case arises when considering as t-conorm S the dual of the t-norm T, i.e. the t-conorm defined by $S(x,y) = 1 - T(1 - x, 1 - y)$. For the sake of simplicity, in that case the (T, S)-Ferrers property is simply referred to as the T-Ferrers property. It then reads as follows

$$T(Q(a,b), Q(c,d)) + T(1 - Q(a,d), 1 - Q(c,b)) \leq 1.$$

The case $T = T_{\mathbf{M}}$ was studied in detail in [6], mainly in terms of α-cuts of fuzzy relations. Clearly, if $T_1 \leq T_2$, then any fuzzy relation that is T_2-Ferrers is also T_1-Ferrers.

Definition 3. [7] *Consider a t-norm T. A binary fuzzy relation Q on A is called a T-biorder if it holds that*

$$Q \circ_T Q^d \circ_T Q \subseteq Q, \tag{6}$$

or equivalently
$$T(Q(a,b), 1 - Q(c,b), Q(c,d)) \leq Q(a,d)$$

for any $a, b, c, d \in A$.

Again, if $T_1 \leq T_2$, then any T_2-biorder is also a T_1-biorder.

2.2   The Equivalence: Rotation-Invariant T-norms

The equivalence between the notions of T-Ferrers relations and T-biorders has been studied before by one of present authors [4], and also earlier by Fodor [7]. The following theorem further generalizes these results and characterizes all t-norms for which both notions coincide.

Theorem 1. *Consider a t-norm T. Then the following statements are equivalent:*

(i) *Any T-Ferrers relation is a T-biorder.*
(ii) *Any T-biorder is a T-Ferrers relation.*
(iii) *T is rotation-invariant:*

$$(\forall (x, y, z) \in [0,1]^3)(T(x, y) \leq z \Leftrightarrow T(y, 1 - z) \leq 1 - x). \tag{7}$$

The equivalence of (i) and (ii) was proven in [4] for $T = T_{\mathbf{L}}$ in the setting of strongly complete fuzzy relations (although the proof did not make use of this assumption). In [7] the same equivalence was shown for a continuous t-norm T that has a contrapositive residual implication, i.e. for which the operation I_T defined by

$$I_T(x, y) = \sup\{z \mid T(x, z) \le y\}$$

satisfies $I_T(x, y) = I_T(1 - y, 1 - x)$. However, careful reading of the paper shows that the continuity assumption is not really invoked and that essentially a rotation-invariant t-norm is used. Indeed, rotation-invariant t-norms are nothing else but left-continuous t-norms that have a contrapositive residual implication (see also [5]). Fodor [9] has also shown that rotation-invariance characterizes the class of t-norms T for which the residual implication and S-implication coincide, i.e. for which $I_T(x, y) = 1 - T(x, 1 - y)$. What the above theorem adds to the existing knowledge is the fact that both notions (*T*-Ferrers relations and *T*-biorders) are only related (and then necessarily equivalent) in case of a rotation-invariant t-norm.

Rotation-invariance can be stated equivalently as

$$(\forall (x, y, z) \in [0, 1]^3)(T(x, y) > z \Leftrightarrow T(y, 1 - z) > 1 - x),$$

and implies in particular that $T(x, y) > 0 \Leftrightarrow x + y > 1$, i.e. the lower-left triangle of the unit square constitutes the zero-divisors of T. This excludes t-norms without zero-divisors (such as $T_{\mathbf{P}}$), nilpotent t-norms other than $T_{\mathbf{L}}$ and the drastic product $T_{\mathbf{D}}$. Note that the equivalence $\Leftrightarrow$ in (7) can be replaced by $\Rightarrow$ or $\Leftarrow$ without any problem.

Example 1. The name *rotation invariance* was adopted from Jenei [11]. Several of his papers are concerned with the construction of rotation-invariant t-norms [13,14,15]. A parametric family of solutions is the Jenei family [12], containing as extreme members the Łukasiewicz t-norm $T_{\mathbf{L}}$ and the nilpotent minimum $T_{\mathbf{nM}}$ of Fodor [9] (defined by $T_{\mathbf{nM}}(x, y) = 0$ if $x + y \le 1$ and $T_{\mathbf{nM}}(x, y) = T_{\mathbf{M}}(x, y)$ elsewhere). $T_{\mathbf{nM}}$ clearly is the greatest rotation-invariant t-norm. This family can easily be described by means of the ordinal sum construction of t-norms [16]. Consider the t-norm T_λ with as only summand the Łukasiewicz t-norm $T_{\mathbf{L}}$ on the interval $[\lambda, 1 - \lambda]$ with $\lambda \in [0, 0.5]$. The Jenei t-norm $T_\lambda^{\mathbf{J}}$ is then defined as follows:

$$T_\lambda^{\mathbf{J}}(x, y) = \begin{cases} T_\lambda(x, y) , & \text{if } x + y > 1 \\ 0 , & \text{if } x + y \le 1 \end{cases}.$$

Note that $T_0^{\mathbf{J}} = T_{\mathbf{L}}$ and $T_{0.5}^{\mathbf{J}} = T_{\mathbf{nM}}$.

Since $T_{\mathbf{nM}} \le T_{\mathbf{M}}$, it follows immediately that any $T_{\mathbf{M}}$-Ferrers relation is a $T_{\mathbf{nM}}$-biorder and that any $T_{\mathbf{M}}$-biorder is a $T_{\mathbf{nM}}$-Ferrers relation. This result can be strengthened considerably.

Proposition 2.

(i) $T_{\mathbf{nM}}$ *is the largest t-norm* T *such that any* $T_{\mathbf{M}}$*-Ferrers relation is a* T*-biorder.*
(ii) $T_{\mathbf{nM}}$ *is the largest t-norm* T *such that any* $T_{\mathbf{M}}$*-biorder is a* T*-Ferrers relation.*

3   Completeness Issues

3.1   Completeness of Reflexive T-Ferrers Relations

In this subsection, we investigate the kind of completeness exhibited by reflexive T-Ferrers relations. Recall that in the crisp case a reflexive Ferrers relation is always complete. The following simple proposition expresses that a minimal type of completeness is always guaranteed. The subsequent theorems characterize the weakly and strongly complete cases.

Proposition 3. *Consider a t-norm T. For any reflexive T-Ferrers relation it holds that*
$$\max(R(a,b), R(b,a)) > 0$$
for any $a, b \in A$.

Theorem 2. *Consider a t-norm T. The following equivalence holds:*

(i) *Any reflexive T-Ferrers relation is* **weakly complete**.
(ii) *The t-norm T satisfies*

$$(\forall(x,y) \in [0,1]^2)(T(x,y) = 0 \Rightarrow x + y \le 1),$$

i.e. the zero-divisors of T are located in the lower-left triangle of the unit square only.

This theorem applies to all rotation-invariant t-norms. In particular, it applies to $T_{\mathbf{L}}$, and hence to any $T \ge T_{\mathbf{L}}$, but not to all nilpotent t-norms. Theorem 2 also applies to t-norms without zero-divisors, although in that case the following much stronger result holds.

Theorem 3. *Consider a t-norm T. The following equivalence holds:*

(i) *Any reflexive T-Ferrers relation is* **strongly complete**.
(ii) *The t-norm T has no zero-divisors.*

This theorem applies for instance to $T_{\mathbf{M}}$ and to all strict t-norms.

3.2   Completeness of Reflexive T-biorders

In this subsection, we repeat the same study for reflexive T-biorders. The first observation is in favour of the T-biorder definition as it holds for any t-norm T.

Theorem 4. *Consider a t-norm T. It then holds that any reflexive T-biorder is weakly complete.*

However, our enthusiasm is tempered by the following negative result implying that there exists no t-norm that can always guarantee strong completeness, which is in disfavour of the T-biorder definition.

Proposition 4. *Not all reflexive $T_{\mathbf{M}}$-biorders are strongly complete.*

4   Large versus Strict Preference Relations

4.1 T-Ferrers Relations

In the classical case, the following equivalence holds.

Proposition 5. *Consider a reflexive binary relation R with corresponding preference structure (P, I, J). Then the following equivalence holds:*

$$R \text{ is Ferrers} \Leftrightarrow J = \emptyset \text{ and } P \text{ is Ferrers}.$$

Rephrasing a result from [4] in current terminology, we obtain the following theorem.

Theorem 5. *Consider a reflexive binary fuzzy relation R with corresponding AFPS (P, I, J) generated by means of $i = T_{\mathbf{L}}$. Then the following equivalence holds, for any t-norm T:*

$$R \text{ is weakly complete and } T\text{-Ferrers} \Leftrightarrow J = \emptyset \text{ and } P \text{ is } T\text{-Ferrers}.$$

In case of a t-norm that satisfies condition (ii) of Theorem 2, the weak completeness condition of R can be omitted from the previous theorem. The condition $J = \emptyset$ is of course a crisp interpretation of the statement 'without incomparability'. Just as there exist different completeness conditions, we could imagine other 'emptiness conditions' on J. Indeed, there is no reason why with a milder emptiness condition (such as $P + I = R$ is weakly complete) and using a generator different from $T_{\mathbf{L}}$ it would not be possible to obtain results similar to Theorem 5. Anyway, such results cannot pertain to a t-norm T without zero-divisors, as in that case, the T-Ferrers property of R implies its strong completeness. However, in that case, $J = \emptyset$ and the generator i becomes irrelevant. For such t-norms, Theorem 5 is the best we can get.

4.2 T-biorders

Of course, Proposition 5 can be stated equivalently as follows.

Proposition 6. *Consider a reflexive binary relation R with corresponding preference structure (P, I, J). Then the following equivalence holds:*

$$R \text{ is a biorder} \Leftrightarrow J = \emptyset \text{ and } P \text{ is a biorder}.$$

It turns out that this proposition can be generalized for exactly those t-norms for which the notions of T-Ferrers relations and T-biorders coincide (Theorem 1).

Theorem 6. *Consider a t-norm T. Then the following statements are equivalent:*

(i) *For any reflexive R with corresponding AFPS (P, I, J) generated by means of $i = T_{\mathbf{L}}$ it holds that if R is a T-biorder, then $J = \emptyset$ and P is a T-biorder.*

(ii) *For any reflexive R with corresponding AFPS (P, I, J) generated by means of $i = T_{\mathbf{L}}$ it holds that if $J = \emptyset$ and P is a T-biorder, then R is a T-biorder.*

(iii) *T is rotation-invariant.*

The fact that Theorem 5 holds for any t-norm T, as opposed to the above theorem, is an argument in favour of the T-Ferrers point of view.

Acknowledgement

The research reported on in this paper was partially supported by project MCYT BFM2001-3515.

References

1. B. De Baets and J. Fodor, *Twenty years of fuzzy preference structures (1978-1997)*, JORBEL **37** (1997), 61–82.
2. B. De Baets and J. Fodor, *Generator triplets of additive fuzzy preference structures*, Proc. Sixth Internat. Workshop on Relational Methods in Computer Science (Tilburg, The Netherlands), 2001, pp. 306–315.
3. B. De Baets and B. Van de Walle, *Weak and strong fuzzy interval orders*, Fuzzy Sets and Systems **79** (1996), 213–225.
4. B. De Baets, B. Van de Walle and E. Kerre, *Fuzzy preference structures without incomparability*, Fuzzy Sets and Systems **76** (1995), 333–348.
5. K. Demirli and B. De Baets, *Basic properties of implicators in a residual framework*, Tatra Mt. Math. Publ. **16** (1999), 31–46.
6. J.-P. Doignon, B. Monjardet, M. Roubens and Ph. Vincke, *Biorder families, valued relations, and preference modelling*, J. Math. Psych. **30** (1986), 435–480.
7. J. Fodor, *Traces of binary fuzzy relations*, Fuzzy Sets and Systems **50** (1992), 331–341.
8. J. Fodor, *A new look at fuzzy connectives*, Fuzzy Sets and Systems **57** (1993), 141–148.
9. J. Fodor, *Contrapositive symmetry of fuzzy implications*, Fuzzy Sets and Systems **69** (1995), 141–156.
10. J. Fodor and M. Roubens, *Fuzzy Preference Modelling and Multicriteria Decision Support*, Kluwer Academic Publishers, 1994.
11. S. Jenei, *Geometry of left-continuous triangular norms with strong induced negations*, JORBEL **98** (1998), 5–16.
12. S. Jenei, *New family of triangular norms via contrapositive symmetrization of residuated implications*, Fuzzy Sets and Systems **110** (2000), 157–174.
13. S. Jenei, *Structure of left-continuous triangular norms with strong induced negations. (I) Rotation construction*, J. Appl. Non-Classical Logics **10** (2000), 83–92.
14. S. Jenei, *Structure of left-continuous triangular norms with strong induced negations. (II) Rotation-annihilation construction*, J. Appl. Non-Classical Logics **11** (2001), 351–366.
15. S. Jenei, *Structure of left-continuous triangular norms with strong induced negations. (III) Construction and decomposition*, Fuzzy Sets and Systems **128** (2002), 197–208.
16. E.-P. Klement, R. Mesiar and E. Pap, *Triangular Norms*, Trends in Logic, Studia Logica Library, Vol. **8**, Kluwer Academic Publishers, Dordrecht, 2000.
17. B. Monjardet, *Axiomatiques et propriétés des quasi-ordres*, Math. Sci. Hum. **63** (1978), 51–82.
18. M. Roubens and Ph. Vincke, *Preference modelling*, Lecture Notes in Economics and Mathematical Systems, Vol. **76**, 1995.

Sugeno Integrals for the Modelling of Noise Annoyance Aggregation

Andy Verkeyn[1], Dick Botteldooren[1], Bernard De Baets[2], and Guy De Tré[3]

[1] Dept. of Information Technology, Ghent University
Sint-Pietersnieuwstraat 41, B-9000 Gent, Belgium
{andy.verkeyn,dick.botteldooren}@rug.ac.be
[2] Dept. of Applied Mathematics, Biometrics and Process Control, Ghent University
Coupure links 653, B-9000 Gent, Belgium
bernard.debaets@rug.ac.be
[3] Dept. of Telecommunications and Information Processing, Ghent University
Sint-Pietersnieuwstraat 41, B-9000 Gent, Belgium
guy.detre@rug.ac.be

Abstract. This paper investigates the use of the Sugeno integral for modelling the unconscious aggregation performed by people when trying to rate the discomfort of their living environment. This general annoyance rating is modelled based on known annoyance caused by a number of sources or activities. The approach is illustrated on data of a Flemish survey.

1 Introduction

Indicators for the impact of environmental stressors are often subjective in nature, such as annoyance caused by noise, odor and light. Information about these indicators is commonly obtained through social surveys, in which people are simply asked for their global judgement on noise, odor and light. The research reported on in this paper tries to model this global judgement by aggregating the impact of several sources or activities (e.g. road traffic, industry, ...).

The problem will be regarded as a multi-criteria decision making problem. In this context, the aim is to choose the best alternative for the objective criterion based on the given evaluations of predictive criteria [9]. Let X be the finite set of n criteria $\{x_1, \ldots, x_n\}$, and denote the satisfaction of a criterion x_i by $f(x_i)$. The problem is then to aggregate the satisfaction of the individual criteria to obtain the overall satisfaction of the objective criterion y,

$$D(y) = G(f(x_1), f(x_2), \ldots, f(x_n)) \tag{1}$$

where G is an aggregation function.

It should be stressed that the general annoyance judgement is not a conscious aggregation process. The subjects are not explicitly asked to perform a multi-criteria analysis. Yet, there is evidence suggesting that the fundamental annoyance emotion is always related to a particular source. This implies that subjects unconsciously aggregate when asked to rate their general annoyance [2].

T. Bilgiç et al. (Eds.): IFSA 2003, LNAI 2715, pp. 277–284, 2003.

2   Methodology

2.1   Mathematical Background

Before discussing the model itself, we recall the mathematical concepts used. The most important concept is the Sugeno integral that is defined with respect to a fuzzy measure. We assume X is a finite set of n elements $\{x_1, x_2, \ldots, x_n\}$. The powerset of X will be denoted as $\mathcal{P}(X)$. A and B are elements of $\mathcal{P}(X)$. $|A|$ is the cardinality of set A.

Definition 1 (Fuzzy measure [13]). *A fuzzy measure on X is a set function* $\mu : \mathcal{P}(X) \to [0, 1]$ *satisfying*

(i) *Boundary condition:* $\mu(\emptyset) = 0$

(ii) *Normalization:* $\mu(X) = 1$

(iii) *Monotonicity:* $A \subseteq B \Rightarrow \mu(A) \leq \mu(B)$

Instead of enumerating all 2^n elements of $\mathcal{P}(X)$, a fuzzy measure can also be defined using a relationship. Such a defining relationship describes how to combine the fuzzy measure value of two sets to obtain the fuzzy measure value of the union of those sets. With this approach, the values for the singletons $\{x_i\}$ for all $i \in \{1, \ldots, n\}$ are sufficient to define the complete fuzzy measure. The prototypical example of a fuzzy measure is a simple additive or probability measure. This measure has the property

$$\mu(A \cup B) = \mu(A) + \mu(B) \quad \text{if } A \cap B = \emptyset \ . \tag{2}$$

Due to the normalization requirement of fuzzy measures, one must take care that in this case $\mu(X) = \sum_{i=1}^{n} \mu(\{x_i\}) = 1$. Another example of a fuzzy measure is a possibility measure

$$\mu(A \cup B) = \max(\mu(A), \mu(B)) \ . \tag{3}$$

It is easy to see that this fuzzy measure satisfies the normalization requirement if at least one of the values $\mu(\{x_i\})$ equals 1 for $i \in \{1, \ldots, n\}$.

The number of elements that must be specified can also be reduced by using alternative representations for μ (e.g. [10]). One such representation is obtained through the possibilistic Möbius transform.

Definition 2 (Possibilistic Möbius transform [12]). *The possibilistic Möbius transform of a fuzzy measure μ on X is the set function* $m^\vee : \mathcal{P}(X) \to [0, 1]$ *defined by*

$$m^\vee(A) = \begin{cases} \mu(A) & \text{if } \mu(A) > \max_{B \subset A} \mu(B) \\ 0 & \text{otherwise.} \end{cases} \tag{4}$$

Definition 3 (Possibilistic Zeta transform [12]). *The possibilistic Zeta transform of a set function m on X is the set function* $Z_m^\vee : \mathcal{P}(X) \to [0, 1]$ *defined by*

$$Z_m^\vee(A) = \max_{B \subseteq A} m^\vee(B) \ . \tag{5}$$

The possibilisitic Möbius and Zeta transforms are each others inverse, so when the possibilistic Zeta transform is applied to $m^\vee$ the original fuzzy measure μ is recovered. However, not every set function $m^\vee$ is the possibilistic Möbius representation of a fuzzy measure, but there exist necessary and sufficient conditions as shown in Theorem 1.

Theorem 1. [5] *A set function* $m^\vee : \mathcal{P}(X) \to [0,1]$ *is the possibilistic Möbius transform of a fuzzy measure* μ *on* X *if and only if*

(i) *Boundary condition:* $m^\vee(\emptyset) = 0$

(ii) *Normalization:* $\max_{A \subseteq X} m^\vee(A) = 1$

(iii) *Monotonicity:* $m^\vee(A) \leq \max_{B \subset A} m^\vee(B) \Rightarrow m^\vee(A) = 0$

A special class of fuzzy measures that can be easily specified using the possibilistic Möbius representation are k-maxitive measures.

Definition 4 (k-Maxitive measure [12]). *Let* $k \in \{1, \ldots, n\}$. *A fuzzy measure* μ *on* X *is called k-maxitive if its possibilistic Möbius transform satisfies* $m^\vee(A) = 0$ *whenever* $|A| > k$ *and there exists at least one subset B of X such that* $|B| = k$ *and* $m^\vee(B) \neq 0$.

One easily verifies that possibility measures are in fact 1-maxitive measures.

As classical integrals are defined w.r.t. classical measures and fuzzy measures are extensions of such measures, the notion of fuzzy integrals was also introduced. One such fuzzy integral is called the Sugeno integral.

Definition 5 (Sugeno integral [13]). *Consider a fuzzy measure* μ *on* X *and a mapping* $f : X \to [0,1]$. *The discrete Sugeno integral of f w.r.t.* μ *is defined as*

$$\mathcal{S}_\mu(f) = \max_{i=1}^{n} \min \left(\mu(H_{(i)}), f(x_{(i)}) \right) , \tag{6}$$

where $f(x_{(1)}) \leq \ldots \leq f(x_{(n)})$ *and* $H_{(i)} = \{x_{(i)}, \ldots, x_{(n)}\}$.

The Sugeno integral w.r.t. a k-maxitive measure will be shortly referred to as a k-maxitive Sugeno integral. A k-maxitive Sugeno integral $\mathcal{S}_\mu(f)$ can be written as a function of the possibilistic Möbius representation $m^\vee$ of μ [11]:

$$\mathcal{S}_\mu(f) = \max_{A \subseteq X} \min \left(m^\vee(A), \bigwedge_{i \in A} f(x_i) \right) . \tag{7}$$

It is important to note that (7) does not require the reordering of the arguments.

Some properties of Sugeno integrals w.r.t. a fuzzy measure μ that are useful in this paper are shown below [6].

(i) Compensating behavior: $\min_{i=1}^{n} f(x_i) \leq \mathcal{S}_\mu(f) \leq \max_{i=1}^{n} f(x_i)$

(ii) Monotonicity w.r.t. integrand: $f \leq f' \Rightarrow \mathcal{S}_\mu(f) \leq \mathcal{S}_\mu(f')$

2.2 Annoyance Aggregation

The cognitive process underlying the general noise annoyance rating [1] is approached as a multi-criteria decision making problem. Each source of annoyance will be considered as a predictive criterion, while total annoyance is the objective criterion. Hence, X will be the set of all annoyance sources. The level of annoyance is expressed as a predefined number $m \in \mathbb{N}$ of linguistic labels. These labels will be represented as the set $L = \{0 = l_1 < l_2 < \ldots < l_m = 1\}$, where 0 means the lowest possible level of annoyance. Using the previous notations, the evaluation function f is then a mapping $X \to L \subseteq [0, 1]$.

In [14] it has been shown that Choquet integrals w.r.t. a possibility measure can predict the global aggregated annoyance quite well. However, the approach taken there is difficult to extend to other measures. Because of this successful use of possibility measures μ, the simple representation of such measures (and k-maxitive extensions with $k > 1$) as their possibilistic Möbius transform $m^\vee$ and the natural expression of a Sugeno integral in function of $m^\vee$, k-maxitive Sugeno integrals also seem good model candidates. While the Choquet integral requires to interpret the scale as a continuum, Sugeno allows to work on an ordinal scale [6]. In this paper, an underlying cardinal annoyance scale is assumed. The strongest component model [2], which is the best known crisp model, simply takes the maximum of its arguments. In [5] the k-maxitive Sugeno integral was linked to a weighted maximum operation. So, from the viewpoint of the "crisp world", this integral also seems a logical choice.

Hence, in our multi-criteria decision model, the k-maxitive Sugeno integral will be used as aggregator in (1), $D(y) = \mathcal{S}_\mu(f)$ where $D(y)$ is the evaluation of the total annoyance y. Considering the elements of L as midpoints of the intervals on our cardinal scale in $[0, 1]$, we can express $D(y)$ in function of the input categories L. The category $l \in L$ is chosen as the classification result if $D(y)$ falls into the interval around l. In the following we will assume the shorthand notation $d = D(y)$ and the value of l will be denoted as d'.

The fuzzy measure used in the Sugeno integral expresses the importance of the subsets of criteria. In the remaining part of this paper, μ will always denote a k-maxitive measure represented by its possibilistic Möbius transform $m^\vee$. We will also restrict ourselves to 1-maxitive and 2-maxitive measures, otherwise the number of parameters would become too large.

2.3 Genetic Optimization

To find the optimal fuzzy measure μ, the classification performance of the model on a dataset obtained from a social survey is maximized. Because of the non-linearities and the complex solution landscape, a genetic algorithm is applied for this optimization [8]. A genetic algorithm makes a distinction between genotype, the internal representation of an individual in the genetic algorithm, and its phenotype, how the individual looks like in its external context.

A genome represents the genotype of a k-maxitive measure. For each $A \in \mathcal{P}(X)$ the genome contains a single real-valued gene in the interval $[0, 1]$. Because we restrict ourselves to 2-maxitive measures, we have genes g_r corresponding to a singleton $\{x_r\}$ and g_{pq} corresponding to a doubleton $\{x_p, x_q\}$ with $p \neq q$, and $p, q, r \in \{1, \ldots, n\}$. The phenotype of the maxitive measure is calculated using the following formulas.

$$m^\vee(\{x_r\}) = g_r \tag{8}$$

$$m^\vee(\{x_p, x_q\}) = \begin{cases} \bar{m} + (1 - \bar{m})g_{pq} & \text{if } g_{pq} \neq 0 \\ 0 & \text{if } g_{pq} = 0 \end{cases} \tag{9}$$

where $\bar{m} = \max(m^\vee(\{x_p\}), m^\vee(\{x_q\}))$.

To ensure the normalization requirement, the phenotype is divided by the maximum of all $m^\vee$ values. The normalized phenotype is then coded back into the genotype. This guarantees only normalized individuals in the population, which eases sensible exchange of genes. Using the above procedure, the genetic algorithm only explores meaningful fuzzy measures that obey the conditions of Theorem 1. Furthermore, the genetic algorithm uses a uniform crossover and self-adaptive mutation operator [8]. The population consists of 50 individuals.

The dataset consists of N records of the form $(f(x_1), f(x_2), \ldots, f(x_n), d^*)$ where d^* denotes the reported evaluation of the total annoyance. The goal of the genetic algorithm is to minimize an error function E, which equalizes the impact of different frequencies of occurrence of d^* values in the dataset. This function is defined as

$$E = \sum_{z=1}^{N} \frac{(d_z - d_z^*)^2}{q_{d_z^*}} + \sum_{z=1, d_z' \neq d_z^*}^{N} \frac{\alpha}{q_{d_z^*}} \tag{10}$$

where $q_i = \text{Prob}(d^* = l_i)$ for $i \in \{1, \ldots, m\}$ and α is an experimentally determined, additional penalty for each wrong prediction.

However, for comparison with other (crisp) models, the performance of the model after optimization of the fuzzy measure on the Flemish dataset will be expressed as the weighted percentage of correct predictions.

3 Survey Data

3.1 General Description

The Sugeno-based annoyance aggregation model has been tested using data collected in a social survey. The survey was conducted with 3200 subjects in Flanders, Belgium. The general topic of the survey was the influence of odor, noise and too much light on the quality of the living environment. The number of data records left after removal of incomplete survey records are shown in Table 1. The questions of importance for this study are the general annoyance question (for noise, odor and light) and the questions concerning annoyance by particular sources or activities. The formulation of all questions is in overall agreement with the ICBEN recommendation put forward in [7]. The subjects are asked to answer the questions using a five point scale labeled "helemaal niet gehinderd"; "een

beetje gehinderd"; "tamelijk gehinderd"; "ernstig gehinderd"; "extreem gehinderd" (which could be translated as "not at all", "slightly", "fairly", "strongly",
"extremely"). It was shown that these selected terms divide the annoyance scale
in almost equidistant way. The general annoyance question appears a few pages
before the question concerning annoyance by particular sources. As a small pre-
study learned that the majority of subjects tend to fill in the written survey from
beginning to end without ever returning to previously answered questions, subjects do not see the detailed list of sources while answering the general annoyance
question.

It is also important to note the unequally distributed frequencies of the annoyance levels. Fortunately, the highest levels occur less often. The relative occurrences of each level of annoyance as well as the number of sources are also
listed in Table 1. The remaining part of this paper will deal only with the noise
annoyance data.

Table 1. The number of complete records N, number of sources n and the
relative occurrence of each total annoyance level (in %) for each environmental
stressor in the dataset.

Stressor	N	n	l_1	l_2	l_3	l_4	l_5
Noise	2661	21	35.59	35.67	18.19	8.57	1.99
Odor	2719	23	54.69	27.73	11.51	4.93	1.14
Light	2845	12	85.69	9.67	3.30	1.16	0.18

3.2 Data Analysis

In view of the model, the data contains three kinds of inconsistencies, in the sense
that some data records can never be correctly classified by the model. A first type
of inconsistency stems from the compensating behavior of the Sugeno integral.
If the reported global annoyance is lower/higher than the minimum/maximum
of the annoyance of any of the sources, the data record will not be correctly classified. A second kind of inconsistency (called *doubt* in [4]) occurs when $f_p = f_q$
for $p \neq q$, $p, q \in N$ where f_i means the combination of annoyance levels for all
sources for record i, but with $d_p^* \neq d_q^*$. Lastly, there is also a problem if $f_p < f_q$
for $p, q \in N$ but $d_p^* > d_q^*$ (called *reversed preference* in [4]) because of the monotonicity property of Sugeno integrals. After removing the smallest number of
data records to get rid of all inconsistencies of the first and second type (keeping
2141 out of the 2661 records), the maximum performance that can be achieved
has been calculated as being 78.99 %. The third type of inconsistencies was left
untouched because it is more difficult to remove (due to its interaction with
the second type) while retaining the maximum number of consistent records.

Therefore, the upper limit performance will be lower than the cited maximum performance of 78.99 %.

4   Conclusions

The performance of the model on all 2661 records, compared to those previously obtained for the strongest component model [2] and the Choquet-based model [14], are summarized in Table 2.

As can be seen, the Sugeno integral is also a good model for the aggregation of noise annoyance. 2-maxitive measures perform slightly better (but not significantly) than 1-maxitive measures. This indicates that annoyance by two sources can accumulate to a stronger effect on the aggregation than both sources alone.

In Fig. 1 the classified general annoyance is compared to the reported general annoyance. The percentages are scaled by the number of occurrences of each label of reported general annoyance, so that the sum of each row equals 100. The area of the bubbles is proportional to the percentage. The classification error shows a systematic overestimation. This fact was also observed for other models, including the strongest component model [2].

Table 2. Weighted performance of several models.

Model	Performance
Strongest component	55.50
1-maxitive Choquet integral	61.28
1-maxitive Sugeno integral	60.94
2-maxitive Sugeno integral	61.37

References

1. Botteldooren, D., Verkeyn, A.: An iterative fuzzy model for cognitive processes involved in environment quality judgement. Proc. IEEE World Congress on Computational Intelligence (2002) 1057–1062
2. Botteldooren, D., Verkeyn, A.: Fuzzy models for accumulation of reported community noise annoyance from combined sources. J. Acoust. Soc. Am. **112** (2002) 1496–1508
3. Cao-Van, K., De Baets, B.: A decomposition of k-additive Choquet and k-maxitive Sugeno integrals. Int. J. Uncertainty, Fuzziness and Knowledge-based Systems **9** (2001) 127–143
4. Cao-Van, K., De Baets, B.: Consistent representations of rankings. European J. Oper. Research (submitted)
5. De Baets, B., Tsiporkova, E.: Basic assignments and commonality functions for confidence measures. Proc. 7th Int. Conf. on Information Proc. and Management of Uncertainty in Knowledge-based Systems Vol 3 (1998) 1382–1389

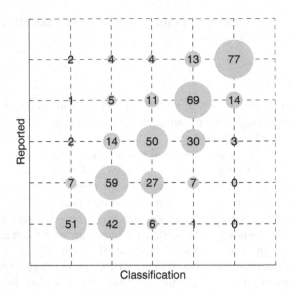

Fig. 1. Percentage of each classified annoyance label weighted by the number of occurrences of each reported label

6. Dubois, D., Marichal, J. L., Prade, H., Roubens, M., Sabbadin, R.: The use of the discrete Sugeno integral in decision making: a survey. Int. J. Uncertainty, Fuzziness and Knowledge-based Systems 9 (2001) 539–561
7. Fields, J. M., De Jong, R. G., Gjestland, T., Flindell, I. H., Job, R. F. S., Kurra, S., Lercher, P., Vallet, M., Yano, T., Guski, R., Felscher-Suhr, U., Schumer, R.: Standardized general-purpose noise reaction questions for community noise surveys: research and a recommendation. J. Sound and Vibration **242** (2001) 641–679
8. Fogel, D. B.: Evolutionary computing: Towards a new philosophy of machine intelligence, 2nd edition. IEEE Press. (2000)
9. Grabisch, M.: Fuzzy integral in multi-criteria decision making. Fuzzy Sets and Systems **69** (1995) 279–298
10. Grabisch, M.: Alternative representations of discrete fuzzy measures for decision making. Int. J. Uncertainty, Fuzziness and Knowledge-based Systems **5** (1997) 587–607
11. Marichal, J.-L.: Aggregation operators for multicriteria decision aid. Ph.D. Thesis. University of Liège. Belgium. (1998)
12. Mesiar, R.: Generalizations of k-order additive discrete fuzzy measures. Fuzzy Sets and Systems. **102** (1999) 423–428
13. Sugeno, M.: Theory of fuzzy integrals and its applications. Ph.D. Thesis. Tokyo Institute of Technology. Japan. (1974)
14. Verkeyn, A., Botteldooren, D., De Baets, B., De Tré, G.: Modelling annoyance aggregation with Choquet integrals. Proc. Eurofuse Workshop on Information Systems (2002) 259–264

On Separability of Intuitionistic Fuzzy Sets

Krassimir T. Atanassov[1], Janusz Kacprzyk[2], Eulalia Szmidt[2], and
Ljudmila P. Todorova[1]

[1] CLBME-Bulgarian Academy of Sciences, Acad. G. Bonchev Str., 105 Bl.,
Sofia-1113, Bulgaria
lpt@clbme.bas.bg and krat@argo.bas.bg
[2] Systems Research Institute, Polish Academy of Sciences
ul. Newelska 6, 01–447 Warsaw, Poland
kacprzyk@ibspan.waw.pl

Abstract. Issues related to the separability of two intuitionistic fuzzy
sets (IFSs; see [1]) are considered. The obtained results are relevant for
decision making and classification. An example illustrating how the proposed solutions may improve decision making is presented.
Keywords: intuitionistic fuzzy sets, separability, decision making, classification.

1 Introduction

The concept of separability is very important for the different kinds of topology,
including the fuzzy topology (see, e.g., [6]). Operators, analogous to the topological operators "closure" and "interior" were defined over intuitionistic fuzzy sets
(IFSs; see [1]) twenty years ago, but their topological properties had not been
studied at all. Here a first step in this direction of development of the theory of
IFSs has been made, introducing the concept of separability of two IFSs. Some
assertions related to this concept were formulated, and one of their applications
was discussed.

2 Short Remarks on Intuitionistic Fuzzy Sets

Let a set E be fixed. An intuitionistic fuzzy set (IFS) A in E is an object of the
following form:
$$A = \{\langle x, \mu_A(x), \nu_A(x)\rangle | x \in E\},$$
where functions $\mu_A : E \to [0,1]$ and $\nu_A : E \to [0,1]$ define the degree of membership and the degree of non-membership of the element $x \in E$, respectively,
and for every $x \in E$:
$$0 \leq \mu_A(x) + \nu_A(x) \leq 1.$$
Let for every $x \in E$:
$$\pi_A(x) = 1 - \mu_A(x) - \nu_A(x).$$
Therefore, function π determines the degree of uncertainty.

T. Bilgiç et al. (Eds.): IFSA 2003, LNAI 2715, pp. 285–292, 2003.

Let a universe E be given. One of the geometrical interpretations of the IFSs uses figure F in Fig. 1.

For every two IFSs A and B a lot of relations, operations and operators are defined (see, e.g. [1]). The necessity for the next research relations and operators are the following.

$$A \subset B \; iff \; (\forall x \in E)(\mu_A(x) \leq \mu_B(x) \& \nu_A(x) \geq \nu_B(x));$$
$$A \supset B \; iff \; B \subset A;$$
$$A = B \; iff \; (\forall x \in E)(\mu_A(x) = \mu_B(x) \& \nu_A(x) = \nu_B(x));$$

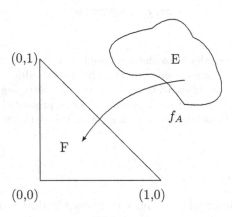

(0,1)

E

f_A

F

(0,0)   (1,0)

Figure 1

The operators that we shall use are of three following types: modal, tolopogical and level.

Let A be an IFS and let $\alpha, \beta \in [0,1]$.

Below, we shall use the following modal operators (see, e.g., [1]), that are extensions of the ordanary modal logic operators "necessity" and "possibility":

$$F_{\alpha,\beta}(A) = \{\langle x, \mu_A(x) + \alpha.\pi_A(x), \nu_A(x) + \beta.\pi_A(x) \rangle | x \in E\}, \; \text{where } \alpha + \beta \leq 1;$$
$$G_{\alpha,\beta}(A) = \{\langle x, \alpha.\mu_A(x), \beta.\nu_A(x) \rangle | x \in E\}.$$
$$H_{\alpha,\beta}(A) = \{\langle x, \alpha.\mu_A(x), \nu_A(x) + \beta.\pi_A(x) \rangle | x \in E\},$$
$$H^*_{\alpha,\beta}(A) = \{\langle x, \alpha.\mu_A(x), \nu_A(x) + \beta.(1 - \alpha.\mu_A(x) - \nu_A(x)) \rangle | x \in E\},$$
$$J_{\alpha,\beta}(A) = \{\langle x, \mu_A(x) + \alpha.\pi_A(x), \beta.\nu_A(x) \rangle | x \in E\},$$
$$J^*_{\alpha,\beta}(A) = \{\langle x, \mu_A(x) + \alpha.(1 - \mu_A(x) - \beta.\nu_A(x)), \beta.\nu_A(x) \rangle | x \in E\};$$

the following topological operators (they are analogous of topological operators "closure" and "intersection", respectively:

$$C(A) = \{\langle x, \sup_{y \in E} \mu_A(y), \inf_{y \in E} \nu_A(y) \rangle | x \in E\},$$

$$I(A) = \{\langle x, \inf_{y \in E} \mu_A(y), \sup_{y \in E} \nu_A(y) \rangle | x \in E\};$$

and the following level operators:

$$P_{\alpha,\beta}(A) = \{\langle x, \max(\alpha, \mu_A(x)), \min(\beta, \nu_A(x))\rangle | x \in E\},$$
$$Q_{\alpha,\beta}(A) = \{\langle x, \min(\alpha, \mu_A(x)), \max(\beta, \nu_A(x))\rangle | x \in E\},$$

for $\alpha, \beta \in [0,1]$ and $\alpha + \beta \leq 1$.

Let the IFS A over the universe E be called *proper*, if there exists at least one $x \in E$, for which $\pi_A(x) > 0$.

3   Main Results on Separability of Intuitionistic Fuzzy Sets

Up to now only the following two assertions related to separability of IFSs are introduced.

Theorem 1 [1]: Let A, B be two proper IFSs, for which there exist $y, z \in E$, so that $\mu_A(y) > 0$ and $\nu_B(z) > 0$. If $C(A) \subset I(B)$, then there are real numbers $\alpha, \beta, \gamma, \delta \in [0,1]$, such that $J_{\alpha,\beta}(A) \subset H_{\gamma,\delta}(B)$.

It is easy to show that the opposite is not always true.

Theorem 2 [1]: For every two IFSs A and B, $C(A) \subset I(B)$, iff there exist two real numbers $\alpha, \beta \in [0,1]$, so that $\alpha + \beta \leq 1$ and $P_{\alpha,\beta}(A) \subset Q_{\alpha,\beta}(B)$.

Now, we shall introduce definitions related to concept separability.

Definition 1: The two IFSs A and B, such that $A \subset B$, are *"strongly separable"* if and only if $C(A) \subset I(B)$.

The geometrical interpretation of these two sets is given in Fig. 2.

Therefore, we can obtain the following equivalent condition:

Definition 1': The two IFSs A and B such that $A \subset B$ are *"strongly separable"* if and only if

$$\sup_{y \in E} \mu_A(y) \leq \inf_{y \in E} \mu_B(y) \quad \text{and} \quad \inf_{y \in E} \nu_A(y) \geq \sup_{y \in E} \nu_B(y).$$

Of course, if A and B are proper IFSs, and

$$\sup_{y \in E} \mu_A(y) > 0, \ \sup_{y \in E} \nu_B(y) > 0, \ \inf_{y \in E} \mu_B(y) < 1 \ \inf_{y \in E} \nu_A(y) < 1,$$

then,

$$\sup_{y \in E} \mu_A(y) < 1, \ \sup_{y \in E} \nu_B(y) < 1, \ \inf_{y \in E} \mu_B(y) > 0, \ \inf_{y \in E} \nu_A(y) > 0.$$

Definition 2: The two IFSs A and B are *"weakly separable"* if and only if

$$\sup_{y \in E} \mu_A(y) \geq \inf_{y \in E} \mu_B(y) \quad \text{and} \quad \inf_{y \in E} \nu_A(y) \geq \sup_{y \in E} \nu_B(y).$$

$$\text{or} \ \sup_{y \in E} \mu_A(y) \leq \inf_{y \in E} \mu_B(y) \quad \text{and} \quad \inf_{y \in E} \nu_A(y) \leq \sup_{y \in E} \nu_B(y).$$

The geometrical interpretations of these two sets are given in Figs. 3 and 4.

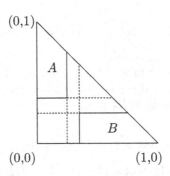

Figure 2

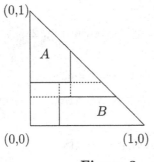

Figure 3

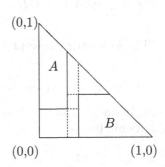

Figure 4

Definition 3: The two IFSs A and B are *"separable"* if and only if they are at least weakly separable.

Definition 4: The two IFSs A and B are *"non-separable"* if and only if they are not separable.

Therefore, we can obtain the following equivalent condition:

Definition 4': The two IFSs A and B are *"non-separable"* if and only if

$$\sup_{y\in E}\mu_A(y) > \inf_{y\in E}\mu_B(y) \ \text{ and } \ \inf_{y\in E}\nu_A(y) < \sup_{y\in E}\nu_B(y). \tag{1}$$

The geometrical interpretation of these two sets is given in Fig. 5.

Theorem 3: For each two non-separable proper IFSs A and B there exist real numbers $\alpha, \beta, \gamma, \delta \in [0,1]$, so that the IFSs $H^*_{\alpha,\beta}(A)$ and $J^*_{\gamma,\delta}(B)$ are separable.

There are different ways for the four real numbers to be determined. Below we shall discuss one (maybe the simplest) of them. The others require more information about the forms and parameters of both IFSs.

Proof: Let

$$a = \sup_{y\in E}\mu_A(y), \tag{1a}$$

$$b = \inf_{y\in E}\nu_A(y), \tag{1b}$$

$$c - \inf_{y\in E}\mu_B(y), \tag{1c}$$

$$d = \sup_{y\in E}\nu_B(y). \tag{1d}$$

Therefore $a, b, c, d \in [0,1]$, $a + b \leq 1$, and $c + d \leq 1$.

We shall discuss the case $a > 0$ and $d > 0$, because the case $a = 0$ and $d = 0$, i.e., for each $x \in E$: $\mu_A(x) = \nu_B(x) = 0$ is obvious.

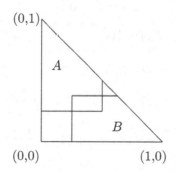

(0,1)

A

B

(0,0)   (1,0)

Figure 5

Let

$$\alpha = \frac{a+c}{2a}, \tag{2a}$$

$$\beta = \frac{d-b}{2-a-2b-c}, \tag{2b}$$

$$\gamma = \frac{b+d}{2d}, \tag{2c}$$

$$\delta = \frac{a-c}{2-b-2c-d}. \tag{2d}$$

We can immediately see that A and B are non-separable by condition, i.e. from (1) it follows that $a > c$, and $d > b$ and hence

$$2 - a - 2b - c \geq 1 - b - c > 1 - a - b \geq 0,$$

i.e.,

$$2 - a - 2b - c > 0,$$

and

$$2 - b - 2c - d \geq 1 - b - c > 1 - a - b \geq 0,$$

i.e.,

$$2 - b - 2c - d > 0.$$

Now, we see that

$$C(H^*_{\frac{a+c}{2a}, \frac{d-b}{2-a-2b-c}}(A)) =$$

$$= C(\{\langle x, \frac{a+c}{2a}\cdot\mu_A(x),$$

$$\nu_A(x) + \frac{d-b}{2-a-2b-c}\cdot(1 - \frac{a+c}{2a}\cdot\mu_A(x) - \nu_A(x)))\rangle | x \in E\})$$

$$= \{\langle x, \sup_{y \in E}(\frac{a+c}{2a}\cdot\mu_A(y)), \inf_{y \in E}(\nu_A(y) +$$

$$+ \frac{d-b}{2-a-2b-c}\cdot(1 - \frac{a+c}{2a}\cdot\mu_A(y) - \nu_A(y))))\rangle | x \in E\})$$

$$\subset \{\langle x, \sup_{y \in E}(\frac{a+c}{2}), \inf_{y \in E}(b + \frac{d-b}{2-a-2b-c}\cdot(1 - \frac{a+c}{2} - b)))\rangle | x \in E\})$$

$$= \{\langle x, \frac{a+c}{2}, b + \frac{d-b}{2}\rangle | x \in E\}) = \{\langle x, \frac{a+c}{2}, \frac{b+d}{2}\rangle | x \in E\}) =$$

$$= \{\langle x, c + \frac{a-c}{2}, \frac{b+d}{2}\rangle\rangle | x \in E\}$$

$$= \{\langle x, \inf_{y \in E}(c + \frac{a-c}{2-b-2c-d}\cdot(1 - c - \frac{b+d}{2})), \sup_{y \in E}(\frac{b+d}{2d}\cdot d)\rangle | x \in E\}$$

$$= \{\langle x, \inf_{y \in E}(\mu_B(y) + \frac{a-c}{2-b-2c-d}\cdot(1 - \mu_B(y) - \frac{b+d}{2d}\cdot\nu_B(y))),$$

$$\sup_{y \in E}(\frac{b+d}{2d}\cdot\nu_B(y)))\rangle | x \in E\}$$

$$\subset I(\{\langle x, \mu_B(x) + \frac{a-c}{2-b-2c-d}\cdot(1 - \mu_B(x) - \frac{b+d}{2d}\cdot\nu_B(x)),$$

$$\frac{b+d}{2d}\cdot\nu_B(x)\rangle | x \in E\})$$

$$= I(J^*_{\frac{a-c}{2-b-2c-d}, \frac{b+d}{2d}}(B)),$$

i.e.,

$$C(H^*_{\frac{a+c}{2a}, \frac{d-b}{2-a-2b-c}}(A)) \subset I(J^*_{\frac{a-c}{2-b-2c-d}, \frac{b+d}{2d}}(B)).$$

Therefore, the IFSs $H^*_{\frac{a+c}{2a}, \frac{d-b}{2-a-2b-c}}(A)$ and $J^*_{\frac{a-c}{2-b-2c-d}, \frac{b+d}{2d}}(A)$ are separable.

The discussed condition for separability can be used in procedures related to decision making and, in particular, in classification.

4 An Example

Different ethiologic factors (polytrauma, isolated craniocerebral trauma, sepsis, thoracic trauma, pneumonia, cardiac arrest and cardiopulmonary resuscitation and others) can generate Acute Respiratory Failure (ARF), that imposes the need of mechanical ventilation. The rate of the patients, who recieve ventilatory support in circumstance of critical care, vary between 20% and 60% according to their clinical features. The mechanical ventilation is life-saving procedure with the risk of potential complications and some unfavourable physiologic effects [4,5]. They are the most important risk factors, leading to development of nosocomial pneumonias (critical ill patients) [2,3]. The risk is several times higher and it grows proportionally to the duration of the breathing support. That is why in such cases it is critically significant that the mechanical ventilation has been withdrawn and patient being extubated in the earliest possible moment, when able to maintain spontaneous breathing.

Now, on the basis of the above discussed research we will construct an algorithm for a decision making in medicine, related to the mechanical ventilation.

On the basis of retrospective reasearch of patients we can determine the values of parameters a, b, c, d for two patient classes: patients in the day of intubation (set A) and patients on the day of weaning from long-term mechanical ventilation (set B). The algorithm is the following:

1. The values of $\alpha, \beta, \gamma, \delta$ from (2a)–(2d) are calculated on the basis of a, b, c, d from (1a)–(1d);
2. For a new patient the parametes $\mu^i_A, \nu^i_A, \mu^i_B, \nu^i_B$ are determined for the day of intubation;
3. For the same patient the parametes $\mu^c_A, \nu^c_A, \mu^c_B, \nu^c_B$ are determined for the current day;
4. The values of $\alpha', \beta', \gamma', \delta'$ are determined on the basis of the previous parameters values;
5. If $\alpha' \geq \alpha$, $\beta' \leq \beta$, $\gamma' \geq \gamma$ and $\delta' \leq \delta$, then the procedure for weaning from long-term mechanical ventilation can start for the patient. In the opposite case we must return to point **3**.

This procedure gives more detailed estimations for the patient's status and therefore, we can perform a more precise decision making.

5   Conclusion

As we noted at the beginning, the present paper is one of the first steps in direction of "topologization" of the theory of IFSs. In next authors' works the above ideas will be applied to decision making where the IFS tools have proven to be very suitable (see, e.g., [7,8,9]).

6   Aknowlegements

The authors are very grateful to the referees for thire valuable remarks.

References

1. Atanassov K., Intuitionistic Fuzzy Sets, Springer Physica-Verlag, Berlin, 1999.
2. Brochard L., A. Rauss, S. Benito, G. Conti, J. Mancebo, N. Rekik, A. Gasparetto, F. Lemaire. Comparison of three methods of gradual withdrawal from ventilatory support during weaning from mechanical ventilation. Am J Respir Crit Care Med 1994, Vol. 150, 896-903.
3. Fagon J.Y., J. Chastre, A. Novara, P. Medioni, C. Gibert. Respective influences of initial severity of illness and of mechanical ventilation in the development of ICU acquired pneumonia (abstract). Am Rev Respir Dis 1991, Vol. 143, A493.
4. Hillman D.R. Physiological aspects of intermittent positive pressure ventilation. Anaesth Intensive Care 1986; Vol. 14, 226-35.;
5. Lin E.S., Oh T.E. Which mode of ventilation? In: Dobb G, ed. Intensive Care: Developments and Controversies. Baillere's Clinical Anaesthesiology. London: WB Saunders, 1991, 441-73.
6. Liu Y.-M., M.-K. Luo, Fuzzy Topology, World Scientific, Singapore, 1997.
7. Szmidt E., J. Kacprzyk, Intuitionistic fuzzy relations and measures of consensus. In: Bouchon-Meunier B., Gutierrez-Rios J., Magdalena L. and Yager R.R. (Eds.): Technologies for Contructing Intelligent Systems 2. Tools. The series "Studies in Fuzziness and Springer - Verlag Heidelberg, 2002, 261-275.
8. Szmidt E., J. Kacprzyk, Analysis of Agreement in a Group of Experts via Distances Between Intuitionistic Fuzzy Preferences. In: Proc. of the Ninth International Conference IPMU 2002, Annecy, France, 1-5 July, 2002, 1859-1865.
9. Szmidt E., J. Kacprzyk, Evaluation of Agreement in a Group of Experts via Distances Between Intuitionistic Fuzzy Sets. In: IS'2002 - 1st. Int. IEEE Symposium: Intelligent Systems, Varna, Bulgaria, September 10-12, 2002, IEEE (Catalog Number 02EX499), pp. 166-170.

Calculating Limit Decisions in Factoring Using a Fuzzy Decision Model Based on Interactions between Goals

Rudolf Felix

FLS Fuzzy Logik Systeme GmbH
Joseph-von-Fraunhofer Straße 20, 44 227 Dortmund, Germany,
Tel. +49 231 9 700 921, Fax. +49 231 9700 929,
felix@fuzzy.de

Abstract. A new application of a model of interactions between goals based on fuzzy relations in the field of factoring has been developed. In contrast to other approaches in multiple attribute decision making, the interactive structure of goals for each decision situation is represented and modeled explicitly. A real world application, a system for limit decisions for factoring is presented.

1. Introduction

Factoring is the sale of accounts receivable or, in other words, it is the purchasing of commercial invoices (debt instruments) from a business at a discount. When making factoring decisions, banks consider a certain number of criteria, weighed according to the factor's established targets. Said multiple interacting decision criteria serve the purpose of evaluating and narrowing the probable risk of default concerning each of the client's customers.

In such a real world application the understanding of decision making is much more flexible than the strict understanding of many existing decision making models, even if they use fuzzy set theory as framework. As discussed in section 4 and in (Felix 1995), the models are not flexible enough to reflect the tension between interacting goals in a way human decision makers do. In contrast to this, the decision making model, upon which the application presented in section 6 is based, is applicable to such complex decision making situations as arise when limit decisions are calculated in the field of factoring. The key issue of a flexible decision making model must be an adequate model of interaction between decision making goals. In the application presented here, goals are criteria which are used for finding the appropriate limit decisions.

2. Basic Definitions

Before we define interactions between goals, we introduce the notion of the positive impact set and the negative impact set of a goal. A more detailed discussion can be found in (Felix 1991 and Felix 1994).

T. Bilgiç et al. (Eds.): IFSA 2003, LNAI 2715, pp. 293-302, 2003.

Def. 1)
Let A be a non-empty and finite set of potential alternatives, G a non-empty and finite set of goals, $A \cap G = \emptyset$, $a \in A$, $g \in G$, $\delta \in (0,1]$. For each goal g we define the two fuzzy sets S_g and D_g each from A into $[0, 1]$ by:

1.   *positive impact function of the goal g*

$$S_g(a) := \begin{cases} \delta, & a \text{ affects positively } g \text{ with degree } \delta \\ 0, & \text{else} \end{cases}$$

2.   *negative impact function of the goal g*

$$D_g(a) := \begin{cases} \delta, & a \text{ affects negatively } g \text{ with degree } \delta \\ 0, & \text{else} \end{cases}$$

Def. 2)
Let S_g and D_g be defined as in Def. 1). S_g is called the *positive impact set of g* and D_g the *negative impact set of g*.

The set S_g contains alternatives with a positive impact on the goal g and δ is the degree of the positive impact. The set D_g contains alternatives with a negative impact on the goal g and δ is the degree of the negative impact.

Def. 3)
Let A be a finite non-empty set of alternatives. Let $\mathcal{P}(A)$ be the set of all fuzzy subsets of A. Let $X, Y \in \mathcal{P}(A)$, x and y the membership functions of X and Y respectively. The *fuzzy inclusion* I is defined as follows:

$$\mathrm{I}: \mathcal{P}(A) \times \mathcal{P}(A) \to [0,1]$$

$$\mathrm{I}(X,Y) =: \begin{cases} \dfrac{\sum\limits_{a \in A} \min(x(a), y(a))}{\sum\limits_{a \in A} x(a)}, & \text{for } X \neq \emptyset \\ \\ 1, & \text{for } X = \emptyset \end{cases}$$

with $x(a) \in X$ and $y(a) \in Y$.

The *fuzzy non-inclusion* N is defined as:

$$\mathrm{N}: \mathcal{P}(A) \times \mathcal{P}(A) \to [0,1]$$

$$\mathrm{N}(X,Y) := 1 - \mathrm{I}(X,Y)$$

The inclusions indicate the existence of interaction between two goals. The higher the degree of inclusion between the positive impact sets of two goals, the more cooperative the interaction between them. The higher the degree of inclusion between the positive impact set of one goal and the negative impact set of the second, the more competitive the interaction. The non-inclusions are evaluated in a similar way. The higher the degree of non-inclusion between the positive impact sets of two goals, the less cooperative the interaction between them. The higher the degree of non-inclusion between the positive impact set of one goal and the negative impact set of the second, the less competitive the relationship.

Note that the pair (S_g, D_g) represents the whole known impact of alternatives on the goal g. Dubois and Prade (1992) show that for (S_g, D_g) the so-called twofold fuzzy sets can be taken. Then S_g is the set of alternatives which more or less certainly satisfy the goal g. D_g is the fuzzy set of alternatives which are rather less possible, tolerable according to the decision maker.

3. Interactions between Goals

Based on the inclusion and non-inclusion defined above, 8 basic types of interaction between goals are defined. The interactions cover the whole spectrum from a very high confluence (analogy) between goals to a strict competition (trade-off). The independence of goals and the case of an unspecified dependence are also considered.

Def. 4)
Let S_{g_1}, D_1, S_{g_2} and D_{g_2} be fuzzy sets given by the corresponding membership functions as defined in Def. 2). For simplicity we write S_1 instead of S_{g_1} etc.. Let $g_1, g_2 \in G$ where G is a set of goals.

The types of interaction between two goals are defined as relations which are fuzzy subsets of $G \times G$ as follows:

1. g_1 is independent of g_2: $\Leftrightarrow$

$$IS - INDEPENDEN \quad T - OF\,(g_1, g_2) :=$$
$$\min\,(N(S_1, S_2), N(S_1, D_2), N(S_2, D_1), N(D_1, D_2))$$

2. g_1 assists g_2: $\Leftrightarrow$

$$ASSISTS(g_1, g_2) := \min(I(S_1 S_2), N(S_1, D_2))$$

3. g_1 cooperates with g_2: $\Leftrightarrow$

$$COOPERATES - WITH(g_1, g_2) := \min(I(S_1, S_2), N(S_1, D_2), N(S_2, D_1))$$

4. g_1 is analogous to g_2: $\Leftrightarrow$

$$IS - ANALOGOUS \quad - TO\,(g_1, g_2) :=$$

$$\min\,(I(S_1, S_2), N(S_1, D_1), N(S_2, D_1), I(D_1, D_2))$$

5. g_1 hinders g_2: $\Leftrightarrow$

$$HINDERS(g_1, g_2) := \min(N(S_1, S_2), I(S_1, D_2))$$

6. g_1 competes with g_2: $\Leftrightarrow$

$$COMPETES - WITH(g_1, g_2) := \min(N(S_1, S_2), I(S_1, D_2), I(S_2, D_1))$$

7. g_1 is in trade-off to g_2: $\Leftrightarrow$

$$IS \_ IN \_ TRADE \_ OFF\,(g_1, g_2) :=$$

$$\min\,(N(S_1, S_2), I(S_1, D_2), I(S_2, D_1), N(D_1, D_2))$$

8. g_1 is unspecified dependent from g_2: $\Leftrightarrow$

$$IS - UNSPECIFIE\ D - DEPENDENT \quad - FROM\,(g_1, g_2) :=$$

$$\min\,(I(S_1, S_2), I(S_1, D_2), I(S_2, D_1), I(D_1, D_2))$$

Please note that interactions between goals have a subsumption relation in the sense of Figure 1.

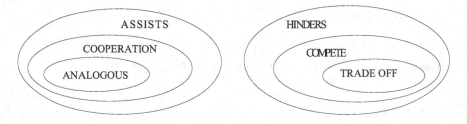

Figure 1: *Subsumption relation of the types of interaction between goals.*

Furthermore, there is a duality relation

$$
\begin{aligned}
\text{assists} &\leftrightarrow \text{hinders} \\
\text{cooperates} &\leftrightarrow \text{competes} \\
\text{analogous} &\leftrightarrow \text{trade off}
\end{aligned}
$$

which corresponds to the common sense understanding of the respective types of interaction between goals.

The interactions between goals are substantial for an adequate modeling of the decision making process because they reflect the way the goals depend on each other and describe the pros and cons of the decision alternatives, with respect to the goals. Together with information about goal priorities, the types of interaction between goals are the basic aggregation guidelines for the decision maker. For example, for cooperative goals a conjunctive aggregation is appropriate. If the goals are rather competitive, then an aggregation based on an exclusive disjunction is appropriate.

4. Types of Interaction Imply Way of Aggregation

The observation, that cooperative goals imply conjunctive aggregation and conflicting goals rather lead to exclusive disjunctive aggregation, is easily to understand from the intuitive point of view.

The fact that the types of interaction between goals are defined as fuzzy relations based on both positive and negative impacts of alternatives on the goals provides for the information about the confluence and competition between the goals: The negative impact functions reflect the negative aspects of the decision alternatives with respect to each goal and, compared with other approaches, represent additional information which enables to distinguish the non-presence of confluence between two goals from an effective competition between them.

Figure 2 shows two different representative situations which can be distinguished appropriately only if besides the positive impact additionally the negative impact of decision alternatives on goals is represented.

In case the goals were represented only by the positive impact of alternatives on them, *situation A* and *situation B* could not be distinguished and a disjunctive aggregation would be the recommended in both cases.

However, in *situation B* a decision set $S_1 \cup S_2$ would not be appropriate because of the conflicts indicated by C_{12} and C_{21}. In this situation the set $(S_1 / D_2) \cup (S_2 / D_1)$ could be recommended in case that the priorities of both goals are similar (where X / Y is defined as the difference between the sets X and Y, that means X / Y = X $\cap \overline{Y}$, where X and Y are fuzzy sets). In case that one of the two goals, for instance goal 1, is significantly more important than the other one, the appropriate decision set would be S_1. The aggregation used in that case had not to be a disjunction, but an <u>exclusive</u> disjunction between S_1 and S_2 with emphasis on S_1.

This very important aspect, can easily be integrated into decision making models by analyzing the types of interaction between goals as defined in Def.4.). The information about the interaction between goals in connection with goal priorities is used in order to define general decision rules which describe the way of aggregation.

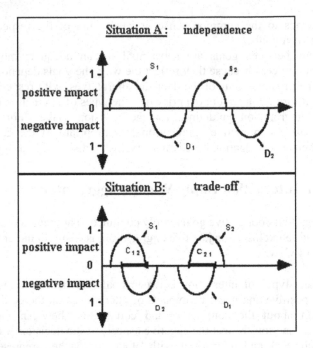

Figure 2: *Distinguishing independence and trade-off based on positive and negative impact functions of goals*

For conflicting goals, for instance, the following decision rule which deduces the appropriate decision set is given:

if $(g_1$ is in trade-off to g_2) and $(g_1$ is significantly more important then g_2) then S_1

Another rule for conflicting goals is the following:

if $(g_1$ is in trade-off to g_2) and $(g_1$ is insignificantly more important then g_2) then S_1 / D_2

Note that both rules use priority information which in case of a conflictive interaction between goals is substantial for a correct decision. Note also that the priority information can only by adequately used if the knowledge about whether or not the goals are conflictive is explicitly modeled (Felix 2000).

5. Related Approaches, a Brief Comparison

Since fuzzy set theory has been suggested as a suitable conceptual framework of decision making (Bellmann and Zadeh 1970), two directions in the field of fuzzy decision making can be observed. The first direction reflects the fuzzification of established approaches like linear programming or dynamic programming (Zimmermann 1991). The second direction is based on the assumption that the process of decision making can be modeled by axiomatically specified aggregation operators (Dubois and Prade 1984). None of the related approaches sufficiently addresses one of the most important aspects of decision making, namely the explicit and non-hard-wired modeling of the interaction between goals. Related approaches like (Dubois and Prade 1984), (Biswal 1992) either require a very restricted way of describing the goals or postulate that decision making shall be performed based on a few, very general mathematical characteristics of aggregation like commutativity or associativity. Other approaches are based on fixed (hard-wired) hierarchies of goals (Saaty 1992) or on the modeling of the decision situations as probabilistic or possibilistic graphs (Freeling 1984). In contrast to that, human decision makers usually proceed in a different way. They concentrate on which goals are positively or negatively affected by which alternatives. Furthermore, they evaluate this information in order to infer how the goals interact with each other and ask for the actual priorities of the goals. In the sense that the decision making approach presented in this contribution explicitly refers to the interaction between goals, it significantly differs from other related approaches (Felix 1995). The importance of modeling interactions between goals is also shown in (Felix, Kühlen, Albersmann 1996) based on a very complex optimization in the field of production planning in the automotive industry.

6. Application to Factoring

For making its limit decisions, the factor follows a set of criteria. The decision alternatives consist of limit levels expressed by certain percentages of the amount requested from the factor. Examples of decision criteria are explained in detail below. During knowledge acquisition, which took place before designing and implementing the system, a lot of criteria have been discussed with experts of the bank. A subset of criteria was finally identified as relevant and implemented. The system did not take into account the easy to evaluate "knockout criteria" as these were to be covered by superior functions of the general factoring system. It rather concentrates on some soft criteria, which are the most important ones.

6.1 Examples of the Handling of Decision Criteria

Let us explain in an example-oriented way, which kind of criteria has been considered.

1. Statistical Information Index SI
This index bears values of a specific SI-interval, ranging between SI_1 and SI_2 . There is, however, a subinterval of $[SI_1 , SI_2]$ where the decision is made only in interaction with other criteria.

2. Integration in the Market
This value indicates in which way the company has been active in the market and what is the corresponding risk of default. Experience shows that the relation

between life span and risk of default is not inversely proportional, but that this relation shows a non-linear behavior, for instance depending on relations to other companies.

3. Public Information Gained from Mass Media, Phone Calls, Staff, etc.
This kind of information has to be entered into the factoring system by means of a special qualitatively densified data statements.

4. Finance Information
This information concerns aspects like credit granting, utilization and over-drawing as well as receivables. The factor's staff will express the information obtained using qualitatively scaled values. The values depend on experience of interaction with other criteria.

There are still more criteria. Their common characteristics are that they interact with each other and that they are rather qualitative than quantitative. Every criterion is weighted individually and has its specific priority. The way priorities are used is explained in Section 4.

6.2   Impact Sets

The decision criteria usually possess a value domain, for instance the value high, neutral or low as presented in Figure 3. The decision alternatives reflect a granularity of percentage of the limit requested by the client. For simplicity we assume that the granularity be 25%. That means that when the limit requested is L, the limit decision equals a percentage $< 100\%$ of the requested amount L. For each criterion C and its corresponding value (as in the example below, where the values are high, neutral and low) the impact sets are defined according to Definition 1. For instance, for the value "high" of the criterion C, the entry "++" will have, according to Def. 1, as δ of $S_{C\,high}$ a value close to 1. Analogously, for the value "low" of the criterion C, the entry "-- " will have as δ of $D_{C\,high}$ a value close to 1. It is important to note that both positive and negative information is needed in order to express the impact. This is reflected by the set $S_{C\,high}$ for positive and by the set $D_{C\,high}$ for negative impacts.

Impact sets for the criterion C and its values

limit granted in % of L	high	neutral	low
100 %	++	0	--
75 %	++	+	--
50 %	+	++	--
25 %	--	+	--
0 %	--	-	++

Figure 3: *Impact sets of the criterion C*

7. Real World Applications

The application presented in this paper is part of an internet-based software system used by a factoring bank. The quality of the system's recommended decisions is equal to the decision quality achieved by senior consultants of the bank. An important aspect of the system is its standardization of the general quality of limit decisions.

8. Conclusions

The importance of an appropriate analysis of decision situations modeling types of interaction between goals in decision making has been discussed. The explicit representation of both positive and negative impacts of decision alternatives on goals provides for the ability of adequate modeling of confluences and conflicts between goals and for an appropriate way of aggregation.

The decision model discussed has been successfully applied to the field of factoring. The organization of the impact sets turned out to be efficient and quick. The system is integrated in an internet-based system interface used directly by the bank's clients.

9. References

(Bellmann, Zadeh 1970)   Bellman, R. E. and Zadeh, L. A., " Decision Making in a Fuzzy Environment", *Management Sciences, 17*, 141-164, 1970.

(Biswal 1992)   Biswal, M. P., "Fuzzy programming technique to solve multi-objective geometric programming problems", *Fuzzy Sets and Systems, Vol.51,* 67-71, 1992.

(Dubois, Prade 1984)   Dubois, D. and Prade, H., "Criteria Aggregation and Ranking of Alternatives in the Framework of Fuzzy Set Theory", *Studies in Management Sciences, Volume 20*, 209-240, 1984.

(Dubois, Prade 1992)   Dubois, D. and Prade, H., "Fuzzy Sets and Possibility Theory: Some Applications to Inference and Decision Processes", *Fuzzy Logic - Theorie und Praxis,*ed B Reusch: Springer-Verlag, 66-83, 1992.

(Felix 1991)   Felix, R., "Entscheidungen bei qualitativen Zielen", *PhD Dissertation,* University of Dortmund, Department of Computer Sciences, 1991.

(Felix 1994)   Felix, R., "Relationships between goals in multiple attribute decision making", *Fuzzy Sets and Systems, Vol.67,* 47-52, 1994.

(Felix 1995)   Felix, R., "Fuzzy decision making based on relationships between goals compared with the analytic hierarchy process", *Proceedings of the Sixth International Fuzzy Systems Association World Congress, Vol.II,* Sao Paulo, Brasil, 253-256, 1995.

(Felix, Kühlen Albersmann 1996)   Felix, R., Kühlen, J. and Albersmann, R., "The Optimiza-tion of Capacity Utilization with a Fuzzy Decision Support

Model", *Proceedings of the fifth IEEE International Con-ference on Fuzzy Systems*, New Orleans, USA, 997-1001, 1996.

(Felix 2000)   Felix, R., "On fuzzy relationships between goals in decision analysis", *Proceedings of the 3rd International Workshop on Preferences and Decisions*, Trento, Italy, 37-41, 2000.

(Felix 2001)   Felix, R., "Fuzzy decision making with interacting goals applied to cross-selling decisions in the field of   private customer banking", *Proceedings of the 10th IEEE International Conference on Fuzzy Systems*, Melbourne, Australia, Vol. ?, 2001.

(Freeling 1984)   Freeling, A. N. S., "Possibilities Versus Fuzzy Probabilities - Two Alternative Decision Aids", *Studies in Management Sciences, Volume 20, Fuzzy Sets and Decision Analysis*, 67-82, 1984.

(FLS 1997)   FLS, "FuzzyDecisionDesk, User Manual for Version 1.7", *FLS Fuzzy Logik Systeme GmbH*, Dortmund, 1997.

(Saaty 1980)   Saaty, T. L., "The Analytic Hierarchy Process", *Mc Graw-Hill*, 1980.

(Zimmermann 1991)   Zimmermann, H. J., "Fuzzy Set Theory and its Applica-tions", *Kluver/Nijhoff*, 1991.

Fuzzy Models of Rainfall–Discharge Dynamics

Hilde Vernieuwe[1], Olga Georgieva[2], Bernard De Baets[1],
Valentijn R.N. Pauwels[3], and Niko E.C. Verhoest[3]

[1] Department of Applied Mathematics, Biometrics and Process Control
Ghent University, Coupure links 653, 9000 Gent, Belgium
{Hilde.Vernieuwe,Bernard.DeBaets}@rug.ac.be
[2] Institute of Control and System Research
Bulgarian Academy of Sciences, P.O.Box 79, 1113 Sofia, Bulgaria
ogeorgieva@iusi.bas.bg
[3] Laboratory of Hydrology and Water Management
Ghent University, Coupure links 653, 9000 Gent, Belgium
{Niko.Verhoest,Valentijn.Pauwels}@rug.ac.be

Abstract. Three different methods for building Takagi–Sugeno models relating rainfall to catchment discharge are tested on the Zwalm catchment. They correspond to the following identification methods: Grid Partitioning (GP), Subtractive Clustering (SC), and Gustafson-Kessel clustering (GK). The models are parametrized on a one-year identification data set and tested against the complete five-year data set. Although these models show a similar behaviour, resulting in comparable values of the Nash and Suttcliffe criterion and the root mean square error, the best values are obtained for the models generated using the GK method.

1 Introduction

For engineering purposes, simple models describing the relationship between rainfall and catchment discharge are commonly used. These models are frequently developed conceptually, meaning that the complex physical reality is simplified using specific hypotheses and assumptions [4]. Such models do not require numerous parameters but are still able to generate predictions with an acceptable accuracy. The parametrization of such models can be based on catchment characteristics [14,18] or can be determined from rainfall and discharge time series [16]. Further developments have led to lumped conceptual models, such as the Probability Distributed Model [13] and the Xinanjiang model [22].

In order to improve the description of the rainfall-discharge behavior of a catchment, physically-based models have been developed, with the objective to describe the interaction between the atmosphere and the land-surface. This has led to the development of Soil-Vegetation-Atmosphere Transfer Schemes (SVATS). The major drawback of SVATS is the need for a large volume of input data, such as topographic, soil, and land cover maps, soil and vegetation parameters, and meteorological forcing data, which renders these models less applicable for engineering purposes.

T. Bilgiç et al. (Eds.): IFSA 2003, LNAI 2715, pp. 303–310, 2003.

304   Hilde Vernieuwe et al.

Over the last decade, fuzzy rule-based models have been introduced in hydrological studies as a powerful alternative modelling tool. With respect to rainfall-discharge prediction some models using fuzzy rules have been reported. See and Openshaw [19] use a combination of a hybrid neural network, an autoregressive moving average model, and a simple fuzzy rule-based model for discharge forecasting. Chang and Chen [5] suggested a fusion of a neural network and fuzzy arithmetic in a counterpropagation fuzzy-neural network for real time streamflow prediction. Hundecha et al. [10] developed fuzzy rule-based routines simulating different processes involved in the generation of discharge from precipitation inputs, and incorporated them in the modular conceptual physical model of Bergstrom [3]. Finally, Xiong et al. [21] used a Takagi-Sugeno model in a flood forecasting study, combining the forecasts of five different rainfall-discharge models. In this paper, we demonstrate the use of Takagi-Sugeno (TS) models for predicting discharge from rainfall time series.

2   Study Area and Data Used

The models developed in this paper have been applied to the catchment of the river Zwalm in Belgium. Troch et al. [20] give a general overview of the soil, vegetative, and topographic conditions of the catchment. The average yearly rainfall is 775 mm and is distributed evenly throughout the year. The annual evaporation is approximately 450 mm.

The data set used consists of hourly precipitation values (obtained through disaggregation of daily observations) and hourly measured discharge values for 5 years: 1994–1998. Pauwels et al. describe in detail this precipitation disaggregation algorithm. The identification data set used to train (build) the models consists of the data for 1994 only. The whole data set was then used to check the capability of all models. The discharge records show a high temporal variability, and include extremely high and low values. Since the hourly precipitation records were obtained using daily observations, the model performance was evaluated using daily averages of the simulated and observed discharge values.

3   Takagi-Sugeno Models for the Zwalm Catchment

3.1   Introduction

The catchment discharge at any time step in the discrete time domain is a function of previous rainfall and discharge values, as well as of the meteorological, topographical, and soil and vegetative conditions of the catchment. This dependency could be expressed by a storage function determined by the nature of the hydrological system being examined. For the river catchment it is a nonlinear and time varying function, which is difficult to define. In order to bypass this problem, models with minimal data requirements, and for which the prediction is relying on available rainfall data, are suggested. In this paper, first order TS models with rules of the form:

$$R_i : \text{ IF } P(k) \text{ is } A_i \text{ AND } Q(k) \text{ is } B_i \text{ THEN } Q_i(k+1) = a_i P(k) + b_i Q(k) + d_i$$

are developed, where $P(k)$ and $Q(k)$ are the precipitation and modeled discharge values at time step k, respectively. Three different identification methods are applied: grid partitioning (GP), subtractive clustering (SC) and Gustafson-Kessel (GK) clustering. In order to make a reliable comparison among the methods, only the previous time step was used.

The model performance is examined by means of the following indices:

(i) The criterion of Nash and Suttcliffe [15] (NS), commonly used in hydrological studies and comparable to the Variance Accounted For (VAF), compares the sum of squares of model errors with the sum of squares of errors when "no model" is present:

$$\text{NS} = 1 - \frac{\sum\limits_{k=1}^{N}(Q_m(k) - Q_{obs}(k))^2}{\sum\limits_{k=1}^{N}(Q_{obs}(k) - \overline{Q_{obs}})^2} \tag{1}$$

where Q_m is the simulated discharge, Q_{obs} is the observed discharge and $\overline{Q_{obs}}$ denotes the mean of the observed data. The optimal value of NS is 1, meaning a perfect match of the model. A value of zero indicates that the model predictions are as good as that of a "no-knowledge" model continuously simulating the mean of the observed signal [4]. Negative values indicate that the model is performing worse than this "no-knowledge" model [4].

(ii) The Root Mean Square Error (RMSE) given by:

$$\text{RMSE} = \sqrt{\frac{\sum\limits_{k=1}^{N}(Q_{obs}(k) - Q_m(k))^2}{N}} \tag{2}$$

3.2 Takagi-Sugeno Models Based on Grid Partitioning

The fuzzy rule-base for the Takagi-Sugeno models based on grid partitioning is determined using the neural network based ANFIS [11] in MATLAB. The grid partitioning method was used to initialize the membership functions. The parameters of the membership functions were calculated on the identification data set through backpropagation while the consequent parameters were calculated using a linear least squares method. Precipitation and discharge values were partitioned into membership functions. Doing so, different models were built, using triangular; trapezoidal; generalized bell-shaped and Gaussian membership functions, as well as a number of membership functions varying between 2 and 5 for each of the antecedent variables. The different types of membership functions are provided in MATLAB as "trimf"; "trapmf"; "gbellmf"; "gaussmf" and "gauss2mf". The latter is a combination of two gaussian membership functions determining the shape of the left and right curves. The models were then trained using the "anfis"-function in MATLAB with the training epoch number set to 100. A validation data set consisting of the data for 1995, was used to prevent overfitting, i.e. this data set was used to determine when the training of these models should be stopped. Each model was then used to perform a baseline run of the identification data. In order to find the optimal model, the NS and the RMSE values were calculated on these simulations. Table 1 lists the best values of these indices obtained for the different types of membership functions and also shows the corresponding number of membership functions. When the trapezoidal membership functions were used, the total firing strength was found to be zero, when more than 3 membership functions were used for the discharge variable, resulting in an error message from MATLAB, so the training could not be performed. Therefore, these combinations were excluded for the trapezoidal membership functions. It can be seen that the "gauss2mf" membership function results in the best values for both the Nash and Sutcliffe index and the RMSE. Since this model consists of 3 membership functions for precipitation and 5 membership functions for discharge, the rainfall-discharge prediction can be performed using 15 rules.

Table 1. Optimal values of the performance indices for the different types of membership functions

	NS	RMSE	Number of membership functions	
	[-]	[m^3/s]	Precipitation	Discharge
trimf	0.40	1.44	5	2
trapmf	0.37	1.48	5	2
gbellmf	0.39	1.45	2	2
gaussmf	0.41	1.43	4	5
gauss2mf	0.42	1.42	3	5

3.3 Takagi-Sugeno Models Based on Subtractive Clustering

The fuzzy rule-base for the Takagi-Sugeno models based on subtractive clustering is also determined by the ANFIS-tool. The initial model parameters were set using subtractive clustering [6]. The same procedure as for the GP model was performed for fine tuning. In order to find the optimal model, built with subtractive clustering, the parameters of the subtractive clustering algorithm were varied between 0.5 and 2 for η and 0.1 and 1 for r_a, $\bar{\varepsilon}$ and $\underline{\varepsilon}$ with steps of 0.1. For each combination of these parameters, a model was built and trained and the baseline run of the identification data was done. According to the values of the performance indices for these simulations, an optimal parameter combination was sought. The optimal values for η and r_a are 1 and 0.1 respectively. The same best values of the performance indices are found for values of $\bar{\varepsilon}$ from 0.2 to 1 and for values of $\underline{\varepsilon}$ from 0.2 to 1 when $\bar{\varepsilon}$ was 0.2 or 0.3, and 0.2 and 0.3 when $\bar{\varepsilon}$ had values from 0.4 to 1. The optimal values found for η and r_a and 0.5 for $\bar{\varepsilon}$ and 0.2 for $\underline{\varepsilon}$ were used to build the model, since these values are close to the ones that are recommended by Chiu [6] . This model resulted in 2 clusters.

3.4   Takagi-Sugeno Fuzzy Models Based on the Gustafson-Kessel Clustering Method

These models were built using the Gustafson-Kessel [8] clustering algorithm in order to define the multidimensional membership functions for P and Q. The consequent parameters were estimated using a global least squares method [1]. The termination criterion of the clustering algorithm was set to 0.001 and the fuzziness exponent used was set to 2. When a nearly singular covariance matrix was found, the technique proposed in [2] was used to overcome this problem. In order to find the number of clusters that result in the best model, this number was varied between 2 and 15. For each number of clusters, a new model was built and the NS index and RMSE were computed for the baseline runs on the identification data. This procedure was repeated 30 times since its result is quite sensitive to the initialisation of the partition matrix. This resulted in two datasets containing the values of the performance indices in different groups according to the number of clusters used.

Table 2. Results of the nonparametric comparisons of the control group (2 clusters) to the other groups

cl. vs. cl.	Q_{NS}	Q_{RMSE}	cl. vs. cl.	Q_{NS}	Q_{RMSE}	cl. vs. cl.	Q_{NS}	Q_{RMSE}
2 vs. 3	0.487	-0.487	2 vs. 8	3.889	-3.889	2 vs.12	4.703	-4.703
2 vs. 4	3.036	-3.036	2 vs. 9	3.966	-3.955	2 vs. 13	6.398	-6.398
2 vs. 5	10.555	-10.555	2 vs. 10	3.472	-3.472	2 vs. 14	5.841	-5.841
2 vs. 6	9.910	-9.910	2 vs. 11	3.950	-3.950	2 vs. 15	6.441	-6.441
2 vs. 7	6.588	-6.588						

On these obtained values, statistical tests were carried out in order to determine the best number of clusters. For each group, the Kolmogorov-Smirnov test was performed to check whether the data was normally distributed. Since this was not the case, non-parametric tests had to be used for further statistical analysis. First, the Kruskal–Wallis test [12] was used in order to test for differences among the groups H_0: *the values of the performance indices are the same for all numbers of clusters*. The result of this test has a value of 234.451 for both performance indices, which is much larger than $\chi_{13}^{2^{-1}}(0.95) = 23.685$. Meaning that H_0 should be rejected, :so significant differences among the groups exist. Subsequent to the Kruskal–Wallis test a non-parametric analysis was performed to seek one-tailed significant differences between one group and each of the other groups. The test applied was the nonparametric comparison of a control group to other groups following the procedure suggested in [7] and [9]. As the performance indices for the group of 2 clusters show better values than the other groups, H_0: *the values of the performance indices for the models with 2 clusters are better than the others* was chosen. The obtained Q-values of this test are listed in Table 2. Note that the values of Q_{RMSE} are the opposite of the values of Q_{NS}. This is explained by the fact that the ranking of the values of the RMSE are the oppposite of the ranking of the values of the Nash and Suttcliffe index. All Q-values are positive for the NS index and negative for the RMSE, indicating that H_0 should be accepted and that the optimal number of clusters found is 2.

As the positioning of fuzzy clusters (and thus the corresponding model) depends on the initial values, a new model is not guaranteed to perform at least as good as the best model in the 30 models, in fact it could perform even worse than all of these 30 models. In order to avoid this, models that show values at least as good as the mean obtained value of the performance indices are chosen from the 30 available models to perform the baseline runs. The averaged result of these 23 simulations is then further used.

4   Simulation Results and Discussion

Table 3 lists the values of the performance indices for the three models and for a linear regression (LR). Values are given for the baseline runs on the identification data set as well as for the baseline runs on the whole data set. This table shows that, although the values of the performance indices do not differ much, the best values are obtained for the average of the simulations of the GK models. Figure 1 shows a comparison between the daily averages of the observations and the daily averages of simulation results for the identification data set (year 1994) and Figure 2 for year 1996. The experimental value of the high peak at day 243 is 18.9m^3/s while the modeled value is 14.9m^3/s The simulation results for the other years show a similar behaviour of the GK models. From both figures, it can be seen that peaks are relatively well predicted whereas the baseflow is modeled slightly higher than the observed one. This overestimation of the baseflow probably causes the relatively low value of the NS index and the relatively high value of the root mean square error.

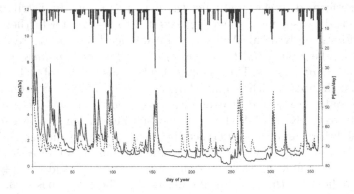

Fig. 1. Results of the GK models for the identification data set. The observations are in solid lines and the simulations are in dashed lines.

Table 3. Values of the performance indies for the identification and complete data sets

Performance Index		Identification Data				Complete Dataset			
		GP	SC	GK	LR	GP	SC	GK	LR
NS [-]	hourly	0.42	0.41	0.44	0.29	0.36	0.38	0.43	0.28
	daily	0.48	0.45	0.49	0.33	0.42	0.40	0.47	0.30
RMSE [m^3/s]	hourly	1.42	1.44	1.39	1.57	1.46	1.44	1.38	1.35
	daily	1.20	1.23	1.18	1.55	1.25	1.27	1.20	1.37

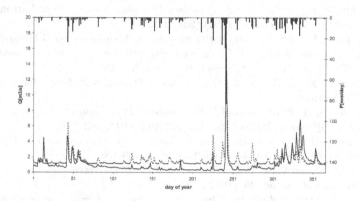

Fig. 2. Results of the GK models for 1996. The observations are in solid lines and the simulations are in dashed lines.

5 Conclusions and Further Research

Three different identification methods were used for building Takagi-Sugeno models. It was found that the different parametrization methods did not lead to major differences in model performance, which can be attributed to the overestimation of the baseflow. The best values of the performance indices are found for the models based on the Gustafson-Kessel clustering method.

Future work will focus on improving the baseflow simulation. Therefore the possibility of projecting the multidimensional clusters on the axes of the antecedent variables will be investigated.

Acknowledgment

O. Georgieva was supported by a NATO Science Fellowship during her stay at Ghent University. This research was also funded under the Research Fund of the Ghent University. Valentijn Pauwels is a postdoctoral researcher supported by the Foundation for Scientific Research of the Flemish Community (FWO-Vlaanderen).

References

1. R. Babuška, *Fuzzy Modeling for Control*, International series in intelligent technologies, Kluwer Academic Publishers, Boston, USA, 1998.
2. R. Babuška, P.J. van der Veen and U. Kaymak, *Improved covariance estimation for Gustafson-Kessel clustering*, Proc. Fuzz-IEEE (Honolulu, Hawaii, USA), 2002.
3. S. Bergström, *The HBV model*, Computer Models of Watershed Hydrology (V. P. Singh, ed.), Water Resources Publications, 1995, pp. 443–476.
4. K.J. Beven, *Rainfall-Runoff Modelling, The Primer*, John Wiley and Sons, Ltd., Chichester, UK, 2000.
5. F.J. Chang and Y.C. Chen, *A counterpropagation fuzzy-neural network modeling approach to real time streamflow prediction*, J. Hydrol. **245** (2001), 153–164.
6. S.L. Chiu, *Fuzzy model identification based on cluster estimation*, Journal of Intelligent and Fuzzy Systems **2** (1994), 267–278.
7. O.J. Dunn, *Multiple contrasts using rank sums*, Technometrics **6** (1964), 241–252.
8. D. Gustafson and W. Kessel, *Fuzzy clustering with a fuzzy covariance matrix*, Proc. IEEE CDC (San Diego, CA, USA), 1979, pp. 761–766.
9. M. Hollander and D.A. Wolfe, *Nonparametric Statistical Methods*, John Wiley, New York, USA, 1973.
10. Y. Hundecha, A. Bárdossy and H.W. Theisen, *Development of a fuzzy logic-based rainfall-runoff model*, Hydrol. Sc. Journal **46** (2001), 363–376.
11. J.S.R. Jang, *ANFIS: adaptive network-based fuzzy inference system*, IEEE Trans. on Systems, Man and Cybernetics **23** (1993), 665–685.
12. W.H. Kruskal and W.A. Wallis, *Use of ranks in one-criterion analysis of variance*, J. Amer. Statist. Assoc. **47** (1952), 583–621.
13. R.J. Moore, *The probability-distributed principle and runoff production at point and basin scales*, Hydrological Sciences Journal **30** (1985), 273–297.
14. J.E. Nash, *A unit hydrograph study, with particular reference to british catchments*, Proc. Inst. Civil Eng. **17** (1960), 299–327.
15. J.E. Nash and J.V. Sutcliffe, *River flow forecasting through conceptual models part I - a discussion of principles*, J. Hydrol. **10** (1970), 282–290.
16. T. O'Donnell, *Instantaneous unit hydrograph derivation by harmonic analysis*, IAHS Resour. Res. **51** (1960), 546–557.
17. V.R.N. Pauwels, N.E.C. Verhoest and F.P. De Troch, *A meta-hillslope model based on an analytical solution to the linearized Boussinesq-equation for temporally variable recharge rates*, Water Resour. Res., **38** (2002), no. 12, 1297, doi:1029/2001WR000714.
18. I. Rodríguez-Iturbe and J. Valdés, *The geomorphologic structure of hydrologic response*, Water Resour. Res. **15** (1979), 1409–1420.
19. L. See and S. Openshaw, *A hybrid multi-model approach to river level forecasting*, Hydrol. Sci. Journal **45** (2000), 523–536.
20. P.A. Troch, F.P. De Troch and W. Brutsaert, *Effective water table depth to describe initial conditions prior to storm rainfall in humid regions*, Water Resour. Res. **29** (1993), 427–434.
21. L. Xiong, A.Y. Shamseldin and K.M. O'Connor, *A non-linear combination of the forecasts of rainfall-runoff models by the first-order Takagi-Sugeno fuzzy system*, J. Hydrol. **245** (2001), 196–217.
22. R.J. Zhao and X.R. Liu, *The Xinanjiang model*, Computer Models of Watershed Hydrology (V.P. Singh, ed.), Water Resources Publications, 1995, pp. 215–232.

A Rule-Based Method to Aggregate Criteria with Different Relevance

Gerardo Canfora and Luigi Troiano

RCOST — Research Centre on Software Technology
Department of Engineering - University of Sannio
Viale Traiano - 82100 Benevento, Italy
{canfora, troiano}@unisannio.it

Abstract. Multiple Criteria Decision Making requires that different relevance assigned to criteria be considered. Most approaches introduced in literature are dependent upon an aggregation model. We propose a rule based approach that is independent of the aggregation model and highlights the uncertainty deriving from the decision maker confidence on the adopted criteria.

1 Introduction

In Multiple Criteria Decision Making (MCDM) it is essential to model the different relevance assigned to criteria. It is intuitive that an aggregation model should consider the way factors influence decision making in reason of the different relevance assigned to them. The more relevant a criterion is, the more it should affect the aggregated result. We can describe criterion C_i's relevance by means of *importance* $v_i \in [0, 1]$. If C_i is relevant, importance is $v_i = 1$. On the opposite side, if C_i is irrelevant, importance is $v_i = 0$. We assume that the aggregation result is computed by application of function F defined as

$$F = F\left((a_1, v_1), \ldots, (a_n, v_n)\right) \tag{1}$$

The problem of finding aggregation models able to deal with the different relevance of criteria has been widely considered in literature. Relevant contributions are Dubois and Prade [1,2,3], Fodor and Roubens [5,6], Sanchez [7], and Yager [10,12,14]. Weighted Average is the simplest form of aggregation that takes into account the relevance of criteria

$$F_{WA} = \frac{\sum\limits_{i=1}^{n} v_i a_i}{\sum\limits_{i=1}^{n} v_i} \tag{2}$$

Weighted Average exhibits two properties:

No effect of irrelevant criteria. Irrelevant criteria ($v_i = 0$) do not affect the result of aggregation, no matter how many they are.

T. Bilgiç et al. (Eds.): IFSA 2003, LNAI 2715, pp. 311–318, 2003.

$$\frac{\sum\limits_{i=1}^{n} v_i a_i}{\sum\limits_{i=1}^{n} v_i} = \frac{\sum\limits_{v_i \neq 0} v_i a_i}{\sum\limits_{v_i \neq 0} v_i} \qquad (3)$$

Invariance of result from co-variation of importance. If we co-variate the importance of criteria, the result of aggregation does not change. Indeed, let us co-variate all importance levels v_i by coefficient k as

$$v_i' = k \cdot v_i \qquad (4)$$

Importance levels v_i are modified without any effect on the relationship between two criteria

$$\nu_{i,j} = \frac{v_i'}{v_j'} = \frac{v_i}{v_j} \qquad (5)$$

The result of aggregation is not affected by such a variation of importance, as shown by

$$\frac{\sum\limits_{i=1}^{n} v_i' a_i}{\sum\limits_{i=1}^{n} v_i'} = \frac{\sum\limits_{i=1}^{n} k v_i a_i}{\sum\limits_{i=1}^{n} k v_i} = \frac{k \sum\limits_{i=1}^{n} v_i a_i}{k \sum\limits_{i=1}^{n} v_i} = \frac{\sum\limits_{i=1}^{n} v_i a_i}{\sum\limits_{i=1}^{n} v_i} \qquad (6)$$

It is interesting to notice that Weighted Average is not defined when all criteria are absolutely irrelevant ($v_i = 0, \forall i$). Intuitively, this is in accordance to impossibility of determining an aggregate result when no relevant criterion is available. In case of boolean importance ($v_i \in \{0, 1\}$), Weighted Average becomes the Arithmetic Mean of all $m \leq n$ relevant criteria

$$\frac{\sum\limits_{i=1}^{n} v_i a_i}{\sum\limits_{i=1}^{n} v_i} = \frac{\sum\limits_{j=1}^{m} a_j}{m} \qquad (7)$$

Criteria relevance has been studied also for fuzzy aggregation schemas. Dubois and Prade [3] consider criteria relevance within the Max and Min aggregations. Yager [13] discusses Weighted Min and Max aggregations. Yager [14] shows how take into account criteria relevance when Ordered Weighted Averaging (OWA) [11] aggregation is adopted. Torra [9] proposes Weighted OWA operators.

Incorporating the criteria relevance is done in order to reduce the effect of the elements which have low importance. The most common approach is to consider criteria relevance coupled to the aggregation schema. The underlying idea is that if a criterion is not relevant, then the decision should not be affected at all by the judgment it gains. In this paper we propose a model based on logical rules that is independent of the aggregation schema adopted. The remind of this paper is dedicated to describe this approach.

2 Rule-Based Approach

Some notable examples of aggregation are

$$M(a_1, \ldots a_k) = \frac{1}{k} \sum_{i=1}^{k} a_i \qquad \text{Arithmetic Mean}$$

$$M(a_1, \ldots a_k) = \min_i [a_i] \qquad \text{Min}$$

$$M(a_1, \ldots a_k) = \max_i [a_i] \qquad \text{Max}$$

$$M(a_1, \ldots a_k) = \left(\frac{1}{k} \sum_{i=1}^{k} \frac{1}{a_i} \right)^{-1} \qquad \text{Harmonic Mean} \tag{8}$$

$$M(a_1, \ldots a_k) = \left(\frac{1}{k} \sum_{i=1}^{k} a_i^{\lambda} \right)^{\frac{1}{\lambda}} \qquad \text{Generalized Mean}$$

$$M(a_1, \ldots a_k) = \frac{1}{k} \sum_{i=1}^{k} w_i a_{i(p)} \qquad \text{Ordered Weighted Averaging}$$

All these aggregations do not deal with criteria relevance. Independently from the aggregation chosen, when a criterion is irrelevant, it should be not considered at all; otherwise that criterion must be took into account. We can describe this reasoning using the following set of propositions

$$p_{i,1} : \textit{if } \text{imp}(C_i) \textit{ is high}$$
$$\textit{then aggregation should consider } C_i.$$
$$p_{i,2} : \textit{if } \text{imp}(C_i) \textit{ is low}$$
$$\textit{then aggregation can ignore } C_i.$$

where $\text{imp}(C_i)$ is the relevance associated to criterion C_i. Ignoring a criterion means that aggregation result is independent of it: the result does not change no matter which value is given to that criterion. Aggregation is then restricted to the resulting subset of criteria. We can formalize that as follows. Let $N = \{1, \ldots, n\}$ be the index set of criteria $C_1, \ldots, C_n$, and $M_{A|B}$ the generic aggregation element such that A is the index subset of criteria surely considered by aggregation M, while B is the index subset of criteria that are certainly not considered. Consequently, $A \cap B = \varnothing$ and $A \cup B \subseteq N$. Rules $p_{i,1}$ and $p_{i,2}$ can be formalized as

$$r_{i,1}: \text{imp}(C_i) \underline{\text{ is high }} \Rightarrow M_{A|B} = M_{A \cup \{i\}|B}$$

$$r_{i,2}: \text{imp}(C_i) \underline{\text{ is low }} \Rightarrow M_{A|B} = M_{A|B \cup \{i\}} \tag{9}$$

$$\forall i, A, B | A \cup B \cup \{i\} \subseteq N, A \cap B = \varnothing, i \notin A \cup B$$

An efficient way for evaluating these rules is by applying of Sugeno-Takagi's method [8]. If v_i is the measure of criteria relevance, then $\tau_{i,1} = v_i$ and $\tau_{i,2} = 1 - v_i$ (so that $\tau_{i,1} + \tau_{i,2} = 1$) are respectively the firing levels for $r_{i,1}$ and $r_{i,2}$. Therefore

$$M_{A|B} = \frac{\tau_{i,1} M_{A \cup \{i\}|B} + \tau_{i,2} M_{A|B \cup \{i\}}}{\tau_{i,1} + \tau_{i,2}} =$$

$$= \tau_{i,1} M_{A \cup \{i\}|B} + \tau_{i,2} M_{A|B \cup \{i\}} \tag{10}$$

$$i \notin A \cup B$$

Moving from $i = 1$ to n, we get

$$F = \sum_{\substack{A \subseteq N \\ A \neq \varnothing}} \tau_A M_A + \tau_\varnothing M_\varnothing \tag{11}$$

where

$$M_A = M_{A|N-A}$$

$$\tau_A = \prod_{i=1}^{n} \tau_i \tag{12}$$

$$\tau_i = \begin{cases} \tau_{i,1} = v_i & i \in A \\ \tau_{i,2} = 1 - v_i & i \in N - A \end{cases}$$

When the criteria subset A, used to compute the aggregation element M_A, is not empty, the result is a number resulting from the particular aggregation chosen. Therefore $\sum_{A \neq \varnothing} \tau_A M_A$ is still a numeric value. But what can we say about $M_\varnothing$? Nothing! It represents aggregation when all criteria are irrelevant. We can assimilate $M_\varnothing$ to total ignorance. It would be not useful to force this concept in a number. The risk would be to bias the result. It is more significant to treat it symbolically. Then, we put

$$M_\varnothing = \mathbf{I} \tag{13}$$

We will refer to $\mathbf{I}$ as *undeterminate element*. Sum (11) has to be intended algebraically: it is a linear combination of $\mathbf{I}$ and the *numeric component* F_Z that qualifies the aggregation. To determine such a combination we can look at aggregation result by another point of view. If all criteria are irrelevant ($v_i = 0, \forall i$), the aggregation result cannot be determined, otherwise the result coincides with the aggregation of relevant criteria. This can be formally described by the following proposition

$p_{\mathbf{I},1}$: *if* imp(C_i) *is low* $\forall i \in N$ *then* $F = \mathbf{I}$
$p_{\mathbf{I},2}$: *otherwise* $F = F_Z$

so that

$$F = \tau_{1,2} \tau_{2,2} \cdots \tau_{n,2} \cdot \mathbf{I} + (1 - \tau_{1,2} \tau_{2,2} \cdots \tau_{n,2}) \cdot F_Z =$$
$$= \xi_J \cdot \mathbf{I} + (1 - \xi_J) \cdot F_Z \tag{14}$$

We will call the coefficient $\xi_J = \tau_{1,2} \tau_{2,2} \cdots \tau_{n,2}$ the *indeterminateness*; it represents the vagueness of aggregation deriving from how much chosen criteria are relevant for evaluation. In addition, we will call the coefficient $\zeta_J = (1 - \xi_J)$ the *determinateness*, so that we can write more conveniently

$$F = \xi_J \cdot \mathbf{I} + \zeta_J \cdot F_Z \tag{15}$$

We will call it *number with indeterminateness*. By equaling Eq.(11) with Eq.(15), we get

$$F_Z = \frac{1}{1 - \xi_J} \sum_{\substack{A \subseteq N \\ A \neq \emptyset}} \tau_A M_A \tag{16}$$

The numeric component F_Z keeps all the qualitative information by which we can assess each alternative. Another way of looking at Eq.(11) is as result of the following set of rules

$$p_A : \text{ if } \text{imp}(C_i) \text{ is high } \forall i \in A \subseteq N, A \neq \varnothing$$
$$\text{and } \text{imp}(C_i) \text{ is low } \forall i \notin A \subseteq N, A \neq \varnothing$$
$$\text{then } F = M_A$$
$$p_I : \text{ if } \text{imp}(C_i) \text{ is low } \forall i \in N \text{ then } F = \mathbf{I}$$

Indeed, if all criteria indexed by A are relevant to aggregation, while criteria not indexed by A are irrelevant, the result of aggregation F should coincide with M_A. When importance is boolean ($v_i \in \{0,1\}$), only one rule is true: that rule assigns the appropriate aggregation M_A to F.

Example 1. Let $C = \{C_1, C_2, C_3\}$ be a set of criteria. Then $N = \{1, 2, 3\}$. We suppose that C_1 and C_3 are fully relevant ($v_1, v_3 = 1$) while C_2 is irrelevant ($v_2 = 0$). Firing levels are computed according to Eq.(12). All rules discarding C_1 have null firing levels in reason of $\tau_{1,2} = 1 - v_1 = 0$, as well as all rules that ignore C_3 ($\tau_{3,2} = 1 - v_3 = 0$); all rules considering C_2 have the firing level which is null due to $\tau_{2,1} = v_2 = 0$. Therefore, only the rule $p_{\{1,3\}}$ has a no-null firing level.

$$\tau_{\{1,3\}} = \tau_{1,1} \cdot \tau_{2,2} \cdot \tau_{3,1} = v_1 \cdot (1 - v_2) \cdot v_3 = 1$$

Then the result of aggregation is

$$F = M_{\{1,3\}}$$

as expected.

Aggregation (11) is computed as the mean of rules weighted by their firing levels. This is pointed out by the following example.

Example 2. Let $v_1 = 1$, $v_2 = 0.4$ and $v_3 = 0.7$. It is easy to verify that

$$\tau_{\{1,2,3\}} = v_1 \cdot v_2 \cdot v_3 = 0.28$$
$$\tau_{\{1,2\}} = v_1 \cdot v_2 \cdot (1 - v_3) = 0.12$$
$$\tau_{\{1,3\}} = v_1 \cdot (1 - v_2) \cdot v_3 = 0.42$$
$$\tau_{\{1\}} = v_1 \cdot (1 - v_2) \cdot (1 - v_3) = 0.18$$
$$\tau_{A-\{1\}} = 0 \ \forall A \subseteq N$$

Then

$$F = 0.28 \cdot M_{\{1,2,3\}} + 0.12 \cdot M_{\{1,2\}} + 0.42 \cdot M_{\{1,3\}} + 0.18 \cdot M_{\{1\}}$$

Let us assign $C_1 \leftarrow 0.6$, $C_2 \leftarrow 0.4$ and $C_3 \leftarrow 0.8$. When we adopt the Arithmetic Mean as aggregator, we get

$$F_{Ave} = 0.28 \frac{0.6 + 0.4 + 0.8}{3} + 0.12 \frac{0.6 + 0.4}{2} + 0.42 \frac{0.6 + 0.8}{2} + 0.18 \frac{0.6}{1} = 0.63$$

This is in accordance with the fact that criterion C_1 is relevant ($v_1 = 1$) and neutral to the Arithmetic Mean ($C_1 \leftarrow 0.6$), and that criterion C_3 is more relevant than C_2 ($v_3 = 0.7 > v_2 = 0.4$). As expected, the resulting value 0.63 is closer to C_1 and C_3, than to C_2.

Example 1 has shown that irrelevant criteria have no effect on the aggregated result. This is because $\tau_{j,1} = v_j = 0$ for at least one $j \in A$ makes null any product τ_A related to an aggregation element M_A considering criterion C_j. This means that all aggregation elements M_A, where one or more criteria indexed by A are irrelevant, do not affect the sum expressed by Eq.(11). More in general, it is easy to verify that the aggregation result does not change, no matter how many irrelevant criteria are took into account, as described by the following example.

Example 3. Next to criteria C_1, C_2, C_3, let us consider the irrelevant criterion C_4 ($v_4 = 0$). Sum (11) can be re-written as

$$F(C_1, C_2, C_3, C_4) = \sum_{A \in P(\{1,2,3,4\})} \tau_A M_A =$$

$$= \sum_{A \cup \{4\}} \tau_{A \cup \{4\}} M_{A \cup \{4\}} + \sum_{A - \{4\}} \tau_{A-\{4\}} M_{A-\{4\}}$$

where $P(\{1, 2, 3, 4\})$ is the power set of $\{1, 2, 3, 4\}$. The firing level associated to $M_{\{1,2,3,4\}}$ is $\tau_{\{1,2,3,4\}} = v_1 \cdot v_2 \cdot v_3 \cdot v_4 = 0$. This happens for any aggregation element $M_{A \cup \{4\}}$ that considers criterion C_4. Then

$$F(C_1, C_2, C_3, C_4) = \sum_{A - \{4\}} \tau_{A-\{4\}} M_{A-\{4\}}$$

Moreover

$$\tau_{A-\{4\}} = (1 - v_4) \cdot \prod_{i=1}^{3} \tau_i = \prod_{i=1}^{3} \tau_i$$

Therefore

$$F(C_1, C_2, C_3, C_4) = \sum_{A \in P(\{1,2,3\})} \tau_A M_A = F(C_1, C_2, C_3)$$

2.1   Semantic Differences with Weighted Average

It is interesting to compare the proposed approach to Weighted Average, when the Arithmetic Mean is selected as aggregation. If criteria importance is boolean ($v_i \in \{0, 1\}$), aggregation (11) coincides with the Arithmetic Mean. Indeed

$$F = M_A = \frac{1}{m} \sum_{i \in A} a_i \tag{17}$$

$$A = \{i \in N | v_i = 1\}, \ m = \text{card}(A)$$

However their behavior is different. Such a difference derives, over a formal plane, from a different logic used to express Weighted Average. Eq.(2) can be seen as result of Sugeno-Takagi's method to the following set of rules

$$p_{C_i} : \textit{if } \text{imp}(C_i) \text{ is high } F = C_i$$

The different formal structure leads to different semantics associated to the proposed model. Unlike the Weighted Average, the proposed model has not invariance of result from co-variation of importance as enunciated above. Such a dissimilarity derives from the different structure of result. The result we get by Eq.(11) expresses the linear combination of numeric component F_Z with the undetermined element $\mathbf{I}$, meaning that a part of the aggregated assessment is unknown because of incompleteness of relevant criteria. If we increase the relevance of criteria by importance co-variation, indeterminateness ξ becomes lower, denoting a lower level of ignorance about the result, and an higher significativeness of the numeric component F_Z. Also F_Z changes. Such a change means that the aggregated evaluation of each alternative depends on the relevance of criteria, then on the significativeness of result: the more criteria are relevant, the lower indeterminateness, the higher the significativeness is. This semantics suggests that relevance of criteria should be considered to a better validation of results. Weighted averaging does not brings this into prominence.

Example 4. Let consider a case with three criteria whose value and relevance are shown in Table 1. If we consider relevance values v_i or v'_i, Weighted Average

Table 1. An example of aggregation with three criteria.

	a_i	v_i	v'_i
C_1	0.4	0.3	0.6
C_2	0.6	0.2	0.4
C_3	0.8	0.4	0.8

remains the same. Indeed the result in both cases is $WA = 0.6222$. While our method produces two different values. When we consider importance values v_i, we get $F = 0.4184$. When importance is v'_i, instead we get $F' = 0.6032$. The different value derives from a redistribution of uncertainty that is not proportional. Indeed, we have indeterminateness $\xi = 0.336$ in the the first case; whilst indeterminateness is $\xi' = 0.048$ in the second case. Such a change reduces the level of uncertainty and increases the significativeness of result.

3   Conclusions

In this paper we proposed a model based on logic rules to deal with relevance of criteria in MCDM problems. The proposed approach is independent of the specific aggregation operator. Fuzzy aggregation operators can be adopted. The paper focused on Weighted Average, the simplest aggregation operator known to deal with different relevance of criteria. The proposed model was compared to Weighted Average, showing analogies and differences. In particular a different semantics emerged: the result and its significativeness depends on the relevance of criteria. Our model is similar to Yager's model [14] in that both use logical rules to take the relevance of criteria into account. However, Yager's model uses the rules to modify the judgments associated with criteria, whereas in our case rules govern the aggregation process. In this paper, we have used Sugeno-Takagi [8] inferential calculus to derive firing levels of logical rules. However, the approach proposed is more general, as other methods of firing rules can be applied [4]. We plan to investigate this issue in future research. Our long term goal is to apply the proposed model to Decision Support Systems to aggregate information bringing into evidence the uncertainty that derives from the confidence on adopted criteria.

References

1. D. Dubois and H. Prade. Criteria aggregation and ranking of alternatives in the framework of fuzzy set theory. *TIMS/Studies in the Management Sciences*, (20):209–240, 1984.
2. D. Dubois and H. Prade. A review of fuzzy sets aggregation connectives. *Information Sciences*, (36):85–121, 1985.
3. D. Dubois and H. Prade. Weighted minimum and maximum operations in fuzzy set theory. *Information Sciences*, (39):205–210, 1986.
4. D. Dubois and H. Prade. What are fuzzy rules and how to use them. *Fuzzy Sets and Systems*, 84(2):169–189, 1996.
5. J. Fodor and M. Roubens. Aggregation and scoring procedures in multicriteria decision-making methods. In *Proceedings of the IEEE International Conference on Fuzzy Systems*, pages 1261–1267, San Diego, 1992.
6. J. Fodor and M. Roubens. *Fuzzy Preference Modelling and Multi-Criteria Decision Support*, volume 14 of *D*. Kluwer, Dordrect.
7. E. Sanchez. Importance in knowledge systems. *Information Systems*, (14):455–464, 1989.
8. M. Sugeno and T. Takagi. *A new approach to design of fuzzy controllers*, pages 325–334. Plenum Press, New York, wang, p. p. edition, 1983.
9. V. Torra. The weighted owa operator. *Int. J. of Intel. Systems*, (12):153–166, 1997.
10. R. Yager. Multiple objective decision making using fuzzy sets. *Int. J. of Man-Machine Studies*, (9):375–382, 1977.
11. R. Yager. On ordered weighted averaging aggregation operators in multi-criteria decision making. *IEEE Trans. on Systems, Man, and Cybernetics*, 18(1):183–190, 1988.
12. R. Yager. On the inclusion of importances in multi-criteria decision making in the fuzzy set framework. *Int. J. of Expert Systems: Research and Application*, (5):211–228, 1992.
13. R. Yager. On weighted median aggregation. *Int. J. of Uncertainty, Fuziness and Knowledge-based Systems*, (1):101–113, 1994.
14. R. Yager. Criteria importances in owa aggregation: An application of fuzzy modelling. In *Proc. FUZZ-IEEE 1997*, volume III, pages 1677–1682. IEEE, 1997.

Imprecise Modelling Using Gradual Rules and Its Application to the Classification of Time Series

Sylvie Galichet [1,2], Didier Dubois [1], Henri Prade [1]

[1] Institut de Recherche en Informatique de Toulouse (IRIT), Université Paul Sabatier,
118, route de Narbonne, 31062 Toulouse Cedex 4
{galichet, dubois, prade}@irit.fr
[2] LISTIC, Université de Savoie,
41, avenue de la Plaine, BP 806, 74016 Annecy Cedex

Abstract. This paper presents an alternative to precise analytical modelling, by means of imprecise interpolative models. The model specification is based on gradual rules that express constraints that govern the interpolation mechanism. The modelling strategy is applied to the classification of time series. In this context, it is shown that good recognition performance can be obtained with models that are highly imprecise.

1 Introduction

Nowadays, most automated applications are based on models of the systems under consideration. In this framework, a precise representation is often used even when this representation is based on a fuzzy rule base. When uncertainties are explicitly dealt with, they affect parameters of analytical models, leading to probabilistic or interval-based processing. The main objective of this paper is to propose an alternative to these analytical approaches by investigating the interest of gradual rules [1] for developing imprecise models. Actually, the proposed strategy relies on a set-valued interpolative reasoning. When specifying representation with gradual rules, there is no need to choose a parameterized function for the interpolator. An imprecise model is directly obtained from the constraints expressed by the rules. What is supposed to be known, in a precise or in an imprecise way, is the behaviour of the system at some reference points, the problem being to interpolate between these points. Figure 1 illustrates our view of an imprecise model in a case where the points on which interpolation is based are imprecise. In order to avoid the choice of an analytical model (linear, polynomial, piecewise linear, cubic spline, ...), an imprecise model which includes different possible precise interpolators is preferred. So, we are no longer looking for a function, only for a relation linking input variables to output variables. In the one-dimensional input case (as in figure 1) considered in this paper, this relation is represented by its graph Γ defined on the Cartesian product $X \times Z$ (where X is the input domain, and Z the output domain). A similar approach, recently proposed in [8], also considers the design of uncertain fuzzy models in the setting of the approximation of multi-valued mappings called "ambiguous functions".

T. Bilgiç et al. (Eds.): IFSA 2003, LNAI 2715, pp. 319-327, 2003.

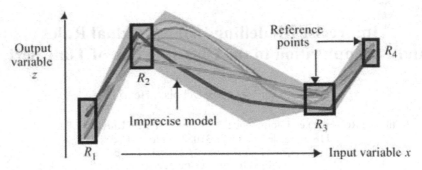

Fig. 1. Imprecise interpolation

The paper, after some brief background on gradual rules, discusses the process of building an imprecise model, constrained by precise reference points. The case of interval-valued reference points is omitted for the sake of brevity (see [3] for details). The proposed modelling strategy is then used for dealing with the classification of time series. Moreover, imprecise reference points are introduced in this applied context.

2 Interpolation and Gradual Rules

The idea of imprecise modelling suggested above is based on constraints to be satisfied, namely the postulate that the results of the interpolation should agree with the reference points. These constraints define the graph Γ of the corresponding relation on $X \times Z$. We first consider the case of precise reference points P_i with coordinates (x_i, z_i), $i = 1, ..., n$. Then the relation Γ should at least satisfy $\Gamma(x_i, z_i) = 1$ and $\forall z \neq z_i \in Z$, $\Gamma(x_i, z) = 0$ for $i = 1, ..., n$. Without any further constraint on the nature of the interpolation, we only have: $\forall x \neq x_i \in X$, $\forall z \in Z$, $\Gamma(x, z) = 1$. So each interpolation point induces the constraint "If $x = x_i$ then $z = z_i$", represented by $(x = x_i) \rightarrow (z = z_i)$ where $\rightarrow$ is material implication. The relation Γ is thus obtained as the conjunction:

$$\Gamma(x, z) = \wedge_{i = 1, ..., n} (x = x_i) \rightarrow (z = z_i). \tag{1}$$

This relation is extremely imprecise since there is no constraint at all outside the interpolation points. The absence of a choice of a precise type of interpolation function should be alleviated by the use of fuzzy rules in order to express additional constraints in the vicinity of the interpolation points. The idea is to use rules of the form "the closer x is to x_i, the closer z is to z_i". The extension to gradual rules of equation (1) provides the following expression for the graph Γ:

$$\Gamma(x, z) = \min_{i = 1, ..., n} \mu_{\text{close to } xi}(x) \rightarrow \mu_{\text{close to } zi}(z) \tag{2}$$

where $\rightarrow$ represents the Rescher-Gaines implication ($a \rightarrow b = 1$ if $a \leq b$ and $a \rightarrow b = 0$ if $a > b$), and $\mu_{\text{close to } xi}(x)$ is the degree of truth of the proposition "x is close to x_i". We now have to define what is meant by "close to". Let A_i denote the fuzzy set of values close to x_i. It is natural to set $\mu_{A_i}(x) = 1$ if $x = x_i$ and to assume that the membership degree to A_i decreases on each side of x_i with the distance to x_i. The simplest solution consists in choosing triangular fuzzy sets with a support denoted by $[x_i^-, x_i^+]$. In a similar way, the closeness to z_i will be modelled by a triangular fuzzy set B_i with modal value z_i and support $[z_i^-, z_i^+]$. Then the interpolation relation only depends on $4n$ parameters $x_i^-, x_i^+, z_i^-, z_i^+$ for n interpolation points. The purpose of the next section is to study criteria for choosing these parameters. In order to simplify the analysis of the interpolation relation, we further assume that at most two rules are simultaneously fired at each point of the input domain (i.e., $x_i \leq x_{i+1}^-$ and $x_i^+ \leq x_{i+1}$, $i = 1, ..., n-1$, which implies for triangular membership functions that the coverage degree is such that $\forall x \in X, \Sigma_{i=1,...,n} \mu_{A_i}(x) \leq 1$).

3 Interpolative Crisp Graph

3.1 Coverage and Consistency of the Rule Base

It is generally expected that a set of parallel fuzzy rules should cover any possible input, which means that each possible input value x should fire at least one rule. In other words, the A_i's should lead to a nonzero degree everywhere in X, i.e.:

$$x_{i+1}^- < x_i^+. \tag{3}$$

Then, we might think of using a strong fuzzy partition both for X and Z (i.e. with a coverage degree exactly equal to 1 for any $x \in X$ and $z \in Z$), with triangular fuzzy sets A_i and B_i. However, in this case, as shown in [1], gradual rules lead to a precise and linear interpolation, as pictured in figure 2 with 3 interpolation points. This is not what we are looking for.

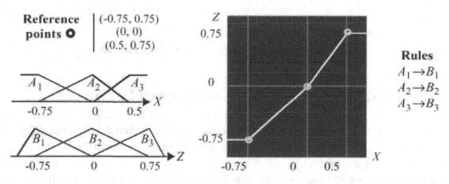

Fig. 2. Linear interpolation

In order to tune the fuzzy set parameters, let us study the generic case of figure 3 corresponding to a pair of gradual rules $A_i \to B_i$ and $A_{i+1} \to B_{i+1}$. The area in dark grey delimited by points M_k, $k = 1, ..., 6$ corresponds to the interpolation graph obtained for the parameters $x_i^-, x_i^+ z_i^-, z_i^+, i=1, 2$. The modification of z_i^+ into z'_i^+ and z_{i+1}^- into z'_{i+1}^- yields an important change in the shape of the area, depicted in light grey in figure 3. Figure 3 makes it clear that the interpolation graph should be connected in order to guarantee that any feasible input is associated with an output value. This means that there is no conflict when two rules are simultaneously fired, i.e. when

$x \in [x_{i+1}^-, x_i^+]$. From a geometric point of view, this amounts to locating the point M_5 in figure 3 under the segment M_1M_2, which is analytically expressed by the inequality:

$$(z_{i+1}^- - z_i)(x_i^+ - x_i) \le (z_i^+ - z_i)(x_{i+1}^- - x_i). \tag{4}$$

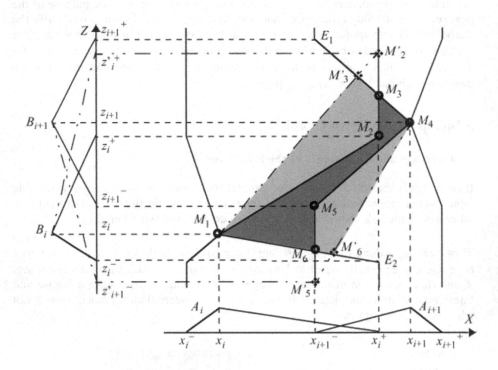

Fig. 3. Graph corresponding to a pair of gradual rules

In a similar way, the point M_2 should be above the segment M_5M_4, i.e.:

$$(z_i^+ - z_{i+1})(x_{i+1}^- - x_{i+1}) \le (z_{i+1}^- - z_{i+1})(x_i^+ - x_{i+1}). \tag{5}$$

Conditions (4) and (5) are necessary for obtaining a connected interpolation graph. They are also sufficient as far as point M_6 is located under segment M_1M_2 (i.e., if $z_i^- < z_i < z_i^+$) and M_3 is above M_4M_5 (i.e., if $z_{i+1}^- < z_{i+1} < z_{i+1}^+$). In fact, conditions (4) and (5) are a particular case of coherence conditions for a set of fuzzy rules, which were previously established in [2]. Note that due to equation (2), the output fuzzy sets can be quite imprecise (as shown in fig. 3), while the resulting model may remain rather precise (as seen on fig. 2).

3.2 Shaping the Interpolation Areas

The above analysis has shown the need for controlling the shape of the interpolation graph. Obviously in figure 3, the light grey area seems to be better than the dark grey area whose shape is too complicated and hard to justify. However it remains to clarify the conditions which determine the exact shape of the interpolation graph between two interpolation points. From a geometric point of view, the deletion of the vertical segment M_2M_3 requires that the point M_2 be located above M_3, which after computation of the z coordinate of M_3 leads to:

$$(z_i^+ - z_{i+1})(x_{i+1}^- - x_{i+1}) \leq (z_{i+1}^+ - z_{i+1})(x_i^+ - x_{i+1}). \qquad (6)$$

In a similar way the constraint that M_5 be under M_6 is expressed by:

$$(z_{i+1}^- - z_i)(x_i^+ - x_i) \leq (z_i^- - z_i)(x_{i+1}^- - x_i). \qquad (7)$$

Keeping the hypothesis of the coverage of X by at most two rules everywhere in the input domain, the satisfaction of (6) and (7) requires that:

$$z_{i+1} \leq z_i^+ \quad \text{and} \quad z_{i+1}^- \leq z_i \qquad (8)$$

and thus the coverage of Z by the B_i's should be sufficient (more precisely $\forall z \in Z$, $\Sigma_{i=1,...,n} \mu_{Bi}(z) \geq 1$), which also ensures the consistency of the rules, since (4) and (5) then hold. Although the coordinates of points M'_3 and M'_6 can be expressed in terms of the considered parameters, it appears simpler to delimit the interpolation graph area by the points M_1, M_3, M_4 and M_6, which corresponds to changing (6) and (7) into equalities. Then, the x-coordinates of M_3 and M_6, namely x_i^+ and x_{i+1}^-, can be explicitly chosen. For a given choice of the x-parameters, the definition of the rules requires the determination of $2n$ z-parameters, which are linked together by $2(n-1)$ equations. Thus, two parameters remain free. A simple way of building the partition of Z is then to prescribe the values on the boundaries, namely z_1^- and z_n^+, and to solve the system of equations obtained by the equality constraints associated with (6) and (7). Figure 4 pictures the interpolation graph which is obtained with the interpolation points of figure 2. The partitioning of X is obtained by cutting the intervals $[x_i, x_{i+1}]$ into three equal parts, i.e. $x_{i+1}^- - x_i = x_i^+ - x_{i+1}^- = x_{i+1} - x_i^+$, as illustrated by the parameters which define the fuzzy sets A_i. The extreme values z_1^- and z_3^+ are also predefined (see figure 4). Lastly, the other parameters are obtained by solving the system of equations derived from (6) and (7).

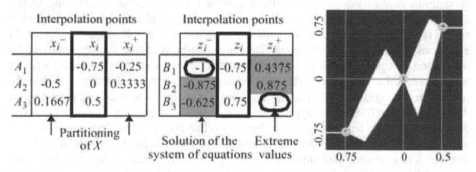

Fig. 4. Construction of a piecewise quadrangle-shaped graph

4 Classification of Time Series

Our purpose is now to illustrate how the imprecise models previously introduced can be used to classify time series. The proposed experiment deals with the «Control-

Chart» database, freely available from the UCI Data Archive [9]. It is a 6-class problem, with 100 examples of each class, a prototype of each class being presented in figure 5. Given an unlabeled time series, the aim of the classification is to decide the class to which it belongs. The idea of the proposed methodology consists in developing an imprecise model of each class. Then, the time series to classify will be assigned to the class whose model has maximal adequacy with the temporal signal under consideration. The imprecise models are specified using gradual rules as advocated in the previous section.

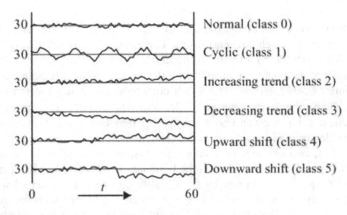

Fig. 5. One example of each class

In figure 6, ten examples of class 5 are plotted simultaneously. It clearly shows that the reference points are no more precise. In this context, triangular membership functions are replaced by trapezoïdal membership functions whose cores delimit the rectangular areas associated with the imprecise reference points. According to this slight modification, the graph plotted in figure 6(a) is obtained from two gradual rules. It can be shown that all considerations on the graph shape expressed by equations (6) and (7) are still valid. The model so-built can be further improved by truncating the upper and lower parts of the quadrangle-shaped graph. An easy strategy to implement the truncation consists in adding a new rule that directly translates the interval-based constraint "If $t \in [a_1, a_2]$ then $z \in [b_1, b_2]$" where a_1, a_2, b_1 and b_2 are defined in figure 6. Such an approach results in the final graph on figure 6(b). Figure 7 presents the implemented models for two other classes. The first one associated with class 3 is based on reference points whose imprecision is only relative to the output. Using strong partitions with triangular input membership functions and trapezoïdal outputs, imprecise linear interpolation is obtained. Concerning cyclic time series (figure 7 (b)), the non-monotonic underlying behavior induces some difficulties in the modelling process. Actually, the right handside and the left handside of the reference points must be handled in different ways. It means that two distinct fuzzy subsets are required for correctly dealing with each reference point. In this framework, the imprecise model of figure 7(b) is composed of 9 gradual rules. Imprecise models are built for the six classes so that the graphs include all the points in the training time series (10 for each class). The classification of a time series, given as a collection of points (t_i, z_i), $t_i = i = 1, ..., 60$, is then carried out according to its adequacy with the class models. The latter is determined from the number of points of the time series under consideration that belong to the model graphs, i.e:

$$N_j = \Sigma_{i=1, ..., 60} \Gamma_j(t_i, z_i), j = 1, ..., 6 \qquad (9)$$

where Γ_j denotes the model graph of the jth class. The final decision then consists in assigning the time series, supposedly unlabeled, to the class that maximizes N_j, $j = 1$, ..., 6. Applying this strategy for the classification of the 600 available examples, perfect classification is obtained, i.e. the error rate is null for the training examples but also for the test time series. This result is better than the one obtained with other approaches to the same problem [5], [7].

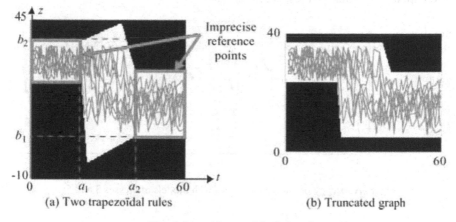

(a) Two trapezoïdal rules

(b) Truncated graph

Fig. 6. Imprecise model of class 5

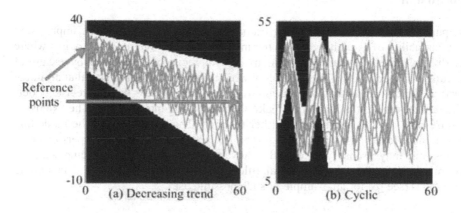

(a) Decreasing trend

(b) Cyclic

Fig. 7. Imprecise models for classes 3 (a) and 1 (b)

It is however important to be cautious about this good performance. Indeed, the discrimation between some classes is not robust. This point is illustrated by figure 8 in the case of classes 3 and 5 which are difficult to differentiate. The adequacy between the 100 time series of class 3 and models of classes 3 and 5 is plotted. It can be stated that, for many time series, the difference between both obtained scores is small, which means that a slight modification of the models would probably result in different final decisions. Actually, an important overlap between the range of both models induces a loss of discrimination power of the adequacy index. In this framework, one may think of improving the robustness of the classification by refining the imprecise models. One possible strategy is then to introduce some membership degrees in the quadrangle areas while keeping their support unchanged. This is the topic of on-going work [4]. While our approach is very insensitive to noise, it is sensitive to outliers. This problem could be addressed by the use of membership degrees as well. Lastly it is clear that the approach should be augmented with a signal processing step that would calibrate the data with respect to the rules.

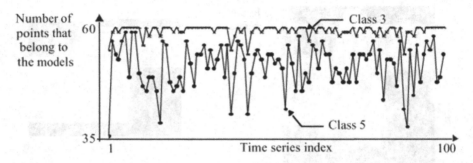

Fig. 8. Adequacy of class 3 examples with models of classes 3 and 5

Conclusion

This paper has proposed a modelling framework which is faithful to the imprecision or the variability of available data. In the intervals between interpolation points where it is difficult to specify an analytical model, imprecision is captured by means of quadrangle-shaped areas. The reader may wonder whether rules are useful at all in the approach. On this issue, the contribution of the paper is to show that imprecise models do have rule representations which make them local and interpretable. The extension to multi input rules is a matter of further research. The application of the modelling methodology for classifying time series has exhibited interesting performance. Besides, one may probably take further advanvage of the easy interfacing with the user, provided by the use of gradual rules, for specifying queries in data mining applications (see [6] for an example of such possible use).

References

[1] D. Dubois, H. Prade, "Gradual inference rules in approximate reasoning", Information Sciences, Vol. 61, no 1-2, pp. 103-122, 1992.
[2] D. Dubois, H. Prade, L. Ughetto, "Checking the Coherence and Redundancy of Fuzzy Knowledge Bases", IEEE Trans. on Fuzzy Systems, Vol. 5, no 3, pp. 398-417, 1997.

[3] Galichet S., Dubois D., Prade H., "Imprecise specification of ill-known functions using gradual rules", Proc. of the 2nd Int. Workshop on Hybrid Methods for Adaptive Systems (EUNITE'02), Albufeira, Portugal, pp. 512-520, Sept. 2002.
[4] Galichet S., Dubois D., Prade H., "Fuzzy interpolation and level 2 gradual rules", submitted to EUSFLAT Conference, Zittau, Germany, Sept. 2003.
[5] E. Keogh, S. Kasetty, "On the need for time series data mining benchmarks: a survey and empirical demonstration", Proc. of the 8th ACM SIGKDD Int. Conf. on Knowledge Discovery and Data Mining, Edmonton, Canada, pp. 102-111, July 2002.
[6] E. Keogh, H. Hochheiser, B. Shneiderman, "An augmented visual query mechanism for finding patterns in time series data", Proc. of the 5th Int. Conf. on Flexible Query Answering Systems (FQAS 2002), Copenhagen, Denmark, pp. 240-250, Oct. 2002.
[7] A. Nanopoulos, R. Alcock, Y. Manolopoulos, "Feature-based classification of time-series data", Int. Journal of Computer Research, Vol. 10, no 3, 2001.
[8] A. Sala, P. Albertos, "Inference error minimisation: fuzzy modelling of ambiguous functions", Fuzzy Sets and Systems, Vol. 121, no 1, pp. 95-111, 2001.
[9] The UCI KDD Archive (http://kdd.ics.uci.edu/), University of California, Irvine.

A Semi-supervised Clustering Algorithm for Data Exploration

Abdelhamid Bouchachia[1] and Witold Pedrycz[2]

[1] University of Klagenfurt, Dept. of Informatics-Systems
Universitätsstrasse 65, A-9020 Klagenfurt, Austria
hamid@isys.uni-klu.ac.at
[2] Dept. Electrical and Computer Engineering, University of Alberta
Edmonton, T6G 2V4, Canada
pedrycz@ee.ualberta.ca

Abstract. This paper is concerned with clustering of data that is partly labelled. It discusses a semi-supervised clustering algorithm based on a modified fuzzy C-Means (FCM) objective function. Semi-supervised clustering finds its application in different situations where data is neither entirely nor accurately labelled. The novelty of this approach is the fact that it takes into consideration the structure of the data and the available knowledge (labels) of patterns. The objective function consists of two components. The first concerns the unsupervised clustering while the second keeps the relationship between classes (available labels) and the clusters generated by the first component. The balance between the two components is tuned by a scaling factor. The algorithm is experimentally evaluated.

1 Introduction

One of the most interesting techniques in pattern recognition, data mining and knowledge discovery is clustering. It aims at finding the hidden structure that underlies a given collection of data points. The task consists of partitioning the data set into containers (clusters), such that similar data points (feature vectors) are grouped into the same container. Similar to the machine learning paradigm, to group data points into containers, two well-known modes can be applied; namely, supervised which is concerned with assigning labelled data points to prespecified groups and unsupervised which aims at assigning unlabelled data points to clusters using some similarity measure, i.e. distance-based, density-based, etc. These modes look extremes in the sense that the former requires complete knowledge on the data being analyzed, while the later uses no knowledge at all. Acquiring knowledge (labelling) on the data points is always an expensive and error-prone task that requires time and human effort. In many situations, the data is neither perfectly nor completely labelled. Thus, we may attempt to benefit from the available knowledge (labels of data points) to cluster those unlabelled data points. This form of combining labelled and unlabelled data to generate the structure of the whole data set is known as *semi-(or partially) supervised clustering* (see Fig. 1). Thus, semi-supervised clustering can be defined as the process of clustering data taking into account available knowledge about data. Labelled data points are used to guide the process of grouping and at the same time to boost the accuracy of unsupervised clustering. The

T. Bilgiç et al. (Eds.): IFSA 2003, LNAI 2715, pp. 328–337, 2003.

goal is then to relate clusters belonging to the same class. Any semi-supervised algorithm should be able to detect such a structure of classes where clusters of the same class are not necessarily adjacent.

Figure 1. Clustering spectrum with respect to knowledge usage

Several semi-supervised approaches have been explored, such as: k-means based seeding [1], fuzzy c-means based seeding [2](where a seed is simply the mean of the labelled data and serve to initialize prototypes), probabilistic clustering [5, 7], genetic algorithms [4], and support vector machine [6]. More relevant to our case is the approach investigated in [8] where a modified version of the FCM algorithm was proposed to handle the problem of partial supervision. The objective function was extended to include a second term that models the relationship between classes and clusters. In this objective function, labelled and unlabelled data are identified by means of a boolean vector $b = [b_k]$, k=1, 2, ..., N, where N is the size of the data set. $b_k = 1$ if pattern x_k is labelled, 0 otherwise. Likewise, the membership values of the labelled patterns are arranged in a matrix $F = [f_{ik}]$, where i=1, 2, ..., C (the number of clusters) and k=1, 2, ..., N. The objective function is:

$$J(U,V) = \sum_{i=1}^{C}\sum_{k=1}^{N} u_{ik}^m \|x_k - v_i\|^2 + \alpha \sum_{i=1}^{C}\sum_{k=1}^{N} (u_{ik} - f_{ik}\, b_k)^m \|x_k - v_i\|^2 \quad (1)$$

where U and V designate the partition matrix and prototypes respectively and the superscript m is the degree of fuzziness associated with the partition matrix ($m > 1$). This modified version of the FCM assumes that the number of classes is predetermined (coded in the matrix F), hence the number of clusters equals the number of classes. The parameter α is a scaling factor to maintain the balance between the supervised and unsupervised components of the objective function.

The approach suggested here overcomes the limitation observed in [8]. Its point of departure is to avoid the assumption that the number of clusters determined by the clustering algorithm should be the same as the number of classes reflected by data labels. In many real-world situations the available labels do not reflect the whole structure of the data. Hence, this work deals with such situations.

The paper is organized as follows. Section 2 introduces the details of our semi-supervised clustering algorithm. Section 3 discusses the analysis of the algorithm according to various aspects such as: the discrimination power, the behavior of the algorithm, and the relationship between clusters and classes. Section 4 concludes the paper.

2   Semi-supervised Clustering Model

The algorithm suggested here relies on FCM [3]. We will extend the objective function of FCM to capture the hidden and the visible data structures. The hidden data structure will be discovered using the FCM objective function as the first term of our objective function. The second term will take into account the visible data structure reflected by the available labels. Thus, the objective function becomes:

$$J(U,V) = \sum_{i=1}^{C} \sum_{k=1}^{N} u_{ik}^{m} \left\| x_{\mathrm{k}} - v_i \right\|^2 + \alpha \sum_{i=1}^{C} \sum_{k=1}^{N} (u_{ik} - \tilde{u}_{ik})^m \left\| x_{\mathrm{k}} - v_i \right\|^2 \quad (2)$$

such that:
$$\sum_{i=1}^{C} u_{ik} = 1 \ \forall \, k, \quad 0 < \sum_{k=1}^{N} u_{ik} < N \ \forall \, i \quad (3)$$

It is natural to assume that a class can be partitioned into several clusters. Let P be the number of classes (labels). Then $c \geq P$, i.e. each class j contains a number of clusters C_j, hence: $\sum_{j=1}^{P} Cj = C$. The terms $\tilde{u}_{ik}$ of the matrix $\tilde{U}$ are iteratively computed as follows:

$$\tilde{u}_{jk}^{(it)} = \tilde{u}_{jk}^{(it-1)} - \beta \frac{\partial Q(F, \tilde{U})}{\partial \tilde{u}_{jk}} \quad (4)$$

where:
$$Q(F, \tilde{U}) = \sum_{i=1}^{P} \sum_{k=1}^{N} \left(f_{ik} - \sum_{j \in \pi_i} \tilde{u}_{jk} \right)^2, \qquad \tilde{u}_{jk} \in [0,1] \quad (5)$$

In Eq. 5, $F = [f_{ik}]$ is a PXN binary matrix such that $f_{ik} = 1$ if pattern x_k belongs to class i, 0 otherwise. π_i is the set of clusters belonging to class i. The way we specify the set π_i will be discussed below. In Eq. 4, β (> 0) represents a learning rate. The minimization of Q (Eq. 5) which is function of:

$$\gamma_{ik} = f_{ik} - \sum_{j \in \pi_i} \tilde{u}_{jk} \quad (6)$$

aims at minimizing the difference between the assigned membership and the sum of all membership degrees for the same point with respect to all clusters involved within the same class. Eq. 4 optimizes the amounts $\tilde{u}_{ik}$ exploiting the computed difference and the learning rate β with the overall goal of reducing the difference between the actual membership grade u_{ik} of a pattern k to a cluster i and the evolving membership $\tilde{u}_{ik}$ expressed in the second term of Eq. 2. The resulting matrix $\tilde{U}$ is used to compute the second term in Eq. 2, that is the difference between U and $\tilde{U}$. The algorithm consists of two minimizations $Min\{J(U,V)\}$ and $Min\{Q(F, \tilde{U})\}$. Let us sketch the way the two performance indices interact:

1. Initialization: At the beginning of the clustering process, the matrix $\tilde{U}$(containing the terms $\tilde{u}_{ik}$) is initialized with the actual partition matrix U.

2. Optimization: Iterate the following two steps until local minima:
   (a) Optimize Eq. 4 using Eq. 5 (Optimization of $\tilde{U}$).
   (b) Minimize Eq. 2 using the output of step 2a.

For the sake of simplicity, we set the fuzziness degree m in Eq. 2 to 2. Then, to minimize Eq. 2 such that the condition in Eq. 3 is satisfied, we partially differentiate $J(U,V)$ with respect to partition matrix U and the prototypes v_i. Applying Lagrangian multiplier for each k=1, 2, ..., N, we have:

$$J(U,V,\lambda) = \sum_{i=1}^{C} u_{ik}^2 \|x_k - v_i\|^2 + \alpha \sum_{i=1}^{C} (u_{ik} - \tilde{u}_{ik})^2 \|x_k - v_i\|^2 - \lambda(\sum_{i=1}^{C} u_{ik} - 1)$$

By setting $\frac{\partial J(U,V,\lambda)}{\partial u_{st}} = 0$ for a given point t and a cluster s, we get:

$$2u_{st} \|x_t - v_s\|^2 + 2\alpha(u_{st} - \tilde{u}_{st}) \|x_t - v_s\|^2 - \lambda = 0, \quad \text{leading to:}$$

$$u_{st} = \frac{2\alpha\tilde{u}_{st} \|x_t - v_s\|^2 + \lambda}{2(1+\alpha) \|x_t - v_s\|^2} = \frac{\alpha\tilde{u}_{st}}{(1+\alpha)} + \frac{\lambda}{2(1+\alpha) \|x_t - v_s\|^2} \tag{7}$$

By setting $\frac{\partial J(U,V,\lambda)}{\partial \lambda_t} = 0$, we have $\sum_{i=1}^{C} u_{it} = 1$, leading to:

$$\lambda \sum_{i=1}^{C} \frac{1}{2(1+\alpha) \|x_t - v_i\|^2} + \frac{\alpha}{1+\alpha} \sum_{i=1}^{C} \tilde{u}_{it} = 1$$

$$\lambda = \frac{1 - \frac{\alpha}{1+\alpha} \sum_{i-1}^{C} \tilde{u}_{it}}{\sum_{i=1}^{C} \frac{1}{2(1+\alpha)\|x_t - v_i\|^2}}$$

Substituting the expression of λ in Eq. 7, we get:

$$u_{st} = \frac{\alpha\tilde{u}_{st}}{(1+\alpha)} + \frac{1 - \frac{\alpha}{(1+\alpha)} \sum_{i=1}^{C} \tilde{u}_{it}}{\sum_{i=1}^{C} \frac{1}{2(1+\alpha)\|x_t - v_i\|^2}} \cdot \frac{1}{2(1+\alpha) \|x_t - v_s\|^2}$$

Then, u_{st} becomes:

$$u_{st} = \frac{\alpha\tilde{u}_{st}}{(1+\alpha)} + \frac{1 - \frac{\alpha}{(1+\alpha)} \sum_{i=1}^{C} \tilde{u}_{it}}{\sum_{i=1}^{C} \frac{\|x_t - v_s\|^2}{\|x_t - v_i\|^2}} \tag{8}$$

Then by setting $\frac{\partial J(U,V,\lambda)}{\partial v_s} = 0$ (here 0 is the null vector), we will have:

$$-2 \sum_{k=1}^{N} u_{sk}^2 (x_k - v_s) - 2\alpha \sum_{k=1}^{N} (u_{sk} - \tilde{u}_{sk})^2 (x_k - v_s) = 0$$

Thus, v_s will be expressed as follows:

$$v_s = \frac{\sum\limits_{k=1}^{N} (u_{sk}^2 + \alpha(u_{sk} - \tilde{u}_{sk})^2)x_k}{\sum\limits_{k=1}^{N} (u_{sk}^2 + \alpha(u_{sk} - \tilde{u}_{sk})^2)} \tag{9}$$

Clearly Eqs.8 and 9 depend on the optimal value of $\tilde{u}_{st}$, which is obtained by differentiating Q in Eq. 5.

$$\frac{\partial Q(F, \tilde{U})}{\partial \tilde{u}_{st}} = -2\sum_{i=1}^{p} \left(f_{it} - \sum_{j \in \pi_i} \tilde{u}_{jt} \right) * \begin{cases} 1 & \text{if } s \in \pi_i \\ 0 & \text{otherwise} \end{cases}$$

Thus, the learning rule in Eq. 4 transforms into:

$$\tilde{u}_{st}^{(it)} = \tilde{u}_{st}^{(it-1)} + 2\beta \sum_{i=1}^{p} \left(f_{it} - \sum_{j \in \pi_i} \tilde{u}_{jt}^{(it-1)} \right) * \begin{cases} 1 & \text{if } s \in \pi_i \\ 0 & \text{otherwise} \end{cases} \tag{10}$$

The clustering process will involve the steps in Alg.1.The expression γ_{ik} in Eq. 6 which

---

Algorithm 1 :   The semi-supervised clustering algorithm

---

1. Apply the standard FCM on the whole data set (both labelled and unlabelled data points) to get the partition matrix $U^{(0)}$.
2. Determine the set π_i of each class using the notion of dominance explained below.
3. Compute the mapping matrix $M_{(PXC)}$ that relates classes to involved clusters:$M(i, j) = 1$ if cluster i is in class j, 0 otherwise.
4. Initialize $\tilde{U}^{(0)}$ with $U^{(0)}$ and set $it = 1$.
repeat
    repeat
        a. Compute $\tilde{U}^{((it)}$ using Eq. 10 (i.e., Minimize $Q(F, \tilde{U})$)
    until $\|\tilde{U}^{(it)} - \tilde{U}^{(it-1)}\| < \tau$ where τ is a small threshold
    repeat
        b. Compute $V^{(it)}$ using Eq. 9
        c. Compute $U^{(it)}$ using Eq. 8 (steps b. and c. correspond to minimizing $J(U, V)$)
    until $\|U^{(it)} - U^{(it-1)}\| < \epsilon$ where ϵ is a small threshold
    d. Compute the mapping matrix $M^{(it)}$
until $M^{(it)} = M^{(it-1)}$ or $it = MaxIter$

---

aims at minimizing the difference between the hard membership degree (0,1) of a pattern k to a class i and the sum of membership degrees of k to all clusters belonging to class i. By fixing the number of clusters for each class, we can test the desired combination of clusters and then find the optimal one. The desired combination is specified by means of π_i (see Eq. 10). It is a part of the initialization process. Now we need to know π_i, or more specifically the amount:

$$\psi_i = \sum_{j \in \pi_i} \tilde{u}_{jk} \tag{11}$$

To compute ψ_i, we use the mapping matrix M, and a corresponding matrix P that specifies the number of patterns from each class in each cluster. A row in P corresponds to a class index, and a column corresponds to a cluster index. A cell $P(i, j)$ represents the number of patterns of class i that appear in cluster j. A pattern k from class i belongs to cluster j if the membership degree of k to j is the highest. The membership degrees are provided in the partition matrix U. Thus, each class will be joined with a list of clusters, and the distribution of its patterns over this list of clusters. To force the algorithm to respect a given combination of clusters, we use the mapping matrix M, and the corresponding matrix P. After sorting P in an ascending order, each row of M contains the list of clusters ordered by dominance. Now we consider the number of clusters per class specified in the requested combination. The set π_i of ψ_i contains those dominant clusters in class i.

3 Analysis of the Algorithm

To analyze this algorithm, a synthetic data set is used. It is generated according to some statistical characteristics, namely a mean and a covariance (μ, Σ). Two classes K_1 consisting of one cluster and K_2 of two clusters are used in this experiment (see Fig 2). Each cluster consists of 100 data points.

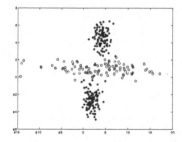

Figure 2. The synthetic data set (o: class 1, *: class 2)

As explained in Section 2, the algorithm consists of a two-step optimization process. The process of optimizing J, requiring the optimization of Q, is repeated iteratively for a certain number of iterations ($MaxIter$). For the sake of illustration, let $\alpha = 0.5$, $\beta = 0.06$ and $MaxIter = 20$ (the clustering process converges at most after 20 iterations). Fig. 3 illustrates the optimization of the performance indices. J and Q decrease as the clustering process progresses from one iteration to the next.

Then we will be interested in discriminating classes from each other. We will show how the scaling parameter α affects the linkage between clusters and classes in a way to produce homogeneous (or pure) clusters that contain data points from the same class only. To discuss the power of discrimination of the algorithm, we need to explain briefly the way we identify misclassified points in clusters. We determine them by looking at the shared clusters then finding the class that corresponds to the highest cardinality of

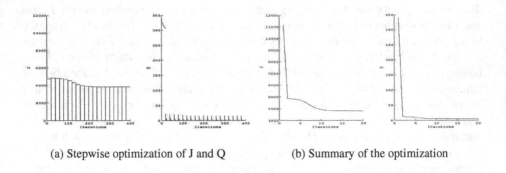

(a) Stepwise optimization of J and Q   (b) Summary of the optimization

Figure 3. Optimization of J and Q

the given cluster. All points that are with different labels (i.e., in different classes) in the same cluster are considered misclassified (or noisy). Now, if we cluster the data into 6 clusters (4 for K_1, and 2 for K_2), set the scaling factor α to various values: 0 (that is equivalent to the standard FCM), then to 0.3, 0.5, 0.7, 0.9, 1, and 2 respectively, and set $\beta = 0.06$ and $MaxIter = 20$, we obtain the results shown in Fig. 4.When $\alpha = 0$

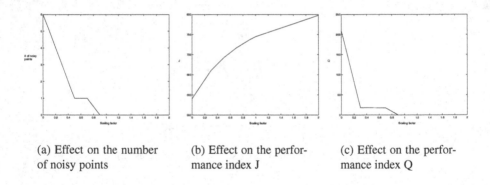

(a) Effect on the number   (b) Effect on the perfor-   (c) Effect on the perfor-
of noisy points   mance index J   mance index Q

Figure 4. Effect of the scaling factor

corresponding to FCM, 6 points are misclassified. As the scaling factor α increases, the number of misclassified points decreases until each of them is assigned to one of the clusters that belongs to the class having the same label. This means that the number of misclassified points is inversely proportional to α (see Fig. 4a). In other words, by assigning higher values to α, we are placing more confidence on the accuracy of the data labels. Furthermore, the first performance index J is proportional to that of α. The performance index J increases as α increases (Fig. 4b). Due to the fact that there is an external force (component 2 of Eq. 2), which aims at attracting patterns to clusters whose prototypes become more central to the data with the same label, the distance between natural centers (i.e., those generated by the standard FCM) and those generated by combining both supervised and unsupervised components of Eq. 2 increases (Fig. 4b). This force relates a pattern with a given label to a cluster that is not necessarily close to it. The algorithm tends to bring data points with the same label into the same container. Furthermore, the algorithm aims at discovering the real structure of the data. Non-adjacent clusters of the same class become labelled uniformly (with the same label). In contrast, the performance index Q decreases as the scaling factor α increases (Fig. 4c). As explained before, Q achieves its minimum when for all points k, the terms γ_{ik} (see Eq. 6) reach their minimum . The membership degree of each point of a given class should be maximally spread on the clusters included in the same class. In other words, each point joins a pure cluster inside a class. Hence, reducing the number of misclassified patterns implies a decrease of Q. On the other hand, we can use the expression ψ_i (Eq. 11 which is part of Eq. 5) to illustrate the evolution of clusters in terms of discriminating patterns of each class. In fact this expression aims at summing up the membership degree of a pattern k spread over all clusters j belonging to class i. A heterogeneous cluster is involved in the computation of ψ_i as many times as the number of distinct labels contained in that cluster, i.e., if a cluster contains two different labels, it is considered twice: for computing the ψ_i values of patterns of K_1 and the ψ_i values of patterns of K_2. Also one can use the matrix U in Eq. 11 instead of $\tilde{U}$, hence:

$$\zeta_i = \sum_{j \in \pi_i} u_{jk}.$$

Figure 5 illustrates the evolution of the overlap between classes during clustering. One can notice how the configuration of histograms changes as α increases. Figs. 5b and 5d ($\alpha = 0.9$) show that ψ_i and ζ_i values of patterns from K_1 are much higher than those of the same patterns with respect to K_2. Patterns of K_2 have higher values with respect to K_2 than with respect to K_1. Hence, patterns of classes can be easily distinguished when $\alpha = 0.9$, i.e., when we have pure clusters. As we increase α, the ψ_i and ζ_i values of patterns of a given class strengthen and those of patterns in other classes weaken. The evolution of clusters, for $\alpha = 0.9$, can be envisioned as shown in Fig. 6a. Furthermore, it is worth mentioning that the algorithm is able to follow a prespecified cluster combination. For example, Fig. 6b shows the contour plot for the combination 2-4 (2 clusters for K_1 and 4 for K_2). Such combination cannot be fulfilled using FCM due to the structure of this data. Besides the use of the performance index Q to find the mapping between classes represented by the matrix F and the clusters represented by the partition matrix U, we can also use a linear regression model to estimate the strength of the relationship between classes and clusters. The linear regression model can be expressed as: $F = A\tilde{U}$ where $F = [f_{ik}]$ is defined as explained earlier. The goal is then to determine the matrix $A(P, C)$ that represents the regression parameters, where P is the number of classes (rows) and C is the number of clusters (columns). A is mathematically derived and is expressed as:

$$A = \left(\sum_{k=1}^{N} F_k \tilde{u}_k^t \right) \left(\sum_{k=1}^{N} \tilde{u}_k \tilde{u}_k^t \right)^{-1} \tag{12}$$

As an illustration, let us use the two combinations (4-2) and (2-4) for two values of the scaling factor α. The results obtained are displayed in Tab.1. It is noticeable that the

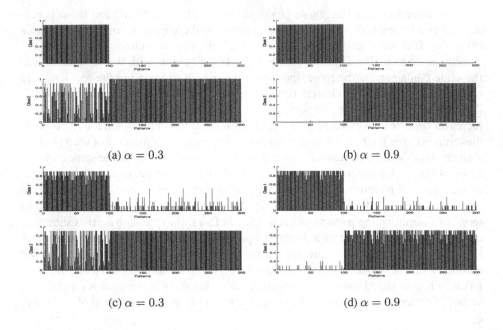

(a) $\alpha = 0.3$ (b) $\alpha = 0.9$

(c) $\alpha = 0.3$ (d) $\alpha = 0.9$

Figure 5. Effect of the scaling factor on the evolution of $\tilde{U}$ and U

first class has strong relationship with the first 4 clusters as provided in the prespecified combination. This relationship is reflected by relatively positive regression parameters. On the other hand, the first class is related to the last 2 clusters negatively. In contrast to this, the second class is negatively related to the first 4 clusters (of the first class) and positively to its corresponding 2 clusters. These observations also apply to the other combinations. We can thus conclude that the strength of the regression parameters typically correspond to the prespecified combination of clusters.

Table 1. Regression parameters

		Combination: (4-2)						Combination: (2-4)					
		C_1	C_2	C_3	C_4	C_5	C_6	C_1	C_2	C_3	C_4	C_5	C_6
$\alpha = 0.3$	K_1	1.0759	1.0843	1.0642	1.1007	-0.1401	-0.1297	1.3390	1.2981	-0.0398	-0.0493	0.0849	0.0981
	K_2	-0.0036	-0.0050	-0.0019	-0.0023	1.0006	1.0006	-0.3390	-0.2981	1.0398	1.0493	0.9151	0.9019
$\alpha = 1$	K_1	1.0000	0.9999	1.0000	1.0001	-0.0030	-0.0029	1.2344	1.2074	-0.0277	-0.0241	-0.0056	-0.0100
	K_2	-0.0025	-0.0034	-0.0012	-0.0015	1.0004	1.0004	-0.2344	-0.2074	1.0277	1.0241	1.0056	1.0100

4   Conclusion

We presented a new approach for performing a semi-supervised clustering. This approach exploits available knowledge about data. The experiments showed that the ap-

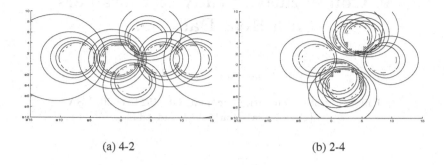

(a) 4-2   (b) 2-4

Figure 6. The evolution of clusters

proach performs very well on different aspects. Some of the aspects described here are the discrimination power, the behavior of the algorithm, and the application of a linear regression model to represent the relationship between clusters and classes. Other aspects such as the detection of mislabelled data, the classification performance and the way to find the optimal cluster combination are studied but not included in this paper.

References

[1] S. Basu, A. Banerjee, and R. Mooney. Semi-supervised clustering by seeding. *Proc. of the Int. Conf. on Machine Learning*, pages 19–26, 2002.
[2] A. Bensaid and J. Bezdek. Partial supervision based on point-prototype clustering algorithms. *The 4th Euro. Cong. on Intelligent Techniques and Soft Computing*, pages 1402–1406, 1996.
[3] J.C. Bezdek. *Pattern Recognition with Fuzzy Objective Function Algorithms*. Plenum, New York, 1981.
[4] A. Demiriz, K. Bennett, and M. Embrechts. Semi-supervised clustering using genetic algorithm. *Intelligent Engineering Systems Through ANN*, pages 809–814, 1999.
[5] B. Jeon and D. Landgrebe. Partially supervised classification using weighted unsupervised clustering. *IEEE Trans. on Geoscience and Remote Sensing*, 37(2):1073–1079, 1999.
[6] R. Klinkenberg. Using labeled and unlabeled data to learn drifting concepts. *Proc. of the Workshop on Learning from Temporal and Spatial Data*, pages 16–24, 2001.
[7] K. Nigam, A. McCallum, S. Thrun, and T. Mitchell. Text classification from labeled and unlabeled documents using em. *Machine Learning*, 39(2/3):103–134, 2000.
[8] W. Pedrycz and J. Waletzky. Fuzzy clustering with partial supervision. *IEEE Trans. on Systems Man and Cybernetics*, B 27(5):787–795, 1997.

A Comparative Study of Classifiers on a Real Data Set

Sofia Visa* and Anca Ralescu

University of Cincinnati, Cincinnati OH 45221-0030, USA
{svisa,aralescu}@ececs.uc.edu

Abstract. This paper investigates the performance of three different classifiers on a common real data set. A discussion about their advantages and limitations is also included. Supervised learning is applied to train the fuzzy sets based, neural network and minimum distance classifiers. Criteria considered as a basis for evaluating the classifiers performance, include the generalization power, the learning curve and ROC points.
keywords: classifier, fuzzy systems, neural networks, minimum distance classifier, ROC, confusion matrix.

1 Introduction

The problem consists in designing and implementing three different classifiers with the goal of comparing their results when applied to a particular data set. The data set consists of verbal descriptions of perceptions of the physical characteristics of a lifting task, whose detailed description can be found in [3]. Two hundred seventeen manual workers were required to imagine various lifting conditions (described verbally) on several aspects (variables) of the lifting task. Seven variables were used to describe the lifting task as follows: **Floor Weight** (FW), *Waist Weight* (W), **Horizontal Distance** (HD), *Twisting Angle* (TA), *Frequency* (F), *Work Duration* (WD), *Vertical Distance* (VD). Their values were assessed by each individual and labeled as *Small* (S), *Medium* (M) or *Big* (B) as illustrated in table 1. The classifier problem is then stated as follows: for each lifting task variable, given label-value correspondences on a subset of the data set, find a model that can capture this correspondence as accurately as possible and which can be used to predict the verbal category for a given numerical description of the input. In [1] a more complete study of the three classifiers on the all seven variables is presented; in this paper we analyzed only the variables with the least (FW) and highest (HD) overlap between classes. The characteristics of data set, high variability between samples within the same category and overlap between categories, pose serious challenges for designing the classifier. To reduce variability in this study, a rational assumption is introduced, to adjust the data as shown in equation (1).

---

* This work was partially supported by a Graduate Fellowship from Ohio Board of Regents.

T. Bilgiç et al. (Eds.): IFSA 2003, LNAI 2715, pp. 338–345, 2003.

$$adj([a,b],L) = \begin{cases} (m_L,b) & \text{if } L = S \\ (a,b) & \text{if } L = M \\ (a,M_L) & \text{if } L = B \end{cases} \tag{1}$$

where $m_L = min_L\{a;[a,b]\}$ and $M_L = max_L\{b;[a,b]\}$.

Table 1. $Between and within - label$ variability (HD)

Label	Value
S	30 - 40
	40 - 55
	28 - 32
M	40 - 65
	50 - 60
	35 - 45
B	33 - 45
	50 - 59
	48 - 63

Table 2. $Adjustment of intervals(HD)$: initial value(1), adjusted value(2)

Label	(1) → (2)
S	30 - 40 → 25 - 40
	40 - 55 → 25 - 55
	28 - 32 → 25 - 32
M	40 - 65 → 40 - 65
	50 - 60 → 50 - 60
	35 - 45 → 35 - 45
B	33 - 45 → 33 - 75
	50 - 59 → 50 - 75
	48 - 63 → 48 - 75

Table 2 illustrates the result of the adjustment for the data in Table 1. From this point onwards the adjusted data is refereed to as the *extended-data*.

2 The Three Classifiers

For all three classifiers the **hold-out** method was used, according to which, half of the data set, selected randomly, is used to model the classifier; the remaining data are used with each classification model for testing purposes.

2.1 Fuzzy System Classifier (FS)

Variability within and overlap of categories can be captured by a fuzzy model [2]. For the current problem the basic mechanism for converting a relative frequency distribution (discrete probability distribution) into fuzzy sets is used [4]. In the following, A denotes a discrete fuzzy set with membership function $\mu_A(x_i)$, $i = 1,...,n$, $\mu_{(k)}$ denotes the values of $\mu_A(x_k)$ arranged in nonincreasing order, and $f_{(k)}$ the frequency distribution on $\{x_1,...,x_n\}$ arranged in nonincreasing order. Then the relation between $\mu_{(k)}$ and $f_{(k)}$ is given by equation 2 [4].

$$\mu_{(k)} = kf_{(k)} + \sum_{i=k+1}^{n} f_{(i)}, \ for \ k = 1,...,n \tag{2}$$

where, $f_{(k)}$ and $\mu_{(k)}$ denote the kth largest value of the frequency distribution and membership function respectively. Figure 1 show the frequency distributions and the fuzzy sets for the variable FW and HD obtained from 2 based on their respective frequencies. The fuzzy sets obtained are used for classification of a given data point as follows. Given the value x, a point or interval value, the following steps are applied:

1. Match x to each label: Calculate the degrees $\mu_S(x)$, $\mu_M(x)$, $\mu_B(x)$.
2. Assign x to a label: Assign x to the label given by 3

$$pred\_l = arg\{max_{L\in\{S,M,B\}}\mu_L(x)\} \tag{3}$$

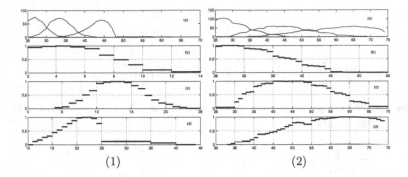

Fig. 1. Frequency distribution (a), and the corresponding fuzzy sets (b,c,d), for the variables FW (1) and HD (2)

2.2   Neural Networks (NN)

For this study a Feed Forward Neural Network(NN) was trained to capture the meaning of the linguistic labels for each variable. One hidden layer of ten neurons, sigmoidal activation function, back-propagation learning algorithm, training error of 0.1, two input values and one output were used. The input values correspond to the endpoints of the interval given as data to be assigned to a linguistic label. The output, corresponding to the class label, is encoded as -1 for **S**$mall$, 0 for **M**$edium$ and $+1$ for **B**ig respectively. After training, for each input data the NN outputs a value in the interval [-1, 1]. A particular test data, the endpoints of an interval, is classified to the label **S, M, B** according to whether the corresponding output is *close enough* to -1, 0, or +1 respectively implemented here by using a treshold γ such that if $NN_V(x)$ denotes the NN output for the variable V, at x, x is assigned the label L_0 if $|NN_V(x) - L_0| \leq \gamma$. Figure 2 shows the NN prediction results for the FW and HD variables. The stars are the NN label prediction (a number between -1 and 1) and the circles are the real class label. For variable HD it can be seen from these that the neural network model distinguishes better between the labels S and M than between M and B; this is due to the higher overlap between the classes M and B than between the classes S and M. Applying the NN to the extended data leads to better performance in classification. In figure 2 (c) and (d) it can be observed that the classes S and B are 'pulled' down and respectively up such that the overlap of the two classes with the class M is greatly reduced; now the classes are better defined. A threshold of 0.5 is used in assigning the labels.

2.3   Minimum Distance Classifier (MDC)

This is a template matching approach in which a prototype of the patterns to be recognized is available. An intuitive approach is to represent each class C_i by its mean (average) vector (prototype vector) p_i as shown in equation 4,

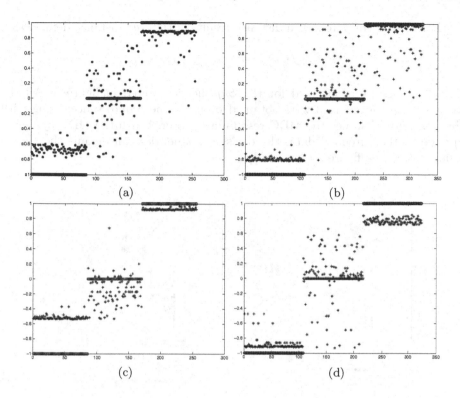

Fig. 2. Neural network classifier for the variables FW (a) and HD (b) and using the extended data (c) and (d), respectivelly

$$p^{(i)} = \frac{1}{\mid C_i \mid} \sum_{x \in C_i} x^{(j)} \tag{4}$$

where $\mid C_i \mid$ denotes the number of data points in class C_i. A data point x is then classified to the class i_0 given by equation 5.

$$x \in C_{i_0} \text{ where } i_0 = arg\{min\, d(x, p_i)\} \tag{5}$$

where d is a distance function defined (Minkowski) in 6.

$$d(i, k) = (\sum_{j=1}^{m} \mid x_{ij} - x_{kj} \mid^r)^{1/r} \text{ where } r \geq 1 \tag{6}$$

Two special cases of the Minkowski metric obtained for two values of r are used here, as follows:

1. For $r = 2$ the Minkowski distance reduces to the usual Euclidean distance

$$d_E(i, k) = (\sum_{j=1}^{d} \mid x_{ij} - x_{kj} \mid^2)^{1/2}$$

2. For $r \to \infty$ the Minkowski distance results in the *Sup* distance defined as

$$d_{Sup}(i, k) = max_{1 \leq j \leq d} \mid x_{ij} - x_{kj} \mid$$

Better results were obtained for the *Sup* metric showed in figure 3. All the four plots in the figure 3 were obtained by using the *Sup* distance; using with the Euclidean distance the MDC results are poorer. When the MDC system is applied to the extended data, the decision boundaries can delimit the classes much better (see figure 3 (c) and (d)).

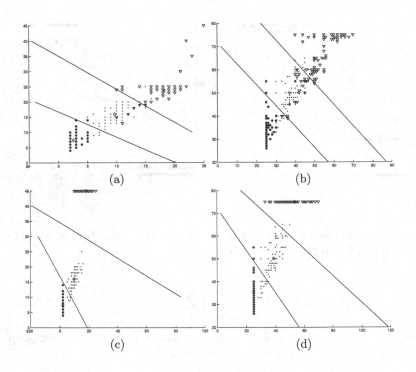

Fig. 3. Minimum distance classifier (Sup distance) for the variables FW (a) and HD (b) and using the extended data (c) and (d), respectivelly

3   Results

To track the errors of classification a simple error model is used:

$error(x) = \begin{cases} 0 \; pred\_l(x) = actual\_l(x) \\ 1 \; otherwise \end{cases}$ and $error(V) = [(\sum_{x \in V} error(x))/|V|] \times 100\%$

where V, is a collection of data points.

3.1   Generalization Power of the Three Classifiers

To train and test the designed classifiers 100 different data sets for each of the two variables were generated randomly. For comparison purpose it is important to train and test all the classifiers on *exactly* the same data points. An average error was calculated from all the data sets. The generalization power of the three

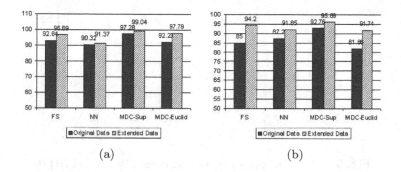

(a)   (b)

Fig. 4. Generalization power for the variables FW (a) and HD (b)

classifiers for the variables FW and HD is plotted in figure 4 along with the values. The results for extended data are significantly better for all the systems.

3.2   ROC (Receiver Operating Characteristic) Points

It is not adequate to characterize the performance of a classifier only by its global error rate. The *confusion matrix* shown in table 3 contains information about actual and predicted classification done by a classification system.

Table 3. The confusion matrix

		Predicted	
		Negative	Positive
Actual	Negative	a	b
	Positive	c	d

The *true positive rate (TP)/false positive rate (FP)* is the proportion of positive/negatives cases that were correctly/incorrectly identified. That is: $TP = \frac{d}{c+d}$, $FP = \frac{b}{a+b}$. A point on the ROC graph is a plot with the false positive rate, FP, on the x-axis and the true positive rate, TP, on the y-axis. The point $(0,1)$ is the perfect classifier: it classifies all positive cases and negative cases correctly; $(0,0)$ represents a classifier that predicts all cases to be negative, while the point $(1,1)$ corresponds to a classifier that predicts every case to be positive and $(1,0)$ is the classifier that is incorrect for all classifications. In many cases, a classifier has a parameter that can be adjusted to increase TP at the cost of an increased FP or decrease FP at the cost of a decrease in TP. Each parameter setting provides a (FP, TP) pairs and a series of such pair can be used to plot an ROC curve. A non-parametric classifier is represented by a single ROC point, corresponding to its (FP, TP) pair. As it can be observed in figure 5 all the classifiers are admis-

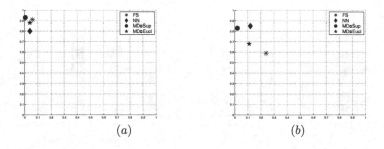

(a) (b)

Fig. 5. The ROC points for the variables FW (a) and HD (b)

sible, in the sense that they corresponding ROC points are in the desired area (above the first diagonal as close as possible to the point $(0, 1)$). To obtain the ROC points, the class M (which overlappes with both the classes S and B) was considered as the positive class and the classes S and B as the negative class. This shows how well the systems recognize the class M over the other two.

3.3 Learning Curves

The learning curve is a plot of classifier predictive power when different size training data sets are used. Given a data set with N elements, $p\%$ of N, $p = 5, 10, 20, \cdots, 90, 95$ are used for training with the remaining $(100 - p)\%$ used for testing. A smooth learning curve with a rapidly increasing slope is preferable. As it can be seen in figure 6, the classifiers analyzed have good learning curves. Even using a data set as small as 2% and test in remaining 98% of data we have generalization power between 78% and 96%. Overall, the MDC with the *Sup* metric has overall best performance followed by the NN and FS; the learning curves for the FS and NN raise sharply when the training set exceeds 5% of the data set for variable HD and behave constant after 20%; for class FW starting with a training set as small as 5% up to a training set of size 80% the learning curves tend to be somewhat flat. This behavior is not surprising: the classes are not much overlapped and are learned by the systems with very few data points; for some classifiers, using a training set of 90% or higher the generalization decreases: the systems overfitts the training data.

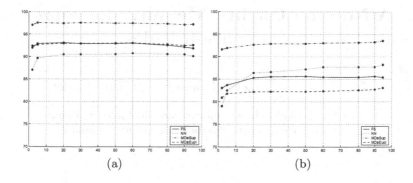

Fig. 6. The learning curves for the variables FW (a) and HD (b)

4 Conclusions

The three approaches considered give similar results in so far as the error rates are concerned, with slightly better results for MDC in conjunction with *Sup* metric. The choice of one of the models over the other may, finally, depend on several factors, including the goal of the modeling, further use of the results, etc. For example, if the user of the model is also interested in being able to differentiate between instances of the same class, then the fuzzy model should be used. If this aspect is not an issue then, perhaps the MDC with *Sup* metric might be selected. If complexity of learning is important, then FS and MDC each reaches a solution in one pass of the data, whereas to train the NN takes thousands of epochs. For the current data the NN system often did not converge for data sets with many contradictory data points. However FS and MDC always produce a model.

References

1. Visa S.: Comparative study of methods for linguistic modeling of numerical data. Master thesis, University of Cincinnati (2002)
2. Zadeh L.: Fuzzy Sets. Information Control **8** (1965) 338–353
3. Yeung S., Genaidy A., Karwowski W., Huston R., Beltran J.: Assessment of manual lifting activities using worker expertise: A comparison of two workers population. Asian Journal of Ergonomics **2** (2001) 11–24
4. Inoue A., Ralescu A.: Generation of Mass Assignment with Nested Focal Elements. Proc. of 18th International Conference of the North American Fuzzy Information Processing Society (NAFIPS-99) (1999) 208–212

A Note on Quality Measures for Fuzzy Association Rules

Didier Dubois[1], Eyke Hüllermeier[2], and Henri Prade[1]

[1] Institut de Recherche en Informatique de Toulouse, France
[2] Informatics Institute, Marburg University, Germany

Abstract. Several approaches generalizing association rules to fuzzy association rules have been proposed so far. While the formal specification of fuzzy associations is more or less straightforward, the evaluation of such rules by means of appropriate quality measures assumes an understanding of the semantic meaning of a fuzzy rule. In this respect, most existing proposals can be considered ad-hoc to some extent. In this paper, we suggest a theoretical basis of fuzzy association rules by generalizing the classification of the data stored in a database into positive, negative, and irrelevant examples of a rule.

1 Introduction

Association rules provide a means for representing dependencies between attributes in databases. Typically, an association involves two sets, A and B, of so-called items (binary features). Then, the intended meaning of a (binary) rule $A \rightharpoonup B$ is that a data record stored in the database that contains the set of items A is likely to contain the items B as well. Example: If a data record is a purchase, the association $\{paper, envelopes\} \rightharpoonup \{stamps\}$ suggests that a purchase containing paper and envelopes is likely to contain stamps as well. Several efficient algorithms for mining association rules in large databases have already been devised [1, 18].

A generalization of binary association rules is motivated by the fact that a database is usually not restricted to binary attributes but also contains attributes with values ranging on (completely) ordered scales, such as cardinal or ordinal attributes. In *quantitative association rules*, attribute values are specified by means of subsets which are typically intervals. Example: "Employees at the age of 30 to 40 have incomes between \$50,000 and \$70,000".

The use of fuzzy sets in connection with association rules – as with data mining in general [17] – has recently been motivated by several authors (e.g. [3, 6, 7, 16]). Moving from set-based (interval-based) to fuzzy associations is formally accomplished by replacing sets (intervals) by fuzzy sets (fuzzy intervals). Still, the evaluation of fuzzy associations through appropriate quality measures, notably the well-known support and confidence measures, is more intricate. Especially, it assumes an understanding of the semantics of a fuzzy rule. In this respect, many existing proposals can be considered ad-hoc to some extent. Here, we suggest a

T. Bilgiç et al. (Eds.): IFSA 2003, LNAI 2715, pp. 346–353, 2003.

theoretical justification of existing measures by generalizing the classification of stored data into positive, negative, and irrelevant examples of a rule.

By way of background, Section 2 reviews classical association rules, and Section 3 gives a brief overview of existing approaches to fuzzy associations. The idea of basing the support and confidence of a fuzzy association on a fuzzy partition of examples is presented in Section 4.

2   Association Rules

Consider a set $\mathcal{A} = \{a_1, \ldots, a_m\}$ of items, and let a transaction (data record) be a subset $T \subseteq \mathcal{A}$. Let $D_X \doteq \{T \in D \mid X \subseteq T\}$ denote the transactions in the database D that contain the items $X \subseteq \mathcal{A}$; the cardinality of this set is $|D_X| = \mathrm{card}(D_X)$. In order to find "interesting" association rules in a database D, a potential rule $A \rightharpoonup B$ is generally rated according to several criteria. For each criterion an appropriate measure is defined, and none of these measures must fall below a certain (user-defined) threshold. In common use are the following measures: A measure of *support* defines the number of transactions in D that contain both A and B:

$$\mathsf{supp}(A \rightharpoonup B) \doteq |D_{A \cup B}|. \tag{1}$$

Support can also be defined by the proportion rather than the absolute number of transactions, in which case (1) is divided by $|D|$. The *confidence* is the proportion of correct applications of the rule:

$$\mathsf{conf}(A \rightharpoonup B) \doteq \frac{|D_{A \cup B}|}{|D_A|}. \tag{2}$$

Further reasonable measures can be considered such as, e.g., the deviation (significance) $\mathsf{int}(A \rightharpoonup B) \doteq |D_{A \cup B}| \cdot |D_A|^{-1} - |D_B| \cdot |D|^{-1}$, expressing that $A \rightharpoonup B$ is interesting only if the occurrence of A does indeed have a positive influence on the occurrence of B. As can be seen, the support measure plays a central role. In fact, all other measures can generally be derived from the support. For example, the confidence of an association $A \rightharpoonup B$ is the support of that association divided by the support of its antecedent, A.

Rather than looking at a transaction T as a subset of items, it can also be seen as a sequence $(x_1, \ldots, x_m)$ of values of binary variables X_i with domain $\mathfrak{D}_{X_i} = \{0, 1\}$, where $x_i = 1$ if the ith item, a_i, is contained in T and $x_i = 0$ otherwise.

Now, let X and Y be quantitative attributes (such as age or income) with completely ordered domains $\mathfrak{D}_X$ and $\mathfrak{D}_Y$, respectively. Without loss of generality we can assume that $\mathfrak{D}_X, \mathfrak{D}_Y \subseteq \mathfrak{R}$, where $\mathfrak{R}$ denotes the set of real numbers. Let x_T and y_T denote, respectively, the values that X and Y take for the transaction T. A quantitative, interval-based association rule $A \rightharpoonup B$ involving the variables X and Y is then of the following form:

$$\text{If } X \in A = [x_1, x_2] \text{ then } Y \in B = [y_1, y_2], \tag{3}$$

where $x_1, x_2 \in \mathfrak{D}_X$ and $y_1, y_2 \in \mathfrak{D}_Y$. This approach can simply be generalized to the case where X and Y are multi-dimensional variables and, hence, A and B hyper-rectangles rather than intervals. Subsequently, we proceed from fixed variables X and Y, and consider the database D as a collection of data points $(x, y) = (x_T, y_T)$, i.e. as a projection of the original database to $\mathfrak{D}_X \times \mathfrak{D}_Y$.

Note that the above quality measures are applicable in the quantitative case as well:

$$\mathsf{supp}(A \rightharpoonup B) = \mathrm{card}\left(\{(x, y) \in D \mid x \in A \wedge y \in B\}\right),$$

$$\mathsf{conf}(A \rightharpoonup B) = \frac{\mathrm{card}\left(\{(x, y) \in D \mid x \in A \wedge y \in B\}\right)}{\mathrm{card}\left(\{(x, y) \in D \mid x \in A\}\right)}.$$

In fact, each interval $A = [x_1, x_2]$ does again define a binary attribute $X_A(x)$ defined by $X_A(x) = 1$ if $x \in A$ and 0 otherwise. In other words, each quantitative attribute X is replaced by k binary attributes X_{A_i} such that $\mathfrak{D}_X \subseteq \bigcup_{i=1}^{k} A_i$.

3   Fuzzy Association Rules

Replacing the sets (intervals) A and B in (3) by fuzzy sets (intervals) leads to fuzzy (quantitative) association rules. Thus, a fuzzy association rule is understood as a rule of the form $A \rightharpoonup B$, where A and B are now fuzzy subsets rather than crisp subsets of the domains $\mathfrak{D}_X$ and $\mathfrak{D}_Y$ of variables X and Y, respectively. In other words, a variable X is now replaced by a number of fuzzy attributes rather than by a number of binary attributes.

The standard approach to generalizing the quality measures for fuzzy association rules is to replace set-theoretic operations, namely Cartesian product and cardinality, by corresponding fuzzy set-theoretic operations:

$$\mathsf{supp}(A \rightharpoonup B) \doteq \sum_{(x,y) \in D} A(x) \otimes B(y), \tag{4}$$

$$\mathsf{conf}(A \rightharpoonup B) \doteq \frac{\sum_{(x,y) \in D} A(x) \otimes B(y)}{\sum_{(x,y) \in D} A(x)}, \tag{5}$$

where $\otimes$ is a t-norm; the usual choice is $\otimes = \min$. Note that the support of $A \rightharpoonup B$ can be expressed by the sum of the *individual supports*, provided by tuples $(x, y) \in D$:

$$\mathsf{supp}_{[x,y]}(A \rightharpoonup B) = A(x) \otimes B(y). \tag{6}$$

According to (6), the tuple (x, y) supports $A \rightharpoonup B$ if both, $x \in A$ and $y \in B$.

Let us mention the possibility of measuring the frequency (support) of a fuzzy itemset $A \cup B$ by a *fuzzy cardinality*, i.e. a fuzzy number, rather than by a single number [4, 9].

The support measure (6) is obviously in line with the conjunction-based approach to modeling fuzzy rules, well-known from Mamdani-like fuzzy control systems. Taking into account the asymmetric nature of a rule, the use of

implication-based fuzzy rules and, hence, of implication operators in place of
conjunctions for the modeling of associations has been proposed by some authors [5, 6, 13]. For example, the following type of measure was suggested in
[13]:

$$\mathsf{supp}_{[x,y]}(A \rightharpoonup B) = A(x) \otimes \big(A(x) \rightsquigarrow B(y)\big). \tag{7}$$

As one advantage of taking the implicative nature of a rule into account, note
that (7) avoids the following questionable property of (4) and (5): Suppose that
attribute A is *perfectly associated* with attribute B, which means that $A(x) =
B(y)$ for all tuples $(x, y) \in D$. Thus, one may find it natural that $A \rightharpoonup B$ has
full confidence. Yet, since $\alpha \otimes \alpha < \alpha$ if $\otimes$ is not idempotent (i.e. $\otimes \neq \min$), (5)
usually yields $\mathsf{conf}(A \rightharpoonup B) < 1$ [14]. (Note that (7) is equivalent to (6) with
minimum t-norm if $\otimes$ is continuous and $\rightsquigarrow$ is the R-implication induced by $\otimes$.)

4 Fuzzy Partitions of Examples

The key idea of the approach as outlined in this section is to provide a sound basis
of fuzzy association rules by generalizing the classification of data into *positive*,
negative, and *irrelevant* examples of a rule. In fact, an set-based association rule
$A \rightharpoonup B$ partitions the database into three types of transactions, namely *positive
examples* $\mathcal{S}_+$ that verify the rule, *negative examples* $\mathcal{S}_-$ that falsify the rule, and
irrelevant examples $\mathcal{S}_\pm$:

$$\mathcal{S}_+ \doteq \{(x,y) \mid x \in A \wedge y \in B\} \tag{8}$$
$$\mathcal{S}_- \doteq \{(x,y) \mid x \in A \wedge y \notin B\} \tag{9}$$
$$\mathcal{S}_\pm \doteq \{(x,y) \mid x \notin A\} \tag{10}$$

The most important quality measures for association rules (support and confidence) are expressed in a natural way in terms of the cardinality of the above
sets. Namely, the support is the number of positive examples, and the confidence
is the number of positive over the number of relevant examples:

$$\mathsf{supp}(A \rightharpoonup B) \doteq |\mathcal{S}_+|, \quad \mathsf{conf}(A \rightharpoonup B) \doteq |\mathcal{S}_+| \cdot \big(|\mathcal{S}_+| + |\mathcal{S}_-|\big)^{-1}$$

The basic question in connection with fuzzy association rules now concerns the
generalization of the partition (8–10). Clearly, if A and B are fuzzy sets rather
than ordinary sets, then $\mathcal{S}_+, \mathcal{S}_-$, and $\mathcal{S}_\pm$ will be fuzzy sets as well. In other words,
a point (x, y) can be a positive (negative) example to some degree, and may also
be irrelevant to some extent. We denote by $\mathcal{S}_+(x, y)$ the degree of membership
of the point (x, y) in the fuzzy set $\mathcal{S}_+$ of positive examples and employ the same
notation for $\mathcal{S}_-$ and $\mathcal{S}_\pm$.

There are different ways to proceed since the logical specification of positive
and negative examples is not unique. In fact, the logical specification of irrelevant
examples via $(x, y) \in \mathcal{S}_\pm \Leftrightarrow \neg\,(x \in A)$ is actually clear, but there are different
options to characterize $\mathcal{S}_+$ and $\mathcal{S}_-$. A straightforward possibility is of course

$$\begin{aligned}
(x,y) \in \mathcal{S}_+ &\doteq (x \in A) \wedge (y \in B), \\
(x,y) \in \mathcal{S}_- &\doteq (x \in A) \wedge \neg(y \in B).
\end{aligned} \tag{11}$$

Still, a viable alternative could be

$$
\begin{aligned}
(x,y) \in \mathcal{S}_+ &\doteq &(x \in A) \wedge (y \in B),\\
(x,y) \in \mathcal{S}_- &\doteq &\neg((x \in A) \Rightarrow (y \in B)),
\end{aligned}
\tag{12}
$$

where $\Rightarrow$ is the standard logical (material) implication. Moreover, referring to (7), one could think of

$$
\begin{aligned}
(x,y) \in \mathcal{S}_+ &\doteq (x \in A) \wedge ((x \in A) \Rightarrow (y \in B)),\\
(x,y) \in \mathcal{S}_- &\doteq (x \in A) \wedge \neg((x \in A) \Rightarrow (y \in B)).
\end{aligned}
\tag{13}
$$

When taking (11) and the standard negation $\alpha \mapsto 1 - \alpha$ as a point of departure, our problem can be specified as follows: Find a generalized conjunction (t-norm) $\otimes$ such that

$$
\mathcal{S}_+(x,y) + \mathcal{S}_-(x,y) + \mathcal{S}_\pm(x,y) = 1
\tag{14}
$$

holds for all $(x,y) \in \mathfrak{D}_X \times \mathfrak{D}_Y$, where

$$
\begin{aligned}
\mathcal{S}_+(x,y) &\doteq A(x) \otimes B(y)\\
\mathcal{S}_-(x,y) &\doteq A(x) \otimes (1 - B(y))\\
\mathcal{S}_\pm(x,y) &\doteq 1 - A(x)
\end{aligned}
\tag{15}
$$

From ALSINA's results in [2] it follows that the only t-norm solving this problem is the product. In fact, ALSINA even solves a somewhat more general problem, seeking solutions $(\otimes, \oplus, n)$ to the functional equation

$$
(\alpha \otimes \beta) \oplus (\alpha \otimes n(\beta)) = \alpha
$$

for all $0 \le \alpha, \beta \le 1$, where $\oplus$ is a t-conorm and $n(\cdot)$ a negation. However, assuming addition (that is the Lukasiewicz t-conorm) as a generalized conjunction is clearly reasonable in our context of data mining, where we are basically interested in generalizing frequency information. Particularly, (14) guarantees that $|\mathcal{S}_+| + |\mathcal{S}_-| + |\mathcal{S}_\pm| = |D|$, which is clearly a reasonable property.

It should be noted that questions of similar type have also been studied, e.g., in fuzzy preference modeling, where the problem is to decompose a weak (valued) preference relation into three parts: strict preference, indifference, and incompatibility [10].

When taking (12) rather than (11) as a point of departure, the problem is to find a generalized conjunction (t-norm) $\otimes$ and a generalized implication operator $\rightsquigarrow$ such that (14) holds with

$$
\begin{aligned}
\mathcal{S}_+(x,y) &\doteq A(x) \otimes B(y)\\
\mathcal{S}_-(x,y) &\doteq 1 - (A(x) \rightsquigarrow B(y))\\
\mathcal{S}_\pm(x,y) &\doteq 1 - A(x)
\end{aligned}
\tag{16}
$$

Note that (14) in conjunction with (16) implies

$$
\alpha \rightsquigarrow \beta = (1 - \alpha) + (\alpha \otimes \beta)
\tag{17}
$$

for all $0 \leq \alpha, \beta \leq 1$ and, hence, suggests a definition of the implication $\leadsto$ in terms of the conjunction $\otimes$. In fact, (17) defines the QL-implication with t-conorm $(\alpha, \beta) \mapsto \min\{1, \alpha + \beta\}$ as a disjunction $(0 \leq (1 - \alpha) + (\alpha \otimes \beta) \leq 1$ always holds since $\alpha \otimes \beta \leq \alpha$ for any t-norm $\otimes$). Here are some examples of standard conjunctions $\otimes$ together with induced implications:

$\otimes$	$\leadsto$
$\min\{\alpha, \beta\}$	$\min\{1, 1 - \alpha + \beta\}$
$\alpha\beta$	$1 - \alpha(1 - \beta)$
$\max\{\alpha + \beta - 1, 0\}$	$\max\{1 - \alpha, \beta\}$

The question concerning the operators $\otimes$ and $\leadsto$ that can be chosen in (16) can be stated as follows: For which t-norms $\otimes$ does (17) define a proper implication operator? Note that the boundary conditions $\alpha \leadsto 1 = 1$ and $0 \leadsto \beta = 1$ do hold for all $0 \leq \alpha, \beta \leq 1$. Apart from that, (17) is obviously increasing in β. Thus, as a major point it remains to guarantee comonotonicity in α. (Of course, apart from that further properties of $\leadsto$ might be required.)

First of all, let us show that indeed not all t-norms are admissible, i.e. there are t-norms $\otimes$ for which (17) is not monotone decreasing in α. In fact, a simple counter-example is the (weakly) drastic product $(\alpha \otimes \beta \doteq \min\{\alpha, \beta\}$ if $\max\{\alpha, \beta\} = 1$ and 0 otherwise), for which (17) becomes

$$\alpha \leadsto \beta = \begin{cases} 1 & \text{if } \beta = 1 \\ \beta & \text{if } \alpha = 1 \\ 1 - \alpha & \text{if } \alpha < 1 \end{cases}.$$

Besides, there are even continuous t-norms that violate the above monotonicity condition. For instance, consider the Hamacher family [12] of t-norms:

$$\alpha \otimes_\gamma \beta \doteq \frac{\alpha\beta}{\gamma + (1 - \gamma)(\alpha + \beta - \alpha\beta)}, \tag{18}$$

where γ is a non-negative parameter. With $\gamma = 10$, (17) yields $0.9 \leadsto 0.5 \approx 0.41 < 0.5 = 1 \leadsto 0.5$. Similar counter-examples can also be constructed for the families of t-norms introduced by YAGER, SCHWEIZER-SKLAR, and DOMBI [15].

Note that the comonotonicity condition

$$(\alpha \leq \alpha') \Rightarrow 1 - \alpha + (\alpha \otimes \beta) \geq 1 - \alpha' + (\alpha' \otimes \beta)$$

is equivalent to

$$(\alpha \leq \alpha') \Rightarrow (\alpha' \otimes \beta) - (\alpha \otimes \beta) \leq \alpha' - \alpha. \tag{19}$$

Thus, it follows that a t-norm $\otimes$ is admissible in (17) if it is a so-called *copula*. In fact, the following result is stated as a theorem in [19]: A t-norm $\otimes$ is a copula iff (19) holds. A related result concerns continuous Archimedean t-norms in particular and shows that such t-norms are admissible if and only if their additive generator is convex.

For many parameterized families of t-norms, the latter result makes it easy to check whether or not a parameter is admissible. For instance, $\gamma \leq 1$ is necessary for the Hamacher family (18).

As a direct consequence of the above results one can prove

Proposition 1: The Lukasiewicz t-norm $\otimes_L : (\alpha, \beta) \mapsto \max\{\alpha + \beta - 1, 0\}$ is the smallest t-norm admissible in (17).

Proposition 2: For the family of Frank t-norms [11], parameterized through $\rho > 0$ according to

$$\otimes_\rho : (\alpha, \beta) \mapsto \begin{cases} \min(\alpha, \beta) & \text{if } \rho = 0 \\ \alpha\beta & \text{if } \rho = 1 \\ \max\{0, 1 - \alpha + \beta\} & \text{if } \rho = \infty \\ \ln_\rho\left(1 + \frac{(\rho^\alpha - 1)(\rho^\beta - 1)}{\rho - 1}\right) & \text{otherwise} \end{cases},$$

(17) is always monotone decreasing in α.

A further interesting result concerns the possibility of combining admissible t-norms into new admissible t-norms.

Proposition 3: The ordinal sum of admissible t-norms is again admissible.

Corollary 4: Each element of the family of t-norms

$$\otimes_\gamma : (\alpha, \beta) \mapsto \frac{\alpha\beta}{\max\{\alpha, \beta, \gamma\}}, \qquad 0 < \gamma \leq 1 \tag{20}$$

introduced by DUBOIS and PRADE [8], is admissible in (17).

Finally, reconsider model (13). In conjunction with (14), we obtain

$$(\alpha \otimes \beta) + (\alpha \otimes \neg\beta) = \alpha$$

with $\alpha = A(x)$ and $\beta = A(x) \rightsquigarrow B(y)$. Thus, we can again refer to ALSINA's result, showing that $\otimes$ should be the product. Apart from that, any implication operator can be used. It should be noted, however, that some implications are unacceptable when imposing further requirements. For example, the reasonable property that $\mathcal{S}_+(x, y)$ is upper-bounded by $B(y)$, especially $\mathcal{S}_+(x, y) = 0$ if $B(y) = 0$, is not satisfied by all operators.

5   Concluding Remarks

The approach outlined in this paper justifies the use of certain fuzzy logical operators in connection with different types of support measures for fuzzy itemsets. Particularly, the t-norm generalizing the logical conjunction should be either the product (and not, as usually, the minimum!) or a so-called copula, depending on how positive and negative negative examples are specified.

Since the membership of a tuple (x, y) in the fuzzy set of *relevant* examples (the complement of $\mathcal{S}_\pm$) is $A(x)$ in any case, our approach also justifies the standard confidence measure (5).

Our results might appear not fully satisfactory since they still permit a rather large class of support measures. Restricting this class further by assuming additional properties is hence an important topic of ongoing research.

References

[1] R. Agrawal and R. Srikant. Fast algorithms for mining association rules. In *Proceedings of the 20th Conference on* VLDB, Santiago, Chile, 1994.

[2] C. Alsina. On a family of connectives for fuzzy sets. *Fuzzy Sets and Systems*, 16:231–235, 1985.

[3] Wai-Ho Au and K.C.C. Chan. An effective algorithm for discovering fuzzy rules in relational databases. In *Proceedings* IEEE *World Congress on Computational Intelligence*, pages 1314 –1319, 1998.

[4] P. Bosc, D. Dubois, O. Pivert, and H. Prade. On fuzzy association rules based on fuzzy cardinalities. In *Proc. IEEE Int. Fuzzy Systems Conference*, Melbourne, 2001.

[5] P. Bosc and O. Pivert. On some fuzzy extensions of association rules. In *Proc. IFSA/NAFIPS-2001*, Vancouver, Canada, 2001.

[6] G. Chen, Q. Wei, and E.E. Kerre. Fuzzy data mining: Discovery of fuzzy generalized association rules. In G. Bordogna and G. Pasi, editors, *Recent Issues on Fuzzy Databases*. Springer-Verlag, 2000.

[7] M. Delgado, D. Sanchez, and M.A. Vila. Acquisition of fuzzy association rules from medical data. In S. Barro and R. Marin, editors, *Fuzzy Logic in Medicine*. Physica Verlag, 2000.

[8] D. Dubois and H. Prade. *Fuzzy Sets and Systems: Theory and Applications*. Academic Press, New York, 1980.

[9] D. Dubois and H. Prade. Fuzzy sets in data summaries – outline of a new approach. In *Proceedings* IPMU-2000, pages 1035–1040, Madrid, Spain, 2000.

[10] J. Fodor and M. Roubens. *Fuzzy Preference Modelling and Multicriteria Decision Support*. Kluwer, 1994.

[11] M.J. Frank. On the simulataneous associativity of $f(x,y)$ and $x + y - f(x,y)$. *Aeq. Math.*, 19:194–226, 1979.

[12] H. Hamacher. *Über logische Aggregationen nichtbinär explizierter Entscheidungskriterien; Ein axiomatischer Beitrag zur normativen Entscheidungstheorie.* R.G. Fischer Verlag, 1978.

[13] E. Hüllermeier. Implication-based fuzzy association rules. In *Proceedings* PKDD–01, pages 241–252, Freiburg, Germany, September 2001.

[14] E. Hüllermeier and J. Beringer. Mining implication-based fuzzy association rules in databases. In B. Bouchon-Meunier, L. Foulloy, and R.R. Yager, editors, *Intelligent Systems for Information Processing: From Representation to Applications*. Elsevier, 2003. To appear.

[15] G.J. Klir and B. Yuan. *Fuzzy Sets and Fuzzy Logic – Theory ad Applications*. Prentice Hall, 1995.

[16] C. Man Kuok, A. Fu, and M. Hon Wong. Mining fuzzy association rules in databases. *SIGMOD Record*, 27:41–46, 1998.

[17] W. Pedrycz. Data mining and fuzzy modeling. In *Proc. of the Biennial Conference of the NAFIPS*, pages 263–267, Berkeley, CA, 1996.

[18] A. Savasere, E. Omiecinski, and S. Navathe. An efficient algorithm for mining association rules in large databases. In VLDB–95, Zurich, 1995.

[19] B. Schweizer and A. Sklar. *Probabilistic Metric Spaces*. North Holland, 1983.

Differentiated Treatment of Missing Values in Fuzzy Clustering

Heiko Timm, Christian Döring, and Rudolf Kruse

Dept. of Knowledge Processing and Language Engineering
Otto-von-Guericke-University of Magdeburg
Universitätsplatz 2, D-39106 Magdeburg, Germany,
{timm,doering,kruse}@iws.cs.uni-magdeburg.de

Abstract. Partially missing datasets are a prevailing problem in data analysis. Since several reasons for missing attribute values can be distinguished, we suggest a differentiated treatment of this common problem. For datasets, in which feature values are missing completely at random, a variety of approaches has been proposed. In other situations, however, the fact that values are missing provides additional information for the classification of the dataset. Since the known approaches cannot exploit this information, we developed an extension of the Gath and Geva algorithm that can utilize it. We introduce a class specific probability for missing values in order to appropriately assign incomplete data points to clusters. Benchmark datasets are used to demonstrate the capability of the presented approach.

Keywords: fuzzy cluster analysis, missing values, class specific probability

1 Introduction

If data analysis methods are applied to practical problems, we often find that datasets contain many missing data elements. A dataset has partially missing data if some attribute values of a feature vector x_j are not observed. An example for an incomplete feature vector is $x_j = (x_{j1}, x_{j2}, ?, x_{j4}, ?)$, which has missing values in the third and fifth attribute. Only the first, second and third attribute are observed. Feature values can be missing for several reasons. Often they are the result of problems or failure in the data collection method. Malfunctioning sensors may collect some but not all features of an event. Questions on a questionnaire remain unanswered, because a person missed the question or refused to answer. Human operators may collect only as many features as they have time for.

Clustering is a technique for classifying data, i.e., to divide a given dataset into a set of classes or *clusters* [2,3,7]. The goal is to assign data points to clusters in such a way that two feature vectors from the same cluster are as similar as possible and two feature vectors from different clusters are as dissimilar as possible. Most approaches for dealing with partially missing datasets in Fuzzy Clustering assume that missing values are *missing completely at random* (MCAR). Such

T. Bilgiç et al. (Eds.): IFSA 2003, LNAI 2715, pp. 354–361, 2003.

missing values behave like a random sample and their probability does not depend on the observed data or the unobserved data [8,9]. Thus, missing values (MCAR) are interpreted as a random reduction of the dataset, which provide no further information for assigning incomplete feature vectors to clusters. However, depending on how the data were collected, the occurrence of a missing value can give a hint which class the incomplete feature vector might belong to. For example, on medical reports some attributes can be left blank, because they are inappropriate for some class of illnesses. Here the person collecting the data felt, that it is unnecessary to record the values of certain attributes. The same kind of missing values can be found as unmarked sections on data sheets, when the options to choose did not apply to the example at hand (or more particular to its class). Another example are missing values due to intentionally unanswered questions (e.g. for income, social status) on questionnaires, when persons are more or less willing to answer depending on their standing (class) within society. In these cases the probability for the occurrence of missing values is class specific. Thus, missing values of this kind should be distinguished and also treated differently from feature values that are missing completely at random. They provide additional information for the classification of the partially missing dataset. After a short introduction to Fuzzy Cluster Analysis, a brief review on approaches for dealing with missing values MCAR is given in section 2. Our approach to fuzzy clustering with missing values occurring with a class dependent probability is presented in section 3.

As already stated, the objective of Fuzzy Clustering methods is to divide a given dataset into a set of clusters based on similarity. In classical cluster analysis each datum must be assigned to exactly one cluster. Fuzzy cluster analysis relaxes this requirement by allowing gradual memberships (membership degrees), thus offering the opportunity to deal with data that belong to more than one cluster at the same time. Most fuzzy clustering algorithms are objective function based: They determine an optimal classification by minimizing an objective function. In objective function based clustering usually each cluster is represented by a *cluster prototype*. This prototype consists of a *cluster center* (whose name already indicates its meaning) and maybe some additional information about the size and the shape of the cluster. The cluster center is an instantiation of the attributes used to describe the domain, just as the data points in the dataset to divide. However, the cluster center is computed by the clustering algorithm and may or may not appear in the dataset. The size and shape parameters determine the extension of the cluster in different directions of the underlying domain.

The degrees of membership to which a given data point belongs to the different clusters are computed from the distances of the data point to the cluster centers w.r.t. the size and the shape of the cluster as stated by the additional prototype information. The closer a data point lies to the center of a cluster (w.r.t. size and shape), the higher is its degree of membership to this cluster. Hence the problem to divide a dataset $X = \{x_1, \ldots, x_n\} \subseteq \mathbb{R}^p$ into c clusters can be stated as the task to minimize the distances of the data points to the cluster centers, since, of course, we want to maximize the degrees of membership. An iterative algorithm is used to solve the classification problem in objective function based clustering: Since the objective function cannot be minimized directly, the cluster prototypes and the membership degrees are alternately optimized.

2   Methods for Missing Values MCAR

The computation of the prototypes and the distance measure in fuzzy clustering algorithms requires to reference the values of all features. Thus the easiest workaround is to remove data points or attributes with missing values from the dataset. Since clustering is then performed on the remaining complete feature vectors only, this approach is called *complete case analysis.* This approach is appropriate if missing values are rare. However, if missing values are frequent, the dataset size may be considerably reduced. Since clustering may then yield unreliable or distorted results, research has focused on imputation methods and on methods to estimate the distance between the cluster prototypes and the incomplete feature vectors [13]. Since imputing missing values during data preprocessing jeopardizes the quality and reliability of the classification results, fuzzy clustering algorithms have been modified to handle missing values MCAR naturally during the iterative clustering process.

Imputing missing values in each iteration of the fuzzy clustering algorithm [10] offers an advantage compared to an imputation during data preprocessing: the current classification (membership degrees and cluster prototypes) can be taken into account for finding better estimates. For instance, missing values can be imputed with the corresponding attribute values of the cluster center to which the corresponding data point has the highest membership degree. Another intuitively motivated approach is to use the weighted mean of all cluster centers [10]. That is, a missing value x_{jk} is computed as $x_{jk} = \frac{\sum_{i=1}^{c} u_{ij}^{m} c_{ik}}{\sum_{i=1}^{c} u_{ij}^{m}}$. x_{jk} is the k-th attribute of datum $\boldsymbol{x}_j$ and c_{ik} is the k-th attribute value of cluster center $\boldsymbol{c}_i$. This formula is also obtained by differentiating the objective function of the fuzzy c-means (FCM) algorithm w.r.t. the missing values if the imputation of missing values is viewed as an optimization problem. Then, the cluster prototypes, membership degrees and missing values are iteratively computed minimizing the objective function [6].

The imputation of missing values introduces new, possibly unreliable information. This results in a higher uncertainty of the classification results. The problem becomes worse, because the imputation of a missing value is influenced by its own imputation in previous iterations. Hence, approaches have been studied that try to reduce the influence of imputation. In the *available case* approach one tries to use all available information of incomplete feature vectors refraining from imputing estimates for missing values [10]. Therefore, the computation formulae for the update of cluster prototypes as well as the formula for the computation of the distance measure have to be modified to use observed feature values only. In the accordingly modified FCM algorithm the feature values of the cluster centers are computed by

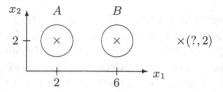

Fig. 1. A dataset with two circular clusters. The centers are marked by a ×. If information on a cluster specific frequency for missing values in the first attribute is available, it can be used to classify the incomplete feature vector $(?, 2)$. Otherwise the membership degree to both clusters should be equal.

$$c_{ik} = \frac{\sum_{j=1}^{n} u_{ij}^m i_{jk} x_{jk}}{\sum_{j=1}^{n} u_{ij}^m i_{jk}}. \qquad (1)$$

i_{jk} is the k-th attribute of the index vector i_j, which indicates whether an attribute value is observed. $i_{jk} = 1$ if the k-th attribute of x_j is observed and $i_{jk} = 0$ otherwise. The distance of an incomplete feature vector to the cluster centers can only be estimated assuming that the analyzed data is clustered. Then, all of the attribute-specific distances behave basically in the same way, such that the attribute-specific distance for an unobserved attribute may be estimated as the mean of the attribute-specific distances for the observed attributes [4]. This leads to the following formula for estimating the euclidian distance in the modification of the FCM algorithm according to the available case approach:

$$d(x_j, c_i) = \frac{p}{\sum_{k=1}^{p} i_{jk}} \sum_{k=1}^{p} i_{jk}(x_{jk} - c_{ik})^2. \qquad (2)$$

3   A Class Specific Probability

In some cases the occurrence of missing values can provide hints for the classification of incomplete feature vectors. The methods presented in section 2 cannot utilize the information on different frequencies for missing values in certain attributes depending on the classes in a dataset. Figure 1 illustrates the assignment of an incomplete feature vector to clusters, whether a class-specific probability for missing values is assumed or not. If we know, for example, that missing values in attribute x_1 occur with a probability of 10% in cluster A and with a probability of 50% in cluster B, we would assign the datum $(?, 2)$ rather to cluster B than to cluster A. In the following we assume that due to the required homogeneity within a cluster, its cluster specific probabilities for missing attribute values apply equally to all data points assigned to it. For treating this special kind of missing values we extend the model of the fuzzy maximum likelihood estimation algorithm (FMLE) presented by Gath and Geva [5]. We further estimate the cluster (and attribute) specific probabilities during the iterative clustering process.

In the FMLE the dataset is interpreted as a realization of c p-dimensional normal distributions, where c is the number of clusters. That is, a datum $\boldsymbol{x}_j$ is created with a prior probability P_i by the normal distribution N_i with the expected value $\boldsymbol{c}_i$ and the covariance matrix $\mathbf{A}_i$. The algorithm computes the classification as a maximum likelihood classifier: it computes P_i and N_i based on the current membership degrees by maximizing the likelihood that the data assigned to a cluster belong to that cluster. The distance computed in the FMLE is inversely proportional to the posterior probability that a datum was created by the probability distribution of the corresponding cluster. Hence a small distance means a high probability and a large distance means a low probability for membership [7].

Class dependent probabilities for missing values $\boldsymbol{mv}_i = \{mv_{i1}, \ldots, mv_{ip}\}$ ($i = 1, \ldots, c$) can be integrated into this model as follows [12]: The data are viewed as a realization of a p-dimensional normal distribution N_i from which the k-th parameter is missing with a probability mv_{ik} and which is chosen with a probability P_i. However, the posterior probability that a datum $\boldsymbol{x}_j$ belongs to class i is difficult to compute if we assume that the decision, which attributes are missing, is made after the creation of a datum. Therefore the model is changed in such a way that first class i is chosen with probability P_i. Then the decision is made which attributes are missing based on the probabilities $\boldsymbol{mv}_i$. And finally the datum is created by the marginal distribution on the space scaffolded by the observable attributes. It is intuitively clear that the posterior probabilities are the same for both models.

This assumption leads to the following posterior probability (likelihood) that a datum $\boldsymbol{x}_j$ with a missing value in the k-th attribute was created by the normal distribution N_i:

$$P_i \cdot mv_{ik} \cdot \prod_{\substack{l=1 \\ l \neq k}}^{p}(1 - mv_{il}) \frac{\exp(-\frac{1}{2}(\boldsymbol{x}_j - \boldsymbol{c}_i)^\top \mathbf{A}^{-1}(\boldsymbol{x}_j - \boldsymbol{c}_i))}{(2\pi)^{p/2}\sqrt{\det(\mathbf{A}_i)}}$$

Because of the special properties of the Gaussian distribution, the needed marginal distribution can be obtained by simply neglecting the dimensions in which values are missing in the evaluation of the expression:

$$\frac{\exp(-\frac{1}{2}(\boldsymbol{x}_j - \boldsymbol{c}_i)^\top \mathbf{A}^{-1}(\boldsymbol{x}_j - \boldsymbol{c}_i))}{(2\pi)^{p/2}\sqrt{\det(\mathbf{A}_i)}}.$$

The distance between a datum $\boldsymbol{x}_j$ and a cluster $\boldsymbol{c}_i$ of the FMLE is inversely proportional to the posterior probability that a datum $\boldsymbol{x}_j$ with a missing value in the k-th attribute was created by the normal distribution N_i. Following the model of Gath and Geva and neglecting the constant factor $(2\pi)^{p/2}$ the distance is therefore computed by:

$$d^2(\boldsymbol{x}_j, (\boldsymbol{c}_i, \mathbf{A}_i, P_i, \boldsymbol{mv}_i)) = \frac{\sqrt{\det(\mathbf{A}_i)}\exp(\frac{1}{2}(\boldsymbol{x}_j - \boldsymbol{c}_i)^\top \mathbf{A}_i^{-1}(\boldsymbol{x}_j - \boldsymbol{c}_i))}{P_i \cdot mv_{ik} \cdot \prod_{\substack{l=1 \\ l \neq k}}^{p}(1 - mv_{il})}$$

Just as in the definition of the posterior probability, attributes in which x_j has missing values are not taken into account for the computation of the term

$$\sqrt{\det(\mathbf{A}_i)} \exp\left(\frac{1}{2}(x_j - c_i)^\top \mathbf{A}^{-1}(x_j - c_i)\right).$$

Because the data belonging to the same class are created based on the same normal distribution regardless of whether they contain missing values or not, the center of the cluster as well as its covariance matrix are computed by neglecting missing values as presented in the available case approach (see section 2, equation 1).

The probabilities P_i and mv_i are estimated for each cluster in each iteration of the extended FMLE algorithm according to:

$$P_i = \frac{\sum_{j=1}^{n} u_{ij}^m}{\sum_{j=1}^{n} \sum_{s=1}^{c} u_{sj}^m}, \tag{3}$$

$$mv_{ik} = \frac{\sum_{j=1}^{n} u_{ij}(1 - i_{jk})}{\sum_{j=1}^{n} u_{ij}}, \tag{4}$$

$i = 1, \ldots, c \; ; \; k = 1, \ldots, p.$

4   Experiments

We tested whether our variant of the FMLE algorithm can exploit class-specifically occurring missing values for classifying the artificially corrupted wine dataset [1]. The wine data contains three different classes with 59, 71, and 48 data points. For the experiment we only used the attributes 7, 10, and 13. Each class was assigned one of the three attributes such that in each class a different attribute contains missing values. Then we deleted values of the assigned attribute in each class with probabilities between 20% and 60%. In addition, we generated missing values MCAR by randomly choosing and deleting feature values in the entire dataset with a probability of 5%. To be able to compare the classification results of our approach, we chose the available case modified version of the FMLE algorithm as the baseline clustering method. Comparability of the results was further ensured by initializing both variants of the FMLE algorithm with a fuzzy c-means clustering of the partially missing wine dataset. We obtained the initial clustering with the available case modified FCM algorithm.

Figure 2 shows the number of classification errors in relation to the probability with which we added class specifically missing values. Both variants of the FMLE show almost equal performance for class specific probabilities of 20% for missing values in one attribute. For higher probabilities, though, the number of classification errors increases when clustering the wine data with the available case variant of the FMLE algorithm. In contrast to this, the number of misclassifications is almost constant for the classifications yielded with the modified version of FMLE algorithm as presented in section 3. While classification performance degrades for the available case approach due to the information loss caused by frequent missing values, our approach is able to compensate by taking into account the that missing values are actually missing with class specific probabilities.

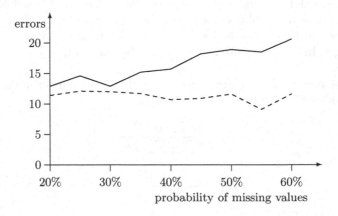

Fig. 2. Number of misclassified data points, FMLE, 3 clusters, wine dataset. Straight line: available case approach, dotted line: model with class specific probabilities.

5   Conclusion

Before analyzing a given partially missing dataset, the causes for missing data elements should be investigated first. In many cases it is impossible to make assumptions on the occurrence of missing values that support an unsupervised classification of the incomplete data. Then, the missing values can be treated as *missing completely at random* and easily dealt with in Fuzzy Cluster Analysis. In other situations, however, information why data are missing is available and the occurrence of missing values provides additional information to classify partially missing datasets. Our presented approach uses class specific probabilities for missing values to benefit from the additional information in these cases. The results of our experiments show that our approach is capable of exploiting classification relevant information contained in the occurrence of missing values. Since the approach is well-defined, it should lead to reliable results whenever class specific probabilities for missing values can be assumed. In the future we want to apply our clustering method to data analysis problems in customer relationship management, e.g. segmenting user profiles.

References

1. Aeberhard, S., Coomans, D., and de Vel, O.: Comparison of Classifiers in High Dimensional Settings. Tech Rep. 92—02, Dept. of Computer Science and Dept. of Mathematics and Statistics, James Cook University of North Queensland, 1992.

2. Bezdek, J.C. and Pal, S.K. (eds.): Fuzzy Models for Pattern Recognition: methods that search for structures in data. IEEE Press, Piscataway, 1992.

3. Bezdek, J.C., Keller, J., Krishnapuram, R., and Pal, N.R.: Fuzzy Models and Algorithms for Pattern Recognition and Image Processing. Kluwer, Boston, London, 1999.

4. Dixon, J.K.: Pattern Recognition with partly missing data. IEEE Transactions on Systems, Man, and Cybernetics, 9(6), 617—621, 1979.

5. Gath, I. and Geva, A.B.: Unsupervised Optimal Fuzzy Clustering. IEEE Transactions on Pattern Analysis and Machine Intelligence, 11, 773—781, 1989.

6. Hathaway, R.J. and Bezdek J.C.: Fuzzy c-Means Clustering of Incomplete Data. IEEE Trans. on Systems, Man, and Cybernetics - Part B, 31(5), 735—744, 2001.

7. Höppner, F., Klawonn, F., Kruse, R. and Runkler, T.: Fuzzy Cluster Analysis, Wiley, Chichester, New York, 1999.

8. Little, R.J.A. und Rubin, D.A.: Statistical analysis with missing data. John Wiley and Sons, New York, 1987.

9. Schafer, J.L.: Analysis of Incomplete Multivariate Data, Chapman & Hall, London, 1997.

10. Timm, H. and Klawonn, F.: Classification of Data with Missing Values. Proc. 6th European Congress on Intelligent Techniques and Soft Computing (EUFIT '98), 1304—1308, Aachen, Deutschland, 1998.

11. Timm, H. and Kruse, R.: Fuzzy Cluster Analysis with Missing Values. Proc. 17th International Conf. of the North American Fuzzy Information Processing Society (NAFIPS98), 242—246, Pensacola, FL, USA, 1998.

12. Timm, H. and Klawonn, F.: Different Approaches for Fuzzy Cluster Analysis with Missing Values, Proceedings of 7th European Congress on Intelligent Techniques & Soft Computing, Aachen, Germany, 1999.

13. Timm, H. Döring, C. and Kruse, R.: Fuzzy Cluster Analysis of Partially Missing Datasets. Proc. of the European Symposium on Intelligent Technologies, Hybrid Systems and Their Implementation on Smart Adaptive Systems (EUNITE 2002), Albufeira, Portugal, 2002.

Mining Multi-level Diagnostic Process Rules from Clinical Databases Using Rough Sets and Medical Diagnostic Model

Shusaku Tsumoto

Department of Medical Informatics, Shimane Medical University, School of Medicine,
89-1 Enya-cho Izumo City, Shimane 693-8501 Japan
tsumoto@computer.org

Abstract. One of the most important problems on rule induction methods is that they cannot extract rules, which plausibly represent experts'
decision processes. On one hand, rule induction methods induce probabilistic rules, the description length of which is too short, compared with
the experts' rules. In this paper, the characteristics of experts' rules are
closely examined and a new approach to extract plausible rules is introduced, which consists of the following three procedures. First, the
characterization of decision attributes (given classes) is extracted from
databases and the concept hierarchy for given classes is calculated. Second, based on the hierarchy, rules for each hierarchical level are induced
from data. Then, for each given class, rules for all the hierarchical levels
are integrated into one rule. The proposed method was evaluated on a
medical database, the experimental results of which show that induced
rules correctly represent experts' decision processes.

1 Introduction

One of the most important problems in data mining is that extracted rules are
not easy for domain experts to interpret. One of its reasons is that conventional
rule induction methods[7] cannot extract rules, which plausibly represent experts' decision processes[9]: the description length of induced rules is too short,
compared with the experts' rules. For example, rule induction methods, including AQ15[3] and PRIMEROSE[9], induce the following common rule for muscle
contraction headache from databases on differential diagnosis of headache:

> [*location = whole*] ∧[Jolt Headache = *no*] ∧[Tenderness of M1 = *yes*]
> → muscle contraction headache.

This rule is shorter than the following rule given by medical experts.

[Jolt Headache = *no*]
∧([Tenderness of M0 = *yes*] ∨[Tenderness of M1 = *yes*] ∨[Tenderness of M2 = *yes*])
∧[Tenderness of B1 = *no*] ∧[Tenderness of B2 = *no*] ∧[Tenderness of B3 = *no*]
∧[Tenderness of C1 = *no*] ∧[Tenderness of C2 = *no*] ∧[Tenderness of C3 = *no*]
∧[Tenderness of C4 = *no*]
   → muscle contraction headache

where [Tenderness of B1 = *no*] and [Tenderness of C1 = *no*] are added.

T. Bilgiç et al. (Eds.): IFSA 2003, LNAI 2715, pp. 362–369, 2003.

One of the main reasons why rules are short is that these patterns are generated only by one criteria, such as high accuracy or high information gain. The comparative studies[9,10] suggest that experts should acquire rules not only by one criteria but by the usage of several measures. Those characteristics of medical experts' rules are fully examined not by comparing between those rules for the same class, but by comparing experts' rules with those for another class[9].

For example, the classification rule for muscle contraction headache given in Section 1 is very similar to the following classification rule for disease of cervical spine:

[Jolt Headache = no]
$\wedge$([Tenderness of M0 = yes] $\vee$[Tenderness of M1 = yes] $\vee$[Tenderness of M2 = yes])
$\wedge$([Tenderness of B1 = yes] $\vee$[Tenderness of B2 = yes] $\vee$[Tenderness of B3 = yes]
    $\vee$[Tenderness of C1 = yes] $\vee$[Tenderness of C2 = yes] $\vee$[Tenderness of C3 = yes]
    $\vee$[Tenderness of C4 = yes])
        $\rightarrow$ disease of cervical spine

The differences between these two rules are attribute-value pairs, from tenderness of B1 to C4. Thus, these two rules are composed of the following three blocks:

$$A_1 \wedge A_2 \wedge \neg A_3 \rightarrow muscle\ contraction\ headache$$
$$A_1 \wedge A_2 \wedge A_3 \rightarrow disease\ of\ cervical\ spine,$$

where A_1, A_2 and A_3 are given as the following formulae:
A_1 = [Jolt Headache = no], A_2 = [Tenderness of M0 = yes] $\vee$ [Tenderness of $M1$ = yes] $\vee$ [Tenderness of M2 = yes], and A_3 = [Tenderness of C1 = no] $\wedge$ [Tenderness of C2 = no] $\wedge$ [Tenderness of C3 = no] $\wedge$ [Tenderness of C4 = no]. The first two blocks (A_1 and A_2) and the third one (A_3) represent the different types of differential diagnosis. The first one A_1 shows the discrimination between muscular type and vascular type of headache. Then, the second part shows that between headache caused by neck and head muscles. Finally, the third formula A_3 is used to make a differential diagnosis between muscle contraction headache and disease of cervical spine. Thus, medical experts first select several diagnostic candidates, which are very similar to each other, from many diseases and then make a final diagnosis from those candidates.

In this paper, the characteristics of experts' rules are closely examined and a new approach to extract plausible rules is introduced, which consists of the following three procedures. First, the characterization of each decision attribute (a given class), a list of attribute-value pairs the supporting set of which covers all the samples of the class, is extracted from databases and the classes are classified into several groups with respect to the characterization. Then, two kinds of sub-rules, rules discriminating between each group and rules classifying each class in the group are induced. Finally, those two parts are integrated into one rule for each decision attribute.

The paper is organized as follows. Section 2 and 3 ntroduces rough sets and a characterization set. Section 4 gives an algorithm for rule induction. Section 5 shows experimental results. Finally, Section 6 concludes this paper.

2   Rough Set Theory and Probabilistic Rules

In the following sections, we use the following notations introduced by Grzymala-Busse and Skowron[8], which are based on rough set theory[4].

Let U denote a nonempty, finite set called the universe and A denote a nonempty, finite set of attributes, i.e., $a : U \to V_a$ for $a \in A$, where V_a is called the domain of a, respectively. Then, a decision table is defined as an information system, $A = (U, A \cup \{d\})$.

The atomic formulae over $B \subseteq A \cup \{d\}$ and V are expressions of the form $[a = v]$, called descriptors over B, where $a \in B$ and $v \in V_a$. The set $F(B,V)$ of formulas over B is the least set containing all atomic formulas over B and closed with respect to disjunction, conjunction and negation. For each $f \in F(B,V)$, f_A denote the meaning of f in A, i.e., the set of all objects in U with property f, defined inductively as follows.

1. If f is of the form $[a = v]$ then, $f_A = \{s \in U | a(s) = v\}$
2. $(f \wedge g)_A = f_A \cap g_A; (f \vee g)_A = f_A \vee g_A; (\neg f)_A = U - f_a$

By the use of the framework above, classification accuracy and coverage, or true positive rate is defined as follows.

Definition 1. *Let R and D denote a formula in $F(B,V)$ and a set of objects which belong to a decision d. Classification accuracy and coverage(true positive rate) for $R \to d$ is defined as:*

$$\alpha_R(D) = \frac{|R_A \cap D|}{|R_A|} (= P(D|R)), \ and \ \kappa_R(D) = \frac{|R_A \cap D|}{|D|} (= P(R|D)),$$

where $|S|$, $\alpha_R(D)$, $\kappa_R(D)$ and $P(S)$ denote the cardinality of a set S, a classification accuracy of R as to classification of D and coverage (a true positive rate of R to D), and probability of S, respectively.

According to the definitions, probabilistic rules with high accuracy and coverage are defined as:

$$R \overset{\alpha,\kappa}{\to} d \ s.t. \ R = \vee_i R_i = \vee \wedge_j [a_j = v_k], \ \alpha_{R_i}(D) \geq \delta_\alpha \ and \ \kappa_{R_i}(D) \geq \delta_\kappa,$$

where δ_α and δ_κ denote given thresholds for accuracy and coverage, respectively.

3   Characterization Sets

In order to model medical reasoning, a statistical measure, coverage plays an important role in modeling, which is a conditional probability of a condition (R) under the decision $D(P(R|D))$. Let us define a characterization set of D, denoted by $L(D)$ as a set, each element of which is an elementary attribute-value pair R with coverage being larger than a given threshold, δ_κ. That is,

Definition 2. *Let R denote a formula in $F(B,V)$. Characterization sets of a target concept (D) is defined as:*

$$L_{\delta_\kappa}(D) = \{R|\kappa_R(D) \geq \delta_\kappa\}$$

Then, three types of relations between characterization sets are defined as follows:

$$
\begin{aligned}
&\text{Independent type: } L_{\delta_\kappa}(D_i) \cap L_{\delta_\kappa}(D_j) = \phi, \\
&\text{Boundary type: } \quad L_{\delta_\kappa}(D_i) \cap L_{\delta_\kappa}(D_j) \neq \phi, \text{ and} \\
&\text{Positive type: } \quad\, L_{\delta_\kappa}(D_i) \subseteq L_{\delta_\kappa}(D_j).
\end{aligned}
$$

All three definitions correspond to the negative region, boundary region, and positive region, respectively, if a set of the whole elementary attribute-value pairs will be taken as the universe of discourse. We consider the special case of characterization sets in which the thresholds of coverage is equal to 1.0. That is,

$$L_{1.0}(D) = \{R_i|\kappa_{R_i}(D) = 1.0\}$$

Then, we have several interesting characteristics.

Theorem 1. *Let R_i and R_j be the formulae in $F(B,V)$ and let $A(R_i)$ denote a set whose elements are the attribute-value pairs of the form $[a,v]$ included in R_i. If $A(R_i) \subseteq A(R_j)$, then we represent this relation as: $R_i \preceq R_j$.*
Let R_i and R_j two formulae in $L_{1.0}(D)$ such that $R_i \preceq R_j$. Then,

$$\alpha_{R_i} \leq \alpha_{R_j}.$$

Thus, when we collect the formulae whose values of coverage are equal to 1.0, the sequence of conjunctive formulae corresponds to the sequence of increasing chain of accuracies.

Since $\kappa_R(D) = 1.0$ means that the meaning of R covers all the samples of D, its complement $U - R_A$, that is, $\neg R$ do not cover any samples of D. Especially, when R consists of the formulae with the same attributes, it can be viewed as the generation of the coarsest partitions. Thus,

Theorem 2. *Let R be a formula in $L_{1.0}(D)$ such that $R = \vee_j [a_i = v_j]$. Then, R and $\neg R$ gives the coarsest partition for a_i, whose R includes D.* $\square$

From the propositions 1 and 2, the next theorem holds.

Theorem 3. *Let A consist of $\{a_1, a_2, \cdots, a_n\}$ and R_i be a formula in $L_{1.0}(D)$ such that $R_i = \vee_j [a_i = v_j]$. Then, a sequence of a conjunctive formula $F(k) = \wedge_{i=1}^{k} R_i$ gives a sequence which increases the accuracy.* $\square$

4 Rule Induction with Grouping

As discussed above, when the coverage of R for a target concept D is equal to 1.0, R is a necessity condition of D. That is, a proposition $D \rightarrow R$ holds and its contrapositive $\neg R \to \neg D$ holds. Thus, if R is not observed, D cannot be a candidate of a target concept. Thus, if two target concepts have a common formula R whose coverage is equal to 1.0, then $\neg R$ supports the negation of two concepts, which means these two concepts belong to the same group. Furthermore, if two target concepts have similar formulae $R_i, R_j \in L_{1.0}(D)$, they are very close to each other with respect to the negation of two concepts. In this case, the attribute-value pairs in the intersection of $L_{1.0}(D_i)$ and $L_{1.0}(D_j)$ give a characterization set of the concept that unifies D_i and D_j, D_k. Then, compared with D_k and other target concepts, classification rules for D_k can be obtained. When we have a sequence of grouping, classification rules for a given target concepts are defined as a sequence of subrules.

From these ideas, a rule induction algorithm with grouping target concepts can be described as follows. First, this algorithm first calculates $L_{1.0}(D_i)$ for $\{D_1, D_2, \cdots, D_k\}$. From the list of characterization sets, it calculates the intersection between $L_{1.0}(D_i)$ and $L_{1.0}(D_j)$ and stores it into L_{id}. Second, the procedure calculates the similarity (matching number)of the intersections and sorts L_{id} with respect of the similarities. Then, the algorithm chooses one intersection ($D_i \cap D_j$) with maximum similarity (highest matching number) and group D_i and D_j into a concept DD_i. These procedures will be continued until all the grouping is considered (Fig. 1). Finally, from the list of grouping, rules are extracted (Fig. 2).

5   Experimental Results

The above rule induction algorithm was implemented in PRIMEROSE4.5 (Probabilistic Rule Induction Method based on Rough Sets Ver 5.0), and was applied to databases on differential diagnosis of headache, meningitis and cerebrovascular diseases (CVD), whose precise information is given in Table 1. The thresh-

Table 1. Information about Databases

Domain	Samples	Classes	Attributes
Headache	52119	45	147
CVD	7620	22	285
Meningitis	141	4	41

olds, δ_α and δ_κ were set to 0.75 and 0.5, respectively. Also, the threshold for grouping is set to 0.8.[1] This system was compared with PRIMEROSE4.5[10], PRIMEROSE[9] C4.5[6], CN2[1], AQ15[3] with respect to the following points:

---

[1] These values are given by medical experts as good thresholds for rules in these three domains.

```
procedure Grouping ;
  var inputs
    L_c : List; /* A list of Characterization Sets */
    L_id : List; /* A list of Intersection */ L_s : List; /* A list of Similarity */
  var outputs
    L_gr : List; /* A list of Grouping */
  var
    k : integer; L_g, L_gr : List;
  begin
    L_g := {} ; k := n /* n: A number of Target Concepts*/
    Sort L_s with respect to similarities;
      Take a set of (D_i, D_j), L_max with maximum similarity values;
      k:= k+1;
      forall (D_i, D_j) ∈ L_max do
        begin
          Group D_i and D_j into D_k;
            L_c := L_c - {(D_i, L_{1.0}(D_i))}; L_c := L_c - {(D_j, L_{1.0}(D_j))};
              L_c := L_c + {(D_k, L_{1.0}(D_k))}; Update L_id for DD_k; Update L_s;
              L_gr := ( Grouping for L_c, L_id, and L_s) ;
              L_g := L_g + {{(D_k, D_i, D_j), L_g}};
        end
    return  L_g;
  end {Grouping}
```

Fig. 1. An Algorithm for Grouping

length of rules, similarities between induced rules and expert's rules and performance of rules. Length was measured by the number of attribute-value pairs
used in an induced rule and Jaccard's coefficient was adopted as a similarity
measure[2]. Concerning the performance of rules, ten-fold cross-validation was
applied to estimate classification accuracy.

Table 2 shows the experimental results, which suggest that PRIMEROSE5
outperforms PRIMEROSE4.5 (two-level) and the other four rule induction methods and induces rules very similar to medical experts' ones.

6 Conclusion

In this paper, the characteristics of experts' rules are closely examined, whose
empirical results suggest that grouping of diseases ais very important to realize
automated acquisition of medical knowledge from clinical databases. Thus, we
focus on the role of coverage in focusing mechanisms and propose an algorithm
for grouping of diseases by using this measure. The above example shows that
rule induction with this grouping generates rules, which are similar to medical
experts' rules and they suggest that our proposed method should capture medical
experts' reasoning.

```
procedure RuleInduction ;
  var inputs
    L_c : List; /* A list of Characterization Sets */
    L_id : List; /* A list of Intersection */
    L_g : List; /* A list of grouping: {{(D_{n+1},D_i,D_j),{(DD_{n+2},.)...}}} */
              /* n: A number of Target Concepts */
  var
    Q, L_r : List;
  begin
    Q := L_g; L_r := {};
    if (Q ≠ ∅) then do
      begin
        Q := Q − first(Q); L_r := Rule Induction (L_c, L_id, Q);
      end
    (DD_k, D_i, D_j) := first(Q);
    if (D_i ∈ L_c and D_j ∈ L_c) then do
      begin
        Induce a Rule r which discriminate between D_i and D_j;
        r = {R_i → D_i, R_j → D_j};
      end
    else do
      begin
        Search for L_{1.0}(D_i) from L_c; Search for L_{1.0}(D_j) from L_c;
          if (i < j) then do
            begin
              r(D_i) := ∨_{R_l∈L_{1.0}(D_j)}¬R_l → ¬D_j; r(D_j) := ∧_{R_l∈L_{1.0}(D_j)}R_l → D_j;
            end
          r := {r(D_i), r(D_j)};
      end
    return L_r := {r, L_r} ;
  end {Rule Induction}
```

Fig. 2. An Algorithm for Rule Induction

Acknowledgments

This work was supported by the Grant-in-Aid for Scientific Research (13131208) on Priority Areas (No.759) "Implementation of Active Mining in the Era of Information Flood" by the Ministry of Education, Science, Culture, Sports, Science and Technology of Japan.

References

1. Clark, P. and Niblett, T., The CN2 Induction Algorithm. *Machine Learning*, 3, 261-283, 1989.
2. Everitt, B. S., *Cluster Analysis*, 3rd Edition, John Wiley & Son, London, 1996.

Table 2. Experimental Results

Method	Length	Similarity	Accuracy
Headache			
PRIMEROSE5.0	8.8 ± 0.27	0.95 ± 0.08	95.2 ± 2.7%
PRIMEROSE4.5	7.3 ± 0.35	0.74 ± 0.05	88.3 ± 3.6%
Experts	9.1 ± 0.33	1.00 ± 0.00	98.0 ± 1.9%
PRIMEROSE	5.3 ± 0.35	0.54 ± 0.05	88.3 ± 3.6%
C4.5	4.9 ± 0.39	0.53 ± 0.10	85.8 ± 1.9%
CN2	4.8 ± 0.34	0.51 ± 0.08	87.0 ± 3.1%
AQ15	4.7 ± 0.35	0.51 ± 0.09	86.2 ± 2.9%
Meningitis			
PRIMEROSE5.0	2.6 ± 0.19	0.91 ± 0.08	82.0 ± 3.7%
PRIMEROSE4.5	2.8 ± 0.45	0.72 ± 0.25	81.1 ± 2.5%
Experts	3.1 ± 0.32	1.00 ± 0.00	85.0 ± 1.9%
PRIMEROSE	1.8 ± 0.45	0.64 ± 0.25	72.1 ± 2.5%
C4.5	1.9 ± 0.47	0.63 ± 0.20	73.8 ± 2.3%
CN2	1.8 ± 0.54	0.62 ± 0.36	75.0 ± 3.5%
AQ15	1.7 ± 0.44	0.65 ± 0.19	74.7 ± 3.3%
CVD			
PRIMEROSE5.0	7.6 ± 0.37	0.89 ± 0.05	74.3 ± 3.2%
PRIMEROSE4.5	5.9 ± 0.35	0.71 ± 0.05	72.3 ± 3.1%
Experts	8.5 ± 0.43	1.00 ± 0.00	82.9 ± 2.8%
PRIMEROSE	4.3 ± 0.35	0.69 ± 0.05	74.3 ± 3.1%
C4.5	4.0 ± 0.49	0.65 ± 0.09	69.7 ± 2.9%
CN2	4.1 ± 0.44	0.64 ± 0.10	68.7 ± 3.4%
AQ15	4.2 ± 0.47	0.68 ± 0.08	68.9 ± 2.3%

3. Michalski, R. S., Mozetic, I., Hong, J., and Lavrac, N., The Multi-Purpose Incremental Learning System AQ15 and its Testing Application to Three Medical Domains, in *Proceedings of the fifth National Conference on Artificial Intelligence*, 1041-1045, AAAI Press, Menlo Park, 1986.
4. Pawlak, Z., *Rough Sets*. Kluwer Academic Publishers, Dordrecht, 1991.
5. Polkowski, L. and Skowron, A.: Rough mereology: a new paradigm for approximate reasoning. Intern. J. Approx. Reasoning **15**, 333–365, 1996.
6. Quinlan, J.R., *C4.5 - Programs for Machine Learning*, Morgan Kaufmann, Palo Alto, 1993.
7. *Readings in Machine Learning*, (Shavlik, J. W. and Dietterich, T.G., eds.) Morgan Kaufmann, Palo Alto, 1990.
8. Skowron, A. and Grzymala-Busse, J. From rough set theory to evidence theory. In: Yager, R., Fedrizzi, M. and Kacprzyk, J.(eds.) *Advances in the Dempster-Shafer Theory of Evidence*, pp.193-236, John Wiley & Sons, New York, 1994.
9. Tsumoto, S., Automated Induction of Medical Expert System Rules from Clinical Databases based on Rough Set Theory. *Information Sciences* **112**, 67-84, 1998.
10. Tsumoto,S. Extraction of Hierarchical Decision Rules from Clinical Databases using Rough Sets. *Information Sciences*, 2003 (in print)

Rough Sets and Information Granulation

James F. Peters[1], Andrzej Skowron[2], Piotr Synak[3], and Sheela Ramanna[1]

[1] Department of Electrical and Computer Engineering, University of Manitoba
Winnipeg, Manitoba R3T 5V6, Canada
{jfpeters,ramanna}@ee.umanitoba.ca
[2] Institute of Mathematics, Warsaw University
Banacha 2, 02-097 Warsaw, Poland
skowron@mimuw.edu.pl
[3] Polish-Japanese Institute of Information Technology
Koszykowa 86, 02-008 Warsaw, Poland
synak@pjwstk.edu.pl

Abstract. In this paper, the study of the evolution of approximation space theory and its applications is considered in the context of rough sets introduced by Zdzisław Pawlak and information granulation as well as computing with words formulated by Lotfi Zadeh. Central to this evolution is the rough-mereological approach to approximation of information granules. This approach is built on the inclusion relation to be a part to a degree, which generalises the rough set and fuzzy set approaches. An illustration of information granulation of relational structures is given. The contribution of this paper is a comprehensive view of the notion of information granule approximation, approximation spaces in the context of rough sets and the role of such spaces in the calculi of information granules.

Keywords: Approximation spaces, calculus of granules, information granulation, rough mereology, rough sets.

1 Introduction: Rough Set Approach to Concept Approximation

One of the basic concepts of rough set theory [7] is the indiscernibility relation defined by means of information about objects. The indiscernibility relation is used to define set approximations [6,7]. There have been reported several generalisations of the rough set approach based, e.g., on approximation spaces defined by tolerance and similarity relation, or a family of indiscernibility relations (for references see the papers and bibliography in [5,12]). Rough set approximations have also been generalised for preference relations and rough-fuzzy hybridisations (see, e.g., [21]). Let us consider a generalised approximation spaces introduced in [18]. It is defined by $AS = (U, \mathcal{N}, \nu)$ where U is a set of *objects* (universe), $\mathcal{N}$ is an *uncertainty function* defined on U with values in the powerset $P(U)$ of U (e.g. $\mathcal{N}(x)$ can be interpreted as a *neighbourhood* of x), and ν is an *inclusion function* defined on the Cartesian product $P(U) \times P(U)$ with values in the interval $[0, 1]$

T. Bilgiç et al. (Eds.): IFSA 2003, LNAI 2715, pp. 370–377, 2003.
© Springer-Verlag Berlin Heidelberg 2003

(or more generally, in a partially ordered set), measuring the degree of inclusion of sets. In the sequel $\nu_p(X,Y)$ denotes $\nu(X,Y) \geq p$ for $p \in [0,1]$. The lower AS_* and upper AS^* approximation operations can be defined in AS by

$$AS_*(X) = \{x \in U : \nu(\mathcal{N}(x), X) = 1\}, \tag{1}$$

$$AS^*(X) = \{x \in U : \nu(\mathcal{N}(x), X) > 0\}. \tag{2}$$

The neighbourhood of an object x can be defined by the indiscernibility relation IND. If IND is an equivalence relation then we have $\mathcal{N}(x) = [x]_{IND}$. In the case where IND is a tolerance (similarity) relation $\tau \subseteq U \times U$, we take $\mathcal{N}(x) = \{y \in U : x\tau y\}$, i.e., $N(x)$ is equal to the tolerance class of τ defined by x. The standard inclusion function is defined by $\nu(X,Y) = \frac{|X \cap Y|}{|X|}$ if X is non-empty and by $\nu(X,Y) = 1$ otherwise. In addition, parameterisation of $\mathcal{N}$ and ν in an AS makes it possible to calibrate (tune) approximation operations, and leads to a form of learning in rough neural networks designed in the context of approximation spaces (see, e.g., [16,10]). For applications it is important to have some constructive definitions of $\mathcal{N}$ and ν. The approach based on inclusion functions has been generalised to the *rough mereological approach* (see, e.g., [11,13,14]). The inclusion relation $x\mu_r y$ with the intended meaning x *is a part of* y *to a degree* r has been taken as the basic notion of the rough mereology which is a generalisation of the Leśniewski mereology [1].

We consider two sources of information granulation. The first source is related to inductive reasoning and the second source arises due to object indiscernibility.

As a result of inductive reasoning one cannot define inclusion degrees of object neighbourhoods directly into the target concepts but only into some patterns relevant to such concepts (e.g. left hand sides of decision rules) (see [16,23]). Such degrees together with degrees of inclusion of patterns in target concepts make it possible to define outputs of information granules, called classifiers, for new objects.

In case of indiscernibility of objects it may be necessary, for example, to consider more general structures of the uncertainty function. The values of an uncertainty function may belong to a more complex set than $P(U)$ (see, e.g., [2,4]). We will illustrate this case using an example of a relational structure granulation.

2   Rough-Mereological Approach to Approximation of Information Granules

Rough mereology offers a methodology for synthesis and analysis of objects in the distributed environments of intelligent agents, in particular, for synthesis of objects satisfying a given specification to a satisfactory degree, or for control in such complex environments. Moreover, rough mereology has been recently used for developing foundations of the *information granule calculus*, an attempt towards formalisation of the paradigm of computing with words based on perception, recently formulated by Lotfi Zadeh [24,25,26].

The rough mereological approach is built on the basis of the inclusion relation *to be a part to a degree* and generalises the rough set and fuzzy set approaches (see. e.g., [11,13,14,15]). Such a relation is called *rough inclusion*. This relation can be used to define other basic concepts like closeness of information granules, their semantics, indiscernibility and discernibility of objects, information granule approximation and approximation spaces, perception structure of information granules as well as the notion of ontology approximation. For details the reader is referred to [4]. The rough inclusion relations together with operations for construction of new information granules from already existing ones create a core of a calculus of information granules.[1] A distributive multi-agent framework makes it possible to create a relevant computational model for a calculus of information granules. Agents (information sources) provide us with information granules that must be transformed, analysed and built into structures that support problem solving. In such a computational model, approximation spaces play an important role because information granules received by agents must be approximated (to be understandable by them) before they can be transformed (see, e.g., [13,19,4,10]).

Developing calculi of information granules for approximate reasoning is a challenge important for many applications including control of autonomous vehicles [22] and line-crawling robots [8], web mining and spatio-temporal data mining [16], design automation, sensor fusion [9], approximation neuron design [10,4], creation of approximate views of relational databases, and, in general, for embedding in intelligent systems the ability to reason with words as well as perception based reasoning [24,25,26]. Some steps towards this direction have been taken. Methods for construction of approximate reasoning schemes (*AR*-schemes) have been developed. Such *AR*-schemes are information granules that are clusters of exact constructions (derivations). Reasoning with *AR*-schemes makes it possible to obtain results satisfying a given specification up to a satisfactory degree (i.e. not necessarily exactly) (see e.g. [13,4,16,17]). Methods based on hybridisation of rough sets with fuzzy sets, neural networks, evolutionary approach or case based reasoning are especially valid in inducing *AR*-schemes.

Let us note that inducing relevant calculi of information granules includes also such complex tasks like discovery of relevant operations on information granules or rough inclusion measures. This is closely related to problems of perception and reasoning based on perception [26].

Using rough inclusions, one can generalise the approximation operations for sets of objects, known in rough set theory, to arbitrary information granules. The approach is based on the following reasoning:

Assume $G = \{g_t\}_t$ is a given family of information granules, g is a given granule, and p, q are inclusion degrees such that $p < q$ (let us recall that inclusion degrees are partially ordered by a relation $\leq$). One can consider two

---

[1] Note, the rough inclusion relations should be extended on newly constructed information granules.

kinds of approximation of granule g by a family of granules G. The (G, q)-*lower approximation* of g is defined by[2]

$$LOW_{G,q}(g) = Make\_granule(\{g_t : \nu(g_t, g) \geq q\}). \tag{3}$$

The (G, p)-*upper approximation* of g is defined by

$$UPP_{G,p}(g) = Make\_granule(\{g_t : \nu(g_t, g) > p\}). \tag{4}$$

The definition of a generalised approximation space defined in section 1 is a special case of the notion of information granule approximation. The presented approach can be generalised to approximation spaces in inductive reasoning.

3 Illustrative Example: Granulation of Relational Structures

In this section we present an illustrative example of information granulation. Among the basic concepts, to be used here, is a relational structure M of a given signature Sig with a domain Dom and a language L of signature Sig. The neighbourhood (uncertainty) function is then any function $\mathcal{N} : Dom \rightarrow P^\omega(Dom)$ where

- $P^\omega(Dom) = \bigcup_{k \in \omega} P^k(Dom)$.
- $P^0(Dom) = Dom$ and $P^{k+1}(Dom) = P(P^k(Dom))$) for any non-negative integer k.

To explain this concept let us consider an information system $\mathbb{A} = (U, A)$ as an example of relational structure. A neighbourhood function $\mathcal{N}_\mathbb{A}$ of $\mathbb{A}$ is defined by $\mathcal{N}_\mathbb{A}(x) = [x]_{IND(A)}$ for $x \in U = Dom$ where $[x]_{IND(A)}$ denotes the A-indiscernibility class of x. Hence, the neighbourhood function forms basic granules of knowledge about the universe U. Let us consider a case where the values of neighbourhood function are from $P^2(Dom)$. Assume that together with an information system $\mathbb{A}$ there is also given a similarity relation τ defined on vectors of attribute values. This relation can be extended to objects. An object $y \in U$ is similar to a given object $x \in U$ if the attribute value vector on x is τ-similar to the attribute value vector on y. Now, consider a neighbourhood function defined by $\mathcal{N}_{\mathbb{A},\tau}(x) = \{[y]_{IND(A)} : x\tau y\}$.

Neighbourhood functions cause a necessity of further granulation. Let us consider granulation of a relational structure M by neighbourhood functions. We would like to show that due to the relational structure granulation we obtain new information granules of more complex structure and in the consequence more general neighbourhood functions than those discussed above. Hence, basic

---

[2] *Make_granule* operation is a fusion operation of collections of information granules. A typical example of *Make_granule* is set theoretical union used in rough set theory. Another example of *Make_granule* operation is realised by classifiers.

granules of knowledge about the universe corresponding to objects become more complex.

Assume that a relational structure M and a neighbourhood function $\mathcal{N}$ are given. The aim is to define a new relational structure $M_\mathcal{N}$ called the $\mathcal{N}$–granulation of M.[3] This is done by granulation of all components of M by means of $\mathcal{N}$. Let us assume $M_\mathcal{N} \supseteq P(Dom)$. We present examples showing that such domain $M_\mathcal{N}$ should consist of elements of $P^\omega(Dom)$.

Let us consider a binary relation $r \subseteq Dom \times Dom$. There are numerous possible ways to define a $\mathcal{N}$–granulation $r_\mathcal{N}$ of relation r – the choice depends on applications. Let us list some possible examples:

$$r_\mathcal{N}(\mathcal{N}(x), \mathcal{N}(y)) \text{ iff } \mathcal{N}(x) \times \mathcal{N}(y) \subseteq r \tag{5}$$
$$r_\mathcal{N}(\mathcal{N}(x), \mathcal{N}(y)) \text{ iff } (\mathcal{N}(x) \times \mathcal{N}(y)) \cap r \neq \emptyset$$
$$r_\mathcal{N}(\mathcal{N}(x), \mathcal{N}(y)) \text{ iff } card((\mathcal{N}(x) \times \mathcal{N}(y)) \cap r) \geq s \cdot card(\mathcal{N}(x))$$

where $s \in [0, 1]$ is a threshold.

In this way some patterns are created for pairs of objects. Such patterns can be used for approximation of a target concept (or concept on an intermediate level) over objects composed from pairs (x, y). Certainly, to induce approximations of high quality it is necessary to search for relevant patterns for concept approximation expressible in a given language. This problem is discussed, e.g., in [20].

Let us consider an exemplary degree structure $D = ([0, 1], \leq)$ and its granulation $D_{\mathcal{N}_0} = (P([0, 1]), \leq_{\mathcal{N}_0})$ by means of an uncertainty function $\mathcal{N}_0 : [0, 1] \to P([0, 1])$ defined by $\mathcal{N}_0(x) = \{y \in [0, 1] : [10^k x] = [10^k y]\}$, for some integer k, where for $X, Y \subseteq [0, 1]$ we assume $X \leq_{\mathcal{N}_0} Y$ iff $\forall x \in X, \forall y \in Y \ x \leq y$. Let $\{X_s, X_m, X_l\}$ be a partition of $[0, 1]$ satisfying $x < y < z$ for any $x \in X_s$, $y \in X_m$, $z \in X_l$. Let $AS_0 = ([0, 1], \mathcal{N}_0, \nu)$ be an approximation space with the standard inclusion function ν. We denote by S, M, L the lower approximations of X_s, X_m, X_l in AS_0, respectively, and by S–M, M–L the boundary regions between X_s, X_m and X_m, X_l, respectively. Moreover, we assume $S, M, L \neq \emptyset$. In this way we obtain restriction of $D_{\mathcal{N}_0}$ to the structure $(Deg, \leq_{\mathcal{N}_0})$, where $Deg = \{S, S$–M, M, M–$L, L\}$. Now, for a given (multi-sorted) structure $(U, P(U), [0, 1], \leq, \mathcal{N}_0, \nu)$, where $\nu : P(U) \times P(U) \to [0, 1]$ is an inclusion function, we can define its $\mathcal{N}_0$–granulation by

$$(U, P(U), Deg, \leq_{\mathcal{N}_0}, \{\nu_d\}_{d \in Deg}) \tag{6}$$

where $Deg = \{S, S$–M, M, M–$L, L\}$ and $\nu_d(X, Y)$ iff $\nu_p(X, Y)$ for some p, d', such that $p \in d'$ and $d \leq_{\mathcal{N}_0} d'$.

---

[3] In general, granulation is defined using the uncertainty function and the inclusion function from a given approximation space. For simplicity, we restrict our initial examples to $\mathcal{N}$–granulation only.

Now, let us consider a function f from M and some possible $\mathcal{N}$-granulations $f_{\mathcal{N}}$ of f.[4]

$$f_{\mathcal{N}}(\mathcal{N}(x), \mathcal{N}(y)) = \{\mathcal{N}(z) : z = f(x', y') \text{ for some } x' \in \mathcal{N}(x), y' \in \mathcal{N}(y)\} \quad (7)$$

$$f_{\mathcal{N}}(\mathcal{N}(x), \mathcal{N}(y)) = \bigcup_{x' \in \mathcal{N}(x), y' \in \mathcal{N}(y)} \{\mathcal{N}(z) : z = f(x', y')\}$$

$$f_{\mathcal{N}}(\mathcal{N}(x), \mathcal{N}(y)) = \bigcup_{x' \in \mathcal{N}(x), y' \in \mathcal{N}(y)} \{\mathcal{N}(z) : card\,(\mathcal{N}(z)) \geq s \text{ and } z = f(x', y')\}$$

where $s \in [0, 1]$ is a threshold.

The values of $f_{\mathcal{N}}$ can be treated as generators for patterns used for the target concept approximation. An example of a pattern language can be obtained by considering the results of set theoretical operations on neighbourhoods. Observe that the values of the function $f_{\mathcal{N}}$ are in $P^2(Dom)$. Hence, one could extend the neighbourhood function and the relation granulation on this more complex domain. Certainly this process can be continued and more complex patterns can be generated. On the other hand it is also necessary to bound the depth of exploration of $P^\omega(Dom)$. This can be done by using the rough set approach. For example, after generation of patterns from $P^2(Dom)$ one should, in a sense, reduce them to $P(Dom)$ by considering some operations from $P^2(Dom)$ into $P(Dom)$ returning the relevant patterns for the target concept approximation. Such reduction is necessary especially if the target concepts are elements of the family $P(Dom)$.

In general, relational structures are granulated by means of a given approximation space. Thus, we use in granulation both the uncertainty and the inclusion functions (see, e.g., (6)). Observe, that approximation spaces are also (multi-sorted) relational structures. Hence, they can be also granulated in searching for relevant approximation spaces for concept approximations. Let us consider an example of granulation of a given approximation space $AS = (U, P(U), [0, 1], \leq, \mathcal{N}, \nu)$ (in most of the cases we use simplified notation $(U, \mathcal{N}, \nu)$). The granulation of AS is defined by using AS itself. The resulting approximation space is then $AS' = (P(U), \mathcal{N}_p, \nu')$ where

1. for some $p \in [0, 1]$ the uncertainty function $\mathcal{N}_p : P(U) \to P^2(U)$ is defined by $\mathcal{N}_p(X) = \{Y \in P(U) : \nu_p(X, Y) \text{ and } \nu_p(Y, X)\}$, i.e., $\mathcal{N}_p(X)$ is a cluster of sets close to X to degree at least p;
2. for $\mathcal{X}, \mathcal{Y} \in P^2(U)$ and $q \in [0, 1]$ we assume $\nu'_q(\mathcal{X}, \mathcal{Y})$ iff for any $X \in \mathcal{X}$ there exists $Y \in \mathcal{Y}$ such that $\nu_q(X, Y)$.

One can consider another granulation of the inclusion functions assuming for $\mathcal{X}, \mathcal{Y} \in P^2(U)$ and $q \in [0, 1]$

$$\nu'_q(\mathcal{X}, \mathcal{Y}) \text{ iff } \frac{card((AS_p)_*(\mathcal{X}) \cap (AS_p)_*(\mathcal{Y}))}{card((AS_p)_*(\mathcal{X}))} > q$$

---

[4] For simplicity, we assume f has two arguments and values of $\mathcal{N}$ are in $P(Dom)$.

where $AS_p = (P(U), \mathcal{N}_p, \nu)$ is an approximation space with the standard inclusion function $\nu : P^2(U) \times P^2(U) \rightarrow [0,1]$.

In multi-agent setting [16,17,19,20] each agent is equipped with its own relational structure and approximation spaces located in input ports. The approximation spaces are used for filtering (approximating) information granules sent by other agents. Such agents are performing operations on approximated information granules and sending the results to other agents, checking relationships between approximated information granules, or using such granules in negotiations with other agents. Parameterised approximation spaces are analogous to weights in classical neurons. Agents are performing operations on information granules (approximating concepts) rather than on numbers. This analogy has been used as a starting point for the rough-neuro computing paradigm [4].

4   Conclusions

We have discussed approximation spaces developed and investigated in the context of rough set theory and the role of such spaces in the calculi of information granules. Generalised approximation spaces have been briefly considered. These spaces include an uncertainty function as well an inclusion function defined relative to a set of objects called a universe. The basic idea of a calculus of granules rooted in a rough mereological approach has been presented in the context of information granulation approximation. It has also been suggested how one might create patterns that can be used to approximate target concepts. Pattern generators have also been considered. This research provides promising avenues for the study of pattern generation and the discovery of patterns for target concept approximation.

Acknowledgements

The research of A. Skowron has been supported by the State Committee for Scientific Research of the Republic of Poland (KBN) research grant 8 T11C 025 19 and by the Wallenberg Foundation grant. The research of J.F. Peters and S. Ramanna has been supported by the Natural Sciences and Engineering Research Council of Canada (NSERC) research grant 185986 and 194376, respectively.

References

1. Leśniewski, S.: Grundzüge eines neuen Systems der Grundlagen der Mathematik. *Fundamenta Mathematicae* **14** (1929) 1–81
2. Lin, T.Y., Yao, Y.Y., Zadeh, L.A. (eds.): *Data Mining, Rough Sets and Granular Computing*. Physica-Verlag, Heidelberg (2002)
3. Mitchell, T.M.: *Machine Learning*. Mc Graw-Hill, Portland (1997)
4. Pal, S.K., Polkowski, L., Skowron, A. (eds.): *Rough-Neuro Computing: Techniques for Computing with Words*. Springer-Verlag, Berlin (2003) (to appear)
5. Pal, S.K., Skowron, A. (eds.): *Rough Fuzzy Hybridization: A New Trend in Decision–Making*. Springer-Verlag, Singapore (1999)

6. Pawlak, Z.: Rough sets. *International Journal of Computer and Information Sciences* **11** (1982) 341–356
7. Pawlak, Z.: *Rough Sets. Theoretical Aspects of Reasoning about Data.* Kluwer Academic Publishers, Dordrecht (1991)
8. Peters, J.F., Ahn, T.C., Degtyaryov, V., Borkowski, M., Ramanna, S.: Autonomous Robotic Systems: Soft Computing and Hard Computing Methodologies and Applications. In: Zhou, C., Maravall, D., Ruan, D. (eds.), *Fusion of Soft Computing and Hard Computing for Autonomous Robotic Systems.* Physica-Verlag, Heidelberg (2003) 141–164
9. Peters, J.F., Ramanna, S., Borkowski, M., Skowron, A., Suraj, Z.: Sensor, filter and fusion models with rough Petri nets, *Fundamenta Informaticae* **47**(3-4) (2001) 307–323
10. Peters, J.F., Skowron, A., Stepaniuk, J., Ramanna, S.: Towards an ontology of approximate reason. *Fundamenta Informaticae* **51**(1-2) (2002) 157–173
11. Polkowski, L., Skowron, A.: Rough mereology: a new paradigm for approximate reasoning. *International J. Approximate Reasoning* **15**(4) (1996) 333–365
12. Polkowski, L., Skowron, A. (eds.): *Rough Sets in Knowledge Discovery* **1-2**. Physica-Verlag, Heidelberg (1998)
13. Polkowski, L., Skowron, A.: Towards adaptive calculus of granules. In: [27], (1999) 201–227
14. Polkowski, L., Skowron, A.: Rough mereological calculi of granules: A rough set approach to computation. *Computational Intelligence* **17**(3) (2001) 472–492
15. Polkowski, L., Skowron, A.: Rough-neuro computing. *Lecture Notes in Artificial Intelligence* **2005**, Springer-Verlag, Berlin (2002) 57–64
16. Skowron, A.: Toward intelligent systems: Calculi of information granules. *Bulletin of the International Rough Set Society* **5**(1-2) (2001) 9–30
17. Skowron, A., Approximate reasoning by agents in distributed environments. In: [28] (2001) 28–39
18. Skowron, A., Stepaniuk, J.: Tolerance approximation spaces. *Fundamenta Informaticae* **27** (1996) 245–253
19. Skowron, A., Stepaniuk, J.: Information granules: Towards foundations of granular computing. *International Journal of Intelligent Systems* **16**(1) (2001) 57–86
20. Skowron, A., Stepaniuk, J.: Information granules and rough-neuro computing. To appear in [4]
21. Słowiński, R., Greco, S., Matarazzo, B.: Rough set analysis of preference-ordered data. LNAI **2475**, Springer-Verlag, Heidelberg (2002) 44–59
22. WITAS. available at http://www.ida.liu.se/ext/witas/eng.html. *Project web page*
23. Wróblewski, J.: *Adaptive Methods of Object Classification.* Ph.D. Thesis, Warsaw University (2002) (in Polish)
24. Zadeh, L.A.: Fuzzy logic = computing with words. *IEEE Trans. on Fuzzy Systems* **4** (1996) 103–111
25. Zadeh, L.A.: Toward a theory of fuzzy information granulation and its certainty in human reasoning and fuzzy logic. *Fuzzy Sets and Systems* **90** (1997) 111–127
26. Zadeh, L.A.: A new direction in AI: Toward a computational theory of perceptions. *AI Magazine* **22**(1) (2001) 73–84
27. Zadeh, L.A., Kacprzyk, J. (eds.): *Computing with Words in Information/Intelligent Systems* **1-2**, Physica-Verlag, Heidelberg (1999)
28. Zhong, N., Liu, J., Ohsuga, S., Bradshaw, J. (eds.): *Intelligent agent technology: Research and development*, 2nd Asia-Pacific Conf. on IAT, Maebashi City (2001)

Indiscernibility-Based Clustering: Rough Clustering

Shoji Hirano and Shusaku Tsumoto

Department of Medical Informatics, Shimane Medical University, School of Medicine
89-1 Enya-cho, Izumo, Shimane 693-8501, Japan
hirano@ieee.org, tsumoto@computer.org

Abstract. This paper presents a new indiscernibility-based clustering method called rough clustering, that works on relative proximity. Our method lies its basis on iterative refinement of N binary classifications, where N denotes the number of objects. First, for each of N objects, an equivalence relation that classifies all the other objects into two classes, similar and dissimilar, is assigned by referring to their relative proximity. Next, for each pair of the objects, we count the number of binary classifications in which the pair is included in the same class. We call this number as indiscernibility degree. If the indiscernibility degree of a pair is larger than a user-defined threshold value, we modify the equivalence relations so that all of them commonly classify the pair into the same class. This process is repeated until class assignment becomes stable. Consequently, we obtain the clustering result that follows given level of granularity without using geometric measures.

keywords: clustering, rough sets, indiscernibility

1 Introduction

Clustering is characterized as a task to find the best partition of objects that maximizes internal cohesion and simultaneously maximizes external isolation. In order to produce high quality clusters, most of the widely-used clustering methods such as k-means, FCM, and BIRCH [1], employ quality measures that are associated with centroids of clusters. For example, internal cohesion of a cluster can be measured as a sum of differences from objects in the cluster to their centroid, and it can be further used as a part of the total quality measure for assessing a clustering result. Such centroid-based methods work well on a data set in which proximity of objects satisfies the natures of distance, that are, positivity ($d(x,y) \geq 0$), identity ($d(x,y) = 0$ iff $x = y$), symmetry ($d(x,y) = d(y,x)$), and triangular inequality ($d(x,z) \leq d(x,y) + d(y,z)$) for any objects x, y and z. However, they have a potential weakness in handling relative proximity. Relative proximity is a class of proximity measures that is suitable for representing subjective similarity or dissimilarity such as likeness of persons. It may not satisfy the triangular inequality because proximity $d(x,z)$ of x and z is allowed

T. Bilgiç et al. (Eds.): IFSA 2003, LNAI 2715, pp. 378–386, 2003.

to be independent of y. Usually, centroid c of objects x, y and z is expected to be in the convex full of them. However, if we use relative proximity, it can be out of their convex full because proximity between c and other objects can be far larger (if we use dissimilarity as proximity) or smaller (if we use similarity) than $d(x, y)$, $d(y, z)$ and $d(y, z)$. Namely, a centroid does not hold its geometric properties there. Thus another measure should be used for evaluating quality of the clusters.

This paper presents a new indiscernibility-based clustering method called rough clustering, that works on relative proximity. The main benefit of this method is that it can be applicable to proximity measures that do not satisfy the triangular inequality. Besides it works on a proximity matrix – thus does not require direct access to the original data values.

2   Rough Clustering

The basic and straightforward approach of rough sets [2] to clustering is to assign equivalence relations on attribute domain and group up objects according to the derived indiscernibility relations [3]. Table 1 shows a simple example. Suppose we have four objects $U = \{x_1, x_2, x_3, x_4\}$ in two-dimensional attribute domain $\Lambda = \{a_1, a_2\}$ as shown in Table 1(a). First, in order to evaluate indiscernibility

Table 1. An example of indiscernibility-based clustering

(a)		
	a_1	a_2
x_1	0.1	0.1
x_2	0.2	0.2
x_3	0.7	0.5
x_4	0.8	0.8

(b)					
	a_1	a_2	U/R_{a_1}	U/R_{a_2}	C
x_1	S	S	0	0	c_1
x_2	S	S	0	0	c_1
x_3	L	M	1	1	c_2
x_4	L	L	1	2	c_3

of objects, we discretize continuous attribute values according to proper criteria, such as discerptibility in their distribution or human decisions. Let us assume that values of a_1 and a_2 are discretized into three levels S(mall), M(edium), and L(arge) as shown in a_1 and a_2 in Table 1(b). Then the following equivalence relations can be defined.

$$U/R_{a_1} = \{\{x_1, x_2\}, \{x_3, x_4\}\},$$
$$U/R_{a_2} = \{\{x_1, x_2\}, \{x_3\}, \{x_4\}\}. \tag{1}$$

Hence using a family of equivalence relations $\mathbf{R} = \{R_{a_1}, R_{a_2}\}$ we obtain classification of U by $\mathbf{R}$ as

$$U/\mathbf{R} = \{\{x_1, x_2\}, \{x_3\}, \{x_4\}\}. \tag{2}$$

Table 2. Dissimilarity matrix for objects in Table 1(1)

	x_1	x_2	x_3	x_4
x_1	0	0.14	0.72	0.99
x_2	0.14	0	0.58	0.84
x_3	0.72	0.58	0	0.31
x_4	0.99	0.84	0.31	0

Table 3. An example of the proposed indiscernibility-based clustering. Note: 0 represents class 'similar' and 1 represents class 'dissimilar'

	U/R_1	U/R_2	U/R_3	U/R_4	C
x_1	0	0	1	1	c_1
x_2	0	0	1	1	c_1
x_3	1	1	0	0	c_2
x_4	1	1	0	0	c_2

Using a set of categories C, this can be rewritten as $U/\mathbf{R} = C = \{c_1, c_2, c_3\}$, where $c_1 = \{x_1, x_2\}$, $c_2 = \{x_2\}$, and $c_3 = \{x_4\}$. If we discretize a_2 into two levels Small and Large, we obtain

$$U/R_{a_2} = \{\{x_1, x_2\}, \{x_3, x_4\}\}, \tag{3}$$

and hence

$$U/\mathbf{R} = \{\{x_1, x_2\}, \{x_3, x_4\}\}. \tag{4}$$

This example shows that, from the standpoint of rough sets, clustering can be regarded as a process to find a set of knowledge, namely a set of equivalence relations on attribute domain, that classifies objects with an appropriate level of granularity. If the knowledge is too fine, a lot of meaninglessly small categories would be obtained. On the contrary, if it is too coarse, only small number of categories with less interesting information would be obtained.

The main difference between existing rough sets-based clustering techniques and our method is that our method defines an equivalence relation not on each attribute but on each objects. That is to say, for N objects, we have N equivalence relations regardless of the number of attributes. Each of the N equivalence relations performs binary classification based on the relative proximity between the object to which the relation is assigned and other objects. Let us show an example using objects in Table 1(a). First, we calculate dissimilarity matrix of the objects as shown in Table 2. For simplicity, we here used the Euclidean distance as a measure of dissimilarity. Next, for object x_1, we assign an equivalence relation R_1 that groups up objects similar to x_1 as follows.

$$U/R_1 = \{\{x_1, x_2\}, \{x_3, x_4\}\}, \tag{5}$$

Here we regard that object $x_j (1 \leq j \leq 4)$ is similar to x_1 if dissimilarity of x_1 to x_j, say $d(x_1, x_j)$ is smaller than 0.5. Note that R_1 evaluates only dissimilarity from x_1 to each of the other objects and does not evaluate dissimilarity between other objects, e.g., $d(x_2, x_3)$ between x_2 and x_3. In the same way, we obtain R_2, R_3, and R_4 as shown in Table 3. Finally, according to the four binary classifications U/R_1, U/R_2, U/R_3 and U/R_4, we obtain a set of categories C. In this example there are two categories c_1 and c_2, where $c_1 = \{x_1, x_2\}$ and

$c_2 = \{x_3, x_4\}$. This can be rewritten in the same way to the previous case as $\mathbf{R} = \{R_1, R_2, R_3, R_4\}$ and $U/\mathbf{R} = \{\{x_1, x_2\}, \{x_3, x_4\}\}$.

The main advantage of this approach is that it requires only proximity matrix of objects and the proximity matrix is not necessarily required to satisfy the triangular inequality. However, the indiscernibility relation derived using these 'initial' equivalence relations usually tends to split the objects into many equivalence classes, i.e., it gives a lot of fine categories. This is because no global relationship among the equivalence relations is considered and thus slightly different equivalence relations may discriminate closely located objects. Therefore, after assigning initial equivalence relations, we refine them so that the resultant indiscernibility relation gives adequately coarse classification to the objects. The refinement is iteratively performed several times by evaluating the *indiscernibility degree* of objects, which reflects global relationships among the equivalence relations. The indiscernibility degree is derived as a ratio of equivalence relations that commonly regard the two objects as indiscernible. When the degree is larger than a predefined threshold value, the two objects are considered to be indiscernible, and all the equivalence relations are refined to classify them into the same category. The refinement process is iterated several times, because the new candidates to be merged could appear on the refined set of equivalence relations.

2.1   Assignment of Initial Equivalence Relations

Let $U = \{x_1, x_2, ..., x_N\}$ be the set of objects we are interested in. An equivalence relation R_i for object x_i is defined by

$$U/R_i = \{P_i, \ U - P_i\}, \tag{6}$$

where

$$P_i = \{x_j | \ d(x_i, x_j) \le Th_{di}\}, \quad \forall x_j \in U. \tag{7}$$

$d(x_i, x_j)$ denotes dissimilarity between objects x_i and x_j, and Th_{di} denotes an upper threshold value of dissimilarity for object x_i. Equivalence relation R_i classifies U into two categories: P_i containing objects similar to x_i and $U - P_i$ containing objects dissimilar to x_i. When $d(x_i, x_j)$ is smaller than Th_{di}, object x_j is considered to be indiscernible to x_i.

Definition of dissimilarity measure $d(x_i, x_j)$ is arbitrary. If all the attribute values are numerical, ordered and independent of each other, conventional Euclidean distance is a reasonable choice since it has been successfully applied to many areas and its mathematical properties are well investigated. More generally, any type of dissimilarity measure can be used regardless of whether or not triangular inequality is satisfied among objects. Threshold of dissimilarity Th_{di} for object x_i is automatically determined based on the denseness of objects. Due to limitation of the pages, we omit description of the process of threshold determination. Details are available in Ref. [4].

2.2   Refinement of Initial Equivalence Relations

Suppose we are interested in two objects, x_i and x_j. In indiscernibility-based classification, they are classified into different categories regardless of other relations if there is at least one equivalence relation that has an ability to discern them. In other words, the two objects are classified into the same category only when all of the equivalence relations commonly regard them as indiscernible objects. This strict property is not acceptable in clustering because it will cause generation of a lot of meaninglessly small categories, especially when no global associations between the equivalence relations is taken into account. We consider that objects should be classified into the same category when most of, not all of, the equivalence relations commonly regard them as indiscernible. In the second stage, we perform global optimization of initial equivalence relations so that they give adequately coarse classification to the objects. The global similarity of objects is represented by a newly introduced measure, *indiscernibility degree*. Our method takes a threshold value of the indiscernibility degree as an input and associates it to the user-defined granularity of the categories. Given the threshold value, we iteratively refine the initial equivalence relations in order to produce categories that meet the given level of granularity.

Now let us assume $U = \{x_1, x_2, x_3, x_4, x_5\}$ and classifications of U by $\mathbf{R} = \{R_1, R_2, R_3, R_4, R_5\}$ is given as follows.

$$U/R_1 = \{\{x_1, x_2, x_3\}, \{x_4, x_5\}\},$$
$$U/R_2 = \{\{x_1, x_2, x_3\}, \{x_4, x_5\}\},$$
$$U/R_3 = \{\{x_2, x_3, x_4\}, \{x_1, x_5\}\},$$
$$U/R_4 = \{\{x_1, x_2, x_3, x_4\}, \{x_5\}\},$$
$$U/R_5 = \{\{x_4, x_5\}, \{x_1, x_2, x_3\}\}. \tag{8}$$

This example contains three types of equivalence relations: $R_1 (= R_2 = R_5)$, R_3 and R_4. Since each of them gives slightly different classification to U, classification of U by the family of equivalence relations $\mathbf{R}$, $U/\mathbf{R}$, contains four very small, almost independent categories.

$$U/\mathbf{R} = \{\{x_1\}, \{x_2, x_3\}, \{x_4\}, \{x_5\}\}. \tag{9}$$

In the following we present a method to reduce variety of equivalence relations and to obtain coarser categories.

First, we define an *indiscernibility degree*, $\gamma(x_i, x_j)$, of two objects x_i and x_j as follows.

$$\gamma(x_i, x_j) = \frac{\sum_{k=1}^{|U|} \delta_k^{indis}(x_i, x_j)}{\sum_{k=1}^{|U|} \delta_k^{indis}(x_i, x_j) + \sum_{k=1}^{|U|} \delta_k^{dis}(x_i, x_j)}, \tag{10}$$

where

$$\delta_k^{indis}(x_i, x_j) = \begin{cases} 1, & \text{if } (x_i \in [x_k]_{R_k} \wedge x_j \in [x_k]_{R_k}) \\ 0, & \text{otherwise.} \end{cases} \tag{11}$$

Table 4. Degree γ for objects in Eq. (8).

	x_1	x_2	x_3	x_4	x_5
x_1	3/3	3/4	3/4	1/5	0/4
x_2		4/4	4/4	2/5	0/5
x_3			4/4	2/5	0/5
x_4				3/3	1/3
x_5					1/1

Table 5. Degree γ after the first refinement.

	x_1	x_2	x_3	x_4	x_5
x_1	3/3	3/4	3/4	2/4	1/5
x_2		4/4	4/4	3/4	0/5
x_3			4/4	3/4	0/5
x_4				3/3	1/5
x_5					1/1

Table 6. Degree γ after the second refinement.

	x_1	x_2	x_3	x_4	x_5
x_1	4/4	4/4	4/4	4/4	0/5
x_2		4/4	4/4	4/4	0/5
x_3			4/4	4/4	0/5
x_4				4/4	0/5
x_5					1/1

and

$$\delta_k^{dis}(x_i, x_j) = \begin{cases} 1, \text{ if } (x_i \in [x_k]_{R_k} \wedge x_j \notin [x_k]_{R_k}) \text{ or} \\ \quad \text{ if } (x_i \notin [x_k]_{R_k} \wedge x_j \in [x_k]_{R_k}) \\ 0, \text{ otherwise.} \end{cases} \tag{12}$$

Equation (11) shows that $\delta_k^{indis}(x_i, x_j)$ takes 1 only when equivalence relation R_k regards x_i and x_j as indiscernible objects under the condition that both of them are in the same equivalence class to x_k. Equation (12) shows that $\delta_k^{dis}(x_i, x_j)$ takes 1 only when R_k regards x_i and x_j as discernible objects under the condition that either of them is in the same class to x_k. By summing up $\delta_k^{indis}(x_i, x_j)$ and $\delta_k^{dis}(x_i, x_j)$ for all $k(1 \leq k \leq |U|)$ as Equation (10), we obtain the percentage of equivalence relations that regard x_i and x_j as indiscernible objects. Note that in Equation (11) we excluded the case when x_i and x_j are indiscernible but both are not in the same class to x_k. This is to exclude the case where R_k does not significantly put weights on discerning x_i and x_j. As mentioned in Section 2.1, P_k for R_k is determined by focusing on similar objects rather than dissimilar objects. This means that when both of x_i and x_j are far dissimilar to x_k, their dissimilarity is not significant for x_k in determining dissimilarity threshold Th_{dk}. Thus we only count the number of equivalence relations that certainly evaluate dissimilarity of x_i and x_j. Indiscernibility degrees for every pairs in U are tabulated in Table 4. Note that the indiscernibility degree of object x_i to itself, $\gamma(x_i, x_i)$, will always be 1.

From its definition the higher $\gamma(x_i, x_j)$ represents that x_i and x_j are commonly regarded as indiscernible objects by large number of the equivalence relations. Therefore, if an equivalence relation R_l discerns the objects that have high γ value, we consider that it represents excessively fine classification knowledge and refine it according to the following procedure (note that R_l is rewritten as R_i below for the purpose of generalization).

Let $R_i \in \mathbf{R}$ be an initial equivalence relation on U. A refined equivalence relation $R_i' \in \mathbf{R}'$ of R_i is defined as

$$U/R_i' = \{P_i', \ U - P_i'\}, \tag{13}$$

where P_i' denotes a set of objects represented by

$$P_i' = \{x_j | \gamma(x_i, x_j) \geq T_h\}, \quad \forall x_j \in U. \tag{14}$$

and T_h denotes the lower threshold value of indiscernibility degree above which x_i and x_j are regarded as indiscernible objects. It represents that when $\gamma(x_i, x_j)$ is larger than T_h, R_i is modified to include x_j into the class of x_i.

Suppose we are given $Th = 3/5$ for the case in Equation (8). For R_1 we obtain the refined relation R'_1 as

$$U/R'_1 = \{\{x_1, x_2, x_3\}, \{x_4, x_5\}\}, \tag{15}$$

because, according to Table 4, $\gamma(x_1, x_1) = 1 \geq T_h = 3/5$, $\gamma(x_1, x_2) = 3/4 \geq 3/5$, $\gamma(x_1, x_3) = 3/4 \geq 3/5$, $\gamma(x_1, x_4) = 1/5 \leq 3/5$, $\gamma(x_1, x_5) = 0/5 \leq 3/5$ hold. In the same way, the rest of the refined equivalence relations is obtained as follows.

$$U/R'_2 = \{\{x_1, x_2, x_3\}, \{x_4, x_5\}\},$$
$$U/R'_3 = \{\{x_1, x_2, x_3\}, \{x_4, x_5\}\},$$
$$U/R'_4 = \{\{x_4\}, \{x_1, x_2, x_3, x_5\}\},$$
$$U/R'_5 = \{\{x_5\}, \{x_1, x_2, x_3, x_4\}\}. \tag{16}$$

Then we obtain classification of U by the refined family of equivalence relations $\mathbf{R}'$ as follows.

$$U/\mathbf{R}' = \{\{x_1, x_2, x_3\}, \{x_4\}, \{x_5\}\}. \tag{17}$$

In the above example, R_3, R_4 and R_5 are modified so that they include similar objects into the equivalence class of x_3, x_4 and x_5, respectively. Types of the equivalence relations remain to be three, however, the categories become coarser than those in Equation (9) by the refinement.

2.3   Iterative Refinement of Equivalence Relations

It should be noted that the state of the indiscernibility degrees could also be changed after refinement of the equivalence relations, since the degrees are recalculated using the refined family of equivalence relations $\mathbf{R}'$.

Suppose we are given another threshold value $T_h = 2/5$ for the case in Equation (8). According to Table 4, we obtain $\mathbf{R}'$ after the first refinement as follows.

$$U/R'_1 = \{\{x_1, x_2, x_3\}, \{x_4, x_5\}\},$$
$$U/R'_2 = \{\{x_1, x_2, x_3, x_4\}, \{x_5\}\},$$
$$U/R'_3 = \{\{x_1, x_2, x_3, x_4\}, \{x_5\}\},$$
$$U/R'_4 = \{\{x_2, x_3, x_4\}, \{x_1, x_5\}\},$$
$$U/R'_5 = \{\{x_5\}, \{x_1, x_2, x_3, x_4\}\}. \tag{18}$$

Hence

$$U/\mathbf{R}' = \{\{x_1\}, \{x_2, x_3\}, \{x_4\}, \{x_5\}\}. \tag{19}$$

The categories in $U/\mathbf{R}'$ are exactly the same as those in Equation (9). However, the state of the indiscernibility degrees are not the same because equivalence relations in $\mathbf{R}'$ are different from those in $\mathbf{R}$. Table 5 summarizes the indiscernibility degrees recalculated using $\mathbf{R}'$. In Table 5 it can be observed that the indiscernibility degrees of some pairs of objects, for example $\gamma(x_1, x_4)$, are increased by the refinement and now they exceed the threshold $th = 2/5$. Thus we perform refinement of equivalence relations again using the same T_h and the recalculated γ. Then we obtain

$$U/R_1' = U/R_2' = U/R_3' = U/R_4' = \{\{x_1, x_2, x_3, x_4\}, \{x_5\}\},$$
$$U/R_5' = \{\{x_5\}, \{x_1, x_2, x_3, x_4\}\}. \tag{20}$$

Hence

$$U/\mathbf{R}' = \{\{x_1, x_2, x_3, x_4\}, \{x_5\}\}. \tag{21}$$

After the second refinement, types of the equivalence relations in $\mathbf{R}'$ are reduced to 2, and the number of categories are also reduced from 4 to 2. We further update the state of the indiscernibility degrees according to the equivalence relations after second refinement. The results are tabulated in Table 6. Since no new pairs whose indiscernibility degree exceeds the given threshold appears, we stop the refinement process and obtain the stable categories as in Equation (21).

As shown in this example, refinement of the equivalence relations may change indiscernibility degree of objects. Thus we iterate the refinement process using the same T_h until the categories become stable. Note that each refinement process is performed using the previously 'refined' set of equivalence relations.

3   Conclusions

In this paper, we have presented rough clustering, which clusters objects according to their relative proximity. We introduced the concept of indiscernibility degree, represented by the number of equivalence relations that commonly regard two objects as indiscernible, as a measure of global similarity between the two. By the use of indiscernibility degree and iterative refinement of equivalence relations, the method classified objects according to their global characteristics, regardless of local difference that might produce unacceptably fine categories.

Acknowledgment

This work was supported in part by the Grant-in-Aid for Scientific Research on Priority Area (B)(No.759) by the Ministry of Education, Culture, Science and Technology of Japan.

References

1. P. Berkhin (2002): Survey of Clustering Data Mining Techniques. Accrue Software Research Paper. URL: http://www.accrue.com/products/researchpapers.html.
2. Z. Pawlak (1991): Rough Sets, Theoretical Aspects of Reasoning about Data. Kluwer Academic Publishers, Dordrecht.

3. J. W. Grzymala-Busse and M. Noordeen (1988): "CRS – A Program for Clustering Based on Rough Set Theory," Research report, Department of Computer Science, University of Kansas, TR-88-3, 13.
4. S. Hirano and S. Tsumoto (2003): An Indiscernibility-Based Clustering Method with Iterative Refinement of Equivalence Relations – Rough Clustering –," Journal of Advanced Computational Intelligence and Intelligent Informatics, (in press).

Fuzzy Multiset Space and c-Means Clustering Using Kernels with Application to Information Retrieval

Sadaaki Miyamoto[1] and Kiyotaka Mizutani[2]

[1] Institute of Engineering Mechanics and Systems, University of Tsukuba,
Ibaraki 305-8573, Japan,
miyamoto@esys.tsukuba.ac.jp
[2] Graduate School of Systems and Information Engineering, University of Tsukuba,
Ibaraki 305-8573, Japan,
kiyotaka@odin.esys.tsukuba.ac.jp

Abstract. A space of fuzzy multisets is considered and applied to clustering of documents/terms for information retrieval. Fuzzy c-means clustering algorithms with kernel functions in support vector machines are studied. A numerical example is given.

1 Introduction

Multiset processing is now widely studied with applications to modeling of information processing, such as relational database, information retrieval, new computation paradigms, and so on [2].

Fuzzy multisets (also called fuzzy bag) have been proposed by Yager [11] and operations have been redefined by Miyamoto [7,8]. Since fuzzy multisets admit multiple occurrences of an object with possibly different memberships, it is appropriate for modeling information retrieval process on the web.

Clustering of information items such as documents and keywords are particularly interested in by many researchers from the viewpoint of data summarizing and mining [5,9]. Thus clustering based on fuzzy multiset space should be studied in this regard.

In this paper we study a fuzzy multiset space and fuzzy c-means clustering with application to information retrieval. In particular kernel functions in support vector machines [12] are used. Our study is similar to the method by Girolami [4], but the similarity measure for document clustering is new.

2 Fuzzy Multiset Space

Let X be a finite universal set on which we consider crisp and fuzzy multisets. A crisp multiset K of X is characterized by the count function

$$Cnt_K : X \to \{0, 1, 2, \dots\}.$$

T. Bilgiç et al. (Eds.): IFSA 2003, LNAI 2715, pp. 387–395, 2003.

The meaning of the function is the multiplicity of an object. If $Cnt_K(a) = 5$, then a occurs five times in K.

When $Cnt_M(a) = 2$, $Cnt_M(b) = 3$, and $Cnt_M(c) = 2$, we write $M = \{2/a, 2/b, 3/c\}$, or $M = \{a, a, b, c, b, c, c\}$ like an ordinary set. A multiset K is called finite if $Cnt_K(x) < \infty$ for all $x \in X$. The collection of all finite crisp multisets of X is denoted by $\mathcal{M}(X)$.

Basic relations and operations for crisp multisets are as follows.

(i) inclusion:
$$M \subseteq N \iff Cnt_M(x) \le Cnt_N(x), \ \forall x \in X.$$

(ii) sum:
$$Cnt_{M+N}(x) = Cnt_M(x) + Cnt_N(x).$$

(iii) union and intersection:
$$Cnt_{M \cup N}(x) = Cnt_M(x) \vee Cnt_N(x),$$
$$Cnt_{M \cap N}(x) = Cnt_M(x) \wedge Cnt_N(x).$$

Let $I = [0, 1]$ be the unit interval. A fuzzy multiset A of X is characterized by the count function
$$Cnt_A \colon X \to \mathcal{M}(I)$$

We assume that $Cnt_A(x)$ is finite for all $x \in X$ throughout this paper. Then we can write
$$Cnt_A(x) = \{v_1(x), v_2(x), \ldots, v_l(x)\}. \tag{1}$$

The collection of all fuzzy multisets of X is denoted by $\mathcal{FM}(X)$. Moreover the collection of all finite sequences $(\mu_1, \ldots, \mu_k)$ in which members are in the unit interval $(\mu_j \in I, \ j = 1, \ldots, k)$ is denoted by $\mathcal{SQ}(I)$. A member $(\mu_1, \ldots, \mu_k)$ in $\mathcal{SQ}(I)$ can be identified with the corresponding multiset $\{\mu_1, \ldots, \mu_k\}$.

It has been shown that a standard form is necessary to define fuzzy multiset operations. A transformation $S \colon \mathcal{M}(I) \to \mathcal{SQ}(I)$ which is a simple sorting into decreasing order of the elements is used for this purpose. That is, for $M = \{v_1(x), v_2(x), \ldots, v_l(x)\}$,

$$S(M) = (\nu^1, \nu^2, \ldots, \nu^l)$$

where $\nu^1 \ge \nu^2 \ge \cdots \ge \nu^l$ and

$$\{\nu^1, \nu^2, \ldots, \nu^l\} = \{v_1(x), v_2(x), \ldots, v_l(x)\}$$

as a multiset. Thus, S is the sorting of a given multiset of the unit interval into the decreasing order. Now, basic relations and operations of fuzzy multisets A and B of X are defined as follows.

1.
$$A \subseteq B \iff S(Cnt_A(x)) \le S(Cnt_B(x)), \ \forall x \in X$$

2.
$$Cnt_{A \cup B}(x) = S(Cnt_A(x)) \vee S(Cnt_B(x))$$

3.
$$Cnt_{A \cap B}(x) = S(Cnt_A(x)) \wedge S(Cnt_B(x))$$

where $\vee$ is componentwise maximum for the two sequences.

4.
$$Cnt_{A+B}(x) = S(Cnt_A(x)) | S(Cnt_B(x))$$

where $\vee$ is componentwise minimum for the two sequences.

Note that $A + B$ is defined by the concatenation of two sequences:

$$(\nu^1, \ldots, \nu^l) | (\nu'^1, \ldots, \nu'^{l'}) = (\nu^1, \ldots, \nu^l, \nu'^1, \ldots, \nu'^{l'}).$$

We moreover introduce a *product* operation $A \cdot B$: if $Cnt_A(x) = (\nu^1, \ldots, \nu^l)$ and $Cnt_B(x) = (\nu'^1, \ldots, \nu'^{l'})$, and $l \leq l'$, then

$$Cnt_{A \cdot B}(x) = (\nu^1 \nu'^1, \ldots, \nu^l \nu'^l, 0, \ldots)$$

For a constant $\beta \in [0, 1]$, βA is defined:

$$Cnt_{\beta A}(x) = (\beta \nu^1, \ldots, \beta \nu^l, 0, \ldots)$$

The cardinality and a *norm* of a multiset A are moreover defined:

$$|A| = \sum_{x \in X} Cnt_A(x),$$

$$\|A\| = \sqrt{\sum_{x \in X} Cnt_{A \cdot A}(x)}.$$

3   Algorithms of c-Means Clustering

Among various methods of clustering, fuzzy c-means [3,1,7] have most been employed in various applications. A drawback of fuzzy c-means is that nonlinear cluster boundary cannot be obtained. Recently, the use of kernel functions in support vector machines [12] in crisp c-means has been proposed by Girolami [4] to obtain clusters with nonlinear boundaries where the Euclidean data space is transformed into a high-dimensional feature space.

In order to apply this method to document clustering, we should consider two points: what space should be adopted and what measure should be used. For the first point we employ the fuzzy multiset space and for the second we use a generalization of the cosine coefficient, as this measure is most frequently used in document clustering [10].

Thus, instead of the Euclidean distance we use

$$s(A, B) = \frac{|A \cdot B|}{\|A\| \|B\|}. \tag{2}$$

Accordingly, the objective function for fuzzy c-means is

$$J_s(U, V) = \sum_{i=1}^{c} \sum_{k=1}^{n} (u_{ik})^m s(A_k, V_i), \quad (m > 1) \tag{3}$$

or the entropy-based objective function [6,7] is used:

$$J_e(U, V) = \sum_{i=1}^{c} \sum_{k=1}^{n} \{u_{ik} s(A_k, V_i) - \lambda^{-1} u_{ik} \log u_{ik}\}, \tag{4}$$

where $\lambda > 0$, $\{A_1, \ldots, A_n\}$ are fuzzy multisets to be classified and $V = (V_1, \ldots, V_c)$ are fuzzy multisets as cluster centers. Notice moreover that we consider maximization of the objective functions, as a similarity measure is used instead of a distance.

Since we consider alternate maximization of the above functions, the iterative solutions for (3) are

$$u_{ik} = \frac{s(A_k, V_i)^{\frac{1}{m-1}}}{\sum_{j=1}^{c} s(A_k, V_j)^{\frac{1}{m-1}}}, \tag{5}$$

$$V_i = \frac{\sum_{k=1}^{n} (u_{ik})^m A_k}{\| \sum_{k=1}^{n} (u_{ik})^m A_k \|}. \tag{6}$$

For (4),

$$u_{ik} = \frac{\exp(\lambda s(A_k, V_i))}{\sum_{j=1}^{c} \exp(\lambda s(A_k, V_j))}, \tag{7}$$

$$V_i = \frac{\sum_{k=1}^{n} u_{ik} A_k}{\| \sum_{k=1}^{n} u_{ik} A_k \|}. \tag{8}$$

Kernel-Based Clustering

Kernel functions [12] enable nonlinear classifications. Effectiveness for the above measure with application to information retrieval should thus be studied. We consider a transformation $\Phi: \mathcal{FM}(X) \to H$ where H is a high-dimensional feature space. Instead of (2), we use

$$s_H(\Phi(A), W) = \frac{\langle \Phi(A), W \rangle}{\|\Phi(A)\|_H \|W\|_H} \tag{9}$$

where $\langle x, y \rangle$ is the inner product of H and $\|x\|_H^2 = \langle x, x \rangle$.

The objective functions are

$$J_s(U, W) = \sum_{i=1}^{c} \sum_{k=1}^{n} (u_{ik})^m s_H(\Phi(A_k), W_i), \tag{10}$$

$$J_e(U, W) = \sum_{i=1}^{c} \sum_{k=1}^{n} \{u_{ik} s_H(\Phi(A_k), W_i) \\ - \lambda^{-1} u_{ik} \log u_{ik}\}, \tag{11}$$

instead of (3) and (4), respectively.

The explicit form of $\Phi(x)$ is not known but the inner product is represented by a kernel:

$$K(x,y) = \langle \Phi(x), \Phi(y) \rangle.$$

In particular, the Radial Basis Function(RBF) kernel [12] is used here:

$$K(A,B) = \exp(-C\|A - B\|^2), \quad (C > 0)$$

where

$$\|A - B\|^2 = \|A\|^2 + \|B\|^2 - 2|A \cdot B|.$$

Notice that the right hand side is calculated by the definitions of the norms and the products for fuzzy multisets.

Let us consider (11) as the objective function, as the derivation is similar for the both objectives. The maximizing solution of u_{ik} is essentially the same as (7):

$$u_{ik} = \frac{\exp(\lambda s_H(\Phi(A_k), W_i))}{\sum_{j=1}^{c} \exp(\lambda s_H(\Phi(A_k), W_j))}, \tag{12}$$

whereas the maximization with respect to W_i leads to

$$W_i = \frac{\sum_{k=1}^{n} u_{ik}\Phi(A_k)}{\| \sum_{k=1}^{n} u_{ik}\Phi(A_k)\|}.$$

in which $\Phi(A_k)$ is unknown.

We remark here that to know $s_H(A_k, W_i)$ instead of W_i is sufficient to calculate the solutions. Substituting the above W_i into (9), we have

$$s_H(\Phi(A_k), W_i) = \frac{\sum_\ell u_{i\ell} K_{k\ell}}{\sqrt{K_{kk} \sum_j \sum_\ell u_{ij} u_{i\ell} K_{j\ell}}}, \tag{13}$$

$$(K_{j\ell} = K(A_j, A_\ell)).$$

Thus, the alternate optimization for (11) is to repeat (12) and (13) until convergence.

The optimization for (10) is similar to the above analysis and hence is omitted.

4   A Numerical Example

Table 1 shows fuzzy multisets for 22 documents which have been extracted from a proceeding in a fuzzy systems symposium in Japan. The original documents are in Japanese and we omitted the titles. The documents are represented by $A, B, \ldots, V$; x, y, and z represent three keywords of 'clustering', 'fuzzy', and 'algorithm', respectively, in these documents. The rules for determining the membership values are as follows.

- If a keyword w is in the title, its membership is 1.0;
- if w is in a section title, its membership is 0.5;
- if w is in the abstract, its membership is 0.2;
- else its membership is 0.

A major part of documents are concerning clustering and other documents do not. Fuzzy c-means by (11) and (4) with and without the kernel have been applied. The parameters are $c = 2$, $\lambda = 10.0$, and $C = 0.5$ in the RBF kernel.

Table 1. Fuzzy multisets for 22 documents.

x: 'clustering'; y: 'fuzzy'; z: 'algorithm'
$A = \{(1,.5,.2)/x, (1,.5,.2)/y, (.5,.2,0)/z\}$
$B = \{(1,.5,.2)/x, (1,.5,.2)/y, (.2,0,0)/z\}$
$C = \{(0,0,0)/x, (1,.5,.2)/y, (.5,.2,0)/z\}$
$D = \{(.5,.2,0)/x, (1,.5,.2)/y, (.5,.2,0)/z\}$
$E = \{(.5,.2,0)/x, (.5,.2,0)/y, (.5,.2,0)/z\}$
$F = \{(1,.2,0)/x, (1,.5,.2)/y, (.2,0,0)/z\}$
$G = \{(0,0,0)/x, (1,.2,0)/y, (1,.5,.2)/z\}$
$H = \{(0,0,0)/x, (1,.5,.2)/y, (1,.2,0)/z\}$
$I = \{(.5,.2,0)/x, (1,.5,.2)/y, (0,0,0)/z\}$
$J = \{(1,.5,.2)/x, (1,.5,.2)/y, (.5,.2,0)/z\}$
$K = \{(1,.5,.2)/x, (1,.5,.2)/y, (.2,0,0)/z\}$
$L = \{(0,0,0)/x, (1,.5,.2)/y, (1,.5,.2)/z\}$
$M = \{(1,.5,.2)/x, (.5,.2,0)/y, (.2,0,0)/z\}$
$N = \{(.2,0,0)/x, (.5,.2,0)/y, (.5,.2,0)/z\}$
$O = \{(1,.2,0)/x, (1,.5,.2)/y, (.2,0,0)/z\}$
$P = \{(.2,0,0)/x, (1,.5,.2)/y, (.5,.2,0)/z\}$
$Q = \{(1,.5,.2)/x, (1,.5,.2)/y, (.2,0,0)/z\}$
$R = \{(1,.2,0)/x, (.5,.2,0)/y, (.2,0,0)/z\}$
$S = \{(1,.2,0)/x, (1,.5,.2)/y, (0,0,0)/z\}$
$T = \{(1,.5,.2)/x, (1,.5,.2)/y, (.2,0,0)/z\}$
$U = \{(1,.5,.2)/x, (1,.2,0)/y, (.5,.2,0)/z\}$
$V = \{(1,.2,0)/x, (1,.5,.2)/y, (1,.2,0)/z\}$

Figures 1 and 2 show the memberships for a cluster. The horizontal axis is the norm of the fuzzy multisets and the vertical axis is the membership value. A square $\square$ shows a document concerning 'clustering', while a cross $\times$ implies other document. Hence this cluster implies documents of 'clustering'. There are four documents misclassified to the other cluster in Fig. 1, while three documents are misclassified in Fig. 2.

Moreover separation of the two clusters is improved in Fig. 2. We use a separation measure SM defined by the entropy for showing improvement of separation:

$$SM = \sum_{i=1}^{c} \sum_{k=1}^{n} u_{ik} \log u_{ik}.$$

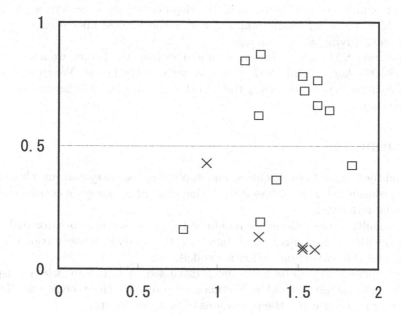

Fig. 1. Result from (4) without a kernel. The horizontal axis is the norm of the fuzzy multisets and the vertical axis is the membership value for a cluster.

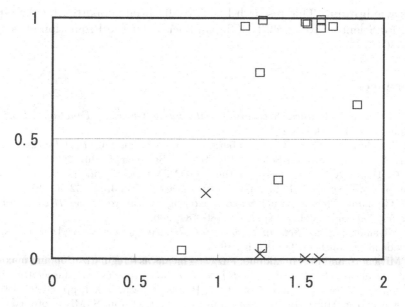

Fig. 2. Result from (11) with the RBF kernel.

Notice that this measure has been used for cluster validation [1]. It is easily seen that the maximum of SM is attained for crisp allocation: $u_{ik} = 0$ or $u_{ik} = 1$ for which $SM = 0.0$, whereas minimum of SM is the fuzziest allocation $u_{ik} = 1/c$ for all i and k giving $SM = -n \log c$.

Calculating SM for the clusters with and without the kernel, we have $SM = -5.31$ with the kernel while $SM = -11.46$ without the kernel. We thus observe improvement of the clusters using the kernel from both the misclassification and the separation measure.

5   Conclusion

Fuzzy multisets have been discussed and application to fuzzy c-means clustering has been considered. Kernel-based clustering algorithm using the cosine correlation has been derived.

Fuzzy multiset operations are redefined using the sorting operator S. When compared with the previous formulation [7,8], the present representation is more compact and the sorting operation is explicit.

The kernel-based approach has been discussed by Girolami, where crisp c-means using the square Euclidean distance are proposed. Here the new similarity measure is proposed using the operations of fuzzy multisets.

Future research include reduction of computation in clustering and application to large number of documents. Furthermore, application to information retrieval of more complex structure including images and multimedia information sources should be studied using the present model.

Acknowledgment This research has partially been supported by the Grant-in-Aid for Scientific Research (C), Japan Society for the Promotion of Science, No.13680475.

References

1. J. C. Bezdek, *Pattern Recognition with Fuzzy Objective Function Algorithms*, Plenum, New York, 1981.
2. C. S. Calude, G. Păun, G. Rozenberg, A. Salomaa, eds., *Multiset Processing*, Lecture Notes in Computer Science, LNCS 2235, Springer, Berlin, 2001.
3. J. C. Dunn, A fuzzy relative of the ISODATA process and its use in detecting compact well-separated clusters, *J. of Cybernetics*, Vol.3, pp. 32–57, 1974.
4. M. Girolami, Mercer kernel based clustering in feature space, *IEEE Trans. on Neural Networks*, Vol.13, No.3, pp. 780–784, 2002.
5. S. Miyamoto, *Fuzzy Sets in Information Retrieval and Cluster Analysis*, Kluwer Academic Publishers, Dordrecht, 1990.
6. S. Miyamoto and M. Mukaidono, Fuzzy c-means as a regularization and maximum entropy approach, *Proc. of the 7th International Fuzzy Systems Association World Congress (IFSA'97)*, June 25-30, 1997, Prague, Chech, Vol.II, pp. 86–92, 1997.
7. S. Miyamoto, Multisets and fuzzy multisets, in Z. Q. Liu, S. Miyamoto, eds., *Soft Computing and Human Centered Machines*, Springer, Tokyo, 2000, pp. 9–34.

8. S. Miyamoto, Fuzzy multisets and their generalizations, in C. S. Calude et al. eds., *Multiset Processing*, LNCS 2235, Springer, Berlin, 2001, pp. 225–235.
9. S. Miyamoto, Information clustering based on fuzzy multisets, *Information Processing and Management* Vol.39, pp. 195–213, 2003.
10. G. Salton, M. J. MacGill, *Introduction to Modern Information Retrieval*, McGraw-Hill, New York, 1983.
11. R. R. Yager, On the theory of bags, *Int. J. General Systems*, Vol. 13, pp. 23–37, 1986.
12. V. Vapnik, *Statistical Learning Theory*, Wiley, New York, 1998.

Using Similarity Measures for Histogram Comparison

Dietrich Van der Weken[1], Mike Nachtegael[1], and Etienne Kerre[1]

Fuzziness & Uncertainty Modelling, Ghent University
Krijgslaan 281 (building S9), 9000 Ghent, Belgium
{dietrich.vanderweken,mike.nachtegael,etienne.kerre}@rug.ac.be
http://fuzzy.rug.ac.be

Abstract. Objective quality measures or measures of comparison are of great importance in the field of image processing. Such measures are needed for the evaluation and the comparison of different algorithms that are designed to solve a similar problem, and consequently they serve as a basis on which one algorithm is preferred above the other. Similarity measures, originally introduced to compare two fuzzy sets, can be applied in different ways to images. In [2] we gave an overview of similarity measures which can be applied straightforward to images. In this paper, we will show how some similarity measures can be applied to normalized histograms of images.

1 Introduction

An important problem in image processing concerns the comparison of images: if different algorithms are applied to an image, we need an objective measure to compare the different output images. Therefore, such measures serve as a basis on which one algorithm is preferred above the other. For example, measures of comparison can be applied to evaluate the performance of different filters for noise reduction [7].

After some preliminaries in Section 2, we will start with a short discussion concerning two statistical measures, the mean square error (MSE) and the peak signal-to-noise ratio (PSNR). Because it is well-known that these classical measures do not always give convincing results we investigated whether similarity measures, originally introduced to compare two fuzzy sets, can be applied successfully to the normalized histograms of images. An overview of the appropriate similarity measures is given in Section 4. Finally, in Section 5, the measures are illustrated with some examples.

2 Preliminaries

2.1 Formal Definition of a Similarity Measure

In the literature a lot of measures can be found to express the similarity or equality between two fuzzy sets. There is no unique definition, but the most

T. Bilgiç et al. (Eds.): IFSA 2003, LNAI 2715, pp. 396–403, 2003.

frequently used one is the following. A *similarity measure* is a fuzzy binary relation in $\mathcal{F}(X)$, with X the universe of grid points of the image, i.e. a mapping $S : \mathcal{F}(X) \times \mathcal{F}(X) \to [0,1]$ satisfying the following properties :

- S is reflexive, i.e. $\forall A : S(A, A) = 1$.
- S is symmetric, i.e. $\forall A, B : S(A, B) = S(A, B)$.
- S is min-transitive, i.e. $\forall A, B, C : S(A, C) \geq \min(S(A, B), S(B, C))$.

2.2 Relevant Properties of Similarity Measures for Application in Image Processing

Not every measure in the literature satisfies the formal definition of a similarity measure given in the previous section. Therefore we will give a larger interpretation to a similarity measure. We will consider a similarity measure purely as a measure to compare two objects from a certain universe, and the properties we will demand depend on the usefulness within the domain of image processing. We consider the following list of relevant properties:

- **Reflexivity** : For two identical images one may expect that the similarity measure has output 1.
- **Symmetry** : Based on our intuition, the output of the similarity measure is expected to be independent of the order in which the two input images are considered. For completeness, we note that one can justify non-symmetric similarity measures by posing that the first (or the second) image is a reference image, with respect to which we want to compare a whole sequence of other images. This implies that we make a distinction between the images based on their role.
- **Reaction to noise** : It is also very interesting to investigate how similarity measures react to noise (e.g. salt & pepper noise or gaussian noise). A good similarity measure should not be affected too much due to the noise (since a noisy image is coming from an original one, it has to be similar to the original image), and to be decreasing with respect to an increasing noise-percentage (a larger value of the similarity measure for a more disturbed image is not in correspondence with our intuition).
- **Reaction to enlightening or darkening** : If one enlightens or darkens an image with a constant value, the similarity measure should return a high value (indeed, one considers almost identical images). One also expects a decreasing behaviour with respect to an increasing enlightening or darkening percentage.
- **Reaction to binary images** : Similarity measures which are applicable to grey-value images should also yield good results for binary images. In particular, also for binary images one may expect that the similarity measure produces a value between 0 and 1, and not only the crisp values 0 or 1.

3   Statistical Measures

The first measure widely used is the mean square error (MSE). For two images A and B, the MSE is defined as follows:

$$MSE(A,B) = \frac{1}{MN} \sum_{(x,y) \in X} |A(x,y) - B(x,y)|^2.$$

A second widely known statistical measure is the peak signal-to-noise ratio (PSNR). The PSNR is in fact a normalization of the MSE and is obtained by dividing the square of the luminance range R of the display device by the MSE and expressing the result in decibels:

$$PSNR(A,B) = 10 \log_{10} \frac{R^2}{MSE(A,B)}.$$

It is well-known that these measures do not always coincide with human perception.

4   Application of Similarity Measures to Histograms

Instead of a straightforward application of similarity measures to images, similarity measures can be applied to the image histograms. The histogram of an image is a chart that shows the distribution of intensities in the image. So the value of the histogram of an image A in the grey value g is equal to the total number of pixels in the image A with grey value g and will be denoted as $h_A(g)$. The histogram of an image can be transformed into a fuzzy set by dividing the values of the histogram by the maximum number of pixels with the same grey value. In this way the most typical grey value has membership degree 1 in the fuzzy set associated with the histogram. So we have the following expression for the membership degree of the grey value g in the fuzzy set Fh_A associated with the histogram of the image A:

$$Fh_A(g) = \frac{h_A(g)}{\max_g h_A(g)}.$$

Now, similarity measures, introduced to express the degree of comparison between two fuzzy sets, can be applied to the fuzzy sets associated with the histogram of the images. A great advantage is that images with different dimensions can be compared and that the calculation is much faster. Next, we give an overview of those similarity measures which can be successfully applied to the fuzzy sets associated with histograms.

In order to decide whether a similarity measure is useful for application to the fuzzy sets associated with the histograms of the images, we again evaluated [6] the 35 different similarity measures with respect to the relevant properties of a similarity measure in image processing. From the total of the 35 similarity measures only the following 8 similarity measures satisfy the relevant properties.

In order to obtain the explicit expressions for the different similarity measures we use the minimum to model the intersection, the maximum to model the union and the standard negator N_s to model the complement.

Let us now take a look at those measures which are applicable to the fuzzy sets associated with the histograms. First of all, 5 similarity measures which are appropriate for straightforward application [1,2], were also found to be appropriate for application to the histograms. If we apply these measures to the fuzzy sets associated with the histograms of the images, we obtain the following expressions for the measures H_1, H_2, H_3, H_4 and H_5, where L is the total number of different grey values :

$$H_1(A, B) = 1 - \left(\frac{1}{L} \sum_g |Fh_A(g) - Fh_B(g)|^r \right)^{\frac{1}{r}} \text{, with } r \in \mathbb{N}\backslash\{0\}$$

$$H_2(A, B) = \frac{|Fh_A^c \cap Fh_B^c|}{|Fh_A^c \cup Fh_B^c|}$$

$$H_3(A, B) = \frac{|Fh_A^c \cap Fh_B^c|}{\max(|Fh_A^c|, |Fh_B^c|)}$$

$$H_4(A, B) = \frac{\min(|Fh_A^c|, |Fh_B^c|)}{|Fh_A^c \cup Fh_B^c|}$$

$$H_5(A, B) = \min \left(\frac{|(Fh_A \triangle Fh_B)^c|}{|(Fh_B \backslash Fh_A)^c|}, \frac{|(Fh_B \triangle Fh_A)^c|}{|(Fh_A \backslash Fh_B)^c|} \right)$$

$$= \frac{|(Fh_A \triangle Fh_B)^c|}{\max(|(Fh_B \backslash Fh_A)^c|, |(Fh_A \backslash Fh_B)^c|)}.$$

The following measures H_6 and H_7 are not suited for straightforward application to the images, although these measures are appropriate for application to the histograms. The first measure [9] is given by the following expression:

$$H_6(A, B) = \frac{\min(|Fh_A^c|, |Fh_B^c|)}{\max(|Fh_A^c|, |Fh_B^c|)}$$

Finally, we consider another similarity measure based on an inclusion measure. Inclusion measures [8] are introduced to express the degree of inclusion between two fuzzy sets. They can be combined using a t-norm in order to construct a similarity measure. Consider an inclusion measure I. Using a t-norm $\mathcal{T}$, we obtain a similarity measure in the following way :

$$S(A, B) = \mathcal{T}(I(A, B), I(B, A)).$$

This expression is based on the crisp expression:

$$A = B \Leftrightarrow A \subseteq B \text{ and } B \subseteq A.$$

There exist a lot of inclusion measures, but using most of these inclusion measures to express similarity yields similarity measures of which at least one of the relevant properties wasn't satisfied [4]. Also, using other t-norms to combine the inclusion measures didn't yield satisfactory results. Only the following inclusion measure is appropriate for constructing a similarity measure:

$$I(A, B) = \frac{\min(|(A \backslash B)^c|, |(B \backslash A)^c|)}{|(B \backslash A)^c|}$$

and we use the minimum to combine the inclusion measure. In this way, we obtain the following similarity measure:

$$H_7(A, B) = \min \left(\begin{array}{c} \frac{\min(|(Fh_A \backslash Fh_B)^c|, |(Fh_B \backslash Fh_A)^c|)}{|(Fh_B \backslash Fh_A)^c|}, \\ \frac{\min(|(Fh_B \backslash Fh_A)^c|, |(Fh_A \backslash Fh_B)^c|)}{|(Fh_A \backslash Fh_B)^c|} \end{array} \right)$$

$$= \frac{\min(|(Fh_A \backslash Fh_B)^c|, |(Fh_B \backslash Fh_A)^c|)}{\max(|(Fh_B \backslash Fh_A)^c|, |(Fh_A \backslash Fh_B)^c|)}.$$

5   Illustration of the Different Similarity Measures

In this section we will illustrate the different similarity measures. More precisely, we will illustrate how the measures react if we increase the noise percentages and the enlightening and darkening. Also the results for completely different images will be shown. The classical measure for comparison of images, the Mean Square Error, is also displayed in the results.

5.1   Reaction to Noise

We add two different percentages of salt & pepper noise to the "cameraman" image. First we add 10 % salt & pepper noise and second we add 35 % of salt & pepper noise. The original image is displayed in Fig. 1 and the noisy images in Fig. 2. The results are shown in Table 1. One can verify that the values are relatively high, and that the similarity values slightly decrease with respect to an increasing noise level. Experiments with gaussian noise yield similar results.

Fig. 1. The original test image "cameraman".

Fig. 2. "Cameraman" with salt & pepper noise. Left: 10 % ; right: 35 % .

	10%	35%
$H_1(A,B)$	0.91023	0.75783
$H_2(A,B)$	0.90348	0.85255
$H_3(A,B)$	0.91109	0.85930
$H_4(A,B)$	0.91183	0.86040
$H_5(A,B)$	0.91175	0.85935
$H_6(A,B)$	0.91951	0.86721
$H_7(A,B)$	0.92005	0.86726
$MSE(A,B)$	1981.94	7071.20

Table 1. Results of the measures applied to $(A = $ original$,B = 10\%$ salt & pepper noise) and $(A = $ original$,B = 35\%$ salt & pepper noise).

5.2   Reaction to Enlightening

In this experiment, we enlighten the "cameraman" image two times (see Fig. 3. The results of the different similarity measures are shown in Table 2. Again, as expected, the similarity values are relatively high and show a decreasing behaviour with respect to a stronger amount of enlightening or darkening.

5.3   Pairs of Different Images

We apply the similarity measures to a pair of totally different images, namely the "cameraman" image and the "Lena" image (see Fig. 4). The results are displayed in Table 3. From this experiment we can conclude that the similarity measures perform better than the MSE. If we compare the results of the "cameraman" image with 35 % salt & pepper noise with the results of the comparison between the "Lena" image and the "cameraman" image, one can observe that the results of most of the measures from the first experiment are larger than the results of the second experiment, although the MSE in the first experiment is larger than the MSE in the second experiment.

Fig. 3. "Cameraman" enlightened. Left: +0.1 ; right: +0.2 .

	+0.1	+0.2
$H_2(A,B)$	0.83377	0.77836
$H_3(A,B)$	0.90935	0.87537
$H_4(A,B)$	0.91689	0.88918
$H_5(A,B)$	0.90938	0.87540
$H_6(A,B)$	0.93638	0.89810
$H_7(A,B)$	0.94358	0.90963
$MSE(A,B)$	613.90	2446.40

Table 2. Results of the measures applied to (A = original,B = enlightened +0.1), (A = original,B = enlightened +0.2).

Fig. 4. "Lena" image and "cameraman" image.

	Lena vs. cameraman
$H_1(A,B)$	0.62093
$H_2(A,B)$	0.66367
$H_3(A,B)$	0.69427
$H_4(A,B)$	0.70775
$H_5(A,B)$	0.70152
$H_6(A,B)$	0.74039
$H_7(A,B)$	0.74655
$MSE(A,B)$	5060.99

Table 3. Results of the measures applied to (A = "Lena",B = "cameraman").

6   Conclusion and Final Remarks

In this paper we have investigated whether similarity measures, originally introduced to express the degree of comparison between two fuzzy sets, can be applied to images. Instead of a straightforward application, we applied the different similarity measures to the fuzzy sets associated with histograms of the images. In this way, 7 similarity measures were found to be appropriate for the comparison of images. All the measures were illustrated with some examples.

References

1. D. Van der Weken, M. Nachtegael, and E.E. Kerre, *The applicability of similarity measures in image processing*. To appear in Proceedings of the 8th International Conference on Intelligent Systems and Computer Sciences (December 4-9, 2000, Moscow, Russia); in Russian.
2. D. Van der Weken, M. Nachtegael, and E.E. Kerre, *An overview of similarity measures for images*. Proceedings of ICASSP'2002 (IEEE International Conference on Acoustics, Speech and Signal Processing), Orlando, United States, 2002, pp. 3317-3320.
3. D. Van der Weken, M. Nachtegael, and E.E. Kerre, *A new similarity measure for image processing*. Journal of Computational Methods in Sciences and Engineering, vol.3, no.2, 2003, pp.17-30.
4. D. Van der Weken, M. Nachtegael, and E.E. Kerre, *Image comparison using inclusion measures*. In preparation.
5. D. Van der Weken, M. Nachtegael, and E.E. Kerre, *Similarity Measures in Image Processing*. Internal Research Report (136 pages); in Dutch.
6. D. Van der Weken, M. Nachtegael, and E.E. Kerre, *Image comparison through application of similarity measures to fuzzy sets associated with the histograms of the images*. Internal Research Report (68 pages); in Dutch.
7. M. Nachtegael, D. Van der Weken, D. Van De Ville, E.E. Kerre, W. Philips, I. Lemahieu, *A comparative study of classical and fuzzy filters for noise reduction*. Proc. of FUZZ-IEEE'2001, pp. 11-14 (CD), 2001.
8. C. Cornelis, C. Vanderdonck, and E.E. Kerre, *Sinha-Dougherty approach to the fuzzification of set inclusion revisited*. Fuzzy Sets and Systems, in press.
9. B. De Baets, and H. De Meyer, *The Frank T-norm family in fuzzy similarity measurement*. Proc. of EUSFLAT'2002, 2nd Conference on the European Society for Fuzzy Logic and Technology, September 5-7, 2001, Leicester (England), pp. 249-252.
10. L.A. Zadeh, *The role of fuzzy logic in the management of uncertainty in expert systems*. Selected papers by L.A. Zadeh, R.R. Yager (editor), New York, Wiley, pp.413-441.

A CHC Evolutionary Algorithm for 3D Image Registration*

Oscar Cordón[1], Sergio Damas[2], and Jose Santamaría[3]

[1] Dept. of Computer Science and A.I. E.T.S. de Ingeniería Informática
ocordon@decsai.ugr.es
[2] Dept. of Software Engineering. E.T.S. de Ingeniería Informática
sdamas@ugr.es
[3] C.S.I.R.C. jsantam@ugr.es
University of Granada. 18071 - Granada (Spain)

Abstract. Image registration has been a very active research area in
the computer vision community. In the last few years, there is an increasing interest on the application of Evolutionary Computation in this
field and several evolutionary approaches has been proposed obtaining
promising results. In this contribution we present an advanced evolutionary algorithm to solve the 3D image registration problem based on
the CHC. The new proposal will be validated using two different shapes
(both synthetic and MRI), considering four different transformations for
each of them and comparing the results with those from ICP and the
usually applied binary coded genetic algorithms.

1 Introduction

Image registration (IR) is a fundamental task in image processing used to finding
a correspondence (or transformation) among two or more pictures taken under
different conditions: at different times, using different sensors, from different
viewpoints, or a combination of them. Evolutionary Computation (EC) [1] uses
computational models of evolutionary processes as key elements in the design and
implementation of computer-based problem solving systems. Genetic algorithms
(GAs), general-purpose search algorithms that use principles inspired by natural
population genetics to evolve solutions to problems, are maybe the most known
evolutionary algorithms.

In the last few years, there is increasing interest on applying EC fundamentals
to IR. Unfortunately, we can find a lack of accuracy when facing this problem and
different contributions fall into simplifications of the problem or, even worse, do
not apply EC concepts in the more suitable way. In this contribution we propose
an advanced evolutionary algorithm to solve the 3D IR problem based on the
CHC [3]. To do so, in section 2 we give some IR basics. Next, we analyze different
approaches to the IR problem from the perspective of EC in section 3. Section

* Research supported by CICYT TIC2002-03276 and by Project "Mejora de Meta-
heurísticas mediante Hibridación y sus Aplicaciones" of the University of Granada.

T. Bilgiç et al. (Eds.): IFSA 2003, LNAI 2715, pp. 404–411, 2003.

4 describes our proposal, which is tested in section 5 over different images and transformations. Finally, in section 6 we present new open lines for future works.

2 Image Registration

IR can be defined as a mapping between two images (I_1 and I_2) both spatially and with respect to intensity:$I_2(x, y, z, t) = g(I_1(f(x, y, z, t)))$ We can usually find situations where intensity difference is inherent to scene changes, and thus intensity transformation estimation given by g is not necessary. In this contribution, we will consider f represents a similarity transformation, i.e. rotation, translation and uniform scaling.

One of the most known algorithms for IR is the Iterative Closest Point (ICP), proposed by Besl and McKay [2], and extended in different papers ([11]):

- The point set P with N_p points p_i from the data shape and the model X, with N_x supporting geometric primitives: points, lines, or triangles is given.
- The iteration is initialized by setting $P_0 = P$, the registration transformation by $q_0 = [1, 0, 0, 0, 0, 0, 0]^t$, and $k = 0$. Next four steps are applied until convergence within a tolerance $\tau > 0$:
    1. Compute the matching between the data (scene) and model points by the closest point assignment rule: $Y_k = C(P_k, X)$
    2. Compute the registration: $f_k(P_0, Y_k)$
    3. Apply the registration: $P_{k+1} = f_k(P_0)$
    4. Terminate iteration when the change in mean square error falls below τ

The algorithm has important drawbacks: (i) it is sensitive to outliers presence; (ii) the initial states for global matching play a basic role for the method success when dealing with important deformations between model and scene points; (iii) the estimation of the initial states is not a trivial task, and (iv) the cost of a local adjustment can be important if a low percentage of occlusion is present.

Hence, the algorithm performance is not good with important transformations. As stated in [11]: "we assume the motion between the two frames is small or approximately known". This is a precondition of the algorithm to get reasonable results.

3 Evolutionary Computation and Image Registration

An exhaustive review of the different approaches to the IR problem from the perspective of EC is out of the scope of our study. Nevertheless, we will mention some of the most important aspects of them in order to achieve a deep understanding of our work.

The first attempts to solve IR using EC can be found in the early eighties. The size of data as well as the number of parameters that are looked for, prevent from an exhaustive search for the solutions. Such an approach based on a GA was proposed in 1984 for the 2D case and applied to angiography images [5]. Since this initial contribution, different authors solved the problem but we can still find important limitations in their approaches:

- The use of a binary coding to solve an inherent real coding problem, with the precision problem depending on a given number of bits in the encoding.
- The kind of GA considered, usually that from the Holland's original proposal [7]. This GA was proposed almost thirty years ago and it suffers from several drawbacks later solved by other components.
- Many contributions only handle images suffering a rigid transformation [9], [6], which is not the case in many real situations where at least a uniform scaling is desirable.

4   A CHC Evolutionary Algorithm for Image Registration

Two different IR algorithms will be proposed using the original binary CHC algorithm and a real-coded version. Bellow, we describe their components.

4.1   The Original Binary-Coded CHC

The key idea of the CHC binary-coded evolutionary algorithm [3] involves the combination of a selection strategy with a very high selective pressure with several components inducing a strong diversity. The main four components of the algorithm are shown as follows:

- An *elitist selection*. The M members of the current population are merged with the offspring population obtained from it and the best M individuals are selected to compose the new population. In case that a parent and an offspring has the same fitness value, the former is preferred to the latter.
- A *highly disruptive crossover*, HUX, which crosses over exactly half of the non-matching alleles, where the bits to be exchanged are chosen at random without replacement. This way, it guarantees that the two offspring are always at the maximum Hamming distance from their two parents, thus proposing the introduction of a high diversity in the new population and lessening the risk of premature convergence.
- An *incest prevention mechanism*. During the reproduction step, each member of the parent (current) population is randomly chosen without replacement and paired for mating. However, not all these couples are allowed to cross over. Before mating, the Hamming distance between the potential parents is calculated and if half this distance does not exceed a difference threshold d, they are not mated and no offspring coming from them is included in the children population. The said threshold is usually initialized to $L/4$ (with L being the chromosome length). If no offspring is in one generation the difference threshold is decremented.
  The effect of this mechanism is that only the more diverse potential parents are mated, but the diversity required by the difference threshold automatically decreases as the population naturally converges.
- A *restart process*, substituting the usual GA mutation, which is only applied when the population has converged. The difference threshold is considered

to measure the stagnation of the search, which happens when it has dropped to zero and several generations have been run without introducing any new individual in the population. Then, the population is reinitialized by considering the best individual as the first chromosome of the new population and generating the remainder M-1 by randomly flipping a percentage (usually a 35%) of its bits.

4.2   The Real-Coded CHC Extension

We have extended the binary-coded CHC to deal with real-coded chromosomes, maintaining its basis as much as possible. Real-coded CHC is based on the same four main components than classical binary-coded CHC. The elitist selection and the restart are exactly the same in both cases. However, there is a need to adapt the incest prevention mechanism (whose operation is guided by the Hamming distance) and to work with a different crossover operator (since HUX is specifically designed for binary chromosomes).

In order to be able to measure the similarity between two parents using the Hamming metric we consider a binary conversion of both. The difference threshold d is then proportionally set up imitating the classical CHC. It is initialized to $\frac{D_{max}}{4}$, where $D_{max} = \sum_{i=1}^{N} param_i \cdot nbits_i$ (with N the number of parameters to be estimated in our IR problem, and $nbits_i$ being the number of bits used for the i-th parameter). So, the crossover is avoided if the distance between both parents is lesser than d. Hence every time no offspring is included in the population after a generation, the difference threshold d is decremented by one and, as usual, the restart is triggered when $d \leq 0$.

On the other hand, the BLX-α crossover [4] is considered to substitute the HUX one. The parameter α allows us to make this crossover as disruptive as desired. This crossover operator is based on obtaining one offspring $H = (h_1, \ldots, h_i, \ldots, h_n)$ from two parents $C_1 = (c_1^1, \ldots, c_n^1)$ and $C_2 = (c_1^2, \ldots, c_n^2)$ by uniformly generating a random value for each gene h_i in the interval $[c_{min} - I \cdot \alpha, c_{max} + I \cdot \alpha]$, with $c_{max} = \max(c_i^1, c_i^2)$, $c_{min} = \min(c_i^1, c_i^2)$, and $I = c_{max} - c_{min}$. Two offspring are generated by applying twice the operator on the two parents.

4.3   Components of the Algorithm

In this section we define some key concepts to understand the CHC adaptation to the IR problem.

Coding Scheme The 3D similarity transformation with uniform scaling is determined fixing seven parameters and they will be the ones we will look for. This is: $(\alpha_x, \alpha_y, \alpha_z, \Delta_x, \Delta_y, \Delta_z, S)$ where α_i are the three Euler angles, Δ_i the three components of the displacement vector and S the uniform scaling factor. Hence, these seven parameters are binary or real coded in the chromosome depending on the algorithm variant applied.

Fitness Function We use a grid data structure in order to improve the efficiency of the closest point assignment computation [10]. Since we are searching in the parameter space and we work on a discrete grid, when defining our fitness function we must prevent cases where scene points are transformed to spatial locations outside the ranges where the grid is defined. Therefore, in such a situation we must assign a low fitness value.

Otherwise, we will try to estimate the parameters of the transformation which transform the scene points and lead us to the maximization of function F in:

$$F = \left[E\left(\frac{N_{insidegrid}}{N_s}\right) * \left(\frac{1}{1 + \sum_{i=0}^{N_s} \| GCP(T(y_i)) \|^2}\right) + \left(\frac{N_{insidegrid}}{N_s}\right) \right]$$

where: E is the floor function, $N_{insidegrid}$ is the number of scene points inside the grid when the registration transformation T to be evaluated is applied to the scene, N_s is the number of scene points, S' is the set of scene points, and $GCP(T(yi))$ is the GCP function [10] (returning the model closest point for every cell center) applied to the result of transforming the y_i scene point.

It can be seen that $F \in [0, 2]$ so that an individual with a fitness value less than 1.0 is the one managing a transformation which is able to put some of the points out from the working grid, while those individuals with $F \geq 1$ correspond to transformations keeping the set of scene points inside the grid. Hence, the larger the F value, the better (since this means that the distance between scene and model set of points is closer to zero).

5   Experiments

We present a number of experiments to study the performance of our proposal. As we will explain below these tests have been carried out under the same conditions since we wanted to extend our conclusions to other possible situations. The results obtained by our two CHC variants, binary *BinCHC* and real *RealCHC*, will be compared against those obtained by a binary-coded GA (*BinGA*) and by the classical *ICP* [11] (see Section 2).

5.1   Model and Scene Images

Our results correspond to a number of registration problems for two different 3D images. All of them have suffered the same four global similarity transformations which must be estimated by the different GAs applied. Because of the amount of data to be managed, it is necessary to carry out a preprocessing step in order to extract a set of feature points to describe the surfaces. Feature points are obtained applying a 3D crest lines edge detector [8] to a superellipsoid and a brain model surfaces, returning 880 and 9108 points respectively. We will refer to each of them as the "piece of cheese" and "Brain" images. Both original and preprocessed images are presented in figure 1.

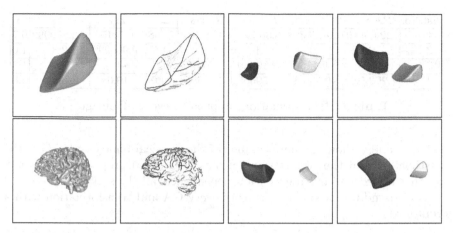

Fig. 1. Original images and their respective crest lines points. The four transformations to be estimated (Table 1) are shown when applied to the first object only (although the same ones will be studied for the "Brain" object as well).

5.2   Parameter Settings

Every method is run for the same fixed time when dealing with the same image (180 seconds in the case of the "piece of cheese" transformations, and 360 seconds when working on the "brain" image). All the runs have been performed on a 2200 MHz. Pentium IV processor.

As regards the number of bits associated to each gene in the binary-coded GAs (*BinGA* and *BinCHC*), large individuals have been considered as we want precise solutions for each transformation parameter. This leads us to define our binary chromosome as a 105 bits structure (fifteen bits for each of the seven parameters). Crossover and mutation probabilities for *BinGA*: $P_c = 0.6$ and $P_m = 0.1$, while $\alpha = 0.5$ in both *BinCHC* and *RealCHC*. Finally, the population size is $M = 100$ individuals in all cases.

5.3   Transformations to Be Estimated

The four transformations considered are stored in table 1 and can be appreciated when being applied to one of the objects in Figure 1. In such table (and from now on), all the different 3D rotations have been expressed in terms of rotation angle ($RAngle°$) and rotation axis ($RAxis_x, RAxis_y, RAxis_z$) to achieve a better understanding of the geometric transformation involved.

5.4   Results

Notice that all statistics in this section are based on a usual error measure in the field of IR, the *Mean Square Error (MSE)*, typically given by: $MSE = \sum_{i=1}^{N} \|f(x_i) - y_i\|^2$ where: f is the estimated registration function, x_i, are the N scene points, and y_i, are the N model points matching the scene ones

Transf.No.	$RAngle°$	$RAxis_x$	$RAxis_y$	$RAxis_z$	Δ_x	Δ_y	Δ_z	S
1	122.699997	0.727393	0.363696	-0.581914	7.568	-15.97	-23.879999	0.7
2	95	0	1	0	-1.5	19.969999	2.8	1.0
3	180.6	-0.11547	0.80829	-0.57735	-7.5	-12	10.8	1.5
4	202.5.0	-0.536895	0.59655	0.59655	24.0	10.6	5.2	2.0

Table 1. Transformations applied to every 3D image

Tables 2 and 3 show the performance of the classical binary-coded GA, the ICP estimation and the two versions of our CHC-based proposal. It is worth noting that in order to avoid execution dependence, all the statistics presented below correspond to five different runs of every GA and transformation with a different seed.

MSE results for the "Piece of cheese" image									
Tr.No.	ICP	BinGA			BinCHC			RealCHC		
		Min	Mean	SDev	Min	Mean	SDev	Min	Mean	SDev
1	2.35	0.0040	1.9440	1.7324	0.0001	0.9661	0.9661	0.0001	0.0001	0
2	4.03	0.0005	3.7900	3.1003	0.0003	1.3822	2.3848	0.0002	0.0003	0
3	0.92	0.0028	2.4979	4.6176	0.0006	0.0199	0.0199	0.0005	5.5520	6.7992
4	5.80	0.0002	5.1755	10.0885	0.0006	0.0017	0.0011	0.0008	0.0012	0.0003

Table 2. MSE corresponding to the four transformations of Table 1 applied to the "Piece of cheese" image (statistics from five different runs).

MSE results for the "Brain" image									
Tr.No.	ICP	BinGA			BinCHC			RealCHC		
		Min	Mean	SDev	Min	Mean	SDev	Min	Mean	SDev
1	43.27	0.0007	1.1237	2.2249	0.0035	0.0053	0.0017	0.0026	0.0034	0.0007
2	32.60	0.0023	3.2665	3.9930	0.0029	0.0450	0.0817	0.0039	0.0050	0.0006
3	87.68	0.0301	25.2653	30.1295	0.0054	0.2795	0.2916	0.0047	0.0061	0.0007
4	234.96	0.0040	0.7891	1.5301	0.0272	0.2436	0.2019	0.0366	0.0444	0.0040

Table 3. MSE corresponding to the four transformations of Table 1 applied to the "Brain" image (statistics from five different runs).

We can see that in both images all the GAs outperform ICP and it is specially important to mention the good and stable performance of our real-coded CHC proposal, with the best mean MSE value in all the experiments but in the third transformation of the "piece of cheese" image. However the binary-coded CHC still achieve better results than the binary-coded GA and ICP.

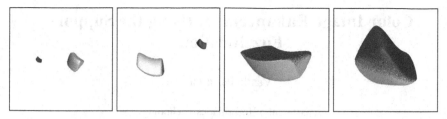

Fig. 2. *ICP*(left) and *RealCHC* (right) results for the first two transformations.

6   Future Works

There is a number of new open lines to be done after this proposal: it is necessary a deep study of the behavior of different EC-based algorithms when dealing with the image registration problem. Likewise, experimentation with noisy images has also to be done. Moreover, new advanced diversity induction mechanisms should be tested in order to study their performance in the IR problem.

References

1. T. Bäck, D.B. Fogel, Z. Michalewicz (Eds.), Handbook of evolutionary computation, IOP Publishing Ltd and Oxford University Press, 1997.
2. P.J. Besl, N.D. McKay, *A method for registration of 3-D shapes*, IEEE Transactions on Pattern Analysis and Machine Intelligence, vol. 14, pp. 239-256, 1992.
3. L.J. Eshelman, The CHC adaptive search algorithm: how to safe search when engaging in non traditional genetic recombination, In Foundations of Genetic Algorithms, G.J.E. Rawlins (Ed.), Morgan Kaufmann, San Mateo, pp. 265-283, 1991.
4. L.J. Eshelman, Real-coded genetic algorithms and interval schemata, In Foundations of Genetic Algorithms 2, L.D. Whitley (Ed.), Morgan Kaufmann Publishers, San Mateo, pp. 187-202, 1993.
5. J.M. Fitzpatrick, J.J. Grefenstette, D. Van Gucht, Image registration by genetic search, In IEEE Southeast Conference, pp. 460-464, Louisville (USA), 1984.
6. R. He, P. A. Narayana, Global optimization of mutual information: application to three-dimensional retrospective registration of magnetic resonance images, Computerized Medical Imaging and Graphics, vol. 26, 277-292, 2002.
7. J.H. Holland, Adaptation in Natural and Artificial Systems, Ann arbor: The University of Michigan Press, 1975, The MIT Press, London, 1992.
8. O. Monga, R. Deriche, G. Malandain, J.P. Cocquerez. Recursive filtering and edge tracking: two primary tools for 3D edge detection, Image and Vision Computing, vol. 9, no. 4, pp. 203-214, 1991.
9. K. Simunic, S. Loncaric, A genetic search-based partial image matching, In 2nd IEEE Intl. Conf. on Intelligent Processing Systems (ICIPS 98), pp. 119-122, Gold Coast, Australia, 1998.
10. S. M. Yamany, M. N. Ahmed, A. A. Farag, *A new genetic-based technique for matching 3D curves and surfaces*, Pattern Recog., vol. 32, pp. 1817-1820, 1999.
11. Z. Zhang, *Iterative point matching for registration of free-form curves and surfaces*, Int. Journal of Computer Vision, vol. 13, no. 2, pp. 119-152, 1994.

Color Image Enhancement Using the Support Fuzzification

Vasile Pătraşcu

Department of Informatics Technology
TAROM Company, Bucharest, Romania
vpatrascu@hotmail.com

Abstract. Simple and efficient methods for image enhancement can be obtained through affine transforms, defined by the logarithmic operations. Generally, a single affine transform is calculated for the whole image. A better quality is possible to obtain if a fuzzy partition is defined on the image support and then, for each element of the partition an affine transform is determined. Finally the enhanced image is computed by summing up in a weight way the images obtained for the fuzzy partition elements.

1 Introduction

The logarithmic models have created a new environment for developing some new methods of image enhancement [1-3], [8]. The logarithmic model presented in this paper is the one developed in [3]. It uses a real and bounded set for gray levels and for colors. Within this structure, image enhancement methods are easily obtained by using an affine transform. Otherwise, such a solution is not sufficient in the case of images with variable brightness and contrast. Better results can be obtained if partitions are defined on the image support and then the pixels are separately processed in each window belonging to the defined partition. The classical partitions frequently lead to the appearance of some discontinuities at the boundaries between these windows. In order to avoid all these drawbacks the classical partitions may be replaced by fuzzy partitions. Their elements will be fuzzy windows and in each of them there will be defined an affine transform induced by parameters using the fuzzy mean and fuzzy variance computed for the pixels that belong (in fuzzy meaning) to the analyzed window. Their calculus uses logarithmical operations. The final image is obtained by summing up in a weight way the images of every fuzzy window as defined before. Within this merging operation, the weights used are membership degrees, which define the fuzzy partition of the image support. Further on, there will be presented the way this logarithmic model of image representation is built. Next, the definition of the fuzzy partition on the image support and then, the determination of the affine transforms for each component (window) of the partition. The last sections comprise experimental results and some conclusions.

T. Bilgiç et al. (Eds.): IFSA 2003, LNAI 2715, pp. 412-419, 2003.
© Springer-Verlag Berlin Heidelberg 2003

2 The Fundamentals of the Logarithmic Model

Further on, there will be a short presentation of the vector space on the gray level set and then, for the vector space of the colors. More details concerning the notions in this section may be found in [3].

2.1 The Vector Space of Gray Levels

Let there be considered the space of gray levels as the set $E = (-1,1)$. In the set of gray levels the addition $\langle + \rangle$ and the multiplication $\langle \times \rangle$ by a real scalar will be defined and then, defining a scalar product $(\cdot | \cdot)_E$ and a norm $\|.\|_E$, a Euclidean space of gray levels will be obtained.

The Addition. The sum of two gray levels, $v_1 \langle + \rangle v_2$ will be defined by:

$$v_1 \langle + \rangle v_2 = \frac{v_1 + v_2}{1 + v_1 v_2}, \forall v_1, v_2 \in E \ . \tag{2.1.1}$$

The neutral element for the addition is $\theta = 0$. Each element $v \in E$ has an opposite $w = -v$. The subtraction operation $\langle - \rangle$ will be defined by:

$$v_1 \langle - \rangle v_2 = \frac{v_1 - v_2}{1 - v_1 v_2}, \forall v_1, v_2 \in E \ . \tag{2.1.2}$$

The Multiplication by a Scalar. The multiplication $\langle \times \rangle$ of a gray level v by a real scalar λ will be defined as:

$$\lambda \langle \times \rangle v = \frac{(1+v)^\lambda - (1-v)^\lambda}{(1+v)^\lambda + (1-v)^\lambda}, \forall v \in E, \forall \lambda \in R \ . \tag{2.1.3}$$

The above operations, the addition $\langle + \rangle$ and the scalar multiplication $\langle \times \rangle$ induce on E a real vector space structure.

The Fundamental Isomorphism. The vector space of gray levels $(E, \langle + \rangle, \langle \times \rangle)$ is isomorphic to the space of real numbers $(R, +, \cdot)$ by the function $\varphi : E \to R$, defined as:

$$\varphi(v) = \frac{1}{2} \ln\left(\frac{1+v}{1-v}\right), \forall v \in E \ . \tag{2.1.4}$$

The isomorphism φ verifies:

$$\varphi(v_1 \langle + \rangle v_2) = \varphi(v_1) + \varphi(v_2), \forall v_1, v_2 \in E \ . \tag{2.1.5}$$

$$\varphi(\lambda \langle \times \rangle v) = \lambda \cdot \varphi(v), \forall \lambda \in R, \forall v \in E \ . \tag{2.1.6}$$

The particular nature of this isomorphism induces the logarithmic character of the mathematical model.

The Euclidean Space of Gray Levels. The scalar product of two gray levels, $(\cdot | \cdot)_E : E \times E \to R$ is defined with respect to the isomorphism from (2.1.4) as:

$$(v_1 | v_2)_E = \varphi(v_1) \cdot \varphi(v_2), \forall v_1, v_2 \in E \ . \qquad (2.1.7)$$

Based on the scalar product $(\cdot | \cdot)_E$ the gray level space becomes a Euclidean space. The norm $\| \cdot \|_E : E \to R^+$ is defined using the scalar product:

$$\| v \|_E = \sqrt{(v | v)_E} = | \varphi(v) |, \forall v \in E \ . \qquad (2.1.8)$$

2.2 The Logarithmic Model for the Color Space

The extension of logarithmic model from gray levels to colors is made in a natural way following the procedure used in [4], [5]. Consider the cube E^3 as the color space. Let be $q \in E^3$ a color having the components r (*red*), g (*green*) and b (*blue*). A color q can be written thus: $q = (r, g, b)$. One can define then the addition, the subtraction and the scalar multiplication operations on the colors set using the formulae (2.1.1), (2.1.2) and (2.1.3).

The Addition.
$$\forall q_1, q_2 \in E^3, \quad q_1 \langle + \rangle q_2 = (r_1 \langle + \rangle r_2, g_1 \langle + \rangle g_2, b_1 \langle + \rangle b_2) \ . \qquad (2.2.1)$$

The Subtraction.
$$\forall q_1, q_2 \in E^3, \quad q_1 \langle - \rangle q_2 = (r_1 \langle - \rangle r_2, g_1 \langle - \rangle g_2, b_1 \langle - \rangle b_2) \ . \qquad (2.2.2)$$

The Scalar Multiplication.
$$\forall \lambda \in R, \forall q \in E^3, \quad \lambda \langle \times \rangle q = (\lambda \langle \times \rangle r, \lambda \langle \times \rangle g, \lambda \langle \times \rangle b) \ . \qquad (2.2.3)$$

The Euclidean Space of the Colors. The color vector space E^3 can be organized as Euclidean space defining the scalar product $(\cdot | \cdot)_{E^3}$ thus: $(\cdot | \cdot)_{E^3} : E^3 \times E^3 \to R$, $\forall q_1, q_2 \in E^3$, $q_1 = (r_1, g_1, b_1)$ and $q_2 = (r_2, g_2, b_2)$ then:

$$(q_1 | q_2)_{E^3} = \varphi(r_1)\varphi(r_2) + \varphi(g_1)\varphi(g_2) + \varphi(b_1)(b_2) \ . \qquad (2.2.4)$$

The norm is obtained using the scalar product (2.2.4):

$$\text{for } \forall q = (r, g, b) \in E^3, \quad \| q \|_{E^3} = \sqrt{\varphi^2(r) + \varphi^2(g) + \varphi^2(b)} \ . \qquad (2.2.5)$$

3 The Fuzzification of the Image Support

A gray level image is described by its intensity function $f : D \to E$ where $D \subset R^2$ is the image support. Without losing generality of the problem, can be considered as image support the rectangle $D = [x_0, x_1] \times [y_0, y_1]$. On the set D a fuzzy partition is built and its elements are called fuzzy windows [7]. The coordinates of a pixel within the support D will be noted (x, y). Let there be $P = \{W_{ij} \mid (i, j) \in [0, m] \times [0, n]\}$ a fuzzy partition of the support D. Consider for $(i, j) \in [0, m] \times [0, n]$ the polynomials $p_{ij} : D \to [0,1]$,

$$p_{ij}(x, y) = C_m^i C_n^j \frac{(x - x_0)^i (x_1 - x)^{m-i}}{(x_1 - x_0)^m} \cdot \frac{(y - y_0)^j (y_1 - y)^{n-j}}{(y_1 - y_0)^n} . \qquad (3.1)$$

where $C_m^i = \dfrac{m!}{i!(m-i)!}$, $C_n^j = \dfrac{n!}{j!(n-j)!}$. The membership degrees of a point $(x, y) \in D$ to the fuzzy window W_{ij} are given by the functions $w_{ij} : D \to [0,1]$ defined by the relation:

$$w_{ij}(x, y) = \frac{\left(p_{ij}(x, y)\right)^\gamma}{\displaystyle\sum_{j=0}^{n} \sum_{i=0}^{m} \left(p_{ij}(x, y)\right)^\gamma} . \qquad (3.2)$$

where $(i, j) \in [0, m] \times [0, n]$. The parameter $\gamma \in (0, \infty)$ has the role of a tuning parameter offering a greater flexibility in building the fuzzy partition P. In other words, γ controls the fuzzification-defuzzification degree of the partition. The membership degrees $w_{ij}(x, y)$ describe the position of the point (x, y) within the support D, namely the upper part or the lower part, the left hand part, the right hand part or in the center of the image. For each window W_{ij} defined by (3.2), the fuzzy cardinality is computed as follows:

$$card(W_{ij}) = \sum_{(x,y) \in D} w_{ij}(x, y) . \qquad (3.3)$$

where $(i, j) \in [0, m] \times [0, n]$. Further on, fuzzy statistics of a gray level image f are used in relation to the window W_{ij}. Thus, the fuzzy mean $\mu_\varphi(f, W_{ij})$ and the fuzzy variance $\sigma_\varphi^2(f, W_{ij})$ of the image f within window W_{ij} are defined by:

$$\mu_\varphi(f, W_{ij}) = \left\langle + \right\rangle_{(x,y) \in D} \left(\frac{w_{ij}(x, y)}{card(W_{ij})} \langle \times \rangle f(x, y) \right) . \qquad (3.4)$$

$$\sigma_\varphi^2(f,W_{ij}) = \sum_{(x,y)\in D} \frac{w_{ij}(x,y)\|f(x,y)\langle-\rangle\mu_\varphi(f,W_{ij})\|_E^2}{card(W_{ij})} . \tag{3.5}$$

where $(i,j)\in[0,m]\times[0,n]$.

4 The Image Enhancement Methods

4.1 The Enhancement Method for Gray Level Image

Let us consider these affine transforms on the images set $F(D,E)$, defined as following: $\psi : F(D,E) \to F(D,E)$, $\forall f \in F(D,E)$

$$\psi(f) = \lambda\langle\times\rangle(f\langle+\rangle\tau) . \tag{4.1.1}$$

where $\lambda \in R$, $\lambda \neq 0$ and $\tau \in E$. The formula (4.1.1) shows that an image can be processed in two steps: a translation with a constant value τ, which leads to a change in the image brightness, then a scalar multiplication by the factor λ - leading to a change in the image contrast. A gray level image can be enhanced using an affine transform [6]. The determination of the two parameters (λ,τ) will be made, so that the new image will have the mean zero and the variance $1/3$. From statistical point of view, this means that the resulted image will be very close to an image with a uniform distribution of the gray levels [6]. On support fuzzification an affine transform will be computed for each element. The fuzzy window W_{ij} will supply a couple of parameters (λ,τ), which reflects the gray level statistics according to the pixels belonging (in fuzzy meaning) to this window. Thus:

$$\lambda_{ij} = \frac{\sigma_u}{\sigma_\varphi(f,W_{ij})} . \tag{4.1.2}$$

$$\tau_{ij} = \langle-\rangle\mu_\varphi(f,W_{ij}) . \tag{4.1.3}$$

where $\sigma_u^2 = 1/3$. Thus the function that transforms the pixels belonging to the fuzzy window W_{ij} takes the following form:

$$\psi_{ij}(f) = \frac{\sigma_u}{\sigma_\varphi(f,W_{ij})}\langle\times\rangle\big(f\langle-\rangle\mu_\varphi(f,W_{ij})\big) . \tag{4.1.4}$$

To obtain the enhanced image the transform ψ_{enh} is built as a sum of the affine transforms ψ_{ij} from (4.1.4), weighted with the degrees of membership w_{ij} :

$$\Psi_{enh}(f) = \sum_{j=0}^{n} \sum_{i=0}^{m} w_{ij} \langle \times \rangle \Psi_{ij}(f) \; . \tag{4.1.5}$$

4.2 The Enhancement Method for Color Images

A color image has three components *red*, *green* and *blue* that are defined by three scalar functions $r : D \rightarrow E$, $g : D \rightarrow E$, $b : D \rightarrow E$. Let there be r_{enh}, g_{enh}, b_{enh} the scalar components of the enhanced image. The image luminosity l is defined by the following function: $l : D \rightarrow E$, $\forall (x,y) \in D$,

$$l(x,y) = \frac{1}{3} \langle \times \rangle \big(r(x,y)\langle + \rangle g(x,y)\langle + \rangle b(x,y)\big) \; . \tag{4.2.1}$$

Using the formulae (3.4) and (3.5) the means $\mu_\varphi(l, W_{ij})$ and the variances $\sigma_\varphi^2(l, W_{ij})$ for a fuzzy window W_{ij} are computed. Using (4.1.4), the affine transform will be defined with these two values. Thus, the enhanced image will be calculated using (4.1.5) with the following functions:

$$r_{enh} = \sum_{j=0}^{n} \sum_{i=0}^{m} \frac{w_{ij} \sigma_u}{\sigma_\varphi(l, W_{ij})} \langle \times \rangle \big(r \langle - \rangle \mu_\varphi(l, W_{ij})\big) \; . \tag{4.2.2}$$

$$g_{enh} = \sum_{j=0}^{n} \sum_{i=0}^{m} \frac{w_{ij} \sigma_u}{\sigma_\varphi(l, W_{ij})} \langle \times \rangle \big(g \langle - \rangle \mu_\varphi(l, W_{ij})\big) \; . \tag{4.2.3}$$

$$b_{enh} = \sum_{j=0}^{n} \sum_{i=0}^{m} \frac{w_{ij} \sigma_u}{\sigma_\varphi(l, W_{ij})} \langle \times \rangle \big(b \langle - \rangle \mu_\varphi(l, W_{ij})\big) \; . \tag{4.2.4}$$

5 Experimental Results

In this section some experimental results will be computed. Thus, for the image "aerial1" shown in Fig. 1 (*left*) was obtained using a (10×10) fuzzy partition the enhanced image shown in Fig. 1 (*right*). In order to compare, there were used two methods: one processing by defining a (3×3) classical partition and another one, which uses only one affine transform for the entire image (i.e. without defining any partition on the image support). In the case of the classical partition shown in Fig. 2 (*right*) the brightness and contrast discontinuities between the partition windows are clearly seen. In the other case, the processing without defining the partition, dark and bright large areas keeping low contrast can be seen in Fig. 2 (*left*).
Two others images "island", and "aerial2" are shown in Figs. 3, 4 (*left*) and their enhanced images in Figs. 3, 4 (*right*).

Fig. 1 The original image "aerial1" (*left*) and the enhanced with fuzzy partition (*right*).

Fig, 2. Enhanced without partition (*left*) and enhanced with classical partition (*right*).

Fig. 3 The original image "island" (*left*) and the enhanced with fuzzy partition (*right*).

Fig. 4. The original image "aerial2" (*left*) and the enhanced with fuzzy partition (*right*).

6 Conclusions

The presented enhancement methods use in the same time elements from the fuzzy set theory with elements from the logarithmic models theory. The calculus is made using logarithmic operations while the image structuring is made using fuzzy partitions. After splitting the image support in fuzzy windows, for each one of them an affine transform will be determined. Thus, there result different transforms depending on the brightness and the contrast of every fuzzy window. The brightness and the contrast dimensions can be found in the fuzzy mean and variance, which were computed for every element of the partition, (i.e. for every window). Using logarithmic operations defined on bounded sets avoids the truncation operations and the fuzzy partition of the image support allows to build affine transforms adjusted to diverse areas of the processed image. The experimental results stand as a proof for the logarithmic model utility and for its rich potential of calculus.

References

1. Jourlin, M., Pinoli, J.C.: Logarithmic Image Processing. The mathematical and physical framework for the representation and processing of transmitted images. Advances in Imaging and Electron Physics, vol. 115, (2001) 129-196
2. Oppenheim, A.V.: Supperposition in a class of non-linear system. Technical Report 432, Research Laboratory of Electronics, M.I.T., Cambridge MA, (1965)
3. Pătraşcu, V.: A mathematical model for logarithmic image processing. PhD thesis, "Politehnica" University of Bucharest, Romania, (2001)
4. Pătraşcu, V., Buzuloiu, V.: A Mathematical Model for Logarithmic Image Processing. Proceedings of the 5th World Multi-Conference on Systemics, Cybernetics and Informatics SCI 2001, vol. 13, Orlando, USA, (2001) 117-122
5. Pătraşcu, V., Buzuloiu, V.: Color Image Enhancement in the Framework of Logarithmic Models. The 8th IEEE International Conference on Telecommunications, Vol. 1, ICT2001, Bucharest, Romania, (2001) 199-204
6. Pătraşcu, V., Buzuloiu, V.: Modelling of Histogram Equalisation with Logarithmic Affine Transforms. Recent Trends in Multimedia Information Processing, (Ed. P. Liatsis), World Scientific Press, 2002, Proceedings of the 9th International Workshop on Systems, Signals and Image Processing, IWSSIP'02, Manchester, United Kingdom, (2002) 312-316
7. Pătraşcu, V.: Logarithmic Image Enhancement Using the Image Support Fuzzification. Symbolic and Numeric Algorithms for Scientific Computing, (Eds. D. Petcu, V. Negru, D. Zaharie, T. Jebeleanu), MIRTON Press, 2002, Proceeding for 4th International Workshop SYNASC 02, Timisoara, Romania, (2002) 253-262
8. Stockham, T.G.: Image processing in the context of visual models. Proceedings IEEE, Vol. 60, no. 7, (1972) 828-842

Lattice Fuzzy Signal Operators and Generalized Image Gradients

Petros Maragos, Vassilis Tzouvaras, and Giorgos Stamou

National Technical University of Athens,
School of Electrical & Computer Engineering,
Zografou, Athens 15773, Greece.
maragos@cs.ntua.gr, tzouvaras@image.ntua.gr, gstam@softlab.ntua.gr

Abstract. In this paper we use concepts from the lattice-based theory of morphological operators and fuzzy sets to develop generalized lattice image operators that are nonlinear convolutions that can be expressed as supremum (resp. infimum) of fuzzy intersection (resp. union) norms. Our emphasis and differences with many previous works is the construction of pairs of fuzzy dilation (sup of fuzzy intersection) and erosion (inf of fuzzy implication) operators that form lattice adjunctions. This guarantees that their composition will be a valid algebraic opening or closing. We have experimented with applying these fuzzy operators to various nonlinear filtering and image analysis tasks, attempting to understand the effect that the type of fuzzy norm and the shape-size of structuring function have on the resulting new image operators. We also present some theoretical and experimental results on using the lattice fuzzy operators, in combination with morphological systems or by themselves, to develop some new edge detection gradients which show improved performance in noise.

1 Introduction

Mathematical morphology (MM) and fuzzy sets share many common theoretical concepts. As an earlier example, the use of min/max to extend the intersection/union of ordinary (crisp) sets to fuzzy sets [14] has also been used to extend the set-theoretic morphological shrink/expand operations on binary images to min/max filtering on graylevel images [11,3]. While the field of morphological image analysis was maturing, several researchers developed various other approaches using fuzzy logic ideas for extending or generalizing the morphological image operations [13,1]. The main ingredients of these approaches have been to (1) map the max-plus structure of Minkowski signal dilation to a sup-T signal convolution, where T is some fuzzy intersection norm, and (2) use duality to map the inf-minus structure of Minkowski signal erosion to a inf-T' convolution, where T' is a dual fuzzy union norm. The main disadvantage of these approaches is that composition of the operators from steps (1) and (2) is not guaranteed to be an algebraic opening or closing. (Openings and closing are the basic morphological smoothing filters.)

T. Bilgiç et al. (Eds.): IFSA 2003, LNAI 2715, pp. 420–427, 2003.

Meanwhile MM was extended using lattice theory [12,4] to more general operators that shared with the standard dilation, erosion, opening and closing only a few algebraic properties. One such fundamental algebraic structure is a pair of erosion/dilation operators that form an *adjunction*. This guarantees the formation of openings and closings.

In a previous work [7,8] some of us used lattice theory to develop generalizations of morphological signal and vector operations based on fuzzy norms. These operations were used in fuzzy dynamical systems to represent the mapping between input and ouput signals (via nonlinear fuzzy-based convolutions) and the mapping between state vectors (via generalized fuzzy-based products of matrices and vectors) as a generalized dilation or erosion acting on signal or vector lattices. In this paper, which is a sequel of [9], we continue this work and apply our general theoretical results from [7] to developing useful nonlinear operators for image/signal analysis based on lattices and fuzzy set operations. From fuzzy set theory [6] we use t-norms and t-conorms to extend intersection and union of crisp sets to signal convolutions. To form openings and closings we use pairs of t-norms and fuzzy implications. (A work similar to our theoretical analysis appeared recently in [2]. Also, some recent work in fuzzy MM includes [5,10].) First, we discuss the theoretical development of the new operators. We also present some results on using the lattice fuzzy operators, in combination with morphological systems or by themselves, to develop some new edge detection gradients which show improved performance in noise. Proofs of our theoretical results will be given in a longer paper.

2 Background: Lattice Morphological Operators

A poset $\mathcal{L}$ is any set equipped with a partial ordering $\leq$. The supremum ($\bigvee$) and infimum ($\bigwedge$) of any subset of $\mathcal{L}$ is its lowest upper bound and greatest lower bound, respectively, induced by the partial order; both are unique if they exist. The algebra $(\mathcal{L}, \vee, \wedge)$ is called a *complete lattice* if the supremum and infimum of any (finite or infinite) collection of its elements exists. An operator ψ on a complete lattice $\mathcal{L}$ is called: *increasing* if it preserves the partial ordering $[f \leq g \implies \psi(f) \leq \psi(g)]$; idempotent if $\psi^2 = \psi$; antiextensive (resp. extensive) if $\psi(f) \leq f$ (resp. $f \leq \psi(f)$). An operator ε (resp. δ) on a complete lattice is called an **erosion** (resp. **dilation**) if it distributes over the infimum (resp. supremum) of any collection of lattice elements; namely $\delta(\bigvee_i f_i) = \bigvee_i \delta(f_i)$ and $\varepsilon(\bigwedge_i f_i) = \bigwedge_i \varepsilon(f_i)$. An operator is called an **opening** (resp. **closing**) if it is increasing, antiextensive (resp. extensive) and idempotent. An operator pair (ε, δ) is called an **adjunction** iff $\delta(f) \leq g \iff f \leq \varepsilon(g)$, $\forall f, g \in \mathcal{L}$. Given a dilation δ, there is a unique erosion $\varepsilon(g) = \bigvee\{f : \delta(f) \leq g\}$ such that (ε, δ) is adjunction, and vice-versa.

Proposition 1 ([12], [4]). *Let (ε, δ) be an adjunction. Then: (i) δ is a dilation and ε is an erosion. (ii) $\delta\varepsilon$ is an opening, and $\varepsilon\delta$ is a closing.*

For lattice-based image/signal processing, the signal space is the collection $\mathcal{L} = \mathbb{V}^{\mathbb{E}}$ of all images/signals $f : \mathbb{E} \to \mathbb{V}$, (a continuous or discrete) domain

$\mathbb{E}$ where $\mathbb{E} = \mathbb{R}^m$ or $\mathbb{Z}^m$, $m = 1, 2, ...,$, and assuming values in $\mathbb{V} \subseteq \overline{\mathbb{R}}$ where $\overline{\mathbb{R}} = \mathbb{R} \cup \{-\infty, \infty\}$. The value set $\mathbb{V}$ must be a complete lattice under the usual ordering $\leq$ of real numbers, with corresponding sup ($\vee$) and inf ($\wedge$) the usual supremum and infimum in $\overline{\mathbb{R}}$. The signal space $\mathcal{L}$ also becomes a complete distributive lattice if we define on it the standard *pointwise* partial ordering $\leq$, supremum $\vee$, and infimum $\wedge$ induced by $\mathbb{V}$.

Proposition 2 ([4]). *The pair* (ε, δ) *is an adjunction on the signal lattice* $\mathbb{V}^{\mathbb{E}}$ *iff for every* $x, y \in \mathbb{E}$ *there exists a scalar adjunction* $(\varepsilon_{y,x}, \delta_{x,y})$ *on* $\mathbb{V}$ *such that*

$$\delta(f)(x) = \bigvee_{y \in \mathbb{E}} \delta_{x,y}(f(y)), \quad \varepsilon(g)(y) = \bigwedge_{x \in \mathbb{E}} \varepsilon_{y,x}(g(x)) \tag{1}$$

3   Lattice Operators Using Fuzzy Norms

In this paper we shall work on the complete signal lattice $\mathcal{L} = \mathbb{V}^{\mathbb{E}}$ where the range of all signals is the complete scalar lattice $\mathbb{V} = [0, 1]$.

The classic *translation-invariant (TI)* dilations and erosions on $\overline{\mathbb{R}}^{\mathbb{E}}$ are built as sup of signal translations of the type $\tau_{y,v}(f)(x) = v + f(x - y)$. In this paper we shall use new translations where the binary operation $a + b$ will be replaced by fuzzy norms. Specifically, we build generalized TI image dilations and erosions by defining the scalar dilations (erosions) $\delta_{x,y}$ ($\varepsilon_{y,x}$) of (1) via some fuzzy intersection (union) between the values of the *image signal* f and a *structuring function* h. First we define fuzzy norms.

A **fuzzy intersection norm**, in short a Tnorm, is a binary operation T: $[0, 1]^2 \rightarrow [0, 1]$ that satisfies the following conditions [6]: For all $a, b, c \in [0, 1]$
F1. $T(a, 1) = a$ and $T(a, 0) = 0$ (boundary conditions).
F2. $T(a, T(b, c)) = T(T(a, b), c)$ (associativity).
F3. $T(a, b) = T(b, a)$ (commutativity).
F4. $b \leq c \Longrightarrow T(a, b) \leq T(a, c)$ (increasing).
For the Tnorm to be a scalar dilation (with respect to any argument) on $\mathbb{V}$, it must also satisfy [7]:
F5. T is a continuous function.
A **fuzzy union norm** [6] is a binary operation U: $[0, 1]^2 \rightarrow [0, 1]$ that satisfies F2-F5 and a dual boundary condition:
F1'. $U(a, 0) = a$ and $U(a, 1) = 1$.
Clearly, U is an erosion on $\mathbb{V}$.

Now the new signal translations on $\mathcal{L} = [0, 1]^{\mathbb{E}}$ are the operators $\tau_{y,v}(f)(x) = T(v, f(x - y))$, where $(y, v) \in \mathbb{E} \times \mathbb{V}$ and $f(x)$ is an arbitrary input signal. A signal operator on is called *translation invariant (TI)* iff it commutes with any such translation. Similarly, we define dual signal translations $\tau'_{y,v}(f)(x) = U(v, f(x - y))$. Consider now two elementary signals, called the *impulse* q and the *dual impulse* q':

$$q(x) \triangleq \begin{cases} 1, & x = 0 \\ 0, & x \neq 0 \end{cases}, \quad q'(x) \triangleq \begin{cases} 0, & x = 0 \\ 1, & x \neq 0 \end{cases}$$

Then every signal f can be represented as a sup of translated impulses or as inf of dual-translated dual impulses:

$$f(x) = \bigvee_y T[f(y), q(x - y)] = \bigwedge_y U[f(y), q'(x - y)]$$

General TI signal dilation and erosion can result, respectively, from the sup-T convolution $\bigcirc_T$ and the inf-U convolution $\bigcirc'_U$ of two signals f and g defined by

$$(f \bigcirc_T g)(x) \triangleq \bigvee_y T[f(y), g(x - y)], \quad (f \bigcirc'_U g)(x) \triangleq \bigwedge_y U[f(y), g(x - y)] \quad (2)$$

The following theorem characterizes all TI signal dilation or erosion operators as nonlinear convolutions of the above type.

Theorem 1. ([7]). *An operator Δ (resp. $\mathcal{E}$) on the signal lattice $[0, 1]^{\mathbb{E}}$ is a translation invariant dilation (resp. erosion) iff it can be represented as the sup-T (resp. inf-U) convolution of the input signal with the operator's (resp. dual) impulse response $h = \Delta(q)$ [resp. $h' = \mathcal{E}(q')$].*

However, the erosion $\mathcal{E}$ of the above theorem may *not* be the adjoint of the dilation Δ. To form an adjunction, we first define a signal **fuzzy dilation** as the previous sup-T convolution

$$\delta(f)(x) \triangleq \bigvee_{y \in \mathbb{E}} T[f(y), h(x - y)] = (f \bigcirc_T h)(x) \quad (3)$$

By recognizing $T[f(y), h(x - y)]$ as the scalar dilations $\delta_{x,y}(f(y))$ in the general decomposition (1) of δ, it follows that the adjoint signal **fuzzy erosion** is

$$\varepsilon(g)(y) \triangleq \bigwedge_{x \in \mathbb{E}} \Omega[g(x), h(x - y)] \quad (4)$$

where Ω represents the adjoint scalar erosions $(\varepsilon_{y,x})$ in (1) and is actually the adjoint of the fuzzy Tnorm:

$$T(v, a) \leq w \iff v \leq \Omega(w, a) \quad (5)$$

Given T we can find its adjoint function Ω by

$$\Omega(w, a) \triangleq \sup\{v \in [0, 1] : T(v, a) \leq w\} \quad (6)$$

The norm T can be interpreted as a logical conjunction, whereas its corresponding adjoint can be interpreted as a logical implication [6].
  Three examples of Tnorms are:

$$\text{Min} : T_1(v, a) = \min(v, a), \quad \text{Product} : T_2(v, a) = v \cdot a$$
$$\text{Yager} : T_3(v, a) = 1 - (1 \wedge [(1 - v)^p + (1 - a)^p]^{1/p}), \; p > 0.$$

The corresponding three adjoint functions are:

$$\Omega_1(w,a) = \begin{cases} w, & w < a \\ 1, & w \geq a \end{cases}, \quad \Omega_2(w,a) = \begin{cases} \min(w/a, 1), & a > 0 \\ 1, & a = 0 \end{cases}$$

$$\Omega_3(w,a) = \begin{cases} 1 - [(1-w)^p - (1-a)^p]^{1/p}, & w < a \\ 1, & w \geq a \end{cases}$$

The generalized **fuzzy opening** α and **fuzzy closing** β are

$$\alpha(f) \triangleq \delta(\varepsilon(f)), \quad \beta(f) \triangleq \varepsilon(\delta(f)) \tag{7}$$

If we define an alternative erosion operator (as an inf-U convolution) by

$$\varepsilon'(f)(y) = \bigwedge_x U[f(x), h(y-x)] \tag{8}$$

where $U(a,b) = 1 - T(1-a, 1-b)$ is a fuzzy union that is dual to T, then $\varepsilon'(f) = 1 - \delta(1-f)$; i.e., this second erosion ε' is the dual of the first dilation δ. Further, the adjoint dilation δ' of ε' is an operator that is dual of the first erosion ε. Many previous works used pairs (ε', δ) which are duality pairs but not adjunctions and hence cannot form openings/closings via compositions.

4   Morphological, Fuzzy, and Hybrid Edge Gradients

In this section we present some theoretical and experimental results on using the lattice fuzzy operators, in combination with morphological systems or by themselves, to develop some new edge gradients based on fuzzy operators.

Morphological Gradient: In the continuous case (images defined on $\mathbb{R}^2$), if $\delta_s(f) = f \oplus sB$ and $\varepsilon_s(f) = f \ominus sB$ are the flat dilation and erosion of f by multiscale disks sB, it is well-known that

$$\lim_{s \downarrow 0} [\delta_s(f)(x) - \varepsilon_s(f)(x)]/(2s) = ||\nabla f(x)|| \tag{9}$$

for a differentiable function f.

Hybrid Gradient: We have proven that a similar result holds if we replace in (9) the morphological dilation with a multiscale fuzzy dilation as follows:

$$\delta_s(f)(x) = \bigvee_y T(f(x-y), h(y/s)) \tag{10}$$

Specifically, assuming that f and T are differentiable, we have proven that

$$\lim_{s \downarrow 0} [\delta_s(f)(x) - f(x)]/s = K \cdot ||\nabla f(x)||, \quad K = \bigvee_y \partial_1 T[f(x), h(y)] \cdot y \tag{11}$$

where ∂_1 denotes partial derivative w.r.t. first argument. Further, for the fuzzy edge gradients, the unit-scale structuring function $h(x)$ must be a unimodal symmetric structuring function with a global maximum at $x = 0$, $h(0) = 1$. Similarly we can define an adjoint multiscale fuzzy erosion ε_s and a corresponding symmetric fuzzy gradient $\lim_{s \downarrow 0} [\delta_s(f)(x) - \varepsilon_s(f)(x)]/(2s)$.

Discrete versions of the above gradients results by replacing the above scale-normalized limits with differences $\delta(f) - \varepsilon(f)$, where δ and ε represent: i) (in the classic morphological case) a morphological dilation and erosion of f by a 3×3-pixel flat structuring element, or ii) (in the hybrid case) a fuzzy dilation and erosion by a small structuring element (at scale $s = 1$). The hybrid (morphological-fuzzy) gradient showed a small improvement over the morphological one.

Fuzzy Gradient: We also propose a new and different type of discrete edge gradient:

$$\text{FuzzyEdge}_{min}(f) = \min[\delta_s(f), 1 - \varepsilon_s(f)] \tag{12}$$

where δ_s and ε_s are the same fuzzy dilation and erosion as above with $s = 1$. A dual type of fuzzy edge gradient, FuzzyEdge_{max}, results when the min is replaced with max. The last two types of edge gradients were inspired by the standard discrete morphological gradient $\delta(f) - \varepsilon(f)$, but to make the gradient operator more consistent with fuzzy set theory we replaced the difference between dilation and erosion with min (or max) of the dilation and the fuzzy complement of the erosion.

In Fig. 1 we present some experimental results illustrating the differences between the classical morphological operators and the generalized lattice-fuzzy operators. Rows 1 and 2 show the dilation, erosion, opening and closing of 1D images. In general, we have observed that, the fuzzy operators are more adaptive and track closer the peaks/valleys of the signal than the corresponding flat morphological operators of the same scale. Similar conclusions were reached during our experiments with 2D images. In Rows 3 and 4, we investigate the performance of the fuzzy operators in edge enhancement-detection. As shown in Fig. 1(h,i), the new fuzzy gradient operators have a quite promising behavior since they yied clean and sharper edges than the morphological gradient. Further, the fuzzy operators are less influenced by noise than the morphological operators. The morphological operators seem to detect more edges in non-noisy images (Fig. 1(n,i)); however, the fuzzy operators may also be designed to detect more edges by optimally tuning the parameter of the Tnorm [9]. The last three rows of Fig. 1 clearly illustrate the superior performance of the fuzzy operators for edge detection in noise. The results are demonstrated by using a test 1D signal with known derivative and edge location. The noisy signal versions are formed by adding 'gray salt and pepper' noise at different SNRs; see Figs. 1(m,n,o). In both the morphological and fuzzy case, the performance, in the presence of noise, is improved with larger size structuring elements.

5   Conclusions

The power but also the difficulty in applying the lattice-based fuzzy operators to image analysis is the huge variety of fuzzy norms and the absence of some

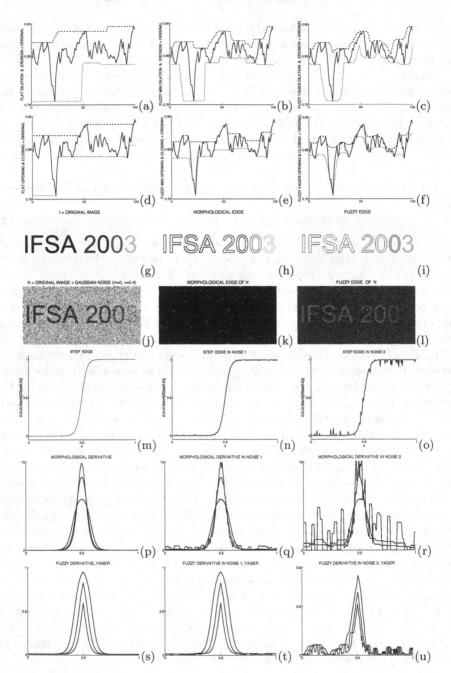

Fig. 1. Rows 1 and 2, left to right: flat, minimum, Yager. Row 1: original signal (solid line), dilation (dashed line), erosion (dotted line). Row 2: closing (dashed line), opening (dotted line). Row 3: Edges in images. Row 4: Edges in noisy images Row 5: step edge function, signal with low noise, signal with high noise. Rows 6,7: performance of morphological and fuzzy edge operators, respectively.

practical experience or systematic ways in selecting them. In general, having a parameter in the Tnorm offers flexibility for the fuzzy operators that use it. The scale (support size) and the shape of the structuring function are also important factors, which can influence the behavior of the fuzzy operators. By tuning these parameters we can find the optimum solution depending on the type of application, such as edge detection in the presence of noise or other feature detection in images with low contrast [9]. In our on-going work we are investigating various methods to design these new operators for various nonlinear filtering and image analysis tasks. Finally, by combining lattice MM and fuzzy set theory, we can create new operators, like the fuzzy edge gradients, that extend and improve the capabilities of the standard morphological operators.

References

1. I. Bloch and H. Maitre, "Fuzzy mathematical morphologies: a comparative study" *Pattern Recognition 28*, vol. 9, pp. 1341–1387, 1995.
2. T. Q. Deng and H.J.A.M. Heijmans, "Grey-Scale Morphology Based on Fuzzy Logic", *J. Math. Imaging and Vision*, 16: 155-171, 2002.
3. V. Goetcherian, "From binary to greytone image processing using fuzzy logic concepts", *Pattern Recognition 12*, pp.7-15, 1980.
4. H. J. A. M. Heijmans, *Morphological Image Operators*, Academic Press, Boston, 1994.
5. E. E. Kerre and M. Nachtegael, "Fuzzy Techniques in Image Processing: Techniques and Applications", in *Studies in Fuzziness and Soft Computing*, Vol. 52, Physica Verlag, 2000.
6. G. J. Klir and B. Yuan, *Fuzzy Sets and Fuzzy Logic: Theory and Applications*, Prentice-Hall, 1995.
7. P. Maragos, G. Stamou and S. Tzafestas, "A Lattice Control Model of Fuzzy Dynamical Systems in State-Space", in *Mathematical Morphology and Its Application to Image and Signal Processing*, J. Goutsias, L. Vincent and D. Bloomberg, Eds, Kluwer Acad. Publ., Boston, 2000, pp. 61–70.
8. P. Maragos and S. Tzafestas, "A Lattice Calculus Unification of Min–Max Control Systems of the Morphological and Fuzzy Type", *Proc. Int'l Symp. on Soft Computing in Engineering Applications (SOFTCOM-98)*, Athens, June 1998.
9. P. Maragos, V. Tzouvaras, and G. Stamou, "Synthesis and Applications of Lattice Image Operators Based On Fuzzy Norms", *Proc. Int'l Conf. Image Processing (ICIP-2001)*, Thessaloniki, Greece, Oct. 2001.
10. M. Nachtegael and E. E. Kerre, "Decomposing and Constructing Fuzzy Morphological Operations Over α-Cuts: Continuous and Discrete Case", *IEEE Trans. Fuzzy Systems*, vol.8, pp.615-626, Oct. 2000.
11. Y. Nakagawa and A. Rosenfeld, "A note on the use of local min and max operations in digital picture processing", *IEEE Trans. Syst., Man, Cybern.*, SMC-8, p.632-635, Aug.1978.
12. J. Serra Ed., *Image Analysis and Mathematical Morphology II: Theoretical Advances*, Acad. Press, NY, 1988.
13. D. Sinha and E. R. Dougherty, "Fuzzy mathematical morphology", *J. Visual Communication and Image Representation*, vol. 3, no. 3, pp. 286–302, 1992.
14. L. A. Zadeh, "Fuzzy Sets", *Information and Control*, vol. 8, pp. 338–353, 1965.

Non-uniform Coders Design for Motion Compression Method by Fuzzy Relational Equations

Hajime Nobuhara and Kaoru Hirota

Department of Computational Intelligence and Systems Science
Tokyo Institute of Technology, 4259 Nagatsuta, Midiri-ku, Yokohama 226-8502, Japan
{nobuhara, hirota}@hrt.dis.titech.ac.jp
http://www.hrt.dis.titech.ac.jp/

Abstract. A motion compression method by max t-norm composite fuzzy relational equations (MCF) is proposed. In the case of MCF, a motion sequence is divided into intra-pictures (I-pictures) and predictive-pictures (P-pictures). The I-pictures and the P-pictures are compressed by using uniform coders and non-uniform coders, respectively. In order to perform an effective compression and reconstruction of the P-pictures, a design method for non-uniform coders is proposed. The proposed design method is based on an overlap level of fuzzy sets and a fuzzy equalization. Through an experiment using 10 P-pictures, it is confirmed that the root means square errors of the proposed method is decreased to 89.4% of the uniform coders, under the condition that the compression rate is 0.0625.

1   Introduction

By normalizing the intensity of an image into $[0, 1]$, it can be treated as a fuzzy relation. Many image processing techniques have been developed in the setting of fuzzy relation calculus, e.g., image compression [1] [2] [3] [5], and digital watermarking [4].

In this paper, a motion compression method by max t-norm composite fuzzy relational equations (MCF) is proposed. In the case of MCF, a motion sequence is decomposed into intra-pictures (I-pictures) and predictive-pictures (P-pictures). The I-picture and P-picture are compressed by using uniform coders and non-uniform coders, respectively. In order to perform an effective compression and reconstruction of P-pictures, a non-uniform coders design method is proposed. The proposed design method is based on an overlap level of fuzzy sets [5] and a fuzzy equalization. Through an experiment using 10 P-pictures, it is confirmed the effectiveness of the proposed method.

T. Bilgiç et al. (Eds.): IFSA 2003, LNAI 2715, pp. 428–435, 2003.

2   A Motion Compression by Fuzzy Relational Equations (MCF)

2.1   An Overview of MCF

A Motion Compression method by Fuzzy relational equations (MCF) treats motion (a sequence of images of size $M \times N$) as fuzzy relations $\mathbf{R} = \{R_k | k = 1, \ldots, S\} \subset F(\mathbf{X} \times \mathbf{Y})$, $\mathbf{X} = \{x_1, \ldots, x_M\}$, $\mathbf{Y} = \{y_1, \ldots, y_N\}$ by normalizing the intensity range of each pixel into $[0, 1]$. Here, the suffix k stands for the sequence of the time. In the MCF, the motion is divided into intra pictures (I-pictures) and predictive pictures (P-pictures) shown in Fig. 1. The I-picture corresponds to the original frame of the motion, and the P-picture is obtained by subtracting the reconstructed images of I-picture and the original frame of the motion. Figures 2 and 3 shows an example of I-picture and a P-picture of a motion. In Fig. 1, the ratio of I-picture and P-picture is 1 : 3, however, there are other possibilities for the selection of the ratio of I-pictures and P-pictures, e.g. 1 : 2, 1 : 4. As it can be seen from Figs. 2 and 3, the information quantity of P-picture is lower than that of I-picture. Thus, the higher ratio of the P-picture of the motion, the better compression rate it is to achieve in the MCF. However, the reconstructed P-picture contains an error appearing during the compression process (predictive error), and it leads to serious degradation of the reconstructed motion. We employ a compromise value of the ratio of the I-pictures and P-pictures.

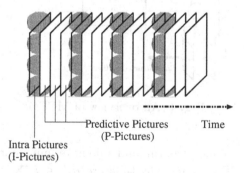

Intra Pictures
(I-Pictures)

Predictive Pictures
(P-Pictures)

Time

Fig. 1. A sequence of images in MCF

The motion is compressed by MCF as shown in Fig. 4, i.e., I-pictures are compressed by U-ICF (Image Compression method based on Fuzzy relational equations, Uniform coders type), while P-pictures are compressed by N-ICF (ICF, Non-uniform coders type).

2.2   U-ICF

The compression of U-ICF is performed by the max t-norm composition of fuzzy relational equations. Other types of fuzzy relational equation, e.g., min s-norm

Fig. 2. Samples of a motion (I-picture)

Fig. 3. Samples of a motion (P-picture)

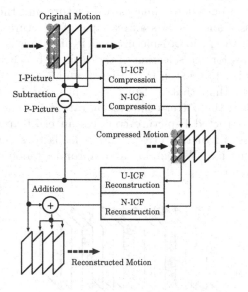

Fig. 4. An overview of MCF

composition, adjoint of max t-norm, and adjoint of min s-norm can be also used for U-ICF. The U-ICF compress the I-picture $R_k^{(I)} \in F(\mathbf{X} \times \mathbf{Y})$ of a motion $\mathbf{R}$ only, where $R_k^{(I)}$ is defined by

$$R_k^{(I)} = R_k. \tag{1}$$

The I-picture is compressed into $G_k^{(I)} \in F(\mathbf{I} \times \mathbf{J})$ by

$$G_k^{(I)}(i,j) = \max_{y \in \mathbf{Y}} \left\{ B_j(y) \ t \ \max_{x \in \mathbf{X}} \left\{ A_i(x) \ t \ R_k^{(I)}(x,y) \right\} \right\}, \tag{2}$$

where t denotes a continuous t-norm, while $A_i \in \mathbf{A} \subset F(\mathbf{X})$, and $B_j \in \mathbf{B} \subset F(\mathbf{Y})$ stand for coders. The coders $\mathbf{A}$ and $\mathbf{B}$ are defined by

$$\mathbf{A} = \{A_1, A_2, \ldots, A_I\}, \tag{3}$$

$$A_i(x_m) = \exp\left(-Sh\left(\frac{iM}{I} - m\right)^2\right),\tag{4}$$

$$(m = 1, 2, \ldots, M),$$

and

$$\mathbf{B} = \{B_1, B_2, \ldots, B_J\},\tag{5}$$

$$B_j(y_n) = \exp\left(-Sh\left(\frac{jN}{J} - n\right)^2\right),\tag{6}$$

$$(n = 1, 2, \ldots, N).$$

where Sh denotes the sharpness of fuzzy sets A_i and B_j. The distribution of coders is uniform, therefore, the image compression process is called U-ICF (image compression and reconstruction based on fuzzy relational equations, uniform coders). The compression rate can be adjusted by the number of fuzzy sets $I \times J$ contained in coders $\mathbf{A}$ and $\mathbf{B}$. Image reconstruction corresponds to an inverse problem under the condition that the compressed image $G_k^{(I)}$ and the coders $\mathbf{A}$ and $\mathbf{B}$ are given. A reconstructed image $\tilde{R}_k^{(I)} \in F(\mathbf{X} \times \mathbf{Y})$ is given by

$$\tilde{R}_k^{(I)}(x, y) = \min_{j \in \mathbf{J}}\left\{B_j(y)\varphi_{t'}\min_{i \in \mathbf{I}}\left\{A_i(x)\varphi_{t'}G_k^{(I)}(i, j)\right\}\right\},\tag{7}$$

where $\varphi_{t'}$ denotes the t-relative pseudo-complement defined by

$$a\varphi_{t'}b = \sup\{c \in [0, 1]|at'c \leq b\},\tag{8}$$

and a t-norm t' must be selected such that

$$\forall a, b \in [0, 1], \quad atb \leq at'b.\tag{9}$$

2.3   N-ICF

The P-picture $R_k^{(P)}(\in F(\mathbf{X} \times \mathbf{Y}))$ is compressed and reconstructed by N-ICF unit shown in Fig. 4, where the P-picture $R_k^{(P)}$ is obtained by

$$\forall(x, y) \quad R_k^{(p)}(x, y) = R_k(x, y) - \tilde{R}_{k-1}^{(I)}(x, y).\tag{10}$$

In the case of N-ICF, the compression and reconstruction are also performed by Eqs. (2), (7)-(9), respectively. However, the distribution of coders $\mathbf{A}$ and $\mathbf{B}$ of N-ICF is different from that of U-ICF. The N-ICF can focus different parts of the image by changing allocation of the fuzzy sets of coders, non-uniformly. Therefore, the N-ICF is useful for the compression/reconstruction of the line image that is often observed in P-pictures.

In order to show the effectiveness of the N-ICF, an example of image compression and reconstruction is presented. In this experiment, a circle image of

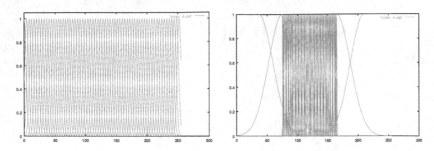

Fig. 5. Uniform Coders **A**   **Fig. 6.** Non-uniform Coders **A**

size $M \times N = 256 \times 256$ shown in Fig. 7, is used as the original image. The U-ICF and N-ICF employ respectively the uniform and non-uniform coders shown in Figs. 5, and 6. The number of fuzzy sets of coders **A** and **B** is $I = 64$ and $J = 64$, respectively, therefore, the compression rate is $\frac{I \times J}{M \times N} = 0.0625$.

Figures 8 and 9 show the reconstructed image of the U-ICF and N-ICF. As can be seen from Figs. 8 and 9, the quality of the reconstructed image of N-ICF is better than that of the U-ICF. With respect to the reconstructed images, the root means square error (RMSE),

$$RMSE = \sqrt{\frac{\sum_{(x,y)\in \mathbf{X}\times\mathbf{Y}}\left(R(x,y) - \tilde{R}(x,y)\right)^2}{|\mathbf{X} \times \mathbf{Y}|}}, \qquad (11)$$

is measured, where R and $\hat{R}$ stand for the original image and reconstructed one, respectively.

Fig. 7. Original Image **Fig. 8.**   Reconstructed **Fig. 9.**   Reconstructed
(Circle)   Image (Uniform)   Images (Non-uniform)

The RMSE of the reconstructed image of U-ICF and N-ICF is 31.86, and 18.40, respectively. This example shows the effectiveness of the N-ICF in the line image compression/reconstruction. However, there is the problem of how to determine the distribution of the coders **A** and **B**. In the next section, a non-uniform coders design for N-ICF is proposed.

3 Non-uniform Coders Design for N-ICF

In the N-ICF, non-symmetric Gaussian coders are defined by

$$\mathbf{A} = \{A_1, A_2, \ldots, A_I\},\tag{12}$$

$$A_i(x_m) = \begin{cases} \exp\left(-Sh_{left}^{(A_i)}\left(A_i^c - m\right)^2\right), & if\ m \le A_i^c, \\ \exp\left(-Sh_{right}^{(A_i)}\left(A_i^c - m\right)^2\right), & if\ m > A_i^c, \end{cases}\tag{13}$$

$$(m = 1, 2, \ldots, M),$$

and

$$\mathbf{B} = \{B_1, B_2, \ldots, B_J\},\tag{14}$$

$$B_j(y_n) = \begin{cases} \exp\left(-Sh_{left}^{(B_j)}\left(B_j^c - n\right)^2\right), & if\ m \le B_j^c, \\ \exp\left(-Sh_{right}^{(B_j)}\left(B_j^c - n\right)^2\right), & if\ m > B_j^c, \end{cases}\tag{15}$$

$$(n = 1, 2, \ldots, N).$$

The distribution of the coders (expressed by $Sh_{left}^{(A_i)}\ Sh_{right}^{(A_i)}$, and A_i^c) is adjusted by the following steps. For simplicity, the coder $\mathbf{A}$ is only considered.

1. Calculate the standard deviation $\sigma(x_m)$ given by

$$\sigma(x_m) = \frac{1}{N}\{\sum_{y\in\mathbf{Y}}(R_k^{(P)}(x_m, y) - E(x_m))^2,\}\tag{16}$$

$$(m = 1,\ldots,M),$$

$$E(x_m) = \frac{1}{N}\sum_{y\in\mathbf{Y}}R_k^{(P)}(x_m, y),\tag{17}$$

$$(m = 1,\ldots,M).$$

2. Calculate the equalization index σ_T/I defined by

$$\sigma_T/I = \frac{1}{I}\sum_{x\in\mathbf{X}}\sigma(x).\tag{18}$$

3. Start from the lower bound of $\mathbf{X}$, i.e., x_1. Proceed toward higher values of $\mathbf{X}$ computing the moving value of the summation,

$$\sum_{x\in\mathbf{X}}\sigma(x).\tag{19}$$

Stop once the value of this summation has reached the value of σ_T/I and record the corresponding value of the argument as π_i (region of fuzzy set $A_i \in F(\mathbf{X})$, see Fig. 10). Determine the center point A_i^c of π_i (see Fig. 11).

4. Initialize the coder $\mathbf{A}$ defined by Eqs. (12) - (13).

5. For the $A_{i-1}, A_i \in \mathbf{A}$, the overlap level $\alpha(A_{i-1}, A_i)$ of A_{i-1} and A_i,

$$\alpha(A_{i-1}, A_i) = \min_{x \in \mathbf{X}_{(i-1)}^{(i)}} \{\max(A_{i-1}(x), A_i(x))\}, \qquad (20)$$

where $\mathbf{X}_{(i-1)}^{(i)} = \{x_m \in \mathbf{X} | a_{i-1} \leq x_m \leq a_i\}$ and a_i corresponds to the center point of A_i, is computed.

5-A) If $\alpha(A_{i-1}, A_i) \geq \alpha_{opt} + \epsilon$, then $Sh_{right}^{(A_i)}$ of A_{i-1} and $Sh_{left}^{(A_i)}$ of A_i are decreased, where α_{opt} denotes the optimal overlap level 0.85 [5].

5-B) If $\alpha(A_{i-1}, A_i) \leq \alpha_{opt} - \epsilon$, then $Sh_{right}^{(A_i)}$ of A_{i-1} and $Sh_{left}^{(A_i)}$ of A_i are increased.

5-C) If $\alpha_{opt} - \epsilon \leq \alpha(A_{i-1}, A_i) \leq \alpha_{opt} + \epsilon$, then the adjustment process of A_{i-1} and A_i is stopped.

6. Repeat step 5 for the successive fuzzy sets in $\mathbf{A}$.

In this method, the first and last fuzzy sets in $\mathbf{A}$ are defined by trapezoidal membership functions shown in Fig. 11.

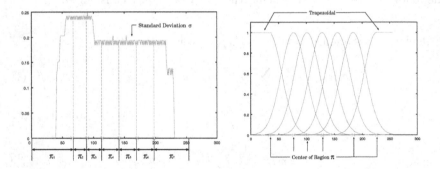

Fig. 10. Standard deviation $\sigma(x)$ of the proposed method

Fig. 11. Example of non-uniform coders $\mathbf{A}$

4   Experimental Comparison

A comparison of non-uniform coders obtained by the proposed method with uniform ones (Eqs. (3)-(6)) is presented. Ten P-pictures (Figs. 12 and 13) of size $M \times N = 384 \times 240$ are compressed into size of $I \times J = 88 \times 60$ by the non-uniform coders and the uniform coders, respectively. The RMSE average of the reconstructed images obtained by the non-uniform and uniform coders is 9.899 and 11.071, respectively.

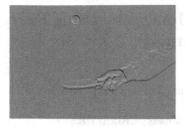 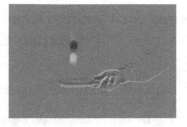

Fig. 12. Test motion samples (Frame 5)   **Fig. 13.** Test motion samples (Frame 10)

5 Conclusions

A motion compression method by fuzzy relational equations (MCF) is presented by extending the still-image compression by fuzzy relational equations [1] [2].

In the case of MCF, a motion sequence is decomposed of intra-pictures (I-pictures) and predictive-pictures (P-pictures). The I-picture and P-picture are compressed by using uniform coders and non-uniform coders, respectively. In order to perform the effective compression and reconstruction of P-pictures, a non-uniform coders design method is proposed. The proposed design method is based on an overlap level of fuzzy sets and a fuzzy equalization. Through an experiment using 10 P-pictures, it is shown that the root means square errors of the proposed method is decreased to 89.4% of the uniform coders, under the condition that the compression rate is 0.0625.

References

1. Hirota, K., and Pedrycz, W.: Fuzzy Relational Compression. IEEE Transactions on Systems, Man, and Cybernetics, Part B, Vol. 29, No. 3, (1999) 407–415
2. Nobuhara, H., Pedrycz, W. and Hirota, K.: Fast Solving Method of Fuzzy Relational Equation and its Application to Image Compression/Reconstruction. IEEE Transaction on Fuzzy Systems, Vol. 8, No. 3, (2000) 325–334
3. Nobuhara, H., Takama, Y., and Hirota, K.: Image Compression/Reconstruction Based on Various Types of Fuzzy Relational Equations. The Transactions of The institute of Electrical Engineers of Japan, Vol. 121-C, No. 6, (2001) 1102-1113
4. Nobuhara, H, Pedrycz, W. and Hirota, K.: A Digital Watermarking Algorithm using Image Compression Method based on Fuzzy Relational Equation. IEEE International Conference on Fuzzy Systems (Fuzz-IEEE 2002), CD-ROM Proceeding, (2002), Hawaii, USA.
5. Nobuhara, H., Pedrycz, W., and Hirota, K.: Fuzzy Relational Image Compression using Non-uniform Coders Designed by Overlap Level of Fuzzy Sets. International Conference on Fuzzy Systems and Knowledge Discovery (FSKD'02), CD-ROM Proceeding, (2002), Singapore.

A Method for Coding/Decoding Images by Using Fuzzy Relation Equations

Ferdinando Di Martino[1], Vincenzo Loia[2], and Salvatore Sessa[3]

[1] Datitalia Processing spa, Research Laboratory CRISALIDE,
Via G. Porzio 4 is. G/5, 80143, Napoli, Italy
dimartino@datitalia.it
[2] Dipartimento di Matematica e Informatica,Università degli Studi di Salerno,
84081 Baronissi (Salerno), Italy
loia@unisa.it
[3] Dipartimento di Costruzioni e Metodi Matematici in Architettura,,
Università di Napoli "Federico II", Via Monteoliveto 3, 80134 Napoli, Italy
sessa@unina.it

Abstract. By using some well known fuzzy relation equations and the theory of the continuous triangular norms, we get lossy compression and decompression of images, here interpreted as two-argument fuzzy matrices.

1 Introduction

Let $t : [0,1]^2 \to [0,1]$ be a triangular norm (for short, t-norm) and by setting, as usually, $t(x,y) = xty$ for all $x,y \in [0,1]$, we know (e.g., see [1]) that if t is assumed continuous, then there is a unique operator $(x \to_t y)$, called residuum of t, defined as $(x \to_t y) = \max\{ z \in [0,1]: (x \, t \, z) \leq y \}$ and it is such that $(x \, t \, z) \leq y$ iff $z \leq (x \to_t y)$. For instance, for the most famous t-norms used in narrow fuzzy logic, we have:

Lukasiewicz t-norm L: $x \, L \, y = \max \{0, x + y - 1\}$, $(x \to_L y) = \min\{1, 1 - x + y\}$

Goguen (or product) t-norm P: $x \, P \, y = x \, y$, $(x \to_P y) = \min\{1, y / x\}$ }

The following properties hold [1], [2] for all $x,y \in [0,1]$:

(α) $\min\{x, y\} = xt(x \to_t y)$,
(αα) $y \leq (x \to_t (xty))$,
(ααα) $(a \to_t x) \leq (a \to_t y)$ if $x \leq y$.

The above properties shall be useful, as already known in literature [3], [4] for solving a system of fuzzy relation equations utilized for coding/decoding images, here seen as binary fuzzy relations. This work continues our stream of investigations already begun in the previous paper [5], where other types of fuzzy relation equations were considered.

T. Bilgiç et al. (Eds.): IFSA 2003, LNAI 2715, pp. 436–441, 2003.

2 Definitions and Results

Let N be the set of natural numbers, h,k,n,m $\in$ N such that k $\leq$ m and h $\leq$ n, I_s = {1...s}, where s $\in$ N. Suppose assigned the fuzzy sets $A_1,...,A_k : I_m \rightarrow [0,1]$ and $B_1,...,B_h : I_n \rightarrow [0,1]$ and the fuzzy relation R : I_m x $I_n \rightarrow [0,1]$. For brevity, we put $A_p(i) = A_{pi}$ for any p $\in$ I_k, i $\in$ I_m, $B_q(j) = B_{qj}$ for any q $\in$ I_h, j $\in$ I_n and R(i,j)=R_{ij} for any i $\in$ I_m, j $\in$ I_n. For any p $\in$ I_k and q $\in$ I_h, we consider the following equation:

$$g_{pq} = \bigcap_{j=1}^{n}\bigcap_{i=1}^{m}[(A_{pi}tB_{qj}) \rightarrow_t R_{ij}] \qquad (1)$$

The resulting fuzzy relation G = [g_{pq}]: I_h x $I_k \rightarrow [0,1]$ is said *"compression of R in terms of the codebooks { $A_1,...,A_k$} and {$B_1,...,B_h$}"*. We note that hk/mn and for comodity of presentation, we give the following notation of the Equation (1):

$$g = \bigcap_{j=1}^{n}\bigcap_{i=1}^{m}[(A_i tB_j) \rightarrow_t R_{ij}] \qquad (2)$$

in which, assuming assigned the fuzzy sets A : $I_m \rightarrow [0,1]$ and B : $I_n \rightarrow [0,1]$ and the number g $\in$ [0,1], we obviously consider R as unknown and we have put, for brevity, A(i) = A_i for any i $\in$ I_m and B(j) = B_j for any j $\in$ I_n. Then we recall the following well-known result (e.g., cfr. [4]):

Theorem 1

If the Equation (2), in the unknown R, has solutions, the fuzzy matrix =Ř : I_m x I_n $\rightarrow [0,1]$ defined pointwise as Ř$_{ij}$ = (A_i t B_j) t g for any i $\in$ I_m and j $\in$ I_n, is the smallest solution of the Equation (2), i.e. Ř $\leq$ R for any solution R.

Proof. If R is a solution of the Equation (2), we have g $\leq$ ((A_i t B_j) $\rightarrow_t R_{ij}$) for all i $\in$ I_m and j $\in$ I_n and using properties (α) and ($\alpha\alpha\alpha$) of Section 1, we deduce :

Ř$_{ij}$ = [(A_i t B_j) t g] $\leq$ (A_i t B_j) t [(A_i t B_j) $\rightarrow_t R_{ij}$] = min { (A_i t B_j), R_{ij}} $\leq R_{ij}$,

i.e. Ř $\leq$ R for any solution R of the Equation (2). So we must only to prove that Ř is a solution of the Equation (2). Indeed, we have by using properties ($\alpha\alpha$) and ($\alpha\alpha\alpha$) of Section 1:

$$g = \bigcap_{j=1}^{n}\bigcap_{i=1}^{m}[(A_i tB_j) \rightarrow_t R_{ij}] \geq \bigcap_{j=1}^{n}\bigcap_{i=1}^{m}[(A_i tB_j) \rightarrow_t ((A_i t B_j) t g)] \geq \bigcap_{j=1}^{n}\bigcap_{i=1}^{m}g = g$$

and then:

$$g = \bigcap_{j=1}^{n}\bigcap_{i=1}^{m}[(A_i tB_j) \rightarrow_t \check{R}_{ij}]$$

i.e. Ř is a solution of the Equation (2). ∎

Returning back to the Equation (1), we can consider it as a system of fuzzy equations in the unknown R and any single equation has the smallest solution $\check{R}_{pq}$, as proved in Theorem 1, whose membership function is $\check{R}_{pqij}=[(A_{pi}tB_{qj})tg_{pq}]$, where $\check{R}_{pqij} = \check{R}_{pq}(i,j)$ for all $p \in I_k, q \in I_h$, $i \in I_m$ and $j \in I_n$. However the following result holds (e.g.,cfr. [4]):

Theorem 2

Let $A_1,...,A_k : I_m \to [0,1]$ and $B_1,...,B_h : I_n \to [0,1]$ be given fuzzy sets together a fuzzy relation G: $I_h \times I_k \to [0,1]$. If the system (1) has solutions R, then the fuzzy relation $\check{R} : I_m \times I_n \to [0,1]$ defined pointwise by

$$\check{R}_{ij} = \bigcup_{p=1}^{k}\bigcup_{q=1}^{h}\check{R}_{pqij}$$

is the smallest solution of the system (1), i.e. $\check{R} \leq R$ for any solution R.

Proof. Since $\check{R}_{pq} \leq R$ for any $p \in I_k$ and $q \in I_h$, we have $\check{R}_{pq} \leq \check{R} \leq R$. Then, using property $(\alpha\alpha\alpha)$ of Section 1, we deduce for any $p \in I_k$ and $q \in I_h$:

$$g_{pq} = \bigcap_{j=1}^{n}\bigcap_{i=1}^{m}[(A_{pi}tB_{qj}) \to_t \check{R}_{pqij}] \leq \bigcap_{j=1}^{n}\bigcap_{i=1}^{m}[(A_{pi}tB_{qj}) \to_t \check{R}_{ij}]$$

$$\leq \bigcap_{j=1}^{n}\bigcap_{i=1}^{m}[(A_{pi}tB_{qj}) \to_t R_{ij}] = g_{pq}$$

i.e. $\check{R}$ is a solution of the system (1).   ∎

Note that $\check{R} \leq R$, i.e. the membership values of $\check{R}$ are not greater of the respective values of R. The fuzzy relation $\check{R}$ is called *"decompression of R in terms of the codebooks { $A_1,...,A_k$} and {$B_1,...,B_h$}"*

3 Experimentations

It is quite natural to interpret a gray-scale image of size m x n pixels to be a fuzzy relation R : $I_m \times I_n \to [0,1]$ by normalizing the intensity range m_{ij} of each pixel into [0,1], that is $m_{ij} = R_{ij}/L$ if L is the length of the gray-scale.

Our approach, different from that one used in [6], follows the same ideas of [5]. We use a gray-scale with L = 256 and divide the original image in (possibly square) submatrices of sizes m x n, called blocks on which to apply the compression via a family of h x k codebooks. The membership functions of the codebooks are arbitrary but they may be chosen in suitable way after some simulations. To reduce the computational time we use the same codebooks in each block.

The t-norms used are L and P . Of course, the compression rate is hk/mn but it is may be diminished by reducing the values of h and k.

We have tested our algorithm of compression/decompression on two images. We firstly have considered blocks 3 x 3 over the original images "**PARROT**" (Figure 1a), and "**CAMERA**" (Figure 1b), each compressed to a block 2 x 2 by using the above t-norms (Figures 2,3,6,7), so having a compression rate of 4/9. Afterwards we have divided the original images in blocks 4 x 4, compressed to blocks 2 x 2 (in this case the compression rate is 4/16) using the same t-norms (Figures 4,5,8,9). We have also calculated the PSNR (Peak Signal to Noise Ratio) for each reconstructed image and we observe that it achieves practically the same value for both above t-norms under the same codebooks and the same compression rate. Indeed, concerning the image "**PARROT**", we have 31.37100 (resp. 31.54880) for the image of Fig. 2 (resp. Fig. 3) and 29.26259 (resp. 29.42459) for the image of Fig. 4 (resp. Fig. 5). Concerning the image "**CAMERA**", we have 23.04455 (resp. 22.98141) for the image of Fig. 6 (resp. Fig. 7) and 21.55562 (resp. 21.44756) for the image of Fig. 8 (resp. Fig.9). The whole algorithm has been implemented in JAVA.

Fig. 1a. PARROT

Fig. 1b. CAMERA

Fig. 2. Decompressed image from blocks 3x3 using L (PSNR=31.37100)

Fig. 3. Decompressed image from blocks 3x3 using P (PSNR=31.54880)

440　　Ferdinando Di Martino, Vincenzo Loia, and Salvatore Sessa

Fig. 4.　Decompressed image from
blocks 4x4 using L (PSNR=29.26259)
(PSNR=29.42549)

Fig. 5. Decompressed image from
blocks 4x4 using P

Fig. 6.　Decompressed image from
blocks 3x3 using L (PSNR=23.04455)
(PSNR=22.98141)

Fig. 7. Decompressed image from
blocks 3x3 using P

Fig. 8.　Decompressed image from
blocks 4x4 using L (PSNR=21.55562)
(PSNR=21.44756)

Fig. 9. Decompressed image from
blocks 4x4 using P

4 Conclusion

We have here proposed an algorithm of coding/decoding images based on fuzzy relation equations with continuous triangular norms, continuing our study already begun in [5]. The fuzzy relation equations considered in this paper are, in a certain sense, inverse to those ones considered in [5] and in [6]. We have made other tests, here not inserted for brevity. However we have noted in all our experiments that there is not essential difference by using the Lukasiewicz t-norm L or the product t-norm P in the algorithm, being practically identical the PSNR in the reconstructed images either using L or using P. Of course the PSNR, as proved in the above tests, is more high in the processes with high compression rate, however we like to underline that the quality of the reconstructed images are strongly dependent from the personal perception and this makes very difficult any possible comparison with other existing methods (e.g., cfr. [6]).

References

1. Klement, E.P., Mesiar, R., Pap, E.: Triangular Norms, Kluwer Academic Publishers, Dordrecht (2000).
2. Hajek, P.: Metamathematics of Fuzzy Logic, Kluwer Academic Publishers, Dordrecht (1998).
3. Di Nola, A., Sessa, S., Pedrycz, W., Sanchez, E.: Fuzzy Relation Equations and their Applications to Knowledge Engineering, Kluwer Academic Publishers, Dordrecht (1989).
4. Di Nola, A., Pedrycz, W., Sessa, S.: Fuzzy Relation Equations and Algorithms of Inference Mechanisms in Expert Systems. In: Gupta, M.M.,Kandel, A.., Bandler,W., Kiszka, J.B. (eds.), North Holland, Amsterdam (1985) 355-368.
5. Loia, V., Pedrycz, W., Sessa, S.: Fuzzy Relation Calculus in the Compression and Decompression of Fuzzy Relations. Internat. J. of Image and Graphics, Vol. 2, n. 4, (2002) 617 - 631.
6. Nobuhara, H., Pedrycz, W., Hirota, K.: Fast Solving Method of Fuzzy Relational Equations and its Application to Lossy Image Compression/Decompression, IEEE Transactions on Fuzzy Systems, Vol. 8, No. 3, (2000) 325-334.

Embedded Fuzzy Control System in an Open Computerized Numerical Control: A Technology Transfer Case-Study

Rodolfo E. Haber[1,2]*, José R. Alique[1], Angel Alique[1], Ramón Uribe-Etxebarria[3], Javier Hernández[3]

[1]Instituto de Automática Industrial (CSIC).
km. 22,800 N-III, La Poveda. 28500 Madrid.
SPAIN.
rhaber@iai.csic.es
[2] Escuela Técnica Superior
Universidad Autónoma de Madrid.
Ciudad Universitaria de Cantoblanco.
Ctra. de Colmenar Viejo, km 15. 28049 Madrid
SPAIN.
[3]Centro Tecnológico IDEKO. Arriaga kalea, 2. P.O. Box: 80
E-20870 Elgoibar (Gipuzkoa). SPAIN

Abstract. In order to improve machining efficiency, the current study is focused on the design and implementation of an intelligent controller in an open computerized numerical control (CNC) system. The main advantages of the present approach include an embedded fuzzy controller in an open CNC to deal with a real life industrial process, a simple computational procedure to fulfill the time requirements and an application without restrictions concerning sensor cost (sensorless application), wiring and synchronization with CNC. The integration process, design steps and results of applying an embedded fuzzy control system are shown through the example of real machining operations.

1 Introduction

Nowadays, manufacturing has more stringent productivity and profitability requirements that can be satisfied only if production systems are highly automated and extremely flexible. One of the main activities the manufacturing industry has to deal with is machining, a process that includes operations that range from rough milling to finishing. There is a number of angles from which to view the optimization of the machining process, angles where minimum production cost, maximum productivity and maximum profit are significant factors [1].

There are also various different ways of implementing machining process optimization. The implementation on which we will focus here attains optimal goals

---

* Corresponding author

T. Bilgiç et al. (Eds.): IFSA 2003, LNAI 2715, pp. 442–449, 2003.

via automatic control of the machining process. In the incessant pursuit of better performance, newer approaches have been tested [2,3]. The results of these tests, however, have not lived up to expectations, because all these approaches have the indispensable design requisite of an accurate (traditional) process model.

The results reported in this paper gives a picture of the present status of a rapidly emerging area that has been under detailed investigation at the Institute of Industrial Automation (IAI). The achievements revealed that improved performance and process optimization could be attained using knowledge-based systems [4]. In fact such achievements encouraged further work in this field, leading in later years to outstanding results in the design and implementation of real-time fuzzy control systems with different self-tuning strategies, as well as the stability analysis of those control systems [5].

The goal of this paper is to show the results of an example of the technology transfer (TT) of a fuzzy controller into the field of machine-tool automation. IAI's technological albeit still theoretical work in this field was transferred to a machine-tool manufacturing company, which planned to include the new control functions as optional equipment in future catalogues, with the help of a CNC provider. The TT funding instrument was PETRI, a program supported by the Spanish administration which affords an opportunity to transfer technology between public institutions and industries. The prime contractor is the public institution, whose goal is to validate the industrial relevance of its scientific results. IAI was the prime contractor, and its partners were a technological center, IDEKO, and SORALUCE S.C.L., a machine-tool manufacturer.

In order to improve machining efficiency, the current study focuses on the design and implementation of an intelligent controller in an open-CNC system. Fuzzy logic (FL) is selected from all the available techniques because it has proven useful in control and industrial engineering as a highly practical optimizing tool. To the best of our knowledge, the main advantage of the present approach is that it includes: (i) an embedded fuzzy controller in an open CNC to deal with a real-life industrial process, (ii) a simple computational procedure to fulfill the time requirements, and (iii) no restrictions in terms of sensor cost (It is a sensorless application), wiring or synchronization with the CNC. The results of the fuzzy-control strategy in actual industrial tests show higher machining efficiency.

This paper is organized as follows. In Section 2 we present a brief study of the machining process, explaining why it is considered a complex process and setting up the milling process as a case study. In Section 3 we design the fuzzy controller to optimize the milling process. Next, in Section 4 we describe how the fuzzy controller can be embedded in open CNCs, and we discuss the key design and programming stages. In Section 5 we share the experimental results. Finally we give some concluding remarks.

2 The Machining Process

The machining process, also known as the metal removal process, is widely used in manufacturing. It consists of four basic types of operations: turning, drilling, milling, and grinding, performed by different machine tools. One of the most complex of the four operations is the milling process [6].

In order to improve machining efficiency, the development of open control systems in the numerical control (NC) kernel offers more facilities for using CNC internal signals (e.g., digital drive signals) without the need to install more sensors. Indeed, cutting processes show significant effects on drive signals such as actual drive current and drive torque. Therefore, main-spindle drive current can be used to optimize cutting speed [7]. Signal behavior is complex because of specific design considerations such as star-triangle switching in drive configuration. Nevertheless, correcting the current offsets during the signal-processing stage (before entering the fuzzy algorithm) solves the problem.

The most relevant factors in terms of automation were sifted out from among the enormous quantity of variables and parameters involved in the machine tool and the machining process. After a preliminary study we selected the spatial position of the cutting tool, considering the Cartesian coordinate axes (x, y, z) [mm], spindle speed (s) [rpm], relative feed speed between tool and worktable (f, feed rate) [mm/min], cutting power invested in removing metal chips from the workpiece (P_c)[kW], current consumed in the main spindle during the removal of metal chips (I_S) [A], radial depth of cut (a, cutting depth) [mm] and cutting-tool diameter (d) [mm].

In order to evaluate system performance, we need to select certain suitable performance indices. This work dealt essentially with rough milling, so the main index was the metal-removal rate (MRR).

3 Fuzzy Approach to Controlling Cutting Force

The design of a fuzzy-logic controller (FLC) as detailed in [4] is introduced in this section. The input variables included in the error vector e are the current-consumed error (ΔI_S in ampere) and the change in current-consumed error ($\Delta^2 I_S$ in ampere). The manipulated (action) variable we selected is the feed-rate increment (Δf in percentage of the initial value programmed into the CNC), whereas the spindle speed is considered constant and preset by the operator. The control scheme is depicted in Fig. 1.

The controller output is inferred by means of the compositional rule. The Sup-Product compositional operator was selected for the compositional rule of inference. For instance, applying the T_2 norm (product) and applying S_1 s-norm (maximum) yields (1).

$$\mu(\Delta I_S, \Delta^2 I_S, \Delta f) = \overset{m \times n}{\underset{i=1}{S_1}} \left[T_2 \left[\mu_{\Delta I_{S_i}}(\Delta I_S), \mu_{\Delta^2 I_{S_i}}(\Delta^2 I_S), \mu_{\Delta f_i}(\Delta f) \right] \right] \qquad (1)$$

where T_2 represents the algebraic-product operation and S_1 represents the union operation (max), $m \times n = 9$ rules.

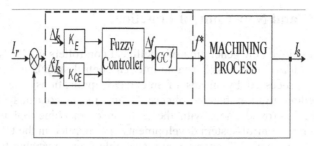

Fig. 1. Fuzzy-control scheme for the machining process.

The crisp controller output, used to change the machine-table feed rate, is obtained by defuzzification employing the centre of area (COA) method defined as

$$\Delta f = \frac{\sum_i \mu_R(\Delta fi) \cdot \Delta fi}{\sum_i \mu_R(\Delta fi)} \ . \tag{2}$$

The strategy used to compute f determines what type of fuzzy regulator is to be used. In this case it is a PD regulator:

$$f*(k) = f_0 + \Delta f*(k) . \tag{3}$$

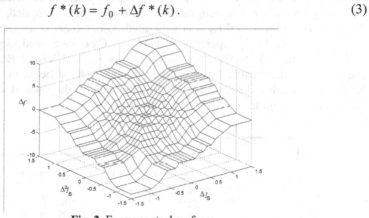

Fig. 2. Fuzzy control surface

Feedrate values (f) were generated on-line by the controller and fed in with the set point for the spindle current (I_{Sr}) and measured value (I_S) provided by the CNC from the internal spindle-drive signal, as is explained in section 4. The static input-output mapping can be represented by the nonlinear control surface shown in Fig. 2, considering that $\Delta I_S \in \left[-1.5, 1.5\right]$, $\Delta^2 I_S \in \left[-1.5, 1.5\right]$ and $\Delta f \in \left[-10, 10\right]$.

4 Open CNC and New Control Functions

The way to embed the new generation of intelligent controllers in CNCs is now ready to be used. Nowadays open CNCs enable internal control signals to be gathered and mathematically processed by means of integrated applications such as additional functions embedded right in the control core. By this means control systems' original features can be improved upon, with the addition of machine-tool manufacturers' know-how and new control-system developments incorporated in the CNC.

At present two levels of opening are available from machine-tool and CNC manufacturers: opening of Man/Machine Communication (MMC) and opening of the CNC's Numerical Control Kernel (NCK). The first category is the one responsible for interaction with the user (e.g., office applications can be developed using the DDE Dynamic Data Exchange protocol). Use of a bus (e.g., Multi-Point Interface bus) enables communication with low-level data. The more important level of opening from the control-system viewpoint, can be found in the NCK, where real-time critical tasks (e.g., axes control) are scheduled and performed.

Nowadays the integration of any control system in the open CNC is a complex task that requires the use of various software utilities, technologies and development tools. Three classical technologies can be used: a software-developing tool (C++, Visual Basic), an open- and real-time CNC and communications technology.

The general outline of the embedded control system is depicted in Figure 3. First the fuzzy controller was programmed in C/C++ and compiled, and as a result a dynamic link library (DLL) was generated. Other tools were used as well. A Sun Workstation, the UNIX operating system and C++ were used to program the NCK [8]. A PC, the WINDOWS 9X operating system and Visual Basic were used for programming the MMC. Inter-module communications between the MMC and the NCK was established through DDE. Finally, and for the sake of simplicity, the user interface was programmed in Visual Basic.

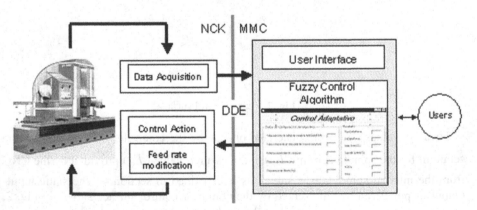

Fig. 3. Diagram of the control system.

The application was developed on the basis of a Sinumerik 840D CNC. The process of integrating the software application into the NCK involved a series of steps. The development was run at a workstation, including edition, compiling and code linking. After that, the file was transferred to the PC, where OEM software ran debugging routines. Finally, the code was copied on a PCMCIA card and inserted into the CNC [13].

An experimental control internal data-acquisition system was developed and used to record the control internal spindle-drive signal. The system enables a selected drive signal to be recorded and provides stored data on the hard drive of the SINUMERIK 840D user interface PC (MMC). Internal data-acquisition software was developed to obtain control information. The maximum sampling frequency could not be any longer than 500 Hz, defined by the servosystems' control cycle. Therefore the acquisition-signal software works at the same time for which servosystems are configured in the CNC (i.e., 2 ms.). The software consists of a data-acquisition module in the NCK that records the selected data into an internal buffer and an MMC background task that receives the completed measurement and stores it in the hard drive of the user-interface PC. Data transfer is performed by splitting the recorded data into a number of fragments, due to limitations in the Sinumerik 840D's file systems. The signals to be recorded (in our case the current consumed at the spindle) are configured using specially added machine data. Recording can be started and stopped either manually through an MMC application or under the NC program's control.

5 Industrial Test. Evaluation

Tests were carried out in the SL400 SORALUCE machining center, which was equipped with a SIEMENS open-CNC Sinumerik 840D. The cutting tool was a Sandvik D38 R218.19 - 2538.35 - 37.070HA (020/943) face mill 25 mm in diameter with four inserts of the SEKN 12 04 AZ (SEK 43A, GC-A p25) type (see Fig. 4b). The workpiece material used for testing was F114 quality steel. The maximum depth of cut was 20 mm, the nominal spindle speed was s_0 =850 rpm, and the nominal feedrate, f_0 =300mm/min. The actual dimensions of the profile were 334x486 (mm). The profile is depicted in Fig. 4a.

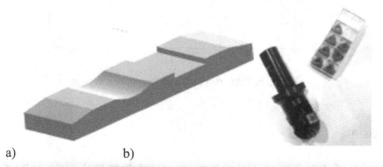

a)   b)

Fig. 4. a) Workpiece for industrial tests b) Tool used for industrial tests

We used control internal spindle-drive signal. The set point was estimated according to constraint given by the power available at the spindle motor, material and the tool characteristics. The results of applying two tests with different set points to mechanize an irregular profile (see Fig. 4b) are depicted in Fig. 5. The improvement in the efficiency is achieved through the reduction in machining time (10%), yet the overshoot was relatively bigger (23.7%). From the technological viewpoint, this overshoot is allowable for rough milling operations. It is important to note that in some approaches the goal of the controller design is to limit the percent overshoot to 20% [3].

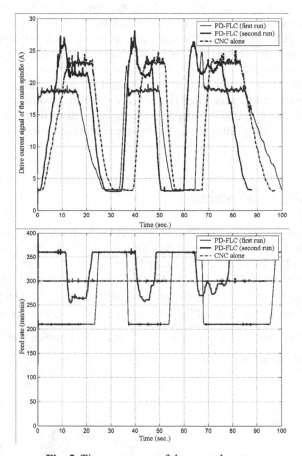

Fig. 5. Time responses of the control system

Conclusion

The results of transferring technology to a machine-tool manufacturer through cooperation with a technological center show the effectiveness of a fuzzy control system for dealing with the nonlinear behavior of the machining process. The embedded fuzzy controller is able to work using only internal CNC signals. Moreover, it can run in parallel with other applications without any synchronization problems. However, future work is necessary to refine the fuzzy controller's performance and so to improve the transient response.

All this work, from hypothesis to implementation and experimentation, including design and analysis, is done following the classical patterns. Our focus at all times lay on the practical implementation, so that the results we have presented in this work are for a real life industrial plant. The results show that internal CNC signals can double as an intelligent, sensorless control system. Actual industrial tests show a higher machining efficiency: in-process time is reduced by 10% and total estimated savings provided by installing the system are about 78%.

References

1. Koren Y.: Control of machine tools. Journal of Manufacturing Science and Engineering 119 (1997) 749-755

2. Lauderbaugh L.K., Ulsoy A. G.: Model reference adaptive force control in milling. ASME Journal Engineering of Industry 111 (1989) 13-21

3. Rober S.J., Shin Y.C., Nwokah O.D.I.: A digital robust controller for cutting force control in the end milling process. Journal of Dynamic Systems, Measurement and Control 119 (1997) 146-152

4. Haber R. E., Haber R.H., Alique A., Ros S.: Application of knowledge based systems for supervision and control of machining processes, in: S. K. Chang, ed., Handbook of Software Engineering and Knowledge Engineering, Vol. 2, (World Scientific Publishing, Singapore) pp. 673-710 (2002)

5. Schmitt-Braess G., Haber R.E., Haber R.H., Alique A.: Multivariable circle criterion: Stable fuzzy control of a milling process, in: Proceedings of the IEEE International Conference on Control Applications, Glasgow, UK, (2002) 385-390

6. Liu Y., Zhuo L., Wang C.: Intelligent adaptive control in milling process. International Journal of Computer Integrated Manufacturing 12(5) (1999) 453-460

7. Kim T,, Kim P.: Adaptive cutting force control for a machining centre by using indirect cutting force measurements. International Journal of Machine Tool Manufacturing 36(8) (1996) 925-937

8. Siemens AG: Sinumerik 840D, OEM-package NCK, software release 4, user's manual, Erlangen (1999)

Activation of Trapezoidal Fuzzy Subsets with Different Inference Methods

Anis Sakly and Mohamed Benrejeb

LA.R.A., Ecole Nationale d'Ingénieurs de Tunis
BP 37, Belvédère 1002 Tunis, TUNISIA
anis.sakly@isetso.rnu.tn, mohamed.benrejeb@enit.rnu.tn

Abstract. Center of gravity defuzzification method requires large amount of computation and memory space caused by fuzzy set discretization. All the same for mean of maxima one. In practice, it is proposed a method simplifying the problem of integration without loss of information by defining the practical methods. In this paper, we propose an approach permitting the pre-calculus of the center of gravity and the mean of maxima of a trapezoidal fuzzy subset when individually inferred.

1 Introduction

One important task to be performed in the design of the inference system of a fuzzy logic controller is to select the implication operator used for this sake. Many authors have presented and analyzed several implication operators in the specialized literature such as: implication functions [10], t-norms [7], [9], Force-implications [6] and a wide range of other kinds of implication [8].

In [6], the fuzzy implication operators were classified into two different families according to the extension of the Boolean Logic they perform.

A lot of works have been developed to select the adequate implication operator based on various criteria. In [4] are presented basic properties for robust implication operators in fuzzy control. The word "robust" is used in the sense of good average behavior in different applications and in combination with different defuzzification methods. In another study [3], it is said that S-implications, class of implications functions, used with the practical form of the center of gravity defuzzification method can produce a non-zero static error in steady state.

This paper is set up as follows: the next section introduces the different implication operators selected in this study; section 3 introduces the principle of practical defuzzification methods, in section 4 we present some propositions permitting to pre-calculate the center of gravity and the mean of maxima of a trapezoidal subset when individually inferred with different implication classes; and finally a conclusion is pointed out in section 5.

T. Bilgiç et al. (Eds.): IFSA 2003, LNAI 2715, pp. 450–457, 2003.
© Springer-Verlag Berlin Heidelberg 2003

2 Fuzzy Implication Operators Selected

In this section we present the fuzzy operators selected to work as implication operators in fuzzy control to carry out our study.
Before doing so, we recall the structure of the different classes of implication operators.

2.1 Implication Functions

The implication functions are the most well known ones that extend the boolean implications. They are classified into tow families [10], [11].

Strong implications (S-implications).
Corresponding to the definition of implication in the classical boolean logic: $A \rightarrow B \equiv N(A) \cup B$. They present the form: $I(x, y) = S(N(x), y)$, being S a t-conorm and N a negation function.

Residual implications (R-implications).
Obtained by residuation of a t-norm T as follows:
$I(x, y) = Sup \{c \in [0,1] / T(x, c) \leq y\}$.

2.2 T-implications

These implications are the t-norms, such as:
$I(x, y) = T(x, y)$, where T is a t-norm.

2.3 Force-Implications

There are two different groups of Force-implications depending on the way in which they are built.

Force-implications based on indistinguishability operators. $I(x, y) = T(x, E(x, y))$, where T is a t-norm and E is an indistinguishability operator, $E(x,y)=T'(I'(x,y),I'(y,x))$ with T' being a t-norm and I' is an implication function.

Force-implications based on distances. $I(x, y) = T(x, 1-d(x, y))$, where T is a t-norm and d is a distance.

2.4   Other Implication Operators

There are many implication operators that do not belong to any of these mentioned above. Although they are not fuzzy implication functions, we will add to our study other operators that generalize the boolean implication.

In Table 1 are summarized several fuzzy implication operators.

3   Practical Defuzzification Methods

The defuzzification interface is the component of the FLC that combines the fuzzy information contained in the individual fuzzy sets inferred and translate it to a crisp control action. It may operate into two ways [4]; mode A and mode B. In mode B; Defuzzification first and Aggregation after; we avoid the computation of the final fuzzy set by considering the contribution of each rule individually, obtaining the final control action by taking a calculation of concrete crisp characteristic value associated to each of them via a weighted sum. That's why this second mode is said practical method.

There is a large group of defuzzification methods in mode B in the specific literature. In this work, we have selected the two generally used, the practical center of gravity and the practical mean of maxima methods given by:

$$\text{OUTPUT} = \frac{\sum_{i=1}^{N_B} \alpha_i \cdot c'_i}{\sum_{i=1}^{N_B} \alpha_i} \tag{1}$$

where N_B is the number of output fuzzy subsets and c'_i is a crisp value corresponding either to the center of gravity g'_i or to the mean of maxima m'_i of the output fuzzy subset i individually inferred with a firing strength α_i.

4   Activation of Trapezoidal Fuzzy Subsets

Let us consider a rule "if x is **A** then y is **B**" where **B** is a trapezoidal fuzzy subset. We will calculate the abscissas of the center of gravity and the mean of maxima of the inference resultant subset **B'**, in function of the firing strength $\alpha = A(x_0)$, for every implication studied.

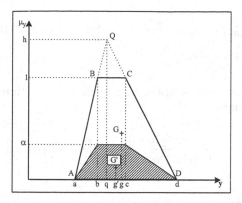

Fig. 1. Activation of a trapezoidal fuzzy subset

4.1 Center of Gravity

According to Fig. 1, let ABCD the trapezoidal output subset **B** where a, b, c and d are respectively the abscissas of its summits A, B, C and D. Let also AQD a triangle where Q is the intersection of the straight lines (AB) and (CD). We note q and h respectively the abscissa and the ordinate of the summit Q. So, we have:

$$q = \frac{a.c - b.d}{(a + c) - (b + d)} \quad \text{and} \quad h = \frac{q - a}{b - a} = \frac{d - q}{d - c}$$

We will express the abscissa **g'** of the center of gravity G' of **B'** in function of the abscissa **g** of the center of gravity G of the triangle AQD.

Remark 1. Excepted of Drastic t-norm, all T-implications and R-implications conserve the support S of the subset **B**. S-implications lead to a widening of the support of **B** over the whole of universe of discourse.

Proposition 1. For T-implications and R-implications, we demonstrate that:

$$g' = g + \delta_0.f(\alpha, h). \tag{2}$$

Where $\delta_0 = g - q$ and $f(\alpha, h)$ is a function depending on α and h.

Proposition 2. For S-implications, we demonstrate that:

$$g' = \frac{(1-\alpha).U_d.\delta_1 + \alpha.S.f_2(\alpha, h).[g + \delta_0.f_1(\alpha, h)]}{(1-\alpha).U_d + \alpha.S.f_2(\alpha, h)}. \tag{3}$$

Where $U_d = L_{sup} - L_{inf}$, $\quad \delta_1 = \dfrac{L_{sup} - L_{inf}}{2}$, $S = d - a$, $f_1(\alpha, h)$ and $\quad f_2(\alpha, h)$ are functions depending on α and h.

Corollary 1. If the universe of discourse is symmetrical with respect to zero, then we have for S-implications:

$$g' = \frac{\alpha.S.f_2(\alpha, h)}{2(1-\alpha).L + \alpha.S.f_2(\alpha, h)} \cdot [g + \delta_0.f_1(\alpha, h)]. \tag{4}$$

With $L = L_{sup} = -L_{inf}$.

Table 2 summarizes the expressions of $f(\alpha, h)$ for some T-implications, R-implications and some other same-behavior implications.

Corollary 2. In particular, if the subset B is triangular in shape then B=C=Q and we have h=1. Therefore, in relations (2), (3) and (4) the functions $f(\alpha, h)$, $f_1(\alpha, h)$ and $f_2(\alpha, h)$ become respectively $f(\alpha)$, $f_1(\alpha)$ and $f_2(\alpha)$ which are obtained by replacing h by 1. Thus, we obtain the results presented in [3].

Remark 2. For Force-implications based on indistinguishability operators whose implication functions are R-implications selected in this study, the expressions of **g'** are identical to form (2). All the same for Gaine and Yager implications.

Remark 3. For Force-implications based on distances or based on indistinguishability operators whose implication functions are S-implications selected in this study, the expressions of **g'** are similar to form (3) and (4) when $\alpha > \dfrac{1}{2}$ and g'=0 when $\alpha \le \dfrac{1}{2}$. All the same for Zadeh implication. In particular, for the implication I_{21}, g'=0 for $\alpha \le \dfrac{\sqrt{5}-1}{2}$.

4.2 Mean of Maxima

There are two different classes of implications for this characteristic value:
-   those conserving the core K of the subset **B** ($K = c - b$),
-   those yielding to a widening of the core K of the subset **B**.

Among the first class, we have Kleene, Reichenbach and Larsen implications. And among the second class, we have all R-implications and Mamdani implication.

Proposition 3. For the first class, the abscissa m' of the mean of maxima M' of the subset B' is:

$$m' = \frac{b+c}{2} \, . \tag{5}$$

Proposition 4. For the second class, we demonstrate that the abscissa m' of the mean of maxima is a convex combination by α of the abscissa of the mean of maxima of the core K and the abscissa of the mean of maxima of the support S, and we have:

$$m' = \alpha . \frac{b+c}{2} + (1-\alpha) . \frac{a+d}{2} \, . \tag{6}$$

Corollary 3. In particular, if the subset B is triangular in shape then B=C. Therefore, in relations (5) and (6) we replace c by b and we obtain the results presented in [2].

Remark 4. Particularly, for Dubois implication, we have:

$$m' = \frac{a+d}{2} \, . \tag{7}$$

Remark 5. For Force-implications based on distances or based on indistinguishability operators whose implication functions are S-implications selected in this study, the expressions of **m'** are approximately similar to form (6) when $\alpha > \frac{1}{2}$.

5 Conclusion

In this work, we proposed expressions permitting to pre-calculate the center of gravity and the mean of maxima of a trapezoidal subset when individually inferred with different implication classes which are used in practical defuzzification methods.

This mathematical approach permits the advance of the comparative study between different classes of fuzzy implications in term of static and dynamic behaviours of the fuzzy controller based on.

References

1. Bouchon-Meunier B.: La logique floue et ses applications. Addison-Wesley, France, S.A. (1995)
2. Castro J.L., Castro J.J. and Zurita J.M.: The influence of Implication functions in Fuzzy Logic with Defuzzification First and Aggregation After. 7th International Fuzzy Systems Association World Congress (IFSA) , Vol. III. Praga (Czech) (1997) 237-242
3. Chenaina T. and Jilani J.: An Analytical Approach for Choosing the Right Fuzzy Implication Based on Performance Criteria for the Fuzzy Control. The 9th IEEE International Conference on Fuzzy Systems. Santa Monica, USA (May 2000)

4. Cordon O., Herrera F. and Peregrin A.: Searching for Basic Properties Obtaining Robust Implication Operators in Fuzzy Control. Fuzzy Sets and Systems, 111. (2000) 237-251
5. Dubois D. and Prade H.: Logique floue, interpolation et commande. RAIRO-APII-JESA, vol. 30. (1996) 607-644
6. Dujet Ch. and Vincent N.: Force Implication: A new approach to human reasoning. Fuzzy sets and Systems, 69. (1995) 53-63
7. Gupta M.M. and Qi J.: Theory of T-norms and Fuzzy Inference Methods. Fuzzy Sets and Systems, 40. (1991) 431-450
8. Kiszka J., Kochanska M. and Sliwinska D.: The Influence of Some Fuzzy Implication Operators on the Accuracy of a Fuzzy Model-Parts I and II. Fuzzy Sets and Systems, 15. (1985) 111-128, 223-240
9. Mizumoto M.: Pictorial Representations of Fuzzy Connectives, Part I: cases of t-norms, t-conorms and averaging operators. Fuzzy Sets and Systems. (1989) 217-242
10. Trillas E. and Valverde L.: On Implication an Indistinguishability in the Setting of Fuzzy Logic: J. Kacpryzk, R.R. Yager, Eds., Management Decision Support Systems Using Fuzzy Sets and Possibility Theory (Verlag TUV Rheinland, Koln, 1985) 198-212
11. Trillas E.: On a Mathematical Model for Indicative Conditionals. Proc. 6th IEEE International Conf. On Fuzzy Systems. (1997) 3-10

Table 1. Implication operators selected

T-norms	S-implications	R-implications
$I_1(x,y) = Min(x,y)$ Mamdani	$I_7(x,y) = Max(1-x,y)$ Kleene	$I_{13}(x,y) = \begin{cases} 1 & \text{if } x \le y \\ y & \text{otherwise} \end{cases}$ Gödel
$I_2(x,y) = x \cdot y$ Larsen	$I_8(x,y) = 1 - x + x \cdot y$ Reichenbach	$I_{14}(x,y) = \begin{cases} Min(1, y/x) & \text{if } x \ne 0 \\ 1 & \text{otherwise} \end{cases}$ Goguen
$I_3(x,y) = Max(0, x+y-1)$ T-Lukasiewicz (bounded norm)	$I_9(x,y) = Min(1, 1-x+y)$ Lukasiewicz	I_9
$I_4(x,y) = \dfrac{x \cdot y}{x + y - x \cdot y}$ T-Hamacher	$I_{10}(x,y) = \dfrac{(1-x)+y-2 \cdot (1-x) \cdot y}{1-(1-x) \cdot y}$	$I_{15}(x,y) = \begin{cases} 1 & \text{if } x \le y \\ \dfrac{x \cdot y}{x - y + x \cdot y} & \text{otherwise} \end{cases}$
$I_5(x,y) = \dfrac{x \cdot y}{1+(1-x) \cdot (1-y)}$ T-Einstein	$I_{11}(x,y) = \dfrac{(1-x)+y}{1+(1-x) \cdot y}$	$I_{16}(x,y) = \begin{cases} 1 & \text{if } x \le y \\ \dfrac{y \cdot (2-x)}{x + y - x \cdot y} & \text{otherwise} \end{cases}$
$I_6(x,y) = \begin{cases} x & \text{if } y = 1 \\ y & \text{if } x = 1 \\ 0 & \text{otherwise} \end{cases}$ T-Weber (drastic t-norm)	$I_{12}(x,y) = \begin{cases} 1-x & \text{if } y = 0 \\ y & \text{if } x = 1 \\ 1 & \text{otherwise} \end{cases}$ Dubois	$I(x,y) = \begin{cases} y & \text{if } x = 1 \\ 1 & \text{otherwise} \end{cases}$ not exploitable

Force-implications	Other implications
$I_{17}(x, y)=\text{Min }(x, E_{\text{Gödel}}(x, y))$ $E_{\text{Gödel}}(x, y) = \begin{cases} 1 & \text{if } x = y \\ \text{Min}(x, y) & \text{otherwise} \end{cases}$	$I_{22}(x,y)=\text{Max}(1-x, \text{Min}(x,y))$ Zadeh
$I_{18}(x, y)=\text{Min }(x, E_{\text{Goguen}}(x, y))$ $E_{\text{Goguen}}(x, y) = \text{Min }(1, \dfrac{\text{Min }(x, y)}{\text{Max }(x, y)})$	$I_{23}(x, y) = \begin{cases} 1 & \text{if } x \le y \\ 0 & \text{otherwise} \end{cases}$ Gaines
$I_{19}(x, y)=\text{Min }(x, E_{\text{Kleene}}(x, y))$ $E_{\text{Kleene}}(x, y) = \text{Min }(\text{Max }(1-x, y),$ $\text{Max }(1-y, x))$	$I_{24}(x, y) = y^{x}$ Yager
$I_{20}(x, y) = \text{Min}(x, 1-\lvert x - y\rvert)$	
$I_{21}(x, y) = \text{Min}(x, 1-\lvert x - y\rvert^{2})$	

Table 2. Expressions of f (α, h) for some T-implications, R-implications and some other same-behavior implications

Implication	$f(\alpha, h)$
I_1 (Mamdani)	$\dfrac{(\alpha - h)^{2}}{h.(2.h - \alpha)}$
I_2 (Larsen)	$\dfrac{(1 - h)^{2}}{h.(2.h - 1)}$
I_3 (T-Lukasiewicz)	$\dfrac{\alpha^{2} + (2.h - 3).\alpha + h^{2} - 4.h + 3}{h.(\alpha + 2.h - 2)}$
I_{13} (Gödel)	$\dfrac{-2.\alpha^{3} + (2.h + 3).\alpha^{2} - 4.h.\alpha + h^{2}}{h.(\alpha^{2} - 2.\alpha + 2.h)}$
I_{14} (Goguen)	$\dfrac{(\alpha - h)^{2}}{h.(2.h - \alpha)}$
I_{18}	$\dfrac{(\alpha^{2} - h)^{2}}{h.(2.h - \alpha^{2})}$
I_{23} (Gaines)	$\dfrac{(h - 3\alpha)}{2.h}$

Design and Simulation of a Fuzzy Substrate Feeding Controller for an Industrial Scale Fed-Batch Baker Yeast Fermentor

Cihan Karakuzu[1], Sıtkı Öztürk[1], Mustafa Türker[2]

[1]Department of Electronics & Telecommunications Engineering, Faculty of Engineering,
University of Kocaeli, Kocaeli, İzmit, Turkey
{cihankk, sozturk}@kou.edu.tr
[2]Pakmaya, PO. Box 149, 41001, Kocaeli, İzmit, Turkey
mustafat@pakmaya.com.tr

Abstract. Conventional control systems can not give satisfactory results in fermentation systems due to process non-linearity and long delay time,. This paper presents design and simulation a fuzzy controller for industrial fed-batch baker's yeast fermentation system in order to maximize the cell-mass production and to minimize ethanol formation. Designed fuzzy controller determines an optimal substrate feeding strategy for an industrial scale fed-batch fermentor relating to status of estimated specific growth rate, elapsed time and ethanol concentration. The proposed controller uses an error in specific growth rate (e), fermentation time (t) and concentration of ethanol (Ce) as controller inputs and produces molasses feeding rate (F) as control output. The controller has been tested on a simulated fed-batch industrial scaled fermenter and resulted in higher productivity than the conventional controller.

1   Introduction

Many different products have been produced by culturing yeast. Baker yeast is the one of these products. *Saccharomyses cerevisia* known as baker yeast is produced using molasses as substrate by means of growing up microorganism. In controlled environment, the main energy source for growing the culture is sucrose in molasses. Energy production, maintenance and growth reaction of the yeast cells can be expressed by following macroscopic chemical reaction.

$$CH_2O + Y_{n/s} NH_3 + Y_{o/s} O_2 \rightarrow Y_{x/s} CH_{1.83}O_{0.56}N_{0.17} + \tag{1}$$

$$Y_{e/s} CH_3O_{0.5} + Y_{c/s} CO_2 + Y_{w/s} H_2O$$

CH_2O, $CH_{1.83}O_{0.56}N_{0.17}$ and $CH_3O_{0.5}$ are respectively sucrose, biomass and ethanol in C-mol unit in Eq. (1). Yi/j s are stoichiometric yield coefficients. Ethanol, which is not desired, is a by product in the process. Ethanol formation takes places either by overfeeding sucrose or oxygen limitation in process. Total receiving sucrose into the fermentor should be oxidised completely by yeast cells for maximum yield productivity during fermentation. This is possible by supplying reasonable molasses feed rate to the system. The main objective is maximum biomass production without ethanol during fermentation. Hence, amount of ethanol in the fermentor should be

T. Bilgiç et al. (Eds.): IFSA 2003, LNAI 2715, pp. 458–465, 2003.
© Springer-Verlag Berlin Heidelberg 2003

under control by controlling feeding rate. Fed-batch process is generally employed by the baker's yeast industry. Fed-batch baker yeast production process is shown in Fig. 1. Essential nutrients and molasses are fed incremental to the fermentor at an predetermined rate during the growth period [1]. Because a predetermined and fixed feeding profile has been used in industry, manual manipulations are needed in some unexpected status. It is therefore desirable, to incorporate a intelligent controller into the feeding systems if possible. In this paper, a fuzzy controller is proposed to determine molasses feeding rate for keeping desired set point of specific growth rate of yeast cells. Designed fuzzy controller is tested on simulation model of the process.

Fig. 1. Block structure of baker yeast production process

2   Process Model

Process modelling for control is one of essential areas of research and industrial applications in biotechnology. Processes based on microorganisms are the most complex in all field of process engineering, and their modelling is considered a difficult task [2]. As Shimizu [3] pointed out, the control system development for fermentation is not easy due to: lack of accurate models describing cell growth and product formation, the nonlinear nature of the bioprocess, the slow process response and a deficiency of reliable on-line sensors for the quantification of key state variables.

Model will be given in this section, is include cell (kinetic) and reactor (dynamic) models.

2.1 Yeast Cell (Kinetic) Model

Microorganism is not still modelled exactly because of relating to many parameters, but modelling studies have gone on. Yeast cell model, which will be given, was based on simple mechanistic model of Sonnleither and Kappeli [4]. The model equations are shown in Table 1. Eq.(2) and Eq.(3) are different from the original model $(1-e^{-t/td})$ and $K_i/(K_i+C_e)$ terms in Eq. (2) and Eq. (3) were added the original model. Pham *et al.* [5] observed that glucose uptake was under the maximum value ($q_{s,max}$) during first an hour after inoculum and shown necessity adding a delay time term to the equation. The other uptake term added to the model, $K_i/(K_i+C_e)$, supports compatibility of model data and real data measured industrial fermentor, especially point of view changing specific oxygen consumption (q_o) and ethanol concentration (C_e). we have observed that the updated model gave closer the real result than the original, in our simulation studies based on data measured from real industrial process. Especially, accuracy of the updated model simulation result have been seen on ethanol production/consumption and specific oxygen consumption kinetic variables (see Fig. 2).

Table 1. Kinetic model of baker yeast cell

$$q_s = q_{s,max} \frac{C_s}{K_s + C_s}(1-e^{-t/td}) \tag{2}$$

$$q_{o,lim} = q_{o,max} \frac{C_o}{K_o + C_o} \frac{K_i}{K_i + C_e} \tag{3}$$

$$q_{s,lim} = \frac{\mu_{cr}}{Y_{x/s}^{ox}} \tag{4}$$

$$q_{s,ox} = min\left(\begin{matrix} q_s \\ q_o / Y_{o/s} \end{matrix}\right) \tag{5}$$

$$q_{s,red} = q_s - q_{s,ox} \tag{6}$$

$$q_{e,up} = q_{e,max} \frac{C_e}{K_e + C_e} \frac{K_i}{K_i + C_s} \tag{7}$$

$$q_{e,ox} = min\left(\begin{matrix} q_{e,up} \\ (q_{o,lim} - Y_{o/s}^{ox}) Y_{o/e} \end{matrix}\right) \tag{8}$$

$$q_{e,pr} = Y_{e/s} q_{s,red} \tag{9}$$

$$\mu = Y_{x/s}^{ox} \cdot q_{s,ox} + Y_{x/s}^{red} \cdot q_{s,red} + Y_{x/e}^{ox} \cdot q_{e,ox} \tag{10}$$

$$q_c = Y_{c/s}^{ox} \cdot q_{s,ox} + Y_{c/s}^{red} \cdot q_{s,red} + Y_{c/e}^{ox} \cdot q_{e,ox} \tag{11}$$

$$q_o = Y_{o/s}^{ox} \cdot q_{s,ox} + Y_{o/e}^{ox} \cdot q_{e,ox} \tag{12}$$

$$RQ = \frac{q_c}{q_o} . \tag{13}$$

2.2 Reactor (Dynamic) Model

This model describes concentrations of the main compound which are dynamically directed by manipulation of input variables and initial conditions. For the simplicity, mixing of both of liquid and gas phases were perfect, is assumed. Hence, model includes basic differential equations. Because main objectives of this paper is reaction taking place in liquid phase, only liquid phase reactor model is given in Table 2 in Eq. 14-18.

Table 2. Reactor model of fed-batch baker yeast fermentation

$$\frac{dCx}{dt} = \mu.Cx - \frac{F}{V}Cx \tag{14}$$

$$\frac{dCs}{dt} = -\left(\frac{\mu}{Y_{x/s}^{ox}} + \frac{q_{e,pr}}{Y_{e/s}} + q_m\right)Cx + \frac{F}{V}(So - Cs) \tag{15}$$

$$\frac{dCe}{dt} = (q_{e,pr} - q_{e,ox}).Cx - \frac{F}{V}Ce \tag{16}$$

$$\frac{dCo}{dt} = -q_o Cx - \frac{F}{V}Co + k_{OL}a(Co^* - Co) \tag{17}$$

$$\frac{dV}{dt} = F \tag{18}$$

Yeast cell and reactor models were constituted by means of a Matlab 6.0 Simulink software to use in all simulation. All of model parameters are taken from [1] and [4], except Ki=2.514 gr/L, $Y_{x/s}^{ox}$ =0.585 gr/gr, td=2 h, So=325gr/L and $k_{OL}a$=500 h^{-1}. Comparison of the process model simulation results and data measured from industrial scale (100m^3) fermentor are shown in Fig. 2. Process model simulation results are satisfactory based on measures in real production media as shown in Fig. 2.

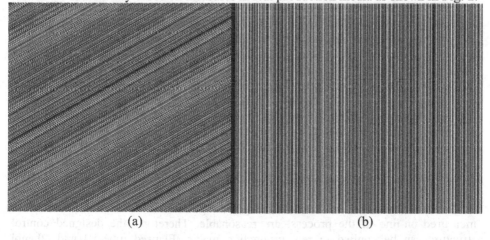

(a)   (b)

Fig. 2. (a) Comparison of data from industrial scale fermentor (*symbols*) and simulation results (*lines*) by the process model for key state variables (C_x *yeast concentration, C_e ethanol concentration, C_s glucose concentration, C_o dissolved oxygen concentration*). **(b)** Comparison of data from industrial scale fermentor simulation model verification for specific oxygen consumption rate, q_o, (o) and respiratory quotient, RQ, (□) with simulation results (dotted line q_o, line RQ).

3 Fuzzy Controller Design for an Industrial Scale Fermentor

In this paper, process control structure is proposed based on fuzzy logic. The proposed block diagram of closed loop control based on simulation model is shown as in Fig. 3. The controller determines molasses feeding rate for keeping desired set point of specific growth rate (μ_{set}) thus minimising ethanol formation to reach maximum biomass yield at the end of fermentation.

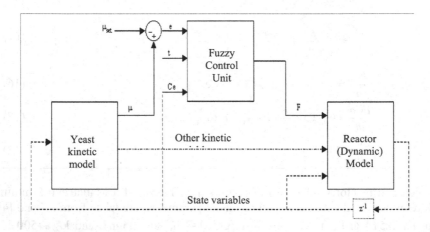

Fig. 3. Block diagram of proposed closed loop control for fed-batch baker's yeast process

The first step is to determine input/output variable in controller design. The error between specific growth rate (μ) and set point (μ_{set}) was chosen as main input variable of the fuzzy controller. The other input variables of the controller are elapsed time (t) and ethanol concentration (Ce). The output variable of the controller was chosen as molasses feed rate (F). Membership functions of the inputs and the output are shown in Fig. 4(a). All of membership functions of the inputs and the output except in last of right and left sides are respectively generalised bell and gauss functions. In the last of right sides membership functions are S-shaped curve membership functions. In the last of left sides membership functions are Z-shaped curve membership functions. Fuzzy control unit in Fig. 3 employs the Mamdani fuzzy inference system using min and max for T-norm and T-conorm operators, respectively. Centroid of area defuzzification schema was used for obtaining a crisp output. The controller rule base tables are shown in Fig. 4(b).

From the point of view control engineering, to choose variables, which are measured on-line in the process, are reasonable. Therefore, the designed control structure can be applied in real production media. Elapsed time (t) and ethanol concentration (Ce) are on-line measured in industrial production process. The other controller input variable (μ), is not on-line measured, but can be estimated from on-line measurable variables. Our studies [6,7] about μ and biomass estimation based on data given in real production media by neural network have shown that this problem could be solved. That is, on-line μ estimation is possible by using soft-sensor such as neural networks. Also, F can be measured on-line in process, hence, applying the proposed control structure in this paper to the industrial production process will be possible.

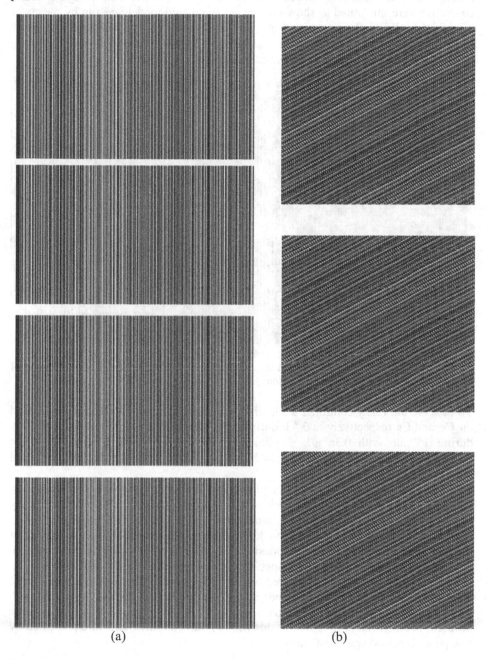

(a)   (b)

Fig. 4. (a) Membership functions of inputs and output variables **(b)** Fuzzy controller rule base table

4 Simulation Results and Conclusions

After yeast cell (kinetic) model, reactor (dynamic) model and designed fuzzy controller were integrated as shown in Fig. 3; simulation model was run by means of Matlab 6.0 Simulink software. Changes of control input determined by fuzzy controller and μ controlled variable during fermentation are given in Fig 5a. Sugar, ethanol and dissolved oxygen concentrations are shown in Fig. 5b.

Fig. 5. (a) Changing of specific growth rate, μ, and molasses feeding rate, F, **(b)** Changing of concentrations during overall fermentation

To observe the fuzzy controller's adaptive behaviour, dummy distortions were added on Ce and Cs respectively at 3rd h during 15 min with 2 g/L amplitude and at 6rd h during 15 min with 0.36 g/L amplitude as shown in Fig. 5(a), response of the controller was observed as shown in Fig. 5(b). Fuzzy controller has done needed manipulation on molasses feeding.

As a result; this paper presents an adaptive and robust control structure, which can applied to fed-batch baker yeast industrial production process. The proposed fuzzy controller has resulted in higher productivity than the conventional controller. Conventional controller follows a predetermined and fixed feeding profile. Hence, it doesn't behave adaptively under undesired key state conditions during fermentation. But, controller designed in this paper determines feeding rate by itself as well as behaves adaptively under undesired media conditions. Moreover, feeding curve determined by the controller does not include sharp changing when comparing with the ones presented in [8,9,10]. Sharp changing in feeding rate may cause some problem in industrial production. In this meaner, the controller structure presented by this paper is robust and satisfactory.

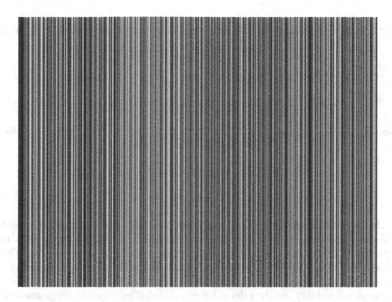

Fig. 6. Controller's adaptive response under dummy distortions on Ce and Cs

References

1. Besli, N., Türker, M., Gul, E.: Design and simulation of a fuzzy controller for fed-batch yeast fermentation, Bioprocess Engineering, Vol.13, (1995) 141-148
2. Kurtanjek Z.: Principal component ANN for modelling and control of baker's yeast production, Journal of Biotechnology, Vol. 65, (1998) 23-35
3. Shimizu H., Miura K., Shioya S., Suga K.: An owerview on the control system design of bioreaktors, Advanced Biochemical Engineering Biotechnology, Vol. 50, (1993) 65-84
4. Sonnleitner B. and Kappeli O.: Growth of *Saccharomyces cerevisiae* is controlled by its Limited Respiratory Capacity: Formulation and Verification of a Hypothesis, Biotechnology and Bioengineering, Vol. 28, (1986) 79-84
5. Bich Pham, H. T., Larsson, G., Enfors, S.-O.: Growth and Energy Metabolism in Aerobic Fed-Batch Cultures of *Saccharomyces cerevisiae*: Simulation and Model Verification, Biotechnology and Bioengineering, Vol. 60,Nno. 4, (1998) 474-482
6. Karakuzu, C., Türker, M., Öztürk, S.: Biomass estimation in industrial baker yeast fermentation using artificial neural networks, 10th Signal Processing and Communication Applications (SIU'2002) Proceedings, Pamukkale,Turkey, (2002) 260-265 (in Turkish)
7. Karakuzu, C., Türker, M.: Biomass prediction in industrial baker yeast fermentation with neural network, V. National Chemical Engineering Congress, Ankara-Turkey Abstracts Proceeding , (2002) PSM04 (in Turkish)
8. Shi Z. and Shimizu K.: Neuro-Fuzzy Control of Bioreaktor Systems with Pattern Recognition, Journal of Fermentation and Bioengineering, Vol. 74, No. 1, 39-45 (1992)
9. Park Y. S., Shi Z. P., et. al.: Application of Fuzzy Reasoning to Control of glucose and Ethanol Concentrations in Baker's Yeast Culture, Applied Microbiology and Biotechnology, Vol. 38, 649-655 (1993)
10. Ye K., Jin S. and Shimizu K., "On the Development of an intelligent control system for Recombinant cell Culture", Int. J. of Intelligent Systems, Vol. 13, 539-560 (1998)

Fuzzy Supervisor for Combining Sliding Mode Control and H∞ Control

Najib Essounbouli, Abdelaziz Hamzaoui, and Noureddine Manamanni

Laboratoire d'Automatique et de Microélectronique, Faculté des Sciences, B.P. 1039,
51687 Reims cedex 2, France
{n.essounbouli, a.hamzaoui}@iut-troyes.univ-reims.fr,
noureddine.manamanni@univ-reims.fr

Abstract: In this paper a fuzzy supervisor for combining H∞ techniques and sliding mode to control a nonlinear uncertain and disturbed SISO system is presented. The plant is approximated by two adaptive fuzzy systems. The process is controlled by a combination of a sliding mode controller, and an H∞ controller according to a weighting factor calculated by the fuzzy supervisor. The global stability and the robustness of the closed-loop system are guaranteed using the Lyapunov approach. An illustration example is presented to show the efficiency of the proposed method.

1. Introduction

Many complex industrial processes whose accurate mathematical models are not available or difficult to formulate. Fuzzy logic provides a good solution for these problems by incorporating linguistic information from human experts. Since, the physical plants are time-varying and uncertain, Wang utilized the general error dynamics of adaptive control to design an adaptive fuzzy controller. In the case of indirect control, the plant model is approximated by adaptive fuzzy systems, and the updating laws are deduced from the stability analysis in the Lyapunov' sense [1][2]. Yet robustness is not guaranteed due to external disturbances. Several approaches using H∞ techniques are presented in the literature [3][4][5]. An H∞ supervisor, calculated from a Riccati-like equation is introduced to maintain the tracking performance and guarantee the robustness by attenuating the effect of both external disturbances and approximation errors on the tracking error to a given level. Nevertheless, these approaches can not give a good time response due to the limited choice of the attenuation level.

Sliding-mode control is a robust alternative SISO system [6][7][8]. However, it has a serious drawback: chattering phenomenon. Many solutions inspired from the boundary-layer approach have been presented in the literature [9][10][11][12], which imposed a compromise between performances and the initial values of the control signal.

We propose a fuzzy supervisor combining H∞ techniques and sliding-mode to control a nonlinear uncertain and perturbed SISO system. Two adaptive fuzzy systems are used to approximate the process, we then divide the state space in fuzzy control

T. Bilgiç et al. (Eds.): IFSA 2003, LNAI 2715, pp. 466–473, 2003.

regions. When the system is far from the sliding surface, sliding mode is applied. In the opposite case, H∞ control is used. For the other cases, the plant is governed by a combination of the two signals according to fuzzy rule basis of the fuzzy supervisor. A simulation example is presented.

This paper is organized as follows: section 2 presents the problem statement. Section 3 gives the H∞ controller. Section 4 formulates the sliding-mode controller problem. Section 5 introduces the fuzzy supervisor and the proposed approach. A pendulum tracking control example is given in section 6 to illustrate the proposed approach.

2. Problem Statement

Consider a Single-input single-output (SISO) nth-order nonlinear system described by the following differential equations:

$$x^{(n)} = f(\underline{x}) + g(\underline{x})u + d \qquad (1)$$

$$y = x$$

Where $f(\underline{x})$ and $g(\underline{x})$ are nonlinear unknown continuous bounded functions, u and y denote, respectively, the input and the output of the system, and d represents the external disturbances assumed to be unknown but bounded. It should be noted that more general classes of nonlinear control problem ban be transformed into this structure [3][6]. Let $\underline{x} = [x, \dot{x}, ..., x^{(n-1)}]^T$ be the state vector of the plant assumed to be available for measurement.

The control objective is to design a control law such that the plant output y follows a bounded reference signal y_r under the constraint that both the stability and the robustness of the closed-loop system are guaranteed.

3. H∞ Control Design

The system model being unknown and disturbed, on can use the following modified feedback control [5]:

$$u_\infty = \hat{g}_1^{-1}(\underline{x})[-\hat{f}_1(\underline{x}) + y_r^{(n)} + K^T E - u_h] \qquad (2)$$

Where $\hat{f}_1(\underline{x}) = \theta_{f1}^T \Psi(\underline{x})$ and $\hat{g}_1(\underline{x}) = \theta_{g1}^T \Psi(\underline{x})$ are two adaptive fuzzy systems approximating the unknown functions $f(\underline{x})$ and $g(\underline{x})$, respectively, as given in [1]. $E = [e, \dot{e}, ..., e^{(n-1)}]^T$, with $e = y_r - y$. $K = [k_n, k_{n-1}, ..., k_1]^T$ represents the dynamic error coefficient vector calculated such that $H(s) = s^n + k_1 s^{n-1} + ... + k_n$ be Hurwitz. u_h attenuates the effect of both the approximation errors and the external disturbances.

The tracking error dynamic equation resulting from (2) can be written as:

$$\dot{E} = AE + B[(\hat{f}_l(\underline{x}) - f(\underline{x})) + (\hat{g}_l(\underline{x}) - g(\underline{x}))u_\infty + u_h - d] \tag{3}$$

where $A = \begin{bmatrix} 0 & 1 & \cdots & 0 \\ \vdots & & \ddots & \vdots \\ 0 & 0 & \cdots & 1 \\ -k_n & \cdots & \cdots & -k_1 \end{bmatrix} \quad B = \begin{bmatrix} 0 \\ \vdots \\ 0 \\ 1 \end{bmatrix}$

A a stable matrix is associated with the following algebraic Riccati equation:

$$A^T P + PA + Q - 2PB\left(\frac{1}{r} - \frac{1}{2\rho^2}\right)B^T P = 0 \tag{4}$$

where Q is a positive definite matrix given by the designer, and ρ is the desired attenuation level. A unique definite solution $P^T = P$ exists if and only if $2\rho^2 \geq r$.

According to the universal approximation theorem [1], there exist optimal approximation parameters θ_{fl}^* and θ_{gl}^* such that $\hat{f}_l(\underline{x}/\theta_{fl}^*)$ and $\hat{g}_l(\underline{x}/\theta_{gl}^*)$ can, respectively, approximate $f(\underline{x})$ and $g(\underline{x})$ as closely as possible. The minimum approximation error is defined as:

$$w_l = (\hat{f}_l(\underline{x}/\theta_{fl}) - f(\underline{x})) + (\hat{g}_l(\underline{x}/\theta_{gl}) - g(\underline{x}))u_\infty \tag{5}$$

Hence, tracking error dynamic can be rewritten as:

$$\dot{E} = AE + B[\Phi_{fl}^T \Psi(\underline{x}) + \Phi_{gl}^T \Psi(\underline{x})u_\infty + u_h + w_l - d] \tag{6}$$

Where $\Phi_{fl} = \theta_{fl} - \theta_{fl}^*$, $\Phi_{gl} = \theta_{gl} - \theta_{gl}^*$.

To determine the adaptation laws for θ_{fl} and θ_{gl}, and the control signal u_h, we consider the following Lyapunov function:

$$V_l = \frac{1}{2}E^T PE + \frac{1}{2\gamma_{fl}}\Phi_{fl}^T\Phi_{fl} + \frac{1}{2\gamma_{gl}}\Phi_{gl}^T\Phi_{gl} \tag{7}$$

Where the positive constants γ_{fl} and γ_{gl} represent the learning coefficients.
Deriving V_l along the error trajectory (7) and using the Riccati equation give:

$$\dot{V}_l = -\frac{1}{2}E^T QE - \frac{1}{\rho^2}E^T PBB^T PE + E^T PB\left(\frac{1}{r}B^T PE + u_h + w_l - d\right) \tag{8}$$

$$+ \frac{1}{\gamma_{fl}}\Phi_{fl}^T\left(\dot{\theta}_{fl} + \gamma_{fl}E^T PB\Psi(\underline{x})\right) + \frac{1}{\gamma_{gl}}\Phi_{gl}^T\left(\dot{\theta}_{gl} + \gamma_{gl}E^T PB\Psi(\underline{x})u_\infty\right)$$

Then using the following control and adaptation laws:

$$u_h = -\frac{1}{r}B^T PE \tag{9}$$

$$\dot{\theta}_{f1} = -\gamma_{f1}E^T PB\Psi(\underline{x}) \tag{10}$$

$$\dot{\theta}_{g1} = -\gamma_{g1}E^T PB\Psi(\underline{x})u_\infty \tag{11}$$

We obtain:

$$\dot{V}_1 \leq -\frac{1}{2}E^T QE + \frac{\rho^2}{2}(w_1 - d)^2 \tag{12}$$

Integrating the above inequality from $t=0$ to $t=T$ and using the fact that $V_1(T) \geq 0$ yields:

$$\int_0^T E^T QE\, dt \leq E^T(0)PE(0) + \frac{1}{\gamma_{f1}}\Phi_{f1}^T(0)\Phi_{f1}(0)$$

$$+ \frac{1}{\gamma_{f1}}\Phi_{f1}^T(0)\Phi_{f1}(0) + \rho^2 \int_0^T (w_1 - d)^2\, dt$$

This inequality is our H∞ criterion guarantying stability and robustness .

4. Sliding-Mode Control Design

To determine the control law, we consider the following sliding surface:

$$S(\underline{x}) = -\lambda_1 e - \lambda_2 \dot{e} - ... - \lambda_{n-1}e^{(n-2)} - e^{(n-1)} \tag{13}$$

The factors λ_1 are calculated such that $H(s) = s^{n-1} + \lambda_{n-1}s^{n-2} + ... + \lambda_1$ is Hurwitz. $f(\underline{x})$ and $g(\underline{x})$ being unknown, we use the following control law [12]:

$$u_{SMC} = \hat{g}_2^{-1}(\underline{x})[-\hat{f}_2(\underline{x}) + y_r^{(n)} + \sum_{i=1}^{n-1}\lambda_i e^{(i)} - D.sign(S)] \tag{14}$$

Where $\hat{f}_2(\underline{x}) = \theta_{f2}^T\Psi(\underline{x})$ and $\hat{g}_2(\underline{x}) = \theta_{g2}^T\Psi(\underline{x})$ are two adaptive fuzzy systems approximating $f(\underline{x})$ and $g(\underline{x})$ respectively. D a positive constant will be determined later.

To determine the adaptation laws for θ_{f1} and θ_{g2}, and the constant D, consider the following Lyapunov function:

$$V_2 = \frac{1}{2}S^2(\underline{x}) + \frac{1}{2\gamma_{f2}}\Phi_{f2}^T\Phi_{f2} + \frac{1}{2\gamma_{g2}}\Phi_{g2}^T\Phi_{g2} \tag{15}$$

where $\Phi_{f2} = \theta_{f2}^* - \theta_{f2}$, $\Phi_{g2} = \theta_{g2}^* - \theta_{g2}$.

Deriving (14) and (17), and after some manipulations we obtain:

$$\dot{V}_2 = S(\underline{x})(w_2 + d - D\,sign(S)) - \frac{1}{\gamma_{f2}}\Phi_{f2}^T(\dot\theta_{f2} - \gamma_{f2}S(\underline{x})\Psi(\underline{x})) \tag{16}$$

$$-\frac{1}{\gamma_{g2}}\Phi_{g2}^T(\dot\theta_{g2} - \gamma_{g2}S(\underline{x})\Psi(\underline{x})u_{SMC})$$

If we choose the following adaptation laws:

$$\dot\theta_{f2} = \gamma_{f2}S(\underline{x})\Psi(\underline{x}) \tag{17}$$

$$\dot\theta_{g2} = \gamma_{g2}S(\underline{x})\Psi(\underline{x})u_{SMC} \tag{18}$$

We obtain:

$$\dot{V}_2 = S(\underline{x})(w_2 + d - D\,sign(S)) \tag{19}$$

To maintain $\dot{V}_2 \le 0$, we choose:

$$D \ge \left| 1 - \frac{\max(g(\underline{x})) - \min(g(\underline{x}))}{\hat{g}_2(\underline{x})} \right| |\max(f(\underline{x})) - \min(f(\underline{x}))| \tag{20}$$

$$+ \hat{g}_2(\underline{x})|\max(g(\underline{x})) - \min(g(\underline{x}))| \left| -\hat{f}_2(\underline{x}) + \sum_{i=1}^{n-1}\lambda_i e^{(i)} + y_r^{(n)} \right| + |d| \Big] + \eta$$

Therefore, the closed-loop system global stability is guaranteed. Furthermore, the transition condition is satisfied which implies the sliding surface is attractive.

5. Fuzzy Supervisor

Having presented the two controllers (H∞ and sliding mode), and proving their stability and robustness by Lyapunov's approach, the next task is to propose a control combining the advantages of the two approaches. In fact, if one considers a control combining "intelligently" the two signals, where the sliding mode is used only in the approaching phase and the H∞ control activated only in the sliding phase. To ensure the continuity of the control signal and to maintain the tracking performance, the transition between the two controllers is done by a combination of these two signals. This combination can be considered as a weighted sum of these two signals. The weighting factor is deduced from a fuzzy system which inputs are the tracking error and its time derivative. We thus propose to partition the state space in several regions using a Takagi-Sugeno-type fuzzy system, and to use the product inference, the algebraic sum for the combination of the different fuzzy rules, and the singleton for the consequent-part. Hence, the fuzzy system output can be written as:

$$\alpha = \frac{\sum_{j=1}^{m} \alpha_j \prod_{i=1}^{n-1} \mu_{H_i^j}(e^{(i)})}{\sum_{j=1}^{m} \prod_{i=1}^{n-1} \mu_{H_i^j}(e^{(i)})} \tag{21}$$

Where H_i^j is a fuzzy set, $\mu_{H_i^j}$ is the membership degree of $e^{(i)}$ to H_i^j, α_I is a singleton, and m of the number of fuzzy rules used.

Hence, the global control law applied to the system is given by:

$$u = (1 - \alpha)u_{SMC} + \alpha u_\infty \tag{22}$$

To prove the global stability of the closed-loop system governed by the control law (27), the following theorem is applied:

Theorem [13]:
Consider a combined fuzzy logic control system as described above.
-    If there exist a positive definite, continuously differentiable, and radially unbounded scalar function V,
-    If every fuzzy subsystem gives a negative definite $\dot{V}$ in the active region of the corresponding fuzzy rule,
-    If the weighted-sum deffuzification method is used, which for any output u of the fuzzy logic supervisor lies between u_{SMC} and u_∞ such that $\min(u_{SMC}, u_\infty) \le u \le \max(u_{SMC}, u_\infty)$

Then according to Lyapunov's theorem, the global stability of the closed-loop system is guaranteed □.

6. Simulation Example

It is well-known that the inverted pendulum represents a class of nonlinear control problem that can be described as the task of balancing a pole on a moving cart.

The control objective is to apply a suitable effort to force the system to track a desired trajectory $y_r(t) = (\pi/30)\sin(t)$. To construct the adaptive fuzzy systems approximating the plant, we select five gaussian membership functions, with centers $-\pi/6$, $-\pi/12$, 0, $\pi/12$, $\pi/6$ and standard deviation $\pi/24$, for both x_1 and x_2 to cover the whole universe of discourse. Using all the possible combination, we obtain 25 fuzzy rules. To update the adjustable parameters, we fix $\gamma_{f1}=\gamma_{f2}=100$ and $\gamma_{g1}=\gamma_{g2}=5$. To calculate the control law (2), we chose $K^T=[5\ 1]$ and $Q=diag(15,15)$, $r=2\rho^2$, and $\rho=0.1$. For the sliding mode control (15), we fix $\lambda_1=5$, and $D=15$. To simplify the computation, only three membership functions for the tracking error e and its time derivative $\dot{e}$, and only five singletons for the output α are used. The structure and the fuzzy rule basis of the fuzzy supervisor are depicted in figure 1.

The simulation results are given by figures 2 and 3. Figure 2 shows good tracking performances, and a fast dynamic response. Figure 3 gives the applied effort using the proposed approach, and the corresponding transient response. We remark that the plant output reaches the desired trajectory using the proposed method at the same time as for the sliding-mode case. Furthermore the initial value of the applied effort is less than the applied effort using sliding-mode.

7. Conclusion

In this work, a fuzzy combination of an H∞ and a sliding-mode controllers for a nonlinear uncertain and perturbed SISO system is presented. The plant model is approximated by two adaptive fuzzy systems. A fuzzy supervisor is used to combine the efficiency and robustness of the sliding-mode and H∞ techniques. The resulting control law leads to a fast dynamic response without chattering. The global stability and the robustness are guaranteed using Lyapunov's theory.

References

[1] Wang, L.-X. (1994) Adaptive fuzzy systems and control, Englewood Cliffs: Prentice-Hall, New Jersey.
[2] Wang, L.-X. (1996) Stable adaptive fuzzy controllers with application to inverted pendulum tracking. IEEE Trans. on Syst., Man and Cybern., vol. 26, pp. 677-691.
[3] Chen B.-S., C.-H. Lee, Y.-C. Chang, (1996) H∞ tracking design of uncertain nonlinear SISO systems: Adaptive fuzzy approache. IEEE Trans. Fuzzy Syst., vol. 4, pp. 32.
[4] Hamzaoui, A., J. Zaytoon, A. Elkari (2000) Adaptive fuzzy control for uncertain nonlinear systems. In Proceeding of IFAC Workshop on Control Optimization 2000, Saint-Petersburg, Russia, pp. -137-141.
[5] Essounbouli, N., A. Hamzaoui, J. Zaytoon (2002) A supervisory robust adaptive fuzzy controller. In Proceeding of 15th IFAC World Congress on Automatic and Control 2002, Barcelona, Spain.
[6] Slotine, J.J.E., W. Li (1991) Applied nonlinear control. Prentice-Hall. Inc.
[7] Yoo, B., W. Ham (1998) Adaptive fuzzy sliding mode control of nonlinear system. IEEE Trans. on Fuzzy Syst., vol. 6, pp. 315-321.
[8] Chang, W., J.B. Park, Y.H. Joob, G. Chen (2002) Design of robust fuzzy-model-based controller with sliding mode control for SISO nonlinear systems. Fuzzy Sets and Syst., vol. 125, pp.1-22.
[9] Kim, S.-W., J.-J. Lee (1995) Design of a fuzzy controller with fuzzy sliding surface. Fuzzy Sets and Syst., vol. 71, pp. 359-367.
[10] Li, H.X., H.B. Gatland, A. W. Green (1997) Fuzzy variable structure control. IEEE Trans. On Syst., Man and Cybern., vol. 27, pp. 306-312.
[11] Wang, J., A.B. Rad, P.T. Chan (2001) Indirect adaptive fuzzy sliding mode control: Part I: fuzzy switching. Fuzzy Sets and Syst., vol. 122, pp. 21-30.
[12] Hamzaoui, A., N. Essounbouli, J. Zaytoon (2003) Fuzzy sliding mode control for uncertain SISO systems. To appear in Proceedings of IFAC International Conference on Intelligent Control Systems and Signal Processing, ICONS 2003, Portugal.
[13] Wang, L.K., F.H.F. Leung, P.JK.S. Tam (2001) A fuzzy sliding controller for nonlinear systems. IEEE Trans. on Indus. Elect., vol. 48, pp. 32-37.

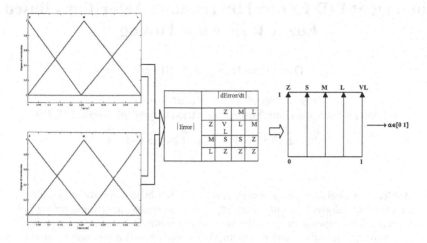

Fig. 1. The structure of the proposed fuzzy supervisor

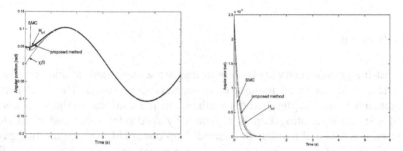

Fig. 2. The angular position using the three control laws The tracking error using the three control laws

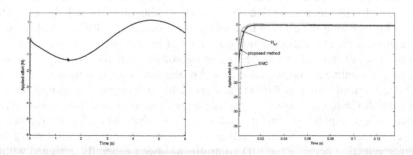

Fig. 3. The applied effort using the proposed method The applied effort using the three control laws (Zoom)

Intelligent PID Control by Immune Algorithms Based Fuzzy Rule Auto-Tuning

Dong Hwa Kim and Jin Ill Park

Dept. of I&C, Hanbat National University,
16-1 San Duckmyong-Dong Yusong Gu Daejon City Seoul, Korea, 305-719.
kimdh@hanbat.ac.kr, Hompage: ial.hanbat.ac.kr
Tel: +82-42-821-1170, Fax: +82-42-821-1164

Abstract. A goal of intelligent control design is to obtain a controller based on input/output information only. Recently a genetic based fuzzy logic tuning is emerged. This paper focuses on the immune algorithms based fuzzy rule tuning. The immune algorithm (IA) is a method that holds promise for such a control system design using the mechanics of antibody against antigen. This algorithm is tried on the boiler level control of power plant which is typically non-linear and multi-variable with multiple control objectives.

1. Introduction

Proportional-Integral-derivative (PID) controllers are widely used in industrial control fields because of the reduced number of parameters to be tuned. The most popular design approach is the Ziegler-Nichols method. On the other hand, the level control system of a steam generating unit is regulates by feedwater valve and main steam valve control position and combustion control loop controls CO/O_2 of the flue gas by manipulating the fuel and air inputs to the steam generating unit. When a coal-fired drum-type boiler efficient dynamic control becomes a very complicated problem. The dynamics of two-phase fluid inside the boiler drum and the nonlinear behavior of the drum water level itself produce many interactive phenomena which ultimately effect the total combustion control scheme of the plant [1, 2]. Therefore, a new concept has been presented with an assumption that CO/O_2 of the flue gas will be controlled separately by adjusting the air/fuel ratio of the plant based on a PID controller. In the present approach the interactions among drum water level, steam flow to a HP turbine, and drum pressure have been studied more rigorously than the existing techniques using a simple nonlinear multivariable model of the steam generating unit [1]. While conventional controls such as PID controllers yield an acceptable response, they do not have the flexibility necessary to provide a good performance over a wide region of operation in complex system or multivariable system. Also, the PID controller cannot effectively meet the requirements of both the set-point-following and disturbance rejection because the PID controller has been generally designed without consideration of disturbance rejection. A highly tuning experience is also required for actual plant since the PID controller is usually poorly tuned [2].

T. Bilgiç et al. (Eds.): IFSA 2003, LNAI 2715, pp. 474-482, 2003.
© Springer-Verlag Berlin Heidelberg 2003

A goal of modem intelligent control design is to obtain a controller based on input/output information only. The immune algorithm (IA) is a method that holds promise for such a control system design or tuning of PID controller.

The IA is an intelligent technique based on the mechanics of natural immunity on human body. The tuning approaches based on the immune algorithm have been suggested for the PID controller [6-8]. In this paper, intelligent PID control has been suggested by immune algorithms based fuzzy rule tuning. The goal of controller is to have an optimal fuzzy rule for control on reference change between 100% and 120 % of the nominal operating point.

2. The Level Control System in Nonlinear System

2.1 The Non-linear Model

The mode is based on the boiler-turbine plant. The boiler is oil-fired and the rated power is 160 MW. Although the model is of low order, it is capable of illustrating some of the complex dynamics associated with the real plant. The dynamics for the system are given by

$$\frac{dh}{dt} = \frac{1}{\partial \delta}\left[\frac{f_e - f_l}{3600\,v_t} - \frac{\partial \delta}{\partial p}\right]\frac{dp}{dt}, \quad \frac{dp}{dt} = \frac{(f_e - f)\left[\dfrac{\delta}{\partial \delta \big/ \partial h}\right] - (f_e h_e - f_l h_l)}{3600 v_t \left[\dfrac{\delta \dfrac{\partial \delta}{\partial p}}{\dfrac{\partial \delta}{\partial h}} + \dfrac{144}{778}\right]}, \quad (1)$$

$$f_e = f_c + f_s, \ f_c = c_{qc}\sqrt{\delta_c(p_c - p)}, \ f_s = \frac{19}{30}c_{vs}\sqrt{\delta_s(p_s - p)},$$

$$c_{qs} = \frac{1}{\sqrt{\dfrac{1}{c_{vc}^2} + c_{pc}^2}}, \ c_{vs} = y_s^3 c_{vs}^{\max}, \ c_{vc} = y_c^3 c_{vc}^{\max}.$$

Where the subscription c means condensate and s stands for steam.

2.2 The Linear Model

The performance of the control systems designed with the IA will be compared to the linear quadratic regulator designed from the linearized model. A linearized model based on the nominal data is obtained from a truncated Taylor series expansion of the non-linear equations. The linear dynamics are of the form

$$\frac{dx}{dt} = Ax + Bu, \ y = Cx + Du. \quad (2)$$

Where, system matrices are

$$A = \begin{bmatrix} -1.9 \times 10^{-3} & 0 \\ -1.0 \times 10^{-3} & 0 \end{bmatrix}, B = \begin{bmatrix} 1.8716 & -0.8295 \\ 0.9905 & -0.4822 \end{bmatrix}, C = \begin{bmatrix} 1.0000 & 0 \\ 8.2400 & -15.3110 \end{bmatrix}.$$

The above equations describe the deaerator model for controller design. These equations are linearized around an equilibrium point (set point) to obtain the linear description of the system. The linearization is dependent on five variables: The pressure in the deaerator, the enthalpy in the deaerator, the feedwater level, the condensate valve position, and the steam valve position. For every equilibrium point, there is a set of these five variables and hence a resulting linear system.

Where x, u, and y are, respectively, state, control, and output vectors. The deaerator is in general a second order, multivariable, and highly non-linear system.

3. Dynamic of the Immune Algorithm for Optimal Controller

3.1 The Response of Immune System

The immune system has two types of response: primary and secondary. The primary response is reaction when the immune system encounters the antigen for the first time. At this point the immune system learns about the antigen, thus preparing the body for any further invasion from that antigen. This learning mechanism creates the immune system's memory.

The secondary response occurs when the same antigen encountered again. This has response characterized by a more rapid and more abundant production of antibody resulting from the priming of the B-cells in the primary response.

3.2 Antibodies in Immune System

In the AIS the antibodies blind to infectious agents and then either destroy these antigens themselves attract help from other components of the immune system. Antibody is

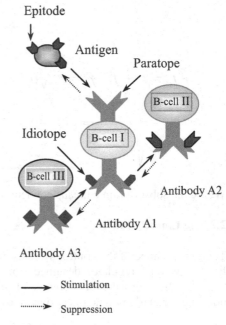

Fig. 1. Structure of idiotypic on Jerne network.

actually three-dimensional Y shaped molecules which consist of two types of protein chain: light and heavy. It also possesses two paratopes which represents the pattern it will use to match the antigen. The regions on the molecules that the paratopes can attach are so-called epitopes.

3.3 Interaction Between Antibodies

Describing the interaction among antibodies is important to understand dynamic characteristics of immune system. For the ease of understanding, Consider the two antibodies that respond to the antigens A1 and A2, respectively. These antigens stimulate the antibodies, consequently the concentration of antibody A1 and A2 increases. However, if there is no interaction between antibody A1 and antibody A2, these antibodies will have the same concentrations. Suppose that the idiotope of antibody A1 and the paratope of antibody A2 are the same. This means that antibody A2 is stimulated by antibody A1, and oppositely antibody A1 is suppressed by antibody A2 as Fig. 1. In this case, unlike the previous case, antibody A2 will have higher concentration than antibody A1. As a result, antibody A2 is more likely to be selected.

This means that antibody A2 has higher priority over antibody A1 in this situation. As we know from this description, the interaction among the antibodies acts based on the principle of a priority adjustment mechanism.

3.4 Dynamics of Immune System

In the immune system, the level to which a B cell is stimulated relates partly to how well its antibody binds the antigen. We take into account both the strength of the match between the antibody and the antigen and the B cell object's affinity to the other B cells as well as its enmity. Therefore, generally the concentration of i-*th* antibody, which is denoted by δ_i, is calculated as follows [3]:

$$\frac{dS_i(t)}{dt} = \left(\alpha \sum_{j=1}^{N} m_{ji} \delta_j(t) - \alpha \sum_{k=1}^{N} m_{ik} \delta_k(t) + \beta m_i - \gamma_i \right) \delta_i(t) \quad (3a)$$

$$\frac{d\delta_i(t)}{dt} = \frac{1}{1 + \exp\left(0.5 - \frac{dS_i(t)}{dt} \right)} \quad (3b)$$

where in Eq. (1), N is the number of antibodies, and α and β are positive constants. m_{ji} denotes affinities between antibody j and antibody i (i.e. the degree of interaction), m_i represents affinities between the detected antigens and antibody i, respectively.

4. The Immune Algorithm Optimal Control System Design

4.1 Characteristics of Boiler-Turbine Control System

The goal of controller is to have an optimal control for reference change between 100% and 120 % of the nominal operating point. A pressure demand and a power demand change between 100% and 120% of the nominal operating point is given by

$$R_1 = \alpha + (\beta - \alpha)\Gamma(t - 200) + (129.6 - \beta)\Gamma(t - 600) + (\alpha - 129.6)$$

$$\Gamma(t - 1000) + (\alpha - \beta)\Gamma(t - 1400) \ at \ R_2 = 66.65, \ R_3 = 0.$$

$$(4)$$

The power demand is changed by

$$R_2 = 66.65 + (85.063 - 66.65\Gamma(t - 200) + (105.8 - 85.063)\Gamma \ (t - 600)$$

$$+ (85.063 - 105.8)\Gamma(t - 1000) + (66.65 - 85.063)\Gamma(t - 1400)$$

$$at \ R_1 = \alpha, \ R_3 = 0, \ R_1 = Pressure \, demand, \ R_2 = Power \, demand,$$

$$\alpha = 108, \ \beta = 118.8.$$

$$(5)$$

4.2 Optimal Controller Design

In linear system $f(x, u) = Ax + Bu + C = 0$, control vector for optimal control is given by

$$u = -R^{-1}B^T (AQ^{-1}A^T + BR^{-1}B^T)^{-1}C.$$

$$(6)$$

The goal of the controller is to track step demands in power and pressure. To achieve this the following performance index is used to be minimized:

$$PI_t = PI_1 + PI_2, \ PI_1 = \int_{t_1}^{t_2} \left[(Y - R_{ref})^T Q(Y - R_{ref}) + u^T Ru \right] dt \quad (8)$$

$$PI_2 = \sum_{i=1}^{3} e_i(t_2), \ e_i(t) = Y_i(t) - R_i(t)$$

This control purpose is implemented by the following procedures for optimal tuning of the PID controller tuning based on immune algorithm.

4.3 The Structure of Coding for Optimal Control

[step 1] Initionalization and Recognize pattern of reference as antigen: The immune system recognizes the invasion of an antigen, which corresponds to input data or disturbances in the optimization problem.

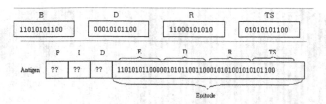

Fig. 2. Coding structure for antigen.

Code the selected E(PI1), D(PI2), R(R), and TS(Q) with binary and string for response specification of reference model as the following Fig. 2.

[step 2] Product of antibody from memory cell:

The immune system produces the antibodies which were effective to kill the antigen in the past, from memory cells. This is implemented by recalling a past successful solution.

I_j^P : the P value of affinity in antibody j, I_j^I : the I value of affinity in antibody j

I_j^D : the D value of affinity in antibody j, P_j : the value of paratope in antibody j

A : the value of epitope in antigen, $\otimes$: exclusive or operator

$\oplus$: mutation and crossover,   j : the length of antibody from 1,

cut_delta : Positive constant,

$$I_{P,I,D}^{new\_k} = F^{new}(\oplus(F(m_j - m_{j+1})\alpha I_k)),\qquad(9)$$

$$F^{new}(x) = \begin{cases} \text{Pr}esent\ value: if\ x \geq A_c \\ \text{Pr}evious\ value: if\ x \prec A_c \end{cases},\quad \alpha = \begin{cases} 1\ if\ |I_j - I_{j+1}| \geq A_{delta} \\ 0\ if\ |I_j - I_{j+1}| \geq A_{delta} \end{cases}$$

$$\oplus(x) = \begin{cases} \oplus(x)\ if\ x = 1 \\ I_j\quad if\ x = 0 \end{cases},\quad F(x,k) = \begin{cases} 1,\ j\ if\ x \geq 0: Stimulation \\ 0,\ j+1\ x \prec 0: Suppression \end{cases}$$

$$m_j = P_j \otimes A$$

[step 3] Initialize antibody group (MCELL) for parameter P=0-1, I=1-10, D=0-10 of the given condition to the desired response of plant.

[step 4] Calculation of affinity between antibodies: The affinities obtained by Eq. (10) and $m_j = P_j \otimes A$ for searching the optimal solution. Arrange with the number of order of affinity value. Select randomly the number of antibody, 25 among the number of MCELL, 100 and calculate affinity, α between both antibodies.

[step 5] Stimulation of antibody: To capture to the unknown antigen, new lymphocytes are produced in the bone marrow in place of the antibody eliminated in step 5. This procedure can generate a diversity of antibodies by a genetic reproduction operator such as mutation or crossover. These genetic operators are expected to be more efficient than generation of antibodies.

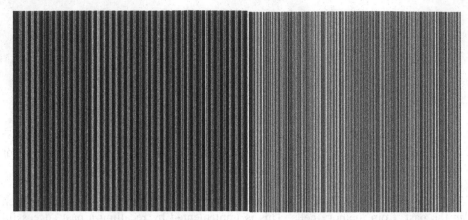

Fig. 3. Fuzzy rule tuning on alpha 0.6 in memory cell of immune network

Fig. 4. Level response fuzzy rule tuned on alpha 0.6.

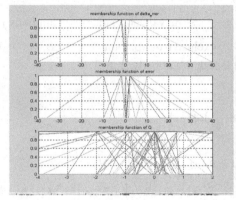

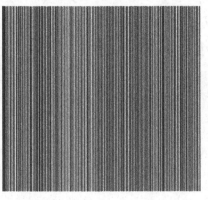

Fig. 5. Fuzzy rule tuning on alpha 0.75 in memory cell of immune network.

Fig. 6. Level response using fuzzy rule tuning on alpha 0.75.

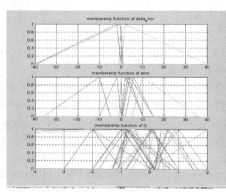

Fig. 7. Fuzzy rule tuning on alpha 0.98 in memory cell of immune network.

Fig. 8. Level response using fuzzy rule tuning on alpha 0.98.

5. Simulations and Discussions

5.1 Simulation Results on Alpha 0.6

To confirm some effect of affinity between alpha and fuzzy rules, this paper used binary code for calculation of fuzzy rules. Figs. 8-10 show fuzzy rule tuning results, level response, and error on alpha 0.6, respectively.

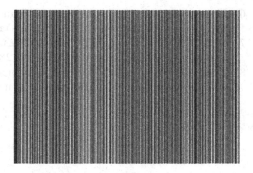

Fig. 9. Error using fuzzy rule tuning on alpha 0.98.

6. Conclusions

The linearization in boiler level system is dependent on five variables: The pressure in the boiler, the enthalpy in the boiler, the feedwater valve position, the main steam valve position.
In this nonlinear system, the PID controller cannot effectively meet the requirements of both the set-point-following and disturbance rejection because the PID controller has been generally designed without consideration of disturbance rejection. A highly tuning experience is also required for actual plant since the PID controller is usually poorly tuned.

This paper focuses on boiler level control of power plants with non-linear and multi-variable using immune algorithm based fuzzy rule tuning. In this paper both methods of binary representation and arithmetic representation in memory cell of immune algorithm is used for alpha calculation of fuzzy rule for optimal response. Figs. represents fuzzy rule tuning results, level response, and error on alpha 0.75, respectively.

References

1. Robert D., et al (1995): Boiler-Turbine control design using a genetic algorithm, Int. J. Man-Machine Studies, Vol. 7, pp. 1-13, 1974.
2. A. Homaifar and E. Mccormick: Simultaneous design of membership functions and rule sets for fuzzy controllers using genetic algorithms, IEEE Trans. Fuzzy Systems, Vol. 3 (1995) 129-139.
3. R. Ketata, D. De Geest and A. Titli, Fuzzy controller: design, evaluation, parallel and hierarchical combination with a PID controller," Fuzzy Sets and Systems, Vol. 71 (1995) 113-129.
4. D. H. Kim, "Intelligent tuning of the two Degrees-of-Freedom Proportional-Integral-Derivative controller on the Distributed control system for steam temperature of thermal power plant," KIEE international Transaction on SC, Vol. 2-D (2002) 78-91.
5. J. D. Farmer, N. H. Packard and A. S. Perelson: The immune system, adaptation, and machine   learning, Physica. D 22 (1986) 187-204.

6. Kazuyuki Mori and Makoto Tsukiyama: Immune algorithm with searching diversity and its application to resource allocation problem, Trans. JIEE, Vol. 113-C (1993).
7. Dong Hwa Kim: Tuning of a PID controller using an artificial immune network model and fuzzy set, IFSA2001Vancouver (2001) July 28.
8. Dong Hwa Kim, Hong Won Pyo: Auto-tuning of reference model based PID controller using immune algorithm, IEEE 2002 international Congress on Evolutionary Computation, Hawaii (2002) May 12-17.

Implementation and Applications of a Constrained Multi-objective Optimization Method

Hossein S. Zadeh

School of Business Information Technology
Royal Melbourne Institute of Technology
GPO Box 2497V
Melbourne 3001, Australia

Abstract. This paper starts by introducing a very flexible aerospace structure and defines the problem of attitude control ($\frac{\pi}{2}$ rad rotation) as well as vibration suppression of the structure. A nominal fuzzy logic controller had been designed and used as a basis for design of a constrained optimized fuzzy controller with multiple objectives. Optimization algorithm and procedure are explained.
Furthermore this paper proposes an innovative application of the optimization and the optimized controller in the field of rover goal-seeking. Rover goal-seeking problem, and use of the optimization methodology are explained.

1 Introduction

Through the introduction of satellite TV channels, cellular telephones, GSM mobile phones, and more recently proliferation of GPS navigation systems, the last few decades have seen a rapid growth in the use of space-based communication satellites. Recent satellites use progressively higher transmission power in order to reduce ground segment size and costs. However increased transmission power also brings about higher power consumption and in case of satellites this translates to larger solar panels. The increased size of the solar panels in conjunction with very low force working environments have resulted in structures which are extremely flexible and have very low frequency structural vibration modes [1]. The structures are usually so flexible that they cannot withstand their own weight under "$1g$" conditions (e.g. on Earth). These structures also exhibit very low damping, especially in the absence of aerodynamic drag (aerodynamic damping). Artist impression of the proposed Colombus spacecraft shown on Figure 1 is an example of a three axis stabilized spacecraft with numerous very low frequency modes.

The requirement to minimize the volume occupied by the satellite at launch leads to use of mechanisms to unfold the vehicle's solar panels and other appendages from their stowed position after launch. The jointed structures required to do this usually possess some degree of backlash as well as nonlinear damping within the joints. Presence of Backlash and damping can be prominent in a

T. Bilgiç et al. (Eds.): IFSA 2003, LNAI 2715, pp. 483–491, 2003.

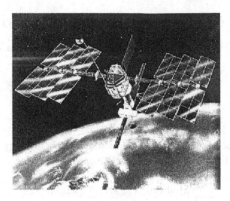

Fig. 1. Artist impression of Colombus radar satellite

multi-jointed structure. The combination of a very flexible structure (with non-linearities), precise pointing requirements, and the difficulty of reproducing the space environment to verify system behavior using conventional test techniques, has made design of control algorithms and investigation of system behavior for the attitude control of flexible spacecraft a challenging area. This is reflected in the large body of literature covering dynamic modeling and control of flexible space structures.

Some notable uses of the classical control techniques for control of attitude and slew motion of a spacecraft were by Hughes [2,3] and Dwyer [4,5]. Hughes analyzed the controls/structure interaction for Canadian Communications Technology Satellite. Use of classical control theory enabled him to incorporate modes of vibration of each of the flexible solar arrays as separate transfer functions in the control system block diagram. This early work was restricted to looking at the effect of including flexible body dynamics on an existing controller utilizing only a single pitch feedback loop, rather than designing control laws to compensate for the flexible body motion. This early work of Hughes, although limited, is of interest as it marks the start of what was to become an immense effort by researchers around the world to develop controllers of ever increasing complexity to overcome the problems of robustness and performance encountered with the rapidly enlarging space structure designs of the seventies and eighties.

Wood [6] applied classical Bode analysis to the design of an attitude control system for the Solar Heliospheric Observatory (SOHO) satellite. Two control schemes were analyzed, the first being the incorporation of a high order roll off term in the attitude feedback loop, below the frequency for the first flexible mode. This technique is known as gain stabilization and has the disadvantage that significant phase lags are introduced to the system. In addition, the presence of low frequency modes requires that a very sharp roll off filter be used to achieve the necessary gain margin for the first mode. If the frequency for the first mode is lower than expected, then the reduction in gain margin could destabilize the system.

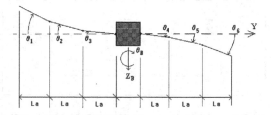

Fig. 2. Schematic of the model used for simulation

Scott [7] designed a number of modern controllers, namely ℓ_1 and $\mathcal{H}_\infty$, and compared their performance to that of more traditional methods. In his research, Scott found that some modern control methods (namely ℓ_1) could produce far superior settling time and robustness to noise.

Zadeh and Scott [8] compared findings of Scott [7] to a nominal Fuzzy Logic controller. The fuzzy logic controller was shown to produce system response much better than that of the $\mathcal{H}_\infty$ controller, but could not outperform the ℓ_1 controller.

2 Problem Definition

Configuration modeled is similar to the Optus class B satellite. The model has three solar panels on each side of a central hub (Figure 2). Derivation of equations of motion is explained in detail by Riseborough [1] and Zadeh [8,9]. The problem is defined as rotating the central hub along the axis passing through the central hub and parallel to the solar panel joints (perpendicular to the page and shown as θ_B on Figure 2). The time taken for the rotation is to be minimized with objectives of minimal vibration induced in the solar panels, and minimal overshoot, and with maximum torque requirement of 1 $N.m$ as the constraint.

3 Fuzzy Logic Controller

Figure 3 depicts block diagram of the system in conjunction with a fuzzy controller. In order to eliminate instability and to give the fuzzy controller a 'prediction' capability, a PD fuzzy controller [10] was designed. The two input signals to the controller are error in angle of rotation (between a reference input and system output) and angular speed of the main bus. These are then used to produce a control signal (torque) with fuzzy logic relations/rules [11, 12].

Universe of discourse (domain of all possible input parameter values) of both inputs was divided using five membership functions. Membership functions for the two inputs are shaped as evenly-spaced unilateral triangles. Fuzzy logic rules are expressed in the form of simple and intuitive if-then statements [12].

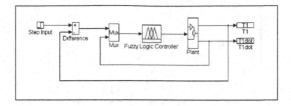

Fig. 3. Block diagram of the plant in relation to the fuzzy logic controller.

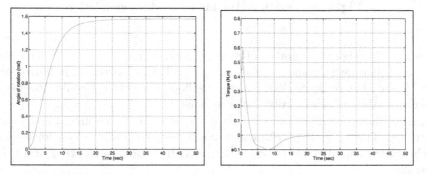

Fig. 4. System response (of the nominal fuzzy logic controller) to the step input. Actuator commands is shown on the right hand pane.

For output, seven evenly-spaced unilateral-triangular membership functions were used. Universe of discourse was set to '$\pm 1 N.m$' which is the maximum available torque.

Defuzzification method used was the simplest available method–'center of gravity'. Defuzzification process produces controller output which in this case is fed back to the plant to close the control loop (Figure 3).

Response of the system and actuator commands to a step input of '$\frac{\pi}{2} rad$' are shown on Figure 4. As is evident from the Figures, the system does not have any overshoot and settles in about 20 seconds. Note that the actuator command is well within the boundaries of the available torque ($\pm 1 N.m$), ensuring that the actuators will not saturate. Actuator saturation could excite higher modes of vibration of the structure.

4   Optimization

Classical optimization theory is well developed for deterministic problems and those of low to moderate dimensions, limiting its application in contemporary problems. There is an optimization method that has risen from statistical physics and has earned popularity in recent years. Known as simulated annealing, it has found diverse applications [13].

For detailed explanation of the optimization procedure see Zadeh and Wood [14]. For completeness however, a small excerpt from the paper [14] is

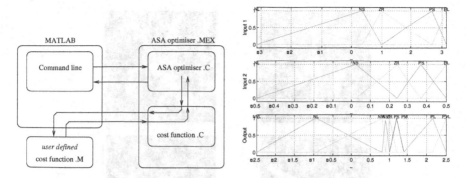

Fig. 5. Schematic of MATLAB-ASA interface. Membership function distribution of the optimized controller is shown on the right pane.

repeated herein: ASA software [15] was interfaced with Matlab© by Drack, Zadeh, and Wharington [13] (left pane on Figure 5). ASA source code was modified slightly and a wrapper function was written to form a library. The library provides access to the ASA engine from within the Matlab command line interface. A cost function library was written in 'C' language. The ASA code, during search, calls the cost function library repeatedly. A second MEX file represents a gateway between this cost function to the user defined cost function written in Matlab 'M' script.

The optimization library was used to minimize the cost function by modifying membership functions of the controller. MFs of the optimized controller are plotted on right pane of Figure 5.

5   Results Visualization

The system of Figure 3 was simulated for 25 seconds. At each 20ms interval (i.e. 50Hz), tracking error as well vibration in individual solar panel was calculated and recorded. The simulation was run twice; once for each of the nominal and the optimized controllers.

Results of the simulation as well as snapshots of the simulation software used are reproduced on Figure 6 (originally presented in another paper [16]).

6   Goal-Seeking in an Unknown Environment

Exploration of planetary surfaces and operation in unknown and (possibly) rough surfaces have been strong motivations for research in autonomous navigation of mobile robots in recent years.

NASA's commitment to the Mars Landing Mission has further fueled a new wave of research activities focusing on real-time autonomous navigation of a

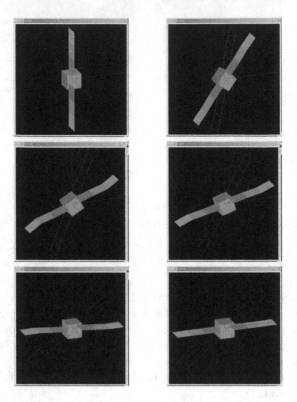

Fig. 6. Snapshots of the satellite model undergoing 90 degree rotation, sequence from left to right, top to bottom. Reproduced from [16].

rover. Carnegie Melon University (CMU) [17, 18] and NASA's Jet Propulsion Lab (JPL) have been two major contributors to this field [19, 20, 21].

The route taken by researchers at CMU uses a mathematical model based on weighted sum of the roll, pitch, and roughness of the map cells between the rover and the goal, incorporating the certainty of each the values [17].

The method used by JPL researchers uses a rule-based (fuzzy logic) Traversability Index [22, 23]. The concept of Traversability Index has recently been expanded to Local, Regional, and Global Traversability Indices [24, 25].

Seraji developed a navigation strategy for a mobile robot traversing a challenging terrain [25]. Developing the navigation strategy, Seraji introduced concept of Traverse-Local, Traverse-Regional, and Traverse-Global navigation behaviors.

The author proposes (and has started working on) a research on the use of visual sensors on the rover to incrementally constructs a model of its environment, constantly updating it as more of the environment is revealed as time progresses. This is an innovative departure from the current state of research, where simulation of a goal-seeking robot has been performed using (at least partial) *a priori* knowledge of the terrain.

The structure of the model is a weighted grid, where the weighting factors are a local, regional and global traversability indices—providing a general measure of the ease at which the rover may move through the terrain. The traversability indices is determined through the use of a set of fuzzy rule bases, taking into consideration factors such as terrain type, roughness and slope. Based on this, potentially incomplete, model of the environment, a modified A^* and simulated annealing algorithm is used to calculate an optimal path from the current rover's position to the goal. Calculating the optimal path, consideration is given to other important factors such as the rover's remaining battery power and the proximity of a 'recharging station'. This is an optimal path based on current knowledge of the environment, and the state of the rover, for a particular point in time. A fuzzy logic controller then takes the calculated path and determines the rover's next movement. The optimal path calculated is constantly updated as new information about the environment is brought to light.

7 Conclusion

A nominal fuzzy controller was designed to perform a certain satellite maneuver. A constrained multi-objective optimization method is developed and used to optimize distribution of MFs in order to minimize the time taken to perform the maneuver and to reduce long term vibrations induced in the structure.

Furthermore an innovative application of the optimization and the optimized controller in the field of rover goal-seeking is introduced. Rover goal-seeking problem, and use of the optimization methodology are explained. A number of important, such as the rover's remaining battery power and proximity of re-charging stations, are also introduced.

8 Acknowledgment

The author would like to thank Professor Bill Martin and Associate Professor Kevin Adams for their support of this and other publications by the author. The support has been whole-heartedly and ongoing. It included, but certainly not limited, to financial support.

References

[1] Riseborough, P.: Dynamics and Control of Spacecraft With Solar Array Joint Nonlinearities. Ph.D. Thesis, RMIT, Melbourne, Australia (1993)
[2] Hughes, P.C.: Attitude dynamics of a three-axis stabilized satellite with a large flexible solar array. Journal of the Astronautical Sciences (1972)
[3] Hughes, P.C.: Flexibility considerations for the pitch attitude control of the communications technology satellite. C.A.S.I. Transactions **5** (1972) 1–4

[4] Dwyer, T.A.W.: Automatic decoupling of flexible spacecraft slewing manoeuvres. In: American Control Conference. (1986) 1529–1534

[5] Dwyer, T.A.W.: Exact nonlinear control of spacecraft slewing manoeuvres with internal momentum transfer. Journal of Guidance and Control 9 (1986) 240–246

[6] Wood, T.D.: Design concepts for the soho spacecraft allowing for structural flexibility. Technical report, British Aerospace (Space Systems) (1990)

[7] Scott, C.N., Wood, L.A.: Optimal robust tracking subject to disturbances, noise, plant uncertainty and performance constraints. Journal of Guidance, Control, and Dynamics (1998)

[8] Zadeh, H.S., Scott, C., Wood, L.A.: Control law comparison for attitude control of a spacecraft. In Zayegh, D.A., ed.: Proceedings of the Third International Conference on Modelling and Simulation. Volume 1., Victoria University of Technology (1997) 200–205

[9] Zadeh, H.S., Wood, L.A.: Optimal active vibraion control of a highly flexible structure. In: Second Australasian Congress on Applied Mechanics. (1999)

[10] Chen, Y.Y., Yen, C.C.: PD-type vs. PID-type fuzzy controllers. In: Proceedings of TENCON'92 IEEE Region 10 Conference. Volume 1., Melbourne, Australia (1992) 341–345

[11] Wang, H.B., Mo, J.P.T., Chen, N.: Fuzzy position control of pneumatic cylinder with two state solenoid valves. In: Control 95 Preprints. Volume 2., Melbourne, Australia (1995) 387–391

[12] MathWorks Inc.: Fuzzy Logic Toolbox for use with Matlab. The MathWorks Inc. (1994)

[13] Drack, L., Zadeh, H.S., Wharington, J.M., Herszberg, I., Wood, L.A.: Optimal design using simulated annealing in Matlab. In: Second Australasian Congress on Applied Mechanics. (1999)

[14] Zadeh, H.S., Wood, L.A.: Membership function optimisation of a fuzzy logic controller for a non-linear system. In: The International Conference on Intelligent Technologies, Bangkok, Thailand (2000)

[15] Ingber, A.: Adaptive Simulated Annealing (ASA) Global optimization C-code. Lester Ingber Research, Chicago IL, "http://www.ingber.com/#ASA-CODE" (1993)

[16] Zadeh, H.S.: Maneuver simulation of a non-linear system using membership function optimization of a fuzzy logic controller. In: 2002 IEEE Aerospace Conference, Big Sky, Montana, USA (2002)

[17] Singh, S., et al.: Recent progress in local and global traversability for planetary rovers. In: IEEE International Conference on Robotics and Automation. Volume 2., San Francisco (2000) 1194–1200

[18] Kelly, A., Stentz, A.: Rough terrain autonomous mobility. Journal of Autonomous Robots 5 (1998)

[19] Seraji, H., Howard, A., Tunstel, E.: Safe navigation on hazardous terrain. In: Proceedings of IEEE International Conference on Robotics and Automation. Volume 3., Seoul (Korea) (2001) 3084–3091

[20] Howard, A., Seraji, H.: Vision-based terrain characterization and traversability assessment. Journal of Robotic Systems 18 (2001)

[21] Seraji, H., Howard, A., Tunstel, E.: Terrain-based navigation of mobile robots: A fuzzy logic approach. In: Proceedings of International Symposium on AI, Robotics, and Automation in Space, Montreal (Canada) (2001)

[22] Seraji, H.: Traversability index: A new concept for planetary rovers. In: Proceedings of IEEE International Conference on Robotics and Automation. Volume 3., Detroit (1999) 2006–2013

[23] Seraji, H.: Fuzzy traversability index: A new concept for terrain-based navigation. Journal of Robotic Systems **17** (2000) 75–91
[24] Seraji, H.: Traversability indices for multi-scale terrain assessment. In: Proceedings of International Symposium on Intelligent Technologies, Bangkok, Thailand (2001)
[25] Seraji, H.: Terrain-based robot navigation using multi-scale traversability indices. In: Proceedings of IEEE Fuzzy Systems Conference, Melbourne, Australia (2001)

System Modelling Using Fuzzy Numbers

Petr Hušek, Renata Pytelková

Department of Control Engineering
Faculty of Electrical Engineering
Czech Technical University in Prague
Technická 2
166 27 Prague 6, Czech Republic
husek,dvorako@fel.cvut.cz

Abstract. In this paper the problem of computing stability margin of uncertain continuous-time linear time invariant systems with coefficients described by fuzzy numbers is solved. All the coefficients are considered as variable intervals parametrized by the degree of confidence in the corresponding model. The lower the confidence level is the larger interval the coefficient can vary in. In the paper the common confidence level for all the coefficient is considered. Two cases are considered: linear and polynomic dependency of coefficients of characteristic polynomial on system parameters. The results are derived based on parameterization of lower and upper bounds of the intervals and using the Kharitonov's and the Hurwitz criterion.

1 Introduction

When creating a mathematical model of a dynamic system the uncertainty is always presented. It is never possible to characterize and describe all the phenomena influencing the system. Either they are too complicated or unknown. Moreover, we cannot determine the parameter values of a system exactly, e.g. the parameters are not admissible to be measured or the values are changing in time. Instead of exact measurements we often use statistic or heuristic knowledge given by an expert. From the latter it seems to be natural that we describe parameters of a system as intervals with different degree of confidence that the parameters lie in those intervals.

If the confidence in validity of a parameter is low the length of the interval the true value of the parameter can lie is bigger. If the confidence is high the considered length of the corresponding interval is smaller. In [3] description of such type of uncertainty was suggested. It is based on theory of fuzzy numbers pioneered by Klir [8]. It represents a different view on model parametric uncertainty in comparison with classical parametric robust control theory where the coefficients vary in intervals of constant length ([2]). More precisely, the uncertain parameter q is represented by a fuzzy number, $\tilde{q}$, with membership function $\alpha = \mu_{\tilde{q}}(q) \in [0,1]$. The maximum confidence level $\alpha = 1$ corresponds to precise knowledge of the parameter q

T. Bilgiç et al. (Eds.): IFSA 2003, LNAI 2715, pp. 492–499, 2003.

($q = \ker(\widetilde{q})$), the minimum confidence level $\alpha = 0$ represents maximum uncertainty ($q \in \mathrm{supp}(\widetilde{q})$).

The controller designed for an uncertain system is usually not required to behave optimally for all possible parameter variations. It suffices if the system behaves optimally in one operating point for nominal values of the parameters and preserves stability under perturbations considered in uncertain system. Taking into account the uncertainty the important question is, how far in some sense we can go from the nominal system preserving stability. Naturally, it is supposed that the nominal system is stable. Such a problem is referred in the literature as a stability margin determination. Solving this problem is generally very complicated task. Only for linear parameter dependency there exist efficient and numerically reliable algorithms ([2]). In cases where the coefficients are multilinear or even polynomical functions of the uncertain parameters the solution can be obtained using time consuming numerical algorithms ([6]).

In this paper we are focusing on the finding a minimum confidence level which is desired to preserve stability of a linear systems with coefficients described by fuzzy numbers. The common confidence level for all the coefficients is supposed. It means that all the coefficients c_i are characterized by means of fuzzy numbers with membership functions $\alpha = \mu_{\widetilde{c}_i}(c_i)$. Naturally at first the confidence level α has to be specified and then the coefficient interval is determined by the α-cut $[c_i]_\alpha$. If $\alpha=1$ (the maximum confidence level - the system works in normal operating conditions) all the coefficients c_i take any value (crisp or interval) within the cores of $\widetilde{c}_i$'s $c_i = \ker(\widetilde{c}_i)$). If $\alpha=0$ (the minimum confidence level) all the coefficients c_i are intervals equal to the supports of $\widetilde{c}_i$'s $c_i \in \mathrm{supp}(\widetilde{c}_i)$.

2 Fuzzy Linear Systems

In the previous section a generalization of systems with parametric uncertainty was given. Systems which coefficients depend on parameters described by fuzzy numbers are called *extended* or *fuzzy linear systems* ([3]).

Generally, an extended continuous-time system with single input and single output is described by the differential equation

$$f(y(t),\dots,y^{(n)}(t),u(t),\dots,u^{(m)}(t),\widetilde{\mathbf{q}}) = 0 \tag{1}$$

where $\widetilde{\mathbf{q}} = (\widetilde{q}_1,\dots,\widetilde{q}_r), q_i \in \widetilde{P}(\Re), i = 1,\dots,r$ is a vector of fuzzy numbers and $\widetilde{P}(\Re)$ denotes the set of all possible fuzzy sets with real universe of discourse.

In the linear case, the extended system is described by linear differential equation

$$y^{(n)}(t) + a_{n-1}(\widetilde{\mathbf{q}})y^{(n-1)}(t) + \dots + a_0(\widetilde{\mathbf{q}})y(t) = b_m(\widetilde{\mathbf{q}})u^{(m)}(t) + \dots + b_0(\widetilde{\mathbf{q}})u(t) \tag{2}$$

where $a_i(\cdot), b_j(\cdot)$ are fuzzy numbers.

The transfer function of the extended system (2) can be defined as

$$\widetilde{G}(s,\widetilde{\mathbf{q}}) = \frac{Y(s)}{U(s)} = \frac{b_m(\widetilde{\mathbf{q}})s^m + b_{m-1}(\widetilde{\mathbf{q}})s^{m-1} + \cdots + b_0(\widetilde{\mathbf{q}})}{s^n + a_{n-1}(\widetilde{\mathbf{q}})s^{n-1} + \cdots + a_0(\widetilde{\mathbf{q}})}. \tag{3}$$

If a degree of confidence in the coefficients is given, the extended system (3) can be represented as an interval system described by the transfer function

$$[\widetilde{G}(s,\widetilde{\mathbf{q}})]_\alpha = \frac{b_m(\mathbf{q}_\alpha)s^m + b_{m-1}(\mathbf{q}_\alpha)s^{m-1} + \cdots + b_0(\mathbf{q}_\alpha)}{s^n + a_{n-1}(\mathbf{q}_\alpha)s^{n-1} + \cdots + a_0(\mathbf{q}_\alpha)}. \tag{4}$$

where $\mathbf{q}_\alpha = (q_{1\alpha},\ldots,q_{r\alpha}), q_{i\alpha} \subset [q_{i\alpha}^-, q_{i\alpha}^+], i = 1,\ldots,r$ is a vector of intervals corresponding to the α-cuts of the parameters $\mathbf{q}_\alpha = [\widetilde{\mathbf{q}}]_\alpha$.

In order to deal with the problems related to extended systems and to apply the robust control and analysis techniques it is useful to describe the parameters expressed by the α-cuts as intervals with varying endpoints. The endpoints are functions of the confidence level α. If convex membership functions are used then we can always find nondecreasing and nonincreasing functions $q_i^-(\cdot), q_i^+(\cdot)$, respectively, such that $[\widetilde{q}_i]_\alpha = [q_i^-(\alpha_i), q_i^+(\alpha_i)] \approx q_i(\alpha_i)$, satisfying $q_i^-(0) = q_i^-$, $q_i^+(0) = q_i^+$ and $q_i^-(1) = q_i^+(1) = q_i^0$ which is the nominal value (or one of those) of the parameter q_i.

For example, triangular membership function with characteristic points $(1,4,5)$ can be represented by means of functions of endpoints as $[3\alpha+1, -\alpha+5]$, $\alpha \in [0,1]$. The framework for computation with such type of coefficient was found in fuzzy mathematics [8].

3 Problem Formulation

Assume that the characteristic polynomial of the extended system (3) using the parameterization of the endpoints described above is obtained as

$$p(s,\alpha) = a_n(\alpha)s^n + a_{n-1}(\alpha)s^{n-1} + \cdots + a_0(\alpha). \tag{5}$$

where $a_j(\alpha)$, $j = 0,\ldots,n$ are varying intervals and $\alpha = (\alpha_0,\ldots,\alpha_n)$ is a vector of degrees of confidence. Such a structure is obtained even if multilinear or polynomic dependency of the coefficients on system parameters is considered.

Definition 1. *The system with characteristic polynomial (5) is said to have a stability margin $\overline{\alpha} = (\overline{\alpha}_0,\ldots,\overline{\alpha}_n)$ if for $\alpha_0 = \overline{\alpha}_0,\ldots,\alpha_n = \overline{\alpha}_n$ the extended system is on the limit of stability.*

In this paper two cases will considered. Firstly the extended system with linear dependency of coefficients of characteristic polynomials on system parameters is studied. In such a case a method based on zero exclusion theorem is used. Then a more general case of polynomic dependency of system parameters is solved. Here the

Hurwitz stability criterion was used. Common confidence $\alpha \in \Re$ for all the coefficients $a_j(\alpha), j = 0, \ldots, n$ is assumed.

4 Linear Parameter Dependency

Bondia and Picó ([3], [4]) used the Argoun stability test [1] to solve the problem. This test corresponds to a graphical representation of Kharitonov's theorem [9] in frequency domain. The significant drawback of the method is that for polynomials of higher order with coefficients depending on a parameter the computation of its roots has to be performed using iterative numerical algorithm. In this paper an algorithm based on zero exclusion theorem is introduced.

Firstly let us recall the Kharitonov's theorem [9]. Consider the set $P(s)$ of real polynomials of degree n composed of the members

$$\delta(s) = \delta_0 + \delta_1 s + \delta_2 s^2 + \delta_3 s^3 + \delta_4 s^4 + \cdots + \delta_n s^n \qquad (6)$$

where the coefficients lie within given ranges,

$$\delta_0 \in [x_0, y_0], \delta_1 \in [x_1, y_1], \cdots, \delta_n \in [x_n, y_n]. \qquad (7)$$

Assume that the degree remains invariant over the set, i.e. $0 \notin [x_n, y_n]$. Such a set of polynomials is called a real *interval* family and we loosely refer to it as an interval polynomial. Kharitonov's theorem provides a surprisingly simple necessary and sufficient condition for the Hurwitz stability of the entire family.

Theorem 1 (Kharitonov's theorem [9]). *Every polynomial in the family P(s) is Hurwitz if and only if the following four extreme polynomials are Hurwitz:*

$$
\begin{aligned}
K^1(s) &= x_0 + x_1 s + y_2 s^2 + y_3 s^3 + x_4 s^4 + x_4 s^4 + y_3 s^3 + \cdots \\
K^2(s) &= x_0 + y_1 s + y_2 s^2 + x_3 s^3 + x_4 s^4 + y_4 s^4 + y_3 s^3 + \cdots \\
K^3(s) &= y_0 + x_1 s + x_2 s^2 + y_3 s^3 + y_4 s^4 + x_4 s^4 + x_3 s^3 + \cdots \\
K^4(s) &= y_0 + y_1 s + x_2 s^2 + x_3 s^3 + y_4 s^4 + y_4 s^4 + x_3 s^3 + \cdots
\end{aligned}
\qquad (8) \blacklozenge
$$

Now we apply the Kharitonov's theorem on the polynomial (5) expressed as

$$p(s, \alpha) = [a_n^-(\alpha), a_n^+(\alpha)]s^n + [a_{n-1}^-(\alpha), a_{n-1}^+(\alpha)]s^{n-1} + \cdots + [a_0^-(\alpha), a_0^+(\alpha)]. \qquad (9)$$

Associate with the polynomial (9) the four Kharitonov's polynomials

$$
\begin{aligned}
K^1(s, \alpha) &= a_0^-(\alpha) + a_1^-(\alpha)s + a_2^+(\alpha)s^2 + a_3^+(\alpha)s^2 + a_4^-(\alpha)s^2 + a_5^-(\alpha)s^2 + \cdots \\
K^2(s, \alpha) &= a_0^-(\alpha) + a_1^+(\alpha)s + a_2^+(\alpha)s^2 + a_3^-(\alpha)s^2 + a_4^-(\alpha)s^2 + a_5^+(\alpha)s^2 + \cdots \\
K^3(s, \alpha) &= a_0^+(\alpha) + a_1^-(\alpha)s + a_2^-(\alpha)s^2 + a_3^+(\alpha)s^2 + a_4^+(\alpha)s^2 + a_5^-(\alpha)s^2 + \cdots \\
K^4(s, \alpha) &= a_0^+(\alpha) + a_1^+(\alpha)s + a_2^-(\alpha)s^2 + a_3^-(\alpha)s^2 + a_4^+(\alpha)s^2 + a_5^+(\alpha)s^2 + \cdots
\end{aligned}
\qquad (10)
$$

Theorem 2 (Stability margin of extended polynomial). *Suppose that the four Kharitonov's polynomials (10) are stable for* $\alpha = 1$. *Denote by* $\overline{\alpha}^k, k = 1,2,3,4$ *the stability margin of the polynomial* $K^k(s,\alpha), k = 1,2,3,4$, *i.e.*

$$\overline{\alpha}^k = \sup\{\alpha \in \Re, \alpha \le 1 : K^k(s,\alpha) \text{ is not Hurwitz}\}, k = 1,2,3,4.. \tag{11}$$

Then the stability margin $\overline{\alpha}$ *of the extended polynomial (9) can be determined as*

$$\overline{\alpha}^k = \max\{\overline{\alpha}^k, k = 1,2,3,4\}.. \tag{12} \blacklozenge$$

The question is how to determine the particular stability margins $\overline{\alpha}^k, k = 1,2,3,4$, i.e., stability margins of uncertain polynomials with linear dependency of their coefficients on confidence level α.

There are two main approaches dealing with parametric robust stability problems, called algebraic and geometric. The former is based on arithmetic manipulations with coefficients of an uncertain polynomial using Routh or Hurwitz stability test. The latter examines the image set of a polynomial evaluated in the points lying on the boundary of stability region and using zero exclusion principle and it will be used in this section.

Suppose that $\delta(s, \mathbf{p})$ denotes a polynomial whose coefficients depend continuously on the parameter vector $\mathbf{p} \in \Re^l$ which varies in a set $\Omega \subset \Re^l$ and thus generates the family of polynomials $\mathbf{p} \in \Re^l$

$$\Delta(s) := \{\delta(s, \mathbf{p}) : \mathbf{p} \in \Omega\}. \tag{13}$$

Let S denotes an arbitrary stability region and ∂S its boundary.

Theorem 3 (Zero exclusion principle [3]). *Assume that the family of polynomials (13) is of constant degree, contains at least one stable polynomial, and* Ω *is pathwise connected. Then the entire family is stable if and only if*

$$0 \notin \Delta(s^*) \text{ for all } s^* \in \partial S. \tag{14} \blacklozenge$$

The image set $\Delta(s^*)$ is called the value set. The consequence of the zero exclusion principle is that a polynomial $\delta^*(s,\mathbf{p}) \in \Delta(s,\mathbf{p})$ satisfying $\delta^*(s^*,\mathbf{p}^*) = 0$ for some $s^* \in \partial S$ and $\mathbf{p}^* \in \Omega$ is on the limit of stability. In other words, if some $s^* \in \partial S$ is the root of polynomial $\delta^*(s,\mathbf{p}) \in \Delta(s,\mathbf{p})$ for some $\mathbf{p} = \mathbf{p}^*$ then this polynomial is on the limit of stability and $\delta^*(s,\mathbf{p}) = 0$ for some $\mathbf{p} = \mathbf{p}^*$. The value of parameter $\mathbf{p}^*$ is a possible candidate for determination of stability margin. In multidimensional case finding all the solutions of the equation $\delta^*(s,\mathbf{p}) = 0$ is very complicated task. However, if the parameter $\mathbf{p}$ is scalar the task becomes much easier. More specifically, consider a polynomial $\delta(s,\mathbf{p}) = 0$ whose coefficients depend linearly on the vector of uncertain parameters $\mathbf{p} = [p_1, p_2, ..., p_l]$. In such cases we may write without loss of generality that

$$\delta(s,\mathbf{p}) = a_1(s)p_1 + \cdots + a_l(s)p_l + b(s) .$$ (15)◆

where $a_i(s), i = 1,\ldots,l$ and $b(s)$ are real polynomials and the parameters p_i are real. For scalar parameter case we can simplify to

$$\delta(s,p) = a_1(s)p_1 + b(s) .$$ (16)

Let s^* denote a point on the stability boundary ∂S. For $s^* \in \partial S$ be a root of $\delta(s,p)$ we must have

$$\delta(s^*,p) = 0 .$$ (17)

The equation (17) is generally algebraic and its solving has to be performed using complex numerical algorithms. However, the equation (17) can be solved for particular $s^* \in \partial S$. Two cases may occur depending on whether s^* is real or complex. Suppose that $s^* = s_r$ where s_r is real. We obtain the single equation

$$a_1(s_r)p = -b(s_r) .$$ (18)

If $s^* = s_c$ where s_c is complex, we will use the following notation:

$$a_1(s_c)p = a_{1r}(s_c)p + j a_{1i}(s_c) \\ b(s_c) = b_r(s_c)p + j b_i(s_c)$$. (19)

The equation (17) then yields

$$\begin{bmatrix} a_{1r}(s_c) \\ a_{1i}(s_c) \end{bmatrix} p = \begin{bmatrix} -b_{1r}(s_c) \\ -b_{1i}(s_c) \end{bmatrix} .$$ (20)

Now we face the problem of determination of the solutions of (18) and (21) that can be generally written as

$$A(s^*)p = b(s^*) .$$ (21)

For each $s^* \in \partial S$ the solution $p = p^*$ of (21) minimizing $\|Ap^* - b\|$ in the sense of least squares, where $\| \; \|$ denotes the Euclidean norm, is of interest. Such a solution is obtained using matrix pseudoinverse as

$$p^* = (A^T A)^{-1} A^T b .$$ (22)

Denote by $P^* = \{p_m^*\}$ the set of all the solutions of (22) satisfying $\|Ap_m^* - b\| < \varepsilon$ where ε is chosen accuracy. Then the stability margin $\overline{\alpha}^\delta$ of polynomial (16) equals

$$\overline{\alpha}^\delta = \max\{p_m^* : p_m^* \le 1\}.$$ (23)

The procedure described above is applied to the stability margins of four Kharitonov's polynomials (10).

5 Polynomic Parameter Dependency

In this case the algebraic approach and Hurwitz stability criterion will be used to find the solution of (11).

Theorem 4 (Hurwitz stability criterion [3]). *Let*

$$\delta(s, p) = a_n(p)s^n + a_{n-1}(p)s^{n-1} + \cdots + a_0(p).$$ (24)

be an uncertain polynomial with coefficients $a_i(p), i = 0,\ldots,n$ *being continuous functions defined for* $p \in Q$, *Q is a real interval. Denote by P family of polynomials* $\delta(s, p), p \in Q$.

Then the family of polynomials (24) is asymptotically stable if and only if
1. there exists a stable polynomial $\delta(s, p) \in P$,
2. $\det \mathbf{H}_n(p) \neq 0$ *for all* $p \in Q$.

where $\mathbf{H}_n(p)$ *(n×n) is the Hurwitz matrix of n-th order associated with* $\delta(s,p)$. ♦

The Hurwitz criterion can be modified for computation of stability margin of uncertain polynomial.

Theorem 5 (Stability margin using Hurwitz stability criterion). *Let us suppose that the polynomial (5) is stable for* $\alpha = 1$, *i.e. the polynomials* $K^k(s,1), k = 1,2,3,4$ *are all stable. Let* $\mathbf{H}_n^k(\alpha)$ *be the n-th order Hurwitz matrix associated with the polynomial* $K^k(s,\alpha), k = 1,2,3,4$. *Then the stability margins* $\overline{\alpha}^k, k = 1,2,3,4$ *can be determined as*

$$\overline{\alpha}^k = \max\{\alpha \in \Re : \alpha \leq 1, \det \mathbf{H}_n^k(\alpha) = 0\}, k = 1,2,3,4.$$ (25)♦

In order to evaluate the determinants $\det \mathbf{H}_n^k(\alpha), k = 1,2,3,4$ very efficient and numerically reliable method based on interpolation-evaluation algorithms using fast Fourier transform algorithm [5], [7]. For determination of real roots of a polynomial lying in the interval [0,1] an efficient algorithm based on Newton-Raphson iterative procedure was developed.

6 Conclusion

In the paper two algorithms for determination of confidence level needed to preserve Hurwitz stability of an extended systems were presented. The common confidence level α for all the uncertain parameters was considered. To derive the result the representation of varying intervals as fuzzy numbers was used. Linear and polynomic parameter dependencies were taken into account. The former is solved using value set concept, the latter is solved via algebraic approach. Very efficient and numerically reliable algorithms were derived and used for computation the determinant of polynomial matrix and determination of real roots of a polynomial. The algorithms are based on FFT algorithm and Newton-Raphson iterative procedure respectively.

Acknowledgements

This work has been supported by the research program No. J04/98:212300013 "Decision Making and Control for Manufacturing" of the Czech Technical University in Prague (sponsored by the Ministry of Education of the Czech Republic) and the project GACR 102/01/1347 (sponsored by the Grant Agency of the Czech Republic).

References

1. Argoun, M.B.: Frequency Domain Conditions for the Stability of Perturbed Polynomials. IEEE Transactions on Automatic Control (1987) No.10, pp.913–916.
2. Bhattacharyya, S.P., Chapellat, H., Keel, L.H.: Robust Control: The Parametric Approach. Prentice-Hall, Inc. (1995).
3. Bondia, J., Picó, J.: Analysis of Systems with Variable Parametric Uncertainty Using Fuzzy Functions. In: Proceedings of European Control Conference ECC'99, Karlsruhe, Germany (1999).
4. Bondia, J., Picó, J.: Application of Functional Intervals to the Stability Analysis of Fuzzy Linear Systems. In: Proceedings of the 9th IFSA World Congress and 20th NAFIPS International Conference, Vancouver, Canada (2001) pp.1337–1342.
5. Dvoráková, R., Hušek, P.: Comments on computing extreme values in "Stability Issues on Takagi-Sugeno Fuzzy Model – Parametric Approach". IEEE Transactions on Fuzzy Systems (2001) Vol. 9, No. 1, pp.221–222.
6. Gaston de, R.R., Safonov, M.G.: Exact Calculation of the Multiloop Stability Margin. IEEE Transactions on Automatic Control (1988) Vol. 33, pp.156–171.
7. Hušek, P., Pytelková, R., Elizondo, C.: Robust Stability of Polynomials with Polynomic Structure of Coefficients. In: Proc. of 5th World Multiconference on Systemics, Cybernetics and Informatics (SCI 2001), Orlando, Florida (2001).
8. Klir, G.J., Folger T.: Fuzzy Sets, Uncertainty and Information. Prentice Hall, Englewood Cliffs (1988).
9. Харитонов, В.Л.: Об Асимптотической Устойчивости Положения Равновесия Семейства Систем Линейных Дифференциальных Уравнений. Дифференциальные Уравнения (1978) ТОМ 14, Hо. 11, с.2086–2088.

Fuzzy Adaptive Sliding Mode Control for a Class of Uncertain Nonlinear MIMO Systems with Application to a 2DOF Twin Propeller

Aria Alasti[1], Hamid Bolandhemat[2] and Navid Dadkhah Tehrani[3]

[1] Department of Mechanical Engineering
Sharif University of Technology
P.O. Box: 11365-9363, Azadi Ave., Tehran, Iran
aalasti@sharif.edu
[2] Department of Aerospace Engineering
Sharif University of Technology
bolandhemat@mehr.sharif.edu
[3] Department of Aerospace Engineering
Sharif University of Technology
n.dadkhah@mehr.sharif.edu

Abstract. A practical design method is presented which used the fuzzy logic advantages in adaptation of sliding mode control. The combined Fuzzy Adaptive Sliding Control (FASC) is designed in such a way to enhance satisfactory sliding performance and robustness with good level of chattering alleviation. The design approach is valid for a class of nonlinear uncertain MIMO systems. This control algorithm does not require the system model. A supervisory term is appended to the controller to assure the stability of fuzzy sliding mode control through Lyapunov theory. The design approach has been applied to a 2DOF twin propeler system with large uncertainty. Simulation results verified effectiveness of proposed method.

1 Introduction

Conventional controllers in tracking problems are well suited only when the control effort can be generated based on a relatively accurate analytical model [1]. In the real world, in the presence of uncertainties, undeterministic phenomenon and unpredictable disturbances, an accurate mathematical model is not available or difficult to formulate, hence, many of these control techniques fail to track the desired motion satisfactorily. Conventional sliding mode control (SMC), based on the theory of variable structure systems (VSS), has been widely applied in tracking problem of uncertain systems [2-4]. In a VSC system, a dicontinuos control law is designed to derive the system states toward a specific sliding surface. As the sliding surface is hit, the nth order tracking problem converts to a first order stabilization problem. In sliding mode , the system response is governed by the surface equation and therefore the robustness to uncertainties and disturbances is achieved [2],[4]. However, SMC suffers from some important drawbacks. First in SMC design procedure the upper and

T. Bilgiç et al. (Eds.): IFSA 2003, LNAI 2715, pp. 500-507, 2003.

lower bound of the uncertain parameters must be estimated. In most cases, such estimation is very difficult and furthermore, it may be usually done conservatively and thus implementation cost becomes too large. Secondly, undesired phenomenon of chattering due to high frequency switching may be occured [1-3]. Note that in practical implementations, chattering may damage system components such as actuators. Recently, combinations of fuzzy logic theory and sliding mode control method have achieved superior performance. For instance in [5-8], the fuzzy adaptive sliding mode controllers have been developed. However there is a main problem existing in their design. Despite of robustness and stability of controllers, the proposed structure of the fuzzy system leads to utilize control energy inefficiently. In this paper we study the output-tracking problem of multi-input multi-output nonlinear systems with uncertainties when the upper bounds on the norm of uncertainties may not be easily known. The proposed control law contains a fuzzy controller and a supervisory control term. The proposed algorithm is associated with effective adaptive laws, which update the controller parameters in an online manner. Differing from the other approaches, the proposed method dosn't need to approximate the unknown system dynamic and only tends to mimic the logic of sliding mode control action . Based on this fact the control rules and membership functions used in this method are very simple, which are chosen such that the proposed fuzzy sliding controller utilizes energy more efficiently and alleviate chattering phenomenon properly.

2 Problem Description

Consider an MIMO nonlinear system governed by [2], [7]:
$$\mathbf{H}(\xi)\ddot{\xi} + \mathbf{C}(\xi,\dot{\xi})\dot{\xi} + \mathbf{F}(g,\xi,\dot{\xi}) + \mathbf{T}_d = u \qquad (1)$$
Where

1) $\xi \in \mathbf{R}^n$ is the coordinate vector,

2) $\mathbf{H}(\xi) \in \mathbf{R}^{n \times n}$ is the inertia matrix,

3) $\mathbf{T}_d \in \mathbf{R}^n$ is the vector of generalized disturbances or unmodeled dynamics,

4) $u \in \mathbf{R}^n$ is the scaled applied torques or forces,

5) $\mathbf{F}(g,\xi,\dot{\xi}) \in \mathbf{R}^n$ is the vector of gravitational and frictional effects.

Equation (1) may be considered as the equation of motion of a rigid satellite, if ξ is chosen as the Gibbs vector. It may also represent a general n-link robot arm, if ξ be the joint angular position vector of the robot [2],[7],[8]. In our case study this equation expresses the behaviour of a twin propeller system where, the vector ξ consists the angle of attack (α) and sideslip angle (β). The controller design problem is defined as follows: given the desired trajectories $\xi_d, \dot{\xi}_d, \ddot{\xi}_d$, with some or all system parameters being unknown, derive a control law $u(t)$ such that the position vector ξ and the velocity vector $\dot{\xi}$ track the desired trajectories effectively.

3 SISO Fuzzy Sliding Mode Controller (FSMC)

Consider $s(\xi, \dot{\xi}) = 0$ represents the sliding hyperplane in state space. The proposed SISO fuzzy sliding mode controller, uses s and its discrete change Δs as inputs. To employ the same membership functions on both inputs, s and Δs are normalized. The selected membership functions on the inputs are "Negative", "Zero", "Positive" for s and "Negative" and "Positive" for Δs. Where N and P are s-curves membership functions and Z is Guassian membership function with zero center. Note that parameters of above s-curves and standard deviation of the Gaussian membership function are used as tuning parameters. Based on the concept of SMC method, the fuzzy sliding mode control rules are designed as:

(R1) IF s is "P" AND Δs is "P" THEN u is "Negative Big (NB)".
(R2) IF s is "P" AND Δs is "N" THEN u is "Negative Medium (NM)".
(R3) IF s is "Z" AND Δs is "P" THEN u is "Negative Small (NS)".
(R4) IF s is "Z" AND Δs is "N" THEN u is "Positive Small (PS)".
(R5) IF s is "N" AND Δs is "P" THEN u is "Positive Medium (PM)".
(R6) IF s is "N" AND Δs is "N" THEN u is "Positive Big (PB)".

For more lightening the above rules, they can be explained as follows: for rule R1, condition "s is positive" and "Δs is positive" implies that the system states are above the sliding surface and moves away from it, hence the control action should be large enough to take the system states downward onto the sliding surface. In the case of rule R2, again, the system states are above the sliding surface, but the rate of change of s is negative, and hence the system states are moving toward the sliding surface. Therefore the control action should be sufficient to this motion to be continued. And in the case of rule R3 only a small control effort must be used to maintain the system states on the sliding surface and not too big to cause chattering. Rules R4, R5 and R6 can be similarly described. By employing product inference engine (PIE), singleton fuzzifier (SF) and center average defuzzifier (CAD), output of the SISO fuzzy sliding mode controller will be:

$$u_{\text{fuzz-sli}} = \frac{\sum_{j=1}^{6} \mu^j \theta^j}{\sum_{j=1}^{6} \mu^j} = W^T \theta \qquad (2)$$

Where $\theta = [\theta^1, \theta^2, \theta^3, \theta^4, \theta^5, \theta^6]^T$ is the consequent vector, in which θ^j is the output of the j th rule, μ^j is the membership value of the j th rule and also:

$$W = [W^1, W^2, W^3, W^4, W^5, W^6]^T \text{ with } W^j = \frac{\mu^j}{\sum_{k=1}^{6} \mu^k}; \ j = 1,...,6 \qquad (3)$$

4 MIMO Fuzzy Sliding Mode Controller

In MIMO nonlinear systems to obtain a good performance, the coupling effect must be considered. Hence, a parallel connection of the proposed fuzzy SISO controller [7], [10] is constructed such that the control effort in each channel is generated not only by employing the related state condition but also by using the other states. Therefore the total control law is:

$$u = u_{fuzz-sli} + \tau_s \qquad (4)$$

With

$$u_{fuzz-sli} = [u_{fs1}, ..., u_{fsn}]^T$$

$$u_{fsi} = \sum_{j=1}^{n} u_{ij} = \sum_{j=1}^{n} W_{ij}^{T}\theta \qquad (5)$$

Where u_{ij} is the output of ith SISO fuzzy sliding mode controller excited with the jth state. The additional control term τ_s is called the supervisory control, which will be designed, in next section to satisfy the sliding condition and guarantee the stability of the proposed controller. The next task is to choose an appropriate Lyapunov function and design adaptation laws to adjust the fuzzy network parameters properly.

5 Learning Algorithms and Stability Analysis

Let adjustable parameters of the fuzzy controller be $\Theta = [\sigma_{ij}, t_{ij}, z_{ij}, \theta_{ij}]$, where σ_{ij}'s are Gaussian membership function standard deviations, t_{ij} and z_{ij} are parameters of s-curve membership functions defined on s and Δs respectively, and θ_{ij}'s are consequent vector of the fuzzy subsystems. Now consider the Lyapunov function candidate

$$V = \frac{1}{2} s^T \mathbf{H} s + \frac{1}{2}(\sum_{i=1}^{n} \sum_{j=1}^{n} \rho_{ij}^{-1}\tilde{\sigma}_{ij}^2 + \sum_{i=1}^{n} \sum_{j=1}^{n} \beta_{ij}^{-1}\tilde{t}_{ij}^2 + \sum_{i=1}^{n} \sum_{j=1}^{n} \alpha_{ij}^{-1}\tilde{z}_{ij}^2 +$$

$$\sum_{i=1}^{n} \sum_{j=1}^{n} \gamma_{ij}^{-1}\tilde{\theta}_{ij}^T\tilde{\theta}_{ij}) \qquad (6)$$

Where $\rho_{ij}, \beta_{ij}, \alpha_{ij}, \gamma_{ij}$ are positive constant weights and $\tilde{\sigma}_{ij} = \sigma_{ij}^* - \sigma_{ij}$, $\tilde{t}_{ij} = t_{ij}^* - t_{ij}$, $\tilde{z}_{ij} = z_{ij}^* - z_{ij}$ and $\tilde{\theta}_{ij} = \theta_{ij}^* - \theta_{ij}$ denote the parameter estimation errors. $\Theta^* = [\sigma_{ij}^*, t_{ij}^*, z_{ij}^*, \theta_{ij}^*]$ are the optimal parameters of the fuzzy controller. It can be shown that by using the following adaptation laws with proper parameter projection algorithm [9]:

$$\dot{\sigma}_{ij} = -\rho_{ij} \frac{\partial u_{ij}}{\partial \sigma_{ij}} s_i, \quad \dot{t}_{ij} = -\beta_{ij} \frac{\partial u_{ij}}{\partial t_{ij}} s_i, \quad \dot{z}_{ij} = -\alpha_{ij} \frac{\partial u_{ij}}{\partial \alpha_{ij}} s_i$$

$$\dot{\theta}_{ij} = -\gamma_{ij} \frac{\partial u_{ij}}{\partial \theta_{ij}} s_i = -\gamma_{ij} W_{ij} s_i \qquad (7)$$

$\dot{V}$ will be reduced to [11]:

$$\dot{V} \le s^T \tau_s + s^T v \qquad (8)$$
$$v = \omega - \text{H.O.T} \qquad (9)$$

Where ω is the difference between the proposed fuzzy sliding mode control law when the controller parameters are the optimal ones and a conventional MIMO sliding mode control action in a similar tracking problem, that is:

$$\omega = u_{\text{fuzz-sli}}(s, \Delta s, \Theta^*) - u_{\text{sli}} \qquad (10)$$

And also H.O.T is the higher order terms generated by Taylor series expansion of the MIMO fuzzy sliding mode control output about the controller adjustable parameters Θ. Now in order to complete the fuzzy sliding mode controller design it is necessary to define the supervisory control term τ_S properly to guarantee the negative definitness of $\dot{V}$. Using the power rate reaching law (Boundary layer approach) [2, 4, 7] the τ_s is constructed as:

$$\tau_{s_i} = -\Lambda_i \left(\frac{|s_i|}{e^{\delta_i}}\right)^\kappa \text{sgn}(s_i) \qquad (11)$$

Where $0 < \kappa < 1$ and Λ_i is a positive coefficient. Also δ_i can be evaluated by using the following formula

$$\delta_i = \left(-\frac{1}{(1 + (s_i / \pi_i)^2)} + 1\right) \qquad (12)$$

And π_i must be a proper coefficient.

6 Case Study and Simulation Results

We demonstrate the effectiveness of proposed FSMC by tracking control problem of a twin propeller system with two degrees of freedom, as shown in Fig.1. Thrust forces, which are generated by two independent propellers, are employed to control yaw (α) and pitch (β) angles of the system. To show robustness of the proposed controller against uncertainties and unmodeled dynamics, a number of different masses are used which their distances to the main shaft of the experimental setup can be arbitrarily adjusted, as follows: $m = 25 \times 10^{-3}, 50 \times 10^{-3}, 75 \times 10^{-3} \text{kg}$ and $l = 50, 100, 150, 22 \text{ mm}$. The dynamic equations of motion governing the twin propeller system are derived by Lagrange method. After substituting numerical values of components the equations are of the following form:

$$[(-I_{yz} + mll_0)\cos\beta]\ddot{\alpha} + (I_{zz} + ml^2)\ddot{\beta} + (I_{yy2} - I_{xx} - ml^2)\dot{\alpha}^2 \sin\beta\cos\beta =$$

$$-(\frac{1}{2}m_r gl_r + gml)\sin\beta - \mu_1\dot{\beta} + (0.124u_1 + 0.03u_2)\times 10^6 + l_{d_1} \tag{13}$$

$$[I_{yy1} + I_{yy2} + ml_0 + (ml^2 + I_{xx})\sin^2\beta]\ddot{\alpha} + [(-I_{yz} + mll_0)\cos\beta]\ddot{\beta} +$$

$$2(ml^2 - I_{yy2} + I_{xx})\dot{\alpha}\dot{\beta}\sin\beta\cos\beta + (I_{yz} - mll_0)\dot{\beta}^2\sin\beta = \tag{14}$$

$$-\mu_2\dot{\alpha} + (0.08u_2 - 0.06u_1)\times 10^6 \cos\beta + l_{d_2}$$

Where l_{d_1}, l_{d_2} are external disturbances generated by random function in Matlab between [-1,1]. The numerical values of the system parameters are as in Table 1, which, are demonstrated in system schematic diagram in Fig.1.

Table 1. Parameters of twin propeller system (units are in *kg* and *mm*)

I_{xx}	2325.10	l_0	28
I_{yy2}	53638.3	m_r	0.048
I_{yy1}	32.2	l_r	211.5
I_{zz}	53579.5	μ_1	0.05
I_{yz}	-150.16	μ_2	0.05

Note that, we encounter with a highly nonlinear, uncertain and couple experimental setup. The desired trajectories for α, β are defined as

$$\alpha_d = 0.73(1 - \cos t) \text{ rad} \qquad \beta_d = 0.7(1 - \cos(2t)) \text{ rad} \tag{15}$$

 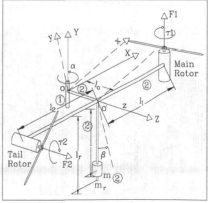

Fig.1. The twin propeller system and its schematic

In control design procedure the sliding surface slopes are chosen as $\lambda = diag(14,10)$. The supervisory control parameters are slected as $\kappa = 0.4$, $\Lambda_1 = 0.345$, $\Lambda_2 = 0.18$ and $\pi_1 = \pi_2 = 0.6$. The sampling period is 0.001sec. Parameters in adaptation laws are $\rho_{ij} = 1.5$, $\beta_{ij} = 1.5$, $\alpha_{ij} = 1.5$, $\gamma_{ij} = 1$. The constraint sets used in the parameter projection algorithm are chosen as $\Omega_{\sigma_{ij}} = [0.02, 0.5]$, $\Omega_{t_{ij}} = \Omega_{z_{ij}} = [0.8, 2.2]$

$\Omega_{t_{ij}} = \Omega_{z_{ij}} = [0.8, 2.2]$ and $\Omega_\theta = \{\theta/\|\theta\| \in [8,35]\}$. Note that to avoid chattering and enhance sliding efficiency, a constraint may be used in the parameter projection algorithm especially to control and adjust proper values for centers of consequent parts of rules (R3) and (R4) in each fuzzy subsystem, which govern the sliding mode motion, and hence, the following constraint set is introduced $\gamma_{\theta_{ij}} = [-0.5, 0.5]$. All

the simulation results demonstrate that the proposed fuzzy controller are robust and acts satisfactorily in each of the above m, l's. For example, simulations results for $m = 75 \times 10^{-3} \, kg$ and $l = 220mm$ are shown in figures 2 to 5.

Fig.2. Desired trajectories (dashed)
and output response (solid)

Fig.3. The supervisory control terms

Fig.4. The control inputs

Fig.5. The s_1, s_2 trajectories

7 Conclusion

In this paper, an adaptive fuzzy sliding mode controller was developed for a large class of nonlinear systems. The proposed controller did not require the accurate model of the system. Control rule base and membership functions of fuzzy controllers were simple and were designed based on the sliding mode control strategy. Using the Lyapunov method, the stability of the closed loop system was guaranteed.

References

1. Isidori, A.: Nonlinear control systems. Springer, New York (1989)
2. Slotine, J. J. E., Li, W.: Applied nonlinear control. Prentice Hall, New Jersey (1991)
3. Utkin, V. I.: Sliding Mode Control and their applications. Moscow Mir (1978)
4. Hung, J. Y., Gao,W., Hung, J. C.: Variable Structure Control: A survey. IEEE Trans. on Industrial Electronics(1993)
5. Yoo, B., Ham, W.: Adaptive FSMC of nonlinear systems. IEEE Trans. on fuzzy systems, Vol. 6, No. 2 (1998)
6. Lin, W. S., Chen, C. S.: Robust Adaptive SMC using fuzzy modeling for a class of uncertain MIMO nonlinear systems. In proc. IEEE int. Conf. Control theory appl., Vol.149, No.3, (2002)
7. Hsu, Y. C., Chen, G.: A Fuzzy Adaptive Variable Structure Controller with Application to Robot Manipulators. IEEE Trans. on systems, Man, and Cybernetics-Part B, Vol. 31,No. 3, (2001)
8. Emami, M.R., Goldenberg, A. A., Turksen, I. B.: A Robust Model-Based Fuzzy Logic Controller for Robot Manipulators. Proc. of the 1998 IEEE International Conference on Robotics and Automation (1998)
9. Wang, Li-Xin.: A Course in Fuzzy systems and Control. Prentice Hall (1997)
10. Gupta, M. M., Kiszka, J. B., Trogan, G. M.: Multivariable Structure of fuzzy control systems. IEEE trans. on syst., Man, Cybern., Vol,SMC-16, pp.638-656 (1986)
11. Lin, W.S., Chen, C. S.: Adaptive fuzzy sliding mode controller design for nonlinear MIMO systems. IEEE conf. proceeding (2000)

Improvement of Second Order Sliding-Mode Controller Applied to Position Control of Induction Motors Using Fuzzy Logic

Mahdi Jalili-Kharaajoo and Hassan Ebrahimirad

Control & Intelligent Processing Center of Excellence,
Department of Electrical & Computer Engineering, University of Tehran, Iran
mahdijalili@ece.ut.ac.ir and h.ebrahimirad@ece.ut.ac.ir

Abstract. In this paper, a fuzzy logic supervised second order sliding mode controller for an indirect vector-controlled induction motor is proposed. A second order sliding mode controller is designed in order to avoid high switching problem and a fuzzy logic controller is employed to enhance convergence speed adaptively. Using the proposed controller, position control of an induction motor is performed. Simulation results show the effectiveness of the proposed controller.

1 Introduction

The induction motor is the motor of choice in many industrial applications due to its reliability, ruggedness, and relatively low cost. Its mechanical reliability is due to the fact that there is no mechanical commutation (i.e., there are no brushes nor commutator to wear out as in a DC motor). Furthermore, it can also be used in volatile environments since no sparks are produced as is the case in the commutator of a DC motor [1]. During the last two decades, Variable Structure Control (VSC) and Sliding Mode Control (SMC) have gained significant interest and are gradually accepted by practicing control engineers [2,3]. There are two main advantages of this approach. Firstly, the dynamic behavior of the system may be tailored by the particular choice of switching functions. Secondly, the closed-loop response becomes totally insensitive to a particular class of uncertainty. In addition, the ability to specify performance directly makes sliding mode control attractive from the design perspective [4]. A phenomenon that usually occurs in SMC is the problem of chattering that can be greatly reduced using higher order sliding controllers [5,6].

In recent years many applications of fuzzy sliding mode control have been introduced [7,8]. The majority of research effort of combining fuzzy logic control and sliding mode control has been spent on how to use fuzzy logic to approximate the control command as a nonlinear function of sliding surface within the boundary layer [7,8]. In this work we will use a fuzzy logic based supervisor to increase the convergence speed of a second order sliding control action applied to position control of an induction motor. Using the proposed fuzzy supervisory control the slope of the sliding surface will be adapted. The performance of the proposed controller is compared with a PID controller in the sense of control effort and robustness of the system in response to mechanical parameter uncertainty and load torque variations.

T. Bilgiç et al. (Eds.): IFSA 2003, LNAI 2715, pp. 508–515, 2003.

2 Second Order Sliding Mode Control Design

The accompanying (sometimes dangerous) vibrations are termed "chattering". The higher the order of an output variable derivative where the high frequency discontinuity first appears, the less visible the vibrations of the variable itself will be. Thus, the remedy to avoid chattering is to move the switching to the higher order derivatives of the control signal. The problem is how to preserve the main feature of sliding modes: exact maintenance of constraints under conditions of uncertainty. Such sliding modes were discovered and termed "higher order sliding modes" (HOSM) [9,10,11]. Consider a nonlinear system

$$\dot{x}(t) = f(t,x) + g(t,x)u(t) \tag{1}$$

where $f(t,x)$ and $g(t,x)$ are smooth uncertain functions and $u(t)$ is the control command. In the design of sliding mode, sliding surface is defined as a function of state variables

$$s = s(x,t) \tag{2}$$

Consider local coordinates $y_1 = s$ and $y_2 = \dot{s}$, after a proper initialization phase, the second order sliding mode control problem is equivalent to the finite time stabilization problem for the uncertain second order system

$$\begin{cases} \dot{y}_1 = y_2 \\ \dot{y}_2 = \varphi(y_1, y_2, x_2, u, t) + \gamma(y_1, y_2, t)v(t) \end{cases} \tag{3}$$

where $v(t) = \dot{u}(t)$.

In the above equations $y_2(t)$ is generally unknown, but $\varphi(t,x)$ and $\gamma(t,x)$ can be bounded as

$$|\varphi(t,x)| \le \Phi, \quad 0 < \Gamma_m < \gamma(t,x) < \Gamma_M, \quad \Phi > 0 \tag{4}$$

Being historically the first known second order sliding controller, that algorithm features twisting around the origin of second order sliding phase plane $y_1 - y_2$. The trajectories perform an infinite number of rotations while converging in finite time. The vibration magnitudes along the axes as well as the rotation times decrease in geometric progression. The control derivative value commutes at each axis crossing, which requires availability of the sign of the sliding-variable time-derivative y_2.

The control algorithm is defined by the following control law, in which the condition on $|u|$ provides for $|u| \le 1$

$$\dot{u}(t) = \begin{cases} -u & if \ \ |u| > 1 \\ -V_m sign(y_1) & if \ \ y_1 y_2 \le 1, |u| \le 1 \\ -V_M sign(y_1) & if \ \ y_1 y_2 > 1, |u| \le 1 \end{cases} \tag{5}$$

The corresponding sufficient conditions for the finite-time convergence to the sliding surface are

$$V_M > V_m, \ V_M > \frac{\Gamma_M V_m + 2\Phi}{\Gamma_m}, \ V_m > \frac{\Phi}{\Gamma_m}, \ V_m > \frac{4\Gamma_M}{s_0} \qquad (6)$$

where $s_0 > |s|$.

3   Position Control of an Induction Motor Using Second Order Sliding Mode

In order to position control in induction motors, the following model is used:

$$J\frac{d^2\theta_m}{d_t^2} + B\frac{d\theta_m}{dt} + T_L = K_t i_{qs} \qquad (7)$$

where

i_{qs} : The q-axis current of stator, w_m : Rotor mechanical speed, T_L : Load torque,

J : Inertia constant, B : Viscous Friction constant, θ_m : Mechanical angle of rotor.

For this purpose, we assume

$$x_1 = \theta^*{}_m - \theta_m, x_2 = \dot\theta_m = w_m \qquad (8)$$

where θ_m^* is the reference position that should be tracked. So, the state space of the model is

$$\begin{cases} \dot{x}_1 = x_2 \\ \dot{x}_2 = -\dfrac{B}{J}x_2 + \dfrac{1}{J}T_L + \dfrac{1}{J}K_t u(t) \end{cases} \qquad (9)$$

where $u(t) = -i_{qs}$ is the control action.

Using (9) we can obtain the equations of the auxiliary system (3) where

$$\varphi(y_1, y_2, x_2, u, t) = (-\frac{B}{J} + C)\left[-\frac{B}{j}x_2 + \frac{1}{J}T_L(t) + \frac{1}{J}K_t u(t)\right] + \frac{1}{J}\frac{dT_L(t)}{dt} \qquad (10)$$

$$\gamma(y_1, y_2, t) = \frac{1}{J}K_t. \qquad (11)$$

The values of B, J, and K_t are uncertain, so $\gamma(.)$ and $\varphi(.)$ have uncertainty.

Since the relative degree of the system is one, the control signal of the main system, $u(t)$, can be obtained by integrating the auxiliary control signal $v(t)$. So, the overall chattering can be significantly reduced.

The system parameters involve the following uncertainties

$$B \le B_{max}, \ J_{min} \le J \le J_{max}, \ K_{t,min} \le K_t \le K_{t,max}, \ |x_2| \le w_{m,max}, \ |T_L(t)| \le T_{L,max},$$

$$\left| \frac{dT_L(t)}{dt} \right| \le dt_{L,max} \tag{12}$$

and the physical limitation of the actuator is

$$-10 \le i_{qs} \le 10 .$$

Using the second order sliding control strategy, the system states slide on the sliding surface

$$s = Cx_1 + x_2 \tag{13}$$

Hence, the dynamic on the sliding surface is

$$\dot{s} = \dot{x}_2 + C\dot{x}_1 = 0 \Rightarrow \dot{x}_2 = -C\dot{x}_1 \tag{14}$$

Using (9) and (14) we come to

$$\dot{x}_2 = -Cx_2 \tag{15}$$

Therefore, x_2 is a first order subsystem with time constant $1/T$ that converges to zero independent of disturbance and uncertainty. The settling time can be improved by increasing the value of C but in this case the sensitivity to uncertainty will be increased. So to achieve a suitable trade-off, at the beginning, C is set to a small value and it is increased after reaching the sliding surface. In the next section the process of increasing C will be derived using a fuzzy logic controller.

4   Improvement of Settling Time Using Fuzzy Logic Controller

Fuzzy logic is mainly introduced to provide tools for dealing with uncertainty [15]. We cannot determine the change of C with respect to the system action precisely. Thus, we will use a fuzzy logic controller to supervise the sliding controller. In order to obtain fuzzy rules of updating the slope of the sliding curve $s = Cx_1 + x_2$, the approximate form of the phase plane around the sliding surface is depicted in Fig. 1. For the operating points A to H we can define the slope change as presented in Table 1. The fuzzy rules are presented in Table 2 and the membership functions are shown in Fig. 2.

5   Simulation Results

In order to investigate the effectiveness of the proposed controller, an induction motor with parameters presented in the following is considered:

3Φ, 208volt, 3hp, 60Hz
$R_s=0.6\Omega$ $R_r=0.4\Omega$ $L_m=0.59$H $L_\sigma=0.0021$H $L_s=0.0061$H $L_r=0.0061$H
$P=2$ $T_{L,max}=2$Nm $B=0.0018673$Nm.sec^2 $J=J_n=0.0117643$Nm.sec/rad

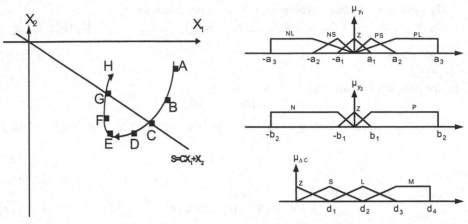

Fig. 1. Phase trajectory around sliding surface.   **Fig. 2. Fuzzy membership functions.**

Table 1: Proper change in *C* for different operating points.

ΔC	$y_2(t)$	$y_1(t)$	Operating Point
Zero	Negative	Positive and large	A
Small	Negative	Positive and small	B
Medium	Negative	Zero	C
Large	Negative	Negative and small	D
Medium	Zero	Negative and large	E
Small	Positive	Negative and small	F
Zero	Positive	Zero	G
Zero	Positive	Positive and small, large	H

Table 2: Fuzzy rules.

		NL	NS	Z	Z	PL
$y_2(t)$	P	M	S	Z	Z	Z
	Z	M	M	M	Z	Z
	N	M	L	M	S	Z

(header: $y_1(t)$)

$w_{m,max}=189$rad/sec $dT_{L,max}=200$Nm/sec $B_{max}=2B$ $J_{min}=0.85J$
$J_{max}=1.75J_n$ $K_{t,min}=0.85$Nm/A $K_{t,max}=1.5$Nm/A

In Figs. 3 and 5 the simulation results for nominal parameters using the second order sliding mode and the proposed fuzzy sliding mode controller for $\theta^*=20°$ and $T_L=0$Nm are shown. The fuzzy controller values are presented in the following:

$a_1=1$ $a_2=5$ $a_3=400$ $b_1=0.5$ $b_2=400$ $d_1=0.05$ $d_2=0.2$ $d_3=0.5$ $d_4=2$

The initial value of C is set to 5. As it can be seen, the proposed fuzzy supervisor is able to improve the performance of the closed-loop system and accelerates the convergence of the system; for instance, the settling time is decrease from 1.2 sec (Fig. 3) to 0.5 sec (Fig. 4). In Fig. 5 angle response of the system under both controllers is compared, clearly the response of proposed fuzzy sliding controller is faster than the other one. In order to investigate further, the performance of the proposed controller is compared with a conventional PID controller. To achieve this goal, we obtain following PI and PID controllers for velocity and position loops, designed to achieve 0.5sec as the settling time
The controller for velocity loop:

$$C_1(s)=3.5+0.215/s$$

The controller for position loop:

$$C_2(s)=9.2+0.145/s+0.04s .$$

The closed-loop response using the designed PI and PID controllers is sketched in Fig. 6. Comparing Figs. 4 and 6, obviously, introduction of proposed fuzzy sliding controller requires less control effort than the case of using PID controller with the same desired performance.

Fig. 7 shows the system response for $\theta^*=20°$ and $T_L=2$Nm where after 1.1sec a load with $T_L=2$Nm is imposed to the system. Clearly, using the proposed controller the disturbance was rejected after 0.1sec with a proper control signal but the PID controller could not reject the disturbance even after 0.5sec. In fact, the PID controller cannot produce a control signal to reject the disturbance in a short time. In another test, to confirm the robustness of the proposed controller in response to the parameter uncertainties, J is increased to $1.75J_n$ (J_n is the nominal inertia constant). The comparison of responses for $\theta^*=20°$ and $\theta^*=30°$ using the proposed fuzzy sliding controller and PID control action are shown in Fig. 8. It is clear that using the proposed fuzzy sliding controller the performance for the two cases, $J=J_n$ and $J=1.75J_n$, is almost the same (8-a). Using the PID controller the settling time of the system for $J=1.75J_n$ was increased and an overshoot was appeared in comparison to the case of $J=J_n$ (8-b). This simulation indicates that the proposed controller has satisfactory robustness in response to the system parameter uncertainty.

6   Conclusion

In this paper, a new fuzzy logic supervised second order sliding mode controller for position control of an induction motor was proposed. A second order sliding mode controller for position control designed and using a suitable fuzzy controller for updating the slope of sliding surface, the convergence speed was improved. The performance of this controller was compared with a conventional PID controller where for same desired performance the former required less control effort and exhibited better robustness in response to system parameter uncertainty.

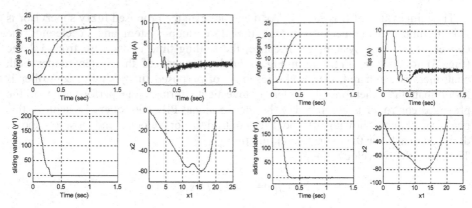

Fig. 3. System responses using the proposed fuzzy second order sliding mode controller with $T_L = 0\text{Nm}$.

Fig. 4. System responses using the second order sliding mode controller with $T_L = 0\text{Nm}$.

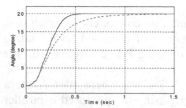

Fig. 5. Angle using the second order sliding controller (—) and proposed fuzzy sliding controller (   ).

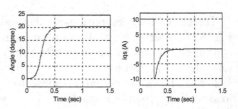

Fig. 6. System responses using the PID controller with $T_L = 0\text{Nm}$.

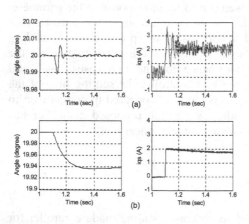

Fig. 7. System responses using (a)- the proposed fuzzy sliding controller (b)- the PID controller with $T_L = 2\text{Nm}$.

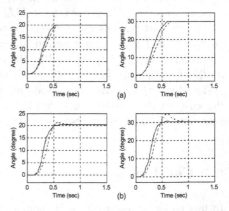

Fig. 8. System responses using (a)- the proposed fuzzy controller (b)- the PID control for $J = J_n$ (—) and $J = 1.75 J_n$ (----) with $T_L = 0\text{Nm}$.

Reference

[1] Leonhard, W., *Control of Electrical Drives*, springer-Verlag (1996).
[2] Utkin, V.I., *Sliding Modes in Control and Optimization*, Springer Verlag (1992).
[3] Edwards, C. and Spurgeon, S.K., *Sliding mode control*. Taylor & Francis Ltd (1998).
[4] Young, D.K., Utkin, V.I. and Özgüner, O., : A control engineer's guide to sliding mode control, *IEEE Transaction on Control System Technology*, vol. 7, No. 3 (1999), 621-629.
[5] Bartolini, G., Ferrara, A., Punta, E. and Usai, E., : Application of a second order sliding mode control to constrained manipulators, *EUCA, IFAC and IEEE European Control Conference*, Bruxelles, Belgium (1997).
[6] Sira-Ramirez, H., : Dynamic second order sliding mode control of the hovercraft vessel,. *IEEE Transaction on Control System Technology*, vol. 10, No. 6 (2002), 880-865.
[7] Kung, C.C. and Lin, S.C., : A Fuzzy-Sliding Mode Controller Design, *IEEE International Conference on Systems Engineering* (1992), 608-611.
[8] Boverie, S., Cerf, P. and Le, J.M. : Fuzzy Sliding Mode Control Application to Idle Speed Control, *IEEE International Conference on Fuzzy Systems* (1994), 974-977.
[9] Bartolini, G., Coccoli, M. and Ferrara, A., : Chattering avoidance by second order sliding mode control, *IEEE Trans. Automatic Control*, vol.43, No. 2 (1998), 497-512.
[10] Levant, A., : Higher order sliding: collection of design tools, *Proc. 4th European control Conference*, Brussels, Belgium (1997).
[11] Bartolini, G., Ferrara, A. and Usai, E., : Output tracking control of uncertain nonlinear second order systems, *Automatica*, vol. 33, No. 12 (1997), 2203-2212.
[12] Jezerink K., Rodic, M. and Sabanovic, A., : Sliding mode application in speed senseless torque control of an induction motor, *Proc.15th IFAC world congress. Barcelona, Spine*, July (2002).
[13] Barreno, F., Gonzalez, H.J. and Shied, B.S., : Speed control of induction motor using a novel fuzzy sliding mode structure, *IEEE Transaction on Fuzzy System*, vol.10, No. 3 (2002), 375-383.
[14] Shyu, k., Lin, F.J., Tollalba, A., Galvan, E. and Franquclo, L.G., : Robust variable structure speed control for induction motor drives, *IEEE Transaction on Aeirospace & Electronic System*, vol. 33, No. 1 (1999), 215-223.
[15] Zadeh, L.A., : Fuzzy sets, *Inf. Control* (1965), 338-353.

Hybrid Electric Vehicles: Application of Fuzzy Clustering for Designing a TSK-based Fuzzy Energy Flow Management Unit

Lucio Ippolito, Pierluigi Siano

Dipartimento di Ingegneria dell'Informazione e Ingegneria Elettrica
Università degli Studi di Salerno
Via Ponte don Melillo, 1
84084 Fisciano (SA) Italy
ippolito@unisa.it

Abstract. Today, the satisfaction of the desire for personal transportation requires developing vehicles that minimize the consequences on the environment and maximize highway and fuel resources. Hybrid electric vehicles (HEVs) could be an answer to this demand. Their use can contribute significantly to reduce their environmental impact, achieving at the same time a rational energy employment. Controlling an HEV requires a lot of experimentations. Experts and training engineers can ensure the good working of the powertrains, but the research of optimality for some criteria combining fuel needs and power requirements is mainly empirical due to the nonlinearity of the driving conditions and vehicle loads. Consequently, in the paper a fuzzy modeling identification approach is applied for modeling the power flow management process. Amongst the various methods for the identification of fuzzy model structure, fuzzy clustering is selected to induce fuzzy rules. With such an approach the fuzzy inference system (FIS) structure is generated from data using fuzzy C-Means (FCM) clustering technique. As model type for the FIS structure a first order Takagi-Sugeno-Kang (TSK) model is considered. From this architecture a fuzzy energy flow management unit based on a TSK-type fuzzy inference is derived. Further, some interesting comparisons and simulations are discussed to prove the validity of the methodology.

1. Introduction

Throughout the world, there is a trend towards the growth of motor vehicle traffic. As a consequence of the rate of growth of the car use, the ever greater environmental nuisance and pollution are causing lethal consequences for humans. Only the adoption of a global strategy, bringing into play both technical innovation and strict regulations, can contribute to reduce emissions from motor vehicles.

In this context hybrid electric vehicles (HEVs) powertrain can offer a sensible improvement of the vehicle environmental impact and a rational energy employment.

The new generation vehicles based on HEV technology are designed by total systems approaches, involving new materials, alternative fuels, and, above all, modern control systems.

T. Bilgiç et al. (Eds.): IFSA 2003, LNAI 2715, pp. 516–525, 2003.
© Springer-Verlag Berlin Heidelberg 2003

Advanced control techniques may have major benefits for the on-board energy flow management. For a HEV, with a parallel configuration, the management of the power drive energy flows requires splitting the instantaneous vehicle power demand between the internal combustion engine, the electric motor and the brake such that these power sources are operated at high efficiency operating points and the related vehicle emissions are minimized.

For the management of power demand the main difficulty is to take into account the non-trivial influence of the main power drive components on the vehicle's performance in terms of emissions and fuel consumption. Sensitivity analysis developed by some authors [1,2,3] has proved that engine produces the greatest quantity (about 80%) of the CO and HC emissions during the first minutes of the driving cycle. Obviously, the fuel economy and the quantity of exhaust emissions generated are extremely influenced also by the driving conditions. Some experimental results have proved the levels of emissions generated, especially the quantity of NOx, increase greatly in correspondence of abrupt engine accelerations [1,2,3].

On this premise, the hybridization strategy must provide to supply a proper fraction of the vehicle's power demand by the electric motor. But at the same time, an excessive usage of the electric source causes a lofty battery pack depletion that could damage the storage modules or shorten their usable life.

The optimal identification of the propulsion hybridization involves therefore a number of objectives to be achieved, which, inherently, have different characteristics. These objectives are in trade-off relations, and with no invariant priority order amongst them. These inherent characteristics of the specific optimization problem suggest as solving methodology the use of the goal programming method [4,5].

On the other hand, even though the goal programming method has proved its effectiveness and robustness, it has revealed a reduced applicability for real-time control applications [5]. This led study to adopt a fuzzy modeling approach for describing the characteristics and behavior of the system using fuzzy reasoning.

More particularly, in the following sections, after an outline of identification of fuzzy models using fuzzy clustering, the application of the methodology to the energy flow management on board a HEV is presented. Simulations are provided to illustrate the performance of the proposed method and, at last, some comparisons are presented in order to draw some conclusions.

2. Theoretic Concepts of Fuzzy Models' Identification Using Fuzzy Clustering

The identification of the existing relations between the input and output variables of the system, expressing them linguistically, is one of the more distinguishing feature of the method.

The fuzzy inference system (FIS) modeling, as that proposed by Takagi-Sugeno-Kang (TSK), is a multimodel approach in which single submodels are combined to describe the global behavior of the system. This partitioning of the input space into regions permits to use a simpler submodel for each of them [6-7].

The description of the system's behavior is obtained generating a set of fuzzy IF-THEN rules, whose in the present paper are extracted by using fuzzy clustering and, more particularly, the fuzzy C-Mean (FCM) clustering technique [6-8].

For multi-input single-output system, the typical TSK model consists of a set of IF-THEN rules. The rules have the following form:

$$R_h : \text{IF } x_1 \text{ is } A_h^1 \text{ and } \ldots \text{ and } x_p \text{ is } A_h^p \text{ THEN } y \text{ is } f_h(x), \quad h = 1, \ldots, n \tag{1}$$

where

$$f_h(x) = a_{0h} + a_{1h}x_1 + a_{2h}x_2 + \cdots + a_{ph}x_p \tag{2}$$

in which $x_{1,\ldots,p}$ are the input variables, y is the output variable, $A_h^{1,\ldots,p}$ are the fuzzy sets, and $f_h(x)$ is a linear function.

The h-th fuzzy rule of the collection is able to describe the local behavior associated to the fuzzy input region characterized by the antecedent of the fuzzy rule.

For any input, $\widetilde{x}$, the inferred value of the TSK model, is calculated as

$$\widetilde{y} = \frac{\sum_{h=1}^n A_h(\widetilde{x}) * f_h(\widetilde{x})}{\sum_{h=1}^n A_h(\widetilde{x})} = \frac{\sum_{h=1}^n \tau_h * f_h(\widetilde{x})}{\sum_{h=1}^n \tau_h} \tag{3}$$

where

$$A_h(\widetilde{x}) = \tau_h = A_h^1(\widetilde{x}_1) \times A_h^2(\widetilde{x}_2) \times \cdots \times A_h^p(\widetilde{x}_p) \tag{4}$$

and τ_h is the level of firing of the h-th rule for the current input $\widetilde{x}$. As the consequent of each rule is linear, its parameters, which minimize the overall error between the TSK fuzzy model and the system being modeled, can be estimated by a recursive least-squares procedure [6].

2.1 Elements about Fuzzy C-Means Clustering Technique

As mentioned above for the extraction of fuzzy IF-THEN rules the proposed method uses the fuzzy C-Mean (FCM) clustering technique.

Through fuzzy clustering, groups of data from a large data set are distilled, obtaining a concise representation of the system's behavior. In this way, clustering becomes the basis of the fuzzy model identification algorithm. Clustering permits the partitioning of the input space into n fuzzy regions, reducing the complexity of the building process of the TSK model.

Applying the FCM algorithm to a set of unlabeled patterns $X=(x_1, x_2, \ldots, x_N)$, $x_i \in R^s$ where N is the number of patterns and S is the dimension of pattern vectors, an optimal fuzzy c-partition and corresponding prototypes are found [9]. The prototypes are selected to minimize the following objective function:

$$F_m(U, W) = \sum_{j=1}^C \sum_{i=1}^N (\mu_{ij})^m d_{ij}^2 \tag{5}$$

subject to the following constraints on U:

$$\mu_{ij} \in [0,1] \quad i = 1, \ldots N \quad j = 1, \ldots C \tag{6}$$

$$\sum_{j=1}^C \mu_{ij} = 1 \quad i = 1, \ldots N \tag{7}$$

$$0 < \sum_{i=1}^{N} \mu_{ij} < N \qquad j = 1,....C \tag{8}$$

where:

N: the number of patterns in X

C: the number of clusters

U: the membership function matrix

μ_{ij} : the value of the membership function of the i-th pattern belonging to the j-th cluster

d_{ij} : the distance, from $\mathbf{x_i}$ to $\mathbf{w}_j^{(t)}$, $d_{ij} = \left\| \mathbf{x_i} - \mathbf{w}_j^{(t)} \right\|$; where $\mathbf{w}_j^{(t)}$ denotes the cluster centre of the j-th cluster for the t-th iteration

$\mathbf{W}$: the cluster centre vector

m: the exponent on μ_{ij} ; to control fuzziness or amount of clusters overlapping

Objective function measures the quality of partitioning that divides a dataset into C clusters by comparing the distance from pattern $\mathbf{x_i}$ to the current candidate cluster centre $\mathbf{w_i}$ with the distance from pattern $\mathbf{x_i}$ to other candidate cluster centres [9].

Equation (5) describes a constrained optimization problem, which can be converted to an unconstrained optimization problem by using the Lagrangian multiplier technique. To minimize the objective function under fuzzy constraints, given a fixed number C, m and ε a small positive constant, the FCM algorithm, starts with a set of initial cluster centres, or arbitrary membership values. Then generate randomly a fuzzy c-partition and set iteration number $t=0$. A two step iterative process works as follows. Given the membership values $\mu_{ij}^{(t)}$, the cluster centre matrix $\mathbf{W}$ is calculated by:

$$w_j^{(t)} = \frac{\sum_{i=1}^{N} (\mu_{ij}^{(t-1)})^m x_i}{\sum_{i=1}^{N} (\mu_{ij}^{(t-1)})^m} \qquad j = 1,....C \tag{9}$$

Given the new cluster centres $\mathbf{W}^{(t)}$, the membership values $\mu_{ij}^{(t)}$ are updated by:

$$\mu_{ij} = \frac{1}{\sum_{l=1}^{C} \left(\frac{d_{ij}}{d_{il}} \right)^{\frac{2}{(m-1)}}} \qquad \begin{array}{l} i = 1,....N \\ j = 1,....C \end{array} \tag{10}$$

where if $d_{ij} = 0 = 0$ then $\mu_{ij} = 1$ and $\mu_{ij} = 0$ for $l \neq j$.

The process stops when $\left\| \mathbf{U}^{(t)} - \mathbf{U}^{(t-1)} \right\| \leq \varepsilon$, or a predefined number of iterations is reached.

To apply the FCM algorithm the definition by the user of several parameters is required: the matrix norm, the fuzziness parameter m, the stopping criterion ε, the maximum number of iterations and the optimal number of clusters.

For such a problem the selection of the optimal number of clusters, fitting a data set is particularly critical, affecting strongly the vehicle performance.

Some authors [10] have proposed quantitative methods for evaluation of clustering results; these methods give a measure of the quality of the resulting partitioning representing a tool at the disposal of the experts in order to evaluate the clustering results. On the other hand, for the specific problem, they are not able to guarantee the respect of the overall vehicle's performance: a good clustering result could correspond to not so good performance of the control system.

From this concept it follows the necessity to carry on an analysis to identify the relationship between number of clusters and vehicle performance. To do this, in the present paper an overall vehicle's performance index (VPI) is defined as the sum of the mean absolute errors for the exhaust gas, the fuel consumption and the batteries' state of charge from their target values. The VPI's calculation, on a prefixed drive cycle, is made increasing the number c of clusters simulation by simulation. Simulated data reveals that, for a given drive cycle, it exists a number of clusters to which it will correspond the best vehicle' performance.

3. Application of Fuzzy Clustering for Designing a TSK-based Fuzzy Energy Flow Management Unit

This section is dedicated to the designing of the energy flow management unit. The model of the control unit, which output is the hybridization degree, is based on the previous described fuzzy model identification approach.

From the before discussed argumentations, the FIS modeling approach consists into identifying the existing relations between the input and output variables of the system. The input/output relations are determined working on a knowledge base, which exemplifies the system's behavior [6-8-11].

Hence, the starting point for the modeling algorithm is to choose the input/output variables and to dispose of a collection of input/output data to which the clustering technique must be applied.

Referring to the choice of the input/output variables, starting from the previous considerations, it can be identified the instantaneous power split between the internal combustion engine and the electric motor as the output variable, while the catalyst temperature, the instantaneous vehicle torque demand, and the state of charge (SOC) of the batteries are selected as input variables.

As far as the data collection generation concerns, for the present application the collection of data is obtained solving the a multiobjective power flows optimisation problem over a standard drive cycle and storing, at each time step, into a database the optimal hybridization degree together with the input variables.

The multiobjective optimization of the energy flows is formulated as in the eqn. (11), assuming as objectives five important terms, which are:

- the instantaneous HC emission
- the instantaneous CO emission
- the instantaneous NOx emission
- the fuel economy FC
- the battery depletion ΔSOC

$$
\begin{cases}
\min_{(P_{ICE},P_{el})\in\Omega} \begin{pmatrix} HC(P_{ICE},P_{el},\Gamma), CO(P_{ICE},P_{el},\Gamma), NO_x(P_{ICE},P_{el},\Gamma), \\ FC(P_{ICE},P_{el},\Gamma), \Delta SOC(P_{ICE},P_{el},\Gamma) \end{pmatrix} \\
P_{ICE}(t) + P_{el}(t) = P_{veh}(t)
\end{cases} \tag{11}
$$

In the previous equation Ω is the solution space and $HC(P_{ice},P_{el},\Gamma)$, $CO(P_{ice},P_{el},\Gamma)$, $NO_x(P_{ice},P_{el},\Gamma)$, $FC(P_{ice},P_{el},\Gamma)$, $\Delta SOC(P_{ice},P_{el},\Gamma)$ are respectively the estimation of the pollutant emissions, fuel consumption and level of batteries discharge referred to the energy flows distribution (P_{ICE}, P_{el}) and the actual power drive state described by the vector Γ.

The problem formalized with eqn. (11) can be solved using the goal attainment method introducing a set of design goals that are associated with the set of design objectives, jointly with a set of under or over attainment factors. The overall problem is then led to the resolution of a standard optimization problem in accordance with the following mathematical formalization:

$$
\min_{\alpha\in\Re,(P_{ice},P_{el})\in\Omega} \alpha
$$

$$
\exists' \begin{cases}
HC(P_{ice},P_{el},\Gamma) - \alpha\,\omega_{HC} \le HC* \\
CO(P_{ice},P_{el},\Gamma) - \alpha\,\omega_{CO} \le CO* \\
NO_x(P_{ice},P_{el},\Gamma) - \alpha\,\omega_{NOx} \le NO_x* \\
FC(P_{ice},P_{el},\Gamma) - \alpha\,\omega_{FC} \le FC* \\
\Delta SOC(P_{ice},P_{el},\Gamma) - \alpha\,\omega_{\Delta SOC} \le \Delta SOC*
\end{cases} \tag{12}
$$

$$
P_{ice}(t) + P_{el}(t) = P_{veh}(t)
$$

where α is the scalar to minimize, $(HC*, CO*, NOx*, FC*, \Delta SOC*)$ is the set of design goals while $(\omega_{HC}, \omega_{CO}, \omega_{NOx}, \omega_{FC}, \omega_{\Delta SOC})$ is the set of degrees of under or over achievement of the goals.

Solving the previous formalized problem over one or more drive cycles the collection of input/output data is generated. As afore said to evaluate the number of clusters to be used, the VPI is defined as the sum of the mean absolute errors between the multiobjective and the fuzzy controlled vehicle's emissions, SOC and fuel consumption computed on the N simulation time steps associated to the selected drive cycle.

$$
VPI = \frac{1}{N}\sum_{i=1}^{N}\left\{ |HC_i - HC_i'| + |CO_i - CO_i'| + |NO_{xi} - NO_{xi}'| + |FC_i - FC_i'| + |SOC_i - SOC_i'| \right\} \tag{13}
$$

For the given definition the minimum value of the VPI corresponds to the best vehicle's performance. Then, applying the FCM method, with the selected number of cluster, to the data collection the cluster centers are determined. They are the prototypical data points that exemplify a characteristic behavior of the system, and therefore, each of them is associated to as a fuzzy rule, with a form defined by eqn. (1). Now, by using the result of clustering technique, it can be assigned to any data $\mathbf{x}$ of X^p a value in Y, being able to generate c fuzzy rules associated to C fuzzy clusters. In such a way, using the fuzzy cluster and the fuzzy rules, it is possible to infer a value $\tilde{y}$ of the output space, using eq. (3).

To calculate the inferred value of the TSK model corresponding to the input $\tilde{x}$, it is necessary to determine $A_h(\tilde{x})$ and $f_h(\tilde{x})$. But, as before mentioned, because each rule is linear, its parameters, which minimize the overall error between the TSK fuzzy model and the system being modeled, are estimated by a linear least-squares regression, in the present case.

Linear least squares regression is by far the most widely used modelling method to fit a model to their data. Used directly, with an appropriate data set, linear least squares regression can be used to fit the data with any function of the same form of $f_h(x)$, in which each explanatory variable in the function is multiplied by an unknown parameter, there is at most one unknown parameter with no corresponding explanatory variable, and all of the individual terms are summed to produce the final function value. With this methodology a fuzzy energy management unit based on the TSK fuzzy model identified above is implemented. The management unit, from the clustering of the data obtained solving the multiobjective problem and using the vehicle's torque demand, the battery's SOC and the catalyst temperature as inputs, is able to furnish the hybridization degree.

4. Simulation Results

In order to evaluate the effectiveness of the proposed methodology for designing of the energy flows management unit, various comparative simulations with other classic approaches have been developed. In particular, for comparing the results furnished by the TSK-based fuzzy energy flows management unit, the multiobjective-based strategy [5] has been considered.

The characteristics of the vehicle used in the simulations are obtained applying the optimal sizing design procedure proposed by the authors in [12,13,14]. The power train characteristics are those reported in [14].

Using Matlab®, the test vehicle, equipped with the multiobjective energy flow management unit, has been simulated over the New European Driving Cycle (NEDC). In order to obtain the partitioning of the input space the FCM method was applied considering five clusters to which are associated the following fuzzy rules:

R_1: *IF* (Torque demand is Low) and (SOC is High) and (Catalyst temperature is Low) *THEN* (Hybridization degree is Out$_5$)

R_2: *IF* (Torque demand is High) and (SOC is Medium) and (Catalyst temperature is High) *THEN* (Hybridization degree is Out$_4$)

R_3: *IF* (Torque demand is VeryHigh) and (SOC is VeryLow) and (Catalyst temperature is VeryHigh) *THEN* (Hybridization degree is Out$_3$)

R_4: *IF* (Torque demand is VeryLow) and (SOC is VeryHigh) and (Catalyst temperature is VeryLow) *THEN* (Hybridization degree is Out$_2$)

R_5: *IF* (Torque demand is Medium) and (SOC is Low) and (Catalyst temperature is Medium) *THEN* (Hybridization degree is Out$_1$)

where Out$_h$ corresponds to the h-th linear function, $f_h(x)$.

The number of cluster was determined equal to five, corresponding to a VPI equal to 1.5×10^{-5}.

For clustering in this application it was used the Euclidean norm as matrix norm, the fuzziness parameter, m, was set to 2.0, the value of ε was set to 10^{-5}, and the maximum number of iterations was set to 100.

For the mapping of each term set on the domain of the corresponding linguistic variable, the Gaussian membership functions were used. On the other hand, to perform some comparative simulations bell-shaped membership functions were used, also. The analysis of the results obtained applying the described methodology to the HEV under test reveals, as it is depicted in fig. 1, that the energy management unit imposes a high degree of hybridization (it uses engine, prevalently) when the torque demand is very high, viceversa, it imposes a low degree of hybridization (it uses electric motor, prevalently) when the catalyst temperature is very low, in order to reduce exhaust emissions.

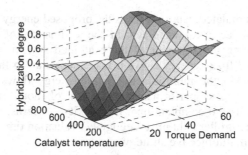

Fig. 1. Hybridization degree surface

Equipping the simulated vehicle with the TSK-based fuzzy energy flows management unit, it allows to obtain the results summarized in table 2.

Table 1. Performance of the TSK-based fuzzy energy flows management unit

Monitored Parameters	Gaussian membership functions	Modified Gaussian membership functions	Bell-shaped membership functions	Multiobjective based control strategy
HC [g/km]	0.2468	0.2509	0.2477	0.2469
CO [g/km]	1.4712	1.4444	1.4415	1.4694
NO_x [g/km]	0.1340	0.1325	0.1446	0.1345
Fuel Consumption [liters/km*100]	3.5781	3.5412	3.5558	3.5933
Final SOC	0.5570	0.5556	0.5558	0.5575

The slight difference in the performance between the controller using standard Gaussian membership functions and those using modified Gaussian membership functions or bell-shaped membership functions is due, mainly, to the assumption of a constant non-zero membership degree at the boundary of the domains.

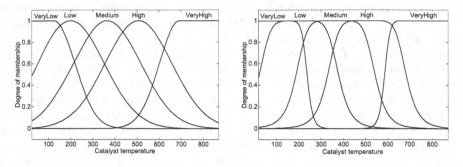

Fig. 2. Modified Gaussian and bell-shaped membership functions

The analysis of the simulated data reveals as the proposed energy flows management algorithm can guarantee great accuracy in the estimation of all the monitored parameters, reproducing the behavior of the examined system over all the driving cycle. The TSK-based fuzzy energy flows management unit, in fact, provides results in great agreement with the multiobjective based energy flows management unit, exhibiting quite the same performance in terms of both emissions and fuel consumption. As far as the computational complexity concerns, the comparison has evidenced the TSK controller exhibits a negligible evaluation time compared with the time required by the multiobjective strategy.

5. Final Remarks

Starting from the consideration that the energy flows management on board a PHEV is an activity involving many general aspects and variables, the paper has dealt with the discussion about the reduced applicability for real-time applications of the multiobjective control strategy, which identifies the instantaneous energy flows distribution. To overcome this limitation, a fuzzy modeling approach has been proposed. The methodology, consisting into identifying the existing relations between the input and output variables of the system, uses a collection of data, acquired experimentally or by laboratory simulations. Then through fuzzy clustering, the input space is partitioned into various fuzzy regions.
Subsequently, by using the result of clustering technique, some fuzzy rules descriptive of the input and output variables relations are induced, obtaining the possibility to infer a value of the output space, corresponding to any combinations of inputs
The methodology has been applied to a parallel HEV for determining the suitable hybridization degree between the electric motor and engine. The analysis of the simulated data has revealed the high degree of accuracy of the TSK-based fuzzy energy flows management unit to identify the optimal power flows distribution. Results suggest to adopt this control system approach for such a class of real-time applications.

6. References

1. W.C. Morchin: Energy management in hybrid electric vehicles. In Proceedings of 17th Digital Avionics Systems Conference, vol. 2, (1998), I41/1-I41/6.
2. S.D. Farrall and R.P. Jones: Energy management in an automotive electric/heat engine hybrid powertrain using fuzzy decision making. In Proceedings of the International Symposium on intelligent control, (1993) 463-468.
3. M.R. Cuddy and K.B. Wipke: Analysis of the Fuel Economy Benefit of Drivetrain Hybridization. In Proceedings of SAE International Congress and Exposition, Detroit, Michigan, (1997).
4. F.W. Gembicki and Y.Y. Haimes. Approach to performance and sensitivity multiobjective optimisation: the goal attainment method IEEE Transaction on Automation and Control, AC-20 (8), vol.6, (1975) 821-830.
5. V. Galdi, L. Ippolito, A. Piccolo and A. Vaccaro: Multiobjective optimization for fuel economy and emissions of HEV using the goal-attainment method. In Proceedings of 18th International Electric Vehicle Symposium, Berlin, (2001).
6. Delgado, M.; Gomez-Skarmeta, A.F.; Martin, F.: A fuzzy clustering-based rapid prototyping for fuzzy rule-based modeling Fuzzy Systems. IEEE Transactions on, Volume: 5 Issue: 2, (1997) 223-23.
7. Zhao, J.; Wertz, V.; Gorez, R.: A fuzzy clustering method for the identification of fuzzy models for dynamic systems. Intelligent Control, 1994, Proceedings of the 1994 IEEE International Symposium on, (1994) 172–177.
8. Chiu, S.L: A cluster estimation method with extension to fuzzy model identification Fuzzy Systems. IEEE World Congress on Computational Intelligence, Proceedings of the Third IEEE Conference on vol.2 (1994) 1240–1245.
9. A.K. Jain, R.C. Dubes: Algorithms for Clustering Data. Prentice-Hall, (1988).
10. G. Gustefson and W. Kessel: Fuzzy clustering with a fuzzy covariance matrix. in Proc. IEEE CDC, , San Diego, (1979) 761-766.
11. Berenji, H.R.; Ruspini, E.H.: Experiments in multiobjective fuzzy control of hybrid automotive engines. Fuzzy Systems Proceedings of the Fifth IEEE International Conference on, Volume: 1, (1996) 681–686.
12. V. Galdi, L. Ippolito, A. Piccolo, and A. Vaccaro: Optimisation of Energy Flow Management in Hybrid Electric Vehicles via Genetic Algorithms. In Proceedings of 2001 IEEE/ASME International Conference on Advanced Intelligent Mechatronics, Como, Italy (2001) 434-439.
13. V. Galdi, L. Ippolito, A. Piccolo and A. Vaccaro: Evaluation of Emissions Influence on Hybrid Electric Vehicles Sizing. In Proceedings of International Conference on Power Electronics, Electrical Drives, Advanced Machines and Power Quality, Ischia, Italy June, (2000).
14. V. Galdi, L. Ippolito, A. Piccolo and A. Vaccaro: A genetic based methodology for Hybrid Electric Vehicle Sizing. Soft Computing, vol 5, issue 6, (2001) 451-457.

A Fuzzy Logic Vision and Control System Embedded with Human Knowledge for Autonomous Vehicle Navigation

Charles E. Kinney and Dean B. Edwards

University of Idaho, Department of Mechanical Engineering, Moscow, Idaho
83844-0902 USA

Abstract. A method for visually navigating an autonomous vehicle in
an unstructured environment with fuzzy logic is summarized in this paper. Fuzzy logic is used with a human embedding method to process
the raw vision data and control the autonomous vehicle. Specifically, the
ability to classify pixels in the image space is discussed as a means of
locating a safe path. Human embedding is used as a method to train
the fuzzy logic system. The trained system is shown able to follow and
recover to paths in the unstructured environment. Additionally, a hierarchical system that fuses the visual information with other sensors is
discussed. The addition of an optical sensor provides the controller with
critical information that helps make intelligent decisions in the presence
of sensor imprecision and uncertainty.

1 Introduction

Autonomous vehicles have been in development for several years. Providing au-
tonomy to robots or vehicles can improve the safety of workers in hazardous
environments as well as assuage the economic burden sustained by companies.
"Possible applications include intelligence service robots for offices, hospitals, and
factory floors; maintenance robots operating in hazardous or accessible areas; do-
mestic robots for cleaning or entertainment; autonomous and semi autonomous
vehicles for help to the disabled and their elderly; and so on[1]."

Another potential application being developed is an autonomous vehicle, log
skidder, used for pulling logs from the area where the timber is cut to another
area where the logs are loaded on a truck. The University of Idaho has been
developing a fuzzy logic controller for a real-time autonomous log skidder over
the last several years[2-9]. The main goal of this project is to enable a log skidder
to recognize paths in the woods and safely navigate them in real-time without
human supervision while pulling logs from one area to the next. "The use of
conventional equipment for thinning and other small-scale operations is usually
not economically justified and can, in fact, cause damage to the soil and trees[2]."
The advent of an autonomous log skidder would provide companies interested in
small-scale timber harvesting an economical and environmentally safe solution.

This type of application necessitates the use of other sensing besides dead
reckoning and enough intelligence for the autonomous agent to perform in a

T. Bilgiç et al. (Eds.): IFSA 2003, LNAI 2715, pp. 526–534, 2003.
© Springer-Verlag Berlin Heidelberg 2003

human-like manner. Global positioning system (G.P.S.) sensors are unreliable due to interference from trees and canyons. Inertial sensors, encoders and other means of determining position of the vehicle accrue error and are unreliable. The imprecise knowledge about the location of the trail and aggregate error from dead reckoning sensors creates the need for other sensing.

An image processing system can potentially fill the void in information. A camera is an inexpensive and effective way to gain more knowledge about the environment quickly. Many robotic vehicles that include visual servoing in the control loop detect edges in the image space to determine the control of the vehicle[10]. Urbie, a joint effort of JPL, iRobot Corp., robotics Institute of Carnegie-Mellon University, and the University of Southern California robotics research laboratory, detects edges for autonomous vision-guided stair climbing[11,12]. Other behaviors detect obstacles in the image field by using stereoscopic vision, light detection and ranging (LIDAR), or acoustic sensors. Recently, researchers at the Jet Propulsion Laboratory have combined LIDAR range information with image texture and color to detect obstacles and classify terrain[13]. Combining sensors and including more information from the image space seems to improve obstacle detection over classical techniques.

A fuzzy logic controller developed at the University of Idaho uses sensors to recognize its environment. Shaft encoders are used to determine the heading of the vehicle. Acoustic sensors are used to determine the location of obstacles and the trail. The fuzzy logic controller uses a hierarchical scheme to gather the sensor data and implement intelligent decisions to control the autonomous vehicle down a trail in the woods. Edwards, Canning, Anderson, and Carlson[2-5,7-9] have shown the viability of fuzzy logic for autonomous vehicle control. To date, the vehicle has been able to maneuver down a 600 foot path unsupervised[4].

This hierarchical fuzzy logic controller is based on a predictor corrector method of trail following. Figure 1 shows the hierarchical structure of the controller. The solid lines represent previous work and the dashed lines represent the work outlined in this paper. The optical trail finding module is a combination of the fuzzy logic vision and control system, hardware, and software. To date, four different modules are combined together with the supervisory module that uses quality factors and a training run to determine the optimal settings of the system[3,4,7]. Quality factors represent the confidence in the module recommendation. Recommendation with a high quality factor are weighted heavier when the supervisory module combines the sensor information. Using quality factors enables the hierarchical control system to deal with uncertainty associated with unconstrained environments like the forest. The dead reckoning module comprises shaft encoders mounted on each track of the vehicle and predicts where the trail should be located. The trail finding, path memorization, and obstacle avoidance modules correct the heading given by the dead reckoning module. The error in the dead reckoning module is cumulative and the corrective modules become more important to the supervisory module as the vehicle travels farther down the trail.

2 Fuzzy Logic Image Processing System

This section explains how pixels in images are identified as belonging to the crisp sets $< Trail >$ and $< Non-Trail >$. The Fuzzy Logic Image Processing System (FLIPS) is divided into three main sections: PREPROCESSING, TRAILFIND-ING, and POSTPROCESSING. The first section reduces the image resolution, adds contrast by adjusting the histogram, and filters the image to reduce noise. The second section is a fuzzy logic system that uses inputs from the image to rank the belonging of the pixels to a fuzzy set $\{Trail\}$. The final section applies a threshold and thins the output image, thus dividing the fuzzy set $\{Trail\}$ into a crisp set $< Trail >$. Figure 2 shows an overview of the fuzzy image processing system. The solid boxes are functions, the lines are

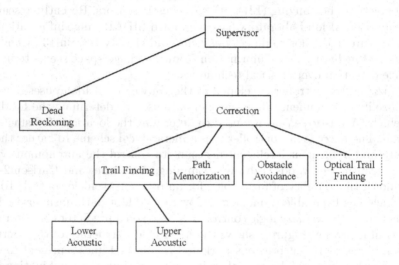

Fig. 1. Hierarchical system overview. The solid lines represent previous work and the dashed lines represent work outlined in this paper

variables, and the images are examples of the processes presented in this chapter. The color images are taken with a CCD camera (480X640 pixels) and are converted into a matrix $[I]$, or $I(i, j, k)$, that is 480X640X3. The indices i and j indicate the spatial location of the pixel, and the last dimension $k = 1,2,3$ represents the color for each pixel. In matrix form, the image is readily analyzed. In this paper, the term "image" represents this matrix format.

2.1 Trail Features

The measurable features of a trail must be defined before the trail can be found[6]. All of the trails are brown and smooth in texture. The shade of brown varies, but the trail remains almost always a shade of brown. The trail also seems to have a relatively smooth texture compared to the surroundings. These two qualities (color and texture) will be measured in order to identify the presence of a trail in images. The surroundings, for the most part, remain green. This simplifies the color definitions. The difference between brown and green only has to be distinguished, not the difference between brown and every other color. An example trail image is shown in Fig. 2.

2.2 PREPROCESSING

PREPROCESSING uses the nearest neighbor interpolation method to reduce the image size; A high resolution isn't required to identify the major features within the image space. Noise in the images are filtered out with a median filter $[M]$. $[M]$ is a 5x5 matrix with each element equaling 1/25, and it averages the values of the pixels surrounding the pixel in question. This averaging process smooths out discontinuities in the image and is accomplished by the convolution of $[M]$ with the image $[I]$. The result is a convolved image with the same size of $[I]$. The histogram is adjusted to 256 different levels with approximately the same number of pixels at each level. This method creates a histogram that is almost flat across the different pixel levels. The final result of PREPROCESSING is a image $[Ip]$. Figure 2 shows an example image after PREPROCESSING. The contrast between the trail and surroundings is more distinct.

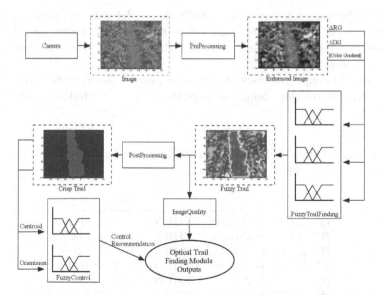

Fig. 2. Optical trail finding module overview. The images are examples of the processes presented in this paper

2.3 TRAILFINDING

The inputs to the TRAILFINDING section of FLIPS are mearsurables of the trail characteristics. Two color input variables are used to distinguish between the color of the trail and the color of the surroundings, the difference between Red and Green ΔRG and the difference between Blue and Green ΔBG.

$$\Delta RG(i,j) = I_p(i,j,1) - I_p(i,j,2) \, . \tag{1}$$

$$\Delta BG(i,j) = I_p(i,j,3) - I_p(i,j,2) \, . \tag{2}$$

Using the difference in these colors seems to be adequate to describe the difference between most shades of brown and green. Most shades of brown have a positive ΔRG and ΔBG while, most shades of Green did not exhibit this trend in the color differences. The other variable used in TRAILFINDING to locate the trail is the squared color gradient ∇C^2. ∇C^2 is the summation of the squared gradient of each color channel[14]. This variable is used to describe the smoothness of the trail in contrast to the high texture of the surroundings.

$$\nabla C(i,j)^2 = \sum_{k=1}^{3} \|\nabla I_p(i,j,k)\|^2 \, . \tag{3}$$

The gradient of an image is commonly used in detecting edges[15]. This input is expected to be beneficial because the surroundings of the trail are filled with many objects, hence, many edges. Smoother regions of images, where the trail is present, possess fewer edges.

The membership functions chosen to represent these inputs were triangular[16] with 50% overlap. Singletons are used for the output sets. The rulebase is shown below :

Table 1. Fuzzy logic rule base for FLIPS. This rule base determines the belonging of each pixel to the fuzzy set {Trail}

IF	ΔRG	is *Positive*	**THEN**	*Trail*	is *High*
IF	ΔRG	is *Zero*	**THEN**	*Trail*	is *Medium*
IF	ΔRG	is *Negative*	**THEN**	*Trail*	is *Low*
IF	ΔBG	is *Positive*	**THEN**	*Trail*	is *High*
IF	ΔBG	is *Zero*	**THEN**	*Trail*	is *Medium*
IF	ΔBG	is *Negative*	**THEN**	*Trail*	is *Low*
IF	∇C^2	is *Low*	**THEN**	*Trail*	is *High*
IF	∇C^2	is *Medium*	**THEN**	*Trail*	is *Medium*
IF	∇C^2	is *High*	**THEN**	*Trail*	is *Low*

Height defuzzification[16] was used to combine the rule base. Height defuzzification is one of the simplest methods to implement and it is very fast to calculate. Figure 2 shows an example of an image after TRAILFINDING. The red areas are trail and the blue areas are non-trail (or dark are trail and light are non-trail if the paper is in black and white).

2.4 POSTPROCESSING

POSTPROCESSING thresholds the fuzzy image to a crisp image by finding a minimum in the linear index of fuzziness . Kaufmann[17,18] introduced the linear index of fuzziness for an image which can be written as

$$\Gamma(T) = \frac{2}{MN} \sum_{x=0}^{L-1} H(x)(min(\mu(T,x), 1 - \mu(T,x))) \ . \tag{4}$$

Here, $H(x)$ is the histogram of the image after TRAILFINDING. T is a potential threshold location, x the location on the histogram, and $\mu(T,x)$ is the membership function that describes the image for the potential threshold location T over the range of histogram values with L different levels.

Thinning the trail is needed because, isolated pixels can be falsely labeled as trail. Only pixels labeled as trail close to other pixels labeled as trail are desired. A summing filter is applied to the binary image. A summing filter sums the pixel values around the pixel of interest. The filter used was 5x5, and each element had a value of one. The convolution of this filter with the image results in a new image, each pixel can range from zero to 25. Pixel locations where the value is at least 15 are labeled as trail and redefined as one. Those pixel locations with a value less than 15 are labeled as non-trail and redefined as zero. FLIPS outputs a binary image $[Tf]$. Where one corresponds to the crisp set $< Trail >$ and zero corresponds to the crisp set $< Non - Trail >$. Figure 2 shows an example image after POSTPROCESSING. Notice that the image is crisp, or binary, and the trail was accurately located.

3 Fuzzy Logic Autonomous Control System

This section describes the Fuzzy Logic Autonomous Control System (FLACS). FLACS uses the output from FLIPS to determine the correct heading adjustment for an autonomous vehicle. FLIPS provides a two-dimensional image $[Tf]$ specifying where the trail is located. The pixels with the value of one indicate location of the trail.

FLACS uses the centroid of the trail in the horizontal direction, Centroid, and the orientation of the trail with respect to horizontal, Orientation, as inputs to a fuzzy logic system that determines the correct heading adjustment for the autonomous vehicle. The orientation of the trail is determined by fitting an ellipse with the same second moments of inertia around the area in the image labeled as trail. The angle in degrees between the major axis of the ellipse and horizontal direction is the orientation of the trail.

The input membership functions are triangular[16] and have 50% overlap with the adjacent sets. Singletons are used for the output set which are placed at 90 degrees, zero degrees, and -90 degrees. This structure was chosen because it is simple and it gives a linear control response until the control saturates[19].

The fuzzy logic rule base for FLACS reflects the mechanics of the vehicle in the environment as well as the judgement of the operator. The rule base used for FLACS is shown below :

Table 2. Fuzzy logic rule base for FLACS. This rule base determines the correct change in heading to keep the autonomous vehicle aligned with the trail

IF *Centroid*	is *Left*	**THEN**	*ΔHeading*	is *Left*
IF *Centroid*	is *Center*	**THEN**	*ΔHeading*	is *Center*
IF *Centroid*	is *Right*	**THEN**	*ΔHeading*	is *Right*
IF *Orientation*	is *Left*	**THEN**	*ΔHeading*	is *Left*
IF *Orientation*	is *Center*	**THEN**	*ΔHeading*	is *Center*
IF *Orientation*	is *Right*	**THEN**	*ΔHeading*	is *Right*

Height defuzzification[16] was used to combine the rule base.

4   Human Embedding

The control recommendations of FLACS were compared to the control recommendations of a human operator for a set of trail images. The membership functions that are used in FLACS were varied until the difference of the operators control recommendations and FLACS control recommendations was minimized for the set of trail images. In this manner, human knowledge was embedded[9] into FLACS to yield the desired behavior.

The simplex method was used to vary the fuzzy parameters in order to reduce the performance index J defined as

$$J = \sum_{n=1}^{N} (H(n) - C(n))^2 \,. \tag{5}$$

Here, J is the summation over N training images of the difference between the human recommendation $H(n)$ and the recommendation from FLACS $C(n)$ squared. When the error is reduced, FLACS is trained to behave like a human. That is, for the same images, the recommendation from FLACS and a human should be very similar, if not identical. The simplex method is a nonlinear optimization technique that does not use differentiation. Therefore, functions that are not continuously differentiable can still be optimized, a shortcoming in the gradient descent method.

5 Experimental

An experiment was conducted at the University of Idaho's Arboretum in order to verify the performance of the optical trail finding module using FLIPS and FLACS. The test measured the ability of an autonomous vehicle using the optical trail finding module to stay on the path. The autonomous vehicle was able to successfully navigate a short trail several times for a total of approximately 100 feet.

6 Conclusions

The goal of this paper was to summarize a fuzzy logic vision and control system that could be used by a hierarchical controller used to navigate an autonomous vehicle down a path in the forest. A Fuzzy Logic Image Processing System, FLIPS, correctly label pixels in an image $< Trail >$ or $< Non - Trail >$. A Fuzzy Logic Autonomous Control System, FLACS, determines the correct control recommendation to safely navigate the vehicle down a path. FLACS was embedded with human knowledge to produce the desired response.

An autonomous vehicle was able to successfully navigate a trail in the woods with FLIPS and FLACS. The feasibility of adding an optical trail finding module into a hierarchical controller has been demonstrated. More experiments, however, need to be conducted to verify the performance improvement that the optical trail finding module will provide the whole system. The experiments conducted were only for the optical trail finding module. The experiments demonstrated the ability of the optical trail finding module to guide the autonomous vehicle down a path. However, the optical trail finding module is not a stand-alone solution to controlling the vehicle and needs to be implemented with the hierarchical control system. Without the obstacle avoidance and dead reckoning modules the autonomous vehicle could collide with trees, people, or other dangerous features. Further, in situations where no or little distinction exists between the trail and its surroundings, such when the ground is covered with snow, the optical trail finding module would be ineffective. However, in these situations the quality would be zero and the hierarchical controller would not use the recommendation from this module.

References

1. Driankov Dimiter, Saffoitti Alessandro: Fuzzy Logic Techniques for Autonomous Vehicle Navigation. Physica-Verlag, New York (2001)
2. Canning John R.: Development of a Fuzzy Logic Guidance System for a MobileRobot. Thesis (August 1999)
3. Carlson Alan C.: Development of a Robust Hierarchical Fuzzy Logic Control System for an Autonomous Vehicle. Thesis (November 2000)
4. Anderson Ian: A Fuzzy Logic Controller That Uses Quality Factors to Blend Behaviors for an Autonomous Forest Vehicle. Thesis (November 2000)

5. Edwards Dean B., Canning John R.: Developing a Radio Controlled Log Skidder With Fuzzy Logic Autonomous Control. Transactions of the ASAE, Vol.38, No. 1 (1995) 243-248
6. Edwards Dean B., Kinney Charles E.: A Fuzzy Logic Trail Finding Algorithm Trained With Human Knowledge. DETC2002/CIE-34396 (September 2002)
7. Alan Carlson, Dean B. Edwards, Michael J. Anderson: Fuzzy Quality Measures for Use in a Hierarchical Fuzzy Controller. Computers and Information in Engineering Conference, Baltimore, Maryland (September 2000)
8. Edwards, D., J. Canning: Fuzzy Control System for an Autonomous Vehicle. Ninth International Conference for the Journal of Mathematics and Computer Modeling, Berkeley, Ca (July 26th-29th, 1993)
9. J.R. Canning, D.B. Edwards: A Method for Embedding Human Expert Knowledge into a Fuzzy Logic Controller. ASME 15th International Computers in Engineering Conference, Boston, MA (September 1995) 1019-1023
10. Boudihir M. Elarbi, Nourine R., Ziou D.: Visual Guidance of Autonomous Vehicle Based on Fuzzy Preception. IEEE International Conference on Intelligent Vehicles (1998)
11. Xiong Yalin, Matthies Larry: Vision Guided Autonomous Stair Climbing. IEEE International Conference on Robotics and Automation (ICRA), San Fransico, California (April 2000)
12. Matthies, Y. Xiong, R. Hogg, D. Zhu, A. Rankin, B. Kennedy: A Portable, Autonomous, Urban Reconnaissance Robot. Sixth International Conference on Intelligent Autonomous Systems, Venice, Italy (July 2000)
13. A Talukder, R. Manduchi, R. Castano, K. Owens, L. Mathies, A.Castano, R. Hogg: Autonomous Terrain Characterisation and Modelling for Dynamic Control of Unmanned Vehicles. IEEE IROS 2002, Switzerland (Sept 30-Oct 2, 2002) Submitted
14. Koschan A.: A Comparitive Study on Color Edge Detection. Proceedings 2nd Asian Conf. On Computer Vision ACCV'95 (1995) 574-578
15. Ziou D., Tabbone S.: Edge Detection Techniques: an Overview. Technical Report, No. 195, Dept. Math and Informatique, Universite de Sherbrooke (1997)
16. J. Yen, R. Langari: Fuzzy Logic: Intelligence, Control, and Information. Prentice Hall (1999)
17. Forero-Vargas M.G., Rojas-Camacho O.: New Formulation in Image Thresholding Using Fuzzy Logic. http://citeseer.nj.nec.com/ (January 2003)
18. Kerre E. E., Nachtegael M.: Fuzzy Techniques in Image Processing. Physica-Verlag, New York (2000)
19. Edwards D.B., Canning J.R.: An Algorithm for Designing Conventional and Fuzzy Logic Control Systems. ASME, 17th International Computers and Information in Engineering, Sacramento, CA (September 1997)

Fuzzy Logic Based Dynamic Localization and Map Updating for Mobile Robots

Mohammad Molhim[1] and Kudret Demirli[2]

[1] Department of Mechanical Engineering, Jordan University of Science and
Technology, Irbid, Jordan,
P.O.Box. 3030
[2] Fuzzy Systems Research Laboratory, Department of Mechanical Engineering,
Concordia University, Montreal, Quebec, Canada H3G 1M8
demirli@me.concordia.ca

Abstract. A fuzzy logic based dynamic localization and map updating
algorithm for mobile robots with ring configuration is introduced. A fuzzy
composite map is constructed based on the sonar readings, obtained
from the robot's environment, and fit to the global map. Then, two sets
of the fuzzy local composite map components are identified. The set of
the matching components and the set of non-matching components. The
former is used to update the current robot's location and the later is
used to update the map of the robot's environment by adding new line
segments. The proposed algorithm is implemented in a real environment
and the results show the effectiveness of the proposed algorithm.

1 Introduction

Mobile robot navigation requires accurate location information. The location
information can be provided on-line for the mobile robot through two types of
devices; proprioceptive and exteroceptive. Odometers are proprioceptive devices
that can provide valid location information only over short distances. Sonar,
laser, and camera are exteroceptive devices that can be used for robot localization based on triangulation or perception.

There are two main localization problems in the literature: One deals with the
situations where the robot has no information about its location in its world.
This is called *Global localization* [6]. The other one deals with the situations
where the robot has a priori information about its location and the robot needs
to update its location while traveling based on the collected sensory information
[7,4].

In this paper we are concerned with the location updating problem. Moreover,
if there is any new changes to the robot's environment, the robot updates its
environment in the form of line segments based on its sonar readings.

There are different localization algorithms in the literature introduced to
solve the location updating problem. In these algorithms the robot constructs a
partial map of its proximity in the form of line segments [1], fuzzy perceptual

T. Bilgiç et al. (Eds.): IFSA 2003, LNAI 2715, pp. 535–543, 2003.

clues [7], and fuzzy line segments [4]. Then, the partial map is matched with global map and if there is a high degree of match the robot's location is updated.

Hoppenot and Colle [5] introduce a localization algorithm that relies on the sonar measurements which are estimated based on the current robot's position obtained by the odometers. These points of impact are then matched to the nearest segment of the robot's environment. The authors use only seven sonar sensors distributed on half ring. The angular and radial uncertainty of sonar readings are not considered.

Demirli and Türkşen [3] introduce a fuzzy logic based static localization algorithm. Their algorithm only estimates the robot's location based on sonar readings while the robot is stationary.

2 Modeling Uncertainty in Sonar Sensors

Sonar sensors, despite their low cost advantage, have some inherent problems that affect their performance. Namely, false reflections, and uncertainty in the orientation and the range of the detected objects. The last two are called *angular and radial uncertainty*, respectively [2].

The angular and radial uncertainty models are presented in [2]. These models are obtained for two sensors detecting the same object (hard wall) and the angle between these two sensors is known.

In Figure 1(a) the angular uncertainty model for sonar readings obtained from two sensors detecting the same object is given as $\Pi_{\Gamma|L}$. In this figure, L_1 is the reading obtained from Sensor 1 and γ_1 is the incidence angle of Sensor 1 with respect to the robot's coordinates. Similarly, L_2 and γ_2 are defined in with respect to Sensor 2. Depending on the values of L_1 and L_2, an appropriate model is selected from this figure. For example, if $L_1 < L_2$, the incidence angle of the sensor with respect to the negative surface normal of the wall is represented by the following trapezoidal possibility distribution:

$$\Pi_{\Gamma|L_1} = [\gamma_1 - 25, \gamma_1 - 11.5, \gamma_1 + 2.5, \gamma_1 + 2.5]$$

As demonstrated experimentally in [2], the radial uncertainty of sonar readings depends on the incidence angle of the sensor and the distance between the sensor and the wall, therefore we obtain radial uncertainty model for each angle belonging to the support of the possibility distribution that represents the angular uncertainty. This is shown in Figure 1(b) for the angle 11.5^o when the sensor is placed at different distances from the wall.

2.1 Fuzzy Local Composite Map

The local composite map consists of sonar data obtained from adjacent sensors with close readings. An example of a local composite map is shown in Figure 2(a), where the map is represented as:

$$\mathcal{M} = \{(L_0, L_1), (L_3, L_4), (L_6, L_7), (L_{11}, L_{12})\}.$$

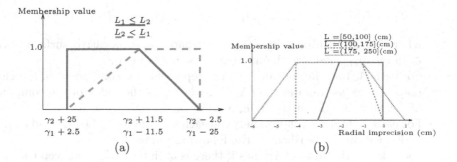

Fig. 1. (a)Angular uncertainty for readings coming from a wall. (b)Reduced radial imprecision when the sensor incidence angle is 11.5°.

This map is simplified as shown in Figure 2(b) and transferred to fuzzy local composite map as shown in Figure 2(c) and represented as:

$$\mathcal{F}=\{\ (\Pi_{A|L_i}, \Pi_{L_i|\lambda})\ \},\ \text{for all } i \in I,$$

where $I = \{i \mid L_i \in \mathcal{S}\}$, $\Pi_{A|L_i}$ is the angular uncertainty associated with L_i and $\Pi_{L_i|\lambda}$ is the radial uncertainty model associated with limits of the core of the angular uncertainty model.

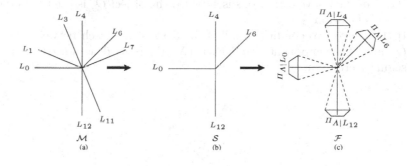

Fig. 2. (a)Local map (b) Simplified local map (c)Fuzzy local map

2.2 Fitting the Fuzzy Local Composite Map to the Global Map

Consider a fuzzy local composite map as defined above, then this map is fit to the global map as follows:

- For each element of $\mathcal{F}$, search the global map of the robot to find the set of objects that have negative surface normals in the support of $\Pi_{\Theta|sL_i}$. This set is represented as

$$A_i = \{O_1, \ldots, O_j, \ldots, O_k\} \forall i \in I$$

where O_j is an object in the global map whose negative surface normal $\in supp\Pi_{\Theta|sL_i}$ and k is the number of elements in A_i.

- For each A_i find the coordinates that represent this component in the global map. These coordinates are (X_i, Y_i), where i is the index of the component of the fuzzy local composite map.
- For each A_i find the relation between each element in A_i, i.e., O_j and (X_i, Y_i). This relation is described by the following parameters:
  - α is a ratio that determines if there is an intersection between the line L (Figure 3), that links the current location of the robot and the (X_i, Y_i), and any global line segment O_j.
  - μ is the ratio that determines if (X_i, Y_i) is within the start and the end points of any line segment that belongs to the global map.
  - (X_E, Y_E) is the intersection point between the line L and the object O_j.
  These ratios are utilized as follows:
  - If $0 \le \mu \le 1$, $0 \le \alpha \le 1$, and $D \le \zeta$, then there is a match between the object O_j and the component A_i, where ζ is a threshold value determined experimentally.
  - If $0 \le \mu \le 1$, and $V \le \zeta$, then there is a match between the object O_j and the component A_i, however, in this case the line L has no intersection with the object O_j and V is the distance between (X_i, Y_i) and the object O_j.
- If one of the above conditions is satisfied the object O_j is used to update the robot's location.
- If the above two conditions fail then there is no match between the object O_j and the component i and then (X_i, Y_i) is used as a seed for new line segment creation.

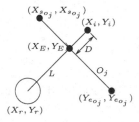

Fig. 3. Local map

2.3 Shortest Distance to an Object

Consider the case in Figure 1 where $L_1 < L_2$, we pick L_1 and rotate $\Pi_{L_1|\gamma}$ by γ to compute the shortest distance between the sensor and the object. Since the incidence angle is not a precise number but represented by a possibility distribution $\Pi_{\Gamma|L_1}$, $\Pi_{L_1|\gamma}$ should be rotated by all the incidence angles γ in the support of $\Pi_{\Gamma|L_1}$. This rotation is shown in Figure 4 for the angles that have a possibility value one.

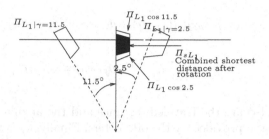

Fig. 4. Conjunctive fusion of two pieces of information to obtain the shortest distance to a wall and corner.

2.4 Fuzzy Triangulation

The shortest distances are used to calculate x and y coordinates with respect to the detected objects. These coordinates are obtained as follows:

$$\Pi_{(X,Y)|tL_i} = \sum_j \pi_j/((X_{O_k}, Y_{O_k}) - sL_i(\cos, \sin)(N_k)), \forall (sL_i)_j \in supp\Pi_{sL_i}$$

$$\Pi_{\Theta|tL_i} = \sum_j \pi_j/(\kappa_{1i} + \lambda_i), \forall (\lambda_i)_j \in supp\Pi_{\Lambda|SL_i}$$

Finally, the robot's coordinates in the global map are obtained by combining all the possibility distributions that represent the x, y and θ coordinates as follows:

$$\Pi_{(X_{so}, Y_{so}, \Theta_{so})} = \bigcap_{i \in I} \Pi_{(X|tL_i, Y|tL_i, \Theta|tL_i)}$$

2.5 Sensor Fusion for Localization

In this paper, the traveled distance of the robot since the previous time-step provided by the odometers is considered to estimate the robot's location. Then, this obtained location is fused with the location information obtained by the sonar to estimate the current robot's loation. The components of current robot's location after the fusion are obtained as follows:

$$\Pi_{(X_f,Y_f,\Theta_f)} = \Pi_{(X_{od},Y_{od},\Theta_{od})} \bigcap \Pi_{(X_{so},Y_{so},\Theta_{so})}$$

where $\Pi_{X_{od}}$, $\Pi_{Y_{od}}$, and Π_Θ are the possibility distributions that represent the components of the robot's location obtained based on the odometers. These components are obtained as follows:

$$\Pi_{(X_{od},Y_{od})} = \sum_{(x_f,y_f)} (\pi_{x_f}, \pi_{y_f}),/((x_f,y_f) + d(\cos,\sin)(\bar{\theta}_f)), \forall (x_f,y_f) \in supp\Pi_{(X_f,Y_f)}$$

$$\Pi_{\Theta_{od}} = \sum_{\theta_f} \pi_{\theta_f}/(\phi + \theta_f), \forall \theta_f \in supp\Pi_{\Theta_f}$$

where d and ϕ are the traveled distance and the angular rotation since the previous time-step provided by the odometers. Finally, $\bar{\theta}_f$ is the crisp value that represents the current location of the robot and it is obtained using the center of gravity method as shown next. Note that in the first localization cycle, i.e., before any movement of the robot, $\Pi_{X_{od}} = \Pi_{X_{so}}$, $\Pi_{Y_{od}} = \Pi_{Y_{so}}$, and $\Pi_{\Theta_{od}} = \Pi_{\Theta_{so}}$.

2.6   Line Segments Extraction

When a component of the fuzzy composite map doesn't fit into the global map, then, this component can be used to initialize a line segment, create a line segment, or to update the parameters of a created line segment. These three cases are explained as follows:

1. **Initialization of a new line segment**
   In this case the component of the fuzzy composite map is used to obtain the starting point and the normal to the surface of a candidate line segment.
2. **Create a new line segment**
   If we have an initialized line segment and a component of the fuzzy composite map that doesn't fit into the global map, then the coordinates of this component is checked with the initialized segment parameters to see if there exists a relation between them. This relation is determined as follows: $\sqrt{(X_{seg} - X_c)^2 + (Y_{seg} - Y_c)^2} \leq \beta_1$ and $\pi_{\Gamma|d_i}(\theta_{seg}) \geq \beta_2$, where β_1 and β_2 are thresholds. If both of these relations are satisfied the coordinates of the component are considered as the ending coordinates of the initialized line segment. This step will complete the parameters of the initialized segment and then we will have a new created segment.
3. **Updating the parameters of a created segment**
   If we have a created line segment and a component of the fuzzy composite map that doesn't fit into the global map, the coordinates of this component can be used to update the parameters of the line segment. Figure 5 shows a line segment and the coordinates of two different components labeled as A and B. The A coordinates are used to update the starting coordinates of the line segment. Similarly, the B coordinates are used to update the ending coordinates of the line segment.

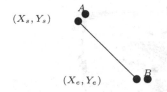

(X_s, Y_s)

(X_e, Y_e)

Fig. 5. Created line segment.

3 Experimental Results

3.1 Localization Experiments

Nomad 200 robot from Nomadic Technologies which has a ring of 16 sonar sensors is used in these experiemnts.

Nomad 200 programmed to follow the path shown in Figure 6(a). The robot had an initial knowledge of its location in the global map. While the robot was moving along its path, it was stopped at different places A, B, C, and D as shown in Figure 6(a) to obtain actual measurements of its position. In addition, the robot's position obtained from the odometers was registered. Then, the error in the robot's position is calculated as the difference between the measured position and the position provided by the odometers. The obtained error versus the localization iterations is plotted by using a dashed line as shown in Figure 6(b). This figure shows that the error obtained by relying only on the odometers is accumulating with time. For example, when the robot is at location A the error provided by the odometers is 73 cm (see Figure 6(b) for robot's path). This error is accumulated over the robot's path until the robot reaches location A. These results support the fact that odometers are not reliable over long paths.

In the second test, the robot follows the same path, however, our proposed localization algorithm is used to continuously update the robot's position while it is moving along its path. The robot was stopped at different places A, B, C, and D as shown in Figure 6(a) so that we can take measurements of its actual position. At the same time we register the position information obtained from the proposed localization algorithm at these locations. We calculate the error in the robot's position. The error versus the number of localization iterations needed to reach each point is plotted and shown by the dashed-dotted line in Figure 6(b). This line shows that the error in the robot's position which is corrected by our localization algorithm is very small and its maximum value is around 2 cm. This is a good range of error when using sonar sensors for localization.

3.2 Map Updating Experiments

In Figure 6(a), one object was added to the robot's environment and while the robot was following its path this object was identified using the sonar readings that did not have matching components in the global map. During this process

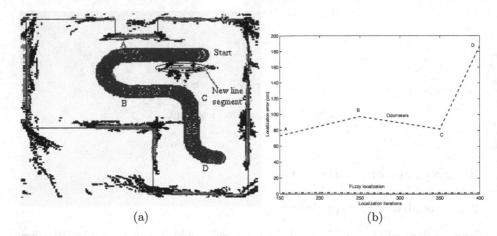

(a)   (b)

Fig. 6. (a)Experimental verification of the proposed localization algorithm.(b) Localization error versus localization iterations.

the number of the line segments in the robot's environment incremented as a result of updating the robot's map by adding this line segment.

4 Conclusions

A fuzzy logic based dynamic localization and map updating algorithm for mobile robots with ring configuration is introduced. The algorithm relies on the sonar readings obtained from the robot's environment. Moreover, it takes into account the uncertainty associated with the sonar readings. This algorithm is implemented on a Nomad 200 mobile robot and satisfactory results are obtained.

References

1. J. Crowley. World modeling and position estimation for a mobile robot using ultrasonic ranging. In *Proceedings of the IEEE International Conference on Robotics and Automation*, pages 674–680, 1989.
2. K. Demirli, M. Molhim, and A. Bulgak. Possibilistic sonar modeling for mobile robots. *International Journal of Uncertainty, Fuzzieness, and Knowledge-based Systems*, 7(2):177, April 1999.
3. K. Demirli and I. Türkşen. Sonar based mobile robot localization by using fuzzy triangulation. *Robotics and Autonomous Systems*, 33(2-3):109–123, December 2000.
4. J. Gasós and A. Martín. Mobile robot localization using fuzzy maps. In *A. Ralescu and T. Martin, editors, Lecture Notes in Artificial Intelligence. Springer Verlag*, 1996.
5. P. Hoppenot and E. Colle. Real time localizing of a low-cost mobile robot with poor ultrasonic data. *Control Engineering Practice*, 6:925–934, 1998.

6. M. Piasecki. Global localization for mobile robots by multiple hypothesis tracking. *Robotics and Autonomous Systems*, 16:93–104, 1995.
7. A. Saffiotti and L. Wesley. Perception-based self localization using fuzzy locations. In *M. van Lambalgen (Ed.) Reasoning with Uncertainty in Robotics. Springer LNCS*, pages 368–386, 1996.

An Overview on Soft Computing in Behavior Based Robotics

Frank Hoffmann

Fakultät Elektrotechnik und Informationstechnik
Universität Dortmund
D-44221 Dortmund (Germany)
hoffmann@esr.e-technik.uni-dortmund.de

Abstract. This paper provides an overview on the contribution of soft computing to the field of behavior based robotics. It discusses the role of pure fuzzy, neuro-fuzzy and genetic fuzzy rule-based systems for behavior architectures and adaptation. It reviews a number of applications of soft computing techniques to autonomous robot navigation and control.

1 Introduction

In many robotic applications, such as autonomous navigation in unstructured environments, it is difficult if not impossible to obtain a precise mathematical model of the robot's interaction with its environment. Even if the dynamics of the robot itself can be described analytically, the environment and its interaction with the robot through sensors and actuators are difficult to capture in a mathematical model. The lack of precise and complete knowledge about the environment limits the applicability of conventional control system design to the domain of autonomous robotics. What is needed are intelligent control and decision making systems with the ability to reason under uncertainty and to learn from experience.

It is unrealistic to assume that any learning algorithm is able to learn a complex robotic task, in reasonable learning time starting from scratch without prior knowledge about the task or the environment. The situation is analogous to software design in which the design process is constrained by the three mutually conflicting constraints cost, time and quality. Optimization of one or two of the objectives, often results in a sacrifice on the third objective. In robot learning the three conflicting objectives are complexity of the task, number of training examples or episodes and prior knowledge. Learning a complex behavior in an unstructured environment without prior knowledge requires a prohibitively long exploration and training phase and therefore creates a serious bottleneck to realistic robotic applications.

Task complexity can be reduced by a divide and conquer approach, which attempts to break down the overall problem into more manageable subtasks. This course is advocated by hierarchical behavior architectures, in that they separate the design or adaptation of primitive behaviors from the task of learning

T. Bilgiç et al. (Eds.): IFSA 2003, LNAI 2715, pp. 544–551, 2003.

a supervisory policy for behavior coordination [3, 12, 14]. The designer biases the learning algorithm towards solutions consistent with the problem specific, prior expert knowledge. Fuzzy control offers a means to integrate explicit domain knowledge in form of linguistic rules that describe the behavioral mapping from perception to action. These rules constitute an initial, sub-optimal behavior that is later refined through experiences gathered from the robot's interaction with the environment [3, 7, 13, 15].

Artificial neural networks and evolutionary algorithms draw inspiration from the capabilities of animals and humans to adapt and learn in dynamic environments under varying conditions, situations and tasks. Fuzzy logic is inspired by the approximate type of reasoning that allows humans to make decisions under uncertain and incomplete information. In the context of the above mentioned trade-offs imposed on robot learning, fuzzy techniques offer a means to sacrifice optimal performance for a reduction in complexity, elimination of unnecessary details and increased robustness of solutions.

Section 2 describes hierarchical approaches and methodologies for fuzzy behavior design and coordination. Section 3 recounts neuro-fuzzy techniques for supervised behavior adaptation. Section 4 discusses the role of evolutionary algorithms for learning primitive fuzzy behaviors and behavior coordination schemes.

2   Fuzzy Behaviors for Robot Control

Behavior coordination architectures can be divided into two categories: arbitration and command fusion schemes. In arbitration, the selected dominant behavior solely controls the robot until the next decision cycle, whereas the motor commands of the suppressed behaviors are completely ignored. The subsumption architecture is a prototypical representative of behavior arbitration [5]. The alternative to arbitration are command fusion approaches, such as dynamical systems [2], which aggregate the control actions of multiple concurrently active behaviors into a consensual decision. Fuzzy rule-based hierarchical architectures offer an alternative approach to robotic behavior coordination [11, 4, 14, 12, 16, 17]. A set of primitive, self-contained behaviors is encoded by fuzzy rule bases that map perceptions to motor commands. Reactive behaviors in isolation are incapable of performing autonomous navigation in complex environments. However, more complex tasks can be accomplished through combination and cooperation among primitive behaviors. A composite behavior is implemented as a supervisory fuzzy controller that activates and deactivates the underlying primitive behaviors according to the current robot's context and goals. Fuzzy behavior coordination is similar to voting, with the main difference that the action selection process is based on fuzzy inference. A fuzzy coordination mechanism offers the advantage that behaviors are active to a certain degree, rather than being either switched on or off. In contrast to mere voting, the weight with which a behavior contributes to the overall decision depends on its current applicability and desirability.

The advantage of fuzzy based behavior fusion turns into a drawback when competing behaviors issue conflicting control commands. In fuzzy fusion the resulting motor command is caused by the weighted average of the decisions proposed by the currently active behaviors. In case of a conflict among active behaviors, this compromise decision might be sub-optimal or even worse than any of the individual commands. What is needed are extensions to mere fuzzy command fusion schemes capable of resolving conflicts among contradicting actions supported by dissenting behaviors.

In context dependent blending of behaviors, originally proposed by Saffiotti et al in [12], higher level supervisory fuzzy rules regulate the activation and deactivation of individual fuzzy behaviors, thereby reintroducing some form of behavior arbitration into fusion. The hierarchical behavior architecture is composed of two distinct layers. On the deliberative level a planner defines a sequence of intermediate goals. At the lower level a fuzzy controller reconciles the abstract goals obtained from the planner, with the constraints and affordances arising from the immediate environmental context. The objective of the coordinating controller is to generate commands that achieve the long-term goals such as reaching a target location while simultaneously satisfying innate goals such as obstacle avoidance. Behavior coordination is achieved by means of supervisory fuzzy rules of the form: **if context then behavior**. The context describes the desirability and applicability of a particular behavior, for example collision avoidance is desirable when obstacles are close. The context also reflects the needs of higher level goals, for example wall-following is applicable when the robot is located in a corridor and desirable if the goal is to navigate to the other end of the corridor.

Yen et al propose an improved defuzzification scheme for fuzzy command fusion in case of conflicting behaviors [16]. Their fusion approach substitutes the center of gravity defuzzification method for multi-modal fuzzy sets by centroid of largest defuzzification. Centroid of largest defuzzification only considers the output fuzzy set with the largest area, whereas fuzzy sets in minor modes are ignored. In a sense, centroid of largest defuzzification becomes similar to majority voting schemes. It guarantees that the final crisp control command is supported by at least one of the active behaviors. The final decision represents the best compromise among those active behaviors that already vote for a sufficiently similar control action, while ignoring minority votes for deviating actions.

Tunstel et al present a fuzzy behavior hierarchy for autonomous navigation and multi-robot control [14]. They distinguish between primitive behaviors that implement a distinct control policy and composite behaviors for behavior coordination. Primitive behaviors interact by means of cooperation and/or competition. Behavior modulation is a process that regulates the activation levels of primitive behaviors, thus determining the impact of a certain behavior on the overall behavior in light of the current situation and goal. The same architecture is applicable not only to multi-behavior coordination but to multi-robot cooperation as well. The authors describe experiments with homogeneous and heterogeneous groups of mobile robots that collectively perform foraging and area coverage. A central agent acquires and processes sensor information and in turn issues directives to individual robots.

Bonarini et al developed a behavior management system for fuzzy behavior coordination [4]. The approach is an extension to the fuzzy behavioral architecture in [12]. Goal-specific strategies are realized by means of conflict resolution among multiple objectives. Behaviors obtain control over the robot according to fuzzy *activation conditions* and *motivations* that reflect the agent's goals and situation. An activation condition, named *cando condition*, is a fuzzy predicate that verifies whether a behavior is applicable in the current context. The fuzzy predicates for motivations, denoted as *want conditions*, are responsible for behavior coordination as they determine whether an applicable behavior is also desirable for achieving the current goal. The cando and wantdo predicates are matched with the information provided by the sensors and an internal world modeler. The world modeler interprets the sensor information and abstracts higher level features described by means of symbolic concepts such as "the door is on the left". A planner module generates goals as symbolic inputs to the behavior coordinator, which are transformed into the corresponding want conditions responsible for selecting and blending actions advanced by primitive behaviors.

Pirjanian et al describe an integration of fuzzy logic with multiple objective decision theory for behavior-based control [11]. The quality of an action-objective pair reflects the desirability of an alternative with regard to a specific goal. The behavior coordination scheme is based on the notion of Pareto-optimality and satisficing solutions. A satisficing solution is the subset of feasible alternatives that achieve each of the objectives to a sufficient degree. The Pareto-optimal subset contains those solutions that can not be further improved on any objective without simultaneously sacrificing the quality of at least one other objective. The multi-objective behavior coordination mechanism first determines the set of feasible actions, second removes inferior feasible solutions to obtain the set of Pareto-optimal actions, third incorporates subjective knowledge such as weights, priorities and goals to find a set of satisficing alternatives and finally determines the most preferred action based on additional criteria. The process systematically condenses the set of possible solutions by incrementally imposing additional decision criteria. Based on the behavioral objectives, the alternative actions are categorized into permissible, Pareto-optimal, satisficing and preferred actions. The authors compare a fuzzy command fusion approach with their method on a mobile robot navigation task, that involves the coordination among three basic behaviors, obstacle avoidance, maintain target heading and move fast forward. The emergent coordinated behaviors are evaluated in terms of safety, velocity, number of successful runs and combined deviation from the ideal values. The results demonstrate that the multiple-objective behavior coordination outperforms the fuzzy approach in terms of safety, velocity, compromise and number of successful runs.

Zhang et al present a fuzzy modular framework for integrating deliberative with reactive strategies [17]. The approach explicitly takes the connection between sensing, planing and control into account. Deliberative behaviors depend on geometric information about the environment and achieve their objectives by means of path planning. Reactive behaviors avoid explicit plans altogether and instead rely on feed-back control to meet their demands. The integration of the beneficial features of deliberative and reactive strategies contributes to solve autonomous navigation tasks in partially unknown environments. The fuzzy control scheme blends deliberative subgoal oriented behaviors with a local obstacle avoidance behavior that copes with unknown objects.

3 Neuro-Fuzzy Systems

From a historic perspective, neuro-fuzzy systems became the first representative of hybridization in soft computing. Neuro-fuzzy systems incorporate the knowledge representation of fuzzy logic with the learning capabilities of artificial neural networks. Both methodologies are concerned with the design of intelligent systems albeit from different directions. The power of neural networks stems from the distributed processing capability of a large number of computationally simple elements. In contrast fuzzy logic is closer related to reasoning on a higher level. Pure fuzzy systems do not possess the capabilities of learning, adaptation or distributed computing that characterize neural networks. On the other hand, neural networks lack the ability to represent knowledge in a manner comprehensible to humans, a key feature of fuzzy rule based systems. Neuro-fuzzy systems bridge the gap between both methodologies, as they synthesize the adaptation mechanisms of neural networks with the symbolic components of fuzzy inference systems, namely membership functions, fuzzy connectives, fuzzy rules and aggregation operators.

Ahrns et al apply neuro-fuzzy control to learn a collision avoidance behavior [1]. Their approach relies on reinforcement learning for behavior adaptation. The learner incrementally adds new fuzzy rules as learning progresses and simultaneously tunes the membership functions of the fuzzy RBF-network.

Godjavec et al present a neuro-fuzzy approach to learn an obstacle avoidance and wall-following behavior on a small size robot [6]. Their scheme allows it to seed an initial behavior with expert rules, which are refined throughout the learning process. During training the robot is controlled either by a human or a previously designed controller. The recorded state-action pairs serve as training examples during supervised learning of neuro-fuzzy control rules. The robot successfully imitates the demonstrated behavior after 1500 iterations.

Ye et al propose a neuro-fuzzy system for supervised and reinforcement based learning of an obstacle avoidance behavior [15]. The scheme follows a two-stage tuning approach, in a first phase supervised learning determines the coarse structure of input-output membership functions. The second reinforcement learning stage fine-tunes the output membership functions. The authors emphasize that the pre-tuned rule-base obtained in the initial supervised learning phase facilitates the subsequent reinforcement learning phase. In conjunction with an improved exploration scheme, the reduced search space complexity accelerates the learning process and leads to more robust control behaviors.

4 Genetic Fuzzy Systems

Evolutionary robotics is concerned with the design of intelligent systems with life-like properties by means of simulated evolution [10]. The basic idea is to automatically synthesize behaviors that enable the robot to perform useful tasks in complex environments. The evolutionary algorithm searches through the space of parameterized controllers that map sensory perceptions to control actions, thus realizing a specific robotic behavior. The evolutionary algorithm maintains and improves a population of candidate behaviors by means of selection, recombination and mutation. A scalar fitness function evaluates the performance of the resulting behavior according to the robot's task or mission. The approaches in evolutionary robotics can be categorized according to the control structures that represent the behavior and the parameters of the controller that undergo adaptation. The following presentation is restricted to evolutionary optimization of fuzzy behaviors. Notice, that combinations of evolutionary with neural techniques even though beyond the scope of this article, play a highly visible role within evolutionary robotics [9, 10]. Neuro-genetic approaches are of particular interest from an ethological perspective as they provide a framework to integrate individual based learning with population based evolutionary adaptation. The evolutionary algorithm either encodes the synaptic weights, the topology or the learning rules of the neural network.

Several authors proposed evolutionary algorithms for learning and tuning of fuzzy controllers for robotic behaviors [3, 7, 8, 13]. In a genetic fuzzy system the evolutionary algorithm evolves a population of parameterize fuzzy controllers. Candidate controllers share the same fuzzy inference mechanism, but establish different input-output mappings according to their genetically encoded knowledge base. The evolutionary algorithm adapts all or part of the components that constitute the knowledge base, namely membership functions, scaling factors and fuzzy rules. The same mechanism is used for the adaptation of primitive behaviors [3, 8] and learning of supervisory fuzzy controllers [4, 7, 13]. The performance of the fuzzy controller while it governs the behavior of the robot is described by a scalar fitness function. Those fuzzy controllers that demonstrate behaviors of higher fitness compared to their competitors are selected as parents for reproduction. Novel candidate behaviors are generated through recombination and mutation of parent fuzzy rules and membership functions. The cycle of selection, reproduction and fitness evaluation progressively leads to improved fuzzy controllers that enable the robot to achieve the behavior design goals in its environment.

In the evolutionary learning of fuzzy rules (ELF) scheme proposed in [3] each chromosome represents a single fuzzy rule rather than an entire rule-base. The population is partitioned into sub-populations, such that fuzzy rules that trigger in similar situations belong to the same sub-population. Rules within one sub-population compete with each other, whereas rules in different sub-populations cooperate to achieve the behavior objectives. A reinforcement scheme, similar to temporal difference learning, distributes the reward among those fuzzy rules that were active during the past training episode. The fitness of a rule is adjusted according to the observed reward and the extent to which the rule contributed to the control actions taken in that episode. Since competition is restricted to sub-populations, fuzzy rules are selected based on the action of the state for which they trigger, rather than the desirability of the state itself. A covering mechanism guarantees that new fuzzy rules are automatically generated for states that are currently not matched by any other rule in the population. ELF has been successfully applied to learn individual robotic behaviors, the coordination of primitive behaviors within a single autonomous agent and cooperation among multiple agents.

Tunstel et al apply genetic programming for off-line identification of supervisory fuzzy rules that coordinate primitive fuzzy behaviors [13]. Composite behaviors are evaluated according to their success in orchestrating the primitive behaviors such that they eventually navigate the robot to its goal location. The fitness is averaged over several trials in different simulated environments in order to obtain robust and reliable behaviors. The generalization capability of the highest scoring behaviors is tested in a more general environment unrelated to the test environments used during evolution. The off-line evolved behaviors are transfered to the physical robot for verification. The robot successfully navigates within close proximity to the goal.

In [7] Hagras et al present a fuzzy classifier systems that utilizes a genetic algorithm for online-learning of a goal seeking and a wall following behavior. The fuzzy classifier system maintains a rule cache to store suitable fuzzy rules, that might be of benefit in future situations and which serve to seed the initial population when a new genetic learning process is invoked. This technique substantially speeds up the learning process, thus making the entire approach feasible for on-line learning of robotic behaviors. In one scenario, the robot acquired the goal seeking and wall following behavior after 96 seconds.

5 Conclusions

Soft computing approaches are preferable over conventional control system design, for problems that are difficult to describe by analytical models. Autonomous robotics is such a domain in which knowledge about the environment is inherently weak and incomplete. Therefore, the features of fuzzy control, neural networks and evolutionary algorithms are of particular benefit to the type of problems emerging in behavior based robotics. The references in the text on fuzzy control, neuro-fuzzy and genetic-fuzzy approaches in robotics do not claim to be complete but rather intend to provide insight into the general utility of soft-computing techniques for behavior based robotics.

Fuzzy behavior hierarchies, neuro-fuzzy and genetic fuzzy system are valuable methodologies for the design and adaptation of complex robotic behaviors. The knowledge representation of fuzzy rule based systems combined with the learning capabilities of neural networks and evolutionary algorithms opens a promising avenue towards more intelligent and robust robotic systems. Soft computing techniques, such as design based on expert knowledge, hierarchical behavior architectures, fuzzy behavior command fusion and evolutionary and neural adaptation contribute to one of the long term goal in robotics, namely that intelligent, autonomous robots demonstrate and acquire complex skills in unstructured real-world environments.

References

[1] I. Ahrns, J. Bruske, G. Hailu, and G. Sommer. Neural fuzzy techniques in sonar-based collision avoidance. In *Soft Computing for Intelligent Robotic Systems*, pages 185–214. Physica, 1998.

[2] P. Althaus, H. I. Christensen, and F. Hoffmann. Using the dynamical system approach to navigate in realistic real-world environments. In *IROS'2001*, 2001.

[3] A. Bonarini. Evolutionary learning of fuzzy rules: competition and cooperation. In W. Pedrycz, editor, *Fuzzy Modelling: Paradigms and Practice*, pages 265–284. Kluwer Academic Press, Norwell, MA, 1996.

[4] A. Bonarini, G. Invernizzi, Th. H. Labella, and M. Matteucci. An architecture to coordinate fuzzy behaviors to control an autonomous robot. *Fuzzy Sets and Systems*, 134(1):101–115, 2003.

[5] R. Brooks. A robust layered control system for a mobile robot. *IEEE Journal of Robotics and Automation*, RA-2(1):14–23, 1986.

[6] J. Godjavec and N. Steele. Neuro-fuzzy control for basic mobile robot behaviors. In *Fuzzy Logic Techniques for Autonomous Vehicle Navigation*, pages 97–117. Springer, 2000.

[7] H.Hagras, V. Callaghan, and M.Colley. Learning fuzzy behaviour co-ordination for autonomous multi-agents online using genetic algorithms and real-time interaction with the environment. In *Fuzzy IEEE*, 2000.

[8] F. Hoffmann. Evolutionary algorithms for fuzzy control system design. *Proceedings of the IEEE*, 89(9):1318–33, September 2001.

[9] J.A. Meyer. Evolutionary approaches to neural control in mobile robots. In *IEEE Int. Conference on Systems, Man and Cybernetics*, 1998.

[10] S. Nolfi and D. Floreano. *Evolutionary Robotics – The Biology, Intelligence, and Technology of Self-Organizing Machines*. MIT Press, 2000.

[11] P. Pirjanian and M. Mataric. A decision theoretic approach to fuzzy behavior coordination. In *CIRA'99*. 1999.

[12] A. Saffiotti, K. Konolige, and E.H. Ruspini. A multivalued-logic approach to integrating planning and control. *Artificial Intelligence*, 76(1-2):481–526, 1995.

[13] E. W. Tunstel. Fuzzy-behavior synthesis, coordination, and evolution in an adaptive behavior hierarchy. In *Fuzzy Logic Techniques for Autonomous Vehicle Navigation*, pages 205–234. Springer, 2000.

[14] E. W. Tunstel, M. A. A. de Oliveira, and S. Berman. Fuzzy behavior hierarchies for multi-robot control. *Int. Journal of Intelligent Systems*, 17:449–470, 2002.

[15] C. Ye, N. H. C. Yung, and D. Wang. A fuzzy controller with supervised learning assisted reinforcement learning algorithm for obstacle avoidance. *IEEE Transactions on Systems, Man and Cybernetics Part B*, 33(1):17–27, 2003.

[16] J. Yen and N. Pfluger. A fuzzy logic based extension to payton and rosenblatt's command fusion method for mobile robot navigation. *IEEE Transactions on Systems, Man and Cybernetics*, 25(6):971–978, 1995.

[17] J. Zhang and A. Knoll. Integrating deliberative and reactive strategies via fuzzy modular control. In *Fuzzy Logic Techniques for Autonomous Robot Navigation*. Springer, 2000.

Asymmetric Redundancy of Tuples in Fuzzy Relational Database

Rolly Intan[1,2] and Masao Mukaidono[1]

[1] Meiji University, Kawasaki-shi, Kanagawa-ken, Japan
[2] Petra Christian University, Surabaya, Indonesia 60236

Abstract. This paper considers and discusses asymmetric redundancy of tuples in fuzzy relational database dealing with *weak fuzzy similarity relation* for scalar domain. Asymmetric redundancy of tuples provides a more comprehensive method compared to the other approaches, especially, when the fuzzy relational data model is used to represent either decision table or more accurate information.

1   Introduction

Fuzzy relational database proposed by Buckles and Petry in 1982 [2], as in classical relational database theory [4], consists of a set of tuples, where t_i represents the i-th tuple and if there are m domains D, then $t_i = (d_{i1}, d_{i2}, ..., d_{im})$. In fuzzy relational database, domains may be discrete scalars as well as discrete numbers taken from either a finite or infinite set. Each component of tuples d_{ij} is not limited to atomic value; instead a subset of D_j, $d_{ij} \subseteq D_j$ $(d_{ij} \neq \emptyset)$ as defined in the following definition.

Definition 1. *[2]. Let $\mathscr{P}(D_j)$ is power set of D_j. A **fuzzy relation**, R, is a subset of the set of cross product*

$$\hat{\mathscr{P}}(D_1) \times \hat{\mathscr{P}}(D_2) \times \cdots \times \hat{\mathscr{P}}(D_m), \quad \text{where } \hat{\mathscr{P}}(D_j) = \mathscr{P}(D_j) - \emptyset.$$

A fuzzy tuple is a member of a fuzzy relation as follows.

Definition 2. *[2]. Let $R \subseteq \hat{\mathscr{P}}(D_1) \times \hat{\mathscr{P}}(D_2) \times \cdots \times \hat{\mathscr{P}}(D_m)$ be a fuzzy relation. A **fuzzy tuple** t (with respect to R) is an element of R.*

Here, an atomic value might be a fuzzy data represented by a fuzzy label as well as a crisp data. The fuzzy relational database of Buckles and Petry relies on the specification of *similarity relation* proposed by Zadeh [13] for each distinct scalar domain in the fuzzy database. A similarity relation, s_j, for a given domain, D_j, maps each pair of elements in the domain to an element in the closed interval [0,1].

Definition 3. *[13] A **similarity relation** is a mapping, $s_j : D_j \times D_j \to [0,1]$, such that for $x, y, z \in D_j$,*

    (a) Reflexivity : $s_j(x,x) = 1$,

    (b) Symmetry : $s_j(x,y) = s_j(y,x)$,

    (c) Max–min transitivity : $s_j(x,z) \geq \max_{y \in D_j} \min[s_j(x,y), s_j(y,z)]$.

T. Bilgiç et al. (Eds.): IFSA 2003, LNAI 2715, pp. 552–559, 2003.

Even though the fuzzy relational database considers components of tuples as set of fuzzy data from the corresponding domains, by applying the concept of similarity classes, it is possible to define a notion of redundancy that is similar to classical relational database theory.

Definition 4. *Two tuples* $t_i = (d_{i1}, d_{i2}, \ldots, d_{im})$ *and* $t_j = (d_{j1}, d_{j2}, \ldots, d_{jm})$, $i \neq j$, *are redundant given* $\alpha = \{\alpha_1, \alpha_2, \ldots, \alpha_m\} \in [0,1]^m$ *corresponding to set of domains, if*

$$\min_{x,y \in d_{ik} \cup d_{jk}} (s_k(x,y)) \geq \alpha_k, \quad \text{for} \quad k = 1, 2, \ldots, m.$$

There is considerable criticism about the use of similarity relation in the fuzzy relational database of Buckles and Petry, especially for the point of max-min transitivity (see e.g., [12]). Max-min transitivity is considered as a very restrictive constraint. Considering this reason, in 1989, Shenoi and Melton [10] extended fuzzy relational database model of Buckles and Petry to deal with *proximity relation* for scalar domain by eliminating max-min transitivity.

However, when we realize that every item of data represented in human's word or phrase, naturally, has different range of meaning where some words have more general meaning than the others, degree of similarity (meaning) of two words may not be symmetric. For simple example, talking about color, the word 'Red' is more general and broader, while 'Crimson' is narrower and more specific. The word 'Red' can cover a wider range of meaning in color than the word 'Crimson'. So the range of meaning in color is different in these two words. In our sentences, it is more correct and common to say that 'Crimson is (like) Red' than to say that 'Red is (like) Crimson'. Moreover, we can say that similarity level of 'Red' given 'Crimson' and similarity level of 'Crimson' given 'Red' are different. Therefore, we consider the relation of similarity between two scalar data may neither necessarily be symmetric nor transitive as represented by *weak similarity relation* (defined in Section 2).

Considering this reason, in this paper, we consider a model of fuzzy relational database dealing with weak similarity relation for scalar domain in which the weak similarity relation is generated by what we called *conditional probability relation*. It can be verified that the weak similarity relation generalizes similarity relation as well as proximity relation. Our primary concern is to define a notion of asymmetric redundancy of tuples generalizing existing concept of redundancy in fuzzy relational database such as proposed by Buckles-Petry.

Fuzzy data representation proposed by Buckles and Petry and its extended work by Shennoi and Melton may be regarded as similarity-based framework as discussed by Chen et al. (1992) [3]. Another representation called possibility-based framework was proposed by Prade and Testemale (1983) [9]. Chen et al. [3] considered a more general model of fuzzy relational database by combining both similarity-based framework and possibility-based framework and proposed a general treatment of its data redundancy by applying Zadeh's extension principle. Related to their work, we are considering extending this paper as a more comprehensive method compared to the existing approaches of data redundancy in its extended version.

2   Weak Similarity Relation

It should be mentioned that max-min transitivity and even symmetry as required
in similarity relation [13] are too strong properties to represent relationships
between two elements of data in real-world application. Although, it is true to
say that if "x is similar to y" then "y is similar to x", but these two statements
might have different degree of similarity depending on their range of meaning.
Hence, we consider conditional probability(-based) relation as a more realistic
relation in representing relationships between two elements of data. A conditional
probability relation is defined in the following definition.

Definition 5. *A **conditional probability relation** is a mapping, $c_j : D_j \times D_j \rightarrow [0,1]$, such that for $x, y \in D_j$,*

$$c_j(x,y) = \mathrm{P}(x \mid y) = \mathrm{P}(y \rightarrow x) = \frac{|x \cap y|}{|y|} \tag{1}$$

*where $c_j(x,y)$ means the degree y supports x or the degree y is similar to x. x
and y are regarded as sets in representing their range of meaning in which $|y|$
means the cardinality of y.*

x has more general or broader meaning than y in a given domain D_j if and only
if $|x| > |y|$. In the definition of conditional probability relations, the probability
values may be estimated based on the semantic relationships between elements
of data by using the epistemological or subjective view of probability theory.
Considering our example in Section 1, it is possible to conclude that,

$$c_{color}(Red, Crimson) > c_{color}(Crimson, Red).$$

The expression means that similarity level of 'Red' given 'Crimson' is greater
than similarity level of 'Crimson' given 'Red'. These conditions can be easily
understood by imaging that every scalar data is represented by a set in terms
of its range of meaning. In our example, size of set 'Red' is bigger than size
of set 'Crimson'. Notion of representing similarity level based on conditional
probability relation implies some properties in the following theorem.

Theorem 1. *Let $c_j(x,y)$ be conditional probability relation of x given y and
$c_j(y,x)$ be conditional probability relation of y given x, such that for $x, y, z \in D_j$,*

(r0) $c_j(x,y) = c_j(y,x) = 1 \Longleftrightarrow x = y$,

(r1) $[c_j(y,x) = 1, c_j(x,y) < 1] \Longleftrightarrow x \subset y$,

(r2) $c_j(x,y) = c_j(y,x) > 0 \Longrightarrow |x| = |y|$,

(r3) $c_j(x,y) < c_j(y,x) \Longrightarrow |x| < |y|$,

(r4) $c_j(x,y) > 0 \Longleftrightarrow c_j(y,x) > 0$,

(r5) $[c_j(x,y) \geq c_j(y,x) > 0, c_j(y,z) \geq c_j(z,y) > 0] \Longrightarrow c_j(x,z) \geq c_j(z,x)$.

From Theorem 1, we can define an interesting mathematical relations, called *weak similarity relation*, based on their constraints represented by axioms in representing similarity level of a fuzzy relation as follows.

Definition 6. *A **weak similarity relation** is a mapping,* $S_j : D_j \times D_j \to [0,1]$, *such that for* $x, y \in D_j$,

      (a) Reflexivity : $S_j(x,x) = 1$,

      (b) Conditional symmetry : if $S_j(x,y) > 0$ then $S_j(y,x) > 0$,

      (c) Conditional transitivity : if $S_j(x,y) \geq S_j(y,x) > 0$ and

          $S_j(y,z) \geq S_j(z,y) > 0$ then $S_j(x,z) \geq S_j(z,x)$.

By definition, weak similarity relation may be considered as a generalization of proximity relation in terms of their symmetric property, where conditional symmetry in the weak similarity relation generalizes symmetric property in the proximity relation.

3 Redundancy of Tuples

In this section, we discuss and define a notion of redundant tuples in the presence of fuzzy relational data model of Buckles and Petry as defined in Definition 1 and 2 in which its distinct scalar domain is dealing with *the weak similarity relation*. First, it is necessary to define similarity of two components of tuples as follows.

Definition 7. *Let* $d_{ij}, d_{kj} \subseteq D_j$ *be two components of tuples, corresponding to tuples,* t_i *and* t_k, *respectively. Degree of similarity between two components of tuples is calculated by a function* $\sigma_j : \hat{\mathscr{P}}(D_j) \times \hat{\mathscr{P}}(D_j) \to [0,1]$ *as given by*

$$\sigma_j(d_{ij}, d_{kj}) = \frac{\sum_{y \in d_{kj}} \sup_{x \in d_{ij}} (S_j(x,y))}{|d_{kj}|}, \tag{2}$$

where $\sigma_j(d_{ij}, d_{kj})$ *represents similarity of* d_{ij} *given* d_{kj}. $|d_{kj}|$ *means cardinality of* d_{kj}.

From (2), it can be easily proved that degree of similarity between two components of tuples also satisfies properties of the weak similarity relation. α-redundancy of tuples is defined as follows.

Definition 8. *Two tuples* $t_i = (d_{i1}, d_{i2}, \ldots, d_{im})$ *and* $t_k = (d_{k1}, d_{k2}, \ldots, d_{km})$, $i \neq k$, *are redundant given* $\alpha = \{\alpha_1, \alpha_2, \ldots, \alpha_m\} \in [0,1]^m$ *corresponding to set of domains, if*

$$\min(\sigma_j(d_{ij}, d_{kj}), \sigma_j(d_{kj}, d_{ij})) \geq \alpha_j, \quad \text{for } j = 1, \ldots, m. \tag{3}$$

Definition 8 gives the same notion of redundancy as proposed by existing concept of redundancy that we may call it *symmetric redundancy of tuples*. Here, it can be said that t_i is a redundant tuple because of t_k, and vice versa. However, we may consider a more general concept of redundancy that we may call it *asymmetric redundancy of tuples*.

3.1 Asymmetric Redundancy of Tuples

The notion of *asymmetric redundancy* of tuples may be considered as a generalization of the existing concept of redundancy of tuples. There are at least two interpretations concerning redundant tuple. First is exerted by a motivation to find *a more accurate information*. For example, there are two domains, *MOUNTAIN* and *HEIGHT* which describe information concerning height of mountains. In general, what we need concerning the height of Mt. Everest as given in Table 1 is the most accurate information. Obviously ⟨Mt. Everest, 8848m⟩ is the most accurate information in which ⟨Mt. Everest, very high⟩ and ⟨Mt. Everest, about 9000 m⟩ are considered as redundant tuples. By this motivation, a notion of redundant tuples is defined as the following.

Table 1. Height of Mountains

MOUNTAIN	HEIGHT
Mt. Everest	very high
Mt. Everest	about 9000 m
Mt. Everest	8848 m

Table 2. HEIGHT $(S_H(x,y))$

$x \setminus y$	VH	A9000	8848m
very high (VH)	1	1	1
about 9000 m (A9000)	0.6	1	0.9
8848 m	0.4	0.6	1

Definition 9. *Tuple* $t_i = (d_{i1}, d_{i2}, ..., d_{im})$ *is* α-*redundant in relation* R *if there is a tuple* $t_k = (d_{k1}, d_{k2}, ..., d_{km})$ *which has more accurate information in all components of tuple than* t_i *with the degree of* α, *where* $\alpha = (\alpha_1, ..., \alpha_m)$ *corresponds to domains* $D_1, ..., D_m$, *whenever*

$$\forall y \in d_{kj}, \exists x \in d_{ij}, \quad (S_j(x,y) \geq S_j(y,x) \text{ and } S_j(x,y) \geq \alpha_j), \qquad (4)$$

or

$$(\sigma_j(d_{ij}, d_{kj}) \geq \sigma_j(d_{kj}, d_{ij}) \text{ and } \sigma_j(d_{ij}, d_{kj}) \geq \alpha_j), \qquad (5)$$

for $j = 1, 2, ..., m$.

Simply, it can be proved that conditions (4) and (5) in Definition 9 satisfy the following theorem.

Theorem 2. *Let* t_i *and* t_k *are two tuples in relation* R. *If* t_i *is* α-*redundant in the presence of* t_k *satisfying condition (4) then it must also satisfy condition (5), not vice versa.*

Here, condition (4) is stricter than (5) in which (4) examines relationship of all element data of given tuple (in case of t_k in Definition 9) that they must satisfy the condition, on the other hand, (5) is based on the average.

Let us consider relation in Table 1. Domain **MOUNTAIN**(M) is crisp domain. On the other hand, domain **HEIGHT**(H) is fuzzy domain. Suppose that similarity level of domain H induced by weak similarity relation is given in Table 2. From Table 2, $S_H(VH, A9000) = 1$, but $S_H(A9000, VH) = 0.6$, etc. Hence, by Theorem 1 (r3), it is concluded that $|VH| > |A9000| > |8848m|$. By arbitrarily given $\alpha = (1, 0.8)$ in which, $\alpha_M = 1$ and $\alpha_H = 0.8$, redundant tuples will be removed from Table 1. By applying (4) or (5) in Definition 9, there are two redundant tuples, $\langle$Mt. Everest, very high$\rangle$ and $\langle$Mt. Everest, about 9000 m$\rangle$, where $\langle$Mt. Everest, 8848m$\rangle$ has the most accurate information.

The second interpretation of redundant tuple is opposite to the first one. Here, a redundant tuple is considered as a tuple covered by other tuple concerning all of their corresponding components of tuple. For example, in decision table, what we need is a more general decision rule. Therefore, in order to reduce the number of decision rules, it is necessary to remove redundant tuples. Simply, for instance given Table 3 shows a relation of *Reproduction of Animals* which is obviously proved that $\langle Horse, Bear \rangle$ is a redundant tuple, for it is covered by $\langle Mammals, Bear \rangle$ in which horse belongs to group of mammals. Based on this interpretation, a notion of redundant tuples is defined as follows.

Table 3. Reproduction of Animals

Description	Reproduction
Horse (h)	Bear (br)
Mammals (m)	Bear (br)
Bird (b)	Egg (eg)

Definition 10. *Tuple $t_i = (d_{i1}, d_{i2}, ..., d_{im})$ is α-redundant in relation R if there is a tuple $t_k = (d_{k1}, d_{k2}, ..., d_{km})$ which covers all information of t_i with the degree of α, where $\alpha = (\alpha_1, ..., \alpha_m)$ corresponds to domains $D_1, ..., D_m$, whenever*

$$\forall x \in d_{ij}, \exists y \in d_{kj}, \quad (S_j(y, x) \geq S_j(x, y) \text{ and } S_j(y, x) \geq \alpha_j) \tag{6}$$

or

$$(\sigma_j(d_{kj}, d_{ij}) \geq \sigma_j(d_{ij}, d_{kj}) \text{ and } \sigma_j(d_{kj}, d_{ij}) \geq \alpha_j) \tag{7}$$

for $j = 1, 2, ..., m$.

Similarly, it can be proved that conditions (6) and (7) in Definition 10 satisfy the following theorem.

Theorem 3. *Let t_i and t_k are two tuples in relation R. If t_i is α-redundant in the presence of t_k satisfying condition (6) then it must also satisfy condition (7), not vice versa.*

In this case, it depends on the application in which redundant tuples can be eliminated either by just removing all redundant tuples (i.e. t_i in Definition 10) or by merging attribute values via the set union operation as discussed in [2,3]. Let us consider a relation scheme ARTIST(NAME, AGE, APTITUDE). An instance of fuzzy relation is given in Table 4 where each tuple represents someone's opinion about the artist who is written in the tuple. Domain NAME is unique which means that every tuple with the same name indicates the same person. Compound data in component of tuple is related by AND operator, for example {*Average, Good*} means *Average* AND *Good*. In other words, all pieces of information involved in the component of tuple are important. Then the problem of redundant tuples should be solved using Definition 10. Let us suppose that similarity of domains AGE and APTITUDE are given in Table 5 and Table

6. From Table 5, similarity level of $Young$ given $[20, 25]$, $S_A(Young, [20, 25])$, is equal to 0.8, on the other hand, similarity level of $[20, 25]$ given $Young$, $S_A([20, 25], Young)$ is equal to 0.4. In that case, $Young$ covers a wider range of meaning in AGE than $[20, 25]$.   Now, we want to remove redundant tuples

Table 4. ARTIST Relation

NAME(N)	AGE(A)	APTITUTE(AP)
John	Young	Good
Tom	Young	{Average,Good}
David	Middle Age	Very Good
Tom	[20,25]	Average
David	About-50	Outstanding

Table 5. AGE $(S_A(x, y))$

$x \setminus y$	Y	[20,25]	MA	A-50
Young (Y)	1.0	0.8	0.3	0.0
[20,25]	0.4	1.0	0.2	0.0
Middle Age (MA)	0.3	0.4	1.0	0.9
A-50	0.0	0.0	0.5	1.0

Table 6. APTITUDE $(S_{AP}(x, y))$

$x \setminus y$	A	G	VG	O
Average (A)	1.0	0.6	0.3	0.0
Good (G)	0.6	1.0	0.8	0.6
Very Good (VG)	0.15	0.4	1.0	0.9
Outstanding (O)	0.0	0.3	0.9	1.0

in Table 4 with arbitrarily given $\alpha = (1.0, 0.0.7, 0.8)$ which corresponds to N, A, and AP, where $\alpha_N = 1.0, \alpha_A = 0.7, \alpha_{AP} = 0.8$. We must set α to 1.0 especially for domain NAME, because domain NAME is crisp domain and each distinct scalar data indicates different person. By applying (6) or (7) in Definition 10, there are two redundant tuples, $\langle Tom, [20, 25], Average \rangle$ and $\langle David, About - 50, Outstanding \rangle$, which are covered by $\langle Tom, Young, \{Average, Good\} \rangle$ and $\langle David, Middle Age, Very Good \rangle$, respectively. Table 7 shows the final result after removing the two redundant tuples.

Table 7. ARTIST Relation (free redundancy)

NAME(N)	AGE(A)	APTITUTE(AP)
John	Young	Good
Tom	Young	{Average,Good}
David	Middle Age	Very Good

Alternatively, compound data in component of tuple is related by OR operator, for example $\{Average, Good\}$ means $Average$ OR $Good$. In other words, not all pieces of information involved in a given component of tuple are important instead we can choice one of them. The problem of redundant tuple is to find the most accurate tuple. Then the problem should use Definition 9 to be solved.

4   Conclusions

In classical relational database, all of its components of tuple are atomic values and each distinct scalar data is regarded to be disjoints. It is clear that the identity relation is used for the treatment of ideal information where an element of a given domain may have no similarity to any other element of the domain; each element may be similar only unto itself. Consequently, a tuple is redundant if it is exactly the same as another tuple. However, in fuzzy relational database, an element of a given domain may have similarity level to other elements of the domain. Moreover, considering the range of meaning, a fuzzy data may cover any other fuzzy data (i.e., 'Red' covers meaning of 'Crimson') with the certain degree of α. Therefore, we define the concept of redundant tuple in fuzzy relational database based on two interpretations as defined in Definition 9 and 10, where components of tuples may not be single value as proposed in the fuzzy relational database model of Buckles and Petry (see Section 1, in which $d_{ij} \subseteq D_j$). Compared to Definition 4 which also defines redundant tuples, Definition 9 and 10 appeal to be more general as a consequence that symmetry is just a special case in the weak similarity relation.

References

1. W. W. Armstrong, Dependency Structures of Database Relationship, *Proceeding of IFIP Congress'74*, (Stockholm, Sweden, August 1974) 580-583.
2. B.P. Buckles, F.E. Petry, A Fuzzy Representation of Data for Relational Database, *Fuzzy Sets and Systems* **5**, (1982), 213-226.
3. G. Chen, J. Vandenbulcle, E. E. Kerre, A General Treatment of Data Redundancy in a Fuzzy Relational Data Model, *Journal of The American Society For Information Science* **43**(4), (1992), 304-311.
4. E.F. Codd, A Relational Model of Data for Large Shared Data Banks, *Communications of The ACM* **13(6)**, (1970), 377-387.
5. D. Dubois, H. Prade, *Fuzzy Sets and Systems: Theory and Applications*, (Academic Press, New York, 1980).
6. R. Intan, M. Mukaidono, Application of Conditional Probability in Constructing Fuzzy Functional Dependency(FFD), *Proceedings of AFSS'00*, (Tsukuba, Japan, 2000), 271-276.
7. R. Intan, M. Mukaidono, Fuzzy Functional Dependency and Its Application to Approximate Querying, *Proceedings of IDEAS'00*, (Yokohama, 2000), 47-54.
8. R. Intan, M. Mukaidono, Conditional Probability Relations in Fuzzy Relational Database, in W. Ziarko and Yiyu Yao, eds., *Proceedings of International Conference on RSCTC'00, LNAI 2005, Springer-Verlag*, (2000), 251-260.
9. H. Prade, C. Testemale, Generalizing Database Relational Algebra for the Treatment of Incomplete/Uncertain Information and Vague Queries, *Proceeding of the 2nd NAFIPS Workshop* (Schenectady, NY, 1983).
10. S. Shenoi, A. Melton, Proximity Relations in The Fuzzy Relational Database Model, *Fuzzy Sets and Systems* **31**, (1989), 285-296.
11. S. Shenoi, A. Melton, L.T. Fan, Functional Dependencies and Normal Forms in The Fuzzy Relational Database Model, *Information Science* **60**, (1992), 1-28.
12. A. Tversky, Features of Similarity, *Psychological Rev.* **84(4)**, (1977), 327-353.
13. L.A. Zadeh, Similarity Relations and Fuzzy Orderings, *Information Science* **3(2)**, (1970), 177-200.

Fuzzy Clustering in Classification Using Weighted Features

Lourenço P.C. Bandeira[1], João M.C. Sousa[1], and Uzay Kaymak[2]

[1] Technical University of Lisbon, Instituto Superior Técnico
Dept. of Mechanical Engineering, GCAR/IDMEC
1049-001 Lisbon, Portugal,
jmsousa@ieee.org
[2] Erasmus University of Rotterdam, Faculty of Economics
P.O. Box 1738, 3000 DR Rotterdam, the Netherlands

Abstract. This paper proposes a fuzzy classification/regression method based on an extension of classical fuzzy clustering algorithms, by weighting the features during cluster estimation. By translating the importance of each feature using weights, the classifier can lead to better results. The proposed method is applied to target selection, where the goal is to maximize profit obtained from the clients. A real-world application shows the effectiveness of the proposed approach.

1 Introduction

Fuzzy modeling combines numerical accuracy with transparency in the form of linguistic rules [9]. One of the most important field of application of fuzzy models is classification, especially in data mining problems. A method that has been used extensively for obtaining fuzzy models is fuzzy clustering, which partition a data set into overlapping groups based on similarity measures. In classification, an object usually belongs to a certain class. However, in some applications it might be useful to give a degree to this classification, combining classification and regression of data. This paper proposes a classification/regression method where the data objects are sorted by real values. This method can be particularly useful in target selection.

Target selection is an important data mining problem from the world of direct marketing. Its goal is to determine the potential customers for a new product by identifying profiles of customers that are known to have shown interest in a product in the past. The key to target selection is maximizing the profits of selling the product, while minimizing the cost of the marketing campaign. In this type of problems, some features can be more relevant than others. Thus, weighting the different features by order of importance can lead to better classifiers. If there is some previous expert knowledge about this interaction between features, it should be looked upon. Therefore, this paper proposes to weight the features of the data based on expert knowledge. These weights are then added to fuzzy clustering techniques. Target selection models can be evaluated using gain charts (also called lift charts), which indicate the advantage obtained by using a derived model for target selection over random selection of targets. In this paper our objective is to maximize the profit that each customer can provide. We also propose a type of gain chart based on real outputs to describe the goal.

T. Bilgiç et al. (Eds.): IFSA 2003, LNAI 2715, pp. 560–567, 2003.

Summarizing, this paper proposes classification by using weighted fuzzy clustering based on the importance of each feature, and an adapted gain chart for classification/regression in target selection problems. The paper starts by presenting briefly fuzzy classification using fuzzy clustering in Section 2. The proposed weighting of features in fuzzy clustering is presented in Section 3. Target selection in direct marketing is presented afterwards in Section 4, where a new type of gain chart is proposed, and feature selection is also briefly discussed. An application is presented in Section 5, where the maximization of donations to a Dutch charity organization is considered. Finally, Section 6 presents the conclusions.

2 Fuzzy Classification

Fuzzy models have gained in popularity in various fields such as control engineering, decision making, classification and data mining [9]. One of the important advantages of fuzzy models is that they combine numerical accuracy with transparency in the form of linguistic rules. Hence, fuzzy models take an intermediate place between numerical and symbolic models. A method that has been extensively used for obtaining fuzzy models is fuzzy clustering. Fuzzy clustering algorithms are unsupervised techniques that partition a data set into overlapping groups based on similarity within the groups and dissimilarity amongst the groups. This paper uses fuzzy models for classifying objects, as described in the following.

Let $\{\mathbf{x}_1, \ldots, \mathbf{x}_N\}$ be a set of N data objects where $\mathbf{x}_k \in \mathbb{R}^n$. The set of data objects can then be represented as a $N \times n$ data matrix $\mathbf{X}$, where n is the number of features (attributes) used to described the data. Note that each object is an instance represented by the vector $\mathbf{x}_k$, which is described by a set of features. The fuzzy clustering algorithm determines a fuzzy partition of $\mathbf{X}$ into C clusters by computing a $N \times C$ partition matrix $\mathbf{U}$ and the C-tuple of corresponding cluster prototypes $\mathbf{V} = \{\mathbf{v}_1, \ldots, \mathbf{v}_C\}$. Often, the cluster prototypes are points in the cluster space, i.e. $\mathbf{v}_i \in \mathbb{R}^n$, but they can also be closed volumes in the clustering space, as in the case of the extended fuzzy c-means algorithm presented in [6]. The elements $u_{ik} \in [0, 1]$ of $\mathbf{U}$ represent the membership of data object $\mathbf{x}_k$ in cluster i.

For the application used in this paper; charity donations (presented in Section 5), the objective is to maximize the number of responders, which is similar to target selection in a commercial environment, see Section 4. However, it is preferable to receive as much donations (positive responses to the direct marketing campaign) contacting as few target supporters as possible (minimizing the mailing costs). This approach has been considered recently [7]. In this paper we are not interested only in maximizing the number of responders, but also in maximizing the donation revenue. Therefore, there is not only the necessity of classification, but it is also required to estimate the donation of each of the identified supporters, by using a regression method. Thus, some new characteristics are added to the problem, as will be explained in Section 5.

3 Weighted Fuzzy Clustering

Many clustering algorithms are available for determining $\mathbf{U}$ and $\mathbf{V}$ iteratively. Fuzzy c-means and the Gustafson–Kessel (GK) clustering algorithms (or variations thereof) are the most popular [2, 4]. The fuzzy clustering algorithm introduced by Gustafson and Kessel introduces feature relevance through covariance matrices. However, this algorithm weight the inputs (features in classification problems) based on the distribution of data. In the algorithm proposed in this paper, the weights are based on the importance of the features directly. In our case, the weights are based on expert knowledge to determine the relevance of the features. A similar approach was proposed recently to handle missing data by fuzzy clustering methods [10], where an index for each data point is considered, in order to deal with missing data. The index has the value 0 when data is missing, and the value 1 when the data is available. This index can be seen as a weight, determining the importance of a certain data object. In our approach, each feature as a weight in the interval [0, 1], where 1 stands for the most important features, and 0 correspond to features that are not relevant at all.

As the curse of dimensionality demands for less features, it is useful to reduce the features as much as possible. However, sometimes an important feature can be disregarded, which leads to poor classification results. Therefore, this paper explores the possibility of weighting the features in the clustering algorithm through expert knowledge, in order to reduce classification errors. Consider a vector of weights $\mathbf{w}$ with n elements, one for each feature:

$$\mathbf{w} = [w_1, w_2, \ldots, w_n], \tag{1}$$

where $w_i \in]0, 1]$, and $i = 1, \ldots, n$. Using this extension, the cluster centers are determined as follows:

$$v_{ij} = \frac{\sum_{k=1}^{N} u_{ik}^m w_j x_{kj}}{\sum_{k=1}^{N} u_{ik}^m w_j}, \quad j = 1, \ldots, n \tag{2}$$

where $i = 1, \ldots, C$, and m is an exponent that determines the fuzziness of the resulting clusters. The weights $\mathbf{w}$ are also used in the computation of the distance. Consider that an $n \times n$ diagonal matrix $\mathbf{W}$ is given by:

$$\mathbf{W} = \begin{bmatrix} w_1 & 0 & \cdots & 0 \\ 0 & w_2 & \cdots & 0 \\ \vdots & \vdots & \ddots & \vdots \\ 0 & 0 & \cdots & w_n \end{bmatrix}. \tag{3}$$

The distance necessary to apply the algorithm is calculated by:

$$d^2(\mathbf{x}_k, \mathbf{v}_i) = (\mathbf{x}_k - \mathbf{v}_i)^T \mathbf{W} (\mathbf{x}_k - \mathbf{v}_i) \tag{4}$$

Note that (4) is a modified version of the Euclidian distance. This paper applies fuzzy classification to target selection in direct marketing, in order to choose the best costumers, as explained in the following.

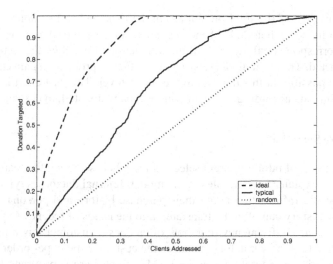

Fig. 1. Example of a gain chart: ideal model (- -), random selection (..), and typical model (–).

4 Target Selection in Direct Marketing

Target selection is an important data mining problem in direct marketing. The main task is to determine the potential customers for a product from a client database, by identifying profiles of customers that are known to have showed interest in a product in the past. Direct marketing is known as the use of existing and new marketing channels to encourage a direct relationship with the customers. Large databases of customers and market data are maintained for this purpose. The customers or clients to be targeted in a specific campaign are selected from the database, given different types of information such as demographic information and information on customer's personal characteristics like profession, age and purchase history.

4.1 Gain Charts

Target selection models can be evaluated in various ways. Often, gain charts (also called lift charts) are used to assess the gain to be expected from the utilization of a target selection model [8]. A gain chart indicates the advantage obtained by using a derived model for target selection over random selection of targets. When a classification problem intends to maximize a revenue, instead of the number of targeted responders, the gain chart presents a different aspect. Note that instead of considering the response percentage as in [8], the vertical axis must now represent the percentage gain over the total amount donated. Thus, this chart presents more explicitly the outcome to be expected from a certain marketing campaign. An example of this gain chart is shown in Fig. 1, where the horizontal axis indicates the percentage of customers that should be mailed to obtain a certain percentage of the total amount. For example, the point (20%, 30%)

indicates that 30% of the total donations can be expected to be captured by the target selection model, when 20% of the customers are selected. A random selection of customers corresponds only to 20% of the total amount raised by the same number of clients contacted. The ideal model (also shown in Fig. 1) produces a gain chart that rises as steeply as possible in this data set to the 100% level. This last chart is obtained by ordering the clients according to the amount of money donated, in descending order.

4.2   Feature Selection

An important step of building target selection models is selecting the features that will be used as the explanatory variables in the model. Internal databases typically contain customer-specific information about their purchase history and personal preferences. The purchase history can often be translated into measures of recency (e.g. how recent is the last purchase?), frequency (e.g. how often does a customer buy a product?) and monetary value (e.g. how much money does the customer spend per order?). It is often assumed in marketing literature that the RFM-variables are appropriate for capturing the specifics of the customer's purchase behavior [1]. From a modeling point-of-view, the RFM features have the advantage that the purchase behavior can be summarized by using a relatively small number of variables, which can be 10 or less for a given data set. Even though the number of available RFM features is typically small, one must still select the most relevant ones for a particular problem, since it is possible that certain variables do not have any explanatory power at all.

When the RFM variables are used, the clustering space can consist of the product-space of RFM features, since the dimensionality of the RFM feature space is often small enough. Fuzzy clustering divides the data into groups with similar properties on the RFM features considered. The clustering results must now be related to the known response behavior of the customers. An application of feature selection in determining the target models using fuzzy algorithms is presented in the next section.

5   Application: Charity Donations

The algorithm proposed in this paper is applied to target selection of a large data base from a Dutch charity organization. Such an organization does not have clients in the usual sense of the word. However, in order to optimize their fund raising results, it must be able to find the supporters who will probably donate more money, to optimize their fund raising results. These targeted supporters are then contacted by mail preferentially in relation to other individuals in the database. A training data set of about 8000 supporters has been collected for modeling purposes. Seven RFM features have been used for characterizing the donation history of the supporters:

1. Number of weeks since last response (TIMELR).
2. Number of months as a supporter (TIMECL).
3. Fraction of mailing responded (FRQRES).
4. Medium time of response (MEDTOR).
5. Average donation amount (AVGDON).

6. Amount of last donation (LSTDON).
7. Average donation per year (ANNDON).

Therefore, each instance $x_k \in X$ can be represented by the expression:

$$x_k = [\text{TIMELR, TIMECL, FRQRES, MEDTOR, AVGDON, LSTDON, ANNDON}] \quad (5)$$

or by any subset of n attributes considered relevant to the problem. In any case, x_k is a vector with n elements, specifying the values of the n chosen attributes. The concept or function to be learned (target concept), denoted by c, is in this case the donation amount of the campaign used to build the target selection models:

$$c = \text{DONAMT} : X \rightarrow \{0, \mathbb{R}^+\} \quad (6)$$

Note that c is both a classification and a regression value. A zero value indicates that the contacted person is not a supporter, and a value different than 0 indicates the expected donation amount. The training examples are described by the ordered pair $<x_k, c(x_k)>$. After the target selection model has been constructed, a score s_k is attributed to each supporter. This score corresponds to the predicted value $c(x_k)$, which is the model prediction of the amount of money that the supporter will donate in the next fund raising campaign. The best supporters, in the sense of donating more money, will be the targets. Usually, the region of interest is composed by 10% up to 50% of the clients, i.e. the region that is selected for mailing purposes in direct marketing [5].

5.1   Comparing the Target Selection Models

In order to compare the performance of the obtained models, the sum squared error is used. Considering that a real number is being maximized, this is one of the most utilized performance measures. This error is computed as the difference between the ideal and the obtained gain charts, for all data points. This measure S can be computed as follows:

$$S = \sum_{k=1}^{N} (gi_k - gc_k)^2 \quad (7)$$

where gi_k are the values obtained for the ideal gain chart, and gc_k are the values obtained for the gain chart computed using the fuzzy model. The model features are selected and weighted, based on the analysis of the sum squared error S. Note that all models have been derived based on the same pseudo-random seed, in order to be comparable. After normalizing the data, a model using the fuzzy c-means has been obtained using all seven RFM variables with $w_i = 1$, for $i = 1, \ldots, n$. The automatic weighting using GK fuzzy clustering has also been derived. Based on expert knowledge and some trial-and-error experiments, the best weight vector obtained for the seven features in (5) is the following:

$$w = [0.10, 0.01, 0.01, 0.01, 0.725, 0.225, 0.05]. \quad (8)$$

The gain chart for the three tested methods is presented in Fig. 2. Note that the proposed weighted method clearly outperforms the other methods.

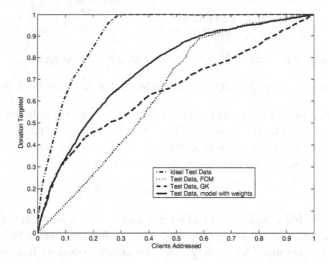

Fig. 2. Gain charts obtained using the seven features: ideal model (-.), c-means (..), weighted c-means(–) and GK model (- -).

By performing some experimentation, analyzing both gain charts and their respective sum squared error S, it was found out that four features presented the most accurate results. So, all the possible combination of target selection models using four features have been tested. The model with the smallest error S, is the one with the following features: TIMELR, AVGDON, LSTDON, ANNDON. Then, based on these four features, one feature was added or excluded, and as so, all models with three and five features have been analyzed, and none had a better result.

Table 1. Comparison of errors using different number of features.

Model	# features	Weights	Train Error	Test Error
c–means	7	no	473.3	490.3
GK	7	yes	166.2	343.9
c–means	7	yes	113.6	155.1
c–means	4	no	128.7	168.8
GK	4	yes	184.9	536.9
c–means	4	yes	113.4	155.8

The sum squared errors obtained in test data are shown in Table 1. Note that the tests using four features are only possible due to the application on expert weights to determine the most important features, as proposed in this paper. Using the seven features, the weighted approach is clearly better than the other approaches. When the four most relevant features are used, the test error S is still clearly better when using the weighted fuzzy clustering algorithm.

6 Conclusions

This paper proposed the weighting of features in fuzzy clustering based on expert's knowledge in classification/regression problems. The method is applied to the maximization of profit in a target selection problem. The application to a charity donations problem shows clearly the advantage of the proposed method.

Future research will deal with the choice of the most relevant features in an automatic way. One possibility is to use logistic regression [3] by testing statistically the parameters obtained for the features. Another possibility is the use of genetic algorithms to optimize the weights.

Acknowledgements

This research is partially supported by the "Programa de Financiamento Plurianual de Unidades de I&D (POCTI), do Quadro Comunitário de Apoio III", FCT, Ministério do Ensino Superior, da Ciência e Tecnologia, Portugal.

References

[1] C. L. Bauer. A direct mail customer purchase model. *Journal of Direct Marketing*, 2:16–24, 1988.
[2] J. C. Bezdek. *Pattern Recognition with Fuzzy Objective Function*. Plenum Press, New York, 1981.
[3] J. S. Cramer. *The LOGIT Model: an introduction for economists*. Edward Arnold, Kent, 1991.
[4] D. E. Gustafson and W. C. Kessel. Fuzzy clustering with a fuzzy covariance matrix. In *Proceedings IEEE CDC*, pages 761–766, San Diego, USA, 1979.
[5] U. Kaymak. Fuzzy target selection using RFM variables. In *Proceedings of joint 9th IFSA World Congress and 20th NAFIPS Int. Conference*, pages 1038–1043, Vancouver, Canada, July 2001.
[6] U. Kaymak and M. Setnes. Fuzzy clustering with volume prototypes and adaptive cluster merging. *IEEE Transactions on Fuzzy Systems*, 10(6):705–712, December 2002.
[7] S. Madeira and J.M. Sousa. Comparison of target selection methods in direct marketing. In *Proceedings of European Symposium on Intelligent Technologies, Hybrid Systems and their Implementation on Smart Adaptive Systems, Eunite'2002*, pages 333–338, Albufeira, Portugal, September 2002.
[8] J. M. Sousa, U. Kaymak, and S. Madeira. A comparative study of fuzzy target selection methods in direct marketing. In *Proceedings of 2002 IEEE World Congress on Computational Intelligence, WCCI'2002, FUZZ-IEEE'2002*, pages 1–6, Paper 1251, Hawaii, USA, May 2002.
[9] J.M. Sousa and U. Kaymak. *Fuzzy Decision Making in Modeling and Control*. World Scientific, Singapore and UK, December 2002.
[10] H. Timm, C. Doring, and R. Kruse. Fuzzy cluster analysis of partially missing data. In *Proceedings of European Symposium on Intelligent Technologies, Hybrid Systems and their Implementation on Smart Adaptive Systems, Eunite'2002*, pages 426–431, Albufeira, Portugal, September 2002.

Data and Cluster Weighting in Target Selection Based on Fuzzy Clustering

Uzay Kaymak

Erasmus University Rotterdam, Faculty of Economics
P. O. Box 1738, 3000 DR, Rotterdam, The Netherlands
u.kaymak@ieee.org

Abstract. We study the construction of probabilistic fuzzy target selection models for direct marketing by using an approach based on fuzzy clustering. Since fuzzy clustering is an unsupervised algorithm, the class labels are not used during the clustering step. However, this may lead to inefficient partitioning of the data space, as the clusters identified need not be well-separated, i.e. homogeneous, in terms of the class labels. Furthermore, the regions of the data space with more high-prospect customers should be explored in larger detail. In this paper, we propose to use a weighting approach to deal with these problems. Additional parameters, i.e. weight factors, associated with data points and the clusters are introduced into the fuzzy clustering algorithm. We derive the optimal update equations for the weighted fuzzy c–means algorithm. A heuristic method for estimating suitable values of the weight factors is also proposed. The benefits of our approach for the target selection models are illustrated by using data from the target selection campaigns of a large charity organization.

1 Introduction

Direct marketing is distinguished from other marketing approaches in that, in direct marketing, one seeks to obtain a direct, individual and measurable response from pre-identified prospective customers. It allows the companies to construct and use detailed customer purchase histories, which tend to be far more useful than traditional predictors based on geo-demographics [1]. An important problem in direct marketing is the identification of prospective customers who are interested in an offer or product. This problem is called *target selection* [2]. The main task in target selection is the determination of prospective customers for a product, given a client database. More precisely, given a customer with a number of features, one has to determine whether he/she will respond to a mailing concerning an offer for a product, either by buying the product or by asking for more information. A customer who responds to the mailing is said to be a *responder*, and a customer who does not respond is said to be a *non-responder*.

Data for constructing target selection models are obtained either from special small-scale mailing campaigns for testing purposes or from the results of previous mailing campaigns. Various methods have been applied to target selection, such as statistical regression [3], neural networks [4, 5], decision and regression trees [6], association rules [7] and fuzzy modelling [8]. The main motivation for using fuzzy modelling is to exploit the linguistic interpretability of fuzzy models, while also obtaining sufficient

T. Bilgiç et al. (Eds.): IFSA 2003, LNAI 2715, pp. 568–575, 2003.
© Springer-Verlag Berlin Heidelberg 2003

numerical accuracy. A fuzzy target selection model that is based on fuzzy c–means clustering is considered in [9]. In this approach, fuzzy clustering is used for determining a fuzzy segmentation of the client base after which the model estimates the probability of response for each customer in a probabilistic fuzzy modelling setting. In this paper, we build on this approach in order to improve the models obtained from this method.

Fuzzy clustering algorithms are unsupervised methods that partition a data set into a number of overlapping groups or clusters based on the distance between the cluster centers and the data points. Since they are unsupervised algorithms, class labels are not used during fuzzy clustering. The clustering solution corresponds to a local optimum of an objective function that is based on the distances between the data points and the cluster prototypes. However, there may be two disadvantages with this solution. First, the clustering algorithm need not achieve the best separation between the classes. Second, the clustering result is influenced by the asymmetry in the distribution of the class labels. In other words, some of the available information regarding the class distribution is not used, which may effect adversely the quality of the final classification model. These two points imply that the result of the clustering algorithm need not correspond to the optimal partition for the target selection problem.

In this paper, we propose a weighting approach for addressing these two points for target selection models obtained by fuzzy clustering. We introduce two sets of weight factors into the clustering algorithm. One set of weight factors is associated with the data points and the other set is associated with the clusters. The distances between the data points and the cluster prototypes are weighted by these weight factors. We derive the optimal update equations for the fuzzy c–means algorithm modified by the weight factors. We also propose a heuristic method for estimating suitable values of the weight factors, given a particular set of clustering data.

The outline of the paper is as follows. Probabilistic fuzzy target selection modelling by using fuzzy clustering is discussed in Section 2. The concept of data and cluster weighting is introduced in Section 3, where also the optimal update equations for the modified fuzzy c–means algorithm are derived. The benefits of our approach for the target selection models are illustrated in Section 4 by using data from the target selection campaigns of a large charity organization. Finally, the conclusions are given in Section 5.

2   Probabilistic Fuzzy Target Selection

Target selection modelling is the generation of customer models (profiles) for a given product by analyzing client data obtained from similar marketing campaigns. These models are used for selecting the customers that are more likely to respond to the direct mailing regarding the product. In this way, the profits for the company can be maximized by reducing the size of the mailing, so that only high-prospect customers are sent an offer instead of the whole client base.

A promising method for obtaining target selection models is based on the use of a probabilistic fuzzy modelling paradigm. These types of models are interpretable linguistically, while the underlying probability distributions for response can also be estimated accurately. Probabilistic fuzzy target selection models can be obtained by fuzzy clustering. In this approach, fuzzy clustering provides a fuzzy partition of the data. Given this partition, conditional probabilities for different classes can be computed by using class label data. Let $\{x_1, \ldots, x_N\}$ be a set of data objects, where $x_j \in \mathbb{R}^M$ is a feature vector that characterizes a customer. A fuzzy clustering algorithm partitions these objects into C overlapping clusters by computing a partition matrix U and C cluster prototypes $\{v_1, \ldots, v_C\}$. Fuzzy c–means (FCM) clustering is known to perform well on target selection problems [9], and so the remainder of the discussion considers only the FCM algorithm. In FCM, the cluster prototypes are also points in the clustering space, i.e. $v_i \in \mathbb{R}^M$, $i = 1, \ldots, C$.

The solution to the FCM algorithm is a local minimum of the objective function

$$J = \sum_{i=1}^{C} \sum_{j=1}^{N} u_{ij}^m \, d_{ij}^2 \, , \tag{1}$$

subject to the constraint $\sum_{i=1}^{C} u_{ij} = 1$, where the exponent m determines the degree of fuzziness of the clusters, and $d_{ij}^2 = (x_j - v_i)^T (x_j - v_i)$ is the squared Euclidian distance between the data vector x_j and the cluster center v_i [10]. The membership of the data vector x_j in cluster i is indicated by u_{ij}.

The probabilistic fuzzy target selection model estimates the conditional probability that a customer is a responder, given the occurrence of fuzzy events that are described by the fuzzy clusters obtained from the FCM algorithm, and the membership of the customer to these fuzzy events. Let ω be the class label belonging to the data vector x and ω_r the class label for the responders. Then, the model is implemented as a zero-order Takagi–Sugeno (TS) system [11], where the rules R_i are of the form

$$R_i \; : \; \text{If } x \text{ is } A_i, \text{ then } \rho_i = \Pr(\omega = \omega_r | A_i), \tag{2}$$

where A_i represents the fuzzy cluster i, and ρ_i is the (singleton) consequent of rule R_i. Each rule thus corresponds to a fuzzy cluster, and its quantifies the conditional probability that the class is ω_r (i.e. customer is a responder) given the fuzzy event described by cluster A_i. The total output s_j of the fuzzy model for vector x_j is then found by Takagi–Sugeno reasoning as

$$s_j = \frac{\sum_{i=1}^{C} u_{ij} \, \rho_i}{\sum_{i=1}^{C} u_{ij}} = \Pr(\omega_j = \omega_r | x_j) \, , \tag{3}$$

with u_{ij} being the membership of data vector x_j in cluster A_i. Note that the denominator of (3) equals 1 due to the probabilistic constraint imposed on the clustering solution. Further, s_j can be interpreted as an estimation of the conditional probability that a customer is a responder, given his/her data vector, which is obtained by interpolation between the conditional probability of different clusters and the membership of the customer in these clusters, as shown in (3).

The model parameters are complete when we calculate the rule consequents. These can be obtained by using probability theory and probability measures of fuzzy events defined by Zadeh [12]. The rule consequents are then given by [13]

$$\rho_i = \Pr(\omega = \omega_r | A_i) = \frac{\Pr(\{\omega = \omega_r\} \cap A_i)}{\Pr(A_i)} = \frac{\sum_{j=1}^{N} \chi_{\omega_r}(\omega_j) \, u_{ij}}{\sum_{j=1}^{N} u_{ij}} \, , \tag{4}$$

where χ_{ω_r} is the characteristic function for the responders. The final model consists of the location of the cluster centers v_i, $i = 1, \ldots, C$ and the corresponding conditional probabilities ρ_i.

3 Data and Cluster Weighting

Since the estimation of the rule consequents is done after determining the fuzzy clusters, the estimation of the conditional probability that a customer is a responder depends on the clustering results. In other words, the location of the clusters should be such that the conditional probabilities can be estimated in a manner that is relevant for the problem analyzed. However, the nature of the FCM algorithm provides no guarantees for obtaining the most efficient partition. In particular, two problems can be distinguished for the target selection method described in Section 2.

1. We seek clusters that separate the responders from the non-responders in an optimal way. However, the FCM algorithm does not consider the class labels during clustering, and so it need not achieve the best separation between the two classes. In other words, the clusters need not be as homogeneous as possible in terms of the class labels.
2. The clustering result is influenced by the asymmetry in the distribution of the two classes. Typically, there are many more non-responders than the responders in a mailing campaign, which will bias the location of the cluster prototypes towards the regions with high concentration of non-responders. The goal of target selection models, however, is to identify the regions in the feature space with relatively high concentrations of responders. Hence, the influence of the responders on the final partition of the FCM algorithm should be larger.

We propose a weighting scheme for dealing with the two issues mentioned above. The main idea is to introduce the weight factors into the objective function that is minimized by the FCM algorithm, so that the optimal partition found by the algorithm is more in line with better separation of the two classes and larger influence of responders. We use two sets of weight factors, σ_j, $j = 1, \ldots, N$, associated with the data points, and τ_i, $i = 1, \ldots, C$, associated with the clusters. The objective function that the FCM algorithm minimizes now becomes

$$J_w = \sum_{i=1}^{C} \sum_{j=1}^{N} u_{ij}^m \, \sigma_j \, \tau_i \, d_{ij}^2 \, , \tag{5}$$

again subject to the constraint $\sum_{i=1}^{C} u_{ij} = 1$. The weight factors are constant during clustering, and they are independent of one another.

The optimal update equations for the alternating iterations are obtained by using the method of Lagrange multipliers [14], similar to the case with the original FCM objective function. Solving for the cluster centers and the membership values, one obtains the update equations

$$v_i = \frac{\sum_{j=1}^{N} u_{ij}^m \, \sigma_j \, x_j}{\sum_{j=1}^{N} u_{ij}^m \, \sigma_j} \, , \tag{6}$$

and

$$u_{ij} = \frac{1}{\sum_{k=1}^{C} \left(\frac{\tau_i d_{ij}^2}{\tau_k d_{kj}^2} \right)^{\frac{1}{m-1}}} . \tag{7}$$

From (7) we see that if the weight τ_i of a cluster is relatively larger than the others, the denominator of (7) will tend to be large, and so the membership will be small. In other words, the important clusters will tend to be more localized, allowing for the investigation of detail. Conversely, less important clusters will have larger membership values, which implies that they will occupy a larger volume. Hence, the region of space that they occupy is explored in less detail. From (6) we see that the location of the centers is now influenced by the weight factors σ_j associated with the data vectors. Relatively more important data points draw the cluster centers more towards themselves. Hence, the clusters will be localized in the regions with high concentrations of important data points.

Various mechanisms for determining the weight factors can be devised. Here, we consider a heuristic approach, which is motivated as follows, inspired by fuzzy exception learning [15]. In target selection, we are interested in identifying the customers that deviate from the average. Customers who are more likely to respond than the average would be selected preferentially, whereas the customers who are more unlikely to respond than the average would definitely not be considered in the mailing. If we explored the data set initially and obtained a target selection model, we would have an idea about the likelihood of response of the customers. We could then use this likelihood to derive a weight factor for each customer. Similarly, we could use this exploratory step to estimate the average response rate in a cluster. Since we are interested in clusters that are homogeneous, we could give higher weights to the relatively homogeneous clusters. Interestingly, these clusters correspond to maximum deviation from the average, and they will be more localized according to (7). In other words, we will be exploring the regions that deviate most from the average in larger detail, whereas regions with average response will be partitioned in a courser way.

To quantify the concept of "deviation from the average", we use a fuzzy set B as shown in Fig. 1, which is a characterization of the linguistic value "Average" as a function of the probability of response. In Fig. 1, p^* stands for the average response rate in the training data. Given the membership function $B(p)$, the deviation from the average is computed as $1 - B(p)$, which is also equal to the weight factor that is used.

The weight factors are thus determined as follows. First, the standard FCM algorithm is applied, and the quantities s_j, $j = 1, \ldots, N$ and ρ_i, $i = 1, \ldots, C$ are determined. The weight factors are then given by

$$\sigma_j = 1 - B(s_j - p^*) , \tag{8}$$

for the data weights, and by

$$\tau_i = 1 - B(\rho_i - p^*) , \tag{9}$$

for the cluster weights. The algorithm for obtaining the target selection model thus looks like as follows.

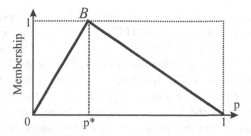

Fig. 1. Definition of the fuzzy set for "Average" as a function of the response rate p.

Algorithm 1. *Choose the number of clusters C, fuzziness parameter m and the termination criterion ϵ.*

1. *Apply standard FCM algorithm with C, m and ϵ.*
2. *From FCM results, compute ρ_i, $i = 1, \ldots, C$ using (4) and s_j, $j = 1, \ldots, N$ using (3).*
3. *Compute the data weights using (8) and the cluster weights using (9).*
4. *Apply weighted FCM algorithm with C, m, ϵ, the data weights and the cluster weights. Initialize the algorithm with the results from step 1.*
5. *From weighted FCM results, compute the new values for ρ_i.*
6. *Record the cluster centers v_i and ρ_i as the target selection model.*

4   Predicting Charity Supporters

In this section, we illustrate the influence of our approach on the target selection results by using data from the target selection campaigns of a large charity organization [4]. The emphasis of this section is to illustrate the benefits of data and cluster weighting, and so we will not deal with the identification of the best model for this problem. In particular, issues regarding feature selection, number of clusters and extensive validation of the models will not be considered. These points have already been considered in [9].

Charity organizations do not have customers in the usual sense of the word. However, they use target selection to trace people who are more likely to donate money in order to optimize their fund raising results. The targeted supporters are then contacted by mail preferentially in relation to other individuals in the database. For this problem, a training data set of about 4000 supporters has been collected for modelling purposes. An independent data set of similar size has been used for validation purposes. In this study, we used three RFM (recency, frequency and monetary value) features that are known to give good results for this data set. These features are

1. Number of weeks since last response (TIMELR).
2. Number of months as a supporter (TIMECL).
3. Fraction of mailings responded (FRQRES).

More details can be found in [4] and [9].

Fuzzy target selection models have been developed with the procedure described in Section 2 and Section 3. Both the standard FCM and the weighted FCM algorithms have been applied with $C = 15$, $m = 2$ and $\epsilon = 0.001$. Figure 2 shows the hit probability charts for both models, for the evaluation data set. Similar results have been obtained from different initializations of the clustering algorithms. A hit probability chart shows what percentage of the mailed customers respond, given a certain mailing size. In general, one tries to maximize the values of response. Hence, a response improvement in a target selection model is visible on a hit probability chart as an increase in the chart values. Considering Fig. 2, the response of the target selection model in the first decile is improved significantly by the weighted FCM, indicating that our approach is suitable to find specific regions with higher probability of response.

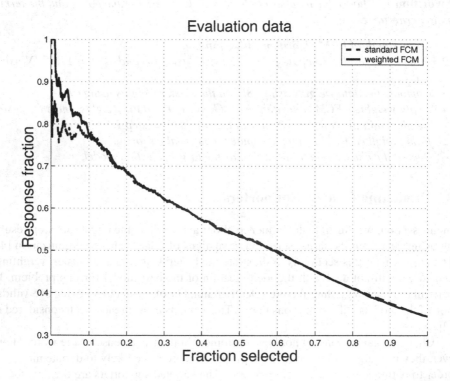

Fig. 2. Hit probability charts from target selection models obtained with the standard FCM algorithm and the weighted FCM algorithm.

5   Conclusions

Probabilistic fuzzy target selection models can be obtained by fuzzy c–means clustering followed by the estimation of conditional probability distributions. A problem with this approach is that the estimation of the conditional probabilities depends on the results of the fuzzy clustering step. We have shown in this paper that weighted FCM can be used to address this issue in order to improve probabilistic fuzzy target selection models. We weigh both the data vectors and the clusters to minimize a weighted objective function for the FCM. The optimal update equations for the modified FCM have been presented. Application to a target selection problem from a large charity organization has illustrated the benefits of our approach for improving the target selection models. At the moment, we present a heuristic approach for determining the weight factors. In future work, we will concentrate on replacing the heuristic method of assigning the weight factors with another approach based on a more detailed analysis of the clustering results.

References

[1] Morwitz, V.G., Schmittlein, D.C.: Testing new direct marketing offerings: the interplay of management judgment and statistical models. Management Science **44** (1998) 610–628
[2] Bult, J.R.: Target Selection for Direct marketing. Ph.D. thesis, Rijksuniversiteit Groningen, Groningen, the Netherlands (1993)
[3] Bult, J.R., Wansbeek, T.J.: Optimal selection for direct mail. Marketing Science **14** (1995) 378–394
[4] Potharst, R., Kaymak, U., Pijls, W.: Neural networks for target selection in direct marketing. In Smith, K., Gupta, J., eds.: Neural Networks in Business: techniques and applications. Idea Group Publishing, London (2002) 89–110
[5] Zahavi, J., Levin, N.: Applying neural computing to target marketing. Journal of Direct Marketing **11** (1997) 5–22
[6] Haughton, D., Oulabi, S.: Direct marketing modeling with CART and CHAID. Journal of Direct Marketing **7** (1993) 16–26
[7] Pijls, W., Potharst, R., Kaymak, U.: Pattern-based target selection applied to fund raising. In Gersten, W., Vanhoof, K., eds.: Data Mining for Marketing Applications, ECML/PKDD-2001 Workshop, Freiburg, Germany (2001) 15–24
[8] Setnes, M., Kaymak, U.: Fuzzy modeling of client preference from large data sets: an application to target selection in direct marketing. IEEE Transactions on Fuzzy Systems **9** (2001) 153–163
[9] Kaymak, U.: Fuzzy target selection using RFM variables. In: Proceedings of Joint 9th IFSA World Congress and 20th NAFIPS International Conference, Vancouver, Canada (2001) 1038–1043
[10] Bezdek, J.C.: Pattern Recognition with Fuzzy Objective Function. Plenum Press, New York (1981)
[11] Takagi, T., Sugeno, M.: Fuzzy identification of systems and its applications to modelling and control. IEEE Transactions on Systems, Man and Cybernetics **15** (1985) 116–132
[12] Zadeh, L.A.: Probability measures of fuzzy events. J. Math. Anal. Appl. **23** (1968) 421–427
[13] van den Berg, J., Kaymak, U., van den Bergh, W.M.: Fuzzy classification using probability-based rule weighting. In: Proceedings of 2002 IEEE International Conference on Fuzzy Systems, Honolulu, Hawaii (2002) 991–996
[14] Höppner, F., Klawonn, F., Kruse, R., Runkler, T.: Fuzzy Cluster Analysis: methods for classification, data analysis and image recognition. Wiley, New York (1999)
[15] van den Bergh, W.M., van den Berg, J., Kaymak, U.: Detecting noise trading using fuzzy exception learning. In: Proceedings of Joint 9th IFSA World Congress and 20th NAFIPS International Conference, Vancouver, Canada (2001) 946–951

On-line Design of Takagi-Sugeno Models

Plamen Angelov[1], Dimitar Filev[2]

[1] Dept of Communications Systems, University of Lancaster, Lancaster, LA1 4YR, UK
P.Angelov@Lancaster.ac.UK
[2] Ford Motor Co., 24500 Glendale av., Detroit, MI 48239, USA
Dfilev@Ford.com

Abstract. An approach to the on-line design of Takagi-Sugeno type fuzzy models is presented in the paper. It combines supervised and unsupervised learning and recursively updates both the model structure and parameters. The rule-base gradually evolves increasing its summarization power. This approach leads to the concept of the evolving Takagi –Sugeno model. Due to the gradual update of the rule structure and parameters, it adapts to the changing data pattern. The requirement for update of the rule-base is based on the spatial proximity and is a quite strong one. As a result, the model evolves to a compact set of fuzzy rules, which adds to the interpretability, a property especially useful in fault detection. Other possible areas of application are adaptive non-linear control, time series forecasting, knowledge extraction, robotics, behavior modeling. The results of application to the on-line modeling the fermentation of *Kluyveromyces lactis* illustrate the efficiency of the approach.

1 Introduction

Takagi-Sugeno (TS) models [5] are widely accepted as a very efficient technique used in solving many practical engineering problems for modeling, classification, and control of complex systems. They have been successfully applied to a variety of problems including nonlinear dynamics approximation, control with multiple operating modes and significant parameter and structure variations. They can be seen as a generalization of the multi-model concept with fuzzily defined regions, in which the nonlinear system is locally linearized. The specific blending of the linear sub-models is the key to the computational efficiency of TS models.

The methods for *off-line* design of TS models based on the data include separate structure and parameter identification [5], [7]. The model structure (fuzzy rules) can be identified by fuzzy clustering of the input/output data. Once the number of fuzzy rules and their focal points are defined, the TS model transforms into a linear model, because the weights of the blending of the local linear models can be fixed. Parameters of the linear sub-models can be obtained by recursive least squares (RLS) method or by a pseudo-inversion [5], [7]. *Off-line* methods, however, suppose that *all* the data is available at the start of the learning process.

As it is well known, however, in many practical problems the object of modeling or the environment can change significantly. Recursive and adaptive techniques are

T. Bilgiç et al. (Eds.): IFSA 2003, LNAI 2715, pp. 576–584, 2003.
© Springer-Verlag Berlin Heidelberg 2003

well developed for the linear models with *fixed structure* [10]. This is not true for the general nonlinear case, however. Moreover, if it is assumed that the structure of the model is also adapting/evolving. Recently, approaches, which treat rule-bases [6] and neural networks [9], [11] with *evolving* structure have been published. They are primarily oriented to control applications and use different mechanism of rules update based on the distance to certain rule center [6], [11] or the error in previous steps [9].

This paper presents an approach to *on-line* design of TS models, which can have evolving structure and can learn recursively in *real-time* from the data. The *on-line* design of TS model can be decomposed into two sub-problems:

i)   *on-line recursive* clustering responsible for the rule base learning;

ii)   *on-line recursive* estimation of the consequent part parameters.

In the present approach, similarly to the mountain and subtractive clustering [1], [7] approaches, the notion of the informative potential of the new data sample is used as a trigger to update (upgrade or modify) the rule-base. It represents the accumulated spatial proximity information, which ensures greater generality of the structural changes and makes possible to disallow the outlays to become rule centers. It also ensures that the rules have more summarization power from the moment of their initialization. In addition, we consider a mechanism of rule-base modification (replacement of less informative rule with a more informative one) ensuring a *gradual* change of the rule-base structure and inheritance of the structural information. The learning could start without *a priori* information and only one data sample, which makes the approach potentially very useful in adaptive control and robotics.

The approach has been applied to the on-line modeling the fermentation of *Kluyveromyces marx.* in lactose oxidation. The results illustrate its efficiency.

2 Off-line Design of Takagi-Sugeno Fuzzy Models

TS rule-based models have fuzzy antecedents and functional consequents [5]:

$\mathfrak{R}_i$: IF (x_1 is $\aleph_{i1}$) AND ... AND (x_n is $\aleph_{in}$)THEN ($y_i = a_{i0} + a_{i1}x_1 + ... + a_{in}x_n$)   (1)

Where $\mathfrak{R}_i$ is the ith fuzzy rule; i=[1, R]; R is the number of fuzzy rules; *x* is the input vector; $x = [x_1, x_2, ..., x_n]^T$; $\aleph_{ij}$ is the antecedent fuzzy sets, j=[1 ,n]; y_i is the output of the ith linear subsystem; a_{il} are its parameters, l=[0,n].

We assume Gaussian-like antecedent fuzzy sets to define the regions in which the local linear sub-models are valid:

$$\mu_{ij} = e^{-\alpha \| x_j - x_i^* \|}$$ (2)

Where $\alpha = 4/r^2$; r defines the spread of the antecedent and the zone of influence of the ith model, i=[1,R]; x_i^* is the focal point of the ith rule antecedent, j=[1,n].

The TS model output is calculated by averaging of individual rules' contributions:

$$y = \sum_{i=1}^{R} \lambda_i y_i = \sum_{i=1}^{R} \lambda_i x_e^T \pi_i = \psi^T \theta$$ (3)

Where y_i is the output of the ith sub-model; $\pi_i = [a_{i0}\ a_{i1}\ a_{i2}\ ...\ a_{in}]^T$, $i=[1, R]$, is the vector of parameters of the ith sub-model; $x_e = [1\ x^T\ ]^T$ is the expanded data vector; $\psi = [\lambda_1 x_e^T, \lambda_2 x_e^T, ..., \lambda_R x_e^T]^T$ is a vector of the inputs, weighted by the normalized firing levels (λ) of the rules; $\theta = [\pi_1^T, \pi_2^T, ..., \pi_R^T]^T$ are parameters of the sub-models.

The degree of firing of each rule is proportional to the contribution of the corresponding linear model to the overall output:

$$\lambda_i = \frac{\tau_i}{\sum_{j=1}^{R}\tau_j}; \quad \tau_i = \mu_{i1}(x_1)\times\mu_{i2}(x_2)\times...\times\mu_{in}(x_n)=\prod_{j=1}^{n}\mu_{ij}(x_j) \tag{4}$$

The *off-line* design of TS models can be decomposed into the following sub-tasks [7]:
  i)   Learning the antecedent part of the fuzzy model (determination of the focal points of the rules, x_i^*; i=[1, R] and spreads, r of the membership functions;
  ii)   Learning the consequent part parameters (a_{ij}; i=[1, R]; j=[0, n]).
First sub-task can be solved by clustering the input-output data space ($z= [x^T;y]^T$).

In a batch-processing learning mode when all the data is available antecedent parameters can be learned by one of the established methods for off-line clustering, such as subtractive clustering [7], FCM [14], Gustafson-Kessel clustering [15], SVM [12].

The second sub-task is transformed into a least squares estimation, when antecedent parameters are defined and fixed. For a given set of input-output data the vector of linear model parameters θ minimizing the locally weighted objective function ensures locally meaningful sub-models [8], [13]:

$$J_L = \sum_{i=1}^{R}\left(Y - X^T\pi_i\right)^T\Lambda_i\left(Y - X^T\pi_i\right) \tag{5}$$

Where matrix X is formed by x_{ek}^T; $X \in R^{TD\times(n+1)}$; matrix Λ_i is a diagonal matrix with $\lambda_i(x_k)$ as its elements in the main diagonal.

The solutions, minimizing individual cost functions can be found by weighted RLS:

$$\hat{\pi}_{ik} = \hat{\pi}_{ik-1}+c_{ik}x_{ek}\lambda_i(x_k)(y_k-x_{ek}^T\hat{\pi}_{ik-1}); \hat{\pi}_0 = 0 \tag{7}$$

$$c_{ik} = c_{ik-1} - \frac{\lambda_i(x_k)\ c_{ik-1}x_{ek}x_{ek}^Tc_{ik-1}}{1+\lambda_i(x_k)\ x_{ek}^Tc_{ik-1}x_{ek}}; k=[1,TD]; c_{i0} = \Omega I \tag{8}$$

The globally optimal cost function, which ensures a better approximation globally for the price of ignoring the local meaning of the sub-models is treated in [13].

3 On-line Design of Takagi-Sugeno Models

The *on-line* learning is under the assumption of continuously collected data. Each new data item can either reinforce the information contained in the previous data or bring enough new information to form a new rule or to modify an existing one. The value of the information they bring is closely related to the information the data collected so far has already posses. Similarly to the subtractive clustering [7] approach the importance of the data is evaluated using its informative potential, measured by the spatial proximity of the data. By differ from [7] we use a Cauchy type function and calculate it recursively. *On-line* design of TS models includes *on-line* clustering and weighted RLS with the assumption of a *gradually evolving* rule-base.

The *on-line* clustering procedure [13] starts with the first data point established as the focal point of the first rule. The antecedent part of the fuzzy rule (1) is formed using the coordinates of this point. Its informative potential is assumed equal to 1. The potential of the next data points is calculated *recursively* by [13]:

$$P_k(z_k) = \frac{k-1}{(k-1)(\vartheta_k + 1) + \sigma_k - 2v_k} \tag{9}$$

Where $P_k(z_k)$ denotes the potential of the data point (z_k) calculated at the moment k; $d_{ik}^j = z_i^j - z_k^j$, denotes projection of the distance between two data points (z_i^j and z_k^j) on the axis z^j (x^j for j=1,2,…,n and on the axis y for j=n+1).

$$\vartheta_k = \sum_{j=1}^{n+1} (z_k^j)^2 ; \ \sigma_k = \sum_{i=1}^{k-1} \sum_{j=1}^{n+1} (z_i^j)^2 ; \ v_k = \sum_{j=1}^{n+1} z_k^j \ \beta_k^j ; \ \beta_k^j = \sum_{i=1}^{k-1} z_i^j \tag{10}$$

Quadratic forms ϑ_k and v_k are calculated from the current data point z_k, while β_k^j and σ_k are recursively updated as follows:

$$\sigma_k = \sigma_{k-1} + \sum_{j=1}^{n+1} (z_{k-1}^j)^2 ; \ \beta_k^j = \beta_{k-1}^j + z_{k-1}^j \tag{11}$$

Each *new* data influences the potentials of the centers of the clusters (z_i^*, i=[1,R]), which are respective to the focal points of the existing rules (x_i^*, i=[1,R]), because the potential depends on the distance to all data points. Potentials of the focal points of the existing clusters are recursively updated taking this effect into account [13]:

$$P_k(z_i^*) = \frac{(k-1)P_{k-1}(z_i^*)}{k-2 + P_{k-1}(z_i^*) + P_{k-1}(z_i^*)\sum_{j=1}^{n+1} (d_{k(k-1)}^j)^2} \tag{12}$$

Where $P_k(z_i^*)$ is the potential of the center of the ith rule at the moment k.

At each step the potentials of the *new* data points are compared to the updated potential of the centers of the existing clusters. If the potential of the new data point is higher than the potential of the existing centers then the new data point is accepted as a new center and a new rule is formed with a focal point based on the projection of this center on the axis x (R:=R+1; $x_R^* = x_k$) [13]. This strong condition is satisfied very rarely. It means that the new data point is more descriptive, has more summarization power than *all* the other data points. The concentration of the data is usually decreasing with the growing number of data. The only exception is when a new region of data space reflecting a new operating regime or new condition appears and a new rule is formed. The outlying data is automatically rejected because its potential is significantly lower due to their distance from the other data. This property of the proposed approach is very useful for fault detection.

If in addition to the previous condition the new data is close to an existing center:

$$\frac{\arg\min_{i=1}^{R} \|z_k - z_i^*\|}{r} + \frac{P_k(z_k)}{\max_{i=1}^{R} P_k(z_i^*)} < 1 \tag{13}$$

then the new data point (z_k) replaces this center ($z_j^* := z_k$).

The proposed on-line clustering approach ensures a gradual upgrade of the rule-base while inheriting the bulk of the rules (all but one of the rules are preserved). Therefore, the normalized firing strengths of the rules (λ_i) will change. Since this effects *all* the data (including the data collected before the moment of the change) the straightforward application of the weighted RLS (7)-(8) is not correct. A proper resetting of the co-variance matrices and parameters initialization is needed at each moment a rule is added to and/or removed from the rule base. We proposed to estimate the co-variance matrices and parameters of the new $(R+1)^{th}$ rule as a weighted average of the respective co-variance and parameters of the remaining R rules [13]. This is possible, since the approach of rule-base upgrade concerns one rule only and all the other R rules remain unchanged. The consequent parameter estimation is based on the weighted RLS.

Parameters of the newly added rule are determined as weighted average of the parameters of the rest R rules by $\hat{\pi}_{R+1k} = \sum_{i=1}^{R} \lambda_i \hat{\pi}_{ik-1}$. Parameters of the other R rules are inherited unchanged ($\pi_{ik} := \pi_{i(k-1)}; i = [1, R]$). When a rule is replaced by another one, which have close center then parameters of all rules are inherited ($\pi_{ik} := \pi_{i(k-1)}; i = [1, R]$). The co-variance matrix of the newly added rule is initialized by $C_{R+1k} = \Omega I$. The co-variance matrices of the rest R rules are inherited ($c_{ik} := c_{i(k-1)}; i = [1, R]$).

The *on-line* design of TS models includes the following phases [13]:

Phase 1: Initialization of the antecedent part of the rules (rule-base structure).
Phase 2: Reading the *next* data sample.
Phase 3: *Recursive* calculation of the informative potential of each new data sample.
Phase 4: *Recursive* up-date of the potentials of old centers.

Phase 5:   Possible update or upgrade of the rule-base *structure*.
Phase 6:   *Recursive* calculation of the consequent parameters.
Phase 7:   Prediction of the output for the next time step by the TS model.
The execution of the algorithm continues for the next time step from stage 2. The

first output to be predicted is $\hat{y}_3$.

4 Experimental Results

On-line modeling of the fermentation of *Kluyveromyces marxianus var. lactis MC 5*
in lactose oxidation [16] is considered as an example. The model of this particular
process includes the dependence between concentrations of the basic substrates: lactose (S) and dissolved oxygen (DO) concentration. There does not exist a general
mathematical model of the microbial synthesis because of the extreme complexity
and variety of living activity of the micro-organisms, although various models of
different parts of the process exist. 6 different fermentation where carried out in aerobic batch cultivation. The cell mass concentration (X_k) is modeled on-line by a TS
model, which evolves to 8 fuzzy rules:

R_1: *IF(S is Extremely High)AND(DO is Very High)* **THEN** $X = a_0^1 + a_1^1 S + a_2^1 DO$

R_2: *IF (S is Very High) AND (DO is Rather High)* **THEN** $X = a_0^2 + a_1^2 S + a_2^2 DO$

R_3: *IF (S is High) AND (DO is Very High)* **THEN**
$$X = a_0^3 + a_1^3 S + a_2^3 DO$$

R_4: *IF (S is Medium) AND (DO is High)* **THEN** $X = a_0^4 + a_1^4 S + a_2^4 DO$

R_5: *IF (S is Rather High) AND (DO is Very Low)* **THEN** $X = a_0^5 + a_1^5 S + a_2^5 DO$

R_6: *IF (S is Rather Low) AND (DO is Rather Low)* **THEN** $X = a_0^6 + a_1^6 S + a_2^6 DO$

R_7: *IF (S is Very Low) AND (DO is Low)* **THEN** $X = a_0^7 + a_1^7 S + a_2^7 DO$

R_8: *IF (S is Low) AND (DO is Extremely)* **THEN** $X = a_0^8 + a_1^8 S + a_2^8 DO$

The centres of the antecedent part are given in Table 1.

Table 1. Centers of the antecedent part of the rules

	S_k, Label	Value, g/l	DO_k, Label	Value, g/l
x_1^*	Extremely High	41.3	Extremely High	5.59
x_2^*	Very High	35	Rather High	2.54
x_3^*	High	32.4	Very High	4.04
x_4^*	Medium	21.4	High	2.80
x_5^*	Rather High	24.0	Very Low	0.13
x_6^*	Rather Low	18.8	Rather Low	0.14
x_7^*	Very Low	10.5	Low	0.18
x_8^*	Low	15.1	Extremely Low	0.07

582 Plamen Angelov and Dimitar Filev

The RMS error in prediction of the cell mass concentration is 2.125 g/l (Fig. 1). The computation time was less than 2 min on 550 MHz PC in Matlab environment. The reason for the computational efficiency of the approach is its non-iterative and recursive nature. This leads to the main advantage of this approach - its applicability for real-time applications. As a shortcoming of this approach one can mention its sub-optimality, which is a result of the decomposition of the identification problem into two-sub-problems. As a result, theoretically better models can be achieved using computationally and time consuming iterative search techniques, like gradient-based search, genetic algorithms etc.

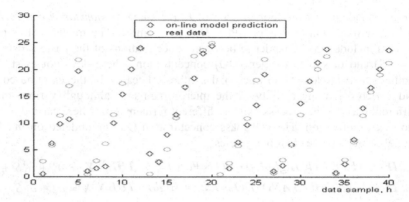

Fig. 1. On-line modeling cell mass concentration (circles - real data; diamonds - TS model)

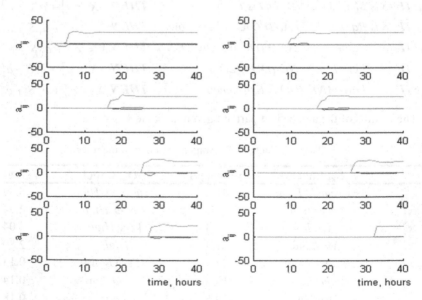

Fig. 2. Consequent parameters evolution ($\pi_i = [a_{i0}\ a_{i1}\ a_{i2}\ \ldots\ a_{in}]^T$, i=[1, R])

5 Concluding Remarks

A computationally effective approach to *on-line* design of TS models is presented in the paper, which does not require re-training of the whole model. It is based on *recursive, non-iterative evolution* of the rule base by unsupervised learning. The adaptive nature of this model in addition to the highly transparent and compact form of fuzzy rules makes them a promising candidate for *on-line* modeling and control of complex processes. The main advantages of the approach are:

✓ it can evolve an existing model when the data pattern changes, while inheriting the rule base;

✓ it can start to learn a process from a single data samples and improve the performance of the model predictions *on-line*;

✓ it is *non-iterative* and *recursive* and hence applicable in real-time.

The results illustrate the viability and efficiency of the approach. The proposed concept has wide implications for many problems, including non-linear adaptive control, fault detection, time-series forecasting, knowledge extraction, behavior-modeling etc. Future implementation in various engineering problems is under consideration.

References

1. Yager, R., Filev, D. : Essentials of Fuzzy Modeling and Control. NY: John Wiley, (1994)
2. Johanson, T. A., Murray-Smith, R.: Operating regime approach to non-linear modeling and control. In: Multiple Model Approaches to Modeling and Control. Murray-Smith, R., Johanson T. A. (eds.). Hants, UK: Taylor Francis (1992) 3-72
3. Jang, J. S. R.: ANFIS: Adaptive network-based fuzzy inference systems, IEEE Transactions on Systems, Man & Cybernetics. 23 (3) (1993) 665-685
4. Angelov, P.: Evolving Rule-based Models: A Tool for Design of Flexible Adaptive Systems. Heidelberg, Germany: Springer-Verlag (2002)
5. Takagi, T., Sugeno, M. : Fuzzy identification of systems and its application to modeling and control. IEEE Trans. on Systems, Man and Cybernetics. 15 (1985) 116-132
6. Filev, D. P.: Rule-base guided adaptation for mode detection in process control. In: 9th IFSA World Congress, Vancouver, BC, Canada (2001) 1068-1073
7. Chiu, S. L.: Fuzzy model identification based on cluster estimation, J. of Intell. & Fuzzy Syst. 2 (1994) 267-278
8. Yen, J., Wang, L., Gillespie, C. W.: Improving the Interpretability of TSK Fuzzy Models by Combining Global and Local Learning. IEEE Trans. on Fuzzy Syst. 6 (1998) 530-537
9. Lin, F.-J., Lin, C.-H., Shen, P.-H.: Self-constructing fuzzy neural network speed controller for permanent-magnet synchronous motor drive. IEEE Tr. Fuzzy Syst. 9 (2001) 751-759
10. Astroem, K. J., Wittenmark, B.: Adaptive Control. Addison Wesley, MA, USA, 1989
11. Kasabov, N. K., Song, Q.: DENFIS: Dynamic Evolving Neural-Fuzzy Inference System and its application for time-series prediction. IEEE Tr. on Fuzzy Syst. 10 (2002) 144-154
12. Vapnik V., The Nature of Statistical Learning Theory. New York: Springer-Verlag (1995)
13. Angelov, P. , D. Filev, "An approach to on-line identification of Takagi-Sugeno fuzzy models. IEEE Trans. on Systems, Man and Cybernetics-B, to appear.
14. Bezdek, J., Cluster Validity with Fuzzy Sets : J of Cybernetics 3(3) (1974) 58-71

15. Gustafson, D. E., Kessel, W. C.: Fuzzy clustering with a fuzzy covariance matrix. IEEE Control and Decision Conference. San Diego, CA, USA (1979) 761-766
16. Angelov, P., Simova, E., Beshkova, D.: Control of cell protein synthesis from Kluyweromyces marxianus var. lactis MC5. Biotech. & Bio Eq. 10(1996) 44-50

Gradient Projection Method and Equality Index in Recurrent Neural Fuzzy Network

Rosangela Ballini[1] and Fernando Gomide[2]

[1] DTE – IE – UNICAMP
ballini@eco.unicamp.br
[2] DCA – FEEC – UNICAMP
13083-970 Campinas, SP, Brazil
gomide@dca.fee.unicamp.br

Abstract. A novel learning algorithm for recurrent neurofuzzy networks is introduced in this paper. The learning algorithm uses the gradient projection method to update the network weights. Moreover, the core of the learning algorithm uses equality index as the performance measure to be optimized. Equality index is especially important because its properties reflect the fuzzy set-based structure of the neural network and nature of learning. The neural network topology is built with fuzzy neuron units and performs neural processing consistent with fuzzy system methodology. Therefore neural processing and learning are fully embodied within fuzzy set theory. The performance recurrent neurofuzzy network is verified via examples of nonlinear system modeling and time series prediction. The results confirm the effectiveness of the neurofuzzy network.

1 Introduction

Recurrent neural networks have received considerable attention during the last years due to its ability to model complex dynamics. This class of neural networks can be classified as either globally or partially recurrent, depending on the structure of the feedback connections. The presence of feedback allows the generation of internal representation and memory capabilities, both essential to process spatio-temporal information [1].

As is widely known, the concept of incorporating fuzzy logic into neural networks has grown into an important research topic in the realm of soft computing. Combinations of these two paradigms, neurofuzzy systems, have been successful in many applications [2], [3]. The neurofuzzy approach joins fuzzy set theory and neural networks into an integrated system to combine the benefits of both. However, a major limitation of most fuzzy neural networks are their restricted application in dynamic mapping domains and problems due to either their feedforward structure, or lack of efficient learning procedures for feedback connections. Different recurrent fuzzy neural networks models to identify nonlinear dynamic systems have been proposed in the literature [4], [5], [6].

The recurrent neurofuzzy network proposed in [6], [11] is a hybrid model, once the network contains logic neurons in the hidden layer and classic neurons

T. Bilgiç et al. (Eds.): IFSA 2003, LNAI 2715, pp. 585–594, 2003.

in the output layer. The learning algorithm is based on reinforcement learning and unconstrained optimization, gradient-based method to update the weights of the second and output layer, respectively.

Differently of the models proposed in [6], [11], the recurrent neural fuzzy network introduced here has a fully recurrent structure whose neural units are all logic neurons processing information throught t-norms and s-norms.

In addition to its structure, this paper suggests a novel learning algorithm for the recurrent neural fuzzy network. The learning algorithm uses the gradient projection method [7] to update the network weights. As the units are logic neurons, yielding a neuron structure in which the weights are within the interval [0, 1], learning becomes a constrained nonlinear optimization problem. The gradient projection method of Rosen projects the gradient in such a way that improves the objective function and meanwhile maintains feasibility.

The core of the learning algorithm uses equality index [8], [9] as the performance measurement to be optimized. Equality indexes are strongly tied with the properties of the fuzzy set theory and logic-based techniques. The neural network recurrent topology is built with fuzzy neuron units and performs neural processing consistent with fuzzy set theory. Therefore neural processing and learning are fully embodied within fuzzy system theory and methodology. This is not the case with most, if not all current approaches developed for both, static and recurrent neural fuzzy networks.

The neural network uses fuzzy neural units modeled through logic operations *and* and *or* processed via t-norms and s-norms, respectively. The fuzzy neuron model is augmented with a nonlinear element placed in series with the previous logical element. This means that each neuron realizes an *on* (rather than an *in*) mapping between unit hypercubes. The incorporation of this nonlinearity changes the numerical characteristics of the neuron, but its essential logic behavior is maintained. The nonlinearities add interpretability because they can be treated as models of linguistic modifiers [6], [9].

The network model implicitly encodes a set of fuzzy if-then rules in its recurrent multilayered topology. The topology suggests a relationship between the network structure and a fuzzy rule-based system. This is especially useful to extract linguistic knowledge after network learning, or to insert preliminary linguistic knowledge for adaptation and tuning to data.

The recurrent neurofuzzy network performance is verified via examples of nonlinear system modeling and compared with classical and alternative models of neural, neural fuzzy networks and recurrent neurofuzzy networks. An application example concerning short term load forecasting is also included to illustrate its potential application in sequence learning. The results confirm the effectiveness of the neural fuzzy network approach.

2   Neural Fuzzy Network Structure

The basic processing units considered here are two fuzzy set-based neurons called logical *and* and *or* neurons, producing mappings $[0,1]^n \rightarrow [0,1]$, [9], [10]. The standard implementation of fuzzy set connectives involves triangular norms, meaning that *and* and *or* operations are implemented through t - and s - norms, respectively. The neural network has a multilayer structure, with three layers (Fig. 1). Recurrence occurs at the second layer neurons. Second layers neurons are T-neurons (*and* neurons with a nonlinear activation function) with output feedback.

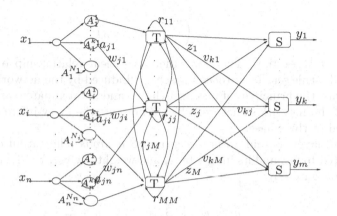

Fig. 1. Recurrent neural fuzzy network structure.

The input layer consists of neurons whose activation functions are membership functions of fuzzy sets that form the input space partition. For each component $x_i(t)$ of a n-dimensional input vector $\mathbf{x}(t)$ there are N_i fuzzy sets $A_i^{k_i}$, $k_i = 1, \ldots, N_i$ whose membership functions are activation functions of corresponding input layer neurons. The variable t is the discrete-time, that is, $t = 1, 2, \ldots$, but it will be omitted in the rest of the paper to simplify notation. Thus, their outputs $a_{ij}, i = 1, \ldots, n$ and $j = 1, \ldots, M$ are the membership degrees of the associated input and M is the number of second layer neurons.

Second layer neurons are constructed with fuzzy set-based neurons called logical *and* that produce mappings $[0, 1]^n \to [0, 1]$, [9], [10]. The standard implementation of fuzzy set connectives involves triangular norms, meaning that operators are implemented through *t-norms* and *s-norms* to yield, when augmented with nonlinear activation functions, a neural structure called *T-neuron* and its dual, named *S-neuron* in which $w_{ij} \in [0, 1]$ weights the input $x_i \in [0, 1]$, and r_{jj} weights the output feedback.

The neural fuzzy architecture considered in this paper has recurrent neurons in the hidden, second layer only. Therefore, the *and* operators have inputs a_{ij} weighted by w_{ij}, feedback connections weighted by r_{jj}, and lateral connections weighted by $r_{jl}, j, l = 1, \ldots, M, j \neq l$. The activation function ψ_{and} is, in general, a nonlinear mapping whose meaning may be associated with a linguistic modifier.

If we assume $a_{jn+1} = q^{-1}z_j$, where q^{-1} is the delay operator and $w_{jn+1} = r_{jl}, l = 1, \ldots, M$ then the net structure imbeds a set of if-then rules $R = \{R_j, j = 1, \ldots, M\}$ of the form:

R_j : If $(x_1$ is $A_1^{k_1}$ with certainty $w_{j1}) \ldots$ and $(x_i$ is $A_i^{k_i}$ with certainty $w_{ji}) \ldots$ and $(x_n$ is $A_n^{k_n}$ with certainty $w_{jn})$ Then z is z_j

where

$$z_j = \psi_{and} \left(\overset{n+M}{\underset{i=1}{\mathbf{T}}} (w_{ji}\, \mathbf{s}\, a_{ji}) \right) \tag{1}$$

and $\psi_{and} : [0,1] \to [0,1]$. Therefore, there is a clear relationship between the network and a rule set. The processing scheme induced by the network structure conforms with the principles of fuzzy set theory and approximate reasoning. The first part of the network structure, which represents the fuzzy inference system, is completed in this phase.

The third layer contains m non-recurrent *S-neurons*. The output of this neuron is denoted by $\widehat{y}_k$, and its inputs by z_j and weights by $v_{kj}, k = 1, \ldots, m$. Each output is computed by

$$y_k = \psi_{or} \left(\overset{M}{\underset{j=1}{\mathbf{S}}} (v_{kj}\, \mathbf{t}\, z_j) \right) \tag{2}$$

where $\psi_{or} : [0,1] \to [0,1]$ is a nonlinear function. Here we assume that ψ_{or} is the logistic function and that ψ_{and} is the identity function.

3   Learning Procedure

Network learning involves three main phases. The first phase uses a convenient modification of the vector quantization approach to granulate the input universes. The next phase simply set network connections and their initial, randomly chosen weights. In the third phase, the network weights are updated via gradient projection learning to maximize an equality index. The essential implementation steps are detailed next. The procedure is a form of supervised learning scheme.

3.1   Generation of Membership Functions

To generate membership functions straightforwardly, we assume triangular functions in the corresponding universes. The simplest form, that is, uniform and complementary partitions, may not be sufficient if a number of patterns is concentrated in certain regions of the pattern space and sparse in others. In these cases non-uniform partitions are more appropriate.

For this purpose, clustering techniques to find the modal values are of interest. In [12] a variation of a self-organized vector quantization neural network is suggested.

In this model, the network weights are cluster centers and the number of output units is the initial number of fuzzy sets of the i-th dimension partition. Initially, the value of this parameter is overestimated, but corrected via learning afterwards. Neurons that win occasionally, as counted by a performance index, are removed. The clustering algorithm details can be found in [6] and [12].

3.2 Weights Initialization

The first and second layers neurons are fully connected. The weights w_{ji}, v_{kj} and r_{jl}, for $i = 1, \ldots, n$; $j, l = 1, \ldots, M$; $k = 1, \ldots, m$ are initializated randomly within $[0, 1]$.

3.3 Determination of Active T Neurons

By construction, for each input pattern $\mathbf{x}$, there are at most two non-zero membership degrees for each of its dimension, and the corresponding membership functions define the active T-neuron neurons. The number of active T-neurons is found combining the indices of the membership functions for which the membership degrees are not zero.

The number of active T-neurons is at most 2^n each time a pattern is presented. Thus the processing time of the network is independent of the number of fuzzy sets of the input space partition [6], [12].

3.4 Fuzzification

This step is straightforward, but note that only membership degrees of active fuzzy sets are computed. This fact occurs due to complementarily of the membership functions. Thus, only $2n$ membership degrees need to be computed.

3.5 Weight Updating

Weight updating is based on gradient projection method suggested by Rosen [7]. The first step computes the network output for a given input $\mathbf{x}$. This corresponds to the fuzzification of the input pattern, and the computation of the outputs for the remaining neurons (using equations (1) and (2)). The learning algorithm uses equality index as the performance measure to be optimized. The equality index criterion is defined as in [9]:

$$Q = \sum_{k=1}^{P} (y_{kj} \equiv \widehat{y}_{kj}) \tag{3}$$

where P is the number of patterns, $\widehat{y}_{kj}$ is the value of the output unit j and y_{kj} is the target value j for the corresponding input vector $\mathbf{x}$. The equality index value is computed as follows:

$$y_{kj} \equiv \widehat{y}_{kj} = \begin{cases} 1 + y_{kj} - \widehat{y}_{kj}, & \text{if } y_{kj} < \widehat{y}_{kj} \\ 1 + \widehat{y}_{kj} - y_{kj}, & \text{if } y_{kj} > \widehat{y}_{kj} \\ 1, & \text{if } y_{kj} = \widehat{y}_{kj} \end{cases} \qquad (4)$$

The sum in (3) is split into three groups, depending upon the relationship between y_{kj} and $\widehat{y}_{kj}$:

$$Q = \sum_{k: \widehat{y}_{kj} > y_{kj}} (y_{kj} \equiv \widehat{y}_{kj}) + \sum_{k: \widehat{y}_{kj} = y_{kj}} (y_{kj} \equiv \widehat{y}_{kj}) + \sum_{k: \widehat{y}_{kj} < y_{kj}} (y_{kj} \equiv \widehat{y}_{kj}) \qquad (5)$$

Since the units are logic neurons, they yield a neuron structure in which $\mathbf{w} \in [0,1]$, where $\mathbf{w} = (w_{11}, \ldots, w_{ji}, v_{11}, \ldots, v_{kj})'$, $i = 1, \ldots, n + M; j = 1, \ldots, M; k = 1, \ldots m$; and $w_{jn+l} = r_{jl}$, $l = 1, \ldots, M$. Therefore, learning becomes a nonlinear optimization problem with linear constraints, summarized as follows:

$$\max \quad Q$$

$$\text{suject to} \begin{cases} \mathbf{w} \leq 1 \\ \mathbf{w} \geq 0 \end{cases} \qquad (6)$$

Given a feasible point $\mathbf{w}$, the steepest ascent direction is the gradient. However, moving along the steepest direction may lead to infeasible points. The gradient projection method of Rosen projects the gradient in such a way that improves the objective function and meanwhile maintains feasibility. Details of the gradient projection can be found in [13]. The network weights are updated as:

$$\Delta \mathbf{w} = \eta(t) \frac{\partial Q}{\partial \mathbf{w}} \qquad (7)$$

where $\eta(t)$ is the learning rate and the index t refers to the iteration step $t = 1, 2, \ldots$. The learning rate can be found using a line search procedure [13]. Here, we adopted dichotomous search method because the procedure uses functional evaluations only.

Let v_{kj} the weight of the output layer for the j-th input of the k-th neuron. The derivative $\partial Q / \partial v_{kj}$ is expressed as

$$\frac{\partial Q}{\partial v_{kj}} = - \sum_{k: \widehat{y}_{kj} > y_{kj}} \frac{\partial \psi_{or}(\cdot)}{\partial v_{kj}} + \sum_{k: \widehat{y}_{kj} < y_{kj}} \frac{\partial \psi_{or}(\cdot)}{\partial v_{kj}} \qquad (8)$$

The hidden units weights w_{ji} are update as follows:

$$\frac{\partial Q}{\partial w_{ji}} = - \left(\sum_{k: \widehat{y}_{kj} > y_{kj}} \frac{\partial \psi_{or}(\cdot)}{\partial v_{kj}} \right) \frac{\partial \psi_{and}(\cdot)}{\partial w_{ji}} + \left(\sum_{k: \widehat{y}_{kj} < y_{kj}} \frac{\partial \psi_{or}(\cdot)}{\partial v_{kj}} \right) \frac{\partial \psi_{and}(\cdot)}{\partial w_{ji}} \qquad (9)$$

where $\psi_{or}(\cdot)$ and $\psi_{and}(\cdot)$ are as in equations (1) and (2), respectively. All the derivatives can be computed when the form of the functions $\psi_{or}(\cdot)$ and $\psi_{and}(\cdot)$ are specified.

We note that the increments are positive when $\widehat{y}_{kj}$ is lower than the required target value and negative when the converse situation holds. Thus all increments are driven by the sign of the difference between $\widehat{y}_{kj}$ and y_{kj}.

4   Simulation Results

To show the performance of the learning algorithm introduced in this paper, we perform a comparison with alternative models suggested in the literature. The examples given here include nonlinear system identification as well as the time series prediction problem.

In all situations, initial weights were generated from a uniform distribution in the range [0, 1]. We choose the *t-norm* as the algebraic product and the probabilistic sum as the *s-norm*. Also, the structure of the RNFN is partially recurrent because here we adopt only output feedback at the hidden layer.

Example 1: The nonlinear plant to be modeled is described by:

$$y_{t+1} = f\left[y_t, y_{t-1}, y_{t-2}, u_t, u_{t-1}\right] = \frac{y_t y_{t-1} y_{t-2} u_{t-1}(y_{t-2} - 1) + u_t}{1 + y_{t-1}^2 + y_{t-2}^2} \qquad (10)$$

where f is assumed to be unknown. During learning, the input u_t was randomly chosen using an uniform distribution in the interval $[-1, 1]$. The RNFN to model this process has only u_t as input. We assume a non-uniform partition with $N_1 = 12$ fuzzy sets, which means 12 hidden layer neurons. The value of the performance index become stable after 1,000 iterations and reaches the value of $Q = 1848.53$. Fig. 2 shows the output of the actual model and of the RNFN when given the following input:

$$u_t = \begin{cases} \sin\left(\frac{2\pi t}{250}\right), & \text{for } t \leq 500 \\ 0.8\sin\left(\frac{2\pi t}{250}\right) + 0.2\sin\left(\frac{2\pi t}{25}\right), & \text{for } t > 500 \end{cases} \qquad (11)$$

Table 1 shows the number of parameters adjusted and number of iterations for neural networks (NN) [14], fuzzy system (FS) [3], neural fuzzy network (NFN) [4], recurrent neural fuzzy network (RNFN) [4] and RNFN models considered in this example. The models are characterized by parameters $[n, p, q, m]$, where n is the number of inputs, p and q are the number of neurons of the first and second hidden layer, respectively, and m is the number of outputs. Note in Table 1 that both, the number of parameters and iterations for the RNFN are smaller than the remaining models. The mean square modeling (test) error was 10^{-3} for the RNFN model.

Example 2: To verify if the proposed RNFN can learn temporal relationships, a short-term load forecasting problem is considered. The main task of short term load forecasting in power systems is to accurately predict the 24 hourly loads

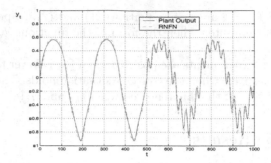

Fig. 2. Plant and recurrent neural fuzzy network outputs.

Table 1. Number of parameters and iterations of the identification models.

Structure	Parameters	Number of Parameters	Iterations
NN [14]	$np + pq + qm$	$[5, 20, 10, 1] = 310$	$10,000$
FS [3]	$3npm$	$[5, 40, -, 1] = 600$	$5,000$
NFN [4]	$2np + mp$	$[5, 16, -, 1] = 176$	$90,000$
RNFN [4]	$3np + mp$	$[2, 16, -, 1] = 112$	$90,000$
RNFN	$p(1 + n + m)$	$[1, 12, -, 1] = 36$	$1,000$

of the next day. The RNFN approach uses the database of an electric utility located at the Southeast region of Brazil.

The training set was composed by hourly loads from January to August of 2001. The input variables for the RNFN model are L_{h-2}, L_{h-1}, which represent the loads at times $h-1$ and $h-2$, respectively. We assume a non-uniform partition with $N_1 = 8$ and $N_2 = 8$ fuzzy sets, which means 64 hidden layer neurons, and 256 parameters to adjust. The values of the performance index become stable after 5,000 iterations when it reaches the value $Q = 4893.74$. Fig. 3 shows the load forecast results for September 12, 2001 whose relative percentual error is below 4.1%. A multilayer neural network with two hidden layers with 12 and 8 neurons, respectively, was trained using the backpropagation algorithm. It took 2,000 iterations to converge. The multilayer neural model achieved a relative percentual error of 7.2%.

5   Conclusions

In this paper we have introduced a new learning algorithm for a class of recurrent neural fuzzy networks. The learning algorithm uses the gradient projection method to update the network weights. The core of the learning algorithm adopts equality index as a performance measure. Equality index is especially important because its properties reflect the fuzzy set-based structure of the neural network and nature of learning. In the recurrent fuzzy neural networks, neurons

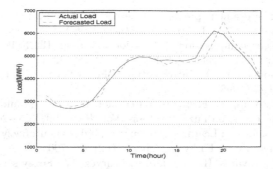

Fig. 3. Load forecast for september 12, 2001.

are computational units that perform operations consistent with fuzzy set theory. The neurons are logic units with synaptic and somatic operations defined by *t-norms* or *s-norms* and nonlinear activation functions. Thus, the designer has higher flexibility to assemble networks to meet application and performance requirements. Simulation results show that the neural fuzzy network achieves high performance in terms of approximation capability, memory requirements, and learning speed.

Acknowledgments

This research was partially supported by the Research Foundation of the State of São Paulo (FAPESP), and the Brazilian National Research Council (CNPq) grant ♯300729/86 − 3.

References

1. Santos, E. P., Von Zuben, F. J.: Efficient Second-Order Learning Algorithms for Discrete-Time Recurrent Neural Networks. In: Medsker, L. R., Jain, L. C. (eds.): Recurrent Neural Networks: Design and Applications, CRC Press (1999) 47-75.
2. Lin, F. J., HwangChang, W. J., Wai, R. J.: A supervisory fuzzy neural network control system for tracking periodic inputs. IEEE Trans. on Fuzzy Systems, Vol. 7 (1999) 41-52.
3. Wang, L. X.: Adaptive Fuzzy Systems and Control, Prentice-Hall (1997).
4. Lee, C. H., Teng, C. C.: Identification and control of dynamic systems using recurrent fuzzy neural networks. IEEE Trans. on Fuzzy Systems, Vol. 8(4) (2000) 349-366.
5. Blanco, A., Delgado, M. Pegalajar, M. C.: Identification of fuzzy dynamic systems using Max-Min recurrent neural networks. IEEE Trans. On Systems, Man and Cybernetics, Vol 26(1) (2001) 451-467.
6. Ballini, R., Soares, S., Gomide, F.: A recurrent neurofuzzy network structure and learning algorithm. In 10th IEEE Int. Conference on Fuzzy Systems, Vol. 3 (2001) 1408-1411.

7. Rosen, J. B.: The gradient projection method for nonlinear programming, Part I, Linear Constraints. SIAM J. Applied Mathematics, Vol. 8 (1960) 514-553.
8. Pedrycz, W.: Neurocomputations in relational systems. IEEE Trans. Pattern Analysis and Machine Intelligence, Vol. 13(3) (1991) 289 - 297.
9. Pedrycz, W., Gomide, F.: An Introduction to Fuzzy Sets: Analysis and Design. MIT Press, Cambridge, (1998).
10. Pedrycz, W., Rocha, A.: Fuzzy-set based models of neuron and knowledge based networks. IEEE Trans. on Fuzzy Systems, Vol. 4(1) (1998) 254-266.
11. Ballini, R., Gomide, F.: Learning in recurrent, hybrid neurofuzzy networks. In 11th IEEE Int. Conference on Fuzzy Systems, (2002) 785-790.
12. Caminhas, W., Tavares, H., Gomide, F. Pedrycz, W.: Fuzzy set based neural networks: structure, learning ands applications. Journal of Advanced Computational Intelligence, Vol. 3(3) (1999) 151-157.
13. Bazaraa, A. S., Shetty, M.: Nonlinear Programming: Theory and Algorithms. John Wiley & Sons, New York (1979).
14. Narendra, K. S., Parthasarathy, K.: Identification and control of dynamical systems using neural networks. IEEE Trans. on Neural Networks, Vol. 1(1) (1990).

Experimental Analysis of Sensory Measurement Imperfection Impact for a Cheese Ripening Fuzzy Model

Irina Ioannou[1], Nathalie Perrot[1], Gilles Mauris[2], Gilles Trystram[3]

[1] UMR Génie Industriel Alimentaire, Equipe REQUALA CEMAGREF,
24 Avenue des Landais, BP 50085, 63172 Aubière cedex,
irina.ioannou@cemagref.fr
[2] LISTIC - Université de Savoie BP 806 74 016 Annecy France cedex,
mauris@univ-savoie.fr
[3] UMR Génie Industriel Alimentaire, ENSIA, France.

Abstract. In the food processes, build tools taking human measurements into account is relevant for the control of the sensory quality of food products. Despite the methodology used to formalize these measurements, these ones are subjected to more imperfections (imprecision, reliability,...). Our aim is to develop tools taking these measurements into account and smoothing the imperfections of these measurements. In this paper, an experimental analysis is led on a fuzzy symbolic model applied to cheese ripening process. This analysis allows to determine the sensory measurements which have the highest impact on the model and to observe the impact of sensory measurements imperfection on the output of the developed fuzzy model.

1. Introduction

Finished food products must meet specific sensory, sanitary and technological requirements. Among these requirements, sensory properties are essential because they condition the choice and the preferences of consumers [1]. As a consequence, it is relevant to develop methods which allow firstly to measure those properties on the manufacturing line and secondly to integrate them in a feedback control of food processing [2]. In this context, the use of the fuzzy symbolic approach for representing human assessments and reasoning in a decision support system is interesting to help people in charge of the process control (operator) [3]. These systems have as inputs sensory measurements made by the operator at line during the manufacturing and/or instrumental measurements. It allows a cooperation between the human operator and the automation system. However, the operator sensory measurements are subjected to higher imperfections than the conventional sensory measurements. The causes are, for example, the place of the measurement (operator influenced by his environment: temperature, humidity, light) or the time of the measurement (pressure due to the need to control, several tasks to do at the same time). From a model developed on an application dealing with the control of a cheese ripening process, we are looking for determining the impact of the imperfections, coming from the expert evaluations, on the fuzzy model output. In this paper, we are interested with two aspects of this kind of

T. Bilgiç et al. (Eds.): IFSA 2003, LNAI 2715, pp. 595-602, 2003.

imperfections: the lack of a sensory measurement and the imprecision of the sensory measurements.

The paper is organized as follows. The second section explains the model developed on the application of cheese ripening. The methodology of the experimental analysis and results are presented in the third section.

2. The Fuzzy Symbolic Approach Applied to Cheese Ripening

Having a mathematical support to represent human assessments and reasoning expressed by linguistic expressions is one of the originate motivation of fuzzy subset theory [4,5]. A fuzzy symbolic description of phenomena consists in affecting degrees (between 0 and 1) to considered terms, e.g. 0.8/High;0.2/Medium;0./Small. The latter can be obtained directly from human assessments or indirectly from the numeric measurements by the concept of fuzzy symbolic measurement, the meaning of terms being encoded in their associated membership functions [6,7]. The interest of the fuzzy symbolic approach is thus the possibility to take both numeric and linguistic assessments into account and also to process further such information, e.g making information fusion (figure 1).

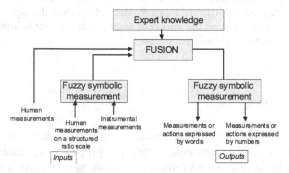

Fig. 1. Principle of building of a model using the fuzzy symbolic approach

This approach was applied to build a decision support system in cheese ripening [8], that is a ageing process lasting about one month. This system was established to inform an operator on the potential drift of the sensory trajectory of the cheese each day in order to help him to control the process. To control the process, the operator uses a variable called ripening degree which represents an indication of the ripening state at a time t of ripening. This variable is assessed on an ordinate scale (from 4 (state A) for a cheese no ripened cheese to 0 (state E) for a ripened cheese) with graduations of 0.25. Despite these graduations, the operators say that their sensitivity on the ripening degree is 0.5. It is why the two sensitivity levels 0.5 and 0.25 are tested for the validation of the model (see fig. 3). The explanatory model of cheese ripening is presented on figure 2. This model takes as input data five variables : the ripening time and four sensory variables : the cheese coat (MS1), its color (MS2), its humidity (MS3) and its consistency (MS4). The output of the model is the ripening degree.

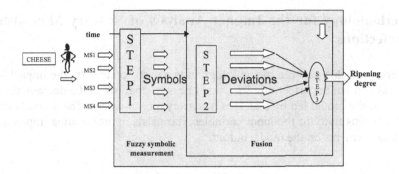

Fig. 2. Principle of the ripening model

The model proposed is composed of three steps. The first step allows to transfer the sensory measurements made on an ordinate scale (from 1 to 6) in a symbolic space (different symbols are defined for each sensory measurement). In the second step, the deviations are calculated for each sensory measurement in comparison with the objective trajectories of the sensory measurement. The third step allows to combine the deviations on each sensory measurement by rules treated in a fuzzy symbolic mathematical frame [7]. The rules used depend on the ripening time, the table 1 presents the rule base for the period from day 8 to day 12, it corresponds to a standard ripening degree C.

Table 1. Rules base for the standard ripening degree C (day 8 to day 12)

Deviations on cheese coat Deviations on cheese consistency	(Very few covered)	(Few covered)	(Medium covered)	(Very covered)
(No proteolysis)	*late (-1)*	*late(-1)*	*late (-0.5)*	
(Few proteolysis)		*late (-0.25)*	Standard state	*early (+0.5)*
(Medium proteolysis)			*early.* *(+1)*	*early (+1)*
(Total proteolysis)			*early (+2)*	*early (+2)*

For example: at day 10 of the ripening time, the cheese coat is measured to 3 and the cheese consistency to 2. The corresponding fuzzy linguistic descriptions are: for cheese coat *medium covered* with a membership degree of one and for cheese consistency: *few proteolysis* with a degree of 1. The standard trajectory of these measurements give the following symbols like standards at day 10: *medium covered* and *few proteolysis*. So, deviations of these both measurements are equal to 0. The table 1 combines these deviations to obtain the ripening degree: the cheese is in the standard state C at day 10.

3. Methodology for the Impact Analysis of Sensory Measurement Imperfections

We propose, in this paper, a methodology in two parts to analyze the impact of sensory measurement imperfections. The first one is an analysis of the degradation of the structure of the model led by a lack of a sensory measurement. The second one is an analysis of sensitivity of the input variables, it consists in propagating imprecision of input measurements on the model output.

3.1 Impact of Lack of a Sensory Measurement

To reproduce the lack of a sensory measurement, an input sensory measurement of the model is deleted. This deletion corresponds to a degradation of the ripening model. It brings information on the importance of each sensory measurement in the structure of the model. The identification of measurements that are more important than others can induce a more attentive assessment by the expert for this measurement : increased vigilance , repetition of the measurement or addition of an information to confirm the sensory measurement.

Method. Six degradations were made, we obtain six damaged models (table2). Each degradation corresponds to the deletion of one sensory measurement for the three first damaged models and two sensory measurements for the three last damaged models. For example, for the first damaged model (MD1), the sensory measurement deleted is the humidity of the cheese (MS3), so the inputs of the model staying are the cheese coat (MS1) and the cheese consistency (MS4). The input, color of the cheese (MS2), is considered like having few impact on the model to find the ripening degree, but it is useful at the end of the ripening to determine a default of color. Therefore, this measurement is deleted for each degradation of the ripening degree model.

Table 2. Composition of the different damaged models

Name of the damaged model	Variables of deleted input	Variables of input staying	Name of he damaged model	Variables of deleted input	Variables of input staying
MD 1	MS3, MS2	MS1,MS4	MD 4	MS3, MS2, MS1	MS4
MD 2	MS1,MS2	MS3, MS4	MD 5	MS3, MS2, MS4	MS1
MD 3	MS4, MS2	MS1, MS3	MD 6	MS1, MS2, MS4	MS3

The deletion of one or several sensory measurements leads to a cancellation of the rule tables of the ripening model, giving the ripening degree. This task is realized with the help of the expert by a reformulation of the original rules. For example, from table 1 shown in part 2, we obtain after the deletion of the measurement cheese consistency the table 3.

Table 3. Rules base for the standard ripening degree C with deletion of the cheese consistency

Deviations on cheese coat	Very few covered	Few covered	Medium covered	Very covered
Ripening degree	*Defect*	*Late (-0.5)*	***Standard***	*Early (+0.5)*

In this way, for the original model, the standard state was defined by a cheese coat *medium covered* and a cheese consistency *few proteolysis*. Now for the damaged model MD1 (cheese consistency deleted), the standard state is defined only by a cheese coat *medium covered.*

The answer of the damaged model (ripening degree) is considered like compatible if it does not differ from the expert answer more than the tolerance thresholds (fixed in function of the precision degree of the human measurements: 0.25 and 0.5 on a scale 0-4). The performance of the model is defined as the number of answers compatible in comparison with the number of cheeses to assess. To compare the performances of the different damaged models with the performance of the original model, they are plotted on a histogram (see fig.2).

Results. On the figure 3, are presented the performances of the different damaged models in comparison with the original model.

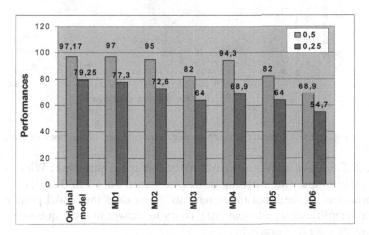

Fig. 3. Performance of the original and the damaged models at two thresholds

A degradation of the structure of the model linked to the cheese humidity (MD1) does not act upon the performances of the model whatever the different thresholds tested. But, when the degradation of the structure of the model is linked to the cheese consistency (MD3), it leads to a loss of 15 % for the two thresholds.

A degradation of the structure of the model linked to the cheese coat (MD2) leads to a decrease of the performances unequalled for the two thresholds. For a threshold of 0.5, 2 % of compatible answers are lost and 8 % to a threshold of 0.25. If the measurement cheese humidity is also deleted (MD4), we notice a loss of 3% of compatible answers to a threshold of 0.5 and 9 % to a threshold of 0.25.

This analysis allows to conclude that the sensory measurement which has the highest impact is the cheese consistency, it is responsible for the 97 % of compatible answers to a threshold of 0.5. Moreover, we can conclude to a strong impact of the cheese coat to give an answer of the model a more refined answer.

To confirm these results, it would be interesting to compare them to the results obtained with others methods of degradation of the model, where the rule tables are not modified such as replacing the missing measurement by the target value of the measurement. For example, for the standard ripening degree C (table 1), the value of the missing measurement (cheese consistency) would be few proteolysis.

3.2 Impact of Sensory Measurement Imprecision

Method. The principle of this analysis is to propagate in the model an imprecision applied on an input variable, the others input variables being fixed. The study is achieved on ripening kinetics, obtained on a pilot able to represent the different ripening dynamics. An example of kinetic is shown on figure 4.

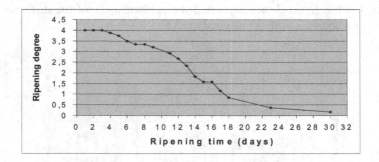

Fig. 4. Evolution of the ripening degree in function of the ripening time

This kind of analysis allows to answer to the following questions : What is the impact on the model output (the ripening degree expressed on a 4 to 0 scale) if we increase the imprecision on the input measurements? How does the model propagate the imprecision (amplification, attenuation)? To try to answer to these questions, let us take the following case like example :

This analysis was made from day 8 until the ripening end. We choose the variable cheese consistency to achieve at first this study because it is the variable the most significant in the model. The others variables are fixed to their objective symbols during the period studied : Cheese colour to yellow, cheese humidity to few humid and cheese coat to medium covered from day 8 to day 16 and to very covered from day 16 until the ripening end. In general, the imprecision of the expert on the measurement cheese consistency is 0.25 (d(input)). Some perturbations can happen like the weakening of the operator due to illness or a particularity of the cheese which leads to a more difficult measurement. These perturbations induce the increasing of the measurement imprecision, for example to 0.5. What is the consequence on the output information given by the model ? The experimentation led, to answer to this question, is explained on figure 5.

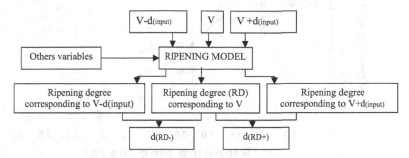

Fig. 5. Method to propagate into the ripening model the measurement imprecision

From a measurement V coming from the kinetic at a time t, we calculate the ripening degrees corresponding to V, to V – d(input) and to V + d(input). Then, we deduce the output deviations (d(RD-) and d(RD+)) which are calculated according to the following equations:

$$\text{Ripening degree} = f(\text{sensory measurements, time}) \qquad (1)$$

With f the inference of the fuzzy symbolic model with the product and bounded sum for respectively the intersection and projection operator.

$$d(RD) = f\,(\text{sensory measurements, time}) - f\,(\text{sensory measurements}+/-\ d(\text{input}), \text{time}) \qquad (2)$$

To analyze the results, we plot the ripening degree in function of the ripening time. Five curves are plotted corresponding to the measurement imprecision (-0.5 ; -0.25 ; 0; 0.25 ; 0.5). Then the critical period is determined (period where d(RD) is strictly superior to 0.5, 0.5 being the expert imprecision on the model output).

Results. On figure 6 are presented the results obtained to analyze the impact of the input measurement imprecision on the model output. The results presented concern the sensitivity of the model to the measurement cheese consistency (MS4), the three others variables being fixed to the following symbols : a color yellow, a humidity few humid and a cheese coat medium covered between day 8 and day16 and very covered from day 16 to the ripening end.

On figure 6, we notice that the d(RD) varies from 0 to 0.67 for d(input) of +/-0.5. It is greater for a d(input) of 0.5 than for a d(input) of 0.25. The critical period is the period from day 17 to 21 for a d(input) of +/-0.5 and does not exist for a d(input) of +/- 0.25. By observation of the propagation of others d(input) like +/-0.75, 1, 1.25 and 1.5, we note that the critical period increases with d(input) until a d(input) of +/-0.75 (day 12 to day 21) and stagnate for the others d(input) tested (+/-1, +/-1.25 and +/-1.5). Thus, precautions must be taken to achieve the measurement cheese consistency in the ripening period day 12 to day 21, which corresponds to the second half of the ripening.

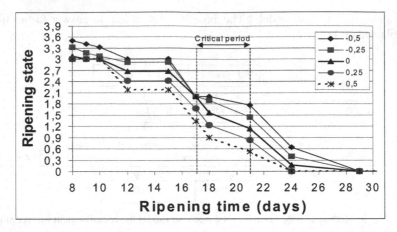

Fig. 6. Variations of the Δoutput in function of the ripening time and of the d(input)

4. Conclusion

The contribution of this paper is to bring information on the impact of imperfections coming from the expert evaluations, on a fuzzy symbolic model describing cheese ripening. Especially, the experimental analysis made allows to indicate the sensory variables that must be measured in priority and also variables for which precautions must be taken to realize a « precise measurement ». In this sense, the cheese consistency is the most important sensory variable. Moreover, we have observed that the sensitivity of the measurements increases during the second half of the ripening period. This study opens an interesting road with regards to the problematic of food process control. Experimentations are in progress to analyse other kinds of imperfections identified in the application like the structure of the model.

References

1. I. Ioannou, N. Perrot and al. (2002). The fuzzy set theory: a helpful tool for the estimation of sensory properties of crusting sausage appearance by a single expert. *Food quality and preference*. 13(7-8): 589-595.
2. N. Perrot, G. Trystram and al. (2000). Feed-back quality control in the baking industry using fuzzy sets. *Journal of food process engineering*. 23: 249-279.
3. V. Davidson, R. B. Brown and al. (1999). Fuzzy control system for peanut roasting. *Journal of food engineering*. 41 : 141-146.
4. L. Zadeh (1965). Fuzzy sets. Information and control, 8: 338-353.
5. L. Zadeh (1999). From computing with numbers to computing with words-From Manipulation of measurements to manipulation of perceptions. *IEEE Transactions on circuits and systems* 45(1): 105-19.
6. L. Zadeh (1971). Quantitative fuzzy semantics. *Inf. Sci.* 3: 159-176.
7. G. Mauris, E. Benoit and al (1996). The aggregation of complementary information via fuzzy sensors. *Measurement*, 17(4): 235-249.
8. N. Perrot, I. Ioannou, and al. (2002). Experiences about operator/technology cooperation using a fuzzy symbolic approach to decision support system design in food processes. *IEEE systems, man and cybernetic society*. TUNISIE.

Generation of Fuzzy Membership Function Using Information Theory Measures and Genetic Algorithm

Masoud Makrehchi, Otman Basir, and Mohamed Kamel

Pattern Analysis and Machine Intelligence Lab, Department of System Design
Engineering, University of Waterloo, Waterloo, Ontario N2L 3G1, Canada
mmakrehc@engmail.uwaterloo.ca, obasir@uwaterloo.ca, mkamel@uwaterloo.ca

Abstract. One of the most challenging issues in fuzzy systems design
is generating suitable membership functions for fuzzy variables. This paper proposes a paradigm of applying an information theoretic model to
generate fuzzy membership functions. After modeling fuzzy membership
function by fuzzy partitions, a genetic algorithm based optimization technique is presented to find sub optimal fuzzy partitions. To generate fuzzy
membership function based on fuzzy partitions, a heuristic criterion is
also defined. Extensive numerical results and evaluation procedure are
provided to demonstrate the effectiveness of the proposed paradigm.

1 Introduction

The fuzzy membership function is a key concept in designing fuzzy systems
(FS). In addition to fuzzy rules construction, generating suitable membership
functions (MF) has been a challenge for three decades. Although there have
been many efforts that aim at automatically generating fuzzy rules, for example using learning from examples [9],[1], and optimization techniques such as
genetic algorithms, simulated annealing and Kalman filtering [4],[12],[10], automatic generation of MF is still based on expert knowledge [11],[8]. Most previous
works on automatic generation of MF have been limited to MF optimization as
a part of FS optimization procedure that considers the optimization of MF and
Fuzzy rule base interdependently. A pioneer attempt for simultaneous optimization of fuzzy rules and MF is that of Homaifar and McCormick [4]. They use
a genetic algorithm (GA) to design the MF's and the fuzzy rule base, simultaneously, within a fuzzy control. This work adapted by other researchers. For
example, in [6] fuzzy rules and MF's are generated at the same time using a
GA. In [5] the authors proposed methods based on evolutionary approaches to
generate flexible, complete, consistent and compact FS. During the optimization of fuzzy rules, fuzzy partitions are optimized to gain better performance.
Cordon *et al* [3] presented an optimization-based method that generates simple
fuzzy rules as well as defining the number of MF's and their parameters to define
triangular-shaped functions, simultaneously. The triangular MF is modeled by

T. Bilgiç et al. (Eds.): IFSA 2003, LNAI 2715, pp. 603–610, 2003.

5 parameters, which are optimized by a GA. The fitness function of the GA is the performance measure of the fuzzy rules in a process control setup.

In this paper, we present a new framework for objectively generating MFs. This framework is guided by information theory (IT) measures such as Shannon entropy and mutual information. The process of generating MFs is achieved through a GA optimization technique. The paper contains six sections. In section two we discuss design issues and factors that are pertinent to the choice MFs. In section three we introduce the MF design problem as a classical optimization formulation. In section 4 we detail our information theory based approach. Experiments to demonstrate effectiveness of the proposed approach are presented in section 5. Section 6 provides concluding remarks.

2   Parameters of Fuzzy Membership Function

Different types of MF have been already proposed [8], for example, trapezoidal, triangular, and sigmoid. Choosing the type of the function has been considered as a subjective procedure [8]. Regardless of its type, an MF can be described by three parameters, namely: support, boundary, and core or prototype (Fig. 1).

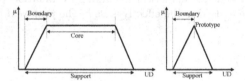

Fig. 1. Parameters of MF in two most popular MF's: trapezoidal and triangular.

If $\mathbf{X}$ is the universe of discourse (UD) and x_i is defined in $\mathbf{X}$;

$$\mathbf{X} = \{ x_1, x_2, \ldots, x_n \} \tag{1}$$

We can define a subset or partition of $\mathbf{X}$ as;

$$X^i = \{ x_1^i, x_2^i, \ldots, x_{k_i}^i \} = [x_L^i, x_H^i] \, , X^i \subset \mathbf{X}, \; 1 \leq i \leq m \tag{2}$$

where $x_L{}^i$ ($x_H{}^i$) is called a lower boundary (an upper boundary). We assume that UD has been partitioned so that all members of $\mathbf{X}$ belong to one or more partitions.

$$\mathbf{X} = \bigcup_{i=1}^{m} X^i = \bigcup_{i=1}^{m} [x_L^i, x_H^i] \tag{3}$$

We define a function on X^i (called membership function) such that;

$$\mu_{X^i} : X^i \subset \mathbf{X} \tag{4}$$

Consequently, parameters of MF can be defined as;

$$\forall x \in X^i; \ Support : \mu_{X^i}(x) > 0, \ Boundary : \mu_{X^i}(x) < 1, \ Core : \mu_{X^i}(x) = 1 \quad (5)$$

Among these three parameters, we are interested in the support which is a partition (a subset) of **X**. Support is a domain in which MF is defined. In this perspective, MF is a mapping from the support to the unit distance (0,1), and assigns a number between 0 and 1 to every member of support. From a fuzzy system's point of view, the UD is partitioned into m overlapped subsets and hence we can express the support of MF as a fuzzy partition of UD. The fuzzy partitions are defined as;

$$X^i \cap X^{i+1} \neq \emptyset, \ 1 \leq i \leq m-1 \quad (6)$$

The formulation assumes overlap between adjacent partitions only.

3   Problem Statement

There are three crucial issues in MF design process, namely: the desired number of partitions over its UD, the shape of MF's in each partition, and UD partitioning including: regions, boundaries, and overlaps between partitions. All these issues or design factors of MF, carry some information about the system.

The number of partitions is highly interdependent with fuzzy rules. Obviously the dimension of fuzzy rule base depends on the number of fuzzy partitions (i.e., the number of MF, each partition is associated to an MF) and the number of inputs and outputs of FS. Consequently, the number of fuzzy partitions influences the number of fuzzy rules and can be finalized during optimization of fuzzy rule base.

The shape of MF (e.g., trapezoidal, triangular or sigmoid) is still a subjective issue and can be determined through a heuristic process. The reason is there is no proven relationship between the degree of fuzziness (the shape of MF) and our information about the environment of fuzzy system. Therefore we can determine the shape of MF using expert knowledge or a learning process that can emulate expert knowledge.

It seems the fuzzy partition (i.e., the support of MF) is more informational part of MF. A fuzzy partition of MF is a part of our information about something in which uncertainty is happening. The problem then is to find a given number of optimum fuzzy partitions that maximizes the information measure of the partitioned UD. This statement addresses a classic optimization problem. Suppose we aim to find m optimum partitions in the UD. From Eq. (2), we have;

$$X^1 = [x_L^1, x_H^1], \quad X^2 = [x_L^2, x_H^2], \quad \ldots, \quad X^m = [x_L^m, x_H^m] \quad (7)$$

Where $x_L^1 = -1$ and $x_H^m = 1$. Since there is an overlap between two adjacent partitions, then we have;

$$Z^{i,i+1} = [x_L^i, x_H^i] \cap [x_L^{i+1}, x_H^{i+1}], \quad x_L^{i+1} < x_H^i, \quad 1 \leq i \leq m-1 \quad (8)$$

where $Z^{i,i+1}$ is the overlap between *ith* and *(i+1)th* partition. We can easily show that;

$$Z^{i,i+1} = [x_L^{i+1}, x_H^i], \; 1 \le i \le m-1 \quad (9)$$

In m -partitioned UD, there are *m-1* overlaps, and based on Eq. (9) we can parameterize these overlaps with *2(m-1)* parameters. If we include the first and the last non-overlapped parameters, x_L^1 and x_H^m , the set of parameters will be;

$$P = \{ \, x_L^1, x_L^2, x_H^1, \dots, x_L^m, x_H^{m-1}, x_H^m \} \quad (10)$$

where P is the complete set of parameters to be optimized. In the next section, we discuss the proposed schema to optimize the set of parameters.

4   Proposed Schema

In this section we establish a relation between IT measures and the parameters of fuzzy partitions (P). After associating the set of parameters to the measure, we can optimize the parameters using an optimization procedure. Although there are some efforts to describe and model the parameters of FS, especially MF, using probabilistic models [7],[2], applying IT measures in optimizing parameters of FS constitutes a new idea.

A genetic algorithm (GA) is used as an optimization tool. To design a GA, we need to set up a series of parameters, including coding and representation schema, and a fitness function. From Eq.(10) we have *2m* parameters to be optimized. To reduce the number of required bits for representing parameters in GA, we use relative parameters (instead of absolute values of parameters). If we express absolute set of parameters (P), as;

$$P = \{ \, p_0, p_1, p_2, \dots, p_{2m-1} \} \quad (11)$$

The relative parameters can be obtained as follows;

$$r_0 = p_0, r_1 = p_1 - p_0, \dots, r_{2m-1} = p_{2m-1} - p_{2m-2} \quad (12)$$

We know that $r_0 = p_0 = -1$, and hence the total number of parameters will be *2m-1*. The final relative parameter set, R, is then as;

$$R = \{ \, r_1, r_2, \dots, r_{2m-1} \} \quad (13)$$

To represent a chromosome binary string, we encode every parameter with q bits. Therefore each parameter lays in the following range; $0 < r_i < 2^q - 1$. It should be noted, to avoid zero overlapping, we don't let the parameters take boundary values (i.e., 0 and 2^{q-1}). The following string shows the encoding parameter set.

$$\underbrace{b_1^1 b_2^1 \dots b_q^1}_{r_1} \; \underbrace{b_1^2 b_2^2 \dots b_q^2}_{r_2} \quad \dots \quad \underbrace{b_1^{2m-1} b_2^{2m-1} \dots b_q^{2m-1}}_{r_{2m-1}} \quad (14)$$

Another important issue in designing a GA is fitness function. One of the main challenges in this formulation has been identifying a set of fitness measures based on IT. That is to attempt to interpret the fitness of a set of parameters as an expression of IT terms. To establish a relation between fuzzy partitions parameterized by R in Eq. (13), we map the set of parameters to the histogram of given data collection (Fig. 2). We define fitness function as the total entropy of partitions whose width are parameterized by vector R. By optimizing parameter set of R to maximize the total entropy of partitioned histogram, the optimum fuzzy partitions associated to the support of MF's are obtained. To generate m membership functions, we need to partition histogram into m joint partitions that cause $m\text{-}1$ overlaps. Every joint partition addresses a joint entropy and each overlap is modelled by mutual information showing interdependency between two entropies. Eq. (15) shows the total entropy that is the fitness function for every set of parameters, R;

$$\mathbf{H} = \sum_{i=1}^{m} H(i) - \sum_{j=1}^{m-1} I(j, j+1) \tag{15}$$

Where $\mathbf{H}$ is the total entropy, $H(i)$ is the entropy of *ith* partition and $I(j,j+1)$ is the mutual information between *jth* and *(j+1)th* partitions.

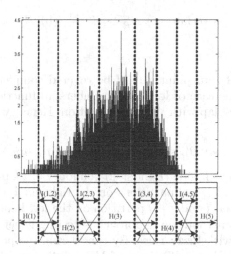

Fig. 2. Fuzzy partitioning of the histogram into m joint partitions and $m\text{-}1$ overlaps.

To define an MF we need at least two parameters among support, boundary and core/prototype. By having the fuzzy partitions, the problem is to assign a value either to the boundary or to the core/prototype. From section 3, we know determining the boundary or core is a subjective issue and related to degree of fuzziness. Although we cannot assign an exact value to the boundary, it is possible to define a suitable range. Suppose X^i and X^{i+1} are two adjacent partitions on $\mathbf{X}$ and d_1 and d_2 are laid on the boundary of the partition X^i. In a *well-defined membership function*, in each partition, we can expect following relation;

$$\forall\, d_1, d_2 \in X^i \wedge (d_1 < d_2) \Rightarrow \mu_{X^i}(d_1) < \mu_{X^i}(d_2) \tag{16}$$

Fig. 3 -a shows a drawback when the right hand boundary of X^i is less than its overlap with the next partition. To compensate this drawback, it is proposed to have another partition that fired by d_1 and d_2 (see Fig. 3-b), as follows;

$$\mu_{X^{i+1}}(d_1) < \mu_{X^{i+1}}(d_2) \tag{17}$$

Eq. (16) and (17) imply that the right side boundary for membership function

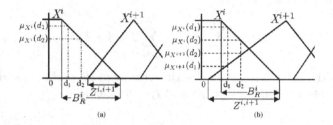

(a)   (b)

Fig. 3. Boundary vs. overlap, (a) $B^i_R > Z^{i,i+1}$, (b) $B^i_R < Z^{i,i+1}$: well-defined boundary.

in the right hand of zero point should be covered completely by overlap with next partition. In the case of left hand membership functions, left side boundary is covered by overlap with previous partition. Hence, we define well-defined boundary as;

$$Right\ hand\ partitions: B^i_R \leq Z^{i,i+1}, \quad Left\ hand\ partitions: B^i_L \leq Z^{i,i-1} \tag{18}$$

where B^i_L (B^i_R) is the left side (right side) boundary of *ith* partition. $Z^{i,i+1}$ is the overlap between i and *(i+1)* partitions on the right hand of the UD and $Z^{i,i-1}$ is the overlap between i and *(i-1)* partition on the left hand of the UD.

To evaluate the method, one approach is testing generated MF's in a complete fuzzy system with fuzzy rules. By considering the performance of the FS as performance of MF, the efficiency of MF is influenced by fuzzy rule base. I this paper, our attempt has been investigating the design process of MF independently from parameters inside FS.

Another approach, that is concerned in this paper, is applying the method on various collections of similar data type. For example, in the case of weather data, we can apply the algorithm on the different collection related to different places or different period of times. The results are compared by a simple distance measure such as Euclidian or Cosine distance. The less variance of distances the better performance of the method.

5   Results

To examine the proposed schema, temperature data of city of Toronto down-loaded from the metrological site of University of Toronto in Mississauga is used as a sample data. The algorithm is applied on two data collections. The first is related to year 2001 as a reference data and another one is to evaluate the result and related to year 2002. Both data collections include more than 8000 data point. To start the procedure, the number of fuzzy partitions should be fixed (e.g., 5 partitions in this research). Fig. 4-a shows the histogram of first data collection and Fig. 4-b is the graph of first data collection versus time after normalization and in Fig. 4-c the resulted MFs has been shown. In Fig. 4-d and 4-e, the histogram and graph of the second data collection (year 2002) are considered. Fig. 4-f shows resulted MFs for the second data collection. Comparing Fig. 4-c and 4-f shows that the two MFs with a negligible difference are almost similar. The results are satisfying because in the MF generation process we switched from a completely subjective problem to a highly objective form. We expect almost the same membership function for various collections unless there is a remarkable change in the statistics of new collection of data. It seems by assuming the input as random variable, in the case of stationary random variable we will have almost a unique MF function over time.

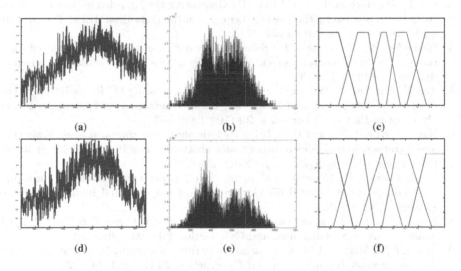

Fig. 4. (a-c) Result of first collection, (d-f) result of second collection.

6   Conclusion

A method based on genetic algorithm optimization and information theory measures for automatic generation of fuzzy membership functions was presented.

The proposed schema that is independent from optimization of fuzzy rules and any other elements inside the fuzzy system. It was considered that among membership function parameters, only support or fuzzy partition could be directly related to information theory measures. The core idea by relating entropy to fuzzy partition of membership function is to change the optimization of the UD partitioning associated to variable X to optimization of partitions of the probability density function of random variable X. The probability density function is approximated by histogram or statistics of X. The results based on proposed evaluation method are satisfying. As a result, the method can be utilized in learning strategies using examples and training data, especially besides fuzzy rules generation may construct a total solution based on learning from example, and by having a plenty of data collections of an individual type, we can generate generic MF for generic data.

References

1. Cano, J.C., Nava, P.A.: A fuzzy method for automatic generation of membership function using fuzzy relations from training examples, Annual Meeting of the North American Fuzzy Information Processing Society. (2002) 158–162
2. Chiu, L. S.: Fuzzy Model Identification Based on Cluster Estimation, Journal of Intelligent and Fuzzy Systems. **2**, (1944) 267–278
3. Cordn, O., Herrera, F., and Villar, P.: Generating the Knowledge Base of a Fuzzy Rule-Based System by the Genetic Learning of the Data Base, IEEE Transactions on Fuzzy Systems. **9** (2001) 667–674
4. Homaifar, A., McCormick, E.: Simultaneous design of membership functions and rule sets for fuzzy controllers using genetic algorithms, IEEE Transaction on Fuzzy Systems. **3** (1995) 129–139
5. Jin, Y., von Seelen, W., and Sendhoff, B.: On Generating FC Fuzzy Rule Systems from Data Using Evolution Strategies, IEEE Transaction on Systems, man and Cybernetics-Part B: Cybernetics. **29** (1999) 829–845
6. Kim, J., Kim, B.M., and Huh, N.C.: Genetic algorithm approach to generate rules and membership functions of fuzzy traffic controller, The 10th IEEE International Conference on Fuzzy Systems. **1** (2001) 525–528
7. Krishnapuram, R.: Generation of membership functions via possibilistic clustering, Proceedings of the Third IEEE Conference on Computational Intelligence, Fuzzy Systems. (1994) 902–908
8. Lee, C.C.: Fuzzy Logic in Control Systems: Fuzzy Logic Controller- PartI,II, IEEE Transactions on Systems, man and Cybernetics. **20** (1990) 404–432
9. Wang, L.X., Mendel, J.M.: Generating fuzzy rules by learning from examples, IEEE Transactions on Systems, Man and Cybernetics. **22** (1992) 1414–1427
10. Simon, D.: Fuzzy membership optimization via the extended Kalman filter, 19th International Conference of the North American Fuzzy Information Processing Society . (2000) 311–315
11. Sugeno, M.: An Introduction Survey of Fuzzy Control, Information Sciences. **36** (1985) 59–83
12. Wong, K., Wong, Y.W.: Combined genetic algorithm/simulated annealing/fuzzy set approach to short-term generation scheduling with take-or-pay fuel contract, IEEE Transactions on Power Systems. **11** (1996) 128–136

Analyzing the Performance of a Multiobjective GA-P Algorithm for Learning Fuzzy Queries in a Machine Learning Environment*

Oscar Cordón[1], Enrique Herrera-Viedma[1], María Luque[1], Félix de Moya[2], and Carmen Zarco[3]

[1] Dept. of Computer Science and A.I. University of Granada.
18071 - Granada (Spain).
{ocordon,viedma,mluque}@decsai.ugr.es
[2] Dept. of Library and Information Science. University of Granada.
18071 - Granada (Spain).
felix@ugr.es
[3] PULEVA Food S.A. Camino de Purchil, 66. 18004 - Granada (Spain).
czarco@puleva.es

Abstract. The fuzzy information retrieval model was proposed some years ago to solve several limitations of the Boolean model without a need of a complete redesign of the information retrieval system. However, the complexity of the fuzzy query language makes it difficult to formulate user queries. Among other proposed approaches to solve this problem, we find the Inductive Query by Example (IQBE) framework, where queries are automatically derived from sets of documents provided by the user. In this work we test the applicability of a multiobjective evolutionary IQBE technique for fuzzy queries in a machine learning environment. To do so, the Cranfield documentary collection is divided into two different document sets, labeled training and test, and the algorithm is run on the former to obtain several queries that are then validated on the latter.

1 Introduction

Information retrieval (IR) may be defined as the problem of the selection of documentary information from storage in response to search questions provided by a user [2]. Information retrieval systems (IRSs) deal with documentary bases containing textual, pictorial or vocal information and process user queries trying to allow the user to access to relevant information in an appropriate time interval.

The fuzzy information retrieval (FIR) model [3] was proposed to overcome several limitations of Boolean IRSs [2], the most extended ones, without a need of a complete redesign. However, the extended Boolean (fuzzy) query structure considered in fuzzy IRSs – weighted, positive or negative terms joined by the AND and OR operators – is difficult to be formulated by non expert users.

* Research supported by CICYT TIC2002-03276 and by Project "Mejora de Meta-heurísticas mediante Hibridación y sus Aplicaciones" of the University of Granada.

T. Bilgiç et al. (Eds.): IFSA 2003, LNAI 2715, pp. 611–619, 2003.
© Springer-Verlag Berlin Heidelberg 2003

The paradigm of Inductive Query by Example (IQBE) [4], where queries describing the information contents of a set of documents provided by a user are automatically derived, has proven to be useful to assist the user in the query formulation process. Focusing on the FIR model, the most known approach is that of Kraft et al. [13], based on genetic programming (GP) [12]. Several other approaches have been proposed based on more advanced evolutionary algorithms (EAs) [1], such as genetic algorithm-programming (GA-P) [11] or simulated annealing-programming, to improve Kraft et al.'s [6,7].

In [9], we proposed a new IQBE algorithm that tackled fuzzy query learning as a multiobjective problem. The algorithm was able to automatically generate several queries with a different trade-off between precision and recall in a single run. To do so, a Pareto-based multiobjective EA scheme [5] was incorporated into the single-objective GA-P IQBE technique proposed in [6].

In this contribution, we design a experimental framework to test the said technique in a machine learning environment. To do so, several queries are selected from the Cranfield collection and the document set is divided into two different subsets, training and test, for each of them. The multiobjective GA-P algorithm is then run on the former sets and the obtained queries are validated on the latter ones to get a view of the real applicatibility of the approach.

The paper is structured as follows. Section 2 is devoted to the preliminaries, including the basis of FIRSs and a short review of IQBE techniques. Then, the multiobjective GA-P proposal is reviewed in Section 3. Section 4 presents the experimental setup designed and the experiments developed. Finally, several concluding remarks are pointed out in Section 5.

2   Preliminaries

2.1   Fuzzy Information Retrieval Systems

FIRSs are constituted of the following three main components:

The documentary data base, that stores the documents and their representations (typically based on index terms in the case of textual documents).

Let D be a set of documents and T be a set of unique and significant terms existing in them. An indexing function $F : D \times T \rightarrow [0,1]$ is defined as a fuzzy relation mapping the degree to which document d belongs to the set of documents "about" the concept(s) represented by term t. By projecting it, a fuzzy set is associated to each document $(d_i = \{< t, \mu_{d_i}(t) > \mid t \in T\};\ \mu_{d_i}(t) = F(d_i,t))$ and term $(t_j = \{< d, \mu_{t_j}(d) > \mid d \in D\};\ \mu_{t_j}(d) = F(d,t_j))$.

In this paper, we will work with Salton's normalized *inverted document frequency* (IDF) [2]: $w_{d,t} = f_{d,t} \cdot log(N/N_t)$; $F(d,t) = \frac{w_{d,t}}{Max_d\, w_{d,t}}$, where $f_{d,t}$ is the frequency of term t in document d, N is the total number of documents and N_t is the number of documents where t appears at least once.

The query subsystem, allowing the users to formulate their queries and presenting the retrieved documents to them. Fuzzy queries are expressed using a query language that is based on weighted terms, where the numerical or linguistic weights represent the "subjective importance" of the selection requirements.

In FIRSs, the query subsystem affords a fuzzy set q defined on the document domain specifying the degree of relevance (the so called *retrieval status value* (RSV)) of each document in the data base with respect to the processed query: $q = \{< d, \mu_q(d) > \mid d \in D\}$; $\mu_q(d) = RSV_q(d)$.

Thus, documents can be ranked according to the membership degrees of relevance before being presented to the user. The retrieved document set can be specified providing an upper bound for the number of retrieved documents or defining a threshold σ for the RSV (the σ-cut of the query response fuzzy set q).

The matching mechanism, that evaluates the degree to which the document representations satisfy the requirements expressed in the query (i.e., the RSV) and retrieves those documents that are judged to be relevant to it.

When using the *importance* interpretation [3], the query weights represent the relative importance of each term in the query. The RSV of each document to a fuzzy query q is then computed as follows [15]. When a single term query is logically connected to another by the AND or OR operators, the relative importance of the single term in the compound query is taken into account by associating a weight to it. To maintain the semantics of the query, this weighting has to take a different form according as the single term queries are ANDed or ORed. Therefore, assuming that A is a fuzzy term with assigned weight w, the following expressions are applied to obtain the fuzzy set associated to the weighted single term queries A_w (*disjunctive queries*) and A^w (*conjunctive ones*):

$$A_w = \{< d, \mu_{A_w}(d) > \mid d \in D\} \quad ; \quad \mu_{A_w}(d) = Min\ (w, \mu_A(d))$$
$$A^w = \{< d, \mu_{A^w}(d) > \mid d \in D\} \quad ; \quad \mu_{A^w}(d) = Max\ (1 - w, \mu_A(d))$$

If the term is negated in the query, a negation function is applied to obtain the corresponding fuzzy set: $\overline{A} = \{< d, \mu_{\overline{A}}(d) > \mid d \in D\}$; $\mu_{\overline{A}}(d) = 1 - \mu_A(d)$.

Finally, the RSV of the compound query is obtained by combining the single weighted term evaluations into a unique fuzzy set as follows:

$$A\ AND\ B = \{< d, \mu_{A\ AND\ B}(d) > \mid d \in D\}\ ;\ \mu_{A\ AND\ B}(d) = Min(\mu_A(d), \mu_B(d))$$
$$A\ OR\ B = \{< d, \mu_{A\ OR\ B}(d) > \mid d \in D\}\ ;\ \mu_{A\ OR\ B}(d) = Max(\mu_A(d), \mu_B(d))$$

2.2   Inductive Query by Example

IQBE was proposed in [4] as "a process in which searchers provide sample documents (examples) and the algorithms induce (or learn) the key concepts in order to find other relevant documents". This way, IQBE is a technique for assisting the users in the query formulation process performed by machine learning methods. It works by taking a set of relevant (and optionally, non relevant documents) provided by a user and applying an off-line learning process to automatically generate a query describing the user's needs from that set. The obtained query can then be run in other IRSs to obtain more relevant documents.

Apart from the IQBE algorithms for the FIR model reviewed in the Introduction, several others have been proposed for the remaining IR models, such as the Boolean [16,8] or the vector space [4] ones.

3 A Multiobjective GA-P Algorithm for Automatically Learning Fuzzy Queries

In [9], we proposed a multiobjective IQBE algorithm to learn fuzzy queries based on the GA-P paradigm whose components are described next.

Coding Scheme: The expressional part (GP part) encodes the query composition – terms and logical operators – and the real-coded coefficient string (GA part) represents the term weights, as shown in Figure 1.

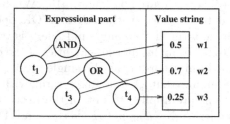

Fig. 1. GA-P individual representing the fuzzy query 0.5 t_1 *AND* (0.7 t_3 *OR* 0.25 t_4)

Fitness Function: The multiobjective GA-P (MOGA-P) algorithm is aimed at jointly optimizing the classical precision and recall criteria [2], as follows:

$$Max\ P = \frac{\sum_d r_d \cdot f_d}{\sum_d f_d} \quad ; \quad Max\ R = \frac{\sum_d r_d \cdot f_d}{\sum_d r_d}$$

with $r_d \in \{0, 1\}$ being the relevance of document d for the user and $f_d \in \{0, 1\}$ being the retrieval of document d in the processing of the current query.

Pareto-Based Multiobjective Selection and Niching Scheme: The Pareto-based multiobjective EA considered is Fonseca and Fleming's Pareto-based MOGA [5]. Each individual is first assigned a rank equal to the number of individuals dominating it plus one (non-dominated individuals receive rank

1) and the population is sorted in ascending order according to that rank. Then, each individual C_i is assigned a fitness value according to its ranking in the population: $f(C_i) = \frac{1}{rank(C_i)}$. Finally, the fitness assignment of each group of individuals with the same rank is averaged among them.

A niching scheme is then applied in the objective space to obtain a well-distributed set of queries with a different trade-off between precision and recall. To do so, Goldberg and Richardson's sharing function [14] is considered:

$$F(C_i) = \frac{f(C_i)}{\sum_{j=1}^{M} Sh(d(C_i, C_j))} \quad ; \quad Sh(x) = \begin{cases} 1 - (\frac{x}{\sigma_{share}})^\gamma, & \text{if } x < \sigma_{share} \\ 0, & \text{otherwise} \end{cases}$$

with M being the population size, σ_{share} being the niche radius and d standing for the Euclidean distance.

Finally, the intermediate population is obtained by Tournament selection [14], i.e., to fill each free place in the new population, t individuals are random selected from the current one and the best adapted of them is chosen.

Genetic Operators: The BLX-α crossover operator [10] is applied twice on the GA part to obtain two offsprings. Michalewicz's non-uniform mutation operator [14] is considered to perform mutation on that part.

The usual GP crossover randomly selecting one edge in each parent and exchanging both subtrees from these edges between the both parents [12] is considered. Each time a mutation is to be made, one of the two following mutation operators (selected at random) is applied: random generation of a new subtree substituting an old one located in a randomly selected edge, and random change of a query term by another one not present in the encoded query, but belonging to any relevant document.

4   Experimental Setup and Experiments Developed

The documentary set used to design our machine learning experimental framework has been the *Cranfield* collection, composed of 1400 documents about Aeronautics. It has been automatically indexed by first extracting the non-stop words, applying a stemming algorithm, thus obtaining a total number of 3857 different indexing terms, and then using the normalized IDF scheme (see Section 2.1) to generate the term weights in the document representations.

Among the 225 queries associated to the Cranfield collection, we have selected those presenting 20 or more relevant documents (queries 1, 2, 23, 73, 157, 220 and 225). The number of relevant documents associated to each of these seven queries are 29, 25, 33, 21, 40, 20 and 25, respectively.

For each one of these queries, the documentary collection has been divided into two different, non overlapped, document sets, training and test, each of them composed of a 50% of both the relevant and non relevant documents. Hence, we represent a retrieval environment where no document retrieved by the learned queries in the test sets has been previously seen by the user.

MOGA-P has been run ten different times on the training document set associated to each query. The parameter values considered are a maximum of 20 nodes for the expression parts, a population size of 800, 50000 evaluations per run, a Tournament size t of 8, 0.8 and 0.2 for the crossover and mutation probabilities in both the GA and the GP parts, a sharing function parameter γ equal to 2, and a niche radius σ_{share} experimentally set to 0.1. The retrieval threshold σ has been set to 0.1 in the FIRS.

Table 1. Statistics of the Pareto sets obtained by the MOGA-P algorithm

$\#q$	$\#p$	$\sigma_{\#p}$	$\#dp$	$\sigma_{\#dp}$	M_2^*	$\sigma_{M_2^*}$	M_3^*	$\sigma_{M_3^*}$
1	110.0	10.5	5.3	0.348	39.901	4.574	0.918	0.044
2	127.4	7.462	4.3	0.202	47.023	3.235	0.895	0.033
23	133.8	5.805	6.9	0.170	52.156	3.004	1.042	0.015
73	93.0	12.435	2.6	0.210	24.893	5.133	0.730	0.041
157	118.9	7.886	7.8	0.310	45.264	3.943	1.066	0.006
220	91.1	6.897	1.9	0.221	18.987	4.395	0.437	0.094
225	98.1	6.243	2.3	0.202	22.931	4.266	0.626	0.083

Table 1 collects several statistics about the ten Pareto sets generated for each query. From left to right, the columns contain the number of non-dominated solutions obtained ($\#p$), the number of different objective vectors (i.e., precision-recall pairs) existing among them ($\#dp$), and the values of two of the usual multiobjective metrics M_2^* and M_3^* [17][1], all of them followed by their respective standard deviation values.

In order to test the real applicability of the algorithm in the machine learning environment, the Pareto sets obtained in the ten runs performed for each query were put together, and the dominated solutions were removed from the unified set. Then, five queries well distributed on the Pareto front were selected from each of the seven unified Pareto sets[2] and run on the corresponding test set once preprocessed (for example, the query terms not existing on the test collection are removed from the query). The results obtained are shown in Table 2, standing Sz for the query size, P and R for the precision and recall values and $\#rr/\#rt$ for the number of relevant and retrieved documents, respectively.

In view of the precision and recall values obtained, the performance of our proposal is very significant[3]. The algorithm is always able to find at least a

---

[1] $M_2^* \in [0, \#p]$ measures the diversity of the solutions found, while M_3^* measures the range to which the Pareto front spreads out in the objective values (in our case, the maximum possible value is $\sqrt{2} = 1.4142$). In both cases, the higher the value, the better the quality of the obtained Pareto set.

[2] An example of such a query ($\#$q1-5) in preorder is: OR AND OR t_{1158}(w=0.298) OR t_{1051}(w=0.518) OR t_{2721}(w=0.957) OR t_{2950}(w=0.838) t_{12}(w=0.970) OR OR t_{1320}(w=0.577) t_{238}(w=0.847) t_{2579}(w=0.737) OR t_{1129}(w=0.701) t_{12}(w=0.329).

[3] The interested reader can refer to [6,7,8,9] to compare the obtained results with those of several other approaches in the same documentary base.

Table 2. Retrieval efficacy of the selected queries on the training and test collections

#q		Sz	P	R	#rr/#rt	Sz	P	R	#rr/#rt
			Training set				Test set		
	1	19	0.304	1.0	14/46	9	0.188	0.4	6/32
	2	19	0.318	1.0	14/44	19	0.111	0.267	4/36
1	3	19	0.591	0.929	13/22	5	0.154	0.133	2/13
	4	19	0.786	0.786	11/14	5	0.143	0.067	1/7
	5	19	1.0	0.643	9/9	15	0.0	0.0	0/3
	1	19	0.273	1.0	12/44	19	0.297	0.846	11/37
	2	19	0.387	1.0	12/31	19	0.216	0.615	8/37
2	3	19	0.579	0.917	11/19	17	0.0	0.0	0/24
	4	19	0.786	0.917	11/14	15	0.059	0.077	1/17
	5	19	1.0	0.667	8/8	17	0.143	0.154	2/14
	1	19	0.232	1.0	16/69	19	0.031	0.118	2/65
	2	19	0.39	1.0	16/41	17	0.208	0.588	10/48
23	3	19	0.591	0.812	13/22	19	0.344	0.647	11/32
	4	19	0.786	0.688	11/14	15	0.455	0.294	5/11
	5	19	1.0	0.625	10/10	19	0.111	0.059	1/9
	1	19	0.692	0.9	9/13	17	0.25	0.455	5/20
	2	19	0.5	1.0	10/20	15	0.208	0.455	5/24
73	3	19	0.526	1.0	10/19	15	0.071	0.091	1/14
	4	19	0.769	1.0	10/13	17	0.062	0.091	1/16
	5	19	1.0	0.9	9/9	17	0.455	0.455	5/11
	1	19	0.299	1.0	20/67	15	0.195	0.8	16/82
	2	19	0.39	0.8	16/41	19	0.119	0.25	5/42
157	3	19	0.593	0.8	16/27	17	0.3	0.3	6/20
	4	19	0.789	0.75	15/19	15	0.25	0.15	3/12
	5	19	1.0	0.5	10/10	15	0.375	0.15	3/8
	1	19	0.833	1.0	10/12	13	0.2	0.1	1/5
	2	19	0.588	1.0	10/17	13	0.167	0.1	1/6
220	3	19	0.588	1.0	10/17	15	0.167	0.1	1/6
	4	17	0.714	1.0	10/14	13	0.6	0.3	3/5
	5	19	1.0	0.9	9/9	19	0.111	0.1	1/9
	1	17	0.324	1.0	12/37	15	0.0	0.0	0/33
	2	17	0.324	1.0	12/37	15	0.0	0.0	0/33
225	3	19	0.579	0.917	11/19	11	0.0	0.0	0/15
	4	19	0.688	0.917	11/16	15	0.0	0.0	0/8
	5	19	1.0	0.917	11/11	15	1.0	0.077	1/1

query retrieving all the relevant documents ($R = 1.0$) provided by the user in the training set. As regards the generalization ability of the learned queries, i.e., their capability to retrieve new relevant documents for the user, it can be seen how it is very satisfactory in the most of the cases. For example, recall levels of 0.4, 0.846, 0.647, 0.455, 0.8 and 0.3 are respectively obtained for queries 1, 2,

618 Oscar Cordón et al.

23, 73, 157, and 220, all of them with appropriate precision values ranging from 0.188 to 0.6. However, we should note that the results obtained in the last query, 225, have not been appropriate as only one of the learned queries has been able to retrieve a relevant document in the test set. In this case, it seems that there is a larger diversity of index terms in the relevant documents for the query, and those index terms existing in the training documents do not describe the test relevant documents.

5   Concluding Remarks

The real applicability of a multiobjective evolutionary IQBE technique for learning fuzzy queries has been tested in a machine learning environment. It has been run on training document sets obtained from the Cranfield collection to derive several queries that have been validated on a different test document set. Very promising results have been achieved for six of the seven Cranfield queries considered in view of the retrieval efficacy obtained.

As future works, we will study other real-like environments based on different training-test partitions of the document collection and will use retrieval measures considering not only the absolute number of relevant and non relevant documents retrieved, but also their relevance order in the retrieved document list.

References

1. Bäck, T.: Evolutionary algorithms in theory and practice. Oxford (1996).
2. Baeza-Yates, R., Ribeiro-Neto, B.: Modern information retrieval. Addison (1999).
3. Bordogna, G., Carrara, P., Pasi, G.: Fuzzy approaches to extend Boolean information retrieval. In: P. Bosc, J. Kacprzyk (Eds.), Fuzziness in database management systems. Physica-Verlag (1995) 231–274.
4. Chen, H., et al.: A machine learning approach to inductive query by examples: an experiment using relevance feedback, ID3, GAs, and SA, Journal of the American Society for Information Science **49:8** (1998) 693–705.
5. Coello, C.A., Van Veldhuizen, D.A., Lamant, G.B.: Evolutionary algorithms for solving multi-objective problems. Kluwer Academic Publishers (2002).
6. Cordón, O., Moya, F., Zarco, C.: A GA-P algorithm to automatically formulate extended Boolean queries for a fuzzy information retrieval system, Mathware & Soft Computing **7:2-3** (2000) 309–322.
7. Cordón, O., Moya, F., Zarco, C.: A new evolutionary algorithm combining simulated annealing and genetic programming for relevance feedback in fuzzy information retrieval systems, Soft Computing **6:5** (2002) 308-319.
8. Cordón, O., Herrera-Viedma, E., Luque, M.: Evolutionary learning of Boolean queries by multiobjective genetic programming. In: Proc. PPSN-VII, Granada, Spain, LNCS 2439. Springer (September, 2002) 710-719.
9. Cordón, O., Moya, F., Zarco, C.: Automatic learning of multiple extended Boolean queries by multiobjective GA-P algorithms. In: V. Loia, M. Nikravesh, L.A. Zadeh (Eds.), Fuzzy Logic and the Internet. Springer (2003), in press.

10. Eshelman, L.J., Schaffer, J.D.: Real-coded genetic algorithms and interval-schemata. In: L.D. Whitley (Ed.), Foundations of Genetic Algorithms 2. Morgan Kaufman (1993) 187–202.
11. Howard, L., D'Angelo, D.: The GA-P: a genetic algorithm and genetic programming hybrid, IEEE Expert **10:3** (1995) 11–15.
12. Koza, J.: Genetic programming. On the programming of computers by means of natural selection. The MIT Press (1992).
13. Kraft, D.H., et al.: Genetic algorithms for query optimization in information retrieval: relevance feedback. In: E. Sanchez, T. Shibata, L.A. Zadeh, Genetic algorithms and fuzzy logic systems. World Scientific (1997) 155–173.
14. Michalewicz, Z.: Genetic algorithms + data structures = evolution programs. Springer (1996).
15. Sanchez, E.: Importance in knowledge systems, Information Systems **14:6** (1989) 455–464.
16. Smith, M.P., Smith, M.: The use of GP to build Boolean queries for text retrieval through relevance feedback, Journal of Information Science **23:6** (1997) 423–431.
17. Zitzler, E., Deb, K., Thiele, L.: Comparison of multiobjective evolutionary algorithms: empirical results, Evolutionary Computation **8:2** (2000) 173–195.

Commutativity as Prior Knowledge in Fuzzy Modeling*

Pablo Carmona[1], Juan L. Castro[2], and José M. Zurita[2]

[1] Universidad de Extremadura, Depto. de Informática,
E. Ingenierías Industriales, Avda. Elvas, s/n, 06017-Badajoz, Spain,
pablo@unex.es
[2] Universidad de Granada, Depto. de Ciencias de la Computación e I.A.,
E.T.S.I. Informática., C/ Daniel Saucedo Aranda, s/n, 18071-Granada, Spain
castro@decsai.ugr.es, zurita@decsai.ugr.es

Abstract. This paper faces with the integration of mathematical properties satisfied by the system as prior knowledge in fuzzy modeling (FM), focusing on the commutativity as a starting point. The underlying idea is to reward the rules in each input fuzzy region that provide good commutativity degrees respecting its complementary —commutatively related— input fuzzy region. With this aim, the similarity between the outputs in both regions will be obtained. The experimental results show the accuracy improvement gained by the proposed method.

Keywords: Fuzzy modeling, prior knowledge, commutativity.

1 Introduction

In FM, the training set is often not enough to attain an accurate and complete description of the system, avoiding to properly model it. A way to attenuate this problem consists in including prior knowledge about the system into FM. One of such knowledge consists in restrictions that the system complies, which can be described as properties satisfied by the system. For example, a system in which small changes in the input cause small changes in the output satisfies a consistency —smoothness— property [3]; a system whose output increases when the input increases satisfies a monotonicity property [9]; foreign exchange markets comply with a symmetry property [1]; etc.

In the literature, two main strategies have been proposed. The first one uses that knowledge to determine the structure of the model [6, 8, 10]. The second one uses the knowledge along the learning process, either by extending the training set with virtual examples on the base of the properties the system obeys [2, 7], or by imposing certain penalty functions that punish the models not complying with the properties [9]. However, for the time being, all these techniques have been used in the neural framework.

We propose the integration of prior knowledge about properties into FM. Unlike the aforementioned techniques, the satisfaction of the properties will be

---

* This work has been supported by Research Project PB98-1379-C02-01

T. Bilgiç et al. (Eds.): IFSA 2003, LNAI 2715, pp. 620–627, 2003.

evaluated directly on the fuzzy model, acting as an additional criterion for the selection of the rules. Although this paper focuses on the commutativity property, an equivalent strategy can be developed for other local properties —such as symmetry— involving only two input regions.

2   Commutativity in a System

The commutativity property has two main advantages that make it appropriate as a first approach to this research topic: its definition is very simple, since it only involves two input variables in its basic form, and it is a local property, only relating two input regions of the system, that will be referred as the *analysed* and the *complementary* regions.

Definition 1. *A system* $f : \mathcal{X}_1 \times \cdots \times \mathcal{X}_n \rightarrow \mathcal{Y}$ *complies with a* bivariate commutativity property *over two input variables* X_{l_1} *and* X_{l_2} *if and only if*

$$f(X_1, \ldots, X_{l_1}, \ldots, X_{l_2}, \ldots, X_n) = f(X_1, \ldots, X_{l_2}, \ldots, X_{l_1}, \ldots, X_n) \ . \quad (1)$$

A multivariate commutativity involving m variables $\S = \{X_{l_1}, \ldots, X_{l_m}\}$ can be defined when any permutation of these variables provides the same output value. Nevertheless, it can be easily proved that this multivariate commutativity can be expressed as the simultaneous fulfillment of two different commutativities involving $(m-1)$ variables in $\S$ and it can be ultimately decomposed as the simultaneous fulfillment of $(m-1)$ bivariate commutativities.[1] Thus, a single value for the multivariate commutativity can be obtained decomposing it into a conjunction of bivariate commutativities and using some aggregation method (e.g., a convex sum).

3   Fuzzy Measures of Commutativity

In this section, two types of measures will be defined taking into account two different elements involved in FM: rule bases and examples.

3.1   Commutativity of a Fuzzy Rule w.r.t. a Fuzzy Rule Base

The rules in a fuzzy model do not appear independently, but they are interacting one another. Besides, determining which pairs of rules must be compared between the input regions involved in an evaluation of the commutativity is not a trivial question, because it depends on the semantics defined over the input variables (i.e., the input fuzzy domains).

Because of that, a pairwise comparison between rules seems not to be appropriate. Instead, the following idea will be used: *the commutativity can be defined as the similarity between the output mapped from two commutatively-related input fuzzy regions.* This will be accomplished with the following definition.

---

[1] E.g., a commutativity over $\{X_1, X_2, X_3, X_4\}$ can be expressed as the simultaneous fulfillment of 2 commutativities over $\{X_1, X_2, X_3\}$ and $\{X_1, X_2, X_4\}$, and thus, as the conjunction of 3 bivariate commutativities, e.g. over $\{X_1, X_2\}$, $\{X_1, X_3\}$, $\{X_1, X_4\}$.

Definition 2 (fuzzy approach — FA). *The bivariate commutativity of a rule*

$$R_{LY}^{i_1...i_{l_1}...i_{l_2}...i_n} : LX_{1,i_1}, \ldots, LX_{l_1,i_{l_1}}, \ldots, LX_{l_2,i_{l_2}}, \ldots, LX_{n,i_n} \to LY$$

over the input variables X_{l_1} and X_{l_2} w.r.t. a rule base $\mathcal{RB}$ is defined as

$$\widetilde{C_{RB}}^{l_1,l_2}(R_{LY}^{i_1...i_n}, \mathcal{RB}) = S(\widehat{LY}_{\mathcal{RB}}(\mathbf{LX}), \widehat{LY}_{\mathcal{RB}}(\mathbf{LX_c})) \tag{2}$$

where $\mathbf{LX}$ is the fuzzy input vector corresponding to the antecedent of the rule, $\mathbf{LX_c} = [LX_{1,i_1}, \ldots, LX_{l_2,i_{l_2}}, \ldots, LX_{l_1,i_{l_1}}, \ldots, LX_{n,i_n}]$ is its commutatively related fuzzy input vector, $\widehat{LY}_{\mathcal{RB}}(\mathbf{LX})$ is the fuzzy output mapped in $\mathcal{RB}$ from $\mathbf{LX}$, and S is a similarity measure between fuzzy sets.

Since the fuzzy output need not to be convex, we selected for the similarity measure S the family of Minkowski-based measures [12] defined as

$$S_r(A, B) = 1 - \left(\int_{\mathcal{X}} |\mu_A(x) - \mu_B(x)|^r \, dx\right)^{\frac{1}{r}}, \quad r \geq 1, \tag{3}$$

which extends the Minkowski metric to the fuzzy environment and deals properly with non-convex fuzzy sets (excepting S_∞) since it involves all the membership values of both fuzzy sets being compared.

3.2   Commutativity of a Fuzzy Rule w.r.t. a Set of Examples

Mainly, the environment where FM is carried out is composed by a training set $E = \{e^1, \ldots, e^m\}$ with examples of the form $e^j = ([x_1^j, \ldots, x_n^j], y^j)$. Then, we must redefine the previous commutativity measure now w.r.t. a set of examples. The simplest strategy consist on translating the examples into a set of rules.

Definition 3. *Given a set of examples E, the rule base $\mathcal{RB}_E$ representing them will be obtained as a set of rules*

$$R_{LY_k}^{i_1...i_n} : LX_{1,i_1}, \ldots, LX_{n,i_n} \to LY_k, \quad k = \arg\max_{j=1...q} \omega(R_{LY_j}^{i_1...i_n})$$

for all $\{i_1, \ldots, i_n\} \in p_1 \times \cdots \times p_n$, where ω is a certainty measure based on E, p_i is the number of values in the input fuzzy domain $\widetilde{\mathcal{X}}_i$, and q is the number of values in the output fuzzy domain $\widetilde{\mathcal{Y}}$.

This method, which takes the rules with maximum certainty degree in every input region of the fuzzy grid,[2] is largely used in the literature (e.g., [5, 11]).

In this paper we use the Mixed Method (MM), recently proposed by the authors in [4]. This method combines the Wang and Mendel's method (WMM) [11] with an extension of the Ishibuchi's rule generation method [5] that deals with

---

[2] If no example covers the region $[LX_{1,i_1}, \ldots, LX_{n,i_n}]$ no rule will be selected; if several rules take the maximum certainty degree, one of them will be selected randomly.

fuzzy consequents. The WMM firstly translates each example into the fuzzy rule best covering it —i.e., with labels having the highest membership degree— and, secondly, once all the examples are processed, selects the rules with maximum certainty degree from (4) among all the conflicting ones:

$$\omega_{WM}(R_{LY}^{i_1\dots i_n}) = \mu_{LX_{1,i_1}}(x_1) \times \dots \times \mu_{LX_{n,i_n}}(x_n) \times \mu_{LY}(y) . \qquad (4)$$

Basically, the MM extends the WMM by adding rules in the input regions where WMM did not identify rules. If there are examples covering one of such fuzzy regions, the rule in this region having the maximum degree from

$$\omega_{Ish}(R_{LY}^{i_1\dots i_n}) = \frac{\beta(R_{LY}^{i_1\dots i_n}) - \bar{\beta}(R_{LY}^{i_1\dots i_n})}{\sum\limits_{k=1}^{q} \beta(R_{LY_k}^{i_1\dots i_n})} \qquad (5)$$

will be added, where q is the number of values in the output fuzzy domain and

$$\beta(R_{LY}^{i_1\dots i_n}) = \sum_{e^j \in E} \mu_{LX_{1,i_1}}(x_1^j) \times \dots \times \mu_{LX_{n,i_n}}(x_n^j) \times \mu_{LY}(y^j)$$

$$\bar{\beta}(R_{LY}^{i_1\dots i_n}) = \sum_{\substack{k=1 \\ LY_k \neq LY}}^{q} \frac{\beta(R_{LY_k}^{i_1\dots i_n})}{q-1} .$$

Once the rule base is obtained, the measure presented in Sect. 3.1 can be directly used by inserting the rule to be evaluated in $\mathcal{RB}_E$ and obtaining its commutativity w.r.t. $\mathcal{RB}_E$.

In the following, we will present a second measure taking advantage of the information available in the training set. This measure takes the examples contained into the complementary region and generates virtual examples [2, 7] by interchanging the values of the two commutative input variables. Then, it estimates the mean error between the expected and the inferred outputs.

Definition 4 (example based approach — EBA). *The bivariate commutativity of a fuzzy rule*

$$R_{LY}^{i_1\dots i_{l_1}\dots i_{l_2}\dots i_n} : LX_{1,i_1}, \dots, LX_{l_1,i_{l_1}}, \dots, LX_{l_2,i_{l_2}}, \dots, LX_{n,i_n} \to LY$$

over the variables X_{l_1} and X_{l_2} w.r.t. a set of examples E is defined as:

$$\dot{C}_E^{\,l_1,l_2}(R_{LY}^{i_1\dots i_n}, E) = 1 - \frac{1}{r} \sum_{e^j \in E^*} \left| y^j - \hat{y}_{\mathcal{RB}_E}(\mathbf{x_c}^j) \right| \qquad (6)$$

where $E^* \subset E$ contains the r examples covering the complementary region:

$$E^* = \{ e^j \in E \mid (\mu_{LX_{l_2,i_{l_2}}}(x_{l_1}^j) > 0) \wedge (\mu_{LX_{l_1,i_{l_1}}}(x_{l_2}^j) > 0) \} , \qquad (7)$$

$\mathbf{x_c}^j = [x_1^j, \dots, x_{l_2}^j, \dots, x_{l_1}^j, \dots, x_n^j]$ *is a vector set up from e^j by interchanging its values in the commutative variables X_{l_1} and X_{l_2}, and $\mathcal{RB}_E$ is the rule base obtained from Definition 3 using the MM.*

The EBA exploits the examples by using them to obtain $\mathcal{RB}_E$ and by integrating them directly into the measurement. As it will be shown below, this can lead to overvalue that information, provoking an overfitting effect.

4   Commutativity into Fuzzy Modeling

Next, we focus on the mechanism to integrate the commutativity as prior knowledge in FM. For clarity, we will select again the MM outlined in the previous section as the underlying method.

Along FM with the MM, it is possible to draw a certainty map $\mathcal{W} \in [0,1]^{n+1}$ associating a certainty degree to each possible rule.[3] Analogously, it is possible to draw a commutative map $\mathcal{C} \in [0,1]^{n+1}$ from the training set through the commutativity measures presented in Sect. 3.2 by inserting each possible rule in the identified rule base $\mathcal{RB}_E$ and measuring its commutativity w.r.t. its complementary region. Then, a suitability map $\mathcal{S} \in [0,1]^{n+1}$ can be obtained using some aggregation mechanism (e.g., a convex sum) that combines the certainty and the commutativity degrees for each rule. Finally, from this suitability map $\mathcal{S}$, the rules with maximum degree in each input fuzzy region can be selected to obtain the final model.

Therefore, the proposed commutativity sensitive FM (CSFM) algorithm will consist in the following four steps:

1. Obtain the certainty map $\mathcal{W}$ from the examples using MM.
2. Obtain the commutativity map $\mathcal{C}$ using a measure of the commutativity of a fuzzy rule w.r.t. a set of examples (Sect. 3.2).
3. Obtain the suitability map $\mathcal{S}$ from $\mathcal{W}$ and $\mathcal{C}$ by means of a convex sum, $\mathcal{S} = \alpha \mathcal{W} + (1 - \alpha)\mathcal{C}$.
4. For each input fuzzy region, select the rule with maximum degree in $\mathcal{S}$.

This simple method will allow to clearly contrast the results of commutativity non-sensitive FM (CNSFM) w.r.t. CSFM in the next section.

5   Experimental Results

In this section, several experiments have been applied to the following three numerical functions with increasing complexity:

$$f_1 : [-1,1] \times [-1,1] \to [-1,1], \quad y = (x_1 + x_2)/2 \tag{8}$$

$$f_2 : [-1,1] \times [-1,1] \to [-1,1], \quad y = 1 - |x_1 - x_2| \tag{9}$$

$$f_3 : [0,0.5] \times [0,0.5] \to [-2,2.4], \ y = x_1 + x_2 - cos(18x_1) - cos(18x_2) \tag{10}$$

whose output surfaces are depicted in Fig. 1. The fuzzy domains are the same for all the functions and are shown in Fig. 2.[4]

The experiments involve five FM methods: two CNSFM methods and two CSFM ones. In the first group, the WMM and the MM described in Sect. 3.2 are

---

[3] The values provided by (4) lie in [0,1] whereas the ones provided by (5) lie in [-1,1]. In order to obtain rules with homogeneous degrees from the MM, (5) will be normalized to [0,1] by the simple transformation $\omega_{Ish_N}(R_{LY}^{i_1...i_n}) = [\omega_{Ish}(R_{LY}^{i_1...i_n}) + 1]/2$.

[4] As each function has its own domain of discourse, the appropriate scaling factors must be applied to each fuzzy domain.

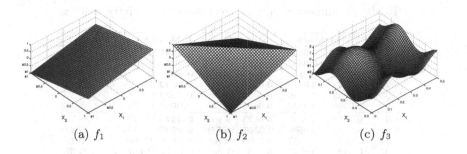

(a) f_1 (b) f_2 (c) f_3

Fig. 1. System surfaces.

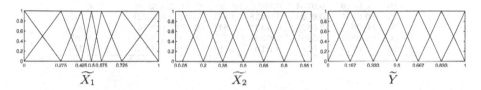

$\widetilde{X_1}$ $\widetilde{X_2}$ $\widetilde{Y}$

Fig. 2. Fuzzy domains.

considered. As MM is the method used to draw the $\mathcal{W}$ in the CSFM, it will allow to establish the contribution of commutativity analysis to FM. The second group corresponds to the FA and EBA proposed in this paper. We consider $\alpha = 0.5$ for combining $\mathcal{W}$ and $\mathcal{C}$, and the similarity measure S_2 for the FA.

Six training set sizes m will be considered: 5, 10, 15, 20, 30, and 50 examples. For each training set size, each method was run 200 times with different randomly generated training sets and the model error was measured using

$$MSE_N = \frac{\sum_{j=1}^{m}[y_N^j - \hat{y}_N(\mathbf{x}^j)]^2}{m} \tag{11}$$

where y_N^j is the normalized output of the j-th example in the test set and $\hat{y}_N(\mathbf{x}^j)$ is the normalized output predicted by the model. The test set consisted in 2500 examples randomly generated from the corresponding function.

Besides, an index of activity for each run of the CSFM methods was worked out as the number of rules it modifies w.r.t. the model identified with MM (i.e., taking into account only the $\mathcal{W}$ map).

Table 1 shows the numerical results, detailing the averaged accuracy indexes —the best highlighted for each training set size— and, in the case of CSFM, their averaged activity indexes (in parenthesis). Last column shows the strength of outperformance between CSFM methods.

CSFM outperforms CNSFM in all the cases, and specially for small training sets (poor informative content). Between CSFM methods, the FA provides in general the best accuracy results. The general poor results of the EBA reveals an overvaluation of the information provided by the examples, unable to guide

Table 1. Numerical results: accuracy and activity indexes

	m	WMM	MM	MM+EBA	MM+FA	FA vs. EBA
	05	0.0826	0.0598	0.0421 (9.32)	**0.0227** (27.00)	85.5%
	10	0.0565	0.0301	0.0193 (10.40)	**0.0095** (22.72)	**103.2%**
f_1	15	0.0391	0.0172	0.0109 (9.46)	**0.0059** (17.11)	84.7%
	20	0.0272	0.0111	0.0065 (8.03)	**0.0042** (12.77)	54.8%
	30	0.0159	0.0059	0.0042 (5.22)	**0.0031** (7.34)	35.5%
	50	0.0069	0.0027	0.0025 (2.23)	**0.0023** (2.73)	8.7%
	05	0.1116	0.0839	0.0623 (9.46)	**0.0446** (27.72)	39.7%
	10	0.0814	0.0505	0.0340 (10.78)	**0.0227** (22.89)	49.8%
f_2	15	0.0578	0.0312	0.0206 (9.70)	0.0147 (17.11)	40.1%
	20	0.0443	0.0221	0.0152 (7.93)	**0.0122** (13.07)	24.6%
	30	0.0273	0.0138	0.0104 (5.55)	**0.0090** (7.97)	15.6%
	50	0.0126	0.0074	**0.0063** (2.50)	0.0065 (3.48)	-3.2%
	05	0.1031	0.0947	0.0810 (9.27)	**0.0761** (27.01)	6.4%
	10	0.0819	0.0704	0.0577 (10.63)	**0.0542** (22.50)	6.5%
f_3	15	0.0651	0.0524	**0.0418** (9.85)	0.0422 (16.72)	-1.0%
	20	0.0543	0.0428	**0.0337** (8.81)	0.0342 (13.14)	-1.5%
	30	0.0391	0.0297	**0.0238** (5.77)	0.0257 (7.58)	-8.0%
	50	0.0244	0.0183	**0.0161** (3.16)	0.0174 (3.86)	**-8.1%**

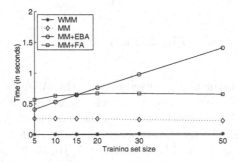

Fig. 3. Computational cost.

the search of the best commutative rule if they are not numerous enough. However, the superiority of the FA diminishes with the complexity of the system, since it determines the commutativity based on a preliminary model $\mathcal{RB}_E$ whose accuracy also diminishes, whereas the EBA obtains the information in the complementary region directly from the examples. Even so, the gain obtained by the FA in f_1 and f_2 is quite higher than the obtained by the EBA in f_3.

Concerning the activity indexes, they seem to be independent both from the system to be identified and from the accuracy obtained, and present similar relative results between CSFM methods, being higher for the FA.

Finally, Fig. 3 presents the results concerning computational cost (only for one function, as all the cases provide similar results) obtained using a Pentium IV 1,7GHz and Matlab. Obviously, the commutativity integration increases the computational cost of FM. However, different behaviors are observed for the different approaches: on one hand, the cost of the EBA depends on the training set size due to the intensive use of the examples it performs, thus being suitable for small training sets, although its low accuracy in such cases limits in general its application; on the other hand, the FA presents the best computational cost in general, since it only depends on the training set size to obtain $\mathcal{W}$ (and $\mathcal{RB}_E$ from it) and the dimensionality of the system affects to obtaining $\mathcal{C}$, but it does not —or slightly— affect to the measurement of each single commutativity.

6   Conclusions

In this paper, a first approach to the integration of prior knowledge about properties the system complies has been developed.

Both proposed approaches outperform the results obtained from the NCSFM. In general, the FA presents the best results, improving to a great extent the model accuracy with an acceptable computational cost. The EBA must be relegated to highly complex systems with large training sets or with few examples and strong computational restrictions.

Independently of the approach being used, the benefits rise with small or low informative training sets, since it is in such cases where the prior knowledge contributes in a decisive way to the FM results. Thus, these are the situations where the ideas proposed in this paper will play a significant role.

References

[1] Y.S. Abu-Mostafa. Financial market applications of learning from hints. In A.P. Refenes, editor, *Neural Networks in the Capital Markets*, chapter 15, pages 221–232. Wiley, 1995.

[2] Y.S. Abu-Mostafa. Hints. *Neural Computation*, 7:639–671, 1995.

[3] P. Carmona, J.L. Castro, and J.M. Zurita. Contradiction sensitive fuzzy model-based adaptive control. *Int. J. Approx. Reas.*, 30(2):107–129, 2002.

[4] P. Carmona, J.L. Castro, and J.M. Zurita. Strategies to identify fuzzy rules directly from certainty degrees: A comparison and a proposal. *IEEE Trans. Fuzzy Syst.*, 2002. (submitted for publication).

[5] H. Ishibuchi, K. Nozaki, N. Yamamoto, and H. Tanaka. Construction of fuzzy classification systems with rectangular fuzzy rules using genetic algorithms. *Fuzzy Sets Syst.*, 65:237–253, 1994.

[6] K. Kohara and T. Ishikawa. Multivariate prediction using prior knowledge and neural heuristics. In *Proc. ISIKNH'94*, pages 179–188, 1994.

[7] P. Niyogi, F. Girosi, and T. Poggio. Incorporating prior information in machine learning by creating virtual examples. *Proc. IEEE*, 86(11):2196–2209, 1998.

[8] C.W. Omlin and C.L. Giles. Training second-order recurrent neural networks using hints. In D. Sleeman and P. Edwards, editors, *Proc. 9th Int. Workshop Machine Learning*, pages 361–366. Morgan Kaufmann, 1992.

[9] J. Sill and Y.S. Abu-Mostafa. Monotonicity hints. *Advances in Neural Information Processing Systems*, 9, 1996.

[10] A. Suddarth and A.D.C. Holden. Symbolic-neural systems and the use of hints for developing complex systems. *Int. J. Man-Mach. Stud.*, 35:291–311, 1991.

[11] L.X. Wang and J.M. Mendel. Generating fuzzy rules by learning from examples. *IEEE Trans. Syst., Man, Cybern.*, 22(6):1415–1427, 1992.

[12] R. Zwick, E. Carlstein, and D.V. Budescu. Measures of similarity among fuzzy concepts: A comparative analysis. *Int. J. Approx. Reas.*, 1:221–242, 1987.

Evolutionary Optimization of Fuzzy Models with Asymmetric RBF Membership Functions Using Simplified Fitness Sharing*

Min-Soeng Kim, Chang-Hyun Kim, and Ju-Jang Lee

Dept. of Electrical Engineering and Computer Science, KAIST,
373-1, Gusung-Dong, Yusung-gu, Taejon, Korea, 305-701
mibella@kaist.ac.kr

Abstract. A new evolutionary optimization scheme for designing a Takagi-Sugeno fuzzy model is proposed in this paper. To achieve better modeling performance, asymmetric RBF membership functions are used. Penalty function is proposed and used in the fitness function to prevent overlapping membership functions in the resulting fuzzy model. The simplified fitness sharing scheme is used to enhance the searching capability of the proposed evolutionary optimization algorithm. Some simulations are performed to show the effectiveness of the proposed algorithm.

1 Introduction

Designing a fuzzy model can be formulated as a search problem in high-dimensional space where each point represents a rule set, membership functions (MFs), and the necessary parameters with the corresponding system's behavior. Given some performance criteria, the performance of the systems forms a hypersurface in the space. Due to the capability of searching irregular and high-dimensional solution space, the evolutionary algorithms (EA's) such as GAs (genetic algorithms) and ESs (evolutionary strategies) have been applied to designing a fuzzy model as in [1], [2], and [3] (find more references and discussions in those papers). Especially, the Takagi-Sugeno (TS)-type fuzzy model [4] has a great advantage due to its representative power and used in many EA-based approaches. Usually, in those fuzzy models, triangular and normal RBF (or Gaussian) MFs are used. In this paper, evolutionary optimization of TS-type fuzzy models with asymmetric RBF MFs are performed to investigate whether using asymmetric RBF MFs are beneficial or not for modeling problem. Penalty function is proposed and used in the fitness function to prevent overlapping MFs in the resulting fuzzy model. From the experimental observation that the search space has the multimodal characteristic, a simplified fitness sharing scheme is used to enhance the searching capability of the proposed evolutionary optimization algorithm. Some simulations are performed to show the effectiveness of the proposed algorithm.

---

* This work is partially supported by KOSEF under the Korea-Japan basic scientific promotion program and by HWRS-ERC

T. Bilgiç et al. (Eds.): IFSA 2003, LNAI 2715, pp. 628–635, 2003.

2 Takagi-Sugeno Fuzzy Model with Asymmetric RBF Membership Functions

In this paper, a TS fuzzy model is used because the TS-type fuzzy model has a great advantage due to its representative power; it is capable of describing a highly nonlinear relations, which is suitable for the problem of modeling. A TS fuzzy model consists of IF-THEN rules where the rule consequents are usually constant values (singletons) or linear functions of the inputs.

$$R_i : \quad \text{IF} \quad x_1 \text{ is } A_{i1} \text{ and } \cdots \text{ and } x_n \text{ is } A_{in} \tag{1}$$
$$\text{then } y_i = c_{i0} + c_{i1}x_1 + \cdots + c_{in}x_n \quad \text{for } i = 1, 2, \cdots, N_R \ .$$

where N_R is the number of rules, $\mathbf{x} = [x_1, x_2, \cdots, x_n]$ is the input vector, y_i is the output of the i-th rule, and c_{ij} are real-valued consequent parameters. A_{ij} are the antecedent fuzzy sets that are characterized by asymmetric RBF MFs $\mu_{A_{ij}}(x_j)$ as shown in the following equation.

$$\mu_{A_{ij}}(x_i) = exp(\frac{-(x_i - w_{ij})^2}{\sigma'_{ij}\,^2}) \ . \tag{2}$$

$$\text{where} \quad \begin{cases} \sigma'_{ij} = \sigma^{lt}_{ij} & \text{if } x_i \leq w_{ij} \\ \sigma'_{ij} = \sigma^{rt}_{ij} & \text{if } x_i > w_{ij} \end{cases} .$$

where w_{ij} represents the center value of the asymmetric RBF MF. σ^{lt}_{ij} and σ^{rt}_{ij} represent width values of MFs for left side and right side, respectively, according to the relative position of input x_i to the center w_{ij}. Given a input sequence, each fuzzy rule produces the degree of memberships for the data. This so-called 'firing strength' of the rule R_i is denoted by τ_i and is calculated as

$$\tau_i = A_{i1}(x_1) \times A_{i2}(x_2) \times \cdots \times A_{in}(x_n) = \prod_{i=1}^{n} \mu_{A_{ij}}(x_i) \ . \tag{3}$$

$\times$ operator in (3) represents '*fuzzy and*' (T-norm) operation and algebraic product is adopted in this paper. Then the fuzzy model output $\hat{y}$ is computed by a defuzzification process as follows:

$$\hat{y} = \frac{\sum_{i=1}^{N_R} \tau_i y_i}{\sum_{i=1}^{N_R} \tau_i} = \frac{\sum_{i=1}^{N_R} \tau_i(c_{i0} + c_{i1}x_1 + \cdots + c_{in}x_n)}{\sum_{i=1}^{N_R} \tau_i} \ . \tag{4}$$

3 Evolutionary Optimization of Fuzzy Models

3.1 Representation

When designing a fuzzy model using EA's, one of the most important considerations is the representation scheme, that is, how to encode a fuzzy model into a chromosome. The proposed algorithm is based on the standard fuzzy model, that is, antecedent parts of each fuzzy rule are obtained by combining each MF of each input variable. If there are N_I input variables and M_i MFs for the i-th input, then the resulting fuzzy model has $N_R = \prod_{i=1}^{N_I} M_i$ fuzzy rules. With this structure, it is only necessary to represent a MF by real values. 2 real values are necessary for the normal RBF MFs, for example, and 3 real values are necessary for asymmetric RBF MFs as in this paper. Then, an individual is denoted as s_k where $k = 1, 2, \cdots N_P$. N_P is the population size. Consequently, a chromosome for a fuzzy model can be represented in the vector form as follows:

$$s_k(t) = [v_{11} \ v_{12} \ \cdots \ v_{1M_1} \ \cdots \ v_{N_I 1} \ \cdots \ v_{N_I M_{N_I}}] \ . \tag{5}$$

where $v_{ij} = [w_{ij}, \sigma_{ij}^{lt}, \sigma_{ij}^{rt}]$ represents the parameters of the j-th membership function for the i-th input variable. The remaining consequent parameters c_{ij} in (1) can be easily obtained by least square methods as shown in [5] not by evolutionary optimization.

3.2   Fitness Function and Fitness Sharing

Defining a proper fitness function is one of the most important issues when using EA's since the fitness function guides the search direction. To evaluate each individual when using the evolutionary approach, a fitness function that is proper to the given problem should be devised. The most general way to define a fitness function is to measure the performance of an individual in terms of the modeling performance, that is, mean-squared-error (MSE):

$$MSE(s_k) = \frac{1}{N_T} \sum_{h}^{N_T} (y_h^d - \hat{y}_h)^2 \ . \tag{6}$$

where y_h^d is the desired output and $\hat{y}_h$ is the model output. If only the MSE is adopted in evaluating individuals, however, then there can be multiple overlapping MFs, which is a typical feature of unconstrained optimization as noticed in [6] and [7]. The overlapping of the MFs results in the following consequences: physical meaning of some fuzzy subsets may be blurred. Also it may be possible that the performance of the resulting fuzzy model becomes worse, since the fact that the overlapping MFs cannot distinguish itself from other MFs means that it lost its freedom of representative capability. Constraints are defined in [6] so that the parameters of fuzzy MFs can vary only within a specific range of their initial values. This approach, however, is somewhat strict in the sense that MFs cannot change its position freely and the multiple overlapping MFs can still be generated according to their initial positions. Another method is to merge similar fuzzy MFs in the process of optimization if necessary. However, merging itself cannot prevent overlapping MFs and the methods needs many MFs on the beginning. Thus, in this paper, a special function which gives penalty for a fuzzy model with multiple overlapping MFs is devised. The proposed penalty function actively calculates the degree of overlapping between neighboring MFs and is defined as the following:

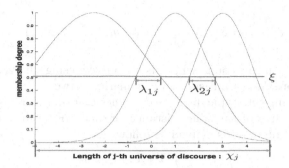

Fig. 1. Definition of the overlapping length in the penalty function.

$$P(s_k) = \frac{1}{N_{ov}^j} \sum_{j=1}^{N_I} \frac{\sum_{i=1}^{N_{ov}^j} \lambda_{ij}}{|\chi_j|} \ . \tag{7}$$

where λ_{ij} is the length of the i-th ($i = 1, \cdots, N_{ov}^j$) overlapping occurrence between two MFs in the j-th input domain. N_{ov}^j represents the total occurrence of overlapping among MFs for the j-th input domain for the individual. χ_j is the length of the j-th input domain. The specific level that constrains the overlapping between two MFs is denoted by ξ as shown in Fig. 1. Using this penalty value with the MSE value, the fitness function is defined as follows:

$$F(s_k) = \frac{1}{MSE(s_k) + \beta \cdot P(s_k)} \ . \tag{8}$$

where β is a design parameter which is used to make a compromise between the MSE and the penalty function. With the proposed fitness function, one can avoid multiple overlapping MFs and the MF itself can be located anywhere on the input domain while not being restricted within the range that is determined by their initial positions. Since the proposed penalty function is not a strict constraint on individuals but an evaluation criteria for individuals, the proposed algorithm can even construct a fuzzy model that has multiple overlapping MFs if it is necessary to solve the given problem, which is not the case when restrictive constraints are used.

By the way, it should be noted that the search space has probably multimodal characteristics from the experimental observation that there are many fuzzy models, which produce almost the same modeling performance, with different set of parameter values. Thus, the fitness sharing scheme is simplified and used in this paper. Originally, fitness sharing modifies the search landscape by reducing the payoff in densely populated regions [8]. In this paper, before performing reproduction, fitness of each individual is changed according to the following equation.

$$F'(s_k) = F(s_k)/m_i \quad \text{where} \quad m_i = \sum_{j=1}^{N_p} sh(d_{ij}) \quad . \tag{9}$$

where m_i is the niche count which is calculated using the sharing function. Sharing function $sh(d_{ij})$ measures the similarity between i-th individual and j-th individual using genotypic distance d_{ij}. Since real values are used to represent each individual, sum of Euclidian distance for each MF between two individuals is used. Sharing function is defined as follows:

$$sh(d_{ij}) = \begin{cases} 1 & \text{if} \quad d_{ij} < \gamma_s \\ 0 & \text{if} \quad \text{otherwise} \end{cases} \quad .$$

where γ_s denotes the threshold value. Since setting the threshold γ_s requires a priori information, we also used adaptive scheme. If the fitness value of best member does not improved during 10 consecutive generations then γ_s is reduced 90% of its current value.

3.3   Evolutionary Operators

Reproduction According to the fitness value of each individual, we first apply a ranking method [9]. After ordering all the individuals in the population according to their fitness value, the upper 30% of the population is used to generate 50% of the new population. The remaining 70% of the population is used to generate 50% of the new population. Eliticism was used to preserve the best individual before applying fitness sharing. For better convergence performance, 5% of the new population is replaced by copying an elite individual into random locations. By speeding up the convergence speed, we need less generations in the evolutionary process. However, since the rapid convergence may result in finding local minima, we maintained the diversity of the population by fitness sharing shown above and also other evolutionary operators as shown in next several paragraphs.

Crossover Crossover is the process of exchanging portions of two 'parent' individuals. An overall probability is assigned to the crossover process, which is the probability that given two parents, the crossover operation will occur. For convenience, we rewrite (5) as $s_k(t) = [p_1 \ p_2 \ \cdots \ p_L]$ where p_i corresponds to a v_{ij} in (5). We have used two types of crossover operation in this paper. The first is a bitwise crossover. When two parents $s_v(t)$ and $s_w(t)$ are selected for the crossover operation, changing point k is selected randomly within the range of an individual and swapping occurs as follows:

$$s_v(t+1) = [v_1 \ v_2 \ \cdots \ v_k \ w_{k+1} \ \cdots w_L] \quad . \tag{10}$$
$$s_w(t+1) = [w_1 \ w_2 \ \cdots \ w_k \ v_{k+1} \ \cdots v_L] \quad .$$

The next crossover is an arithmetic crossover operator, which produces the children using a linear combination of two parents as follows:

$$s_v(t+1) = \alpha \cdot s_v(t) + (1 - \alpha) \cdot s_w(t) \quad . \tag{11}$$
$$s_w(t+1) = \alpha \cdot s_w(t) + (1 - \alpha) \cdot s_v(t) \quad .$$

where the parameter α is generated randomly each time the arithmetic crossover is applied. The arithmetic crossover is applied to the center value and the width value. The range of α to [-0.5, 1.5] is used so that the center of a MF can be located not only inside of the values of two parents, but also outside. This kind of crossover can be thought as a combination of a convex crossover and an affine crossover [10]. Since the center value of a membership function can be very distant from the effective input domain after several arithmetic crossover operations are applied, some boundary values are defined so that the center values can remain within the effective input domain. The boundary values for the center value of membership functions for the i-th input are denoted as w^i_{min} and w^i_{max}, respectively. These values are determined according to the range of each input domain.

Mutation The next operation is mutation. Mutation consists of changing an element's value of an individual at random, often with a constant probability. Mutation is performed column-wise for every center value and every width value of all individuals as follows:

$$\theta(t+1) = \theta(t) + N(0,\delta) \ . \tag{12}$$

where θ is a parameter value: the center or the width of a membership function and $N(\cdot,\cdot)$ represents normal distribution function, which generates a random value around the zero-mean with the variance parameter δ. As a result of the crossover operations and the mutation operations, the width can be less or equal to zero. To prevent this, a lower boundary value σ_{min} for the width is set.

3.4 Summary

Given the data pattern matrix **Z**, the maximum number of generation G_{max}, the population size N_P, the number of input variables N_I, we set the probabilities for the crossover operations and the mutation operations, γ_s, δ, σ_{min} and determine the design parameters β and ξ. According to the range of each input variable, the boundary values w^i_{min} and w^i_{max} are determined. The framework of evolutionary procedure used in this paper is as follows:

- 1) Generate an initial population $P(0) = [s_1(0) \ s_2(0) \ \cdots \ s_{N_P}(0)]$ at random and set $i = 0$.
- 2) Using the parameter values of each individual, construct a fuzzy model and calculate consequent parameters by least square methods for all individuals.
- 3) Evaluate every individual and save the best individual
- 4) Change fitness value of each individual by applying fitness sharing
- 5) Apply evolutionary operators: reproduction, crossover, and mutation to obtain the next population $P(i+1)$
- 6) $i = i+1$, return to step 2) if the G_{max} is not reached or the procedure is terminated

4   Simulation Results

The aim of this section is to find a fuzzy model for a nonlinear dynamic plant using the proposed algorithm. We will consider the second-order nonlinear plant studied in [5], [6], and [11].

$$y(k) = g(y(k-1), y(k-2)) + u(k) \ . \tag{13}$$

where

$$g(y(k-1), y(k-2)) = \frac{y(k-1)y(k-2)(y(k-1) - 0.5)}{1 + y^2(k-1) + y^2(k-2)} \ . \tag{14}$$

The goal is to approximate the nonlinear component $g(y(k-1), y(k-2))$, which is usually called the 'unforced system' in control literature, using a TS fuzzy model. The problem is 2 inputs and 1 output example. As in [6], 400 simulated data points were generated from the plant model (13). With the starting equilibrium state (0,0), 200 samples of training data were obtained using a random input signal $u(k)$ that is uniformly distributed in $[-1.5, 1.5]$. The rest 200 samples of validation data were obtained using a sinusoidal input signal $u(k) = sin(2\pi k/25)$. Parameter settings for the given problem are as follows: The population size N_P was 200. The maximum generation G_{max} was 200. Crossover probability was 0.4. Mutation probability was 0.7. δ was 0.02. β was 0.04 and ξ was 0.7. Also, sharing threshold γ_s was initially set to 1.0. We used 2 MFs for each input, which results in 4 rules. Table 1 shows the simulation results obtained by applying the proposed algorithm. Results are averaged over 10 trials and the average value was inserted for the proposed algorithm and for the algorithm of [5]. In [5], the same penalty functions was used with normal RBF MFs. Thus, we also applied the fitness sharing to check how the performance changes. [11] used symmetric MFs and triangular MFs are used in [6]. In [5] normal RBF MFs are used. From the results, we can conclude that using asymmetric RBF MFs is more beneficial for modeling performance. Also from the Table 1, one can easily see the effectiveness of the proposed evolutionary optimization algorithm. It is shown that using a fitness sharing scheme is effective in finding a better fuzzy model.

5   Conclusion

A new evolutionary optimization scheme for designing a fuzzy model is proposed in this paper. It is shown that using asymmetric RBF membership functions is beneficial to achieve better modeling capability. By using a penalty function in the fitness function, overlapping of adjacent membership functions are prevented. Also to enable better exploration of search space, simplified and adaptive fitness sharing is used. It is shown that using a sharing fitness scheme is helpful to find a better fuzzy model. Simulation results shows that the proposed algorithm is successful to find a fuzzy model for modeling problems.

Table 1. Comparison results for the nonlinear system modeling problem (* results are averaged over 10 trials)

Algorithm	No. of rules	MSE(train)	MSE(validation)
Yen and Wang [11]	25	$2.3e^{-4}$	$4.1e^{-4}$
	20	$6.8e^{-4}$	$2.4e^{-4}$
Setnes and Roubos [6]	4	$1.2e^{-3}$	$4.7e^{-4}$
* Kim and Lee [5]	4	$1.38e^{-4}$	$1.82e^{-4}$
with fitness sharing	4	$1.40e^{-4}$	$1.85e^{-4}$
with adaptive fitness sharing	4	$1.39e^{-4}$	$1.80e^{-4}$
* Proposed	4	$1.34e^{-4}$	$1.62e^{-4}$
with fitness sharing	4	$1.34e^{-4}$	$1.51e^{-4}$
with adaptive fitness sharing	4	$1.34e^{-4}$	$1.53e^{-4}$

References

1. Y. Shi, R. Eberhart and Y. Chen, Implementation of Evolutionary Fuzzy Systems, IEEE Trans. on Fuzzy Syst., vol. 7, No. 2, (1999)109-119
2. Marco Russo, Genetic Fuzzy Learning, IEEE Trans. on Evolutionary Computation., vol.4, No. 3, (2000) 259-273
3. S. J. Kang, H. S. Hwang and K. N. Woo, Evolutionary design of fuzzy rule base for nonlinear system modeling and control, IEEE Trans. Fuzzy Syst., vol. 8, (2000) 37-45
4. T. Takagi and M. Sugeno, Fuzzy identification of systems and its applications to modeling and control, IEEE Trans. Syst., Man, Cybern., vol. 15, (1985) 116-132
5. Min-Soeng Kim, Sun-Gi Hong, and Ju-Jang Lee, Evolutionary design of a fuzzy system for various problems including vision based mobile robot control, Proc. 2002 IEEE/RSJ Intl. Conf. Intelligent Robots and Systems, EPFL, Lausanne, Switzerland, (2002) 1056-1061
6. M. Setnes and H. Roubos, GA-fuzzy modeling and classification: complexity and performance, IEEE Trans. Fuzzy Syst., vol. 8, (2000) 509-522
7. Min-Soeng Kim and Ju-Jang Lee, Evolutionary design of Takagi-Sugeno type fuzzy model for nonlinear system identification and time series prediction, Proc. the Int'l Conf. on Control, Automation and Systems, Oct. Cheju Univ. Korea, (2001) 667-670
8. B. Sareni and L. Krähenbühl, Fitness sharing and niching methods revisited, IEEE Trans. on EC, vol. 2, no. 3, (1998) 97-106
9. Michalewicz, Z.: Genetic Algorithms + Data Structures = Evolution Programs. 3rd edn. Springer-Verlag, Berlin Heidelberg New York (1996)
10. M. Gen and R. Cheng, Genetic algorithms & enginerring optimization, John Wiley & Sons, Inc., (2000)
11. J. Yen and L. Wang, Simplifying fuzzy rule-based models using orthogonal transformation methods, IEEE Trans. Syst., Man, Cybern., pt, B, vol. 29, (1999) 13-24

Fuzzy Multi-objective Optimisation Approach for Rod Shape Design in Long Product Rolling

Victor Oduguwa[1], Rajkumar Roy[1] and Didier Farrugia[2]

[1] Department of Enterprise Integration, School of Industrial and Manufacturing Science,
Cranfield University, Cranfield, Bedford, MK43 0AL, UK
{v.oduguwa and r.roy}@cranfield.ac.uk
[2] Corus R, D and T, Swinden Technology Center,
Rotherham, S60 3AR, UK

Abstract. Fuzzy fitness evaluation within evolutionary algorithms is increasingly being used to manage the fitness associated with genetic individuals as defined by the membership function of the fuzzy set. In most of the approaches reported in the literature, the fitness assignment mechanism is dependent on the defuzzified domain value or the degree of membership of the consequent solution set. This suggests that the trade-off between optimal solutions values and the truth-values (membership function values) of the associated solutions is not fully explored. This can result to a deception problem for search algorithm in cases where fuzzy fitness is used for evaluation. This paper presents a novel approach to deal fuzzy fitness evaluation within multi-objective optimisation framework to address the deceptive problem associated with current fuzzy fitness techniques reported in the literature. The proposed approach is applied to a rod rolling shape optimisation problem for automatic generation of optimal rod shapes.

1   Introduction

In complex engineering processes such as hot rolling of rods, Finite Element (FE) analysis is often used to study the effect of roll design during high temperature rolling. Owing to their discrete nature, FE models have been used to develop a detailed understanding of the rolling process at meso-scale level. Although FE techniques allow an entire rolling sequence to be studied, it is still time consuming (mostly in 3D), to use them as an embedded optimizer for roll design sequences despite improvement in both hardware and software. It is therefore necessary to develop optimisation approaches that are not only based on FE responses, but also on the engineer's qualitative knowledge for solving complex engineering design problems.

Although engineers still use their expert design knowledge to reason about the complex non-linear numerical models of real roll pass design problems, fuzzy sets with If-Then rules can also be used to model the complex roll pass design problem [1]. This form of representation is particularly useful in situations where the behaviours of the underlying phenomenon are not fully understood but its influence on other objects is known, for example the roll shape behaviour in a rolling system de-

T. Bilgiç et al. (Eds.): IFSA 2003, LNAI 2715, pp. 636–643, 2003.

sign. Successful applications of fuzzy logic do exist in the steel manufacturing process for instance fuzzy logic for automatic shape control in the rolling process [2], and in roll pass design [3] have been reported.

1.1 Fuzzy Fitness Evaluation in Evolutionary Algorithms

Genetic Fuzzy System (GFS) is a class of numerous approaches reported in the literature for combining Fuzzy logic and genetic algorithm (GA), where the GA are used to design fuzzy systems. Two possible types of the GFS are considered in this study. The first one is the Fuzzy Genetic Algorithm (FGA), considered as a GA that uses fuzzy logic based techniques or fuzzy tools to improve the GA behaviour by modelling different GA components. The second often incorrectly referred to as FGAs applies GA's within an imprecise environment to produce approximate solutions to fuzzy optimisation problems. GA's are used to manage fuzzy objectives and the fitness associated to the chromosomes is defined by a membership value of the fuzzy set. There are several cases reported in the literature were fuzzy evaluation have been used to assign quality value to every GA solution.

Fuzzy logic was applied within GA to model objectives and constraints in the evaluation function. In electric machine design, fuzzy constraint have also being used to describe the fitness of the machine [4]. Koskimaki et al [5] proposed a genetic algorithm as a substitute for the designer's heuristic optimising process for electric machine design. Fuzzy constraint was used to describe the fitness of the machine. In image analysis, Bhandari et al [6] combined fuzzy based set theoretic measures to evaluate picture quality to evaluate picture quality. A quantitative index was derived and used to evaluate solutions within the GA. Roy [7] developed a GA based approach to identify optimal solutions and then fuzzy logic to qualitatively evaluate the solutions. In most of the approaches reported in the literature, the fitness assignment mechanism is dependent either on the vector of the objective variable or the degree of membership of the consequent solution set. This suggests that the trade-off between optimal solutions values and the truth values (membership function values) of the associated solutions is not fully explored.

1.2   Problem Statement

This section explores the concept of the fuzzy fitness problem space as a two objective optimisation problem, where the two objectives have a conflicting relationship. The first objective is the defuzzified scalar value of the objective value and the second objective is the corresponding membership grade. The membership grade can be viewed as the degree to which the solution scalar value is true; it best represents the information contained in the consequent solution. In fuzzy reasoning of such problems, there can be instances where the same antecedent rules are fired resulting to different consequent fuzzy set solutions. These solutions can results to the same defuzzified objective variables in the underlying domain with different degrees of membership. The example in Figure 1 shows the effect of different truth-values of the 'Average' fuzzy set on the consequent solution set. The defuzzification method adopted is the centroid. The Figure shows the objective values at 450 for two cases with corresponding membership grade of 0.30, and 0.55 respectively.

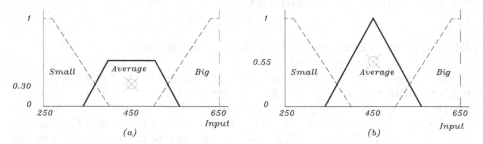

Fig. 1 Effect of different truth values on the fuzzy solution set

If the objective value was used solely as the fitness of the solution, this suggests that solutions with the same objective values are ranked equally regardless of the magnitude of the truth-value. In view of the deceptive problem this poses for evolutionary search algorithms it seems more appropriate that the fitness criterion should have an additional dimension that ranks the solutions on the basis of the truth-value (membership grade) of the solution set as well as the objective values. This paper presents a fuzzy multi-objective optimisation approach to address the deceptive problem associated with current fuzzy fitness techniques reported in the literature.

2   Fuzzy Multi-objective Optimisation (FUMOO): Solution Approach

FUMMO is a multi-objective genetic algorithm based solution approach. One of the most popular multi-objective GA, NSGAII [8] was adopted in this study. FUMOO was developed to explore the trade-off between the membership grade and the objective values of a given set of solutions. In this study, the maximisation of both objectives is sought to address the deceptive problem associated with current fuzzy fitness techniques reported in the literature. The algorithm consists of three main parts: the basic genetic algorithm, fuzzy evaluator and the multi-objective fuzzy fitness assignment. The solution strategy is coded in C++ using the proposed algorithm shown in Figure 2. A brief description of the key features is given as follows.

GA's search by using a population of points and as a result they are effective in finding the global optimum. These features are used to assign values to each decision variables. These values indicate the vector of each variable in the design space. Values of the decision variables are then passed on to the fuzzy evaluator. These input values are fuzzified, and fuzzy IF-THEN rules are applied within the fuzzy inference mechanism. The evaluation of a proposition produces one fuzzy set associated with each model solution variable. For example, in evaluating the following propositions: *If a is Y then S is P, If b is X then S is Q, If c is Z then S is R.*. The consequent fuzzy set P, Q, R is correlated to produce a fuzzy set representing the solution variable S. An appropriate method of defuzzification is used to find a scalar value and the corresponding membership grade that best represents the information contained in the consequent fuzzy set D. Both the scalar value and the membership grade represent the fuzzy fitness of the each individual solution. In order to select the fittest member of the population, each individual is ranked based on the Pareto dominance criteria.

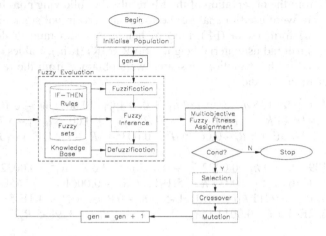

Fig. 2   Multi-Objective Fuzzy Fitness Assignment Algorithm

3   Case Application to Optimisation of Rod Rolling Shape

There are many problems in shape optimization where it is essential to have efficient search in the complex space in order to achieve optimal solution. It is difficult to select a functional form without some prior knowledge of shape statistics. It is therefore necessary to develop approximate approaches that utilise FE formulations and incorporate engineers reasoning qualitative knowledge for automatic generation of optimal rod profiles.

3.1   Experimental Procedure

A single roll pass was modelled using the ABAQUS finite element simulation software. The case study described in this paper deals with the shape optimisation of a single oval to round pass. The geometrical parameters relevant to the present study that affects shape in the rod rolling process can be categorised as: (a) initial thickness (h_1), (b) final thickness (h_2), (c) initial width (w_1), (d) final width (w_2), (e) work roll radius (R_r), and (f) roll gap (R_g). Rolls and stock were drawn as 2D sections using Patran pre-processor software and extruded into 3D. FE analysis was performed using ABAQUS Explicit software. Stock deformation was assumed symmetrical across horizontal and vertical planes through the stock centre therefore a quarter symmetry of the stock was modelled. A design of experiment technique was used to generate the input conditions for the FE runs [9, 10] and 30 runs of different input configurations were proposed for the investigation. In order to approximate the behaviour of the rod shape from the FE runs the input conditions as well as the engineer's interpretation of the FE outputs were used to create the fuzzy sets and the fuzzy rules. This approach was adopted because the experts understood the rod deformation more in terms of these responses than the first order variables. This also allows for reduced number of fuzzy rules. From the observation of the FE results the following non-dimensional design parameters were used to analyse the rod shape: (a) initial stock area/roll pocket area (SAR), (b) form factor (FF): ratio of height to width ratio of deformed stock (h_2/w_2) and (c) roll radius/material height ratio(RRMR) (R/h_1). Values of these design parameters used in the evolutionary search were obtained from the response models shown in equation 1, 2 and 3.

$$SAR = -1.976 + 0.1106\,h_1 - 0.00157\,h_1^2 + 0.184\,w_1 - 0.0012\,w_1^2 - 0.104\,P_r + \qquad (1)$$
$$0.0025\,P_r^2 + 0.0046\,R_r - 2.708\text{E-6}\,R_r^2 + 11.24\text{E-6}\,T + 0.0026\,h_1w_1 - 5.728\text{E-4}$$
$$h_1P_r - 1.0455\text{E-4}\,h_1R_r - 0.002\,w_1P_r - 0.0036\,w_1R_g - 1.207\text{E-4}\,w_1T$$

$$FF = 11.109 - 0.190\,h_1 + 0.0022\,h_1^2 - 0.525\,w_1 + 0.0077\,w_1^2 + 0.0022\,P_r^2 + \qquad (2)$$
$$0.176\,R_g + 0.00561\,R_g^2 - 0.0035\,R_r + 5.191\text{E-6}\,R_r^2 - 0.0061\,T + 9.765\text{E-7}\,T^2 -$$
$$8.722\text{E-4}\,h_1w_1 + 0.0011\,h_1P_r + 7.015\text{E-5}\,h_1R_r - 0.0061\,w_1R_g - 4.1\text{E-5}\,w_1R_g +$$
$$2.532\text{E-4}w_1T - 0.0011\,P_rR_g - 1.695\text{E-4}\,R_gR_r + 4.319\text{E-5}\,R_gT$$

$$RRMR = 6.155 - 0.375h_1 + 0.0056h_1^2 + 0.061R_r + 5.877\text{E-5}\,h_1R_g - 9.319\text{E-4} \qquad (3)$$
$$h_1R_r - 1.267\text{E-5}\,R_gR_r$$

3.2 Fuzzy Sets and Rules

The membership function for each fuzzy variable shows the degree of membership of each value in the variable's fuzzy sets for the range of interest. For example, Figure 3a shows the membership function for stock area (SAR). At a value of 0.77, the degree of membership is 0.375 in the fuzzy set 'Low', a degree of membership 0.1 in the fuzzy set 'Average', and a degree of membership 0 in the fuzzy set 'High'. The membership functions for the SAR, FF and RRMR (Figure 3a, 3b, and 3c respectively) were developed using the sampling space from the design of experiment technique and the fuzzy sets to facilitate the rule development process. The membership function of the output fuzzy set (roundness (Figure 3d)) was developed using the engineer's interpretation of the FE outputs. The FE plot of the rod profiles were classified into 5 sets: *Elliptical (E), Fairly Elliptical (FE), Flat Round (FLTR), Fairly Round (FR), and Round (R)* as shown in Figure 4. The classified profiles were then used to develop the roundness fuzzy set. The number of fuzzy sets for the roundness was agreed by the engineers in order to incorporate their reasoning process.

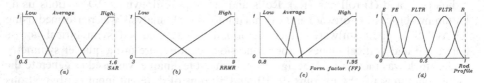

Fig. 3   Membership Functions for Rolling Variables

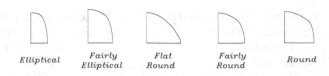

| Elliptical | Fairly Elliptical | Flat Round | Fairly Round | Round |

Fig. 4   Classification of FE Rod Shape Profiles

A rule base that specifies the qualitative relationship between the output parameter (roundness) and the input parameters: initial stock area/roll area (SAR), form factor (FF) and the roll radius/material height ratio (RRMR) is formulated in Table 1. Theses rules were developed by interactive interview with the domain experts. For example rule 1 shows that, if the area ratio is *average*, the RRMR is *low*, and the form factor is *low* then rod profile is predicted as *'round'*.

Table 1: Fuzzy decision table for Roundness

Rule no	SAR	RRMR	FF	Rod profile
1	Ave	L	L	R
2	H	H	L	FLTR
3	H	L	L	FLTR
4	Ave	H	H	FE
5	Ave	H	Ave	FR
6	L	L	H	E
7	Ave	L	H	E
8	L	H	H	E
9	Ave	H	L	R

The compensatory weighted mean operator was used to aggregate the fuzzy sets in the antecedent part of the rule. This also ensures that the cumulative effect of the other rules influences the determination of the rod profile. These fuzzy sets are then converted into a scalar value using the centroid method of defuzzification in the final step of the fuzzy inference cycle. This value was used as the first objective fitness value and the associated membership function value was used as second fitness value in the multi-objective optimisation problem.

3.3 Test Results and Discussion

FUMOO was used to maximise both the defuzzified domain value and the corresponding membership grade. Experiments were carried out using a simple GA for a single objective domain value and the FUMOO on both objectives to illustrate how the FUMOO deals with the deception problem. Results were obtained using the fuzzy evaluation rules shown in Table 1 for both approaches. The performances of the simple GA and FUMOO for different values of crossover and mutation probabilities were first investigated. Ten independent GA runs were performed in each case using a different random initial population. Results are reported for the following parameters: population size of 100 over 300 generations with single point crossover probability of 0.9, mutation probability of 0.02, and tournament selection with size 3. In most of the cases examined seven out of ten runs obtained similar results. Figure 5 shows a plot of the search space for the rod profile problem. Figure 5a shows a plot obtained by performing a random search with 5000 points. During the initial study, the simple GA identified (0.92, 0.58) as the best solution for the defuzzified domain value. This solution shows a good roundness value of 0.92 but the membership grade is rather low at 0.58. This is not acceptable from engineering point of view. In order to obtain good designs in terms of roundness value and membership grade, FUMOO is proposed.

The proposed FUMOO approach uses NSGAII as discussed in section 2. Figure 5b shows the best convergence obtained from the NSGAII result. Selected solutions are shown Table 2 and their variable values for points 1, 2, and 3 indicated in Figure 5b. Results in Table 2 show that the approach can identify design with good roundness value as well as membership grade.

     (a) Qualitative Search Space   (b) NSGAII Solution Plot

Fig. 5   Rod Profile Search Space

Table 2: FUMMO Solutions

Run	Height	Width	Pass Radius	Roll gap	Roll Radius	Temp	Roundness	Membership Grade
1	29.56	21.32	30.1	2.643	260.1	1071	0.7461	0.9902
2	30.12	21.32	30.75	2.653	260.1	1065	0.7813	0.8297
3	29.48	21.32	30.75	2.693	259.8	893.2	0.9219	0.7721

4   Conclusions

Current fuzzy fitness methods can cause deception for genetic algorithms which can result to unrealistic solutions. This paper presents a novel approach to deal with fuzzy fitness evaluation within multi-objective optimisation framework to address this deception problem. The proposed technique was applied to a rod rolling shape evaluation problem. Results obtained demonstrate that the proposed technique can be used to identify optimal solutions.

Acknowledgements

The authors wish to acknowledge the support of the Engineering and Physical Sciences Research Council (EPSRC) and Corus R, D and T, STC, UK.

Reference

1. Shivpuri, R. and Kini, S. *Application of Fuzzy Reasoning Techniques for Roll Pass Design Optimisation, Proceedings of the 39th Mechanical Working and Steel Processing Conference and Symposium on New Metal Forming Processes.* 1998. p. 755-763.
2. Hur, Y. and Rhee, D. *Application of fuzzy logic for automatic shape control in stainless steel rolling process, Joint 9th IFSA World Congress and 20th NAFIPS International Conference.* 2001. p. 251 -256.
3. Pataro, C., D, M, and Helman, H. *Direct determination of sequences of passes for the strip rolling process by means of fuzzy logic rules, Proceedings of the Second International Conference on Intelligent Processing and Manufacturing of Materials. IPMM '99.* 1999. p. 549 -554.
4. Shibata, T. and Fukuda, T. *Intelligent motion planning by genetic algorithm with fuzzy critic, Proceedings of the 1993 Intelligent Symposium on Intelligent Control.* 1993. Chicago, IIIinois. p. 565-570.
5. Koskimaki, E. and Goos, J. *Fuzzy Fitness Function for Electric machine Design by Genetic Algorithm, Proceedings of the Second Nordic Workshop on Genetic Algorithms and Their Applications (2NWGA).* 1996. Vaasa, Finland. p. 237-244.
6. Bhandari, D., Pal, S.K., and Kundu, M.K. *Image enhancement incorporating fuzzy fitness function in genetic algorithms, Proceedings of the Second International Conference on Fuzzy Systems,.* 1993. San Francisco, U.S.A. p. 1408-1413.
7. Roy, R., *Adaptive Search and the Preliminary Design of Gas Turbine Blade Cooling System,* PhD Thesis, 1997, University of Plymouth: Plymouth.
8. Deb, K., Agrawal, S., Pratap, A., and Meyarivan, T. *A Fast Elitist Non-dominated Sorting Genetic Algorithm for Multi-objective Optimization: NSGA-II, Parallel Problem Solving from Nature VI (PPSN-VI).* 2000. p. 849-858.
9. Oduguwa, V. and Roy, R. *Multi-Objective Optimisation of Rolling Rod Product Design using Meta-Modelling Approach, Proceedings of the Genetic and Evolutionary Computation Conference (GECCO-2002).* 2002. New York: Morgan Kaufmann Publishers, San Francisco, USA. p. 1164-1171.
10. Oduguwa, V., Roy, R., Watts, G., Wadsworth, J., and Farrugia, D. *Comparison of Low-Cost Response Surface Models for Rod Rolling Design Problems, 18th National Conference on Manufacturing Research.* 2002. Leeds Metropolitan University, UK: Professional Engineering Publishing. p. 209-213.

A Fuzzy-Based Meta-model for Reasoning about the Number of Software Defects

Marek Reformat

University of Alberta, Edmonton, Canada,
reform@ee.ualberta.ca,
http://www.ee.ualberta.ca/~reform

Abstract. Software maintenance engineers need tools to support their work. To make such tools relevant, the tools should provide engineers with quantitative input, as well as the knowledge needed to understand those factors influencing maintenance activities. This paper proposes a comprehensive meta-model for the prediction of a number of defects; it dwells on evidence theory and a number of fuzzy-based models developed using different techniques applied to different subsets of data. Evidence theory and belief function values assigned to generated models are used for reasoning purposes. The study comprises a detailed case for estimating the number of defects in a medical imaging system.

1 Introduction

Software maintenance tasks account for more than half of the typical software budget [12]. In order to make these activities more efficient, software engineers need to understand relationships between software attributes and maintenance efforts.

Corrective software maintenance is associated with those tasks related to the elimination of software defects. One of the most important aspects of such maintenance is related to the prediction of a number of defects which can be found in a given software system. However, besides a quantitative prediction, managers should also be provided with knowledge of those factors which influence the number of defects, and with confidence levels associated with obtained results and acquired knowledge.

This paper proposes a meta-model, which addresses these needs. Prediction capabilities and extraction of knowledge are two important objectives of the proposed system. A number of fuzzy-based models are built using different subsets of the data. This results in a wide range of representations of relationships existing among data attributes. This variety is exploited by the proposed system, where the outputs of models are combined using elements of evidence theory [9].

2 Rule-Based Models

Rule-based models receive information describing a situation, process that information using a set of rules, and produce a specific response as their result. In

T. Bilgiç et al. (Eds.): IFSA 2003, LNAI 2715, pp. 644–651, 2003.
© Springer-Verlag Berlin Heidelberg 2003

its simplest form, a rule-based model is just a set of IF-THEN statements called rules. Each rule consists of an IF part, called the antecedent, and a THEN part, called the consequent.

In the case when antecedent and consequent parts of IF-THEN rules contain labels of fuzzy sets characterized by appropriate membership functions, the rules are called fuzzy IF-THEN rules. Fuzzy IF-THEN rules are often employed to capture imprecise information. At the same time they can also be used for prediction and estimation purposes.

Rules can be generated on the basis of expert knowledge, or automatically derived from available data using a variety of different approaches and tools. In the paper, fuzzy IF-THEN rules are extracted from fuzzy-neural networks (FNN) [7], which are developed using an evolutionary-based approach. Genetic Algorithms [4] are used to perform a structural optimization of FNN [8]. The FNN is constructed by detecting the most essential connections that shape its architecture. The weights of connections are binary (0-1) for the purpose of this optimization. By using this method, the effort is put on determining the most essential structure of the model. The structure of the network directly identifies the rules.

In the area of software engineering, prediction models supporting maintenance activities have been built for many years. Reliability models predicting a number of defects [3] or fault correction efforts [2] are the most popular models. Complexity and size metrics, testing metrics and process quality data are utilized as predictor variables. A number of techniques and methods, such as regression analysis, neural networks and pattern recognition, to name but a few [5], are applied to build these models. Work has been also reported using rule-based approaches in order to build prediction models in the area of software engineering [1], as well as Bayesian Belief Networks [3].

3   Fuzzy-Based Meta-model for Prediction

3.1   Concept

Fuzzy-based models, developed using different methods and data subsets, represent different relationships existing among data attributes. The approach proposed in this paper takes advantage of that by combining many differently derived models into a single prediction system. The purpose of the approach is to:

- use a number of different fuzzy rule-based models developed by different methods, using different data subsets;
- use a concept of belief functions from evidence theory [9] [11] to represent beliefs in "goodness" of the models;
- use the Transferable Belief Model [10], built on evidence theory, and the beliefs to relate a new data point to a specific category and to provide a certain level of confidence about the result.

One of the main elements of evidence theory is the concept of a basic belief mass (bbm), which represents a degree of belief that something is true. This concept is an essential element of the proposed system. *Bbm* values are assigned to each output of a model. This means that if a model output is activated by a given data point then there is a belief equal to *bbm* value that this data point belongs to a category indicated by the output. At the same time, the belief value $1 - bbm$ is assigned to a statement: it is not known to which category the data point belongs. The *bbm* values of all outputs of models activated by a data point, together with categories identified by these outputs, constitute the input to an inference engine (section 3.3).

3.2   Generation and Validation of Fuzzy Models

The development stage of a meta-model based system is about the construction of fuzzy-based models and assigning belief values to them. Each model is "checked" against all training data points and bbm_T values are generated. They are indicators of "goodness" of outputs of the models. The formula (the Laplace ratio) used for calculations of *bbm* for every output of each model is:

$$bbm_T = \frac{Number\_Classified_T + 1}{Total\_Number\_Points_T + 2}$$

where $Number\_Classified_T$ represents a number of training data points classified properly for a given category, and $Total\_Number\_Points_T$ represents a total number of training data points belonging for this category.

Another set of data, called validation data, is used to "validate goodness" of all models. Bbm_V (where V stands for validation) values are calculated:

$$bbm_V = \frac{Number\_Classified_V + 1}{Total\_Number\_Points_V + 2}$$

Following that, a combination of both *bbm* values is calculated. For this purpose a special formula is proposed which preserves the value of bbm_T in the case when bbm_V is equal to 0.5 (a value of 0.5 indicates 50-50 chances that something is either true or false). The formula is presented below:

$$bbm_{UPDATE} = \frac{bbm_T * bbm_V}{bbm_T * bbm_V + (1 - bbm_T)(1 - bbm_V)}$$

3.3   Inference Engine

The proposed system contains an inference engine and all generated models together with their *bbm* values. The inference engine is an implementation of the following concept: when a new data point activates a number of outputs of models, *bbm* values are "generated". These values indicate the relation of this data point either to the same or to a number of different categories. In the case when a single category is identified by all outputs, the inference engine is not engaged - it is predicted that the data point belongs to the identified category.

In the case when a number of different categories are identified, the Transferable Belief Model [10] is used to derive possibilities that the data point belongs to each of these categories.

Example: Two fuzzy based models have been built to represent relationships between an n-dimensional input and a single-dimensional output with two categories *small* and *large*. *Bbm* values of model outputs are shown in Table 1. When a new

Table 1. Bbm values of model outputs

model output		bbm
first	*small*	0.85
	large	0.75
second	*small*	0.75
	large	0.98

data point is checked against the models, it activates the following outputs: *small* from the first model, *small* and *large* from the second. *Bbm* values associated with the outputs are: 0.85, 0.75 and 0.98. Based on this information Table 2 is created. These

Table 2. Basic bbm

possible outcome	1st model "small"	2nd model "small"	2nd model "large"
0	0	0	0
small	0.85	0.75	0
large	0	0	0.98
small or *large*	0.15	0.25	0.02

values represent input to the TBM. Inside the TBM, the *bbm* values are combined

Table 3. Basic belief masses assigned to each possible output

m(0)	0.94325
m(*small*)	0.01925
m(*large*)	0.03675
m(*small, large*)	0.00075

using the conjunctive combination rule [10]. The results are shown in Table 3.

The next step is to calculate pignistic probabilities. One represents the belief that the data point belongs to the category *small*, another that it belongs to the category *large*. This pignistic probability BetP on the belonging of the data point to one of the two categories is computed from the *bbm* values according to the TBM [10]:

$$BetP(samll) = \frac{m(small) + \frac{m(small,large)}{2}}{1 - m(0)} = 0.3458$$

$$BetP(large) = \frac{m(large) + \frac{m(small",m(large)}{2}}{1 - m(0)} = 0.6542$$

Based on that, it can be said that the new data point belongs to the category *small* with belief of 0.3458, and to the category *large* with belief of 0.6542.

4 Experimental Setup

4.1 Data

The Medical Imaging System (MIS) is a commercial software system consisting of approximately 4500 routines. It is written in Pascal, FORTRAN and PL/M assembly code. In total, there are about 400,000 lines of code. The MIS data set [6] contains a description of 390 software modules which comprise the system. The description is done in terms of 11 software measures (lines of code – LOC, lines of code without comments – CL, number of characters – TCHAR, number of comments – TCOMM, number of comment characters, – MCHAR, number of code characters – DCHAR, program length – N, estimated program length – NE, Jensen's program length - NF, McCage's cyclomatic number - VG, Belady's bandwidth metric – BW) and the number of changes which are made to each module, due to faults discovered during system testing and maintenance.

4.2 Genetic-Based Construction of Models

A standard version of a Genetic Algorithm is applied. A binary string represents the structure of the network. The adjustment of connections is driven by a fitness function, which is one of the objectives of an optimization process. The role of the fitness function is to assess how well the model matches the training data. Three fuzzy sets are defined in the input and the output space: *small*, *medium* and *large*. A network has a number of outputs equal to the number of fuzzy sets in the output space. The optimization process is conducted with limits regarding a number of neurons in a hidden layer (the maximum is five) and a number of inputs to each neuron of a hidden layer (the maximum is five). Three different fitness functions are used as follows:

$$Fit\_Fun_A = \frac{1}{1+Q}$$

$$Fit\_Fun_B = \Pi_{i=1}^{m} \frac{1 + Number\_Classified_i}{Number\_Points_i}$$

$$Fit\_Fun_C = Fit\_Fun_A * Fit\_Fun_B$$

where m is a number of outputs, $Number\_Classified_i$ is the number of properly predicted data points of category i ("properly" means that an activated output matches the fuzzy set label of a training data point), $Number\_Points_i$ represent the total number of data points in category i, Q is a commonly used performance index assuming the form

$$Q = \sum_{k=1}^{N} (\mathbf{F}(k) - \hat{\mathbf{F}})^T (\mathbf{F}(k) - \hat{\mathbf{F}})$$

with N as a number of training data points, $F_i(k)$ as a value of the output i for given training data point k obtained from the model, $\hat{F}_i(k)$ - original value of the output i for given training data point k, and $\mathbf{F}(k) = [F_1(k), F_2(k), ..., F_m(k)]^T$.

5   Number of Software Defects

5.1   Development and Validation of Fuzzy Models

The original data is split into three sets: training set (60 per cent of data points), validation set (20 per cent), and testing set (20 per cent). For the experiment, the testing set does not change. However, a new split into training and validation sets is generated every time a new model is built. Overall, the Genetic Algorithm is used with three different fitness functions and three different splits; this results in nine models. Each output of each model is then "checked" against training data and bbm_T values ar assigned to models' outputs.

In the next step, the models are "checked" against data points from the validation sets. New belief values bbm_V are calculated for each output of each model, and are used to update bbm_T values. The result of this process, bbm_{UPDATE} values, is presented in Table 4.

Table 4. Bbm values of model outputs

model	category: **small**		category: **medium**		category: **large**	
	bbm_T	bbm_{UPDATE}	bbm_T	bbm_{UPDATE}	bbm_T	bbm_{UPDATE}
A_1	0.9077	0.9917	0.1714	0.1145	0.6000	0.3333
A_2	0.9171	**0.9943**	0.1944	0.0215	0.1818	0.1000
A_3	0.9091	0.9896	0.2903	0.0806	0.5455	0.3750
B_1	0.6869	0.8558	0.4688	0.1692	0.8000	**0.9231**
B_2	0.7644	0.9036	0.6923	0.3913	0.6000	0.3333
B_3	0.7772	0.8865	0.4444	0.0678	0.9091	0.8333
C_1	0.7208	0.8771	0.5758	0.4043	0.7000	0.7000
C_2	0.7538	0.8757	0.5641	**0.9119**	0.8333	0.6250
C_3	0.8010	0.9468	0.4242	0.3294	0.7273	0.5714

5.2   Analysis of IF-THEN Rules

In the case of the category *small*, outputs of models developed using $FitFun_A$ have the largest values of bbm. Their values have increased after validation. This shows that the rules extracted from these three models are good candidates for prediction purposes. The rule with the highest value of bbm (bold fonts in Table 4) is presented below:

> **if**   LENGTH_OF_CODE_WITH_COMMENTS is small (LOC_small) &
> LENGTH_OF_CODE_NO_COMMENTS is medium (CL_medium)
> **or**
> LENGTH_OF_CODE_NO_COMMENTS is small (CL_small) &
> NUMBER_OF_COMMENTS is small ($TCOMM\_small$)
> **then**
> **NUMBER_OF_CHANGES is small**

For the categories *medium* and *large*, the situation looks quite different. In this case, only a single rule in each category increased its *bbm* value. All other rules have smaller values of *bbm* after validation. There are two possible explanations either: there are no clear relationships among *medium* and *large* number of defects and data attributes; or there are not enough data points in both categories (the histogram has indicated that 82 per cent of data points belong to a single category *small*). The best rule for the category *medium* is:

if   NUMBER_OF_COMMENTS is small (*TCOMM_small*) &
        JENSEN'S_PROGRAM_LENGTH is medium (*NF_medium*)
or
        NUMBER_COMMENT_CHARACTERS is small (*MCHAR_small*) &
        JENSEN'S_PROGRAM_LENGTH is medium (*NF_medium*)
or
        NUMBER_COMMENT_CHARACTERS is medium (*MCHAR_medium*) &
        ESTIMATED_PROGRAM_LENGTH is large (*NE_large*)
then
        NUMBER_OF_CHANGES is medium

In the case of the category *large* the rule is:

if   NO_CHARACTERS is medium (*TCHAR_medium*) &
        JENSEN'S_PROGRAM_LENGTH is large (*NF_large*)
or
        NUMBER_OF_COMMENTS is medium (*TCOMM_medium*) &
        BELADY'S_BANDWIDTH is small (*BW_small*)
then
        NUMBER_OF_CHANGES is large

5.3   Predicting the Number of Defects

The constructed meta-model estimation system is used to predict a number of defects for testing data (see section 5.1). Two aspects of the results are important - classification rate and confidence levels assigned to obtained classifications. The testing set contains 78 data points. Classification rates for the nine models are in the range from 62.82 per cent for model B_-1 (49 data points properly classified) to the maximum value of 79.49 per cent for all models A (62 properly classified data points). Application of the meta-model prediction system increased the classification rate to 85.90 per cent, which translates to 67 proper classifications.

Not all testing points are unequivocally classified to a single category. Many *medium* and *large* points activate outputs of different categories. In such cases, the TBM is involved. An interesting case is related to a data point, of category *medium*, which has activated several *medium* and *large* outputs. The system has indicated that the point belongs to the category *medium* with belief 0.4219, and to the category *large* with belief 0.5756. As can be seen, there is no "winning" category. A numerical comparison of beliefs indicates that the point should belong to the category *large*, but at the same time similar values of beliefs point to a very low confidence in the prediction.

6 Conclusions

In this study, a new concept is put forward for analyzing software engineering data. This concept applies elements of evidence theory to a combination of fuzzy rule-based models, in order to create an estimation system, which then has the capability to determine confidence levels of the generated predictions. Application of a number of rule-based models developed with different techniques enhances prediction capabilities and allows more comprehensive extraction of knowledge from the data. In this paper, attention has been directed to the prediction of the number of defects in a given software system. Estimation systems, extracted knowledge and results of prediction processes have been described and investigated.

7 Acknowledgment

The author would like to acknowledge the support of the Alberta Software Engineering Research Consortium (ASERC) and the Natural Sciences and Engineering Research Council of Canada (NSERC).

References

1. Chatzoglou, P.D. and Macaulay L.A.: A Rule-Based Approach to Developing Software Development Prediction Models. Automated Software Eng. **5** (1998) 211-243
2. Evanco, W. M.: Prediction Models for Software Fault Correction Effort. Fifth Conf. on Software Maintenance and Reengineering, Lisbon, Portugal (2001)
3. Fenton, N.E., Neil, M.: A Critique of Software Defect Prediction Models, IEEE Trans. on Software Eng. **25** (1999) 675-689
4. Goldberg, D. E.: Genetic Algorithms in Search, Optimization, and Machine Learning, Addison-Wesley, Reading (1989)
5. Jorgensen, M.: Experience with the accuracy of Software Maintenance Task Effort Prediction Models. IEEE Trans. of Software Eng. **21** (1995) 674-681.
6. Munson, J.C. and Khoshgoftaar, T.M.: Software metrics for reliability assessment, In: M.R. Lyu (ed.) Software Reliability Engineering, Computer Society Press, Los Alamitos (1996) 493-529
7. Pedrycz, W. and Gomide, F.: An Introduction to Fuzzy Sets: Analysis and Design. MIT Press (1998)
8. Reformat M., Pedrycz, W., Pizzi, N.: Building a Software Experience Factory using Granular-based Models, to appear in Fuzzy Sets and Systems
9. Shafer, G.. A mathematical Theory of Evidence. Princeton University Press (1976)
10. Smets, Ph. and Kennes, R.: The Transferable Belief Model. Artificial Intelligence, **66** (1994) 191-234
11. Smets, Ph.: Belief Functions. In Non Standard Logics for Automated Reasoning, ed. Smets Ph., Mamdani A., Dubois D. and Prade H. Academic Press, London (1988) 253-286.
12. Smith, D.: Designing maintainable software. New York, Springer (1999)

A Dual Representation of Uncertain Dynamic Spatial Information

Gloria Bordogna[1], Paola Carrara [2] , Sergio Chiesa [1], Stefano Spaccapietra [3]

[1] CNR-IDPA, sezione di Milano, p.zza Cittadella 4, 24129 Bergamo Italy.
gloria.bordogna@idpa.cnr.it
[2] CNR-IREA, sezione di Milano, via Bassini 15, 20131 Milano, Italy.
paola.carrara@irea.cnr.it
[2] Database Lab., Swiss Federal Institute of Technology (EPFL), 1015 Lausanne, Switzerland
stefano.spaccapietra@epfl.ch

Abstract. A dual representation of spatio-temporal phenomena whose observation is affected by uncertainty is proposed in the context of fuzzy set and possibility theory. The concept of fuzzy time validity of a snapshot of a dynamic phenomenon is introduced as well as the concept of possible spatial reference of the phenomenon at a given time instant. Then a mechanism to generate virtual snapshots of the phenomenon, showing its possible spatial reference at consecutive time instants, is described.

1. Introduction

The last decade has seen a blooming of research on spatio-temporal databases, aiming at filling the gap between application requirements and GIS functionality in terms of representation and retrieval of dynamic spatial entities, i.e. spatial entities for which the application requires recording their evolution in space [1-4]. Spatial evolution refers to a change in the spatial position and/or a change of the shape of the entity.

Temporal data modeling in GISs has been approached by defining snapshot models, by introducing time-space compound attributes and spatial objects belonging to time sequences [1]. More recently, unified spatio-temporal models have been proposed in which entities may have both a spatial and a temporal extent [5, 6], the two extents being dealt with as orthogonal dimensions. The temporal extent denotes the period of time in which the observation of a dynamic phenomenon is valid, i.e., the period of time in which the spatial reference of an entity can be considered constant (stable). However, these more powerful approaches did not address the need for representing the uncertainty that may affect spatio-temporal information. Yet temporal uncertainty, for instance, arises in many real applications where temporal knowledge cannot be precisely captured, either because of the scarcity of the observations (so that it is impossible to state when exactly the change occurred), or because the timing of the observation is not exactly known (e.g., a human may observe an event without exactly recording the instant it happened and use therefore uncertain expressions such as *"the accident arrived around 2 pm"*), or because the continuously changing nature of the observed phenomenon makes it impossible to associate a precise time to a precise value (e.g., while observing the

T. Bilgiç et al. (Eds.): IFSA 2003, LNAI 2715, pp. 652-659, 2003.

temperature in a furnace it may not be possible to state exactly at what time temperature has reached 350° Celsius).

The analysis of uncertainty in spatial data has ancient roots [7,8] and the literature on GISs has produced several proposals for managing uncertain spatial information in the context of fuzzy set theory [9,10]. A parallel approach addressing the modeling of uncertain temporal information may be found in [11,12,13]. Only a few proposals aim at managing uncertainty in spatio-temporal information [14,15,16].

In this contribution we propose a fuzzy model for spatial dynamic phenomena that provides a dual representation of uncertain spatio-temporal information dealt with as orthogonal concepts. We start from the idea that dynamic phenomena are generally represented by a sequence of observations (snapshots), each one describing the system status at a precisely known time instant. This time instant is assumed as indicating the time in which the observation can be considered a valid description of the phenomenon. However, since the transition from a given status to another one is generally smooth and occurs with a set of micro variations over a period of time, we try to model this transition by associating with each snapshot a fuzzy interval of validity [12,13,17]. In this way we can have several instants of time associated with a single snapshot, each one having its distinct degree of validity.

An alternative view of a spatio-temporal phenomenon is to account for the lack of knowledge on the spatial reference by associating with each precise time interval or instant an indeterminate snapshot synthesizing the multitudes of the phenomenon's spatial changes occurring during the precise period of time or the possible position at the given time instant [18]. In this way, the degree associated with each spatial grain, i.e., pixel or primitive geometric entity such as polygons, lines etc., states its possibility to be interested by the phenomenon during the whole period of time. These alternative views of an uncertain spatio-temporal phenomenon can be seen as special cases of a more general representation in which both the spatial and the temporal information are represented by means of validity intervals.

An important facility that a spatio-temporal database should provide is the possibility to follow the spatial evolution of a phenomenon over a period of time. This ability, known as temporal navigation or time-travel facility [1,19], is generally provided by showing snapshots corresponding to consecutive time instants. Based on the dual representation we have defined in this paper, we will describe a novel method to realize this facility. The method generates virtual snapshots of the phenomenon that correspond with instants of time in between those of consecutive actual snapshots. The finer is the granularity of the time information the smoother are the changes of the spatial reference, then the better is the description of the phenomenon evolution. By taking into account the uncertainty we achieve a finer timestamping of the phenomenon dynamics. In this paper, section 2 carries an analysis of uncertainty in spatio-temporal information and introduces the dual representation. In section 3 a formalization of the representation of dynamic spatial phenomena is proposed and the evaluation of queries imposing constraints on either the temporal reference or the spatial reference is formalized. Section 4 illustrates a method to realize the time travel facility based on the dual representation.

2. Analysis of Uncertain Spatio-temporal Information

Spatio-temporal information can be affected by uncertainty due to several reasons. In this section we analyze some characteristics of this uncertainty and highlight some relations existing between uncertainty in spatial information and in temporal information.

Spatial *uncertainty* expresses vague and incomplete knowledge on the phenomenon resulting from several sources such as the observation, the phenomenon itself [9,18,20], or its representation.

The introduction of temporal information as a new dimension in the representation of a spatial phenomenon may lead to a further level of uncertainty, depending on the semantics of the temporal information. Basically, there are two semantics for the time variable associated with spatial or non-spatial information [1]. The most common semantics is to consider that the temporal specifications associated to an observation of a phenomenon record the evolution of the phenomenon over a period of time (by timestamping snapshots of the phenomenon) as observed in the real world from the application perspective. This semantic interpretation is known as valid time, as opposed to transaction time, which is used to record when an observation is entered into or deleted from the database. Values for valid time are determined by the application. Values for transaction time are given by the computer system. Attribute values, relationships and objects in a database can have a period of validity associated with them. An additional semantic option is to allow representation of time relationships between phenomena [19].

Uncertainty relative to the time validity of a phenomenon may occur in the case in which there is not a precise knowledge of the instant of observation of the phenomenon, and we can just specify it approximately (*temporal indeterminacy*). This may be caused by the granularity of the time information. A way to represent this information is by means of a fuzzy subset on the time reference domain interpreted as a possibility distribution [12,13,17]. A situation of *temporal vagueness* occurs whenever the definition of the period of validity for the observation of a phenomenon is hampered by heuristic knowledge of the phenomenon dynamics [13]. An example can be the validity period for the minimum surface extent of glaciers detected using satellite images, as there may not be an image available taken at exactly the time the event occurred.

Uncertainty relative to the time relationships between phenomena may occur when one is unable to specify a total order between events or in the case in which one is only able to imprecisely qualify the extent of the time relationship between events. An example of the latter is the specification that an event occurred *long before*, *very long before*, etc., another event.

If we adopt an object-based approach to represent spatio-temporal phenomena, as in [5], uncertainty modeling will rely on the definition of a set of Abstract Spatio-Temporal Data Types for representing phenomena affected by distinct combinations of vagueness and/or uncertainty on the spatio-temporal reference. Nevertheless, for many real applications some scenarios can be disregarded based on the assumption that in these applications indeterminacy of the spatial information is inversely related to the uncertainty of the temporal information. We can then specify either an exact spatial information and be vague on its time validity, or we can have indeterminacy on the spatial information and be precise on the time validity. For example, I can say that I know precisely the position of a taxicab on a map but vague on the time validity corresponding to that position: "the taxi is at the central station *around 12.00*". Alternatively I can be precise on the time information but vague on the spatial reference and say: "the taxi is *around* the Central Station at 12.00". This inverse relationship suggests a dual representation for uncertain spatio-temporal phenomena: on the one side a phenomenon can be represented by precise spatial reference information associated with an indeterminate or vague validity interval. On the other side, it can be represented by summarizing its numerous micro-variations occurring on the spatial domain in a precise time interval through an indeterminate or vague spatial reference [18]. These alternative representations are subsumed by a more general one in which uncertainty is modeled on both the temporal and the spatial reference: for example to represent the information "the taxi is *close* to the Central Station *around 12*". In the next section we will formally introduce this dual representation for uncertain dynamic spatial information.

3. A Fuzzy Dual Representation of Uncertain Dynamic Spatial Information

In this section we formalize the dual representation for dynamic spatial phenomena that is suited to manage the uncertainty relative to either the temporal or the spatial reference. In this context we adopt an object-based representation, i.e. a dynamic spatial object is an entity that changes its spatial reference in time. We first introduce the representation in which we have precise spatial reference and indeterminate or vague time reference. Generally the information that we have on spatio-temporal phenomena consists of sequences of snapshots, maps of the spatial reference of the entity taken at specific instants of time: o_1 at t_1, o_2 at t_2, o_3 at t_3,..., o_n at t_n in which $t_1 < t_2 < ... < t_n$ are instants of time defined on a temporal domain T, and o_i for $i=1,...n$ are spatial objects, i.e., instances of some abstract spatial data type defined on a spatial domain X. From this information we can guess that at each instant of time $t \in [t_{i-1}, t_i]$ between two consecutive time snapshots t_{i-1} and t_i the spatial reference of the dynamic object is "in between" the ones defined by the spatial objects o_{i-1} and o_i; the two spatial objects can be both *valid* to a distinct degree for representing the dynamic object at time t. In fact, when no further information on the phenomenon dynamics is available, one can guess that for t closer to t_{i-1} o_{i-1} would be the most likely spatial manifestation of the dynamic object, while for t closer to t_i, o_i would be the most preferred manifestation. Formally we can represent a dynamic spatial object o_d as a set of pairs (τ_i, o_i):

$$o_d := \{(\tau_1, o_1), (\tau_2, o_2), ..., (\tau_n, o_n)\} \qquad (1)$$

where τ_i is the *fuzzy time validity range* associated with the spatial object o_i. We can consider that o_i is a valid spatial manifestation of the dynamic object o_d during the time interval τ_i. τ_i is defined as a fuzzy subset of the time domain T as follows:

$$\tau_i := \int_{t_k \in T} \mu_i(t_k) / t_k$$

$\mu_i(t_k) \in [0,1]$ can be interpreted as the time validity of o_i at time instant t_k that is dealt with as a continuous function. In the case we do not have any knowledge on the dynamic laws governing the evolution of the phenomenon, we can define the semantics of τ_i by a triangular membership function centered in t_i, the precise instant of observation of o_i with left and right displacements $\mid t_i - t_{i-1} \mid$, $\mid t_{i+1} - t_i \mid$ respectively. This is the usual linear interpolation.

Based on this representation we can easily answer queries imposing temporal constraints on a dynamic phenomenon represented by o_d [13]. Since the phenomenon is represented with a temporal reference described by possibility distributions on T, we can apply one of the several approaches defined in the literature for the evaluation of soft conditions against fuzzy attribute values [23,24]. To answer the query *"What is the possible position of the phenomenon o_d at the precise time t?"* we compute the virtual snapshot $o_d(t)$ defined by a fuzzy subset of the spatial domain X as follows:

$$o_d(t) = \{\tau_1(t) \: / \: o_1, \tau_2(t) \: /o_2, ..., \tau_n(t) \: /o_n\} = \Sigma_{i=1,...,n}(\mu_i(t) \: / \: o_i) = \sigma \qquad (2)$$

We interpret the fuzzy set σ as the possible spatial reference of the phenomenon at time instant t, i.e., $\mu_i(t)$ is interpreted as the possibility that the phenomenon has position o_i at time instant t_i. Based on representation (1) other kinds of temporal constraints can be evaluated.

To answer queries specifying spatial constraints such as *"At what time o_d covers position o?"* we have to evaluate the spatial constraint $C=covers$ against the spatial reference $o_1, ..., o_n$ that is dealt with as a sparse information in this representation. In current GISs, the topological operators such as *covers* are defined as binary operators admitting only two degrees of satisfaction (*False* and *True*) [9]. In general, if we define *covers* as a soft constraint C: $X \times X \rightarrow [0,1]$ we can obtain a fuzzy set $\Sigma_{k=1,...,n}$ $C(o_k, o) / \tau_k$ as a result of its evaluation [25]. We can interpret τ_k as the possible fuzzy time instant (or the vague period) in which the phenomenon covers o to an extent $C(o_k, o) \in [0,1]$.

The dual representation of (1) is defined with a precise time reference, and a fuzzy spatial reference by a set of pairs (t_i, σ_i) :

$$\underline{o_d} := \{ \: (t_1, \sigma_1), (t_2, \sigma_2), ...(t_n, \sigma_n) \} \qquad (3)$$

in which σ_i is the spatial validity of the phenomenon at time instant t_i, or the possible spatial reference of the dynamic phenomenon at time instant t_i. σ_i is defined as a fuzzy subset of the whole spatial domain X as follows:

$$\sigma_i := \int_{o_i \in X} \pi_i(o_k) \: / \: o_k$$

in which the value $\pi_i(o_k) \in [0,1]$ can be interpreted as the possibility that the spatial object o_k is interested by the phenomenon at time instant t_i.

Based on the representation defined in (3) we can answer queries imposing spatial constraints on a dynamic phenomenon. For example, to answer the query *"At what time o_d is located inside region o?"* we compute the following fuzzy set defined on the time domain:

$$o_d(o) = \{ \: \Delta_1 \: / \: t_1, \Delta_2 \: / \: t_2, ... \Delta_n \: / \: t_n \} = \Sigma_{i=1,...,n}(\Delta_i \: / \: t_i) = \tau$$

in which $\Delta_i \in [0,1]$ is the degree of a fuzzy inclusion $\Delta_i = degree(\sigma_i \subseteq_f o)$ computed for example as $\Delta_i = Area(\sigma_i \cap o) / Area(\sigma_i)$ in which $(\sigma_i \cap o)$ is the fuzzy subset of the spatial domain obtained by the intersection of σ_i and o. τ is interpreted as the possible temporal reference of the phenomenon when its position is within o. On the basis of definition (3) other kinds of spatial constraints can be evaluated. : For example one could ask when the phenomenon is *north* of a region of interest, or *close* to a given point on the map.

Based on representation (3), to answer queries specifying a temporal constraint χ such as "*Where is o_d at (after, before, during etc) time t?*" we have to evaluate the constraint χ against the temporal reference of the phenomenon, i.e., $t_1,...,t_n$. If $\chi : T \times T \rightarrow [0,1]$ is a soft constraint we obtain as a result the fuzzy set $\Sigma_{k=1,...n}\chi(t_k,t)/\sigma_k$ in which σ_k can be interpreted as the possible location of the phenomenon when χ is satisfied to the extent $\chi(t_k, t) \in [0,1]$.

To convert the representation formalized in (1) into its dual representation defined in (3), a mapping *Dual:* $\wp(T) \times X \rightarrow T \times \wp(X)$, in which $\wp(A)$ denotes the powerset of fuzzy set on A, can be defined i.e., $Dual(o_d) = \{(t_j , \sigma_j)\}_{t_j \in T} = o_d$. This is obtained by computing $\sigma_j = o_d(t_j)\ \forall t_j \in T$ as defined in formula (2). It can be proved that $Dual(o_d)$ can be transformed back into representation (1) without loss of information by applying the inverse function of *Dual*. It can be noticed that the dual representations defined by (1) and (3) are subsumed by a more general one in which a dynamic spatial phenomenon is represented by a set of pairs (τ_i , σ_i) in which both the temporal reference τ_i and the spatial reference σ_i are fuzzy subsets of the temporal and spatial domain, respectively.

4. A Time Travel Facility for Dynamic Spatial Phenomena

In this section we address the problem of integrating the lack of information between a series of actual snapshots of a phenomenon. In fact, if the interval between two consecutive actual snapshots is not adequate the essential information about the dynamics changes remains undetected. On the basis of representation (1) a set of virtual snapshots $o_d(t_1)$, ..., $o_d(t_n)$ (as defined in (2)) can be generated in between two consecutive actual snapshots $o_d(t_A)$ and $o_d(t_B)$, thus achieving a finer timestamping of the spatial evolution of the phenomenon during an interval of time $[t_A,t_B]$. To map the virtual snapshots we adopt a raster representation of the spatial reference. We partition the range $[0,1]$ of the time validity degrees $\mu_i(t)$ into a set of k subintervals. Then we associate with each of these subsets a grey level through a function *map_grey level* so that intervals get a fainter grey as they are closer to 0 and a darker grey as they are closer to 1. The possible spatial reference of the phenomenon at a time instant t, $o_d(t)$, can then be graphically represented by a grey level region. Each pixel p satisfying the condition $p \subseteq (o_1 \cap ... \cap o_m)$ with $o_1, ...,o_m$ in $o_d(t)$ and $1 \leq m \leq n$ gets a grey level g that can be computed as the maximum of the membership degrees $\mu_1(t),.... ,\mu_m(t)$ associated with $o_1,...,o_m$: $g = map\_grey\ level(max(\mu_1(t), ... , \mu_m(t)))$. Alternatively a weighted average can be used to determine g in the case in which $o_d(t)$ is interpreted as the valid spatial reference of the phenomenon at time t. Figure 1 shows the frames of a temporal sequence of a dynamic phenomenon. At the top we have the sequence constituted by the six actual snapshots that are fully valid at time instants $t_1, ... t_6$. These snapshots are black and white images. At the bottom of figure 1 we have the generated sequence consisting of the actual and the virtual snapshots. The virtual snapshots are grey level images. It can be noticed that the variation of the spatial reference observable in the generated temporal sequence is smoother than the one in the original sequence of actual snapshots.

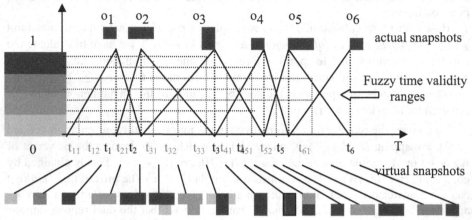

Fig. 1. Timestamping of a dynamic phenomenon through 10 virtual snapshots and 6 actual snapshots

5. Conclusions

The representation of dynamic spatial phenomena has to account for the uncertainty that generally characterizes observations. In this contribution, a dual representation of uncertain spatio-temporal information is proposed in the context of fuzzy set and possibility theory. This is a first step towards the definition of abstract spatio-temporal data types for representing dynamic spatial phenomena in an object-based framework. The time and space dimensions are seen as orthogonal dimensions and the two dual representations can be regarded as special views of a generalized fuzzy representation of spatio-temporal information. This special views may be used to index spatio-temporal data for efficient retrieval.

References

1. T. Abraham, J. F. Roddick, *Survey of Spatio-temporal databases,* Geoinformatica, 3(1), (1999), 61-99.
2. K.K. Al-Taha , R.T. Snodgrass, M.D. Soo, *Bibliography on spatio-temporal databases,* SIGMOD Record, 22 (1993), 56-67.
3. M.J.Egenhofer, R.J.Golledge, *Spatial and Temporal reasoning in geographic information systems,* Oxford University Press, (1998).

4.   Frank ed., *Theories and methods of spatio-temporal reasoning in geographic space*, Lecture Notes in Computer Science 639, Springer-Verlag, (1992).
5.   M.F. Worboys, *A Unified model for spatial and temporal information*, The Computer Journal, 37(1),(1994),27-34.
6.   Parent, S. Spaccapietra, E. Zimanyi, *Spatio-temporal conceptual models: data structures + space + time*, 7th ACM Sym. on Advances in GIS, Kansas City, Kansas, (1999).
7.   M.F. Goodchild, S.Gopal eds, *Accuracy of spatial databases*, Taylor and Francis, London, (1989)
8.   D.H.Maling, *Measurements from maps: principles and methods of cartometry*, Pergamon press, NY, (1989).
9.   P.A.Burrough, A.U.Frank eds, *Geographic Objects with Indeterminate Boundaries*, in GISDATA series, Taylor & Francis, (1996).
10. M.Codd, F.Petry, V.Robinson eds, *Uncertainty in Geographic Information Systems and Spatial data*, special issue of Fuzzy Sets and Systems, 113,(1), (2000), 1-159.
11. D.Dubois, J.Lang, E.Prade, *Timed possibilistic logic*, Fundamenta Informaticae, XV, 4, (1991), 211-234.
12. R. De Caluwe, G. De Tré, B. Van der Cruyssen, F. Devos, P. Maesfranckx, *Time Management in Fuzzy and Uncertain Object-Oriented Databases*, O. Pons, M.A. Vila, J. Kacprzyk (eds.), Knowledge Management in Fuzzy Databases, Studies in Fuzziness and Soft Computing, Physica-Verlag, Heidelberg, Germany, (2000), 67-88.
13. R.R.Yager, *Retrieval from Multimedia databases using Fuzzy temporal Concepts*, ibidem, 261-274.
14. S.Dragicevic, D. J. Marceau, *An application of fuzzy logic reasoning for GIS temporal modeling of dynamic processes*, Fuzzy sets and Systems, 113(1), (2000), 69-80.
15. Ratsiatou, E. Stefanakis, *Spatio-Temporal Multicriteria Decision making under Uncertainty*, in Proc. of the 1st Int. Symposium on Robust statistics and fuzzy techniques in geodesy and GIS, Zurich, (2001).
16. Yazici, Q. Zhu, N. Sun, *Semantic data modeling of spatiotemporal database applications*, Int. J. of Intelligent Systems, 16(7), (2001), 881-904.
17. L.A.Zadeh, *Fuzzy Sets as a Basis for a Theory of Possibility*, Fuzzy Sets and Systems, 1, (1978), 3-28.
18. G.Bordogna, S. Chiesa, *A Fuzzy Object based Data Model for Imperfect Spatial Information Integrating Exact-Objects and Fields*, Int. J. of Uncertainty, Fuzziness and Knowledge-Based Systems, 11(1) (2003), 23-41.
19. S.Spaccapietra, C.Parent, E. Zimanyi, *MurMur: a research agenda on multiple representations*, Proc. of DANTE'99, Kyoto, (1999).
20. P. Fisher, *Sorites paradox and vague geographies*, Fuzzy Sets and Systems, 113(1), (2000), 7-18.
21. L. A. Zadeh, *The concept of a linguistic variable and its application to approximate reasoning, parts I, II*. Information Science, 8, 199-249, (1987), 301-357.
22. R.M. Guting et al., *A foundation for representing and querying moving objects*, FernUniversitaet Hagen, Informatik-report 238, (1998).
23. F.E. Petry, *Fuzzy Databases*, Kluwer Academic Pub., (1996).
24. P.Bosc, and H.Prade, *An Introduction to the Fuzzy Set and Possibility Theory-based Treatment of Flexible Queries and Uncertain and Imprecise Databases*, in Uncertainty Management in Information Systems: from Needs to Solutions, A. Motro and P. Smets eds, Kluwer Academic Pub., (1994), 285-324.
25. G. Bordogna, S. Chiesa, *Fuzzy Querying of Spatial Information*, Proc. of EUROFUSE, Varenna, (2002).

Enabling Fuzzy Object Comparison in Modern Programming Platforms through Reflection*

Fernando Berzal, Juan-Carlos Cubero, Nicolás Marín, and Olga Pons

IDBIS Research Group - Dept. of Computer Science and A. I.,
E.T.S.I.I. - University of Granada, 18071, Granada, Spain
{fberzal|jc.cubero|nicm|opc}@decsai.ugr.es
http://frontdb.ugr.es

Abstract. Comparison plays a determining role in many problems related to object management. Every modern programming language provides a comparison method (e.g. the *equals* method in Java) that allows us to compare two objects of any user-defined class. However, if we develop classes to represent objects imprecisely described using Fuzzy Subset Theory, the built-in method is no longer suitable. Fuzzy object comparison must be handled using a fuzzy equality concept, that is, resemblance relationships instead of classical equality. In this paper we present how to enable modern programming platforms so that classes with imprecise attributes can be developed and fuzzy objects of these classes can be easily compared. In particular, we introduce a *fuzzyEquals* method for a *FuzzyObject* class that can be used in user-defined classes without being overridden.

1 Introduction

The Object-Oriented Data Model[6,1] is one of the most important data paradigms for both programmers and database developers, particularly when they deal with complex and dynamic problems. As was the case with older data models, like the relational one, the object-oriented model is being widely studied in order to make possible the management of imprecise information [3].

The basic criterion used to distinguish objects in a object-oriented context is the Identity equality: *two objects are identical if they are the same object (i.e. if they have the same object identifier).* However, there exist many situations where this measure is very strict and we need to use the concept of Value equality: *two objects are equal if the values of all their attributes are recursively equal.* When we deal with perfect information this concept of equality (and, in general, the usual set of relational comparators) permits us to compare the objects of a given class. However, if the objects have imprecise values in their descriptions, then, this classical notion of equality is no longer valid. When Fuzzy Subset Theory[8] is used to handle object imprecision, fuzzy resemblance measures are acceptable operators in order to perform comparisons.

---

* This work was partially supported by Spanish R&D project TIC2002-04021C0202

T. Bilgiç et al. (Eds.): IFSA 2003, LNAI 2715, pp. 660–667, 2003.

Recently, we have developed a proposal[2] to represent fuzzy information in an object-oriented data model. Our research is mainly motivated by the need for an easy-to-use transparent mechanism to develop applications dealing with fuzzy information. Following our proposal, programmers and designers should be able to directly use new structures developed to store fuzzy information without the need for any special treatment, without altering the underlying programming platform and with the most possible transparency. Our proposal allows the programmer to handle imprecision in an important set of the situations where imprecision can appear in an object-oriented software development effort.

For our proposal to come to fruition, we have developed a set of operators which compute the resemblance between objects of a given class[4]. Following the principles that guided our research, in this paper we describe the way this theoretical approach can be implemented in a modern programming platform, so that programmers can easily design their classes to handle imprecise objects and compare the fuzzy objects of these classes with a minimum of effort. As we will see, we have used the reflection capability that many modern programming languages offer to develop a framework that can be used by user-defined classes in order to compare objects.

This paper is organized as follows: Section 2 is devoted to the presentation of a brief summary of the theoretical aspects of our comparison approach, while Section 3 explains the use of reflection in order to compute the resemblance between objects, and Section 4 analyzes the main features of our solution as implemented in Java and an example program that highlights the main advantages of our approach. Finally, Section 5 contains some concluding remarks and future work to be done.

2   An Approach to Compare Fuzzy Objects

The example illustrated in Figure 1 will help us to introduce the problem that motivates this work. The figure depicts the information about two rooms characterized by their quality, their extension, the floor they are on, and the set of students who attend their lessons in each room. The description of the objects belonging to class *Room* is imprecise due to the following reasons:

- The *quality* is expressed by an imprecise label.
- The attributes *extension* and *floor* can be expressed using a numerical value or an imprecise label.
- The set of students is fuzzy, taking into account the percentage of time each student spends receiving the lessons in each room to compute the membership degrees.

We have to solve the following problems in order to compare two objects of the class Room:

- First, we have to handle resemblance in basic domains (e.g. attributes *quality, extension,* and *floor*).

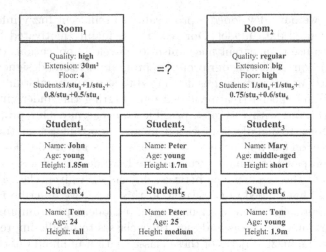

Fig. 1. Comparing imprecise objects: An example.

– Second, we also have to be able to compare fuzzy collections of imprecise objects (e.g. attribute *set of students*).
– Finally, we need to aggregate the resemblance information that we have collected by studying particular attributes and compute a general resemblance opinion for the whole complex objects.

Complex objects will need a recursive computation of their resemblance (i.e. student resemblance must be computed). During this process we have to deal with the possible presence of cycles in the data graph (i.e. it is possible that we have to compute the resemblance of objects o_1 and o_2 in order to compute the resemblance between objects o_1 and o_2).

2.1   Computing the Resemblance between Two Objects

Taking into account the ideas presented in [4], we can define the calculus of the resemblance between two objects of a given class as follows:

Definition 1 (Resemblance between two objects) *Let C be a class whose type is made up by the set of attributes $Str_C = \{a_1, a_2, ..., a_n\}$. Let o_1 and o_2 be two objects of C. We define the resemblance (FE) between the objects o_1 and o_2 as the value returned by the following function:*

$$FE : F_C \times O(F_C) \times O(F_C) \times \mathcal{P}(\mathcal{P}_2(O(F_C))) \times \mathcal{P}(\mathcal{P}_2(O(F_C))) \to [0,1]$$

where F_C is the family of all the classes and $O(F_C)$ is the set of all the class instances. $\mathcal{P}$ stands for the power set and $\mathcal{P}_2$ represents those members of the power set whose cardinality is 2 (i.e., pairs of objects in our context). The calculus of $FE(C, o_1, o_2, \Omega_{visited}, \Omega_{aprox})$[1] involves the following recursive computation:

---

[1] Henceforth, FE stands for $FE(C, o_1, o_2, \Omega_{visited}, \Omega_{aprox})$

(a) If $o_1 = o_2$, then: $FE = 1$

(b) If there exists a resemblance relation S defined in C, then: $FE = \mu_S(o_1, o_2)$. In particular, if o_1 and o_2 are fuzzy sets then:

$$FE = \beth_{FE_\Omega, \otimes}(o_1, o_2) = \otimes(\Theta_{FE_\Omega}(o_2|o_1), \Theta_{FE_\Omega}(o_1|o_2)), \text{ where}$$

$$\Theta_{FE_\Omega}(o|o') = \min_{x \in Spp(o')} \max_{y \in Spp(o)} \{I(\mu_{o'}(x), \mu_o(y)) \otimes FE(C_D, x, y, \Omega_{visited}, \Omega_{approx})\},$$

where C_D stands for the class that is the reference universe of the sets,
I is an implication operator, $Spp(o)$ is the support-set of o, and $\otimes$ is a t-norm.

(c) If $\{o_1, o_2\} \notin \Omega_{visited}$, then

$$FE = V_Q(W, R).$$

where R contains the resemblance values $FE(C_{a_i}, o_1.a_i, o_2.a_i, \Omega_{visited} \cup \{\{o_1, o_2\}\}, \Omega_{approx})$ where defined (C_{a_i} is the domain class of the attribute a_i), W contains the weights for attributes a_i, and V_Q is Vila's aggregation operator[7], which is defined as:

$$o_Q \max_{i:r_{a_i} \in R} \{w_{a_i} \wedge r_{a_i}\} + (1 - o_Q) \min_{i:r_{a_i} \in R} \{r_{a_i} \vee (1 - w_{a_i})\}$$

(d) If $\{o_1, o_2\} \in \Omega_{visited} \wedge \{o_1, o_2\} \notin \Omega_{approx}$, then:

$$FE = FE(C, o_1, o_2, \emptyset, \Omega_{approx} \cup \{\{o_1, o_2\}\})$$

(e) Otherwise, when $\{o_1, o_2\} \in \Omega_{visited} \wedge \{o_1, o_2\} \in \Omega_{approx}$, then FE is undefined.

The above function, in spite of the complexity in its definition, is just a case selection that comprises the different tools that can and should be used to compare pairs of objects:

- There are two basic cases:
  - When the identity equality holds between the objects.
  - When a known defined resemblance relation exists in the class. As a particular example, when we compare two fuzzy sets of objects we can use a generalized resemblance degree[4] which recursively compares the elements in the sets.
- The third case provides a general recursive model which applies the aggregation operator V_Q[7] over recursive calls that compute the resemblance between couples of attribute values. When aggregating, not all the attributes have the same importance (w_{a_i} weights the importance of the attribute a_i). The aggregation is founded on the semantics of a quantifier Q (by using o_Q - orness of Q).
- The fourth and fifth cases use the variables $\Omega_{visited}$ and Ω_{approx} to deal with the existence of cycles:
  - The first time that the couple $\{o_1, o_2\}$ produces a cycle, which is detected because $\{o_1, o_2\}$ is already in $\Omega_{visited}$, then the couple is inserted into Ω_{approx} in order to compute an approximation that focuses only on non-problematic attributes (those which do not lead to cycles).

- If the couple of objects is in Ω_{approx} (i.e., its resemblance is currently being approximated), then we do not calculate any resemblance value, and the function FE is undefined.

The above function is a resemblance relation because the properties of the operators used in each of the basic cases are those of a resemblance relation (see [4] for a more in depth study).

3   Using Reflection in Modern Programming Platforms

Reflection is a feature of many of the modern programming languages (e.g. Java – java.sun.com – and C♯ – www.microsoft.com) . This feature allows an executing program to examine or "introspect" itself, and manipulate internal properties of the program. For example, it is possible for a class to obtain the names of all its members and display them.

Since our final aim is to allow the programmer to perform fuzzy comparisons without having to write specific code for each class he/she writes, we can define a generic *FuzzyObject* class which will serve as base class for any classes whose objects need fuzzy comparison capabilities. We can avoid duplicating code in different classes if we write a generic *fuzzyEquals* method at the *FuzzyObject* class. Taking into account that the *fuzzyEquals* method requires access to the particular object fields, the only way we can implement such a general version of this operator is through reflection.

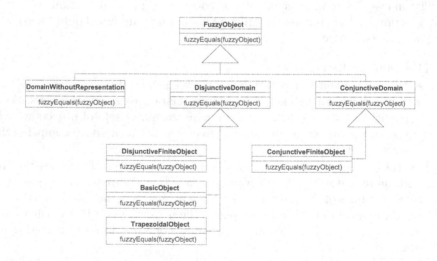

Fig. 2. Our framework.

Just by extending this general *FuzzyObject*, the programmer can define her own classes to represent fuzzy objects. Our framework, as depicted in Figure 2,

also includes some classes to represent common kinds of domains for imprecision, such as linguistic labels without underlying representation (*DomainWithoutRepresentation*), domains where labels are possibility distributions over an
underlying basic domain (*DisjunctiveDomain* and its subclasses to represent labels with finite support-set, basic domain values, and functional representations
of labels with infinite support-set, like trapezoidal ones), and, finally, fuzzy collections of fuzzy objects (*ConjunctiveDomain*). These classes define their proper
fuzzyEquals logic to handle the different cases we discussed in Section 2.

4   Using Java to Experiment with FuzzyEquals

Let us suppose that we want to compare the classrooms already mentioned in a
previous section (see Figure 1). We can define the following class in Java[5] to
represent the classrooms:

```
public class Room extends FuzzyObject
{
    // Instance variables
    public Quality           quality;
    public Extension         extension;
    public Floor             floor;
    public StudentCollection students;

    // Constructor
    public Room (Quality quality,Extension extension,Floor floor,StudentCollection students)  {
        this.quality   = quality;
        this.extension = extension;
        this.floor     = floor;
        this.students  = students;
    }

    // Field importance
    public static float fieldImportance (String fieldName)   {
        String fields[]    = new String[] { "quality", "extension", "floor", "students" };
        float  importance[] = new float[] { 0.5f,      0.8f,        1.0f,    1.0f       };

        for (int i=0; i<fields.length; i++)
            if (fields[i].equals(fieldName))
                return importance[i];

        return 1.0f;
    }
}
```

where the *fieldImportance* method is specified to set the attribute importances (although it could be omitted if the user gives the same importance to
all of them). This overridden method is defined by default in *FuzzyObject* and
used by its *fuzzyEquals* method. The room imprecise attributes can be easily
implemented just by extending the classes provided by our framework without
having to worry about the *fuzzyEquals* implementation:

– The imprecise room quality is an object of a class (*Quality*) which extends
  DomainWithoutRepresentation.
– The extension and floor attributes are both particular cases of *DisjunctiveDomain* and, as such, they can be basic values, trapezoids, or finitely-
  described labels.

– Finally, the set of students is a fuzzy collection of students, where the fuzzy
collection *StudentCollection* inherits from *ConjunctiveFiniteObject* and students are defined as fuzzy objects as follows:

```
public class Student extends FuzzyObject
{
  // Instance variables
  public String name;
  public Age    age;
  public Height height;

  // Constructor
  public Student (String name, Age age, Height height) {
    this.name   = name;
    this.age    = age;
    this.height = height;
  }
  ...
}
```

Using the above classes, we can easily write an application which compares
fuzzy objects, regardless the object graphs they belong to, since our resemblance
function, as defined in Section 2, works well with cyclic structures.

For instance, we can define some students using the following code, which
makes use of linguistic labels:

```
// Label definitions for students
Age young  = new Age ( new Label("young"),        0,  0, 23, 33 );
Age middle = new Age ( new Label("middle-aged"), 23, 33, 44, 48 );
Height shortHeight  = new Height ( new Label("short"),    0,   0, 150, 160);
Height mediumHeight = new Height ( new Label("medium"), 150, 160, 170, 180);
Height tall         = new Height ( new Label("tall"),   170, 180, 300, 300);

//Student definition
Student student1 = new Student ( "John",  young,        new Height(185) );
Student student2 = new Student ( "Peter", young,        new Height(170) );
Student student3 = new Student ( "Mary",  middle,       shortHeight     );
Student student4 = new Student ( "Tom",   new Age(24), tall            );
Student student5 = new Student ( "Peter", new Age(25), mediumHeight    );
Student student6 = new Student ( "Tom",   young,        new Height(190) );
```

Using a similar approach, we create a couple of rooms:

```
// Label definitions for rooms:
// highQuality, mediumQuality, highFloor... (as above)

// Sets of students
Vector vector1 = new Vector();
vector1.add ( new MembershipDegree (1.0f, student1 ) );
vector1.add ( new MembershipDegree (1.0f, student2 ) );
vector1.add ( new MembershipDegree (0.8f, student3 ) );
vector1.add ( new MembershipDegree (0.5f, student4 ) );
StudentCollection set1 = new StudentCollection ( vector1 );
Vector vector2 = new Vector();
vector2.add ( new MembershipDegree (1.0f, student1 ) );
vector2.add ( new MembershipDegree (1.0f, student5 ) );
vector2.add ( new MembershipDegree (0.75f, student3 ) );
vector2.add ( new MembershipDegree (0.6f, student6 ) );
StudentCollection set2 = new StudentCollection ( vector2 );

//Room definitions
Room room1 = new Room ( highQuality,   new Extension(30), new Floor(4), set1 );
Room room2 = new Room ( mediumQuality, big,                  highFloor,    set2 );
```

Once we have some fuzzy objects, we can compare them just by invoking their *fuzzyEquals* method, as in

```
System.out.println ( "room1 fvs. room2 = " + room1.fuzzyEquals(room2) );
```

which returns an approximate value of 0.81. Thus, we have encapsulated fuzzy object comparisons in our framework classes so that programmers can now freely compare imprecisely-described objects without having to code any comparison logic.

5   Conclusions and Further Work

In this paper, we have demonstrated how to implement reusable fuzzy comparison capabilities in modern programming platforms through the use of reflection and theoretical results which help us apply fuzzy techniques in object-oriented models.

Our approach seamlessly deals with the cyclic structures which are common in object graphs and cause problems when computations have to be recursively solved. Moreover, although we have used our own operators, our proposal can be extended to deal with any operator able to compare imprecisely-described objects. We can even parameterize our fuzzy comparison methods to allow for different comparison strategies at run-time.

Our future efforts will be directed towards the refinement of our framework in order to make it efficient enough to deal with the huge datasets commonly used in real-world applications, thus making it available for those interested in using it.

References

1. M. Berler, J. Eastman, D. Jordan, C. Russell, O. Schadow, T. Stanienda, and F. Velez. *The object data standard: ODMG 3.0.* Morgan Kaufmann Publishers, 2000.
2. F. Berzal, N. Marín, O. Pons, and M. A. Vila. *Software Engineering with Fuzzy Theory*, chapter Using Classical Object-Oriented Features to build a FOODBS. Physica Verlag, 2002. To appear.
3. J. Lee, J-Y. Kuo, and N-L. Xue. A note on current approaches to extent fuzzy logic to object oriented modeling. *International Journal of Intelligent Systems*, 16:807–820, 2001.
4. N. Marín, J.M. Medina, O. Pons, D. Sánchez, and M. A. Vila. Complex object comparison in a fuzzy context. *Information and Software Technology*, to appear.
5. B. Spell. *Professional Java Programming.* Wrox Press, 2000.
6. B. Stroustrup. What is object-oriented programming? *IEEE Software*, 1988.
7. M. A. Vila, J. C. Cubero, J. M. Medina, and O. Pons. The generalized selection: an alternative way for the quotient operations in fuzzy relational databases. In B. Bouchon-Meunier, R. Yager, and L. Zadeh, editors, *Fuzzy Logic and Soft Computing*. World Scientific Press, 1995.
8. L. A. Zadeh. Fuzzy sets. *Information and Control*, 8:338–353, 1965.

An XML-based Approach to Processing Imprecise Requirements*

Jonathan Lee, Yong-Yi Fanjiang, Tzung-Jie Chen, and Ying-Yan Lin

Department of Computer Science and Information Engineering,
National Central University, Chungli, Taiwan
{yjlee, yyfanj, jchen, slose}@selab.csie.ncu.edu.tw

Abstract. Fuzzy theory is suitable to capture and analyze the imprecise requirements [2][4], meanwhile, with recent development in XML standards and availability of tools supporting, it has become possible to generate multiple types of document by applying XML documents transformation technology. In this paper, we propose an approach based on the XML schema to defining the transformation rules from imprecise requirements into its corresponding source codes, and constructing the code generator in a semi-automatic manner.

1 Introduction

Generating source codes from requirement specifications is valuable in maintaining consistency between a specification and its implementation, and abating the routine work of writing skeleton source codes. With the recent development in XML standards and availability of tools supporting, it has become possible to generate multiple types of document by applying XML documents transformation technology. Numerous researchers have reported progress towards the successful code generation by means of XML, which can be classified into three categories: (1) constructing user interface [6], (2) generating database [1][7], and (3) transforming UML into codes [8].

In this paper, we propose an approach to constructing a source code generator through the transformation rules by using the APIs generated from the XML schema (an overview of our approach is depicted in Fig. 1). Two kinds of transformation rules schema: organization and structure rules schema are defined and a code generator is constructed in a semi-automatic manner to facilitate the source code generation.

The organization of this paper is as follow. We first briefly describe the FOOM [5] schema based on the XML to model imprecise requirements in the next section. In section 3, the transformation from the FOOM specification to its corresponding codes is discussed. The implementation of FOOM prototype is described in section 4, and concluding remarks are given in section 5.

---

* This research is supported by National Science Council (Taiwan) under grants NSC91-2213-E-008-012.

T. Bilgiç et al. (Eds.): IFSA 2003, LNAI 2715, pp. 668–676, 2003.

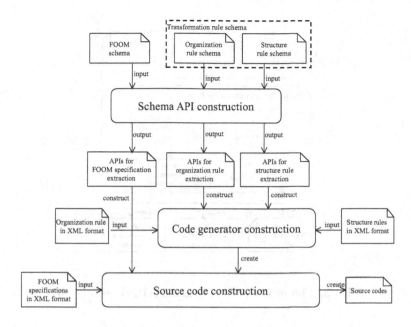

Fig. 1. An overview of the proposed approach.

2 Modeling Imprecise Requirements with XML

In order to model imprecise requirements with XML, we have formulated the FOOM schema [3] based on the key features defined in FOOM: fuzzy set, fuzzy attribute, fuzzy rule, and fuzzy association, by means of the XML schema.

The basic element of the fuzzy term is the fuzzy set described by a membership function. There are two kinds of fuzzy set: discrete and continuous. Both types of fuzzy sets are captured using the XML schema. A fuzzy attribute is mapped to a pair of the element <name> and <fuzzy-type>. The <name> tag can be any string to represent the attribute's name, and the attribute's fuzzy type can be either a <fuzzy-variable> or a <typical-variable> element. Each fuzzy-variable is characterized by a set of <fuzzy-set> elements. Each typical-value includes a set of <t-point> elements, which contain a <t-value> and <t-degree> elements. Using the fuzzy rules is one way to deal with imprecision where a rule's conditional part and/or the conclusion part contains linguistic variables. A fuzzy rule is mapped to a sequence of <if> elements and <then> elements. A fuzzy association is mapped to an association type which is a complex type. The content model of an association is a sequence of <description>, <link-attribute>, <association-end>, <association-end>, <degree-of-participation> and <possibility-degree> elements. The value that a link takes for the link attribute is described by the degree of participation to represent the degree that a link participates in this association. The <possibility-degree> is a confidence level of this fuzzy association, whose value is a fuzzy truth value.

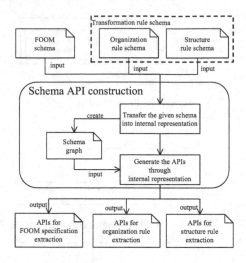

Fig. 2. The process of the schema API construction.

3   Transforming FOOM Specifications into Source Codes

In this section, the transformation rule schema is defined for representing the transformation rules from the FOOM specifications into implementations. The transformation rule schema consists of the organization rule schema and the structure rule schema. The structure rule schema defines rules for generating source codes, and the organization rule describes the structure of codes to be generated using the data specified in the structure rule documents [8]. In our approach, both the organization and structure rule schema are transformed into a set of APIs, and then two rule extractors are built to parse the transformation rules into the object trees. After the object trees are constructed by the rule extractors, a code generator based on object trees is then developed. Finally, the source codes are generated through the code generator.

3.1   Schema API Construction

Fig. 2 shows the detailed processes to generate a set of APIs from the XML schema for content validation and data access in an automatic manner [3]. A schema graph extended from DTD graph [9] with typing information is used to serve as an intermediate representation for the structure of XML schema to bridge the XML schema and APIs for both content validation and data access. Through the APIs generated from the schema graph, the XML documents are parsed into object trees.

Transformation rule is an XML document that describes how to translate requirements specification to source codes. We adopt the description in [8] to define the organization rule schema and structure rule schema, respectively.

Fig. 3. An organization rule example.

In the organization rule schema, the class is composed of the place group, <Repeat>, <Opener> and <Closer>. The place group is consisted of <Place>, and <onExist> element. The <Place> element has an attribute - name to identify where a particle code is generation. The <onExist> element describes a specific action only if the place specified by its attribute - name is processed. The organization rule schema uses the <Repeat> element to specify the repeating generation. For example, Fig. 3 specifies the structure for constructing a particular class. The attribute, constructor, and operation code are produced within class's opener and closer. And <Repeat> element encircles the <fuzzyset> element to represent that fuzzy terms can be repeatedly appeared.

In the structure schema, the <Rule> element contains an attribute - language to identify which language that the transformation rule is specified The transformation component defined between <Generation> element can be a class, an attribute, a constructor, or a operation. Both of them consists of three elements: <Insertion>, <Switch>, and <Selection>. Those elements are the primitive element in the structure rule schema. <Insertion> element is used to simply insert value of text attribute. The <Switch> element provides the conditional insertion through the <Case> element and <Default> element. Finally, the <Selection> element allows a user to select one of several alternative elements encircled by the <Option> elements. Fig. 4 shows a fuzzy variable transformation rule example. Note that the FuzzyJ[1] is denoted as the target language specified in the <Rule> tag.

---

[1] Refer to FuzzyJ toolkits at http://www.iit.nrc.ca/IR_public/fuzzy/fuzzyJToolkit2.html.

```
<Rule name="FuzzyJ" language="FuzzyJ">
  <Generation filename="$fuzzy-variable@ nam e+C lass">
    <Class>
      <Insertion place="visibility" text=" "/>
      <Insertion place="type-keyword" text="class"/>
      <Insertion place="classnam e" text="$fuzzy-variable@ nam e+Class"/>
      <Opener symbol="{ "/>
      <Closer symbol="}"/>
    </Class>
    <Attribute>
      <Insertion place="visibility" text="public"/>
      <Insertion place="static" text="static"/>
      <Insertion place="type" text="FuzzyVariable"/>
      <Insertion place="attNam e" text="$fuzzy-variable@ nam e"/>
    </Attribute>
    <Constructor>
      <Insertion place="visibility" text=" "/>
      <Insertion place="return" text=" "/>
      <Insertion place="cons_nam e" text="$fuzzy-variable@ nam e+Class"/>
      <Opener symbol="{ "/>
      <Closer symbol="}"/>
    </Constructor>
    <Operation>
      <Insertion place="visibility" text="public"/>
      <Insertion place="static" text="static"/>
      <Insertion place="type" text="void"/>
      <Insertion place="op_nam e" text="init"/>
      <Opener symbol="{ "/>
      <Closer symbol="}"/>
    </Operation>

  <FuzzyVariable>
    <Insertion place="fuzzy_varnam e" text="$fuzzy-variable@ nam e"/>
    <Insertion place="assignm ent" text="="/>
    <Insertion place="new_keyword" text="new "/>
    <Insertion place="fuzzy_vartype" text="FuzzyVariable"/>
    <GroupInsertion place="arg_assignm ent" delimiter=",">
      <Term mark="true" text="$fuzzy-variable@ nam e"/>
      <Term mark="false" text="$low er"/>
      <Term mark="false" text="$upper"/>
      <Term mark="true" text="$unit"/>
    </GroupInsertion>
  </FuzzyVariable>
  <FuzzySet>
    <Insertion place="visibility" text=" "/>
    <Insertion place="fuzzy_setnam e" text="FuzzySet"/>
    <Insertion place="fuzzy_setnam e" text="$fuzzy-set@ nam e+Set"/>
    <Insertion place="assignm ent" text="="/>
    <Insertion place="new_keyword" text="new "/>
    <Insertion place="fuzzy_settype" text="FuzzySet"/>
    <Insertion place="arg_assignm ent" text=" "/> <!-- maybe GroupInsertion -->
    <Insertion place="fset_append" text="$fuzzy-variable@ nam e+ addTerm "/>
    <GroupInsertion place="fset_arg" delimiter=",">
      <Term mark="true" text="$fuzzy-set@ nam e"/>
      <Term mark="false" text="$fuzzy-set@ nam e+Set"/>
    </GroupInsertion>
  </FuzzySet>
  <Point>
    <Insertion place="point_append" text="$point> appendSetPoint"/>
    <GroupInsertion place="point_arg" delimiter=",">
      <Term mark="false" text="$f-value"/>
      <Term mark="false" text="$m em bership-degree"/>
    </GroupInsertion>
  </Point>
  </Generation>
</Rule>
```

Fig. 4. A structure rule example.

3.2   Constructing APIs for Code Generator

Fig. 5 illustrates the process of constructing code generator. The procedure of constructing code generator is spilt into two processes: parsing the transformation rule, and building the code generator. The main idea is to traverse each node contained in the organization rule tree by breadth-first manner to lay out the structure of the codes, and to match the particular programming syntax specified in the structure rule tree to build the code generator in FuzzyJ. The details are described below:

Algorithm 1

1. *use the organization rule extractor and structure rule extractor to construct the organization rule tree (ot) and structure rule tree (st) based on the organization rule document and structure rule document.*
2. *Traverse the ot tree in the breadth-first manner*
   (a) *if the visited node is <Place> tag, then generate an output statement.*
      i. *first, use the text specified in this tag's name attribute as keyword to search corresponding st tree for finding a <Insertion> or <GroupInsertion> node which contains a attribute named* place *whose value is equal to the keyword; and then*
      ii. *if the found node is a <Insertion>, then use getText() API defined in this node to get the value of the text attribute and put this value as the output content of this output statement. Otherwise, for all the sub nodes contained in this node, use the getText() API defined in those sub node for each other to get their value of text attribute and put all of those value together as the output content of this output statement.*
   (b) *else if the visited node is <Insertion> tag, then generate an output statement. The output content of this statement is the value of* name *attribute in this node.*
   (c) *else if the visited node is <Opener> or <Closer>, then do the same actions as <Place> tag.*

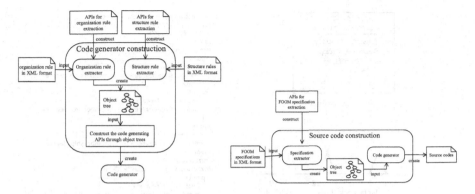

Fig. 5. The process of the code gen- **Fig. 6.** The process of the source erator construction.   code construction.

(d) else if the visited node is <Repeat> tag, then generate a while loop statement. The ending condition of this statement is specified in the lastOccurrence *attribute in this tag.*

3.3   Source Code Construction

Fig. 6 shows the source codes generating processes. Firstly, the FOOM specification is parsed into an object tree by using the specification extractor. Secondly, the code generator extracts the necessary data from the object tree to generate the corresponding source codes.

The *MeetingRegistration* class specified in the meeting schedule problem is chosen an example to demonstrate the results generated by our approach from the requirements specifications to its corresponding FuzzyJ source codes. The *MeetingRegistration* class is a crisp class, with an attribute - status, which is associated with a fuzzy range (see Fig. 7(A) and (B)). The XML document instantiated by the FOOM schema for representing the *MeetingRegistration* class and fuzzy variable - ParticipantStatus is shown in Fig. 7(C) and (D), respectively. Note that the <fuzzy-set> denoted the possible fuzzy terms and their membership function. Fig. 8 shows the FuzzyJ source codes generated from Fig. 7. Two classes are generated. The first class (see Fig. 8(A)) declares the attribute - status's type as *FuzzyValue* which is the fuzzy package defined in the FuzzyJ, and the other one (*participantClass* class) is constructed based on the fuzzy variable - *Participant-Status*. The *participantClass* contains only one attribute - participant-status which is declared as FuzzyVariable, and the static keyword means that this attribute is shared by all the instances instantiated by this class. Furthermore, this class contains an init operation to create a FuzzyVariable instance and all the fuzzy sets declared in the Fig. 7.

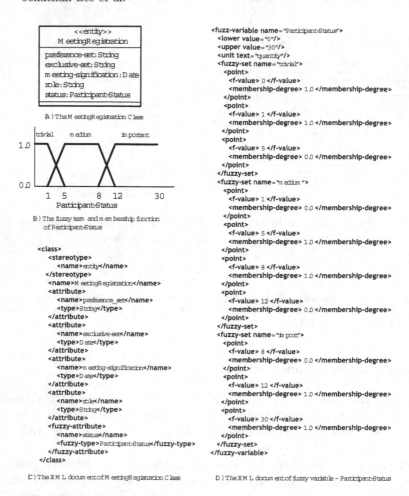

(A) The MeetingRegistration Class

(B) The fuzzy term and membership function of Participant-Status

(C) The XML document of MeetingRegistration Class

(D) The XML document of fuzzy variable – Participant-Status

Fig. 7. An example of FOOM specification - *MeetingRegistration* Class.

4   Implementation

We adopted Java as the programming language for the FOOM prototype. There are four main parts in this prototype. The first part is to generate the content validation and data access APIs according the input schema. The second one is to assist users in using the XML specification constructor to construct the requirements specification into XML format visually. The third part is to construct the source code generator through the transformation rules. And the last part contains the specification extractor and code generator to build the object tree from FOOM specification and generate the source codes.

```
import nrc.fuzzy.*;
class MeetingRegistration {
    public String preference-set;
    public String exclusive-set;
    public Date meeting-signification;
    public String role;
    public FuzzyValue status;

    MeetingRegistration (String r) {
        role = r;
        if (role== "faculty") {status = new FuzzyValue (participant-statusClass. participant-status, "important"); }
        else if (role == "staff") {status = new FuzzyValue (participant-statusClass. participant-status, "medium"); }
        else if (role == "student") {status = new FuzzyValue (participant-statusClass. participant-status, "trivial"); }
        else {status = new FuzzyValue (participant-statusClass. participant-status, "medium"); }
    }
}
```

(A) The generated FuzzyJ source code of MeetingRegistration Class

```
import nrc.fuzzy.*;
class participant-statusClass {
    public static FuzzyVariable participant-status;
    participant-statusClass () {}
    public static void init () {

        participant-status = new FuzzyVariable("participant-status", 0, 30,"quality");

        FuzzySet trivialSet = new FuzzySet();
        trivialSet.appendSetPoint(0,1);
        trivialSet.appendSetPoint(1,1);
        trivialSet.appendSetPoint(5,0);
        participant-status.addTerm("trivial", trivialSet);

        FuzzySet mediumSet = new FuzzySet();
        mediumSet.appendSetPoint(1,0);
        mediumSet.appendSetPoint(5,1);
        mediumSet.appendSetPoint(8,1);
        mediumSet.appendSetPoint(12,0);
        participant-status.addTerm("medium", mediumSet);

        FuzzySet importantSet = new FuzzySet();
        importantSet.appendSetPoint(8,0);
        importantSet.appendSetPoint(12,1);
        importantSet.appendSetPoint(30,1);
        participant-status.addTerm("important", importantSet);
    }
}
```

(B) The generated FuzzyJ source code of Participant-Status Class

Fig. 8. The generated FuzzyJ source code from the Fig. 7.

5 Conclusion

In this paper, we proposed the transformation rule schema based on the XML schema to define the transformation rules from imprecise requirements into its corresponding source codes and constructed the code generator in a semi-automatic manner. Our future research plan will focus on the issues of dynamic behaviors source code generation.

References

1. B. Demuth, H. Hussmann, and S. Obermaier. Experiments with XMI based transformations of software models. In *Workshop on Transformations in UML, ETAPS 2001 Satellite Event, Genova, Ilaly,* April 7th, 2001.
2. J. Lee. *Software Engineering with Computational Intelligence.* Springer-Verlag, Heidelberg, 2003.
3. J. Lee and Y.Y. Fanjiang. Modeling imprecise requirements with XML. (to appear). *Information and Software Technology,* 2003.
4. J. Lee, J. Y. Kuo, and N. L. Xue. A note on current approaches to extending fuzzy logic to object-oriented modeling. *International Journal of Intelligent Systems,* 16(7):807–820, 2001.
5. J. Lee, N.L. Xue, K.H. Hsu, and S.J. Yang. Modeling imprecise requirements with fuzzy objects. *Information Sceinces,* 118:101–119, 1999.
6. A. Mueller, T. Mundt, and W. Lindner. Using XML to semi-automatically derive user interfaces. In *Proceedings of the Second International Workshop on User Interfaces to Data Intensive Systems,* pages 91–95, 2001.
7. E. Nantajeewarawat, V. Wuwongse, S. Thiemjarus, K. Akama, and C. Anutariya. Generating relational database schemas from UML diagrams through XML declarative descriptions. In *Proceedings of the 2nd International Conference on Intelligent Technologies (InTech'2001),* November 2001.

676   Jonathan Lee et al.

8. D.H. Park and S.D. Kim. XML rule based source code generator for UML CASE tool. In *The Eight Asis-Pacific Software Engineering Conference (APSEC 2001)*, pages 53–60, Dec. 4-7 2001.
9. J. Shanmugasundaram, K. Tufte, G. He, C. Zhang, D. DeWitt, and J. Naughton. Relational databases for querying xml documents: limitations and opportunities. In *Proceedings of the 25th International Conference on Very Large Data Bases*, pages 302–314, 1999.

Inducing Fuzzy Concepts through Extended Version Space Learning

Eyke Hüllermeier

Informatics Institute
Marburg University, Germany
eyke@informatik.uni-marburg.de

Abstract. The problem of inducing a concept from a given set of examples has been studied extensively in machine learning during the recent years. In this context, it is usually assumed that concepts are precisely defined, which means that an object either belongs to a concept or not. This assumption is obviously over-simplistic. In fact, most real-world concepts have fuzzy rather than sharp boundaries, an observation that motivates the development of methods for fuzzy concept learning. In this paper, we introduce generic algorithms for inducing fuzzy concepts within the framework of version space learning.

1 Introduction

The acquisition of *concepts* constitutes one of the fundamental learning skills of human beings, and its underlying mechanisms have been investigated thoroughly in cognitive psychology and related fields. Thus, it comes at no surprise that concept learning has also received a great deal of attention in *machine learning*.

A concept is usually identified with its *extension*, that is a subset C of an underlying set $\mathcal{U}$ of objects. For example, C might be the concept "dog" whose extension is the set of dogs presently alive, a subset of all creatures on earth. The goal of (machine) learning is to induce an *intensional* description of a concept from a set of (positive and negative) examples, that is a characterization of a concept in terms of its properties.

It is widely recognized that most natural concepts have non-sharp boundaries [7]. To illustrate, consider concepts like wood, river, lake, hill, street, house, or chair. Obviously, these concepts are *vague* or *fuzzy*, in that one cannot unequivocally say whether or not a certain collection of trees should be called a wood, whether a certain building is really a house, and so on. Rather, one will usually agree only to a certain extent that an object belongs to a concept.

Given the fuzziness of natural concepts, it is a straightforward idea to extend the formal definition of (the extension of) a concept from a subset to a *fuzzy subset* of $\mathcal{U}$. A concept is then characterized by a membership function $\mathcal{U} \to \mathcal{L}$, where $\mathcal{L}$ is a suitable membership scale. Henceforth, we shall not distinguish between a concept and its fuzzy extension, that is, we shall use the same symbol to denote a concept and its membership function. Thus, $C(u)$ is the degree to which object $u \in \mathcal{U}$ is considered an element of the concept C.

T. Bilgiç et al. (Eds.): IFSA 2003, LNAI 2715, pp. 677–684, 2003.

In this paper, we develop generic algorithms for inducing fuzzy concepts within the framework of version space learning. Section 2 gives a specification of the learning problem. Different approaches conceivable for learning fuzzy concepts are briefly discussed in Section 3. In Section 4, we generalize version space learning to the fuzzy case, and in Section 5 we consider the problem of inducing a single consistent (fuzzy) hypothesis for a given set of examples.

2   The Learning Task

We assume that membership degrees are measured on a finite scale

$$\mathcal{L} = \{ \lambda_0, \lambda_1, \ldots, \lambda_m \} \tag{1}$$

that is endowed with a complete order $\preceq$, i.e. $(\mathcal{L}, \preceq)$ is a *chain*. As usual we denote by $\prec$ the strict part of $\preceq$. The λ_i in (1) are such that $\lambda_0 \prec \lambda_1 \prec \ldots \prec \lambda_m$. Thus, $C(u) \prec C(u')$ means that the concept C does more apply to u' than to u or, say, that u' is more prototypical of C. Note that a finite scale (1) is usually sufficient from a practical point of view. In fact, the assumption that membership degrees can be measured precisely in $[0, 1]$ is even unrealistic in most situations.

An object must be distinguished from its formal representation. In machine learning, an object is usually described through a fixed set of attributes (features). Thus, an object $u \in \mathcal{U}$ is identified with a vector $x = x(u) \in \mathcal{X}$, where $\mathcal{X}$ is the feature space. Note that the mapping $\mathcal{U} \to \mathcal{X}$ is not necessarily injective, i.e. different objects might be indistinguishable in terms of the attributes under consideration. Still, we assume that these attributes are sufficient to guarantee a well-defined fuzzy concept C in the form of a mapping $\mathcal{X} \to \mathcal{L}$, which means that $x(u) = x(u')$ implies $C(u) = C(u')$. Consequently, C can be defined as a subset of $\mathcal{X}$.

We consider the problem of learning a fuzzy concept C from given examples in the form of labeled (classified) objects, where the label (class) of an object is its membership degree in C. More precisely: Given a sample

$$S = \{ \langle x_1, \lambda_{x_1} \rangle, \langle x_2, \lambda_{x_2} \rangle, \ldots, \langle x_n, \lambda_{x_n} \rangle \} \subseteq (\mathcal{X} \times \mathcal{L})^n, \tag{2}$$

where λ_x denotes the label (membership degree) of object $x \in \mathcal{X}$, induce a complete *concept description* from these examples, that is a mapping $C^{\mathrm{est}} : \mathcal{X} \to \mathcal{L}$. This mapping is an estimation (approximation) of the true (but unknown) concept C. As an aside, let us mention that we are well aware of the subjectivity and context-dependency of fuzzy concepts in general and, hence, of the "true" concept C, whose existence is taken for granted, in particular.

3   Approaches to Learning Fuzzy Concepts

How can standard machine learning algorithms be used or extended for inducing fuzzy concepts? In principle, passing from ordinary concepts to fuzzy concepts comes down to passing from the problem of *classification* to the one of *regression*: An ordinary concept is identified by an indicator function $\mathcal{X} \rightarrow \{0, 1\}$, whereas a fuzzy concept is defined in terms of a more general membership function $\mathcal{X} \rightarrow \mathcal{L}$. If membership degrees are expressed in terms of real numbers, learning a membership function is a standard regression problem (actually, it is a *constrained* regression problem since $\mathcal{L}$ is usually restricted to the unit interval). If we assume an ordinal scale (1), then we have a problem of *ordinal regression*.

As can be seen, one possibility to address fuzzy concept learning is to proceed from standard regression methods, or – when taking our ordinal setting outlined in Section 2 as a point of departure – to apply methods of ordinal regression. In fact, ordinal regression is well-known in the field of statistics [5] and has recently attracted attention in the machine learning community as well [1, 2, 3, 4].

Yet, we suggest a different alternative, namely to reduce the problem of inducing a fuzzy concept to a set of standard classification problems. This idea is directly related to the level-cut representation of a fuzzy set. In fact, deciding whether $\lambda_x \succeq \lambda$ for an object x and a label $\lambda \in \mathcal{L}$ corresponds to deciding whether x is an element of the λ-cut $C_\lambda =_{\text{def}} \{ x \in \mathcal{X} \mid C(x) \succeq \lambda \}$ of the fuzzy set C under consideration. This suggests learning a fuzzy concept C "levelwise", by learning the "ordinary" concepts C_λ for all $\lambda \in \mathcal{L}$.

This approach seems to be attractive for several reasons, notably the following: Firstly, an indicator function is a rather simple type of function, and hence solving a (binary) classification problem is usually simpler than solving a regression problem. Secondly, the λ-cuts C_λ are *nested* in the sense that $C_{\lambda_j} \subseteq C_{\lambda_i}$ for $i < j$. This property puts a structure on the sequence of binary classification problems associated with the problem of learning a fuzzy concept. In fact, it suggests a kind of "relaxation learning": Start with learning a description D_m of the concept C_{λ_m}, that is of the prototypes of the fuzzy concept C, and induce a description D_{i-1} of $C_{\lambda_{i-1}}$ on the basis of D_i, by relaxing some of the conditions imposed by D_i. Thirdly, the representation of a fuzzy concept in the form of descriptions D_i of λ-cuts C_{λ_i} appears very natural and allows for an intuitive interpretation of an induced model. Especially, the relaxation of D_i reveals the properties that demarcate objects with membership degree λ_i from objects with a lower membership λ_{i-1}. Such information is particularly interesting if membership degrees are interpreted in terms of preference.

4 Generalized Version Space Learning

Let $\mathcal{H}$ denote a fixed *hypothesis space*, where each hypothesis $h \in \mathcal{H}$ is associated with a binary classification function $h : \mathcal{X} \rightarrow \{0, 1\}$. We assume that $\mathcal{H}$ is sufficiently rich to reconstruct the sample (2) correctly: For any C_{λ_j}, $0 \leq j \leq m$, there exists a *consistent* hypothesis $h_j \in \mathcal{H}$, where consistency means that h_j classifies all members of C_{λ_j} as *positive* and all non-members as *negative*:

$$\forall 1 \leq i \leq n : (h_j(x_i) = 1) \Leftrightarrow (x_i \in C_{\lambda_j}).$$

Note that $C_\lambda = \mathcal{X}$ for the smallest membership degree $\lambda = \lambda_0$, hence $\mathcal{H}$ contains the hypothesis $h \equiv 1$.

4.1   Version Space Learning

The hypothesis space $\mathcal{H}$ can be endowed with a partial order relation:

$$h_1 \trianglelefteq h_2 \Leftrightarrow_{\text{def}} \{x \mid h_1(x) = 1\} \subseteq \{x \mid h_2(x) = 1\}. \tag{3}$$

If (3) holds, then h_1 is said to be at least as *specific* as h_2, or h_2 to be at least as *general* as h_1. We denote by $\triangleleft$ the strict part of $\trianglelefteq$, i.e. $h_1 \triangleleft h_2 \Leftrightarrow_{\text{def}}$ $(h_1 \trianglelefteq h_2) \wedge \neg (h_2 \trianglelefteq h_1)$. If $h_1 \triangleleft h_2$ holds, then h_1 is called a specialization of h_2 and h_2 a generalization of h_1. The ordering (3) defines a complete lattice structure with top element $h_\top \equiv 1$ and bottom element $h_\perp \equiv 0$.

In version space learning [6], the lattice structure is exploited in order to efficiently maintain a representation of the complete subset of hypotheses $h \in \mathcal{H}$ that are consistent with a given sample

$$S = \{(x_1, y_1), (x_2, y_2), \ldots, (x_n, y_n)\} \in (\mathcal{X} \times \{0, 1\})^n,$$

i.e. that classify all examples correctly. It is not difficult to prove that this subset is given by the *version space*

$$\mathcal{V} = \mathcal{V}(S) = \langle \mathcal{V}_s, \mathcal{V}_g \rangle =_{\text{def}} \{h \in \mathcal{H} \mid \exists\, h_s \in \mathcal{V}_s, h_g \in \mathcal{V}_g : h_s \trianglelefteq h \trianglelefteq h_g\},$$

where the *specific boundary* $\mathcal{V}_s$ and the *general boundary* $\mathcal{V}_g$ are, respectively, the set of all maximally specific and maximally general consistent hypotheses. More precisely, $\mathcal{V}_s$ consists of all hypotheses $h \in \mathcal{H}$ for which the following holds: h is consistent with S and there is no $h' \in \mathcal{H}$, also consistent with S, and such that $h' \triangleleft h$. Likewise, a hypothesis $h \in \mathcal{H}$ belongs to $\mathcal{V}_g$ if it is consistent with S and if there is no $h' \in \mathcal{H}$ which is also consistent with S and such that $h \triangleleft h'$.

Version space learning basically comes down to maintaining the boundaries $\mathcal{V}_s$ and $\mathcal{V}_g$, which means updating these boundaries for each new observation. The following cases can occur for a hypothesis $h \in \mathcal{V}_s$: (1) The new example is a *false positive*, i.e. it is erroneously classified as positive by h. Therefore, h is too general and must be eliminated from $\mathcal{V}_s$. (2) The new instance is a *false negative* for h. In that case, h is replaced by all its immediate generalizations. Likewise, the following cases are distinguished for a hypothesis $h \in \mathcal{V}_g$: (1) The new example is a *false positive*. Thus, h is replaced by all its immediate specializations. (2) The new instance is a *false negative*, hence h is eliminated from $\mathcal{V}_g$.

Learning starts with $\mathcal{V}_s = \{h_\perp\}$ and $\mathcal{V}_g = \{h_\top\}$, and proceeds by updating the current boundaries for each of the examples (x_i, y_i), $1 \leq i \leq n$. Note that $\mathcal{V}_s = \emptyset$ or $\mathcal{V}_g = \emptyset$ indicates that no consistent hypothesis exists. Different implementations of version space learning have been proposed so far. A key point in version space learning is the (efficient) computation of the immediate specializations and generalizations of a hypothesis. Needless to say, the latter strongly depends on the representation of hypotheses.

Given a new query instance x_0 whose label is to be estimated, the set

$$\{h(x_0) \mid h \in \mathcal{V}(S)\}$$

of predictions made by the consistent hypotheses constitutes the set of *possible* labels. If a single label must be estimated, one may suggest e.g. the majority vote as the most likely candidate.

4.2   Generalized Version Space Learning

Applying version space learning to fuzzy concept induction amounts to constructing a version space for each of the level-cuts C_λ, $\lambda \in \mathcal{L}$. The first step is to replace the original sample (2) by m samples

$$ S_j = \left\{ (x_1, y_1^j), (x_2, y_2^j), \ldots, (x_n, y_n^j) \right\}, $$

$1 \leq j \leq m$, where $y_i^j = 1$ if $\lambda_{x_i} \geq \lambda_j$ and $y_i^j = 0$ otherwise. Note that we do not need to learn a version space for the λ_0-cut of the fuzzy concept C, which is simply given by $\mathcal{X}$. Therefore, it is not necessary to consider a sample S_0.

Let $\mathcal{V}^j =_{\text{def}} \langle \mathcal{V}_s^j, \mathcal{V}_g^j \rangle$ denote the version space constructed for the sample S_j. Given a new query x_0 to be classified, one has to distinguish the following cases for any of the version spaces $\mathcal{V}^j$: (1) $h(x_0) = 0$ for all hypotheses $h \in \mathcal{V}^j$. (2) $h(x_0) = 1$ for all hypotheses $h \in \mathcal{V}^j$. (3) There are hypotheses $h_1, h_2 \in \mathcal{V}^j$ such that $h_1(x_0) = 0$ and $h_2(x_0) = 1$. In the first case, there is a common agreement that x_0 is not an element of the λ_j-cut of C. In the second case, there is an agreement that it belongs to the λ_j-cut, and in the third case the situation is ambivalent. This suggests approximating the true degree of membership of x_0 by the following interval: $\lambda_{x_0}^{\text{est}} = [\lambda_{x_0}^l, \lambda_{x_0}^u]$, where

$$ \lambda_{x_0}^l =_{\text{def}} \max \left\{ \lambda_j \,|\, \forall\, h \in \mathcal{V}^j : h(x_0) = 1 \right\}, $$
$$ \lambda_{x_0}^u =_{\text{def}} \max \left\{ \lambda_j \,|\, \exists\, h \in \mathcal{V}^j : h(x_0) = 1 \right\}. $$

Of course, this definition makes sense only if $\lambda_{x_0}^l \leq \lambda_{x_0}^u$. Fortunately, this property can be guaranteed by the following result.[1]

Theorem 1. For a version space $\mathcal{V} \neq \emptyset$ and a query x_0, let $\mathcal{V}(x_0) =_{\text{def}} \{h(x_0) \,|\, h \in \mathcal{V}\}$. Moreover, define the following ordering on $2^{\{0,1\}}$: $\{0\} \leq \{0,1\} \leq \{1\}$. Then, the following holds true: $\mathcal{V}^j(x_0) \geq \mathcal{V}^k(x_0)$ for all $j < k$.

Example 1. Suppose that an instance x is a tuple $(x_1, x_2) \in \mathfrak{N}^2$ of natural numbers, and that the hypothesis space $\mathcal{H}$ is given by the class of rectangles $[a, b, c, d] =_{\text{def}} \{(u, v) \,|\, a \leq u \leq b, c \leq v \leq d\}$. Since $\mathcal{H}$ is closed under intersection, the specific boundary $\mathcal{V}_s(S_j)$ is then given by a single rectangle, namely the smallest one that covers all examples belonging to the λ_j-cut of the fuzzy concept C. The general boundary $\mathcal{V}_g(S_j)$ is given by the set of rectangles that are maximally extended under the constraint of not covering any example x_i such that $\lambda_{x_i} < \lambda_j$.

5   Learning a Single Hypothesis

Version space learning is usually quite complex from a computational point of view. Thus, one would often be satisfied with finding a single consistent hypothesis. To avoid a more or less arbitrary choice one could try to find a maximally

---

[1] We omit proofs for reasons of space.

specific hypothesis among the consistent ones, which can be seen as a "cautious" strategy. The problem in connection with learning fuzzy concepts is the fact that two hypotheses $h_\jmath \in \mathcal{V}^\jmath$ and $h_k \in \mathcal{V}^k$, whether maximally specific or not, do not necessarily obey the monotonicity condition $C_{\lambda_k}^{\mathrm{est}} \subseteq C_{\lambda_\jmath}^{\mathrm{est}}$ for $\jmath < k$, where $C_{\lambda_k}^{\mathrm{est}}$ is the estimated λ_k-cut of C induced by h_k. In fact, this condition is violated as soon as $h_k \trianglelefteq h_\jmath$ does not hold.

The goal of learning a single hypothesis for a fuzzy concept can hence be defined as follows: Induce a *monotone* sequence

$$h_m \to h_{m-1} \to \ldots \to h_1, \tag{4}$$

of *consistent* hypotheses, where consistency means that $h_\jmath \in \mathcal{V}^\jmath$ ($h_\jmath$ is consistent with $S_\jmath$), and monotonicity that $h_{\jmath+1} \trianglelefteq h_\jmath$ for $m < \jmath \leq 1$. The (hypothetical) fuzzy concept determined by such a sequence is then given by the mapping

$$C^{\mathrm{est}} : x \mapsto \max\{\lambda_\jmath \,|\, 0 \leq \jmath \leq m, \, h_\jmath(x) = 1\},$$

where $h_0 = h_\top \equiv 1$. Henceforth, we shall call a monotone sequence (4) of consistent hypotheses a *consistent compound hypothesis*.

5.1   Searching for Consistent Compound Hypotheses

In order to find a consistent compound hypotheses, we pursue the following idea: We start with a hypothesis $h_m \in \mathcal{V}^m$, that is with an approximation of the λ_m-cut C_{λ_m}. Having constructed a subsequence $h_m \to \ldots \to h_\jmath$, we try to continue this sequence by generalizing the hypothesis $h_\jmath$ appropriately. If this does not work, we have to backtrack, i.e. we have to modify some of the hypotheses in the tail of the current sequence.

Let

$$\mathcal{G}(h) =_{\mathrm{def}} \{h' \in \mathcal{H} \,|\, h \trianglelefteq h'\}$$

denote the set of all generalizations of the hypothesis h. As before, we make the assumption that $\mathcal{V}(S_\jmath) \neq \emptyset$ for all $1 \leq \jmath \leq m$. Now, suppose that $h_\jmath$ is a hypothesis for the concept $C_{\lambda_\jmath}$. In order to find a generalization of $h_\jmath$ that is consistent with the concept $S_{\lambda_{\jmath-1}}$ we consider the set $\mathrm{Cand}(S_{\lambda_{\jmath-1}}, h_\jmath)$ of hypotheses. The latter is defined as the set of maximally specific elements of the set $\mathcal{V}(S_{\lambda_{\jmath-1}}) \cap \mathcal{G}(h_\jmath)$, i.e. those hypotheses h satisfying

$$h \in \mathcal{V}(S_{\lambda_{\jmath-1}}) \cap \mathcal{G}(h_\jmath) \wedge \forall h' \in \mathcal{H} : (h' \triangleleft h) \Rightarrow (h' \notin \mathcal{V}(S_{\lambda_{\jmath-1}}) \cap \mathcal{G}(h_\jmath)). \tag{5}$$

Note that the elements of $\mathrm{Cand}(S_{\lambda_{\jmath-1}}, h_\jmath)$ can be computed by successively generalizing $h_\jmath$ in a minimal way until a hypothesis in $\mathcal{V}(S_{\lambda_{\jmath-1}})$ is obtained. In fact, the computation of $\mathrm{Cand}(S_{\lambda_{\jmath-1}}, h_\jmath)$ corresponds to the computation of a *specific boundary* as in version space learning. The only difference concerns the fact that search does not start at the maximally specific hypothesis $h_\perp$. Therefore, it may happen that $\mathrm{Cand}(S_{\lambda_{\jmath-1}}, h_\jmath) = \emptyset$.

Now, we realize fuzzy concept learning by searching for a monotone sequence (4) in a systematic way. To this end, we define a suitable search tree T, which can be traversed, e.g., in a depth-first manner, as follows: Each node represents a hypothesis $h \in \mathcal{H}$. The root of the tree is the hypothesis $h_\perp$. The nodes at level 1 correspond to the hypotheses $h \in \mathcal{V}_s(S_m)$, i.e. the hypotheses h such that

$$h \in \mathcal{V}(S_m) \wedge \forall h' \in \mathcal{H} : (h' \lhd h) \Rightarrow (h' \notin \mathcal{V}(S_m)).$$

A node of depth $\jmath$ represents a hypothesis $h_{m-\jmath+1}$ for the concept $C_{\lambda_{m-\jmath+1}}$. The successor nodes of a node that represents a hypothesis $h_\imath$ is the set of nodes associated with the hypotheses $\mathrm{Cand}(S_{\lambda_{\jmath-1}}, h_\jmath)$ which are given by (5). The possibility of computing this set (efficiently) is a prerequisite of the learning method.

The following result shows that searching the tree T as defined above will be successful in the sense that a consistent compound hypothesis will be found whenever such a hypothesis exists.

Theorem 2. The tree T as defined above contains a consistent compound hypothesis whenever such a hypothesis exists.

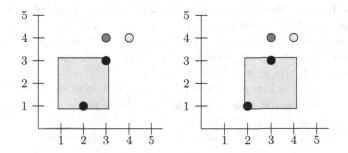

Fig. 1. The left hypothesis $h_3 = [1, 3, 1, 3]$ can be generalized to a consistent hypothesis $h_2 - [0, 3, 1, 4]$. A consistent generalization does not exist for the right hypothesis.

Example 2. If hypotheses are rectangles, as in Example 1, it is easy to see that a monotone subsequence can always be extended. In this case, the above algorithm simply yields the sequence of maximally specific hypotheses $h_\jmath \in \mathcal{V}_s(S_\jmath)$. As already said in Example 1, these hypotheses are unique.

In order to illustrate the need to backtrack, let the hypothesis space $\mathcal{H}$ be reduced to the class of *squares*. Figure 1 shows four observations: Two objects x with $\lambda_x = \lambda_3$ (black points), one object x with $\lambda_x = \lambda_2$ and one object x with $\lambda_x = \lambda_1$ (light point). As can be seen, there are two maximally specific hypotheses consistent with the sample S_3, namely $h_3 = [1, 3, 1, 3]$ and $h'_3 = [2, 4, 1, 3]$. However, the latter cannot be generalized to a hypothesis consistent with S_2. The first hypothesis h_3 can be generalized to the consistent hypothesis $h_2 = [0, 3, 1, 4]$. The latter can in turn be generalized to consistent hypotheses $[0,4,0,4]$ and $[0,4,1,5]$. Thus, the complete search tree contains two consistent compound hypotheses: $[1,3,1,3] \rightarrow [0,3,1,4] \rightarrow [0,4,0,4]$ and $[1,3,1,3] \rightarrow [0,3,1,4] \rightarrow [0,4,1,5]$. Note that neither $[0,4,0,4]$ nor $[0,4,1,5]$ is maximally specific among the hypotheses consistent with the sample S_1. Both hypotheses are, however, maximally specific consistent *generalizations* of $h_2 = [0,3,1,4]$.

6   Summary and Conclusion

We developed generic algorithms for inducing fuzzy concepts within the framework of version space learning. The first method maintains the complete (generalized) version space and suggests interval-valued predictions of membership functions. The second method is computationally less expensive and generates a single consistent hypothesis. This is accomplished by systematically searching the hypothesis space for a *monotone sequence* of consistent hypotheses, each of which approximates a level-cut of the underlying fuzzy concept.

An important point of future work concerns the practical realization of the above algorithms by means of concrete hypothesis spaces and related learning procedures. This has already been done for some selected types of logical concept descriptions, e.g. CNF formulae. For these approaches, we have also carried out some experimental studies. For reasons of space, these experiments are not presented here; an extended version of the paper (technical report) including these results can be requested from the author.

A further aspect of future work is the generalization of the framework developed in this paper to the case where the hypothesis space $\mathcal{H}$ is not guaranteed to contain a compound hypothesis consistent with the sample, i.e. to the case where the learning problem can only be solved in an approximate way.

References

[1] K. Cao-Van and B. De Baets. Impurity measures in ranking problems. Technical report, Ghent University, Belgium, 2001.

[2] E. Frank and M. Hall. A simple approach to ordinal classification. In *Proc. ECML–2001, 12th European Conference on Machine Learning*, pages 145–156, Freiburg, Germany, 2001.

[3] R. Herbrich, T. Graepel, and K. Obermayer. Regression models for ordinal data: A machine learning approach. Technical Report TR 99-3, Department of Computer Science, Technical University of Berlin, Berlin, Germany, 1999.

[4] S. Kramer, G. Widmer, B. Pfahringer, and M. De Groeve. Prediction of ordinal classes using regression trees. *Fundamenat Informaticae*, 34:1–15, 2000.

[5] P. McCullagh and J.A. Nelder. *Generalized Linear Models*. Chapman & Hall, London, 1983.

[6] T.M. Mitchell. *Version spaces: An approach to concept learning*. PhD thesis, Electrical Engineering Department, Stanford University, Stanford, CA, 1979.

[7] A.L. Ralescu and J.F. Baldwin. Concept learning from examples and counter examples. *International Journal of Man-Machine Studies*, 30(3):329–354, 1989.

A Symbolic Approximate Reasoning under Fuzziness

Mazen El-Sayed and Daniel Pacholczyk

Faculty of sciences, University of Angers
2 Boulevard Lavoisier, 49045 ANGERS Cedex 01, France
{elsayed,pacho}@univ-angers.fr

Abstract. We study knowledge-based systems using symbolic many-valued logic. In previous papers we have proposed a symbolic representation of nuanced statements. Firstly, we have introduced a symbolic concept whose role is similar to the role of the membership function within a fuzzy context. Using this concept, we have defined *linguistic modifiers*. In this paper, we propose new deduction rules dealing with nuanced statements. More precisely, we present new *Generalized Modus Ponens rules* within a many-valued context.

1 Introduction

The development of knowledge-based systems is a rapidly expanding field in applied artificial intelligence. The knowledge base is comprised of a database and a rule base. We suppose that the database contains facts representing *nuanced statements*, like "x is m_α A" where m_α and A are labels denoting respectively a nuance and a vague or imprecise term of natural language. The rule base contains rules of the form *"if x is m_α A then y is m_β B"*. Our work presents a symbolic-based model which permits a qualitative management of vagueness in knowledge-based systems. In dealing with vagueness, there are two issues of importance: (1) how to represent vague data, and (2) how to draw inference using vague data. When the imprecise information is evaluated in a *numerical way*, fuzzy logic which is introduced by Zadeh [9], is recognized as a good tool for dealing with aforementioned issues and performing reasoning upon vague knowledge-bases. A second formalism, refers to *multiset theory* and a *symbolic many-valued logic* [6, 8], is used when the imprecise information is evaluated in a *symbolic way*. In previous papers [4, 5], we have proposed a symbolic model to represent nuanced statements. This model is based on multiset theory and a many-valued logic proposed by Pacholczyk [8]. In this paper, our basic contribution has been to propose deduction rules dealing with nuanced information. For that purpose, we propose deduction rules generalizing the *Modus Ponens* rule in a many-valued logic context [8]. The first version of this rule has been proposed in a fuzzy context by Zadeh [9] and has been studied later by various authors [1, 2, 7]. In section 2, we present briefly the basic concepts of the

T. Bilgiç et al. (Eds.): IFSA 2003, LNAI 2715, pp. 685–693, 2003.
© Springer-Verlag Berlin Heidelberg 2003

M-valued predicate logic which forms the backbone of our work. Section 3 introduces briefly the symbolic representation model previously proposed. In section 4, we propose new *Generalized Modus Ponens* rules in which we use only simple statements. In section 5, we propose *Generalized Modus Ponens* rules in more complex situations.

2   M-valued Predicate Logic

Within a multiset context, to a vague term A and a nuance m_α are associated respectively a multiset $\mathbb{A}$ and a symbolic degree τ_α. So, the statement "x is m_α A" means that x belongs to multiset $\mathbb{A}$ with a degree τ_α. The M-valued predicate logic [8] is the logical counterpart of the multiset theory. In this logic, to each multiset $\mathbb{A}$ and a membership degree τ_α are associated a M-valued predicate **A** and a truth degree τ_α−true. In this context, the following equivalence holds: x is m_α A $\Leftrightarrow$ $x \in_\alpha \mathbb{A}$ $\Leftrightarrow$ "x is m_α **A**" is true $\Leftrightarrow$ "x is **A**" is τ_α−true. One supposes that the membership degrees are symbolic degrees which form an ordered set $\mathcal{L}_M = \{\tau_\alpha, \alpha \in [1, M]\}$. This set is provided with the relation of a total order: $\tau_\alpha \leq \tau_\beta \Leftrightarrow \alpha \leq \beta$. We define in $\mathcal{L}_M$ two operators $\wedge$ and $\vee$ and a decreasing involution $\sim$ as follows: $\tau_\alpha \vee \tau_\beta = \tau_{max(\alpha,\beta)}, \tau_\alpha \wedge \tau_\beta = \tau_{min(\alpha,\beta)}$ and $\sim \tau_\alpha = \tau_{M+1-\alpha}$. On this set, an implication $\rightarrow$ and a T-norm T are defined respectively as follows: $\tau_\alpha \rightarrow \tau_\beta = \tau_{min(\beta-\alpha+M,M)}$ and $T(\tau_\alpha, \tau_\beta) = \tau_{max(\beta+\alpha-M,1)}$.

Example 1. For example, by choosing M=9, we can introduce: $\mathcal{L}_9 = \{not\ at\ all,$ *little, enough, fairly, moderately, quite, almost, nearly, completely}*.

3   Representation of Nuanced Statements

Let us suppose that our knowledge base is characterized by a finite number of concepts $\mathcal{C}_i$. A set of terms P_{ik} is associated with each concept $\mathcal{C}_i$, whose respective domain is denoted as X_i. As an example, terms such as *"small"*, *"moderate"* and *"tall"* are associated with the particular concept *"size of men"*. We have assumed that some nuances of natural language must be interpreted as *linguistic modifiers*. In the following, we designate by m_α a linguistic modifier. In previous papers [4, 5], we have proposed a symbolic-based model to represent nuanced statements. In the following, we present a short review of this model. We have proposed firstly a new method to symbolically represent vague terms. In this method, we assume that a domain of a vague term, denoted by X, is not necessarily a numerical scale. This domain is simulated by a *"rule"* (Figure 1) representing an arbitrary set of objects. Our basic idea has been to associate with each multiset P_i a symbolic concept which represents an equivalent to the membership function in fuzzy set theory. For that, we have introduced a new concept, called *"rule"*, which has a geometry similar to a membership L-R function and its role is to illustrate the membership graduality to the multisets.

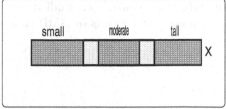

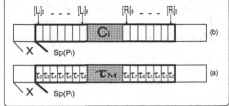

Figure 1: An universe X   **Figure 2:** a *"rule"* associated with P_i

By using the "rule" concept we have defined some linguistic modifiers. We have used two types of linguistic modifiers.

- *Precision modifiers*: They increase or decrease the precision of the basic term. We distinguish two types of precision modifiers: contraction modifiers and dilation modifiers. We use $\mathbb{M}_6 = \{m_k | k \in [1..6]\} = \{exactly, really, \emptyset, more~or~less, approximately, vaguely\}$ which is totally ordered by $j \le k \Leftrightarrow m_j \le m_k$.

- *Translation modifiers*: They operate both a translation and precision variation on the basic term. We use $\mathbb{T}_9 = \{t_k | k \in [1..9]\} = \{extremely~little, very~very~little, very~little, rather~little, \emptyset, rather, very, very~very, extremely\}$ totally ordered by $k \le l \Longleftrightarrow t_k \le t_l$. The multisets $t_k P_i$ cover the domain X.

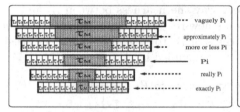

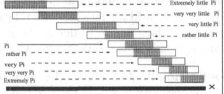

Figure 3: Precision modifiers   **Figure 4:** Translation modifiers

In this paper, we continue to propose our model for managing nuanced statements and we focus to study the exploitation of nuanced statements.

4   Exploitation of Nuanced Statements

In this section, we are interested to propose some generalizations of the Modus Ponens rule within a many-valued context [8]. We notice that the classical Modus Ponens rule has the following form: If we know that $\{If~"x~is~A"~then~"y~is~B"~is~true~and~"x~is~A"~is~true\}$ we conclude that "y is B" is true. In a many-valued context, a generalization of Modus Ponens rule has one of two forms:

F1- If we know that $\{If~"x~is~A"~then~"y~is~B"~is~\tau_\beta\text{-}true~and~"x~is~A'~"~is~\tau_\epsilon\text{-}true\}$ and that $\{A'~is~more~or~less~near~to~A\}$, what can we conclude for "y is B", in other words, to what degree "y is B" is true?

F2- If we know that $\{If~"x~is~A"~then~"y~is~B"~is~\tau_\beta\text{-}true~and~"x~is~A'~"~is~\tau_\epsilon\text{-}true\}$ and that $\{A'~is~more~or~less~near~to~A\}$, can we find a B' such as $\{B'~is~more~or~less~near~to~B\}$ and to what degree "y is B'" is true?

These forms of *Generalized Modus Ponens* (GMP) rule have been studied firstly by Pacholczyk in [8]. In this section, we propose new versions of GMP rule in which we use new relations of nearness.

4.1   First GMP Rule

In Pacholczyk's versions of GMP rule, the concept of nearness binding multisets A and A' is modeled by a similarity relation which is defined as follows:

Definition 1. *Let A and B be two multisets. A is said to be τ_α- similar to B, denoted as $A \approx_\alpha B$, iif: $\forall x | x \in_\gamma A$ and $x \in_\beta B \Rightarrow min\{\tau_\gamma \to \tau_\beta, \tau_\beta \to \tau_\gamma\} \geq \tau_\alpha$.*

This relation is (1) reflexive: $A \approx_M A$, (2) symmetrical: $A \approx_\alpha B \Leftrightarrow B \approx_\alpha A$, and (3) weakly transitive: $\{A \approx_\alpha B, B \approx_\beta C\} \Rightarrow A \approx_\gamma C$ with $\tau_\gamma \geq T(\tau_\alpha, \tau_\beta)$ where T is a T-norm. By using the similarity relation to model the nearness binding between multisets, the inference rule can be interpreted as: {*more the rule and the fact are true*} and {*more A' and A are similar*}, *more the conclusion is true*. In particular, when A' is more precise than A ($A' \subset A$) but they are very weakly similar, any conclusion can be deduced or the conclusion deduced isn't as precise as one can expect. This is due to the fact that the similarity relation isn't able alone to model in a satisfactory way the nearness between A' and A. For that, we add to the similarity relation a new relation called *nearness relation* and which has as role to define the nearness of A' to A when $A' \subset A$. In other words, it indicates the degree to which A' is included in A.

Definition 2. *Let $A \subset B$. A is said to be τ_α-near to B, denoted as $A \sqsubset_\alpha B$, if and only if $\{\forall x \in F(B), x \in_\beta A$ and $x \in_\gamma B \Rightarrow \tau_\alpha \to \tau_\beta \leq \tau_\gamma\}$.*

The nearness relation satisfies the following properties: (1) Reflexivity: $A \sqsubset_M A$, and (2) Weak transitivity: $A \sqsubset_\alpha B$ and $B \sqsubset_\beta C \Rightarrow A \sqsubset_\gamma C$ with $\tau_\gamma \leq min(\tau_\alpha, \tau_\beta)$. In the relation $A \sqsubset_\alpha B$, the less the value of α is, the more A is included in B. Finally, by using similarity and nearness relations, we propose a first *Generalized Modus Ponens* rule.

Proposition 1. *Given the following assumptions:*

1. it is τ_β-true that if "x is A" then "y is B"

2. "x is A'" is τ_ϵ-true with $A' \approx_\alpha A$.

Then, we deduce : "y is B" is τ_δ-true with $\tau_\delta = T(\tau_\beta, T(\tau_\alpha, \tau_\epsilon))$. if A' is such that $A' \sqsubset_{\alpha'} A$, we deduce: "y is B" is τ_δ-true with $\tau_\delta = T(\tau_\beta, \tau_{\alpha'} \longrightarrow \tau_\epsilon)$.

Example 2. Let "really tall" $\approx_8$ "tall" and "really tall" $\sqsubset_8$ "tall". If we have:
- if "x is tall" then "its weight is important" is true,
- "Pascal is really tall" is quite-true,
then we can deduce: "Pascal's weight is really important" is almost-true.

4.2 GMP Rules Using Precision Modifiers

In the previous paragraph we calculate the degree to which the conclusion of the rule is true. In the following, we present two new versions of GMP rule in which the predicate of the conclusion obtained by the deduction process is not B but a new predicate B' which is more or less near to B. More precisely, the new predicate is derived from B by using precision modifiers[1] ($B' = mB$). The first version assumes that the predicates A and A' are more or less similar. In other words, A' may be less precise or more precise than A. The second one assumes that A' is more precise than A.

Proposition 2. *Let the following assumptions:*

1. it is τ_β-true that if "x is A" then "y is B"

2. "x is A'" is τ_ϵ-true with $A' \approx_\alpha A$.

Let $\tau_\theta = T(\tau_\beta, T(\tau_\alpha, \tau_\epsilon))$. If $\tau_\theta > \tau_1$ then there exists a $\tau_{n(\delta)}-$dilation modifier m, with $\tau_\delta \leq T(\tau_\alpha, \tau_\beta)$, such that "y is mB" is $\tau_{\epsilon'}$-true and $\tau_{\epsilon'} = \tau_\delta \longrightarrow \tau_\theta$. Moreover, we have: $B \subset mB$ and $mB \approx_\delta B$.

This proposition prove that if we know that A' is more or less similar to A, without any supplementary information concerning its precision compared to A, the predicate of the conclusion obtained by the deduction process (mB) is less precise than B (i.e. $B \subset mB$) and which is more or less similar to B.

Proposition 3. *Let the following assumptions:*

1. it is τ_β-true that if "x is A" then "y is B"

2. "x is A'" is τ_ϵ-true with $A' \sqsubset_\alpha A$.

Let $\tau_\theta = T(\tau_\beta, \tau_\alpha \longrightarrow \tau_\epsilon)$. If $\tau_\theta > \tau_1$ then there exists a $\tau_{n(\delta)}-$contraction modifier m, with $\tau_\delta \geq \tau_\beta \longrightarrow \tau_\alpha$, such that "y is mB" is $\tau_{\epsilon'}$-true and $\tau_{\epsilon'} = T(\tau_\delta, \tau_\theta)$. Moreover, we have: $mB \sqsubset_\delta B$.

This proposition prove that from a predicate A' which is more or less near to A we obtain a predicate mB which is more or less near to B. More precisely, if A' is more precise than A then mB is more precise than B. The previous propositions (2 and 3) present two general cases in which we consider arbitrary predicates A'. In the following, we present two corollaries representing special cases of propositions 2 and 3 in which we assume that the rule is completely true and that A' is obtained from A by using precision modifiers.

Corollary 1. *Let the following rule and fact:*

1. it is true that if "x is A" then "y is B"

2. "x is $m_k A$" is τ_ϵ-true where m_k is a $\tau_{\gamma_k}-$dilation modifier.

If $T(\sim \tau_{\gamma_k}, \tau_\epsilon) > \tau_1$ then we conclude: "y is $m_k B$" is $\tau_{\epsilon'}$-true, with $\tau_{\epsilon'} =\sim \tau_{\gamma_k} \longrightarrow T(\sim \tau_{\gamma_k}, \tau_\epsilon)$.

---

[1] The definitions of these are presented in appendix A.

Example 3. Given the following data:
- if "x is tall" then "its weight is important" is true,
- "Jo is more or less tall" is moderately-true.
Then, we can deduce: "Jo's weight is more or less important" is moderately-true.

Corollary 2. *Let the following rule and fact:*

1. it is true that if "x is A" then "y is B"

2. "x is $m_k A$" is τ_ϵ-true where m_k is a τ_{γ_k}−contraction modifier.

Then, we conclude that: "y is $m_k B$" is τ_ϵ-true.

Example 4. Given the following data:
- if "x is tall" then "its weight is important" is true,
- "Pascal is really tall" is moderately-true.
Then, we can deduce: "Pascal's weight is really important" is moderately-true.

5   Generalized Production System

In this section, we present some generalizations of Modus Ponens rule in more complex situations. More precisely, we study the problem of reasoning in 4 situations: (1) when the antecedent of the rule is a conjunction of statements, (2) when the antecedent is a disjunction of statements, (3) in presence of propagation of inference (i.e. when the conclusion of the first rule is the antecedent of the second rule, and so on), and (4) when a combination of imprecision is possible (i.e. when we have some rules which have the same statement in the conclusion parts of these rules). So, we present the following 4 propositions representing inference rules in these situations.

Proposition 4 (Antecedent is a conjunction). *Given the following assumptions:*

1. if "x_1 is A_1" and ... and "x_n is A_n" then "y is B" is τ_β-true,

2. for i = 1..n, "x_i is $A_i^{'}$" is τ_{ϵ_i}-true,

3. for i = 1..n, $A_i \approx_{\alpha_i} A_i^{'}$.

Then, we can deduce: "y is B" is τ_δ-true with $\tau_\delta = T(\tau_\beta, T(\tau_{\alpha_1}, \tau_{\epsilon_1})) \wedge ... \wedge T(\tau_\beta, T(\tau_{\alpha_n}, \tau_{\epsilon_n}))$. If, for i = j .. k, the predicates $A_i^{'}$ are such that $A_i^{'} \sqsubset_{\alpha_i^{'}} A_i$, we can deduce: "y is B" is τ_δ-true with $\tau_\delta = \tau_{\delta_1} \wedge ... \wedge \tau_{\delta_n}$ and $\tau_{\delta_i} = T(\tau_{\alpha_i^{'}} \longrightarrow \tau_{\epsilon_i}, \tau_\beta)$ if $i \in [j, k]$ and $\tau_{\delta_i} = T(\tau_\beta, T(\tau_{\alpha_i}, \tau_{\epsilon_i}))$ if not.

Proposition 5 (Antecedent is a disjunction). *Given the following assumptions:*

1. if "x_1 is A_1" or ... or "x_n is A_n" then "y is B" is τ_β-true,

2. for i = 1..k, "x_i is $A_i^{'}$" is τ_{ϵ_i}-true,

3. for i = 1..k, $A_i \approx_{\alpha_i} A_i^{'}$.

Then, we can deduce: "y is B" is τ_δ-true with $\tau_\delta = T(\tau_\beta, T(\tau_{\alpha_1}, \tau_{\epsilon_1})) \vee ... \vee T(\tau_\beta, T(\tau_{\alpha_k}, \tau_{\epsilon_k}))$. If, for $i = j .. L$, the predicates A_i' are such that $A_i' \sqsubset_{\alpha_i'} A_i$, we can deduce: "y is B" is τ_δ-true with $\tau_\delta = \tau_{\delta_1} \vee ... \vee \tau_{\delta_k}$ and $\tau_{\delta_i} = T(\tau_{\alpha_i'} \longrightarrow \tau_{\epsilon_i}, \tau_\beta)$ if $i \in [j, L]$ and $\tau_{\delta_i} = T(\tau_\beta, T(\tau_{\alpha_i}, \tau_{\epsilon_i}))$ if not.

Proposition 6 (Propagation of inference). *Given the following assumptions:*

1. *if "x is A" then "y is B" is τ_β-true,*
2. *if "y is B" then "z is C" is τ_γ-true,*
3. *there exists $\tau_\epsilon > \tau_1$ such that "x is A'" is τ_ϵ-true,*
4. *there exists τ_α such that $A \approx_\alpha A'$.*

Then, we can deduce: "z is C" is τ_δ-true with $\tau_\delta = T(T(\tau_\beta, \tau_\gamma), T(\tau_\alpha, \tau_\epsilon))$. If the predicate A' is such that $A' \sqsubset_{\alpha'} A$, then we can deduce: "z is C" is τ_δ-true with $\tau_\delta = T(T(\tau_\beta, \tau_\gamma), \tau_{\alpha'} \longrightarrow \tau_\epsilon)$.

Proposition 7 (Combination of imprecision). *Given the following assumptions:*

1. *for $i = 1..n$, if "x_i is A_i" then "y is B" is τ_{β_i}-true,*
2. *for $i = 1..n$, "x_i is A_i'" is τ_{ϵ_i}-true,*
3. *for $i = 1..n$, $A_i \approx_{\alpha_i} A_i'$,*

then we can deduce that: "y is B" is τ_δ-true with $\tau_\delta = T(\tau_{\beta_1}, T(\tau_{\alpha_1}, \tau_{\epsilon_1})) \vee ... \vee T(\tau_{\beta_n}, T(\tau_{\alpha_n}, \tau_{\epsilon_n}))$.
If, for $i = j .. k$, the predicates A_i' are such that $A_i' \sqsubset_{\alpha_i'} A_i$, then we can deduce: "y is B" is τ_δ-true with $\tau_\delta = \tau_{\delta_1} \vee ... \vee \tau_{\delta_n}$ and $\tau_{\delta_i} = T(\tau_{\alpha_i'} \longrightarrow \tau_{\epsilon_i}, \tau_{\beta_i})$ if $i \in [j, k]$ and $\tau_{\delta_i} = T(\tau_{\beta_i}, T(\tau_{\alpha_i}, \tau_{\epsilon_i}))$ if not.

We present below an example in which we use the GMP rules presented in this section. In this example, we use index cards written by a doctor after his consultations. From index cards (IC_i) and some rules ($\mathcal{R}_j$), we wish deduce a diagnosis.

Example 5. Let assume that we have the following rules in our base of rules.

$\mathcal{R}_1-$ "If the temperature is high, the patient is ill" is almost true,
$\mathcal{R}_2-$ "If the tension is always high, the patient is ill" is nearly true,
$\mathcal{R}_3-$ "If the temperature is high and the eardrum color is very red, the disease is an otitis" is true,
$\mathcal{R}_4-$ "If fat eating is high, the cholesterol risk is high" is true,
$\mathcal{R}_5-$ "If the cholesterol risk is high, a diet with no fat is recommended" is true.

Let us assume now that we have an index card for a patient:

$\mathcal{F}_1-$ "the temperature is rather high" is nearly true,
$\mathcal{F}_2-$ "the tension is always more or less high" is almost true,

$\mathcal{F}_3-$ "the eardrum color is really very red" is quite true,
$\mathcal{F}_4-$ "the fat eating is very very high" is moderately true.

Using the GMP rules previously presented, we deduce the following diagnosis:

$\mathcal{D}_1-$ "the patient is ill" is almost true,
$\mathcal{D}_2-$ "the disease is an otitis" is almost true,
$\mathcal{D}_3-$ "the cholesterol risk is high" is true,
$\mathcal{D}_4-$ "a diet with no fat is recommended" is true.

6   Conclusion

In this paper, we have proposed a symbolic-based model dealing with nuanced information. This model is inspired from the representation method on fuzzy logic. In previous papers, we have proposed a new representation method of nuanced statements. In this paper, we proposed some deduction rules dealing with nuanced statements and we presented new *Generalized Modus Ponens* rules. In these rules we can use either simple statements or complex statements. Finally, we plan to study: (1) the graduality of inference and we will investigate the forms of graduality satisfied by our GMP rules, and (2) the gradual rule of the form *'the more x is A, the more y is B'.*

References

[1] J. F. Baldwin. A new approach to approximate reasoning using fuzzy logic. *Fuzzy Sets and Systems*, 2:309 – 325, 1979.
[2] B. Bouchon-Meunier and J. Yao. Linguistic modifiers and imprecise categories. *Int. J. of Intelligent Systems*, 7:25–36, 1992.
[3] D. Dubois and H. Prade. Fuzzy sets in approximate reasoning, part 1: Inference with possibility distributions. *Fuzzy Sets and Systems*, 40:143–202, 1991.
[4] M. El-Sayed and D. Pacholczyk. A qualitative reasoning with nuanced information. *8th European Conference on Logics in Artificial Intelligence (JELIA 02), 283 - 295, Italy*, 2002.
[5] M. El-Sayed and D. Pacholczyk. A symbolic approach for handling nuanced information. *IASTED International Conference on Artificial Intelligence and Applications (AIA 02), 285 - 290, Spain*, 2002.
[6] M. De glas. Knoge representation in fuzzy setting. Technical Report 48, LAFORIA, 1989.
[7] L. D. Lascio, A. Gisolfi, and U. C. Garcia. Linguistic hedges and the generalized modus ponens. *Int. Journal of Intelligent Systems*, 14:981–993, 1999.
[8] D. Pacholczyk. *Contribution au traitement logico-symbolique de la connaissance.* PhD thesis, University of Paris VI, 1992.
[9] L. A. Zadeh. A theory of approximate reasoning. *Int. J. Hayes, D. Michie and L. I. Mikulich (eds); Machine Intelligence*, 9:149–194, 1979.

Appendix A: Definitions of Precision Modifiers

We distinguish two types of precision modifiers: contraction modifiers and dilation modifiers. A contraction (resp. dilation) modifier m produces nuanced term mP_i more (resp. less) precise than the basic term P_i. In other words, the "rule" associated with mP_i is smaller (resp. bigger) than that associated with P_i. We define these modifiers in a way that the contraction modifiers contract simultaneously the core and the support of a multiset P_i, and the dilation modifiers dilate them. The amplitude of the modification (contraction or dilation) for a precision modifier m is given by a new parameter denoted as τ_γ. The higher τ_γ, the more important the modification is.

Definition 3. *m is said to be a τ_γ-contraction modifier if, and only if it is defined as follows:*

- *if $P_i = (L_i, C_i, R_i)$ then $mP_i = (L_i', C_i', R_i')$ such that $L_i' \trianglelefteq_M L_i$ and $R_i' \trianglelefteq_M R_i$*
- *$\forall x, x \in_\alpha P_i$ with $\tau_\alpha < \tau_M \Rightarrow x \in_\beta mP_i$ such that $\beta = max(1, \alpha - \gamma + 1)$*

Definition 4. *m is said to be a τ_γ-dilation modifier if, and only if it is defined as follows:*

- *if $P_i = (L_i, C_i, R_i)$ then $mP_i = (L_i', C_i', R_i')$ such that $L_i' \trianglelefteq_M L_i$ and $R_i' \trianglelefteq_M R_i$*
- *$\forall x, x \in_\alpha P_i$ with $\tau_\alpha > \tau_1 \Rightarrow x \in_\beta mP_i$ such that $\beta = min(M, \gamma + \alpha - 1)$*

Making Fuzzy Absolute and Fuzzy Relative Orders of Magnitude Consistent

Didier Dubois, Allel HadjAli, and Henri Prade

IRIT, Université Paul Sabatier
118 Route de Narbonne
31062 Toulouse Cedex 4, France
{dubois, hadjali, prade}@irit.fr

Abstract. Fuzzy absolute and fuzzy relative orders of magnitude models are recalled. Then, the problem of the consistency of these models is addressed. The paper provides conditions under which the relation between quantities estimated in terms of fuzzy absolute labels can be expressed in terms of fuzzy relative orders of magnitude. Conversely, possible estimates in terms of fuzzy absolute labels are obtained from information about fuzzy relative and fuzzy absolute orders of magnitude.

1 Introduction

Orders of magnitude were introduced in qualitative reasoning [8][11] about physical systems to solve ambiguities inherent in the sign algebra (−, 0, +) used. Orders of magnitude knowledge may be absolute or relative. The Absolute Orders of Magnitude (AOM) are based on a partition of the real line finer than the one of the sign-based model. We not only represent the sign of a quantity but also its order of magnitude. Relative Orders of Magnitude (ROM) are crisp relations that qualify the relative position of a quantity with respect to another quantity. The first approach for handling ROM relations was the formal system FOG [7]. FOG is based on three basic relations, which capture "Close to" (Cl), "Negligible with respect to" (Ne) and "Comparable to" (Co). It is described by 32 intuition-based inference rules. Several systems were proposed later to improve FOG both at modeling and reasoning levels [5][1]. Recently in [4], a fuzzy set-based approach has been proposed for handling ROM reasoning. This approach is faithful to the gradual nature of ROM relations and provides a natural interface between numbers and qualitative terms.

Generally, one type of reasoning based on AOM or ROM may be insufficient in many practical applications. Both models are necessary to capture all the relevant information. This is why it would be useful to bridge AOM and ROM. Few attempts have been made in this sense, the most noticeable one is proposed in [10] where constraints under which consistency between AOM and ROM models can hold, are established. Under these conditions, we can determine which ROM relation holds between two quantities described by AOM labels. Conversely, we can provide a possible qualification in terms of absolute qualitative labels for a quantity known to be

T. Bilgiç et al. (Eds.): IFSA 2003, LNAI 2715, pp. 694–701, 2003.

related to an AOM by a given ROM relation. The aim of this paper is to investigate how this consistency notions can be extended when AOM and ROM models are defined in the setting of fuzzy sets. Fuzzy AOM (F-AOM) models considered are the ones based on partitioning the real line into a set of fuzzy intervals [3]. The Fuzzy ROM (F-ROM) relations considered are the ones expressed in terms of Cl or Ne introduced in [4]. The next section presents the consistency problem between AOM and ROM models in the crisp case. Section 3 provides fuzzy counterparts of these orders of magnitude models. In section 4, we formulate the consistency problem in the framework of the proposed F-AOM and F-ROM models.

2 AOM and ROM Models Consistency

AOM Models. The AOM models rely on a partition of 3, each element of the partition standing for a basic qualitative class to which a label is associated. The partition is defined by a set of real landmarks including 0 and generates the universe of description of the AOM models.

Symmetrical absolute partitions are the most interesting ones in practice. The symmetrical AOM model with n positive (negative) qualitative labels is denoted by OM(n) and is referred as the AOM model of granularity n [10]. For example, OM(5) model is based on the following set of landmarks : $\{-\delta, -\beta, -\alpha, -\gamma, 0, \gamma, \alpha, \beta, \delta\}$, with the corresponding labels: NVL = $]-\infty, -\delta]$ (Negative Very Large); NL = $]-\delta, -\beta]$ (Negative Large); NM =$]-\beta, -\alpha]$ (Negative Medium); NS = $]-\alpha, -\gamma]$ (Negative Small); NVS = $]-\gamma, 0[$ (Negative Very Small); [0] = {0}; PVS = $]0, \gamma[$ (Positive Very Small); PS = $[\gamma, \alpha[$ (Positive Small); PM = $[\alpha, \beta[$ (Positive Medium); PL = $[\beta, \delta[$ (Positive Large); PVL = $[\delta, +\infty[$ (Positive Very Large). The resulting absolute partition is shown in Fig.1.

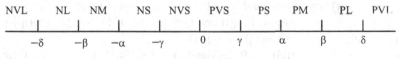

Fig.1 Crisp partition

Qualitative algebras (Q-algebras) over these models and their properties have been discussed in depth, see for instance [9].

ROM Models. ROM models are based on the definition of a set of binary relations r_i which are invariant by homothety, i.e. x r_i y only depends on the ratio x/y, and the axiomatics of which are described by a set of rules. One of the noticeable systems for handling ROM relations, which was introduced to overcome the FOG's problems (particularly, the impossibility of incorporating quantitative information when available) is the system ROM(3) proposed by Dague [1]. ROM(3) extends FOG by adding a new relation Di which stands for "Distant from".

Dague has proposed to define the four relations Ne, Cl, Co and Di in 3 by means of three primitive relations involving a single real parameter k as follows [1]:

Proximity at order k (P_k): x P_k y if and only if $|x - y| \leq k \times \text{Max}(|x|, |y|)$.

Negligibility at order k (N_k): x N_k y if and only if $|x| \leq k \times |y|$.

Distant at order k (D_k): x D_k y if and only if $|x - y| \geq k \times \text{Max}(|x|, |y|)$.

The four ROM relations are then defined as follows, based on the choice of two parameters k_1 and k_2 ($0 < k_1 \leq k_2 \leq 1/2$).

$$Cl \leftrightarrow P_{k_1}, \qquad Co \leftrightarrow P_{1-k_2}, \qquad Ne \leftrightarrow N_{k_1}, \qquad Di \leftrightarrow D_{k_2}.$$

The behavior of these relations are described by means of 15 axioms, from which about 45 useful inference rules have been established [1]. ROM(3) has two degrees of freedom, represented by the parameters k_1 and k_2, which define a relative partition of the real line. This partition concerns the quotients of quantities, for instance, x Ne y if and only if x/y belongs to $[-k_1, k_1]$. For positive numbers, this partition is characterized by the set of relative landmarks: $Q = \{k_1, k_2, 1-k_2 \ 1-k_1\}$.

Consistency of the AOM and ROM models. The notion of consistency between AOM and ROM models proposed in [10], enables us to address the two following problems: i) given two quantities, which are qualified in terms of absolute labels, how to relate them by an appropriate ROM relation; ii) conversely, given two quantities that are related by a given ROM relation, how to characterize them in terms of absolute qualitative labels. The AOM and ROM models considered are the ones previously presented and the ROM(3) model, respectively. Let us first give the definition of consistency as proposed in [10]:

Definition 1. If the quotients between the landmarks of an absolute model OM(n) coincide with the landmarks of a relative partition of ROM(3) then, OM(n) and the ROM(3) models are said to be consistent.

Following this definition, it has been shown in [10] that consistency between OM(5) models and ROM(3) model only hold in the following three cases (for positive numbers):

Case 1: The landmarks of the absolute model and landmarks of relative partition correspond to $(\gamma, \alpha, \beta, \delta) = (\alpha/q, \alpha, \beta, q\beta)$ and $Q = \{k_1, k_2, 1-k_2 \ 1-k_1\} = (\alpha/(q^2\beta), \alpha/(q\beta), \alpha/\beta, 1/q)$ with $q = \sqrt{2}$ and $\beta = (2+\sqrt{2})/2 \cdot \alpha$, respectively.

Case 2: $(\gamma, \alpha, \beta, \delta) = (\alpha/q, \alpha, \beta, q\beta)$ and $Q = \{k_1, k_2, 1-k_2 \ 1-k_1\} = (\alpha/(q^2\beta), \alpha/(q\beta), 1/q, \alpha/\beta)$ with $q \cong 1.75487$ and $\beta = \alpha/(q-1)$.

Case 3: $(\gamma, \alpha, \beta, \delta) = (\beta/q, \alpha, \beta, q\alpha)$ and $Q = \{k_1, k_2, 1-k_2 \ 1-k_1\} = (\beta/(q^2\alpha), 1/q, \beta/(q\alpha), \alpha/\beta)$ with $q \cong 2.32472$ and $\beta = (q-1)\alpha$.

A simple analysis of these results shows that consistency is highly constrained. Indeed, only one degree of freedom is preserved among four of the AOM model, while the resulting ROM(3) model is fully determined. Some examples of ROM $\rightarrow$ AOM and AOM $\rightarrow$ ROM correspondences can be provided as follows (for positive numbers):

- In any case, any very small number is negligible with respect to any very large number: $\forall x \in PVS, \forall y \in PVL$, x Ne y.
- In case 1, any very small number is not close to any medium, large or very large number: $x \in PVS$, x Cl y $\Rightarrow$ y $\in$ PVS or y $\in$ PS.
- In case 2 and 3, any very small or small number is not close to any large or large number: $x \in PVS$ (or PS), x Cl y $\Rightarrow$ y $\in$ PVS, y $\in$ PS or y $\in$ PM.

3 Fuzzy Orders of Magnitude Models

In this section, we introduce the fuzzy counterparts of AOM models presented in section 2 and the modeling of ROM(3) relations in the setting of fuzzy set theory.

F-AOM Models. Dubois and Prade [3] have proposed a fuzzy set-based model for absolute orders of magnitude by partitioning the real numbers into a set of fuzzy intervals. The labels of the absolute partition are now interpreted as overlapping fuzzy intervals as pictured in Fig.2 below.

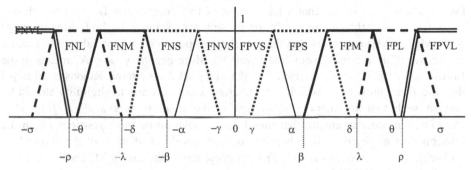

Fig.2 Fuzzy partition

The F-AOM models considered here are based on the fuzzy absolute partition {FNVL, FNL, FNM, FNS, FNVS, FPVS, FPS, FPM, FPL, FPVL}. For instance, the label FPS stands for the fuzzy class of positive small numbers with [γ, δ] as support (i.e. {t, $\mu_{FPS}(t) > 0$}). It has been shown in [3][6] that the closure and the combination tables for the fuzzy intervals of Fig.2, can be obtained by defining the operations over all possible combinations of adjacent intervals in addition to the intervals themselves.

F-ROM Model. A fuzzy set-based approach for handling relative orders of magnitude notions expressed in terms of closeness (Cl) and negligibility (Ne) has been developed in [4]. We have shown that these notions can be modeled by means of fuzzy relations controlled by tolerance parameters. The membership functions of these relations are defined in terms of ratios. The following definitions have been proposed [4]:

Closeness. The idea of relative closeness expresses the approximate equality between two real numbers x and *y* and can be captured by the following:

Definition 2. The closeness relation (Cl) is a reflexive and symmetric fuzzy relation such that

$$\mu_{Cl}(x, y) = \mu_M(x/y),$$

where μ_M is the characteristic function of a fuzzy number close to 1, such that $\mu_M(1) = 1$ (since x is close to x), and $\mu_M(t) = 0$ if t ≤ 0 (since it is required that two numbers that are close should have the same sign). *M* is called a *tolerance parameter.*

Closeness is naturally symmetric i.e., $\mu_{Cl}(x, y) = \mu_{Cl}(y, x)$. It implies that M should satisfy $\mu_M(t) = \mu_M(1/t)$. So, the support S(M) of the fuzzy number M is of the form [1−ε, 1/(1−ε)], with ε ∈ [0, 1[.

Negligibility. We define negligibility from closeness. They can be related in the following way: "x is negligible with respect to y" if and only if "x + y is close to y". It leads to define a fuzzy negligibility relation as follows,

Definition 3. Negligibility is a fuzzy relation defined from a closeness relation Cl based on a fuzzy number *M* via the following:

$$\mu_{Ne}(x, y) = \mu_{Cl}(x+y, y) = \mu_M((x+y) / y).$$

The comparability relation Co can be then defined as x Co y $\Leftrightarrow$ ¬(x Ne y ∨ y Ne x) [4]. Regarding the "Distant" relation, if x Di y means ((x − y Co y) ∨ (y − x Co y)) [10], it amounts to saying that x Di y $\Leftrightarrow$ ¬(x Cl y); moreover x Di y states a difference-based rather than a quotient-based relation w.r.t x and y. In the following, (x, y) ∈ Cl[M] (Ne[M]) expresses that the more or less possible pairs of values for x and y are restricted by the fuzzy set Cl[M] (Ne[M]). More generally "x ∈ A" means in the following that x is fuzzily restricted by the fuzzy set A. We have demonstrated in [4] that the fuzzy number M which parameterizes closeness and negligibility should be chosen such that its support S(M) lies in the validity interval $V = [(\sqrt{5}-1)/2, (\sqrt{5}+1)/2]$ in order to ensure that the closeness relation be more restrictive than the relation "not negligible". In others terms, each level cut of M is of the form $[1-\varepsilon, 1/(1-\varepsilon)]$, with $\varepsilon \in [0, (3-\sqrt{5})/2]$. The parameterized relations Cl[M] and Ne[M] will be referred to as F-ROM relations.

Note that F-ROM reasoning in terms of closeness and negligibility relations has been axiomatized by proposing a set of locally complete inference rules [4]. For example, a rule states: (a, c) ∈ Cl[M] and (b, c) ∈ Ne[N] ⇒ (a+b, c) ∈ Cl[M⊕N   1]

(where ⊕ and   denote the addition and subtraction extended to fuzzy numbers [2]). The tolerance parameters of the deduced relations are manipulated at a symbolic level in agreement with the arithmetic of fuzzy numbers. The above semantic requirements on the tolerance parameters ensure the semantic meaningfulness of the conclusions.

4   Consistency Problem in Fuzzy Case

We now examine the extension of the idea of consistency formulated in section 2 to the fuzzy case. In other terms, knowing the two quantities, which are described in terms of labels of the F-AOM models, we want to be able to establish the corresponding F-ROM relations of type Cl[M] or Ne[M] that link them. This will be referred to as the F-AOM → F-ROM problem. Conversely, given two quantities known to be related by one of the F-ROM relations Cl[M] or Ne[M], and knowing the value of one of the quantities in terms of labels of the fuzzy absolute partition, we want to be able to express the other quantity in terms of these fuzzy labels. This will be referred to as the F-ROM → F-AOM problem. In the following, and for the sake of brevity and clarity we only focus on positive numbers.

F-ROM → F-AOM Problem. Assume that a result of the form (x, y) ∈ R[M] is symbolically obtained using the inference rules mentioned in the last paragraph of section 3. R represents relation Cl or Ne. Knowing the value of one of these quantities in terms of fuzzy absolute labels, under what conditions can we characterize the other quantity in terms of these fuzzy labels? The application of the combination/projection principle on available pieces of information enables us to compute the fuzzy set that restricts the values of the quantity of interest. Then, the obtained fuzzy set will be expressed in terms of the fuzzy absolute labels. Let us consider a few examples:

i) Assume that $x \in FPVS$ and the result $(x, y) \in Cl[M]$ has been obtained. The fuzzy set B which restricts the values of y can be computed in the following way:

$$\mu_B(y) = \sup_x \min(\mu_{FPVS}(x), \mu_{Cl[M]}(x, y)) = \sup_x \min(\mu_{FPVS}(x), \mu_M(x/y))$$
$$= \sup_x \min(\mu_{FPVS}(x), \mu_M(y/x)), \textit{ since M is symmetrical}$$
$$= \mu_{FPVS \otimes M}(y).$$

Then, $B = FPVS \otimes M$ where $\otimes$ denotes the product extended to fuzzy numbers [2]. Now, if the level cuts of the fuzzy number FPVS are of the form $[\underline{x}_{FPVS}, \overline{x}_{FPVS}]$ with $\underline{x}_{FPVS} = 0$ then, the corresponding level cuts of B write: $[\underline{b}, \overline{b}] = [0, \overline{x}_{FPVS}] \cdot [1-\varepsilon, 1/(1-\varepsilon)] = [0, \overline{x}_{FPVS} + \eta \overline{x}_{FPVS}] \subseteq [0, \alpha + \eta\alpha] \subseteq [0, \alpha + (\sqrt{5}-1)/2 \cdot \alpha]$, where α is the upper bound of the support of FPVS (see Fig.2) and $\eta = \varepsilon/(1-\varepsilon) \in [0, (\sqrt{5}-1)/2]$. This means that B is less restrictive than FPVS in the sense of fuzzy sets inclusion (i.e. B $\supseteq$ FPVS) and B $\cap$ FPS $\neq \emptyset$. We can easily check that B $\cap$ FPM $= \emptyset$ if $\alpha \leq (\sqrt{5}-1)/2 \cdot \beta$. Thus, the following result can be proved:

$$\forall\, x \in FPVS, (x, y) \in Cl[M] \Rightarrow y \in FPVS \cup FPS, \text{ if } \alpha \leq (\sqrt{5}-1)/2 \cdot \beta.$$

Similarly, the following F-ROM $\to$ F-AOM correspondences can be established:

ii) If $\delta \leq (\sqrt{5}-1)/2 \cdot \lambda$,

    a) $\forall\, x \in FPS, (x, y) \in Cl[M] \Rightarrow y \in FPVS \cup FPS \cup FPM.$

    b) $\forall\, x \in FPL, (x, y) \in Cl[M] \Rightarrow y \in FPM \cup FPL \cup FPVL.$

iii) If $\theta \leq (\sqrt{5}-1)/2 \cdot \rho$, $\forall\, x \in FPVL, (x, y) \in Cl[M] \Rightarrow y \in FPL \cup FPVL.$

Proof. Let us only consider the proof of (ii-a). By applying the combination/projection principle, the fuzzy set B of values of y can be obtained as follows:

$$\mu_B(y) = \sup_x \min(\mu_{FPS}(x), \mu_{Cl[M]}(x, y)) = \sup_x \min(\mu_{FPS}(x), \mu_M(y/x)) = \mu_{FPS \otimes M}(y).$$

Then, $B = FPS \otimes M$. Now, if $[\underline{x}_{FPS}, \overline{x}_{FPS}]$ represents the level cuts of the fuzzy number FPS, the corresponding level cuts of B can be computed as follows: $[\underline{b}, \overline{b}] = [\underline{x}_{FPS}, \overline{x}_{FPS}] \cdot [1-\varepsilon, 1/(1-\varepsilon)] = [\underline{x}_{FPS} - \varepsilon \underline{x}_{FPS}, \overline{x}_{FPS} + \eta \overline{x}_{FPS}] \subseteq [\gamma - \varepsilon\gamma, \delta + \eta\delta] \subseteq [\gamma - (3-\sqrt{5})/2 \cdot \gamma, \delta + (\sqrt{5}-1)/2 \cdot \delta]$. This means that B $\supseteq$ FPS, FPVS $\cap$ B $\neq \emptyset$ and B $\cap$ FPM $\neq\emptyset$. Now, if $\delta \leq (\sqrt{5}-1)/2 \cdot \lambda$, we have $\forall\, x \in FPS, (x, y) \in Cl[M] \Rightarrow y \in FPVS \cup FPS \cup FPM.$ ◆

In the case where the F-ROM relation that holds between x and y is expressed in terms of Ne, some F-ROM $\to$ F-AOM correspondences can be provided:

iv) Assume that $y \in FPL$ and $(x, y) \in Ne[M]$. The fuzzy set A restricting the values of x can be obtained as follows:

$$\mu_A(x) = \sup_y \min(\mu_{FPL}(y), \mu_{Ne[M]}(x, y)) = \sup_y \min(\mu_{FPL}(y), \mu_M(1+x/y))$$
$$= \sup_y \min(\mu_{FPL}(y), \mu_{M-1}(x/y)), \textit{ since } \mu_M(f^{-1}(t)) = \mu_{f(M)}(t)$$
$$= \mu_{FPL \otimes (M-1)+}(x).$$

Then, $A = FPL \otimes (M \quad 1)^+$ where $(M \quad 1)^+$ is a fuzzy number such that $\mu_{(M \ 1)+}(t) = \mu_{(M \ 1)}(t)$ if $t \geq 0$ and 0 otherwise. Let $[\underline{x}_{FPL}, \overline{x}_{FPL}]$ be the level cut of the fuzzy number FPL, the corresponding level cuts of A are of the form: $[\underline{a}, \overline{a}] = [\underline{x}_{FPL}, \overline{x}_{FPL}] \cdot [0, \eta] = [0, \overline{x}_{FPL} \cdot \eta] \subseteq [0, \sigma\eta] \subseteq [0, (\sqrt{5}-1)/2 \cdot \sigma]$. Now, if $\gamma \leq (\sqrt{5}-1)/2 \cdot \sigma \leq \beta$ then the following result holds:

$$\forall y \in FPL, (x, y) \in Ne[M] \Rightarrow x \in FPVS \cup FPS.$$

v) In the same way, we can prove that if $y \in FPVL$ and $(x, y) \in Ne[M]$ then the fuzzy set restricting the values of x is $A = FPVL \otimes (M \quad 1)^+$. Now, if we consider the finite level cuts $[\underline{x}_{FPVL}, \overline{x}_{FPVL}]$ of the fuzzy number FPVL then, the level cuts of A are of the form $[\underline{a}, \overline{a}] = [0, \overline{x}_{FPVL} \cdot \eta] \subseteq [0, (\sqrt{5}-1)/2 \cdot \overline{x}_{FPVL}]$. Now, if $\gamma \leq (\sqrt{5}-1)/2 \cdot \overline{x}_{FPVL} \leq \lambda$ then the following result also holds:

$$\forall y \in FPVL, (x, y) \in Ne[M] \Rightarrow x \in FPVS \cup FPS \cup FPM.$$

F-AOM → F-ROM Problem. Given two real numbers x and y, which are qualified in terms of fuzzy absolute labels. The question that we want to answer is what F-ROM relation Cl[M] or Ne[M] may hold between x and y? Let us show on some examples how this question can be dealt with:

i) First assume that $x \in FPVS$ and $y \in FPVL$. In this case, the intuitive F-ROM relation that could relate x and y is Ne. Now, we have to compute the fuzzy set that expresses the semantics of this relation, i.e. the tolerance parameter underlying this relation. For this, we first compute the fuzzy set M of values of the function $f(x, y) = 1 + x/y$ using the extension principle in the following way:

$$\forall x \in FPVS, \forall y \in FPVL,$$
$$\sup_{w=1+x/y} \min(\mu_{FPVS}(x), \mu_{FPVL}(y)) = \mu_M(w) = \mu_M(1+x/y),$$

with $M = 1 \oplus FPVS \oslash FPVL$ where $\oslash$ denotes the operation of division extended to fuzzy numbers [2]. Now, the fuzzy number M should satisfy the following semantic requirements stressed in section 3:

$$c_1: \mu_M(1) = 1, \quad c_2: S(M) \subseteq V = [(\sqrt{5}-1)/2, (\sqrt{5}+1)/2].$$

Note that $f(x, y) = 1$ if and only if $x = 0$ or $y \to +\infty$. We can easily check that in both cases $\mu_M(1) = 1$ (since $\mu_{FPVS}(0) = 1$ and $\mu_{FPVL}(y) = 1$ when $y \to +\infty$). To verify the second requirement let us use the level cuts notions. Let $[\underline{x}_{FPVS}, \overline{x}_{FPVS}]$ and $[\underline{x}_{FPVL}, \overline{x}_{FPVL}]$ be the level cuts of the fuzzy numbers FPVS and FPVL respectively then, the level cuts of M writes: $[\underline{m}, \overline{m}] = [\underline{x}_{FPVS}, \overline{x}_{FPVS}] / [\underline{x}_{FPVL}, \overline{x}_{FPVL}] + 1 \subseteq [1, 1 + \overline{x}_{FPVS}/\underline{x}_{FPVL}]$. It is also easy to see that $[\underline{m}, \overline{m}] \subseteq [1, 1+\alpha/\rho]$. Then, the support of M lies in the interval V if $\alpha \leq (\sqrt{5}-1)/2 \cdot \rho$. Thus, the following semantic entailment holds:

$$\forall x \in FPVS, \forall y \in FPVL, (x, y) \in Ne[1 \oplus FPVS \oslash FPVL], \text{ if } \alpha \leq (\sqrt{5}-1)/2 \cdot \rho.$$

Using the same calculus, we can establish other F-AOM → F-ROM correspondences. The most interesting ones are:

ii) $\forall\, x \in$ FPS, $\forall\, y \in$ FPVL, $(x, y) \in$ Ne[1⊕ FPS⊘FPVL] if $\delta \leq (\sqrt{5}-1)/2 \cdot \rho$.

iii) $\forall\, x \in$ FPVS, $\forall\, y \in$ FPL, $(x, y) \in$ Ne[1⊕ FPVS⊘FPL] if $\alpha \leq (\sqrt{5}-1)/2 \cdot \lambda$.

But in the case where $x \in$ FPS and $y \in$ FPL, $\mu_M(1) = 1$ with M = 1⊕ FPS⊘FPL does not hold. Then, no information about the link between x and y can be obtained.

5 Conclusion

In this paper, we have shown that bridging F-AOM into F-ROM relations and conversely is possible. Consistency between F-AOM and F-ROM models is less constrained than in the crisp case. Only some conditions on the bounds on the supports of fuzzy numbers associated to the labels of the partition, are required. Another advantage of the fuzzy semantics for ROM relations is its ability to interface symbolic results expressed in terms of Cl[M] (or Ne[M]) with numerical ill-known values.

References

1. Dague P., Numeric reasoning with relative orders of magnitude, Proc. of the 11th Amer. Assoc. Conf. on AI (AAAI-93), July, 1993, Washington, DC, pp. 541-547.
2. Dubois D., Kerre E., Mesiar R., Prade H. (2000), Fuzzy Intervals Analysis, In: Fundamentals of Fuzzy Sets, The Handbooks of Fuzzy sets Series (Dubois D., Prade H., Eds), Kluwer Academic Publishers, Dordrecht, pp. 483-581.
3. Dubois D., Prade H., Fuzzy Arithmetic in Qualitative Reasoning, in Modeling and Control of Systems, A Blaquière, Ed. Lecture Notes in Control and Information Sciences, 121, Springer-Verlag (1988), pp. 457-467.
4. Hadj Ali A., Dubois D., Prade H., Qualitative reasoning based on fuzzy relative orders of magnitude, IEEE Transactions on Fuzzy Systems, Vol. 11, No 1, February 2003, pp. 9-23.
5. Mavrovouniotis M.L., Stephanopoulos G., Order-of-magnitude reasoning with O[M], Artificial Intelligence in Engineering (1989), Vol. 4, No. 3, pp. 106-114.
6. Parsons S., Interval algebra and order of magnitude reasoning, Applications of Artificial Intelligence in Engineering VI, Oxford (1991), pp. 945-961.
7. Raiman O., Order of magnitude reasoning, Proc. of the 5th Amer. Assoc. Conf. on Artificial Intelligence (AAAI-86), Philadelphia (1986), PA, pp. 100-104.
8. Travé-Massuyès L., Dague P., Guerrin F., Le Raisonnement Qualitatif pour les Sciences de l'Ingénieur, Hermès, Paris (1997).
9. Travé-Massuyès L., Piera N., Order of magnitude models as qualitative algebra, Proc. of the Inter. Joint Conf. on Artificial Intelligence, Detroit (1989), USA, pp. 1261-1266.
10. Travé-Massuyès L., Prats F., Sanchez M., Agell N., Consistent relative and absolute order of magnitude models, Int. Workshop on Qualitative Reasoning, June 9-12, 2002, Spain,
11. Weld D.S., de Kleer J., Readings in Qualitative Reasoning about Physical Systems, Morgan Kaufmann (1990), San Mateo, Ca.

Learning First Order Fuzzy Logic Rules

Henri Prade, Gilles Richard, and Mathieu Serrurier

IRIT - Universite Paul Sabatier
118 route de Narbonne 31062 Toulouse FRANCE
henri.prade@irit.fr, grichard@ifi.edu.vn, serrurier@irit.fr

Abstract. The paper presents an algorithm based on Inductive Logic Programming for inducing first order Horn clauses involving fuzzy predicates from a database. For this, a probabilistic processing of fuzzy function is used, in agreement with the handling of probabilities in first order logic. This technique is illustrated on an experimental application. The interest of learning fuzzy first order logic expressions is emphasized.

Keywords : Inductive Logic Programming, fuzzy rule, data mining, uncertainty

1   Introduction

Learning fuzzy rules is a research topic which has raised a considerable interest in the last decade. At least three main trends of works can be distinguished. First, neuro-fuzzy learning techniques have been developed for tuning fuzzy membership functions in fuzzy rules (which are first obtained by fuzzy clustering methods usually); see [10] for a survey. The fuzzy rules which are thus produced are then used for approximating functions in control problems. Another research line has been investigated with a greater concern for the descriptive power of the fuzzy rules, from a user point of view. Thus, Quinlan's [12] ID3 algorithm has been extended to fuzzy decision trees, involving a fuzzy description of classes and making use of entropy measures (extended to fuzzy sets) for building the fuzzy rules; see [3] for a survey. More recently, the use of fuzzy membership functions has been advocated by several researchers for providing association rules in data mining with a better representation power, e.g. [8].

In these different problems, fuzzy rules are derived, which generally involve *unary* fuzzy properties. Moreover, fuzzy association rules are completed with (usually scalar) confidence and support degrees. So, what is obtained has the limited expression power of propositional-like rules. Inductive Logic Programming (ILP) [11] provides a general framework for learning classical first order logic rules, for which reasonably efficient algorithms have been developed (Progol [9], FOIL [13],[14]...). The aim of this paper is to adapt the ILP approach, restricted on free-function Horn clauses, in order to allow for the learning of rules which may involve fuzzy predicates (which are not necessarily unary) and existential quantifiers, together with confidence evaluation. In order to do that, we take advantage of a probabilistic interpretation of fuzzy membership functions and of Halpern's

T. Bilgiç et al. (Eds.): IFSA 2003, LNAI 2715, pp. 702–709, 2003.
© Springer-Verlag Berlin Heidelberg 2003

view [7] for equipping first order logic with a probabilistic structure. Moreover machineries such as FOIL embed confidence degrees for controlling the learning process. This will enable us to handle the fuzziness of the predicates directly in the computation. The paper is organized as follows : Sections 2 and 3 provide the backgrounds on ILP and probabilistic logic. Section 4 presents our approach to the learning of first order logic rules involving fuzzy predicates. Finally, section 5 illustrates the approach on an example.

2 ILP Review

We first briefly recall the standard definitions and notations. Given a first order language $\mathcal{L}$ with a set of variables Var, we build the set of terms $Term$, atoms $Atoms$ and formulas as usual. The set of ground terms is the Herbrand universe $\mathcal{H}$ and the set of ground atoms or facts is the Herbrand base $\mathcal{B} \subset Atom$. A *literal* l is just an atom a (positive literal) or its negation $\neg a$ (negative literal). A (resp. ground) substitution σ is an application from Var to (resp. $\mathcal{H}$) $Term$ with inductive extension to $Atom$. We denote $Subst$ the set of ground substitutions. A *clause* is a finite disjunction of literals $l_1 \vee \ldots \vee l_n$ also denoted $\{l_1, \ldots, l_n\}$. A Horn clause is a clause with at most one positive literal. A Herbrand interpretation I is just a subset of $\mathcal{B}$: I is the set of true ground atomic formulas and its complementary denotes the set of false ground atomic formulas. Let us denote $\mathcal{I} = 2^{\mathcal{B}}$, the power set of $\mathcal{B}$ i.e. the set of all Herbrand interpretations. We can now proceed with the notion of a logical consequence.

Definition 1 *Given A an atomic formula, $I, \sigma \models A$ means that $\sigma(A) \in I$. As usual, the extension to general formulas F uses compositionality.*
$I \models F$ means $\forall \sigma, I, \sigma \models F$ (we say I is a model of F).
$\models F$ means $\forall I \in \mathcal{I}, I \models F$.
$F \models G$ means all models of F are models of G (G logical consequence of F).

Stated in the general context of first-order logic, given a background theory B and a set of observations E (training set), the task of *induction* is to find a set of formulas H such that :
$$B \cup H \models E \tag{1}$$
where E, B and H here denote sets of clauses. A set of formulas is here, as usual, considered as the conjunction of its elements.
Of course, one may add two natural restrictions :

- $B \not\models E$ since, in such a case, H would not be necessary to explain E.
- $B \cup H \not\models \bot$: this means $B \cup H$ is a consistent theory.

In the setting of relational databases, inductive logic programming (ILP) is often restricted to Horn clauses and function-free formulas, E is just a set of ground facts. Moreover, the set E itself satisfies the previous requirement, but it is generally not considered as an acceptable solution since it gives no information about new facts so it has no predictive ability. Usually, ILP fits with the idea of providing a compression of the information content of E.

There are two general types of algorithms, *top down* and *bottom up* algorithms. *Top down* ones start from the most general clause and specialize it step by step. *Bottom up* procedures start from a fact and generalize it. In our case, we will use the FOIL algorithm [13][14] which is of the *top down* type. The goal of FOIL is to produce rules until all the examples are covered. Rules with conclusion part F, the target predicate, are found in the following way:

1. take $R \rightarrow F$ the most general clause with $R = \top$
2. take the literal L such as the clause $L \wedge R \rightarrow F$ maximizes the gain function
3. $R = L \wedge R$
4. if confidence($R \rightarrow F$)<threshold goto 2
5. return $R \rightarrow F$

The gain function is computed by the formula :

$$gain(L \wedge R \rightarrow F, R \rightarrow F) = n * (log_2(cf(L \wedge R \rightarrow F)) - log_2(cf(R \rightarrow F)))$$

where n is the number of distinct examples covered by $L \wedge R \rightarrow F$. Given a Horn clause $A \rightarrow B$, the confidence $cf(A \rightarrow B) = \frac{P(A \wedge B)}{P(A)}$. The way for computing probabilities of first order logic rules is presented in the next section.

3 Probabilistic Logic

We focus here on the logic of probability as developed by Halpern in [7]. The aim of Halpern's work, following ideas in Bacchus [1], is to design a first-order logic for capturing reasoning about beliefs and statistical information. Here we restrict Halpern's framework to the usual logic programming setting. This means that the domain object is the Herbrand universe $\mathcal{H}$, an interpretation I is just a subset of $\mathcal{B}$. In fact, we just apply Halpern's definitions for attaching probabilities to Horn clauses, without using the associated notion of logical consequence.

First, let us give a meaning to the probability of a non-closed first order formula in a given interpretation. Halpern names type 1 structure the triple $M = \{I, \mathcal{H}, \mu_1\}$ where I is an Herbrand interpretation, $\mathcal{H}$ is the Herbrand universe, and μ_1 is a probabilistic measure over $\mathcal{H}$ (of course, the probability μ_1^n is available over the product domain $\mathcal{H}^n$). Given a type 1 structure M and a non closed formula F with the vector $\overrightarrow{t}$ of n free variables, the meaning of $\mu_1^I(F)$ (abbreviated in $\mu_1(F)$ when I is clear from the context) is the probability that a random vector $\overrightarrow{x}$ on $\mathcal{H}^n$ makes $\sigma[\overrightarrow{t}/\overrightarrow{x}](F)$ true in I. Hence the formal definition :

Definition 1. *Given* $M = \{I, \mathcal{H}, \mu_1\}$, F *a formula with a vector* $\overrightarrow{t}$ *on* n *free variables, the type 1 probability of* F *is :*

$$\mu_1(F) = \mu_1^n\{\overrightarrow{x} \in \mathcal{H}^n | I \models \sigma[\overrightarrow{t}/\overrightarrow{x}](F)\} \tag{2}$$

If μ_1 is a uniform probability over $\mathcal{H}$, it is easy to see that $\mu_1(F)$ is a frequency. Type 1 structures are useful for capturing statistical information but these structures are insufficient for describing probabilities on closed formulas. Indeed, a closed formula has no free variable. So the type 1 probability of a closed formula is 0 or 1 according as this formulas is true or false in I.

Halpern names type 2 structure the pair $M = \{S, \mu_2\}$ where S is a set of Herbrand interpretations, and μ_2 is a probabilistic measure over S. Given a type 2 structure M and a closed formula F, the meaning of $\mu_2(F)$ is the probability that F is true in a randomly chosen interpretation in S.

Definition 2. *Given $M = \{S, \mu_2\}$, F a closed formula, its type 2 probability is :*

$$\mu_2(F) = \mu_2\{I \in S | I \models F\} \tag{3}$$

These structure can be used for describing the belief degree of a ground fact or, in our case, for dealing with membership degrees.

For giving a meaning to probabilities on all rules, we must mixing type 1 and type 2 structures. Halpern names type 3 structure the 4-tuple $M = \{S, \mathcal{H}, \mu_1, \mu_2\}$, where S is a set of interpretations with the same Herbrand's domain $\mathcal{H}$. μ_1 is a probability measure on $\mathcal{H}$ and μ_2 is a probability measure on S. Given a type 3 structure, we could have two definitions of the probability of a formula.

First, given a formula F, it is clear that the value $\mu_1^I(F)$, where I is an interpretation, is a random variable on S with measure μ_2). We can associate the probability $\mu(F)$ with the expectation of $\mu_1^I(F)$ in the measurable space (S, μ_2). So the formal definition follows :

$$\mu(F) = E_{\mu_2, S}(\mu_1^I(F)) \tag{4}$$

thus $\mu(F)$ is just the mean value of $\mu_1(F)$ considered as a random variable over S. $\mu(F)$ has a meaning for all formulas, closed or not.

Besides, given a vector $\vec{t}$ on $\mathcal{H}^n$, $\sigma[\vec{t}/\vec{x}](F)$ (where $\vec{x}$ is the vector of the n free variables in F) is a closed formula, we can compute $\mu_2(\sigma[\vec{t}/\vec{x}](F))$. So, $\mu_2(\sigma[\vec{t}/\vec{x}](F))$ is a random variable on the set $\mathcal{H}^n$ (the set of all substitutions of the n free variables of F) with the measure μ_1. This leads to another definition $\mu(F) = E_{\mu_1, \mathcal{H}^n}(\mu_2(\sigma[\vec{t}/\vec{x}](F)))$ which is in accordance with the previous once since we have :

Proposition 1.

$$E_{\mu_2, S}(\mu_1^I(F)) = E_{\mu_1, \mathcal{H}^n}(\mu_2(\sigma[\vec{t}/\vec{x}](F))) \tag{5}$$

Proof : by variable swapping and using the independence of μ_1 and μ_2.

This shows that the two ways for computing probabilities of formulas turn to be equivalent.

4   Learning First Order Fuzzy Rule with Fuzzy Predicate

For using FOIL algorithm, we must define probabilities on first order formulas from ILP data. ILP data describe only one interpretation under Closed World Assumption (CWA). So we can use type 1 probabilities. We call I_{ILP} the interpretation of $B \wedge E$ under CWA. So given a fact a :

$$I_{ILP}(a) = \begin{cases} \top \text{ if } B \wedge E \models a \\ \\ \bot \text{ otherwise} \end{cases}$$

The domain $\mathcal{H}$ is the Herbrand domain describes by B and E. We take μ_1 as a uniform probability on $\mathcal{H}$. So it's easy to deduce that the confidence in a clause $A \to B$, with $\overrightarrow{t}$ as vector on the n free variables, is :

$$cf(A \to B) = \frac{|\{\overrightarrow{x} \in \mathcal{H}^n | I_{ILP} \models \sigma[\overrightarrow{t}/\overrightarrow{x}](A \wedge B)\}|}{|\{\overrightarrow{x} \in \mathcal{H}^n | I_{ILP} \models \sigma[\overrightarrow{t}/\overrightarrow{x}](A)\}|}$$

where $|\ |$ denotes cardinality.

This formal framework is insufficient for databases associated with fuzzy predicates and membership degrees. More precisely, we consider a first order database K with fuzzy predicates as a set of positive facts labeled by real numbers in $[0, 1]$. For instance in section 5, we shall deal with a database containing facts such as $comfortable(a), 0.5$ and $closeto(a, sea), 0.4$. Thus, K is made of pairs of the form $A, \beta(A)$, where A is a ground fact, and $\beta(A)$ is the fuzzy degree associated with A.

For describing probabilities in this base, we have to introduce fuzzy degrees in Halpern framework. One of the understandings of the membership degree of a fact is to view it as the probability that an instance $\overrightarrow{x}$ be labelled by the fuzzy predicate A (rather than not) by an expert [5][6]. So in our probability framework, if we take an expert as an interpretation, the membership degree is exactly the type 2 probability of the fact. For computing our probabilities we will construct a type 3 structure from K . We have $\mu_2(A) = \beta(A)$ for all fact $A \in K$. We take μ_1 as a uniform probability on $\mathcal{H}$. So any non-fuzzy fact that is true in at least one interpretation $I \in S$ is true in all the interpretations in S and receive 1 as a degree.

By this way, we are not entirely describing S in the type 3 structure, and we can't compute the probability of a conjunction of facts as needed in (5). There are three classical ways for combining membership degrees x and y in a fuzzy set conjunction : $min(x, y), x \times y, max(0, x + y - 1)$. It can be shown that a type 3 structure can be built for each choice of the conjunction.

However, with the above-mentioned likelihood interpretation of membership degrees, min remains a natural choice for the conjunction [4]. In this case S is the set of interpretations corresponding to α-cut of I_{ILP}, where an α-cut of I_{ILP} is the subset of I_{ILP} where facts have a membership degree greater than α. Thus, it turns out that what is computed in (5) is an average value of the confidence measure which would be associated with crisp rules corresponding to the α-cuts of the induced fuzzy rules involving fuzzy predicates.

Example of confidence computation : Let us consider the following database with fuzzy degrees associated with fuzzy predicates A et B :
$K = \{A(a,b),1; A(a,c),0.7; A(b,a),0.2; B(a),0.8\}$.
Any other fact is assumed to be false. The Herbrand domain is $\mathcal{H} = \{a,b,c\}$. We want to compute the confidence of the formula $A(X,Y) \rightarrow B(X)$ in the type 3 structure associated with K. We have :
$$cf(A(X,Y) \rightarrow B(X)) = \frac{\mu(A(X,Y) \wedge B(X))}{\mu(A(X,Y))}$$

$$\begin{aligned}
=_{(5)} & \frac{E_{\mu_1,\mathcal{H}^n}(\mu_2(\sigma[\overrightarrow{t}/\overrightarrow{x}](A(X,Y) \wedge B(X))))}{E_{\mu_1,\mathcal{H}^n}(\mu_2(\sigma[\overrightarrow{t}/\overrightarrow{x}](A(X,Y))))} = \frac{\sum_{\overrightarrow{x} \in \mathcal{H}^n} \mu_2(\sigma[\overrightarrow{t}/\overrightarrow{x}](A(X,Y) \wedge B(X)))}{\sum_{\overrightarrow{x} \in \mathcal{H}^n} \mu_2(\sigma[\overrightarrow{t}/\overrightarrow{x}](A(X,Y)))} \\
& = \frac{\mu_2(A(a,b) \wedge B(a)) + \mu_2(A(a,c) \wedge B(a))}{\mu_2(A(a,b)) + \mu_2(A(a,c)) + \mu_2(A(b,a))} = \frac{min\{1,0.8\} + min\{0.7,0.8\}}{1 + 0.7 + 0.2} = 0.78.
\end{aligned}$$

Given a database with fuzzy predicates, based on a type 3 structure, we can compute confidence degrees for induced formulas, and have a FOIL-like mechanism which handles membership degrees.

5   Illustrative Example

As an illustration of our approach, we have explored a database that can be found on the PRETI platform (http://www.irit.fr/PRETI/). This database describes houses to let for vacations. There are more than 600 houses described in terms of about 25 attributes. A lot of these attributes are about distances between the house and another place (sea, fishing place, swimming pool, ...) or about prices at different periods in a year (June, weekend, school vacations ...). From this database, we can obtain a fuzzy data base by merely changing price and distance information into fuzzy information such as "cheap", "expensive", "far", "not_too_far", "not_far", together with a membership degree.

For instance, "cheap" and "expensive" are represented by the following trapezoids (0,0,800,2000) and (2500,3800,10000,10000). "close_to", "not_too_far" and "far" are represented by the following trapezoids (0,0,5,10), (5,10,30,35) and (30,35,100,100) respectively (trapezoid (a,b,c,d) defines a fuzzy set with support [a,d] and core [b,c]). For discrete attribute membership, degrees have to be directly associated which each possible value.

For example, the fact that the house x_1 lies at 5.5 km from sea, represented by the logical fact $distance(x_1, sea, 5.5)$, is translated into the fuzzy predicate expression $close\_to(x_1, sea)$ with degree 0.8 and $no\_too\_far(x_1, sea)$ with degree 0.2. The table below shows some examples of rules involving fuzzy predicates which are induced by our extended FOIL algorithm applied to the PRETI database. We assume that we use constraints on predicates, such as in $cheap(A, B)$, for making sur that A is an house and B is a time period.

rules	confidence
$expensive(A, B), dishwasher(A), small\_capacity(A) \rightarrow comfortable(A)$	0.91
$expensive(A, C), B = "shop" \rightarrow close\_to(A, B)$	0.91
$area(A, NARBONNAIS), expensive(A, C), high\_capacity(A) \rightarrow close\_to(A, sea)$	0.92
$pets\_accepted(A) \rightarrow washingmachine(A)$	0.9
$small\_capacity(A), cheap(A, June), far(A, sea) \rightarrow not\_comfortable(A)$	0.77

Rules 1, 2, 3 are genuine first order rules which could not be found by a "propositional" method since they exhibit an existential quantifier. As pointed out by the first rule, the algorithm can mix fuzzy predicates and non-fuzzy predicates (e.g. "dishwasher"). The rule 4 is a classical rule. Rules 4 and 5 can be represented in propositional logic, but in our case, propositional variables like $cheap(A, June)$ are produced by the FOIL-like algorithm (which here selects the "June" instantiation).

6   Conclusion

This paper has described a formal framework and a procedure for learning first order rules involving fuzzy and non-fuzzy predicates. A probabilistic view of fuzzy membership degrees has been used. The definition of confidence degrees for rules with fuzzy predicates allows us to easily introduce them in any learning algorithm which uses confidence degrees as a basis for the guiding process. Since the confidence computation is a weigthed version of FOIL's one, it's easy to deduce that the complexity of the algorithm is the same as the FOIL's one. Learning rules with fuzzy predicates has some obvious advantages. A first one is that fuzzy rules can involve fuzzy categories as often used by people. Generally speaking, it is well-known that fuzzy sets defined on subsets of the real line provide a flexible interface with precise numerical values. Moreover, since it's also known that ILP has difficulties for handling real numbers, the use of fuzzy predicates for representing numerically valued predicates (as made with the PRETI database) can provide a valuable improvement .

However the proposed algorithm could still be improved. First, as in data mining, other degrees (like support, etc...) could be computed and used for controlling the learning process. Besides, it has the same limits as FOIL, which cannot find all the "interesting" rules [15]. Moreover, it may be important to figure out if the confidence degree of a rule is brittle with respect to a variation of the range(s) of the predicate(s) involved in the rule; it amounts to know if the confidence would vary with different crisp approximations of the rule, made by different level-cuttings of the fuzzy predicate(s) involved in the fuzzy rule. If this variation is small, then our fuzzy rule would be genuinely robust [2]. Future works will also focus on a formal description of ILP in a fuzzy context.

References

1. F. Bacchus. *Representing and Reasoning With Probabilistic Knowledge*. MIT press, 1990.
2. P. Bosc, D. Dubois, O. Pivert, H. Prade, and M. de Calmes. Fuzzy sumarisation of data using fuzzy cardinalities. In *Proc. Inter. Conf. Information Processing and Management of Uncertainty in Knowledge-based Systems (IPMU 2002)*, pages 1553–1559, Annecy-France, July 2002.
3. B. Bouchon-Meunier and C. Marsala. *Learning fuzzy decision rules*, chapter 4 in. Fuzzy Sets in Approximate Reasoning and Information Systems,(J.C. Bezdek, D. Dubois, H. Prade, eds.), The Handbooks of Fuzzy Sets Series. Kluwer Academic Publishers, 1999, 279-304.
4. D. Dubois, S. Moral, and H. Prade. A semantics for possibility theory based on likelihoods. *J. of Math. Analysis and Applications*, 205:359–380, 1997.
5. D. Dubois and H. Prade. Fuzzy sets, probability and measurement. *Europ. J. of Operational Research*, 40:135–154, 1989.
6. D. Dubois and H. Prade. *Modelling uncertain and vague knowledge in possibility and evidence theories*, pages 303–318. *Uncertainty in Artificial Inelligenge 4*,(D.R. Shachter, T.S. Levitt, L.N. Kanal, J.F. Lemmer, eds.). North-Holland, 1990.
7. J. Halpern. An analysis of first-order logics of probability. *Artificial Intelligence*, 46:310–355, 1990.
8. E. Hüllermeier. Implication-based fuzzy association rules. In L. De Raedt and A. Siebes, editors, *Proceedings PKDD-01, 5th Conference on Principles and Pratice of Knowledge Discovery in Databases*, number 2168 in LNAI, pages 241–252, September 2001.
9. S. Muggleton. Inverse entailment and Progol. *New Generation Computing*, 13:245–286, 1995.
10. D. Nauck and R. Kruse. *Neuro-fuzzy methods in fuzzy rule generation*, chapter 5 in. Fuzzy Sets in Approximate Reasoning and Information Systems,(J.C. Bezdek, D. Dubois, H. Prade, eds.), The Handbooks of Fuzzy Sets Series. Kluwer Academic Publishers, 1999, 305-334.
11. S-H Nienhuys-Cheng and R. de Wolf. *Foundations of Inductive Logic Programming*. Number 1228 in LNAI. Springer, 1997.
12. J. R. Quinlan. Induction of decision trees. *Machine Learning*, 1(1):81–106, 1986.
13. J. R. Quinlan. Learning logical definitions from relations. *Machine Learning*, 5:239–266, 1990.
14. J. R. Quinlan. Knowledge acquisition from structured data. *IEEE Expert*, 6(6):32–37, 1991.
15. B.L. Richards and R.J. Mooney. Learning relations by pathfinding. In *Proc. of the AAAI conference*, pages 50–55, San Jose, 1992. AAAI Press.

An Interactive Fuzzy Satisfying Method for Multiobjective Nonlinear Integer Programming Problems through Genetic Algorithms

Masatoshi Sakawa and Kosuke Kato

Hiroshima University, Higashi-Hiroshima, 739-8527, JAPAN,
Corresponding author: sakawa@msl.sys.hiroshima-u.ac.jp,
http://mandala.msl.sys.hiroshima-u.ac.jp/

Abstract. In this paper, we propose a general-purpose solution method for nonlinear integer programming problems by extending genetic algorithms with double strings for linear ones. After describing the proposed genetic algorithm, its efficiency will be shown through numerical experiments using single-objective nonlinear programming problems. Furthermore, focusing on multiobjective nonlinear integer programming problems, we present an interactive fuzzy satisficing method through the proposed genetic algorithm.

1 Introduction

In general, a variety of actual decsion making situations are formulated as large-scale mathematical programming problems. Mathematical programming problems whose decision variables are discrete can be theoretically solved by dynamic programming techniques. However, it is difficult to obtain strict optimal solutions to large-scale ones because their search areas are vast. In addition, since strict optimal solutions are not always necessary for practical use, efficient approximate solution algorithms have been desired.

Genetic algorithms (GA), proposed by J.H. Holland [1] as an algorithm to simulate the mechanism of natural evolution, have recently attracted considerable attention in a number of fields as a methodology for optimization and learning. As we look at recent applications of genetic algorithms to optimization problems, regardless of single-objective or multiobjective, we can see continuing advances [2,4].

For multiobjective multidimensional 0-1 knapsack problems, M. Sakawa et al. proposed an approximate solution method based on genetic algorithms with double strings (GADS) [7] . As an extension of the GADS, they proposed a genetic algorithm with double strings based on reference solution updating (GADSRSU) [8] for multiobjective general 0-1 programming problems involving both positive coefficients and negative ones. Furthermore, they proposed a genetic algorithm with double strings using linear programming relaxation (GADSLPR) [6] for multiobjective multidimensional integer knapsack problems and a genetic algorithm with double strings using linear programming relaxation based on reference solution updating (GADSLPRRSU) [5] for general integer programming problems.

T. Bilgiç et al. (Eds.): IFSA 2003, LNAI 2715, pp. 710–717, 2003.

These approximate solution methods based on genetic algorithms are designed for linear discrete programming problems, whereas a number of mathematical programming problems in the real world involve integer decision variables, nonlinear objective functions and nonlinear constraint functions. Since there does not exist a general-purpose solution method for such nonlinear integer programming problems like the branch and bound method or GADS for linear ones, a solution method peculiar to each problem has been proposed.

Under these circumstances, in this paper, we propose a genetic algorihtm with double strings as a general-purpose solution method for nonlinear integer programming problems by extending the GADSLPRRSU [5] and attempt to derive satisficing solutions to multiobjective nonlinear integer programming problems through the proposed genetic algorithm.

2   Nonlinear Integer Programming Problems

Generally, nonlinear integer programming problems are formulated as:

$$\left.\begin{array}{ll} \text{minimize} & f(\boldsymbol{x}) \\ \text{subject to} & g_i(\boldsymbol{x}) \leq 0, \ i = 1, \ldots, m \\ & x_j \in \{0, 1, \ldots, \nu_j\}, \ j = 1, \ldots, n \end{array}\right\} \tag{1}$$

where $\boldsymbol{x}$ is an n dimensional integer decision variable vector, $f(\cdot), g_i(\cdot), i = 1, \ldots, m$ are nonlinear functions.

3   Genetic Algorithm for Nonlinear Integer Programming Problems

In this section, we mention the extension of a genetic algorithm with double strings based on reference solution updating (GADSRSU) proposed by M. Sakawa et al. for linear integer programming problems [5] to nonlinear ones.

3.1   Individual Representation

As in [5], the individual representation by double strings shown in Fig. 1 is adopted. In the figure, each of $s(j), j = 1, \ldots, n$ is the index of an element in a solution vector and each of $y_{s(j)} \in \{0, 1, \ldots, \nu_{s(j)}, j = 1, \ldots, n$ is the value of the element, respectively.

Individual **S** :

$s(1)$	$s(2)$	$\cdots$	$s(n)$
$y_{s(1)}$	$y_{s(2)}$	$\cdots$	$y_{s(n)}$

Fig. 1. Double string

712   Masatoshi Sakawa and Kosuke Kato

3.2 Decoding Algorithm

Basically, an individual **S** represented by a double string is decoded to a solution x by the substitution $x_{s(j)} := y_{s(j)}$ [5,6,7,8]. However, since the obtained solution x may be infeasible, some device is needed to make it feasible. Thus, in [5,6,7,8], according to the problem type, decoding algorithms which map any individual to the corresponding feasible solution are proposed. In this paper, we construct a decoding algorithm of double strings for nonlinear integer programming problems using a reference solution x^*, which is a feasible solution and is used as the origin of decoding.

Proposed decoding algorithm using a reference solution
_____

In this algorithm, it is assumed that a feasible solution x^* is obtained in advance. Let n and N be the number of variables and the number of individuals in the population, respectively.

Step 1: If the index of an individual to be decoded is in $\{1, \ldots, \lfloor N/2 \rfloor\}$, go to step 2. Otherwise, go to step 7.
Step 2: Let $j := 1$, $x := \{0, \ldots, 0\}$, $l := 1$.
Step 3: Let $x_{s(j)} := y_{s(j)}$.
Step 4: If $g_i(x) \leq 0, i = 1, \ldots, m$, let $l := j$, $j := j + 1$, and go to step 5. Otherwise, let $j := j + 1$, and go to step 5.
Step 5: If $j \leq n$, go to step 3. Otherwise, go to step 6.
Step 6: If $l > 0$, go to step 7. Otherwise, go to step 8.
Step 7: By substituting $x_{s(j)} := y_{s(j)}, 1 \leq j \leq l$ and $x_{s(j)} := 0, l < j \leq n$, we obtain a feasible solution x corresponding to the individual **S** and stop.
Step 8: Let $j := 1$, $x := x^*$.
Step 9: Let $x_{s(j)} := y_{s(j)}$. If $y_{s(j)} = x^*_{s(j)}$, let $j := j + 1$, and go to step 11. If $y_{s(j)} \neq x^*_{s(j)}$, go to step 10.
Step 10: If $g_i(x) \leq 0, i = 1, \ldots, m$, let $j := j + 1$, and go to step 11. Otherwise, let $x_{s(j)} := x^*_{s(j)}, j := j + 1$, and go to step 11.
Step 11: If $j \leq n$, go to step 9. Otherwise, we obtain a feasible solution x from the individual **S** and stop.

The proposed decoding algorithm can make the correspondence of any individual to a feasible solution. Because solutions obtained the decoding algorithm using a reference solution tend to concentrate around the reference solution, M. Sakawa et al. introduced the reference solution updating procedure [8]. The procedure is also adopted in this paper.

3.3 Usage of Solution to Continuous Relaxation Problem

In order to find an approximate optimal solution with high accuracy in reasonable time, we need some schemes such as the restriction of the search space to a promising region, the generation of individuals near the optimal solution and so forth. From the point of view, M. Sakawa et al. [5] made use of information about an optimal solution to the corresponding continuous relaxation problem in the generation of the initial population and the mutation. Since these operation seem efficient for nonlinear integer programming problems, we adopt them in the present paper.

The continuous relaxation problem, where the integer condition of decision variables is relaxed in (1), is written as:

$$\left.\begin{array}{ll} \text{minimize} & f(\boldsymbol{x}) \\ \text{subject to} & g_i(\boldsymbol{x}) \leq 0, \ i = 1, \ldots, m \\ & 0 \leq x_j \leq \nu_j, \ j = 1, \ldots, n \end{array}\right\}. \tag{2}$$

When this problem is convex, we can obtain a global optimal solution by some convex programming technique, e.g., the sequential quadratic programming. Otherwise, i.e., when it is non-convex, because it is difficult to find a global optimal solution, we search an approximate optimal solution by some approximate solution method such as genetic algorithms or simulated annealing.

The information about an (approximate) optimal solution $\hat{\boldsymbol{x}}$ to the continuous relaxation problem (2) is used in the generation the initial population and the mutation operator.

3.4 Generation of Initial Population

In general, the initial population is generated completely at random. However, it is important to generate more promising initial population so as to obtain a good approximate optimal solution. For this reason, the information about an optimal solution $\hat{\boldsymbol{x}}$ to the continuous relaxation problem is used in this paper. To be more specific, determine $y_{s(j)}, j = 1, \ldots, n$ randomly according to the corresponding Gaussian distribution with mean $\hat{x}_{s(j)}$ and variance σ^2.

The procedure of generation of initial population is summarized as follows.

Generation of initial population

Step 1: Let $r := 1$.
Step 2: If a uniform random number rand() in $[0, 1]$ is less than or equal to the constant R introduced in 3.2, go to step 3. Otherwise, go to step 7.
Step 3: Let $j := 1$.
Step 4: Determine $s(j)$ by a uniform integer random number in $\{1, \ldots, n\}$ so that $s(j) \neq s(j'), j' = 1, \ldots, j - 1$.
Step 5: Determine $y_{s(j)}$ by a Gaussian random number with mean $\hat{x}_{s(j)}$ and variance σ^2, and let $j := j + 1$.
Step 6: If $j > n$, let $r := r + 1$ and go to step 11. Otherwise, go to step 4.
Step 7: Let $j := 1$.
Step 8: Determine $s(j)$ by a uniform integer random number in $\{1, \ldots, n\}$ so that $s(j) \neq s(j'), j' = 1, \ldots, j - 1$.
Step 9: Determine $y_{s(j)}$ by a uniform integer random number in $\{0, 1, \ldots, \nu_{s(j)}\}$, and let $j := j + 1$.
Step 10: If $j > n$, let $r := r + 1$ and go to step 11. Otherwise, go to step 8.
Step 11: If $r > N$, stop. Otherwise, go to step 2.

3.5 Mutation Operator

It is considered that mutation plays the role of local random search in genetic algorithms. A direct extension of mutation for 0-1 problems is to change the value of $g_{s(j)}$ at random in $[0, \nu_{s(j)}]$ uniformly, when mutation occurs at $g_{s(j)}$. The mutation operator is further refined by using information about an optimal solution $\hat{x}$ to the continuous relaxation problem. To be more specific, change $g_{s(j)}, j = 1, \ldots, n$ randomly according to the corresponding Gaussian distribution with mean $\hat{x}_{s(j)}$ and variance τ^2.

The procedures of mutation are summarized as follows.

Mutation

Step 0: Let $r := 1$.

Step 1: Let $j := 1$.

Step 2: If a random number rand() in $[0, 1]$ is less than or equal to the probability of mutation p_m, go to step 3. Otherwise, go to step 4.

Step 3: If another random number rand() in $[0, 1]$ is less than or equal to a constant R, determine $x_{s(j)}$ randomly according to the Gaussian distribution with mean $\hat{x}_{s(j)}$ and variance τ^2, and go to step 4. Otherwise, determine $x_{s(j)}$ randomly according to the uniform distribution in $[0, \nu_j]$, and go to step 4.

Step 4: If $j < n$, let $j := j + 1$ and go to step 2. Otherwise, go to step 5.

Step 5: If $r < N$, let $r := r + 1$ and go to step 1. Otherwise, stop.

3.6 Other Genetic Operators

With respect to other genetic operators, as in [5], we adopt the elitist expected value selection, the partially matched crossover and inversion.

3.7 Proposed Genetic Algorithm

Now we are ready to summarize the genetic algorithm with double strings using continuous relaxation for solving nonlinear integer programming problems.

Computational procedures

Step 0: Determine values of the parameters used in the genetic algorithm: the population size N, the generation gap G, the probability of crossover p_c, the probability of mutation p_m, the probability of inversion p_i, the minimal search generation $I_{\min}$, the maximal search generation $I_{\max} (> I_{\min})$, the scaling constant c_{mult}, the convergence criterion ε, the degree of use of information about solutions to linear programming relaxation problems R, and set the generation counter t at 0.

Step 1: Generate the initial population consisting of N individuals.

Step 2: Decode each individual (genotype) in the current population and calculate its fitness based on the corresponding solution (phenotype).

Step 3: If the termination condition is fulfilled, stop. Otherwise, let $t := t + 1$ and go to step 4.

Step 4: Apply reproduction operator using elitist expected value selection, after performing linear scaling.

Step 5: Apply crossover operator, called PMX (Partially Matched Crossover) for double string.

Step 6: Apply mutation.

Step 7: Apply inversion operator. Go to step 2.

4 An Interactive Fuzzy Satisfying Method for Multiobejctive Nonlinear Integer Programming Problem

In general, multiobjective nonlinear integer programming problems are formulated as:

$$
\left.
\begin{array}{ll}
\text{minimize} & f_l(\boldsymbol{x}),\ l = 1, \ldots, k \\
\text{subject to} & g_i(\boldsymbol{x}) \leq 0,\ i = 1, \ldots, m \\
& x_j \in \{0, 1, \ldots, \nu_j\},\ j = 1, \ldots, n
\end{array}
\right\}.
\tag{3}
$$

In the following part of this paper, the feasible region of (3) is denoted by X.

For the multiobjective nonlinear integer programming problem (3), in the present paper, we introduce fuzzy goals to consider the ambiguity or fuzziness of the decision maker's judgement on objective functions, and incorporate the proposed genetic algorithm into an interactive method to derive a satisficing solution for the decision maker by drawing out local preference information through interactions [3].

Interactive algorithm

Step 1: Calculate the individual minimum of each objective function $f_l(\boldsymbol{x})$, $l = 1, \ldots, k$ under the given constraints in (3) using the proposed genetic algorithm.

Step 2: Ask the decision maker for determining a membership function of each objective function and setting initial reference membership levels $\bar{\mu}_l$, $l = 1, \ldots, k$.

Step 3: For current reference membership levels, solve the corresponding augmented minimax problem (4) using the proposed genetic algorithm to obtain the M-Pareto optimal solution and the membership function value.

$$
\underset{\boldsymbol{x} \in X}{\text{minimize}}\ \max_{l=1,\ldots,k} \left\{ \bar{\mu}_l - \mu_l(f_l(\boldsymbol{x})) + \rho \sum_{i=1}^{k} (\bar{\mu}_i - \mu_i(f_i(\boldsymbol{x}))) \right\}
\tag{4}
$$

Step 4: If the DM is satisfied with the current levels of the M-Pareto optimal solution, stop. Then the current M-Pareto optimal solution is the satisficing solution of the DM. Otherwise, ask the DM to update the current reference membership levels by considering the current values of the membership functions and return to Step 3.

5 Numerical Experiment

In order to show the efficiency of the proposed method, it is applied to the following two-objective nonlinear integer programming problem:

$$
\left.
\begin{aligned}
\text{maximize } & f_1(\boldsymbol{x}) = \prod_{j=1}^{n}[1 - (1 - r_j)^{x_j}] \\
\text{minimize } & f_2(\boldsymbol{x}) = \sum_{j=1}^{n} q_j[x_j + \exp\left(\frac{x_j}{4}\right)] \leq Q \\
\text{subject to } & g_1(\boldsymbol{x}) = \sum_{j=1}^{n} p_j x_j^2 - P \leq 0 \\
& g_2(\boldsymbol{x}) = \sum_{j=1}^{n} w_j x_j \exp\left(\frac{x_j}{4}\right) - W \leq 0 \\
& x_j \in \{1, 2, \dots, \nu_j\}, j = 1, \dots, n
\end{aligned}
\right\}
\tag{5}
$$

where $n = 8$, $\nu_j = 10$, $j = 1, \dots, n$, and $\boldsymbol{r}, \boldsymbol{p}, \boldsymbol{q}, \boldsymbol{w}$ are n dimensional real-valued coefficient row vector, P, Q, W are real-valued constants.

We compare the result of the proposed genetic algorithm (GA) with that of the complete enumeration (CE) in application of the interactive fuzzy satisficing method to (5).

First, according to the first step of the interactive algorithm in the previous section, optimization problems for each objective functions in (5) are solved by the proposed genetic algorithm and the complete enumeration. Table 1 shows results of both methods and the result by the proposed genetic algorithm is the average of ten trials.

Table 1. Minimal values of each individual objective function.

	f_1	f_2
GA	0.690	204.8
CE	0.690	204.8

According to the second step, after membership functions for objective functions are determined by the decision maker based on the result in step 1, initial membership levels are set as $(\bar{\mu}_1, \bar{\mu}_2) = (1.0, 1.0)$.

Then, in the third step, the augmented minimax problem (4) for the initial reference levels is solved by both methods. The obtained solutions are shown at the second and third rows in Table 2. The hypothetical decision maker cannot be satisfied with this solution, particularly, he wants to improve $f_1(\boldsymbol{x})$. at the expense of $f_2(\boldsymbol{x})$. Then, the decision maker update the reference membership levels to $(1.00, 0.80)$. By repetition of such interaction with the decision maker, in this example, a satisficing solution is obtained at the third interaction.

From the table, the proposed genetic algorithm can find a strict optimal solution obtained by the complete enumeration in all trials for each interaction phase. In addition,

Table 2. The whole interaction process.

		μ_1	μ_2	f_1	f_2
1 st	GA	0.636	0.624	0.445	319.0
$\bar{\mu}_1 = 1.00, \bar{\mu}_2 = 1.00$	CE	0.636	0.624	0.445	319.0
2 nd	GA	0.755	0.540	0.526	344.5
$\bar{\mu}_1 = 1.00, \bar{\mu}_2 = 0.80$	CE	0.755	0.540	0.526	344.5
3 rd	GA	0.708	0.557	0.493	339.5
$\bar{\mu}_1 = 0.95, \bar{\mu}_2 = 0.80$	CE	0.708	0.557	0.493	339.5

as to computational time, the proposed genetic algorithm takes only about 5 seconds on average while the complete enumeration does about 400 seconds on average. Therefore, the interactive fuzzy satisficing method for multiobjective nonlinear programming problems based on the proposed genetic algorithm is supposed to be useful.

6 Conclusion

In this paper, a new genetic algorithm with double strings for nonlinear integer programming problems was proposed and an interactive fuzzy satisficing method for multiobjective nonlinear integer programming problems base on the proposed genetic algorithm was presented. Furthermore, the application of the proposed interactive method through the genetic algorithm to an illustrative numerical example demonstrated its feasibility and usefulness.

References

1. Holland, J.H.: Adaptation in Natural and Artificial Systems, University of Michigan Press (1975)
2. Michalewicz, Z.: Genetic Algorithms + Data Structures = Evolution Programs, Springer-Verlag, (1992), Second extended edition, (1994), Third revised and extended edition, (1996)
3. Sakawa, M.: Fuzzy Sets and Interactive Multiobjective Optimization, Plenum Press, New York (1993)
4. Sakawa, M.: Genetic Algorithms and Fuzzy Multiobjective Optimization, Kluwer Academic Publishers, Boston/Dordrecht/London (2002)
5. Sakawa, M., Kato, K.: Integer programming through genetic algorithms with double strings based on reference solution updating, 2000 IEEE International Conference on Industrial Electronics, Control and Instrumentation, (2000) 2915-2920
6. Sakawa, M., Kato, K., Shibano, T., Hirose, K.: Fuzzy multiobjective integer programs through genetic algorithms using double string representation and information about solutions of continuous relaxation problems, Proceedings of 1999 IEEE International Conference on Systems, Man and Cybernetics, (1999) 967-972
7. Sakawa, M., Kato, K., Sunada, H., Shibano, T.: Fuzzy programming for multiobjective 0-1 programming problems through revised genetic algorithms, European Journal of Operational Research **97** (1997) 149–158.
8. Sakawa, M., Kato, K., Ushiro, S., Ooura, K.: Fuzzy programming for general multiobjective 0-1 programming problems through genetic algorithms with double strings, 1999 IEEE International Fuzzy Systems Conference Proceedings **3** (1999) 1522-1527

A Global Optimization Method for Solving Fuzzy Relation Equations

Ş. İlker Birbil[1] and Orhan Feyzioğlu[2]

[1] Erasmus Research Institute of Management, Erasmus University,
Postbus 1738, 3000 DR, Rotterdam, The Netherlands
sibirbil@few.eur.nl
[2] Galatasaray University, Department of Industrial Engineering
Çırağan Caddesi No: 36, 34357, Ortaköy/İstanbul, Turkey
ofeyzioglu@gsu.edu.tr

Abstract. A system of fuzzy relation equations can be reformulated as a global optimization problem. The optimum solution of this new model corresponds to a solution of the system of fuzzy relation equations whenever the solution set of the system is nonempty. Moreover, even if the solution set of the fuzzy relation equations is empty, a solution to the global optimization problem provides a point such that the difference between the right and the left hand side of the fuzzy relation equations is minimized. The new global optimization problem has a nonconvex and nondifferentiable objective function. Therefore, a recent stochastic search approach is applied to solve this new model. The performance of the approach is tested on a set of problems with different dimensions.

Keywords: fuzzy relation equations, $max\text{-}min$ composition, global optimization method, electromagnetism-like mechanism (EM).

1 Introduction

In different areas of fuzzy set theory, it is crucial to solve a system of fuzzy relation equations [1]. For instance, in fuzzy linear programming the system of fuzzy linear equations determines the feasible region over which an optimization problem has to be solved. Fuzzy relation equations also arise in fuzzy control theory, where the fuzzy relations are used to establish the association of the different parameters in the system. In lack of full association, i.e., when the solution set of the fuzzy relation equations system is empty, the least discrepancy in the associations is tried to be computed [2].

Given a coefficient matrix $A \in [0,1]^{m \times n}$ and a right-hand side vector $b \in [0,1]^m$, a typical fuzzy relation equation (FRE) problem is aimed at finding the solution set $S \subseteq [0,1]^n$ such that $x \in S$ if and only if

$$A \circ x = b, \tag{1}$$

where $\circ$ denotes the max-min composition associated with the fuzzy relation [2]. Explicitly, we have

T. Bilgiç et al. (Eds.): IFSA 2003, LNAI 2715, pp. 718–724, 2003.

$$\max\{\min(a_{11}, x_1), \min(a_{12}, x_2), \cdots, \min(a_{1n}, x_n)\} = b_1,$$
$$\max\{\min(a_{21}, x_1), \min(a_{22}, x_2), \cdots, \min(a_{2n}, x_n)\} = b_2,$$

$$\vdots = \vdots$$

$$\max\{\min(a_{m1}, x_1), \min(a_{m2}, x_2), \cdots, \min(a_{mn}, x_n)\} = b_m.$$

If we denote max and min functions by $\vee$ and $\wedge$, respectively, then the above system becomes

$$\bigvee_{j=1}^{n} (a_{ij} \wedge x_j) = b_i, \quad i = 1, 2, \cdots, m. \tag{2}$$

Notice that if we define a vector-valued function $\Phi : [0,1]^n \to \Re^m$ by

$$\Phi(x) = (\phi_1(x), \cdots, \phi_m(x))^T,$$

where $\phi_i : [0,1]^n \to \Re$, $1 \le i \le m$ are given by

$$\phi_i(x) = \bigvee_{j=1}^{n} (a_{ij} \wedge x_j) - b_i, \tag{3}$$

then it is easy to see that $x \in S$ if and only if x satisfies

$$\Phi(x) = 0. \tag{4}$$

However, in certain cases the solution set S can be an empty set, then we are interested in the set of x such that the discrepancy between the left and right hand sides of the system (1) is minimized. Suppose $\pi : [-1,1]^m \to \Re$ is a real-valued *penalty* function

$$\pi(\phi_1(x), \cdots, \phi_m(x)) := \Pi(x), \tag{5}$$

then we want to solve the following optimization problem

$$\min_{x \in [0,1]^n} \Pi(x). \tag{6}$$

In this paper we have selected the penalty function π

$$\pi(y) = \sum_{i=1}^{m} y_i^2,$$

which has lead to the following instance of (6)

$$\min_{x \in [0,1]^n} \Pi(x) = \sum_{i=1}^{m} [\phi_i(x)]^2. \tag{7}$$

There exist certain difficulties in solving (7). As shown in Figure 1, the functions (3) are not differentiable because of the nondifferentiability of the *max* and *min* functions. Moreover, in general the objective function of the optimization problem (7) is not convex. An elementary example of the function Π is illustrated in Figure 2.

As a direct consequence of these difficulties, main assumptions for standard nonlinear optimization techniques can not be applied [3]. Thus, after defining the penalty function, the problem (7) becomes a *global optimization problem*. Thus, it can be solved with various global optimization methods proposed especially in the recent years.

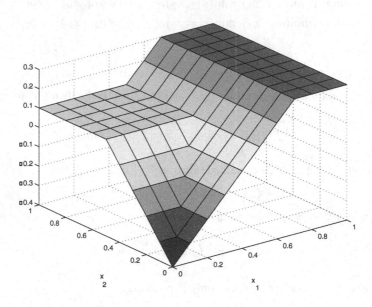

Fig. 1. $\phi(x) = \{(0.5 \wedge x_1) \vee (0.7 \wedge x_2)\} - 0.4$

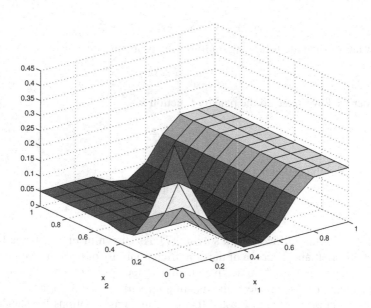

Fig. 2. $f(x) = [\{(0.5 \wedge x_1) \vee 0.7 \wedge x_2\} - 0.4]^2 + [\{(0.3 \wedge x_1) \vee (0.8 \wedge x_2)\} - 0.5]^2$

2   Solution Approach

Electromagnetism-like Mechanism (EM) is a new population-based stochastic search method, which is developed to solve box-constrained global optimization problems [4, 5, 6]. Moreover, it does not require the differentiability nor the convexity of the objective function. Therefore, we have selected EM as our global optimization method to solve the problem (6).

In a recent article, Birbil et. al. conducted thorough computational experiments on a set of problems from the literature and showed that EM has an outstanding performance when compared with some of the other global optimization methods [4]. Also, in a following article, they have shown that the method converges to the vicinity of the global optimum in the limit [5].

This method imitates the behavior of electrically charged particles that are released to the space. A set of points from the feasible region corresponds to the charged particles in the space. The strength of the method lies in the idea of directing sample points toward local optimizers, which point out attractive regions of the feasible region. EM focuses on a special class of optimization problems with bounded variables in the form of:

$$\min \quad f(x)$$
$$\text{s.t.} \quad x \in [l, u],$$

where $[l, u] := \{x \in \Re^n | l_k \le x_k \le u_k, k = 1, \cdots, n\}$. Note that the problem (6) is an instance of the above formulation with $l_k = 0$ and $u_k = 1$ for all $k = 1, \cdots, n$. Algorithm 1 shows the general scheme of EM.

Algorithm 1 : EM
```
1:  INITIALIZE the algorithm.
2:  WHILE Stopping criteria are not satisfied
3:     Apply a LOCAL SEARCH procedure.
4:     Calculate the TOTAL FORCE VECTOR on each point.
5:     MOVE the points to new locations.
6:  END WHILE
```

The procedure INITIALIZE is used to sample p points (*population*) randomly from the feasible region (Step 1). We denote the points by x^i, $i = 1, \cdots, p$. In this paper, each coordinate of a point is uniformly distributed between the corresponding upper bound and lower bound. After a point is sampled from the feasible region, the objective function value for the point, $f(x^i)$ is calculated using the function pointer.

The LOCAL SEARCH procedure (Step 3) is used to gather the local information for the point which has the minimum objective function value denoted by x^{best}. Here, we have utilized a simple searching procedure, which works on a small ball of radius δ around the point. This procedure randomly picks any coordinate and searches for a possible improvement with a random step length up to δ. Moreover, it does not require any gradient information. Though a more complicated local search method may replace this procedure, it has been already studied that even with this trivial procedure EM shows an impressive performance [6].

In Step 4, a charge is assigned to each point of the population like electromagnetic particles. These charges are not constant and change from one iteration to another according to the objective function value of the points. This charge determines the point's power of attraction or repulsion. At each iteration the charge of a point, denoted by q^i, is evaluated as follows

$$q^i = exp(-n \frac{f(x^i) - f(x^{best})}{\sum_{k=1}^{p}(f(x^k) - f(x^{best}))}), \quad i = 1, \cdots, p. \tag{8}$$

Notice that, unlike electrical charges, no signs are attached to the charge of an individual point in Equation (8). Instead, the direction of a particular force between two points is decided after comparing their objective function values. Hence, the TOTAL FORCE VECTOR, F^i exerted on any point i computed by the following:

$$F^i = \sum_{j \neq i}^{p} \begin{cases} (x^j - x^i)\frac{q^i q^j}{\|x^j - x^i\|^2} & if \quad f(x^j) < f(x^i) \\ (x^i - x^j)\frac{q^i q^j}{\|x^j - x^i\|^2} & if \quad f(x^j) \geq f(x^i) \end{cases}, \quad i = 1, \cdots, p. \tag{9}$$

After comparing the objective function values, the direction of the component forces between the point and the other points is selected. Between two points, the one that has a better objective function value attracts the other one. Contrarily, the point with worse objective function value repels the other. Since x^{best} has the minimum objective function value, it acts as an absolute point of attraction, i.e., it attracts all other points in the population.

After evaluating the total force vector F^i, in procedure MOVE of Step 5, the point i is moved in the direction of the total force vector by a random step length:

$$x^i = x^i + \lambda \frac{F^i}{\|F^i\|}(RNG) \quad i = 1, \cdots, p. \tag{10}$$

where the random step length, λ, is uniformly distributed between 0 and 1, and RNG is a vector whose components denote the allowed feasible movement toward the upper bound, u^k, or the lower bound, l^k, for the corresponding dimension. Furthermore, the force exerted on each particle is normalized so that the feasibility can be maintained.

3 Computational Study

In order to test the performance of EM, we have compiled two sets of problems. The first set consists of six problems that have solutions. The six problems in the second set have the same sizes as in the first set but their solution sets are empty [1]. In all problems we set the population size parameter equal to n and the local search parameter δ to 0.005. As a stopping criterion, we have used a maximum number of iterations. In other words, the algorithm is stopped after a maximum number of WHILE loops are exceeded in Algorithm 1. We set this parameter to 25 in all problems.

---

[1] For interested reader, the sets of problems can be downloaded from the following internet address *http://www.few.eur.nl/few/people/sibirbil/files/emfre.h*

Tables 1 and 2 show the results of EM on the first and the second set of problems, respectively. The columns 2 and 3 of the tables give the dimensions of the corresponding fuzzy relation equation problems. Since EM is a stochastic search method, we have taken 25 runs for each problem and reported the average results for the optimum solution and the number of function evaluations in columns 4 and 5, respectively. We did not report the computation times, since they were negligible even for the higher dimensional problems.

Table 1. Problems with a nonempty solution set.

Function No	n	m	Π^*	Evals.
1	5	5	0.0	68
2	5	20	0.0	214
3	10	20	0.008	925
4	15	10	0.001	542
5	15	20	0.009	1146
6	30	30	0.014	2806.2

The figures in the first two rows of the Table 1 show that EM is able to provide a solution when the problem size ($n \times m$) is small. Furthermore, as it is given in the last four rows, EM has also provided relatively good solutions with higher dimensional problems. The number of function evaluations in the last column show that the computational effort invested is less than the computational effort invested in solving a typical global optimization problem of the same size (cf. [7]).

Table 2. Problems with an empty solution set.

Function No.	n	m	Π^*	Evals.	MATLAB
1	5	5	0.160	390	0.303
2	5	20	0.485	639	1.048
3	10	20	1.204	1125	1.642
4	15	10	0.280	1032	0.675
5	15	20	1.101	1720	1.655
6	30	30	1.153	2876	1.783

Table 2 gives the performance of EM with the problems that have an empty solution set. The results show that EM provides solutions, which give relatively small residual errors between the two sides of the fuzzy relation equations. In order to justify the quality of the results, we have also solved the problems with the optimization toolbox of MATLAB using the procedure fmincon. At each run, the starting point is selected randomly from the n-dimensional unit cube. The last column shows the results found by

MATLAB. It is clear that EM outperforms the MATLAB solver in all the problems. This also suggests that EM is able to avoid trapping in the local optima. We can conclude that the selection of a proper global optimization method, like EM, becomes crucial in solving fuzzy relation equations even when the system has an empty solution set.

4   Conclusion

We have proposed to formulate a system of fuzzy relation equations as a global optimization problem. This allows us to minimize the difference between the right and left hand sides of the system. Therefore, this approach provides a solution even if the solution set is empty. Nevertheless, the resulting model is a difficult nonconvex and nondifferentiable optimization problem. In order to solve this problem, we have applied a recent stochastic global optimization method. The performance of the proposed approach is first tested on a set of problems with a nonempty solution set. Moreover, the proposed approach is compared with a regular nonlinear optimization procedure after solving a second set of problems that have an empty solution set.

In this paper we have only dealt with max-min composition. A straightforward application of the proposed method can be used to solve fuzzy relation equations modelled with max-product composition. In our future research, we also intend to utilize the proposed approach for solving fuzzy mathematical programming problems that have a feasible set defined by a system of fuzzy relation equations.

References

[1] Di Nola, A., Sessa, S., Pedrycz, W., Sanchez, E.: Fuzzy Relation Equations and Their Applications to Knowledge Engineering. Kluwer Academic Publishers, Dordrecht, The Netherlands (1989)
[2] Klir, G.J., Folger, T.A.: Fuzzy Sets, Uncertainty, and Information. Prentice-Hall, Englewood Cliffs, New Jersey (1988)
[3] Bertsekas, D.: Nonlinear Programming. Athena Scientific, Belmont, Massachusetts (1995)
[4] Birbil, Ş.İ., Fang, S.C.: An electromagnetism-like mechanism for global optimization. Journal of Global Optimization **25** (**3**) (2003) 263–282
[5] Birbil, Ş.İ., Fang, S.C., Sheu, R.L.: On the convergence of a population-based global optimization algorithm. Journal of Global Optimization (2002, to appear)
[6] Birbil, Ş.İ.: Stochastic Global Optimization Techniques. PhD thesis, North Carolina State University, Raleigh (2002)
[7] Törn, A., Zilinskas, A.: Global Optimization. Springer Verlag, Berlin (1989)

A Study on Fuzzy Random Linear Programming Problems Based on Possibility and Necessity Measures

Hideki Katagiri and Masatoshi Sakawa

Graduate School of Engineering, Hiroshima University, Kagamiyama 1-4-1,
Higashi-Hiroshima, Hiroshima, 739-8527 Japan
{katagiri, sakawa}@msl.sys.hiroshima-u.ac.jp

Abstract. In this paper, we deal with linear programming problems with fuzzy random variable coefficients and propose two decision making models based on possibility and necessity measures. One is the expectation optimization model, which is to maximize the expectation of degrees of possibility or necessity that the objective function value satisfies with a fuzzy goal given by a decision maker. The other is the variance minimization model, which is to minimize the variance of the degree. We show that the formulated problems based on the expectation optimization model and on the variance minimization model are transformed into a linear fractional programming problem and a convex quadratic programming problem, respectively and that optimal solutions of these problems are obtained by using conventional mathematical programming techniques.

1 Introduction

In the classical mathematical programming, the coefficients of objectives or constraints in problems are assumed to be completely known. However, in real systems, they are rather uncertain than constant. In order to deal with the uncertainty, stochastic programming [1] and fuzzy one [2] were considered. They are useful tools for the decision making under stochastic and fuzzy environment, respectively.

Most researches in respect to mathematical programming problems take account of fuzziness or randomness. However, in practice, decision makers are faced with situations where both fuzziness and randomness exist. In the case where some expert estimates parameters with uncertainty in a mathematical programming problem, the values are not always given as random variables or fuzzy sets. For example, in a production planning problem, the demand of some commodity might be dependent on weathers, i.e., fine, cloudy and rainy. If each of weathers occurs randomly and an expert estimates the demand for each weather as an ambiguous value such as a fuzzy number, then the demand is represented with a fuzzy random variable. In recent years, several authors considered linear programming problems including fuzzy random variables [3,4,5,6].

T. Bilgiç et al. (Eds.): IFSA 2003, LNAI 2715, pp. 725–732, 2003.

In this research, we consider linear programming problems where coefficients of the objective function are fuzzy random variables. Since the problems including fuzzy random variables are ill-defined problems, we interpret the problem with some point of view and transform into the deterministic problem. Here, we will propose two models based both on possibilistic programming and on stochastic programming.

Section 2 provides a definition of fuzzy random variables in this research. In Section 3, we formulate a linear programming problem with fuzzy random variable coefficients. Section 4 considers the problem to maximize degrees of possibility and necessity that the objective function value satisfies a fuzzy goal. Since the degree varies randomly due to the randomness of the objective function value, we propose the expectation optimization model to maximize the expectation of the degree. Furthermore, we consider the variance minimization model to minimize the variance of the degree, which is a risk aversion model. It is shown that the problems based on the expectation optimization model and on the variance minimization model are transformed into linear fractional programming problems and a convex quadratic programming problem, respectively. Finally, Section 5 concludes this paper and discusses further research.

2   Fuzzy Random Variable and the Related Concepts

Fuzzy random variable is a useful tool for a decision making under fuzziness and randomness, and it was first defined by Kwakernaak [7]. The mathematical basis was established by Puri and Ralescu [8]. Kruse et al. [9] provides a slightly different definition. In this research, we consider the discrete type of fuzzy random variable defined as follows:

Definition 1. *Let $F(\mathbb{R})$ be the set of fuzzy numbers, $s = \{1, \ldots, S\}$ a set of scenario and p_s the probability that the scenario s occurs. Suppose that $\sum_{s=1}^{S} p_s = 1$. If X is a mapping from $s \rightarrow F(\mathbb{R})$, then X is a fuzzy random variable.*

The above definition of fuzzy random variables corresponds to a special case of those given by Kwakernaak and Puri-Ralesu. The hybrid number introduced by Kaufman et al. [10], which is applied to the various decision making problems, satisfies the nature of the definition above.

There are other concepts to deal with both fuzziness and randomness such as *fuzzy evnet* and *probabilistic set*. *Fuzzy event* is well-known and was introduced by Zadeh [11]. Hirota et al. [12,13] considered *probabilistic set* to treat randomness of membership degree. Here, it seems important to realize the difference between a fuzzy event and a fuzzy random variable. A fuzzy event deals with a situation where the realization of a random variable is a real value but the observed event is fuzzy. Accordingly, in a fuzzy event, the degree of membership for each realization of a random variable is between 0 and 1. On the other hand, a fuzzy random variable is useful to deal with the situations where the realization of a random variable is not constant but fuzzy and for example, it is represented with a fuzzy number.

3   Formulation

In this paper, we consider the following problem:

$$\left.\begin{array}{c} \text{minimize } \tilde{\bar{C}}x \\ \text{subject to } Ax \le b, \ x \ge 0 \end{array}\right\} \tag{1}$$

where $x = (x_1, \ldots, x_n)^t$ is a decision vector and $\tilde{\bar{c}} = (\tilde{\bar{c}}_1, \ldots, \tilde{\bar{c}}_n)$ is a coefficient vector. Let A be an $m \times n$ matrix and b an $m \times 1$ vector. Each $\tilde{\bar{c}}_j$ is a fuzzy random variable with the following membership function:

$$\mu_{\tilde{\bar{c}}_j}(t) = \max\left\{0, 1 - \frac{|t - \bar{c}_j|}{\alpha_j}\right\}, \quad j = 1, \ldots, n \tag{2}$$

where $\bar{c}_j$ denotes a random variable (or a scenario variable) whose realization under the scenario s is c_{js}, and the number of scenarios is S. Let p_s be the probability that the scenario s occurs. It is assumed that $\sum_{s=1}^{S} p_s = 1$ holds. Each α_j denotes the spread of a fuzzy number.

This type of problem is often seen in a real decision making problem. For example, in the agricultural project for maximizing the gross profit, a profit for sale on some commodity per unit often changes in dependence on the crop yields. If the crop yields come under the influence of weathers, then the profit for each weather might be estimated by an expert as an ambiguous value. Since each weather occurs randomly, the profit per unit is represented by a fuzzy random variable. In this case, the objective function in (1) corresponds to the gross profit.

Since the coefficients of objective functions are the symmetric triangular fuzzy random variables, each objective function also becomes the same type of fuzzy random variable $\tilde{\bar{Y}}$ with the following membership function:

$$\mu_{\tilde{\bar{Y}}}(y) = \max\left\{0, \ 1 - \frac{\left|y - \sum\limits_{j=1}^{n} \bar{c}_j x_j\right|}{\sum\limits_{j=1}^{n} \alpha_j x_j}\right\}. \tag{3}$$

Considering the imprecision or fuzziness of the decision maker's judgment, for each objective function of problem (1), we introduce the fuzzy goal $\tilde{G}$ with the membership function expressed as

$$\mu_{\tilde{G}}(y) = \begin{cases} 0, & y > h^0 \\ \dfrac{y - h^0}{h^1 - h^0}, & h^1 \le y \le h^0 \\ 1, & y < h^1. \end{cases} \tag{4}$$

Since problem (1) is not a well-defined problem, we interpret the problem from some view point. It should be reminded that a fuzzy random variable deals with the ambiguity of a random variable, and on the other hand, a fuzzy event treats the vagueness of a random variable. In fuzzy mathematical programming, there are two major approaches. One is possibilistic programming [2], which is a useful tool for the problem where the coefficients of an objective function and/or constraints are fuzzy sets. The other is flexible programming [14], which is valuable for treating the fuzzy constraints and fuzzy goals. Since a fuzzy random variable is related to ambiguity, we propose decision making models using possibility and necessity measures based on possibilistic programming.

4   Fuzzy Random Linear Programming Models

4.1   Fuzzy Random Programming Model Using a Possibility Measure

Using the definition of a possibility measure, we consider the degree of possibility $\Pi_{\tilde{Y}}(\tilde{G})$ that the objective function value satisfies the fuzzy goal $\tilde{G}$ as follows:

$$\Pi_{\tilde{Y}}(\tilde{G}) = \sup_{y} \min \left\{ \mu_{\tilde{Y}}(y), \ \mu_{\tilde{G}}(y) \right\}. \tag{5}$$

Now, we consider the following problem instead of the original problem (1):

$$\left. \begin{array}{l} \text{maximize } \Pi_{\tilde{Y}}(\tilde{G}) \\ \text{subject to } A\boldsymbol{x} \leq \boldsymbol{b}, \ \boldsymbol{x} \geq \boldsymbol{0} \end{array} \right\} \tag{6}$$

The objective of problem (6) is to maximize a degree of possibility with respect to a fuzzy goal.

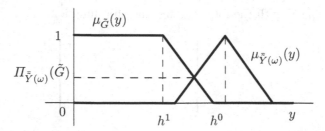

Fig. 1   Degree of possibility with respect to a fuzzy goal

Let

$$h^{\max} = \max_{s} \max_{\boldsymbol{x} \in X} \sum_{j=1}^{n} c_{js} x_j \ \text{ and } \ h^{\min} = \min_{s} \min_{\boldsymbol{x} \in X} \sum_{j=1}^{n} c_{js} x_j$$

where $X \triangleq \{x|Ax \leq b, \; x \geq 0\}$. If $h^0 > h^{max}$ and $h^1 < h^{min}$ are satisfied, by using (3) and (4), it is easy to calculate the degree of possibility as follows:

$$\Pi_{\tilde{Y}}(\tilde{G}) = \frac{\displaystyle\sum_{j=1}^{n}\{\alpha_j - \bar{c}_j\}x_j + h^0}{\displaystyle\sum_{j=1}^{n}\alpha_j x_j - h^1 + h^0}.$$

It should be noted that the degree varies randomly due to the randomness of $\bar{c}_j$ and that the degree for each scenario s becomes a real value. In other words, the degree can be represent with a random variable, and hence problem (6) is regarded as a stochastic programming problem.

Stochastic programming is a powerful tool for various decision making under uncertainty and have been studied by so many researchers. Dantzig [15] introduced the two-stage problem, and Charnes and Cooper [16] considered a chance constrained programming. In stochastic programming, there are four typical models: the expectation optimization model, the variance minimization model, the probability maximization model [16] and the fractile optimization model [17]. In this paper, we formulate the problems based on the expectation optimization model and on the variance minimization model.

1a. Expectation optimization model using a possibility measure

In this subsection, we consider the following problem based on the expectation model, which is to optimize the expectation of the objective function value:

$$\left. \begin{array}{c} \text{maximize } E[\Pi_{\tilde{Y}}(\tilde{G})] = \dfrac{\displaystyle\sum_{s=1}^{S} p_s \sum_{j=1}^{n}\{\alpha_j - \bar{c}_j\}x_j + h^0}{\displaystyle\sum_{j=1}^{n}\alpha_j x_j - h^1 + h^0} \\[2em] \text{subject to } x \in X \end{array} \right\} \qquad (7)$$

Since problem (7) is a linear fractional programming problem, it is transformed into a linear programming problem by using the method of Charnes and Cooper's transformation.

2a. Variance minimization model using a possibility measure

The expectation model is one of the useful decision making models under fuzzy stochastic environments; however, in the obtained solution based on this model, there is a possibility that the degree of possibility corresponding to a certain scenario is fairly small because the variance of the degree of possibility is unconsidered. Therefore, we consider the model to minimize the variances of degrees of possibility.

At first, we try to calculate the variance of the degree of possibility. Based on the nature of variance, it follows that

$$
Var[\Pi_{\tilde{Y}}(\tilde{G})] = \frac{1}{\left(\sum_{j=1}^{n} \alpha_j x_j - h^1 + h^0\right)^2} Var\left[\sum_{j=1}^{n} \bar{c}_j x_j\right].
$$

where Var denotes the variance. Let V be the variance-covariance matrix of $\bar{c}$. Then the problem to minimize the variances of degrees of possibility is formulated as

$$
\left.\begin{array}{c}
\text{minimize} \dfrac{1}{\left(\sum_{j=1}^{n} \alpha_j x_j - h^1 + h^0\right)^2} x^T V x \\[2em]
\text{subejct to } Ax \le b, \ x \ge 0
\end{array}\right\} \tag{8}
$$

The variance-covariance matrix is expressed by

$$
V = \begin{bmatrix}
v_{11} & v_{12} & \cdots & v_{1n} \\
v_{21} & v_{22} & \cdots & v_{2n} \\
\vdots & \vdots & \ddots & \vdots \\
v_{n1} & v_{n2} & \cdots & v_{nn}
\end{bmatrix}
$$

where

$$
v_{jj} = V[\bar{c}_j] = \sum_{s=1}^{S} p_s\{c_{js}\}^2 - \left\{\sum_{s=1}^{S} p_s c_{js}\right\}^2, \ j = 1, \dots, n
$$
$$
v_{jl} = Cov[\bar{c}_j, \bar{c}_l] = E[\bar{c}_j, \bar{c}_l] - E[\bar{c}_j]E[\bar{c}_l], \ j \neq l, \quad l = 1, \dots, n
$$

and

$$
E[\bar{c}_j, \bar{c}_l] = \sum_{s=1}^{S} p_s c_{js} c_{ls}.
$$

Let $1/t = \sum_{j=1}^{n} \alpha_j x_j - h^1 + h^0$ and $y \overset{\triangle}{=} tx$. Then it is easily shown that problem (8) is transformed into the following problem:

$$
\left.\begin{array}{l}
\text{minimize } y^T V y \\[1em]
\text{subject to } Ay \le tb, \ y \ge 0, \ \sum_{j=1}^{n} \alpha_j y_j + (h^0 - h^0)t = 1, \ t > 0.
\end{array}\right\} \tag{9}
$$

Since V is a positive-definite matrix from the nature of variance, it is easily understood that the problem is a convex quadratic programming problem. Accordingly, the global optimum solution of (9) is obtained by conventional quadratic programming techniques.

4.2   Fuzzy Random Programming Model Using a Necessity Measure

In this subsection, we consider the models using a necessity measure, which is useful for the case where a decision maker prefer to the solution with more certainty. The degree of necessity is defined by

$$N_{\tilde{\bar{Y}}}(\tilde{G}) = \inf_{y} \max\left\{1 - \mu_{\tilde{\bar{Y}}}(y),\ \mu_{\tilde{G}}(y)\right\}. \tag{10}$$

1b. Expectation optimization model using a necessity measure

It is easily shown that the problem to maximize the expectation of the degree of necessity is formulated as follows:

$$\left.\begin{array}{c} \text{maximize } E[N_{\tilde{\bar{Y}}}(\tilde{G})] = \dfrac{\displaystyle\sum_{s=1}^{S} p_s \sum_{j=1}^{n}(h_0 - \bar{c}_j x_j)}{\displaystyle\sum_{j=1}^{n}\alpha_j x_j - h^1 + h^0} \\[4pt] \text{subject to } \boldsymbol{x} \in X \end{array}\right\} \tag{11}$$

The above problem is also a linear fractional programming problem and can be solved by a linear programming technique.

2b. Variance minimization model using a necessity measure

It is interesting to note that the variance of the degree of necessity is equal to that of the degree of possibility. The reason is apparent from the nature of variance. Accordingly, the problem to minimize the variance of a degree of necessity is equivalent to (9).

5   Conclusion

In this research, we have investigated linear programming problem with fuzzy random variable coefficients, and proposed two models based on possibilistic programming and stochastic programming. Furthermore, we have shown that the problems based on these models are equivalently transformed into the deterministic problems and that the deterministic problems are easily solved by conventional mathematical programming techniques such as linear programming and quadratic programming techniques. In general, it is often seen that mathematical programming problems under fuzzy stochastic environments are formulated as little complicated problems. In this sense, our models have the advantage that the formulated problems are easily solved by conventional techniques, and it is quiet useful to apply our model to many real decision making problems.

In future, we will try to consider other models such as the probability maximization model and the fractile optimization model.

References

1. Vajda, S.: Probabilistic Programming, Academic Press, (1972).
2. Inuiguchi, M. and Ramik, J.: Possibilistic Linear Programming: A Brief Review of Fuzzy Mathematical Programming and a Comparison with Stochastic Programming in Portfolio Selection Problem, Fuzzy Sets and Systems 111 (2000) 3–28.
3. Wang, G.-Y., Zhong, Q.: Linear Programming with Fuzzy Random Variable Coefficients, Fuzzy Sets and Systems 57 (1993) 295-311
4. Katagiri, H., Ishii, H.: Chance Constrained Bottleneck Spanning Tree Problem with Fuzzy Random Edge Costs, Journal of the Operations Research Society of Japan 43 (2000) 128–137
5. Katagiri, H., Ishii, H.: Linear Programming Problem with Fuzzy Random Constraint, Mathematica Japonica 52 (2000) 123–129.
6. Luhandjula, M.K., Gupta, M.M.: On Fuzzy Stochastic Optimization, Fuzzy Sets and Systems 81 (1996) 47–55.
7. Kwakernaak, H.: Fuzzy Random Variable-1, Definitions and theorems, Information Sciences 15 (1978) 1–29.
8. Puri, M.L., Ralescu, D.A.: Fuzzy Random Variables, Journal of Mathematical Analysis and Applications 114 (1986) 409–422.
9. Kruse, R., Meyer, K.D.: Statistics with Vague Data, D. Reidel Publishing Company (1987).
10. Kaufman, A., Gupta, M.M.: Introduction to Fuzzy Arithmetic: Theory and Applications, Van Nostrand Reinhold Company (1985).
11. Zadeh, L.A., Probability Measure of Fuzzy events, Journal of Mathematical Analysis and Applications 23 (1968) 421–427.
12. Hirota, K., Concepts of Probabilistic Sets, Proceedings of IEEE Conference on Decision and Control (1977) 1361–1366.
13. Cogala, E., Hirota, K.: Probabilistic Set: Fuzzy and Stochastic Approach to Decision, Control and Recognition Processes, Verlag TUV Rheinland, Koln (1986).
14. Zimmermann, H.-J.: Applications of Fuzzy Sets Theory to Mathematical Programming, Information Sciences 36 (1985) 29–58.
15. Dantzig, G.B.: Linear Programming under Uncertainty, Management Science 1 (1955) 197–206.
16. Charnes, A., Cooper, W.W.: Chance Constrained Programming, Management Science 6 (1959) 73–79.
17. Geoffrion, A.M.: Stochastic Programming with Aspiration or Fractile Criteria, Management Science 13 (1967) 672–679.

Author Index

Lecture Notes in Artificial Intelligence (LNAI)

Lecture Notes in Computer Science